[8.1]	i	the imaginary unit, $\sqrt{-1}$
	$\sqrt{-b},\ b>0$	$i\sqrt{b}$
	C	the set of complex numbers
	z	complex number
	$\bar{z}$	the conjugate of z
[8.3]	cis θ	$\cos\theta + i\sin\theta$
[8.4]	z^0	$1 + 0i$ $(z \neq 0 + 0i)$
	z^{-m}	$1/z^m$ $(z \neq 0 + 0i)$
[9.2]	$(x,\ y,\ z)$	the ordered triple of numbers whose first component is x, second component is y, and third component is z
[10.1]	$\begin{bmatrix} a_1 & b_1 & c_1 \\ a_2 & b_2 & c_2 \end{bmatrix}$, etc.	matrix
	$A_{m \times n}$	m by n matrix
	$a_{i,j}$	the element in the ith row and jth column of the matrix A
	A^t	the transpose of the matrix A
	$0_{m \times n}$	the m by n zero matrix
	$-A_{m \times n}$	the negative of $A_{m \times n}$
[10.2]	$I_{n \times n}$	the n by n identity matrix
[10.3]	$A \sim B$	A is equivalent to B (for matrices)
[10.4]	$\begin{vmatrix} a_1 & b_1 & c_1 \\ a_2 & b_2 & c_2 \\ a_3 & b_3 & c_3 \end{vmatrix}$, etc.	determinant
	M_{ij}	the minor of the element a_{ij}
	A_{ij}	the cofactor of a_{ij}
	$\delta(A)$	the determinant of A
[10.6]	A^{-1}	the inverse of A
[11.1]	$s(n)$, or s_n	the nth term of a sequence
[11.2]	S_n	the sum of the first n terms in a sequence
	Σ	the sum
	S_∞	the sum of an infinite sequence
[11.3]	$\lim_{n \to \infty} s_n$	the limit of a sequence
[11.4]	$n!$	n factorial, or factorial n
	$0!$	zero factorial
	$\binom{n}{r}$	$\dfrac{n!}{r!(n-r)!}$
[13.4]	$(r,\ \theta)$	a set of polar coordinates for a point in the plane

Functions and Graphs

Functions and Graphs

Edwin F. Beckenbach

University of California, Los Angeles

Michael D. Grady

Loyola Marymount University

Irving Drooyan

Los Angeles Pierce College

Wadsworth Publishing Company
Belmont, California
A division of Wadsworth, Inc.

Mathematics Editor: Richard Jones
Production: Greg Hubit Bookworks
Cover photo: John N. Drooyan

© 1983 by Wadsworth, Inc. All rights reserved. No part of this book may be reproduced, stored in a retrieval system, or transcribed, in any form or by any means, electronic, mechanical, photocopying, recording, or otherwise, without the prior written permission of the publisher, Wadsworth Publishing Company, Belmont, California 94002, a division of Wadsworth, Inc.

Printed in the United States of America

1 2 3 4 5 6 7 8 9 10—87 86 85 84 83

Library of Congress Cataloging in Publication Data

Beckenbach, Edwin F.
  Functions and graphs.

  Includes index.
  1. Functions. 2. Algebra—Graphic methods.
I. Grady, Michael D., 1946–   II. Drooyan, Irving.
III. Title.
QA331.3.B43 1983   515   82-19983
ISBN 0-534-01180-2

ISBN 0-534-01180-2

Contents

Preliminary Concepts — 1

1 *Equations and Inequalities in One Variable* — 10

1.1 Equivalent Equations; First-Degree Equations — 10
1.2 Second-Degree Equations — 16
1.3 Equations Involving Certain Algebraic Expressions — 21
1.4 Solution of Inequalities — 26
1.5 Equations and Inequalities Involving Absolute Values — 37
Chapter Review — 41

2 *Relations and Functions; Graphs* — 43

2.1 Pairings of Real Numbers — 43
2.2 Function Notation; Composition — 48
2.3 Linear Equations and Inequalities in Two Variables — 52
2.4 Forms of Linear Equations — 60
2.5 Parallel and Perpendicular Lines — 63
2.6 Quadratic Equations and Inequalities in Two Variables — 68
2.7 Inverse Functions — 74
2.8 Special Functions — 79
Chapter Review — 84

3 *Polynomial and Rational Functions* — 88

3.1 Quotients of Polynomials — 88
3.2 Zeros of a Polynomial Function — 94
3.3 Rational Zeros of Polynomial Functions — 100
3.4 Graphing Polynomial Functions — 104
3.5 Rational Functions — 109
Chapter Review — 116

4 Exponential and Logarithmic Functions — 118

- 4.1 Exponential Functions — 118
- 4.2 Logarithmic Functions — 121
- 4.3 Special Logarithms and Powers — 127
- 4.4 Solution of Exponential Equations; Applications — 136
- Chapter Review — 144

Supplemental Exercises for Chapters 1–4 — 147

5 Circular Functions — 150

- 5.1 The Functions Sine and Cosine — 150
- 5.2 Special Function Values of Cosine and Sine — 155
- 5.3 Function Values of $\cos x$ and $\sin x$, for $x \in R$ — 161
- 5.4 Graphs of $y = \cos x$ and $y = \sin x$ — 167
- 5.5 Tangent Function — 176
- 5.6 Other Circular Functions — 182
- 5.7 Inverse Circular Functions — 188
- Chapter Review — 195

6 Trigonometric Functions; Vectors — 198

- 6.1 Angles and Their Measure — 198
- 6.2 Functions of Angles — 204
- 6.3 Right Triangles — 214
- 6.4 Law of Sines — 219
- 6.5 Law of Cosines — 224
- 6.6 Geometric Vectors — 227
- Chapter Review — 233

7 Identities and Conditional Equations — 237

- 7.1 Basic Identities — 237
- 7.2 Special Formulas for the Cosine Function — 242
- 7.3 Special Formulas for the Sine Function — 247
- 7.4 Special Formulas for the Tangent Function — 251
- 7.5 Conditional Equations — 257
- 7.6 Conditional Equations for Multiples — 262
- Chapter Review — 264

8 Complex Numbers — 266

- 8.1 Definitions; Basic Operations — 266
- 8.2 Complex Zeros of Polynomial Functions — 276
- 8.3 Trigonometric Form of Complex Numbers — 279
- 8.4 De Moivre's Theorem—Powers and Roots — 285
- Chapter Review — 289

Supplemental Exercises for Chapters 5–8 — 291

9 Systems of Equations and Inequalities — 293

- 9.1 Systems of Linear Equations in Two Variables — 293
- 9.2 Systems of Linear Equations in Three Variables — 301
- 9.3 Partial Fractions — 307
- 9.4 Systems of Nonlinear Equations — 314
- 9.5 Systems of Inequalities — 324
- 9.6 Convex Sets—Polygonal Regions — 326
- 9.7 Linear Programming — 329
- Chapter Review — 332

10 Matrices and Determinants — 334

- 10.1 Definitions; Matrix Addition — 334
- 10.2 Matrix Multiplication — 340
- 10.3 Solution of Linear Systems by Using Row-Equivalent Matrices — 347
- 10.4 The Determinant Function — 356
- 10.5 Properties of Determinants — 362
- 10.6 The Inverse of a Square Matrix — 367
- 10.7 Solution of Linear Systems Using Matrix Inverses — 374
- 10.8 Cramer's Rule — 378
- Chapter Review — 382

11 Sequences and Series — 384

- 11.1 Sequences — 384
- 11.2 Series — 391
- 11.3 Limits of Sequences and Series — 397
- 11.4 The Binomial Theorem — 403
- 11.5 Mathematical Induction — 411
- Chapter Review — 415

Supplemental Exercises for Chapters 9–11 — 417

12 Conic Sections — 419

- 12.1 Relations Whose Graphs Are Given — 419
- 12.2 Parabolas — 422
- 12.3 Circles and Ellipses — 426
- 12.4 Hyperbolas — 431
- 12.5 More About Graphs of Quadratic Equations — 435
- Chapter Review — 437

13 Additional Topics in Analytic Geometry — 440

- 13.1 Translation of Axes — 440
- 13.2 Rotation of Axes — 445
- 13.3 Parametric Equations — 449
- 13.4 Polar Coordinates — 454
- 13.5 Graphing in Polar Coordinates — 458
- Chapter Review — 462

Supplemental Exercises for Chapters 12–13 — **464**

Appendices

A Reference Outline — 469

B Tables — 481

- Table I   Exponential Functions, Base e — 481
- Table II Common Logarithms — 482
- Table III Natural Logarithms — 484
- Table IV Values of Circular Functions — 485
- Table V   Values of Trigonometric Functions — 489
- Table VI Squares, Square Roots, and Prime Factors — 498

Answers to Odd-Numbered Exercises — **499**

Index — **569**

Preface

Audience	*Functions and Graphs* is designed as a course for the student who has taken one and one-half or two years of high-school algebra but who needs more preparation before undertaking the study of calculus. Upon successful completion of this course, the student can proceed directly to the study of calculus.

Course Use	The text is designed for maximum flexibility. Choosing topics that are appropriate for a given class allows the book to be used in a three- or five-unit quarter, or in a three-unit semester course. Alternatively, the entire book can be used in a five-unit semester course.

Review Material	In order to accelerate the student's progress, the review of algebra has been minimized. However, a comprehensive summary of terminology, notation, properties of real numbers, and basic algebraic properties is included following this preface. This summary, called "Preliminary Concepts," is referred to throughout the text.

Treatments	The function concept along with techniques for graphing the elementary functions plays a central role in this book.

Polynomial and rational functions are introduced in Chapters 2 and 3. The topics related to the theory of equations in Chapter 8 can be considered in conjunction with Sections 3.2 and 3.3 if the instructor wishes to include a more comprehensive treatment of polynomial functions early in the course. However, these topics are not necessary to develop a competency in sketching the graphs of polynomial functions. Exponential and logarithmic functions are treated in Chapter 4. Heavy emphasis is given to the solution of logarithmic and exponential equations and the applications of these functions. The trigonometric functions are treated in Chapters 5 through 7. The circular function approach, which pairs real numbers with real numbers, is used to introduce these functions. This gives early importance to those aspects of trigonometry that students will need in calculus. Complex numbers, including De Moivre's theorem, are introduced in Chapter 8. Systems of equations and matrices are treated in Chapters 9 and 10. Sequences and series, including the binomial theorem, are treated in Chapter 11. An extensive treatment of analytic geometry in the plane including translation and rotation of axes is presented in Chapters 12 and 13.

Exercises   Each exercise set is graded starting with the simpler exercises and ending with the "B" exercises, which are more challenging.

Sets of supplemental exercises, which are designed to bring together concepts from several of the preceding chapters, are included after Chapters 4, 8, 11, and 13. These optional exercises are designed to offer a challenge to the more outstanding student.

Answers are given for odd-numbered exercises for each section in the text, and for all problems in the chapter reviews and supplemental exercises.

Calculations   Provision has been made for the optional use of a hand calculator when working with the exponential, logarithmic, and trigonometric functions. Table IV is graduated in hundredths of a radian, and Table V is graduated in tenths of a degree, so that the entries are compatible with the values typically obtained with a calculator.

Acknowledgment   We sincerely thank Mary-Margaret Grady for her excellent work in preparing the manuscript.

Edwin F. Beckenbach
Michael D. Grady
Irving Drooyan

Preliminary Concepts

The following definitions, notations, and algebraic properties have been introduced in previous courses. They are presented here to provide you with a convenient reference for your work in the present text.

Lowercase letters, such as $a, b, x,$ and y, will be used to denote real numbers both in the following lists and in the text.

Glossary

Set A **set** is simply a collection of objects. Any one of the objects of a collection is called a **member** or an **element** of the set. For example, the collection of numbers 2, 4, 6, 8 is a set denoted $\{2, 4, 6, 8\}$ and the number 2 is a member or an element of that set.

Set equality Two sets A and B are **equal**, $A = B$, if and only if they consist of exactly the same elements. For example, $\{3, 1\} = \{3, 2 - 1\}$.

Subset If every element of a set A is an element of a set B, then A **is a subset of** B. For example, $\{1, 2, 3\}$ is a subset of $\{1, 2, 3, 4\}$.

Union of two sets The **union** of two sets A and B is the set of all elements that belong either to A or to B or to both. For example, for $A = \{2, 4, 6\}$ and $B = \{1, 3, 4, 5\}$, "A union B" is $\{1, 2, 3, 4, 5, 6\}$.

Intersection of two sets The **intersection** of two sets A and B is the set of all elements common to both A and B. For example, for $A = \{1, 2, 3, 4\}$ and $B = \{2, 3, 4, 5\}$, "A intersect B" is $\{2, 3, 4\}$.

Algebraic expressions	Any grouping of constants and variables obtained by applying a finite number of the elementary operations—addition, subtraction, multiplication, division, or the extraction of roots—is called an **algebraic expression**. For example, $$\frac{3x^2 + \sqrt{2x-1}}{3x^3 + 7} \quad \text{and} \quad xy + 3x^2z - \sqrt[5]{z}$$ are algebraic expressions.
Equivalent expressions	If two expressions have equal values for all values of the variables for which both expressions are defined, then the expressions are **equivalent**. For example, $$2(x+3) \quad \text{and} \quad 2x+6$$ are equivalent algebraic expressions.
Power of a, a^n	An expression of the form a^n is called a **power** of a, where a is the **base** of the power and n is the **exponent** of the power. If n is a positive integer and a is a real number, then $$a^n = \underbrace{a \cdot \cdots \cdot a}_{n\text{-factors}}.$$ For example, $$3^2 = 3 \cdot 3 \quad \text{and} \quad x^4 = x \cdot x \cdot x \cdot x.$$
Monomial expression	An algebraic expression of the form cx^n, where n is a nonnegative integer, is called a **monomial** in the variable x. The number c is called the **coefficient** of the monomial, and the number n is called the **degree** of the monomial in x. A constant other than 0 is said to be a monomial of degree 0; no degree is assigned to the special monomial 0. For example, $4x^2$ is a monomial of degree 2, and 4 is a monomial of degree 0.
Terms of an algebraic expression	In any algebraic expression of the form $A + B + C + \cdots$, where $A, B, C, \ldots$ are algebraic expressions, $A, B, C, \ldots$ are called **terms** of the expression. For example, in $x + (y + 3)$, the terms are x and $(y + 3)$; in $x + y + 3$, the terms are x, y, and 3.
Polynomial	A **polynomial** is any algebraic expression that can be written as a sum of terms that are monomials. The **degree of a polynomial** is the same as the degree of its term of highest degree. Since no degree is assigned to the monomial 0, no degree is assigned to the polynomial 0 either. For example, $x^2 + 2x$ is a polynomial of degree 2, and $3x^5 - 2x^2 + 3$ is a polynomial of degree 5.
Standard form of a polynomial	A polynomial of degree n, $n \geq 0$, in x can be represented—when its terms are rearranged, if need be—by an expression in the **standard form** $$a_n x^n + a_{n-1} x^{n-1} + a_{n-2} x^{n-2} + \cdots + a_1 x + a_0 \quad (a_n \neq 0),$$

where it is understood that the a's are the (constant) coefficients of the powers of x in the polynomial. For example, $x^2 + 2x^3 + x + x^4 - 1$ written in standard form is

$$x^4 + 2x^3 + x^2 + x - 1.$$

Leading coefficient

The term $a_n x^n$ is called the **leading term**, and the coefficient a_n is called the **leading coefficient** in a polynomial. It is often convenient to have the leading term of the form x^n. In this case (that is, when $a_n = 1$), the polynomial is said to be **monic**. For example, the leading coefficient of $3x^2 - x + 1$ is 3, and the polynomial $x^2 - 2x + 4$ is monic.

Real polynomial

If the coefficients in a polynomial are real numbers, then the polynomial is called a **polynomial over R**. If the variable is restricted to represent only real numbers, then the polynomial is said to be a **polynomial in the real variable x**. If both the coefficients and the variable are restricted to real values, we say that the polynomial is a **real polynomial**.

Rational expression

A **fraction** is an expression denoting a quotient. If the numerator (dividend) and the denominator (divisor) are polynomials, then the fraction is said to be a **rational expression**. Trivially, any polynomial can be considered as being a rational expression since it is the quotient of itself and 1. For example,

$$\frac{x+2}{x}, \quad x+2 \quad \text{and} \quad \frac{x^2-1}{x^3+1}$$

are rational expressions.

Scientific notation

A number that is expressed as a product of a number between 1 and 10 (excluding 10) and a power of 10 is said to be in **scientific notation**. For example, 7,600,000 expressed in scientific notation is 7.60×10^6, and 0.0000487 expressed in scientific notation is 4.87×10^{-5}.

nth root of a

A number having a as its nth power is called an ***n*th root of a**. For example, 3 is a second (or square) root of 9 since $3^2 = 9$ and the third (or cube) root of -27 is -3 since $(-3)^3 = -27$.

Radical expression

If n is a positive integer and $a^{1/n}$ is a real number, then

$$\sqrt[n]{a} = a^{1/n}.$$

In the symbolism $\sqrt[n]{a}$, a is called the **radicand** and n is the **index** of the radical, and the expression is called a **radical expression** of order n. If no index is shown with a radical expression, as, for example, in the case $\sqrt{a}$, then the index 2 is understood to apply.

a^0

If a is a real number, $a \neq 0$, then
$$a^0 = 1.$$

For example,
$$3^0 = 1 \quad \text{and} \quad (x^2 + 1)^0 = 1.$$

a^{-n}

If a is a real number, $a \neq 0$, and n is a positive integer, then
$$a^{-n} = \frac{1}{a^n}.$$

For example,
$$3^{-2} = \frac{1}{3^2} \quad \text{and} \quad (x^2 + 9)^{-4} = \frac{1}{(x^2 + 9)^4}.$$

$a^{1/n}$

If a is a real number and n is a positive integer, then $a^{1/n}$ is the real number if only one exists and is the positive real number if two exist, such that
$$(a^{1/n})^n = a.$$

For example, $9^{1/2} = 3$ and $(-27)^{1/3} = -3$.

$a^{m/n}$

If n is a positive integer, $a^{1/n}$ is a real number, and m is any integer, then
$$a^{m/n} = (a^{1/n})^m.$$

For example,
$$4^{3/2} = (4^{1/2})^3 = 2^3 = 8.$$

Notation

$\{a, b\}$	the set whose elements are a and b
$A \subset B$	A is a subset of B
$\emptyset$	the null, or empty, set
$A \cup B$	the union of sets A and B
$A \cap B$	the intersection of sets A and B
$\in$	is an element of
$\{x \mid \ldots\}$	the set of all x such that $\ldots$
N	the set of natural numbers; $\{1, 2, 3, \ldots\}$

Preliminary Concepts 5

J	the set of integers; $\{\ldots, -2, -1, 0, 1, 2, \ldots\}$
Q	the set of rational numbers
H	the set of irrational numbers
R	the set of real numbers
C	the set of complex numbers
$a + b$	the sum of a and b
$a \cdot b$	the product of a and b; also written ab, or $(a)(b)$
$-a$	the negative of a; $(-1)a = -a$
$a - b$	the difference b subtracted from a; $a - b = a + (-b)$
$\dfrac{a}{b}$	the quotient a divided by b; $\dfrac{a}{b} = a \cdot \dfrac{1}{b}$
$a < b$	a is less than b
$a > b$	a is greater than b
$a \leq b$	a is less than or equal to b
$a \geq b$	a is greater than or equal to b
$\lvert a \rvert$	the absolute value of a; $$\lvert a \rvert = \begin{cases} a & \text{if } a \geq 0 \\ -a & \text{if } a < 0 \end{cases}$$
$P(x)$	algebraic expression in the variable x

Order Axioms of Real Numbers

O-1 If a is a real number, then exactly one of the following statements is true: a is positive, a is zero, or $-a$ is positive. *Trichotomy law.*

O-2 If a and b are positive real numbers, then $a + b$ is positive and ab is positive. *Closure law for positive numbers.*

O-3 There is a one-to-one correspondence between the real numbers and the points on a geometric line. *Completeness property.*

Geometric Interpretation of Order in the Real Numbers

Certain algebraic statements concerning the order of real numbers can be interpreted geometrically. We summarize some of the more common correspondences in the table on page 6, where in each case $a, b, c \in R$.

Algebraic statement	Geometric statement	Graph
1. a is positive	1. The graph of a lies to the right of the origin.	1.
2. a is negative	2. The graph of a lies to the left of the origin.	2.
3. $a > b$	3. The graph of a lies to the right of the graph of b.	3.
4. $a < b$	4. The graph of a lies to the left of the graph of b.	4.
5. $a < c < b$	5. The graph of c is to the right of the graph of a and to the left of the graph of b.	5.
6. $|a| < c$	6. The graph of a is less than c units from the origin.	6.
7. $|a - b| < c$	7. The graphs of a and b are less than c units from each other.	7.
8. $|a| < |b|$	8. The graph of a is closer to the origin than the graph of b.	8.

Notice that Statement 5, $a < c < b$, is a contraction of the two inequalities $a < b$ and $b < c$, and is read "a is less than b and b is less than c." Similarly, $a < x$ and $x \leq b$ can be written together as $a < x \leq b$.

Axioms for the Real Numbers

1. $a + b$ is a unique real number.   *Closure law for addition.*
2. $(a + b) + c = a + (b + c)$ *Associative law for addition.*
3. For each $a \in R$,   *Additive-identity law.*

$$a + 0 = a.$$

0 is the *additive identity* for R.

4. For each $a \in R$,   *Additive-inverse law.*

$$a + (-a) = 0.$$

$-a$ is the *additive inverse of a*.

5. $a + b = b + a$ *Commutative law for addition.*
6. $a \cdot b$ is a unique real number.   *Closure law for multiplication.*

Preliminary Concepts

7. $(a \cdot b) \cdot c = a \cdot (b \cdot c)$ *Associative law for multiplication.*
8. $a \cdot (b + c) = (a \cdot b) + (a \cdot c)$ *Distributive law.*
9. For each $a \in R$,   *Multiplicative-identity law.*

$$1 \cdot a = a.$$

1 is the *multiplicative identity* in R.

10. $a \cdot b = b \cdot a$ *Commutative law for multiplication.*
11. For each $a \in R$, $a \neq 0$,   *Multiplicative-inverse law.*

$$\frac{1}{a} \cdot a = 1.$$

$1/a$ is the *multiplicative inverse of a.*

Properties of Real Numbers

1. If $a + b = 0$, then $b = -a$ and $a = -b$.
2. If $a \cdot b = 1$, then $a = 1/b$ and $b = 1/a$.
3. $a \cdot b = 0$ if and only if either $a = 0$ or $b = 0$, or both.
4. $-(-a) = a$
5. $\dfrac{-a}{b} = \dfrac{a}{-b} = -\dfrac{a}{b} = -\dfrac{-a}{-b}$, $b \neq 0$
6. $\dfrac{a}{b} = \dfrac{-a}{-b} = -\dfrac{a}{-b} = -\dfrac{-a}{b}$, $b \neq 0$
7. $\dfrac{a}{b} = \dfrac{c}{d}$ if and only if $ad = bc$, $b, d \neq 0$.
8. $\dfrac{ac}{bc} = \dfrac{a}{b}$, $b, c \neq 0$
9. $\dfrac{1}{a} \cdot \dfrac{1}{b} = \dfrac{1}{ab}$, $a, b \neq 0$
10. $\dfrac{a}{b} \cdot \dfrac{c}{d} = \dfrac{ac}{bd}$, $b, d \neq 0$
11. $\dfrac{a}{c} + \dfrac{b}{c} = \dfrac{a + b}{c}$, $c \neq 0$
12. $\dfrac{a}{b} + \dfrac{c}{d} = \dfrac{ad + bc}{bd}$, $b, d \neq 0$
13. $\dfrac{a}{c} - \dfrac{b}{c} = \dfrac{a - b}{c}$, $c \neq 0$

14. $\dfrac{a}{b} - \dfrac{c}{d} = \dfrac{ad - bc}{bd}, \quad b, d \neq 0$

15. $\dfrac{1}{a/b} = \dfrac{b}{a}, \quad a, b \neq 0$

16. $\dfrac{a/b}{c/d} = \dfrac{a}{b} \div \dfrac{c}{d} = \dfrac{a}{b} \times \dfrac{d}{c} = \dfrac{ad}{bc}, \quad b, c, d \neq 0$

17. If $a = b$, then

$$a + c = b + c$$

and

$$ac = bc.$$

18. If $a < b$, then

$$a + c < b + c \quad \text{for every } c \in R,$$
$$a \cdot c < b \cdot c, \quad \text{if } c > 0,$$

and

$$a \cdot c > b \cdot c, \quad \text{if } c < 0.$$

Special Products

1. $a(b_1 + b_2 + \cdots + b_n) = ab_1 + ab_2 + \cdots + ab_n$
2. $(x + a)(x + b) = x^2 + (a + b)x + ab$
3. $(x + a)^2 = x^2 + 2ax + a^2$
4. $(x + a)(x - a) = x^2 - a^2$
5. $(x + a)(x^2 - ax + a^2) = x^3 + a^3$
6. $(x - a)(x^2 + ax + a^2) = x^3 - a^3$

Properties of Exponents

If a and b are real numbers, $b \neq 0$, and m and n are integers, then:

1. $b^m \cdot b^n = b^{m+n}$
2. $\dfrac{b^m}{b^n} = b^{m-n}$
3. $(b^m)^n = b^{mn}$

4. $(ab)^n = a^n b^n, \quad a \neq 0 \text{ if } n = 0$
5. $\left(\dfrac{a}{b}\right)^n = \dfrac{a^n}{b^n}, \quad a \neq 0 \text{ if } n = 0$

These properties are valid if m and n are rational numbers provided a and b are positive.

Properties of Radicals

For real values of a and b for which all the radical expressions denote real numbers,

1. a. $\sqrt[n]{a^n} = a$ (n an odd positive integer)

   b. $\sqrt[n]{a^n} = |a|$ (n an even positive integer)

2. $\sqrt[n]{a^m} = (\sqrt[n]{a})^m$ (n, m integers, $n > 0$)

3. $\sqrt[n]{a} \cdot \sqrt[n]{b} = \sqrt[n]{ab}$ (n a positive integer)

4. $\dfrac{\sqrt[n]{a}}{\sqrt[n]{b}} = \sqrt[n]{\dfrac{a}{b}}$ ($b \neq 0$, n a positive integer)

5. $\sqrt[cn]{a^{cm}} = \sqrt[n]{a^m}$ (m, n, c integers, $n, c > 0$)

1 Equations and Inequalities in One Variable

In this chapter we shall consider real-number solutions of first- and second-degree equations and of equations involving radicals and absolute values. We shall also consider inequalities involving rational expressions and absolute values. Interval notation will be introduced as a form for expressing the solution sets of inequalities.

1.1 Equivalent Equations; First-Degree Equations

We shall use symbols such as $P(x)$, $Q(x)$, and $R(x)$ to represent algebraic expressions in the variable x (see "Preliminary Concepts").

If we replace the variable x in $P(x) = Q(x)$ with a real number for which both members are defined, and if the resulting statement is true, then the number is a **solution** of the equation and is said to **satisfy** the equation. Thus, 2 is a solution of $x + 3 = 5$, because $2 + 3 = 5$ is a true statement. The set of all solutions of an equation is said to be the **solution set** of the equation.

Equations that have the same solution set are called **equivalent equations**. For example, the equations

$$x + 3 = -3 \quad \text{and} \quad x = -6$$

are equivalent, because the solution set of each is $\{-6\}$.

Elementary transformations

To solve an equation, we usually either determine the members of the solution set by inspection or else generate a sequence of equivalent equations until we arrive at one with an obvious solution set. The following theorem, which is a direct consequence of the addition and multiplication laws for real numbers (see "Preliminary Concepts"), is frequently used in generating equivalent equations.

1.1 Equivalent Equations; First-Degree Equations

Theorem 1.1 *If $P(x)$, $Q(x)$, and $R(x)$ are expressions, then for all values of x for which $P(x)$, $Q(x)$, and $R(x)$ are real numbers,*

$$P(x) = Q(x)$$

is equivalent to each of the following equations:

    I. $P(x) + R(x) = Q(x) + R(x)$;

    II. $P(x) \cdot R(x) = Q(x) \cdot R(x)$ for $R(x) \neq 0$.

Observe that, since

$$P(x) - Q(x) = P(x) + [-Q(x)]$$

and

$$\frac{P(x)}{Q(x)} = P(x)\left[\frac{1}{Q(x)}\right], \quad Q(x) \neq 0,$$

the conclusions of Theorem 1.1 are also valid for the operations of subtraction and division.

Example   Solve $\dfrac{x+8}{x} = 5$.

Solution   We use parts of Theorem 1.1 to generate the following sequence of equivalent equations.

$$(x)\frac{x+8}{x} = 5(x)$$

$$x + 8 = 5x \quad (x \neq 0)$$

$$x + 8 - x = 5x - x$$

$$8 = 4x$$

$$2 = x$$

The equation $2 = x$ is equivalent to the original equation for $x \neq 0$. Further, 0 is not a solution of the original equation. Therefore the solution set is $\{2\}$.

Any application of any part of Theorem 1.1 is called an **elementary transformation**. An elementary transformation *always* produces an equivalent equation. Care must be exercised in the application of Part II, however, for we have specifically excluded multiplication or division by zero. For example, to solve the equation

$$\frac{x}{x-3} = \frac{3}{x-3} + 2, \tag{1}$$

we might first multiply each member by $(x - 3)$ to find an equation that is free of fractions. We have

$$(x - 3)\frac{x}{x - 3} = (x - 3)\frac{3}{x - 3} + (x - 3)2,$$

$$x = 3 + 2x - 6, \tag{2}$$

from which

$$x = 3.$$

Thus 3 is a solution of Equation (2). But, upon substituting 3 for x in Equation (1), we have

$$\frac{3}{3 - 3} = \frac{3}{3 - 3} + 2 \quad \text{or} \quad \frac{3}{0} = \frac{3}{0} + 2,$$

and neither member is defined. In obtaining Equation (2), each member of Equation (1) was multiplied by $(x - 3)$; but if x is 3, then $(x - 3)$ is zero, and Theorem 1.1-II is not applicable. Equation (2) is *not* equivalent to Equation (1), and in fact Equation (1) has no solution.

We can always decide whether what we think is a solution of an equation is such in reality by substituting the suggested solution in the original equation and determining whether or not the resulting statement is true. If each equation in a sequence is obtained by means of an elementary transformation, the sole purpose for such checking is to detect arithmetic errors. We shall dispense with checking solution sets in the examples that follow unless we apply what may be a nonelementary transformation—that is, unless we *multiply or divide by an expression that vanishes for some value or values of the variable.*

Solution of a linear equation

The equation

$$ax + b = 0, \tag{3}$$

where $a \neq 0$, is a **first-degree**, or **linear**, **equation**. Any equation that can be reduced to this form by elementary transformations is therefore equivalent to a first-degree equation. We can show that such an equation always has one and only one solution. By Theorem 1.1-I, Equation (3) is equivalent to

$$ax = -b; \tag{4}$$

further, by Theorem 1.1-II, Equation (4) is equivalent to

$$x = -\frac{b}{a}, \tag{5}$$

which, of course, has the unique solution $-b/a$. Since Equations (3), (4), and (5) are equivalent, Equation (3) has the unique solution $-b/a$.

An equation containing more than one variable, or containing symbols such as a, b, and c representing constants, can often be solved for one of the symbols in terms of the remaining symbols by applying elementary transformations until the desired symbol is obtained by itself as one member of an equation.

1.1 Equivalent Equations; First-Degree Equations

Example   Solve $ay = b + y$, for y.

Solution   We generate the following sequence of equivalent equations.
$$ay - y = b$$
$$y(a - 1) = b$$
$$y = \frac{b}{a - 1} \quad (a \neq 1)$$

Exercise 1.1

A Solve.

Examples
a. $6y - 2(2y + 5) = 6(5 + y)$
b. $3x^2 + 2x - 1 = 3x(x - 4)$

Solutions
a. $6y - 4y - 10 = 30 + 6y$
   $2y - 10 = 30 + 6y$
   $-40 = 4y$
   $-10 = y$
   The solution set is $\{-10\}$.

b. $3x^2 + 2x - 1 = 3x^2 - 12x$
   $-1 = -14x$
   $\dfrac{1}{14} = x$
   The solution set is $\{1/14\}$.

1. $3x + 2(8 - x) = 3(x - 2)$
2. $5(x + 1) = 4(x + 2) - 3$
3. $-2[x - (x + 3)] = x + 1$
4. $-[2x - (3 - x)] = 4x - 11$
5. $x^2 - 3 = 1 + (x + 1)(x - 2)$
6. $2 - x - x^2 = 1 - (x - 1)^2$
7. $2[3x - (4x + 1)] = 2x - 3$
8. $4x - (2 + 3x) = x - [2x - (3x - 2)]$
9. $[2x + (3 - x)] - [x - (2 + x)] = x$
10. $x - [(x - 1) - (2x - 2)] = -2x + 1$
11. $\{3 - [x - (2x + 1)] + 1\} = -2x$
12. $x - \{2x - [x + (x - 4)]\} = 2x$

Examples
a. $\dfrac{3x + 4}{5} = \dfrac{7x + 6}{10}$
b. $\dfrac{x + 2}{x + 4} = 2$

Solutions
a. $10\left(\dfrac{3x + 4}{5}\right) = 10\left(\dfrac{7x + 6}{10}\right)$
   $2(3x + 4) = 7x + 6$
   $6x + 8 = 7x + 6$
   $2 = x$
   The solution set is $\{2\}$.

b. $\dfrac{x + 2}{x + 4}(x + 4) = 2(x + 4)$
   $x + 2 = 2(x + 4) \quad (x \neq -4)$
   $x + 2 = 2x + 8$
   $-6 = x$
   The solution set is $\{-6\}$.

13. $\dfrac{x-5}{4} = 1 + \dfrac{x-9}{12}$ 14. $\dfrac{2x+3}{6} = \dfrac{x+1}{9}$ 15. $\dfrac{x}{3} - 1 = \dfrac{2x-3}{3}$

16. $\dfrac{x}{3} + \dfrac{x+1}{2} = 3$ 17. $\dfrac{2+x}{x} = 3$ 18. $\dfrac{x-2}{x-4} = \dfrac{3}{4}$

Example

$$\dfrac{4}{2x+3} + \dfrac{4x}{4x^2-9} = \dfrac{12}{2x+3}$$

Solution

$$\left(\dfrac{4}{2x+3} + \dfrac{4x}{4x^2-9}\right)(2x+3)(2x-3) = \dfrac{12}{2x+3}(2x+3)(2x-3)$$

$$4(2x-3) + 4x = 12(2x-3) \quad (x \neq -3/2, 3/2)$$

$$-12x = -24$$

$$x = 2$$

The solution set is $\{2\}$.

19. $\dfrac{2}{x+1} = \dfrac{x}{x+1} + 1$ 20. $\dfrac{3}{x-2} = \dfrac{1}{2} + \dfrac{2x-7}{2x-4}$

21. $\dfrac{2x+1}{x-4} - \dfrac{4x-3}{2(x-4)} = 0$ 22. $\dfrac{3x}{x+3} + \dfrac{-9x-1}{3(x+3)} = \dfrac{4}{x+3}$

23. $\dfrac{3x}{x-4} = \dfrac{x}{2x-8} + \dfrac{1}{x-4}$ 24. $\dfrac{6x}{3x+2} - \dfrac{4x}{3x+2} = \dfrac{1}{6x+4}$

25. $\dfrac{1}{x-1} + \dfrac{2}{x+1} = \dfrac{3x-1}{x^2-1}$ 26. $\dfrac{4}{x+2} - \dfrac{1}{x} = \dfrac{2x-1}{x^2+2x}$

27. $\dfrac{3}{x+1} + \dfrac{x-4}{x^2-x-2} = \dfrac{-10}{x-2}$ 28. $\dfrac{1}{x} + \dfrac{1}{x^2+x} = \dfrac{3}{x+1}$

29. $\dfrac{2}{x-1} + \dfrac{3x}{x^2-4x+3} = \dfrac{5}{x-3}$ 30. $\dfrac{2x-1}{x-4} + \dfrac{x}{x^2-x-12} = \dfrac{2x-1}{x+3}$

31. $\dfrac{x-3}{x+2} + \dfrac{11x-8}{x^2-2x-8} = \dfrac{x+2}{x-4}$ 32. $\dfrac{2x+1}{2x-3} + \dfrac{-2x+1}{2x^2-5x+3} = \dfrac{x}{x-1}$

Solve for the indicated variable. Leave the result in the form of an equation equivalent to the given equation. Indicate any restrictions on the variables.

Example

Solve $l = a + (n-1)d$, for d.

Solution

Add $-a$ to each member.

$$l - a = (n-1)d$$

1.1 Equivalent Equations; First-Degree Equations

Multiply each member by $\dfrac{1}{n-1}$ $(n \neq 1)$.

$$\frac{l-a}{n-1} = d$$

$$d = \frac{l-a}{n-1} \quad (n \neq 1)$$

33. $S = 2\pi rh$, for r
34. $S = 2\pi rh$, for h
35. $v = k + gt$, for k
36. $v = k + gt$, for t
37. $A = \dfrac{h}{2}(b+c)$, for c
38. $S = \dfrac{a}{1-r}$, for r
39. $l = a + (n-1)d$, for n
40. $\dfrac{1}{r} = \dfrac{1}{r_1} + \dfrac{1}{r_2}$, for r
41. $V = \pi R^2 h - \pi r^2 h$, for r^2
42. $V = \dfrac{4}{3}\pi R^3 - \dfrac{4}{3}\pi r^3$, for R^3

Example Solve $x'(x-3) + 4 = 5(x + x') - 9$, for x.

Solution

$$x'x - 3x' + 4 = 5x + 5x' - 9$$

$$x'x - 5x = 5x' - 9 + 3x' - 4$$

$$x(x' - 5) = 8x' - 13$$

$$x = \frac{8x' - 13}{x' - 5} \quad (x' \neq 5)$$

43. $2y'y + 2x = 0$, for y'
44. $2x + 2y + 2xy' + 2yy' = 0$, for y'
45. $\dfrac{2xy - 2x^2 yy'}{y^4} = 0$, for y'
46. $\dfrac{(2x + 2yy')2y - 2y(x^2 + y^2)}{4y^2} = 1$, for y'
47. $x^2 y' - 3x - 2y^3 y' = 1$, for y'
48. $2xy' - 3y' + x^2 = 0$, for y'
49. $x_1 x_2 - 2x_1 x_3 = x_4$, for x_1
50. $3x_1 x_3 + x_1 x_2 = x_4$, for x_1
51. $\dfrac{y - y_1}{x - x_1} = 6$, for y
52. $\dfrac{y - y_1}{x - x_1} = 2$, for x

B 53. Find a value of k so that $2x - 3 = \dfrac{4+x}{k}$ is equivalent to the equation $2x - 1 = -3$.

54. Find a value of k so that $3x - 1 = k$ is equivalent to the equation $2x + 5 = 1$.

55. For what value of k does $\dfrac{2x^2 - kx + 1}{x + 3} = 2x - 3$ have $\{-1\}$ as its solution set?

56. For what value of k does $\dfrac{2kx^3 - kx^2 + x - 1}{x + 2} = x$ have $\{1\}$ as its solution set?

57. For what value of k does $\dfrac{x - 3}{x + 2} + \dfrac{kx + 8}{x^2 - 2x - 8} = \dfrac{x + 2}{x - 4}$ have $\{x \mid x \neq -2, 4\}$ as its solution set?

58. For what value of k does $\dfrac{kx}{x + 1} + \dfrac{1}{2x + 2} = \dfrac{4}{x + 1}$ have the empty set as its solution set?

1.2 Second-Degree Equations

The equation
$$ax^2 + bx + c = 0 \quad (a \neq 0)$$
is a **second-degree**, or **quadratic**, equation. Any equation that can be reduced to this form by elementary transformations is therefore equivalent to a quadratic equation. We shall designate the form shown above as the **standard form** for such equations.

Solution by factoring

The following theorem, which is a consequence of the properties of real numbers (see "Preliminary Concepts"), will be useful in solving quadratic equations.

Theorem 1.2 *If $P(x)$ and $Q(x)$ are expressions in x, then $P(x)Q(x) = 0$ if and only if both $P(x)$ and $Q(x)$ are defined and*
$$P(x) = 0 \quad \text{or} \quad Q(x) = 0, \quad \text{or both.}$$

Example Solve $x^2 + 2x - 15 = 0$.

Solution The equation
$$x^2 + 2x - 15 = 0$$
is equivalent to
$$(x + 5)(x - 3) = 0.$$
Since $(x + 5)(x - 3) = 0$ is true if and only if
$$x + 5 = 0 \quad \text{or} \quad x - 3 = 0,$$
we can see by inspection that the only values of x that satisfy the original equation are -5 and 3. Hence the solution set is $\{-5, 3\}$.

1.2 Second-Degree Equations

Number of solutions

The solution set of a quadratic equation over the real numbers can contain zero, one, or two members. The equation in the foregoing example has two real solutions. Now consider the equation

$$x^2 + 1 = 0.$$

Since the square of any real number is nonnegative, this equation has *no real solutions*. Next consider the equation

$$x^2 - 2x + 1 = 0.$$

Since $x^2 - 2x + 1 = 0$ is equivalent to

$$(x - 1)^2 = 0,$$

and since the only value of x for which $(x - 1)^2 = 0$ is 1, this is the only member of the solution set of $x^2 - 2x + 1 = 0$. Because we wish to consider every quadratic equation which has any real solution to have two solutions, we say that the solution is of **multiplicity two**; that is, we count such a value twice as a solution.

General solution

If we cannot readily factor a quadratic equation, we must use other methods of solution. Quadratic equations of the form

$$(x - p)^2 = q \quad (q \geq 0)$$

can be solved by observing that $x - p$ must be one of the square roots of q, that is, either

$$x - p = \sqrt{q} \quad \text{or} \quad x - p = -\sqrt{q},$$

and conversely. Thus the solution set of $(x - p)^2 = q$ is $\{p + \sqrt{q}, p - \sqrt{q}\}$. The assumption that $q \geq 0$ was made because at this time we are considering only real numbers as solutions. This method of solving quadratic equations is called **extraction of roots**.

Example

Solve $(x - 3)^2 = 4$.

Solution

By extraction of roots, we have

$$x - 3 = \sqrt{4} \quad \text{or} \quad x - 3 = -\sqrt{4},$$

from which

$$x - 3 = 2 \quad \text{or} \quad x - 3 = -2.$$

Then $x = 5$ or $x = 1$, and the solution set is $\{1, 5\}$.

Quadratic formula

Being able to find solution sets for quadratic equations of the form $(x - p)^2 = q$ enables us to find the solution set of any quadratic equation. Let us first consider the general quadratic equation in standard form

$$ax^2 + bx + c = 0 \quad (a \neq 0), \tag{1}$$

for the special case in which $a = 1$ (in case $a \neq 1$, we can divide both members of Equation (1) by a to obtain an equivalent equation with leading coefficient 1), that is,

$$x^2 + bx + c = 0. \tag{1'}$$

By using a process called completing the square, we can write this equation in the equivalent form

$$(x - p)^2 = q,$$

which we can solve as above. We begin the process of **completing the square** by adding $-c$ to each member of Equation (1'), which yields

$$x^2 + bx = -c. \tag{2}$$

If we then add $(b/2)^2$ to each member of Equation (2), we obtain

$$x^2 + bx + \left(\frac{b}{2}\right)^2 = -c + \left(\frac{b}{2}\right)^2, \tag{3}$$

in which the left-hand member is equal to $(x + b/2)^2$, and we have

$$\left(x + \frac{b}{2}\right)^2 = -c + \frac{b^2}{4}. \tag{4}$$

Since we have performed only elementary transformations, Equation (4) is equivalent to Equation (1'), and we can solve (4) by extraction of roots, obtaining the following theorem, which gives the solutions of the general quadratic equation in terms of the coefficients. The proof is left as an exercise.

Theorem 1.3   If a, b, and $c \in R$, the solutions to $ax^2 + bx + c = 0$ ($a \neq 0$) are

$$x = \frac{-b \pm \sqrt{b^2 - 4ac}}{2a}. \tag{5}$$

Equation (5) is called the **quadratic formula**. The symbol $\pm$ is used to condense the two equations

$$x = \frac{-b + \sqrt{b^2 - 4ac}}{2a} \quad \text{or} \quad x = \frac{-b - \sqrt{b^2 - 4ac}}{2a}$$

into a single equation. We need only substitute the coefficients a, b, and c of a given quadratic equation in the formula to find the solution set for the equation.

Example   Solve $x^2 + 3x - 2 = 0$ by using the quadratic formula.

Solution   Substituting 1 for a, 3 for b, and -2 for c in the quadratic formula and simplifying, we have

$$x = \frac{-3 \pm \sqrt{9 - 4 \cdot 1 \cdot (-2)}}{2 \cdot 1} = \frac{-3 \pm \sqrt{17}}{2}.$$

The solution set is $\left\{ \dfrac{-3 + \sqrt{17}}{2}, \dfrac{-3 - \sqrt{17}}{2} \right\}.$

1.2 Second-Degree Equations

Determination of number of real solutions

An examination of the quadratic formula,

$$x = \frac{-b \pm \sqrt{b^2 - 4ac}}{2a},$$

shows that, if $ax^2 + bx + c = 0$, $a, b, c \in R$, is to have a real solution then $\sqrt{b^2 - 4ac}$ must be real. This, in turn, implies that only those quadratic equations for which $b^2 - 4ac \geq 0$ have real solutions. If $b^2 - 4ac > 0$, there are two unequal real solutions, and if $b^2 - 4ac = 0$, there is one real solution of multiplicity two. The number represented by $b^2 - 4ac$ is called the **discriminant** of the quadratic equation $ax^2 + bx + c = 0$.

Exercise 1.2

A *Solve by factoring.*

Example $\quad x^2 + x = 12$

Solution Write an equivalent equation in standard form and then factor the left-hand member.

$$x^2 + x - 12 = 0$$
$$(x + 4)(x - 3) = 0$$

Determine solutions by inspection, or set each factor equal to zero and solve.

$$x + 4 = 0 \qquad x - 3 = 0$$
$$x = -4 \qquad x = 3$$

The solution set is $\{-4, 3\}$.

1. $x^2 + x = 2$
2. $x^2 = 6 - x$
3. $x^2 - 5x = 0$
4. $x^2 - 5x - 10 = 4$
5. $x^2 - x = -1 + x$
6. $x(2x - 1) = 1$
7. $6x^2 + 5x + 1 = 0$
8. $3x^2 = 16x - 5$
9. $2x^2 - 13x + 6 = 0$
10. $12x^2 - 13x + 3 = 0$
11. $6x^2 + 7x + 2 = 0$
12. $6x^2 + 13x - 2 = 0$
13. $2x^2 = 6 - x$
14. $3x^2 - 5 = 2x$
15. $x(2x + 9) = -9$
16. $(x - 5)(x + 1) = -8$
17. $5 = \dfrac{6}{x^2} - \dfrac{7}{x}$
18. $3 = \dfrac{-2}{x + 2} + \dfrac{2}{x + 5}$

Solve by the extraction of roots.

Example $\quad (x - 2)^2 = 5$

Solution Set $x - 2$ equal to each square root of 5.

$$x - 2 = \sqrt{5} \qquad x - 2 = -\sqrt{5}$$
$$x = 2 + \sqrt{5} \qquad x = 2 - \sqrt{5}$$

The solution set is $\{2 + \sqrt{5}, 2 - \sqrt{5}\}$.

19. $x^2 = 4$
20. $x^2 = 9$
21. $2x^2 = 6$
22. $5x^2 = 8$
23. $x^2 - 4 = 4$
24. $x^2 - 6 = 12$
25. $(x + 1)^2 = 4$
26. $(2x - 3)^2 = 16$
27. $(x - 1)^2 = 3$
28. $(x + 2)^2 = 7$
29. $-4(x - 1)^2 = -1$
30. $-125(x - 3)^2 = -5$

Solve for x by using the quadratic formula.

Example

$x^2 - 5x + 5 = 0$

Solution

Substitute 1 for a, -5 for b, and 5 for c in the quadratic formula and simplify.

$$x = \frac{5 \pm \sqrt{25 - 4 \cdot 1 \cdot 5}}{2 \cdot 1} = \frac{5 \pm \sqrt{5}}{2}$$

The solution set is $\left\{\dfrac{5 + \sqrt{5}}{2}, \dfrac{5 - \sqrt{5}}{2}\right\}$.

31. $x^2 + 8x - 9 = 0$
32. $x^2 - 4x + 4 = 0$
33. $x^2 + 11x + 30 = 0$
34. $4x^2 - 3x - 1 = 0$
35. $3x^2 - 5x - 2 = 0$
36. $5x^2 + 8x = 4$
37. $x^2 - x - 4 = 0$
38. $x^2 - 2x - 2 = 0$
39. $x^2 - 3x - 3 = 0$
40. $3x^2 + x - 7 = 0$
41. $-x^2 + 6x + 13 = 0$
42. $-9x^2 - 12x + 5 = 0$

43. $\dfrac{x}{x^2 + 2x - 2} = 1$
44. $\dfrac{3x + 1}{x^2 + x + 5} = 1$

45. $2x^2 + (x - 1)^2 = 1$
46. $-2x^2 + (2x - 1)^2 = 4$
47. $-(x + 4)^2 + 2(x - 2)^2 = 1$
48. $(2x + 3)^2 - (x - 4)^2 = 2$

49. $\dfrac{2x}{x - 1} - \dfrac{x + 1}{2} = 0$
50. $\dfrac{3}{2x + 1} - \dfrac{2x - 3}{x} = 0$

51. $\dfrac{2x}{x - 1} - \dfrac{3}{x^2 - x} = \dfrac{x + 1}{x}$
52. $\dfrac{3x - 1}{3x + 1} + \dfrac{2x}{2x + 1} = 1$

B Find the value of k so that the given equation has one root of multiplicity two.

53. $x^2 + 2x + k = 0$
54. $x^2 - 4x + k = 0$
55. $x^2 + kx = -9$
56. $x^2 - kx = -49$
57. $kx^2 + 4x + 1 = 0$
58. $kx^2 + 12x + 9 = 0$

Solve for x in terms of k.

59. $kx^2 - 1 = 0$
60. $k^2x^2 - x = 0$
61. $-kx^2 + x = 0$
62. $-kx^2 + 2k = 1$

63. $x^2 + 2kx + k^2 = 0$
64. $2x^2 + kx - 3k^2 = 0$
65. $x^2 - kx - k^2 = 0$
66. $2x^2 - kx - 2k^2 = 0$
67. Show that if $a, b, c \in R$, then the equation $ax^2 + bx + c = 0$ $(a \neq 0)$ can be written equivalently as

$$x = \frac{-b \pm \sqrt{b^2 - 4ac}}{2a}.$$

68. Show that if r_1 and r_2 are roots of the quadratic equation $ax^2 + bx + c = 0$, then $r_1 + r_2 = -b/a$ and $r_1 r_2 = c/a$.

1.3 Equations Involving Certain Algebraic Expressions

In this section we shall discuss some methods of solving equations that may contain expressions that are not polynomials. We first consider equations that contain radical expressions.

Equality of like powers

In order to find solution sets for equations containing radical expressions, we shall need the following result.

Theorem 1.4 *If $U(x)$ and $V(x)$ are expressions in x, then the solution set of*

$$U(x) = V(x)$$

is a subset of the solution set of

$$[U(x)]^n = [V(x)]^n,$$

for each natural number n.

This theorem, which follows simply from the fact that products of equal numbers are equal numbers, permits us to raise both members of an equation to the same natural-number power with the assurance that we do not lose any solutions of the original equation in the process. On the other hand, it does *not* assert that the resulting equation will be equivalent to the original equation, and indeed it will not always be so. The equation

$$[U(x)]^n = [V(x)]^n$$

may have additional solutions (called **extraneous solutions**) that are not solutions of $U(x) = V(x)$. For example, the solution set of $x = 10$ is $\{10\}$, whereas the solution set of $x^2 = 10^2$, or $x^2 = 100$, is $\{10, -10\}$. Thus the equations $x = 10$ and $x^2 = 10^2$ are not equivalent.

Necessity of checking solutions

Because raising both members of an equation to a natural-number power does not always produce an equivalent equation, each solution obtained using this process *must* be substituted for the variable in the original equation to check its validity. An application of Theorem 1.4 is not an elementary transformation.

Example

Find the solution set of $\sqrt[3]{x-1} = -1$.

Solution

We raise each member of $\sqrt[3]{x-1} = -1$ to the third power to obtain
$$(\sqrt[3]{x-1})^3 = (-1)^3,$$
$$x - 1 = -1,$$
which is equivalent to
$$x = 0.$$
Since $\sqrt[3]{0-1} = -1$, a solution of the original equation is 0. Moreover, 0 is the only real solution, since Theorem 1.4 guarantees that the solution set of the equation $\sqrt[3]{x-1} = -1$ is a subset of the solution set of $x = 0$.

Example

Find the solution set of $\sqrt{x+2} + 4 = x$.

Solution

We first write the equivalent equation
$$\sqrt{x+2} = x - 4$$
and then square both members to obtain
$$(\sqrt{x+2})^2 = (x-4)^2,$$
$$x + 2 = x^2 - 8x + 16.$$
This last equation is equivalent to
$$x^2 - 9x + 14 = 0,$$
$$(x-2)(x-7) = 0,$$
which clearly has solutions 2 and 7. Upon replacing x with 2 in the original equation, however, we obtain
$$\sqrt{2+2} + 4 = 2,$$
which is false. Hence, 2 is not a solution of the original equation; it is an extraneous root. On the other hand, 7 does satisfy the original equation, so the solution set is $\{7\}$.

It is sometimes necessary to apply Theorem 1.4 more than once in solving certain equations.

Example   Find the solution set of $\sqrt{x+4} + \sqrt{9-x} = 5$.

Solution   It is helpful if this equation is first transformed so that each member of the equivalent equation contains only one of the radical expressions. Thus, by adding $-\sqrt{9-x}$ to each member, we obtain

$$\sqrt{x+4} = 5 - \sqrt{9-x}.$$

Squaring both members leads to

$$(\sqrt{x+4})^2 = (5 - \sqrt{9-x})^2,$$
$$x + 4 = 25 - 10\sqrt{9-x} + 9 - x,$$
$$2x - 30 = -10\sqrt{9-x},$$
$$x - 15 = -5\sqrt{9-x}.$$

Squaring both members of this equation we obtain

$$(x - 15)^2 = (-5\sqrt{9-x})^2,$$
$$x^2 - 30x + 225 = 25(9 - x),$$
$$x^2 - 5x = 0,$$
$$x(x - 5) = 0.$$

This equation has 0 and 5 as solutions. Both of these satisfy the original equation. Hence, by Theorem 1.4 the solution set is $\{0, 5\}$.

Substitution method   Some equations that are not polynomial equations can nevertheless be solved by means of related polynomial equations. For example, though $y + 2\sqrt{y} - 8 = 0$ is not a polynomial equation, if the variable p is substituted for the radical expression $\sqrt{y}$, we then have $p^2 + 2p - 8 = 0$, which is a polynomial equation in p.

Example   Find the solution set of $y + 2\sqrt{y} - 8 = 0$, $y \geq 0$.

Solution   If we set $p = \sqrt{y}$ and substitute in the given equation, we have

$$p^2 + 2p - 8 = 0,$$
$$(p + 4)(p - 2) = 0,$$

which has 2 and -4 as solutions. The value $p = 2$ leads to $\sqrt{y} = 2$ or $y = 4$. The value $p = -4$ leads to $\sqrt{y} = -4$, which has no solution since $\sqrt{y}$ is always nonnegative when $y \geq 0$. The solution set is therefore $\{4\}$.

The technique of substituting one variable for another—or, more generally, a variable for an expression—is called the **method of substitution**. It is not limited to cases involving radicals, but is useful in any situation in which an equation is polynomial in form. For example, in the equation

$$\left(x + \frac{1}{x}\right)^{-2} + 6\left(x + \frac{1}{x}\right)^{-1} + 8 = 0,$$

we would set

$$p = \left(x + \frac{1}{x}\right)^{-1}$$

to obtain the equation

$$p^2 + 6p + 8 = 0.$$

Similarly, in the equation

$$(y + 3)^{1/2} - 4(y + 3)^{1/4} + 4 = 0,$$

we would set

$$p = (y + 3)^{1/4}$$

to obtain the equation

$$p^2 - 4p + 4 = 0.$$

Exercise 1.3

A Solve by using Theorem 1.4 and check. If there is no solution, so state.

Examples

  a. $\sqrt{x} = 4$
  b. $\sqrt{x + 1} = 9$

Solutions

  a. $(\sqrt{x})^2 = 4^2$
  $x = 16$
  Check: $\sqrt{16} = 4$
  The solution set is {16}.

  b. $(\sqrt{x + 1})^2 = 9^2$
  $x + 1 = 81$
  $x = 80$
  Check: $\sqrt{80 + 1} = 9$
  The solution set is {80}.

1. $\sqrt{x} = 8$
2. $4\sqrt{x} = 9$
3. $\sqrt{y + 8} = 1$
4. $\sqrt{y + 3} = 2$
5. $x - 1 = \sqrt{2x + 1}$
6. $2y + 3 = \sqrt{y + 2}$
7. $\sqrt[3]{2 - y} = 3$
8. $\sqrt[5]{7 - x} = 2$
9. $\sqrt{x}\sqrt{x + 9} = 20$
10. $\sqrt{x + 3}\sqrt{x - 9} = 8$
11. $\sqrt{y^2 - y} = \sqrt{y^2 + 2y - 3}$
12. $x - 2 = \sqrt{2x^2 - 3x + 2}$
13. $\sqrt{y + 4} = \sqrt{y + 20} - 2$
14. $\sqrt{x} + \sqrt{2} = \sqrt{x + 2}$

1.3 Equations Involving Certain Algebraic Expressions

15. $\sqrt{x+4} + \sqrt{x+2} = 2$
16. $\sqrt{x-1} + \sqrt{x+2} = 1$
17. $(5+x)^{1/2} + x^{1/2} = 5$
18. $(y+7)^{1/2} + (y+4)^{1/2} = 3$
19. $(y^2 - 3y + 5)^{1/2} - (y+2)^{1/2} = 0$
20. $(z-3)^{1/2} + (z+5)^{1/2} = 4$

Solve for the indicated variable. Leave the results in the form of an equation. Assume that denominators are not zero.

21. $r = \sqrt{\dfrac{A}{\pi}}$, for A
22. $t = \sqrt{\dfrac{2v}{g}}$, for g
23. $x\sqrt{xy} = 1$, for y
24. $P = \pi\sqrt{\dfrac{l}{g}}$, for g
25. $x = \sqrt{a^2 - y^2}$, for y
26. $y = \dfrac{1}{\sqrt{1-x}}$, for x
27. $R = \sqrt{\dfrac{V}{\pi h} + r^2}$, for V
28. $R = \sqrt[3]{\dfrac{3V}{4\pi} + r^3}$, for V

Solve by the method of substitution.

29. $x + 6\sqrt{x} - 7 = 0$
30. $y - 12y^{1/2} + 35 = 0$
31. $z^{2/3} - 3z^{1/3} - 18 = 0$
32. $x^{2/3} - x^{1/3} - 6 = 0$
33. $\dfrac{1}{y^2} - \dfrac{7}{y} - 18 = 0$
34. $z^{-2} - 5z^{-1} - 14 = 0$
35. $\dfrac{x^2}{(x+1)^2} + \dfrac{x}{x+1} = 30$
36. $\left(1 + \dfrac{1}{y}\right)^2 + 3\left(1 + \dfrac{1}{y}\right) = 40$
37. $\sqrt{x} - 6\sqrt[4]{x} + 8 = 0$
38. $\sqrt{x} - 3\sqrt[4]{x} - 4 = 0$
39. $\sqrt{x-6} + 3\sqrt[4]{x-6} - 18 = 0$
40. $\sqrt[3]{x^2} - 12\sqrt[3]{x} + 20 = 0$
41. $(x+3)^{-2} + 4(x+3)^{-1} = 32$
42. $z^{-4} - 8z^{-2} + 15 = 0$

B *Find all the real solutions of each equation. If there are none, so state.*

43. $\sqrt{5 + \sqrt{x}} = \sqrt{x} - 1$
44. $\sqrt{13 + \sqrt{x}} = \sqrt{x} + 1$
45. $\sqrt{x + \sqrt{x}} = 1 - \sqrt{x}$
46. $\sqrt{\sqrt{x} - x} = \sqrt{x} + 3$
47. $x - x^{2/3} = x^{1/3} - 1$
48. $x^{4/3} + x^{2/3} = -x^2 - 1$
49. $x^{1/2} + x^{1/3} = 4x^{1/6} + 4$
50. $x^{3/4} + 9 = x^{1/2} + 9x^{1/4}$
51. $8x + 4(2x+1)^{2/3} = 9(2x+1)^{1/3} + 5$
52. $2x + (x+1)^{2/3} = 2(x+1)^{1/3} - 1$

1.4 Solution of Inequalities

Sentences such as
$$x + 3 \geq 10 \tag{1}$$
and
$$\frac{-2y - 3}{3} < 5 \tag{2}$$
are called **inequalities**. For appropriate values of the variable, one member of an inequality represents a real number that is less than ($<$), less than or equal to ($\leq$), greater than or equal to ($\geq$), or greater than ($>$) the real number represented by the other member.

For inequalities we shall consider only real-number replacements for the variables. Any element of the replacement set of the variable for which an inequality is valid is called a **solution** of the inequality, and the set of all solutions of an inequality is called its **solution set**.

Elementary transformations

As in the case with equations, we shall solve a given inequality by generating a series of **equivalent inequalities** (inequalities having the same solution set) until we arrive at one for which the solution set is obvious. To do this, we shall need the following theorem applicable to inequalities. The proof of this theorem follows directly from the properties of order for the set of real numbers (see "Preliminary Concepts").

Theorem 1.5 If $P(x)$, $Q(x)$, and $R(x)$ are expressions, then for all values of x for which $P(x)$, $Q(x)$, and $R(x)$ are real numbers,
$$P(x) < Q(x)$$
is equivalent to each of the following statements.

I. $P(x) + R(x) < Q(x) + R(x)$;
II. $P(x) \cdot R(x) < Q(x) \cdot R(x)$ for $R(x) > 0$;
III. $P(x) \cdot R(x) > Q(x) \cdot R(x)$ for $R(x) < 0$.

Similarly,
$$P(x) \leq Q(x)$$
is equivalent to inequalities of the form I–III, with $<$ (or $>$) replaced by $\leq$ (or $\geq$) under the same conditions as above.

The theorem can be interpreted as follows:

I. The addition of the same expression to each member of an inequality produces an equivalent inequality in the same sense.

1.4 Solution of Inequalities

II. If each member of an inequality is multiplied by the same expression representing a *positive* number, the result is an equivalent inequality.

III. If each member of an inequality is multiplied by the same expression representing a *negative* number, and the direction of the inequality is reversed, the result is an equivalent inequality.

Example By Theorem 1.5-I the inequality
$$3x - 4 < -2x - 5$$
is equivalent to
$$3x - 4 + 2x + 4 < -2x - 5 + 2x + 4,$$
$$5x < -1.$$

By Theorem 1.5-II we have
$$(5x)\left(\frac{1}{5}\right) < (-1)\left(\frac{1}{5}\right),$$
$$x < -\frac{1}{5}.$$

Observe that, since
$$P(x) - Q(x) = P(x) + [-Q(x)]$$
and
$$\frac{P(x)}{Q(x)} = P(x)\frac{1}{Q(x)},$$
the conclusions of Theorem 1.5 are also valid for the operations of subtraction and division.

Note that Theorem 1.5 does not permit multiplying (or dividing) by zero, and variables in multipliers and divisors are restricted from values for which the expression vanishes. The result of applying any part of this theorem is an elementary transformation.

Interval notation A notation called **interval notation** is sometimes used to denote sets such as $\{x | -2 < x \leq 5\}$, which can be called *intervals* of real numbers. Such notation involves the use of a parenthesis to denote the exclusion of an endpoint and a bracket to denote the inclusion of an endpoint. Thus,
$$(-2, 5] = \{x | -2 < x \leq 5\}.$$

For an infinite interval such as $\{x | x < 6\}$ we would write $(-\infty, 6)$, where the symbol $-\infty$ denotes the inclusion of all real numbers less than 6 in the interval. Similarly, $\{x | x > 4\}$ can be written as $(4, +\infty)$.

Intervals of real numbers can be graphed on a number line as shown in the table on page 28.

Interval notation	Graph
1. (a, b)	1. open circle at a, open circle at b
2. $[a, b)$	2. closed at a, open at b
3. $(a, b]$	3. open at a, closed at b
4. $[a, b]$	4. closed at a, closed at b
5. $(-\infty, a)$	5. open at a, extending left
6. $(-\infty, a]$	6. closed at a, extending left
7. (a, ∞)	7. open at a, extending right
8. $[a, \infty)$	8. closed at a, extending right

Note that an open circle indicates that the number corresponding to the indicated point is *not* in the set.

The solution set of an inequality frequently takes the form of an interval or the union of several intervals. Thus such a solution set can be written in interval notation and can be expressed graphically.

Solution of a linear inequality

Several applications of Theorem 1.5 can be applied to solve inequalities in the same way that the theorems of equality are applied to solve equations.

Example Find and graph the solution set of $\dfrac{-x + 3}{4} > -\dfrac{2}{3}$.

Solution Multiplying each member by 12 yields
$$3(-x + 3) > -8,$$
$$-3x + 9 > -8.$$

Adding -9 to each member, we have
$$-3x > -17.$$

1.4 Solution of Inequalities

Finally, multiplying each member by $-1/3$, we obtain

$$x < \frac{17}{3},$$

and the solution set is written

$$S = \left\{x \mid x < \frac{17}{3}\right\} = \left(-\infty, \frac{17}{3}\right).$$

The graph of the solution set is shown in the figure below. The open dot on the right-hand endpoint indicates that 17/3 is not a member of the solution set.

Inequalities sometimes appear in a form such as

$$-6 < 3x \leq 15, \tag{3}$$

where an expression is bracketed between two inequality symbols. The solution set of such an inequality is obtained in the same manner as the solution set of any other inequality. In (3) above, each expression may be multiplied by 1/3 to obtain

$$-2 < x \leq 5.$$

The solution set,

$$S = \{x \mid -2 < x \leq 5\} = (-2, 5],$$

is shown on a line graph in Figure 1.1. The open dot at the left-hand endpoint of the interval indicates that -2 *is not* a member of the solution set, whereas the solid dot at the right-hand endpoint shows that 5 *is* a member of the solution set.

Figure 1.1

Solution of quadratic inequalities

As in the case with first-degree inequalities, we can generate equivalent second-degree inequalities by applying any part of Theorem 1.5. Additional procedures are necessary, however, to obtain the solution sets of such inequalities. For example, consider the inequality

$$x^2 + 4x < 5.$$

To determine values of x for which this condition holds, we might first rewrite the inequality equivalently as

$$x^2 + 4x - 5 < 0$$

and then as

$$(x + 5)(x - 1) < 0.$$

It is clear here that only those values of x for which the factors $x + 5$ and $x - 1$ are opposite in sign will be in the solution set. These can be determined algebraically by noting that $(x + 5)(x - 1) < 0$ implies either

Case I: $x + 5 < 0$ and $x - 1 > 0$

or else

Case II: $x + 5 > 0$ and $x - 1 < 0$.

Each of these two cases can be considered separately.

First, for Case I,

imply
$$x + 5 < 0 \quad \text{and} \quad x - 1 > 0$$
$$x < -5 \quad \text{and} \quad x > 1,$$

a condition that is not satisfied by any value of x. The solution set of Case I is $\emptyset$.
The inequalities for Case II,

imply
$$x + 5 > 0 \quad \text{and} \quad x - 1 < 0,$$
$$x > -5 \quad \text{and} \quad x < 1,$$

which lead to the solution set

$$\{x \mid -5 < x < 1\}.$$

Thus the complete solution set is

$$S = \{x \mid -5 < x < 1\} \cup \emptyset = \{x \mid -5 < x < 1\}.$$

Critical numbers

Notice in the foregoing solution process that the numbers -5 and 1 separate the set of real numbers into the three intervals shown in Figure 1.2.

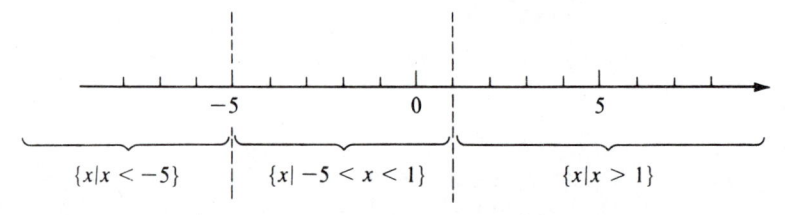

$\{x \mid x < -5\}$ $\qquad$ $\{x \mid -5 < x < 1\}$ $\qquad$ $\{x \mid x > 1\}$

Figure 1.2

Each of these intervals either is or is not a part of the solution set of the inequality $x^2 + 4x < 5$. To determine which, we need only substitute an arbitrarily selected number from each interval and test it in the inequality. Let us use $-6, 0,$ and 2.

$$(-6)^2 + 4(-6) \stackrel{?}{<} 5 \qquad 0^2 + 4(0) \stackrel{?}{<} 5 \qquad 2^2 + 4(2) \stackrel{?}{<} 5$$
$$36 - 24 \stackrel{?}{<} 5 \qquad 0 + 0 \stackrel{?}{<} 5 \qquad 4 + 8 \stackrel{?}{<} 5$$
$$12 \stackrel{?}{<} 5 \qquad 0 \stackrel{?}{<} 5 \qquad 12 \stackrel{?}{<} 5$$

No. $\qquad\qquad$ Yes. $\qquad\qquad$ No.

1.4 Solution of Inequalities

Clearly, the only interval involved that is in the solution set is $\{x \mid -5 < x < 1\}$, and hence this interval is the solution set. The graph is shown in Figure 1.3. If the inequality in this example were $x^2 + 4x \leq 5$, the endpoints on the graph would be shown as closed dots.

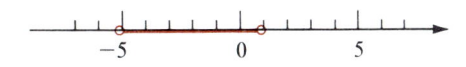

Figure 1.3

In an inequality of the form $Q(x) < 0$ or $Q(x) > 0$, numbers for which either $Q(x) = 0$ or $Q(x)$ is undefined are called **critical numbers**.

Examples   Find the critical numbers for each inequality.

a. $2x^2 - x - 1 > 0$ **b.** $\dfrac{x}{x^2 - 4} < 0$

Solutions   **a.** Factoring the left-hand member of

$$2x^2 - x - 1 = 0,$$

we obtain

$$(2x + 1)(x - 1) = 0,$$

from which

$$x = -\frac{1}{2} \quad \text{or} \quad x = 1.$$

Thus the critical numbers are $-1/2$ and 1.

b. $\dfrac{x}{x^2 - 4} = 0$ for $x = 0$. The expression $\dfrac{x}{x^2 - 4}$ is undefined when $x = \pm 2$. Thus the critical numbers are 0, 2, and -2.

The critical numbers of an inequality separate the set of real numbers into disjoint intervals. Each of these intervals is either contained in or disjoint from the solution set of the inequality. This fact provides us with an alternative method of solving many inequalities that are not linear.

Example   Solve $x^2 - 3x - 4 \geq 0$.

Solution   Factoring the left-hand element yields

$$(x + 1)(x - 4) \geq 0,$$

Solution continued overleaf

and we observe that -1 and 4 are critical numbers because $(x + 1)(x - 4) = 0$ for these values. Hence, we wish to check the intervals shown on the number line.

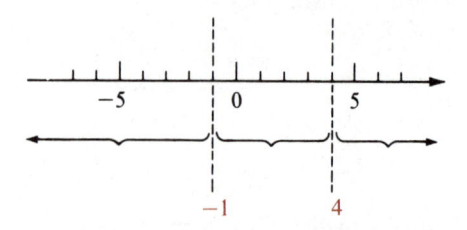

We now substitute an arbitrary value in each interval, say $-2, 0,$ and 5, for the variable in the original inequality.

$(-2)^2 - 3(-2) - 4 \overset{?}{\geq} 0$ $\qquad$ $(0)^2 - 3(0) - 4 \overset{?}{\geq} 0$ $\qquad$ $(5)^2 - 3(5) - 4 \overset{?}{\geq} 0$

Yes. $\qquad\qquad\qquad$ No. $\qquad\qquad\qquad$ Yes.

Hence, the solution set is

$$\{x \mid x \leq -1\} \cup \{x \mid x \geq 4\} = (-\infty, -1] \cup [4, \infty).$$

The graph is shown in the figure below.

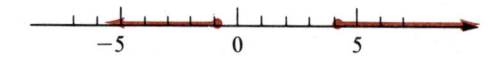

Variables in the denominator

The solution of an inequality involving fractions has to be approached with care if any fraction contains a variable in the denominator. If each member of such an inequality is multiplied by an expression containing the variable, we have to be careful either to distinguish between those values of the variable for which the expression denotes a positive and negative number, respectively, or to make sure that the expression by which we multiply is always positive. However, we can use the notion of critical numbers to avoid these complications.

Example $\qquad$ Solve the inequality $\dfrac{x}{x - 2} \geq 5$.

Solution $\qquad$ We first write the given inequality equivalently as

$$\frac{x}{x - 2} - 5 \geq 0,$$

from which

$$\frac{x - 5(x - 2)}{x - 2} \geq 0,$$

$$\frac{-4x + 10}{x - 2} \geq 0. \tag{4}$$

1.4 Solution of Inequalities

In this case, the critical numbers are $5/2$ and 2, because $\dfrac{-4x+10}{x-2}$ equals zero for $x = 5/2$, and $\dfrac{-4x+10}{x-2}$ is undefined for $x = 2$. Thus we want to check the intervals shown on the number line.

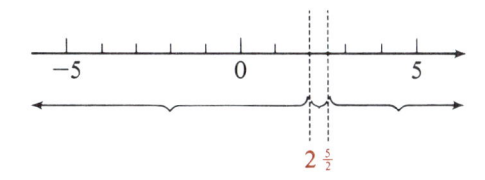

Substituting arbitrary values from each of the three intervals, say 0, 9/4, and 3, for the variable in the original inequality or in (4), we can identify the solution set,

$$\left\{x \mid 2 < x \leq \frac{5}{2}\right\} = \left(2, \frac{5}{2}\right],$$

as we did in the previous example. The graph is shown in the figure above. Note that the left-hand endpoint is an open dot (2 is not a member of the solution set) because the left-hand member of the original inequality is undefined for $x = 2$.

Exercise 1.4

A Solve the inequality. Write the solution set in interval notation and represent it on a line graph.

Examples

a. $\dfrac{3x+1}{5} < x - 2$

b. $-1 < 2x + 3 \leq 8$

Solutions

Generate the following sequences of equivalent inequalities.

a. $3x + 1 < 5x - 10$

$11 < 2x$

$\dfrac{11}{2} < x,\ \left(\dfrac{11}{2}, +\infty\right)$

b. $-4 < 2x \leq 5$

$-2 < x \leq \dfrac{5}{2}$

$\left(-2, \dfrac{5}{2}\right]$

1. $x + 11 > 5$
2. $x - 3 < -2$
3. $3x + 1 \leq 10$
4. $5x - 4 \leq 11$
5. $2x + 7 \geq -5$
6. $3 - x \leq -1$
7. $x + 6 \leq 3x + 1$
8. $2x + 7 \leq x + 4$
9. $1 - 4x > x + 6$
10. $4x \geq -5x - 1$
11. $-(x + 3) \leq -1$
12. $-(2x + 4) \geq 3$
13. $-2 \leq \dfrac{3x + 1}{4}$
14. $1 \leq \dfrac{-2x - 1}{3}$
15. $\dfrac{3x + 4}{2} < 6$
16. $\dfrac{-3x + 2}{4} < 5$
17. $\dfrac{2x - 3}{3} \leq \dfrac{3x}{2}$
18. $\dfrac{3x - 4}{2} \leq \dfrac{-2x}{5}$
19. $-1 \leq x + 5 \leq 6$
20. $0 \leq 2x - 3 < 1$
21. $-4 < \dfrac{3x + 2}{5} \leq -2$
22. $4 \leq \dfrac{3x - 1}{2} \leq 10$
23. $5 \leq \dfrac{x + 1}{-3} \leq 8$
24. $-2 \leq \dfrac{x + 5}{2} \leq 0$

Solve the inequality. Write the solution set in interval notation.

Example $x^2 + x \leq 12$

Solution The given inequality is equivalent to $(x + 4)(x - 3) \leq 0$. Thus the critical numbers are -4 and 3 because $(x + 4)(x - 3) = 0$ for these values. Check the intervals shown on the number line in the figure by substituting an arbitrary value from each interval for the variable in the inequality. Here we use $-5, 0,$ and 5.

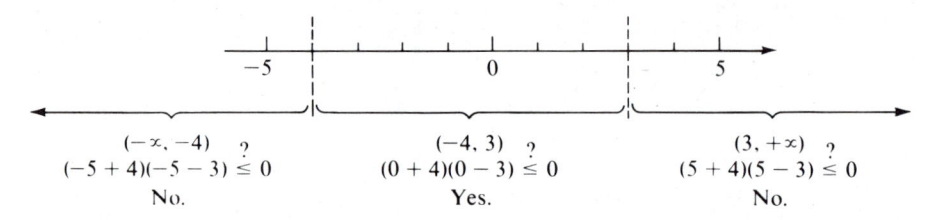

$(-\infty, -4)$? $(-5 + 4)(-5 - 3) \leq 0$ No.

$(-4, 3)$? $(0 + 4)(0 - 3) \leq 0$ Yes.

$(3, +\infty)$? $(5 + 4)(5 - 3) \leq 0$ No.

Next check the critical numbers themselves; here both are solutions. Therefore the solution set is $[-4, 3]$.

25. $(x + 1)(x - 5) > 0$
26. $(x + 6)(x + 2) < 0$
27. $(2x + 1)(3x - 8) \leq 0$
28. $(3x - 1)(3x - 10) \geq 0$
29. $x^2 + x - 6 > 0$
30. $x^2 + 3x - 4 \leq 0$
31. $x^2 - 2x - 8 > 0$
32. $x^2 + x - 12 \geq 0$
33. $x^2 + 4x > 3x + 20$
34. $8x + 8 \geq -x^2 + 2x$
35. $2x^2 - 2 \leq -x^2 + 5x$
36. $3x^2 + 2x < -3x^2 + x + 2$
37. $6x^2 + 6x \geq -2x^2 + 3$
38. $6x^2 - 6 > 5x - 2$
39. $x^2 < 0$
40. $x^2 + 1 \geq 0$

1.4 Solution of Inequalities

Example $x^2 - 2x - 1 \leq 0$

Solution Find the critical numbers by the quadratic formula:
$$x = \frac{2 \pm \sqrt{4+4}}{2} = \frac{2 \pm 2\sqrt{2}}{2} = 1 \pm \sqrt{2}.$$

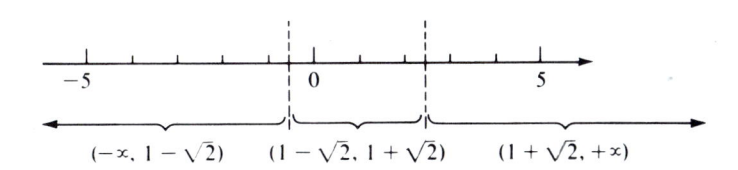

```
         |              |          |
   ──────┼──────────────┼──────────┼──────▶
        -5              0          5
   ◀─────────┬──────────┬──────────┬──────▶
        (-∞, 1-√2)  (1-√2, 1+√2)  (1+√2, +∞)
```

Check each interval by testing an arbitrary point in it, say -4, 1, and 4.

$$(-4)^2 - 2(-4) - 1 \stackrel{?}{\leq} 0 \qquad 1^2 - 2(1) - 1 \stackrel{?}{\leq} 0 \qquad 4^2 - 2(4) - 1 \stackrel{?}{\leq} 0$$

No. Yes. No.

Test the critical numbers themselves; since each makes the expression 0, they are solutions. The solution set is therefore $[1 - \sqrt{2}, 1 + \sqrt{2}]$.

41. $x^2 - 4x + 1 \geq 0$ 42. $x^2 + 4x + 2 \leq 0$
43. $x^2 - x + 4 < 0$ 44. $-x^2 + x + 1 > 0$

Example $\dfrac{x+1}{x} \geq 2$

Solution Write the equivalent inequalities

$$\frac{x+1}{x} - 2 \geq 0$$

$$\frac{x+1-2x}{x} \geq 0$$

$$\frac{1-x}{x} \geq 0$$

The critical numbers are 0 and 1 since $\dfrac{1-x}{x}$ is zero when $x = 1$ and undefined for $x = 0$.

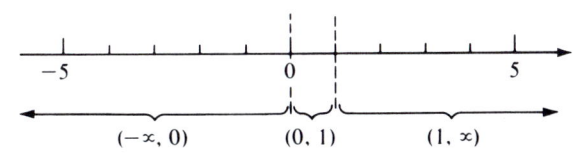

```
         |              |          |
   ──────┼──────────────┼──────────┼──────▶
        -5              0          5
   ◀─────────┬──────────┬──────────┬──────▶
         (-∞, 0)     (0, 1)      (1, ∞)
```

Solution continued overleaf

Check each interval by testing an arbitrary point in it, say $-1, 1/2, 2$.

$$\frac{-1+1}{-1} \stackrel{?}{\geq} 2 \qquad \frac{\frac{1}{2}+1}{\frac{1}{2}} \stackrel{?}{\geq} 2 \qquad \frac{2+1}{2} \stackrel{?}{\geq} 2$$

No.   Yes.   No.

Test the critical numbers; 1 is a solution but 0 is not. Therefore the solution set is $(0, 1]$.

45. $\dfrac{x-1}{x} \leq 3$
46. $\dfrac{x}{x-1} \geq 5$
47. $\dfrac{x+1}{x-1} < 1$
48. $\dfrac{x+1}{x-1} \leq 1$
49. $\dfrac{2}{2+x} - 3 > -2$
50. $\dfrac{x}{3-x} - 3 \leq 6$

B Solve the inequality. Write the solution set in interval notation.

Example

$$\dfrac{(x-1)(x+2)}{x-3} \leq 0$$

Solution

The critical numbers are $1, -2,$ and 3.

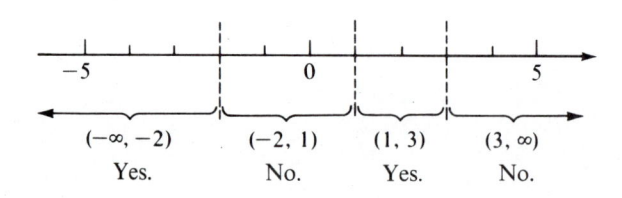

$(-\infty, -2)$ $(-2, 1)$ $(1, 3)$ $(3, \infty)$
Yes. No. Yes. No.

Check each interval by testing an arbitrary point in it, say $-3, 0, 2,$ and 4. Test the critical numbers themselves; 1 and -2 are solutions but 3 is not. Therefore the solution set is $(-\infty, -2] \cup [1, 3)$.

51. $\dfrac{(x+1)}{(x-1)(x-3)} > 0$
52. $\dfrac{(x-1)(x+1)}{(x-5)} < 0$
53. $\dfrac{(x-2)(x+1)}{x} < 0$
54. $\dfrac{x}{(x+2)(x-4)} \geq 0$
55. $(x-1)(x+1)(x+3) \geq 0$
56. $(x+1)(x-4)x < 0$
57. $\dfrac{(x+2)(x-2)}{x(x+4)} > 0$
58. $x(x+2)(x-2)(x-4) \geq 0$

In Exercises 59–64, assume ε is a positive real number. Solve each inequality for x in terms of ε. Write the solution set in interval notation.

Examples

a. $-\varepsilon < 2x + 1$ b. $2x + 1 < \varepsilon$

Solutions

a. $-\varepsilon < 2x + 1$
$-1 - \varepsilon < 2x$
$-\dfrac{1+\varepsilon}{2} < x$

The solution set is $\left(-\dfrac{1+\varepsilon}{2}, \infty\right)$.

b. $2x + 1 < \varepsilon$
$2x < \varepsilon - 1$
$x < \dfrac{\varepsilon - 1}{2}$

The solution set is $\left(-\infty, \dfrac{\varepsilon - 1}{2}\right)$.

59. $-\varepsilon < 4x - 5$
60. $4x - 5 < \varepsilon$
61. $-\varepsilon < 3x + 2 < \varepsilon$
62. $-\varepsilon < 2x + 4 < \varepsilon$
63. $-\varepsilon < -3x + 7 < \varepsilon$
64. $-\varepsilon < -2x + 1 < \varepsilon$

Example $\quad x^2 - 2 < \varepsilon$

Solution

Solve $x^2 - 2 = \varepsilon$ for x in terms of ε to obtain the critical values $-\sqrt{2 + \varepsilon}$ and $\sqrt{2 + \varepsilon}$. These values separate the number line into the three disjoint intervals $(-\infty, -\sqrt{2 + \varepsilon})$, $(-\sqrt{2 + \varepsilon}, \sqrt{2 + \varepsilon})$, and $(\sqrt{2 + \varepsilon}, \infty)$. Check each interval by testing values from each of these intervals, say $-2\sqrt{2 + \varepsilon}, 0,$ and $2\sqrt{2 + \varepsilon}$.

$(-2\sqrt{2+\varepsilon})^2 - 2 \stackrel{?}{<} \varepsilon \qquad (0)^2 - 2 \stackrel{?}{<} \varepsilon \qquad (2\sqrt{2+\varepsilon})^2 - 2 \stackrel{?}{<} \varepsilon$

No.   Yes.   No.

The solution set is $(-\sqrt{2 + \varepsilon}, \sqrt{2 + \varepsilon})$.

65. $x^2 < \varepsilon$
66. $4x^2 < \varepsilon$
67. $x^2 - 1 < \varepsilon$
68. $x^2 - 4 < \varepsilon$

1.5   Equations and Inequalities Involving Absolute Values

The absolute value of a real number x is defined by

$$|x| = \begin{cases} x & \text{if } x \geq 0, \dagger \\ -x & \text{if } x < 0. \end{cases}$$

For example, $|-5| = 5$. Since it is also true that $|5| = 5$, the equation $|x| = 5$ has two solutions, 5 and -5. In general, simple equations involving absolute values are contractions of two equations without absolute values. The same is

† See "Preliminary Concepts."

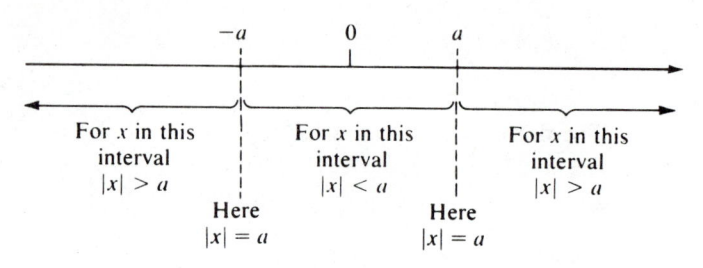

Figure 1.4

true of inequalities. The relationships are indicated in Figure 1.4. To summarize the relationships, note that for $a > 0$,

I. $|x| = a$ is equivalent to $x = a$ or $x = -a$;
II. $|x| < a$ is equivalent to $-a < x$ and $x < a$, i.e., $-a < x < a$;
III. $|x| > a$ is equivalent to $x < -a$ or $x > a$.

To solve equations or inequalities involving absolute value, we can first write two equivalent equations or inequalities without absolute value. We then recall that the everyday word *and* corresponds to set intersection and *or* (the nonexclusive *or*, meaning *one or the other, or both*) corresponds to set union. Our solution set will be the intersection of two sets for type II inequalities, the union of two sets for equations or inequalities of type I or III, respectively.

Example   Solve $|x - 3| = 5$.

Solution   By Statement I above, this equation is equivalent to
$$x - 3 = 5 \quad \text{or} \quad x - 3 = -5.$$
The solution to the first equality is 8; the solution to the second, -2. The solution set is $\{-2, 8\}$.

Example   Solve $|x - 3| < 5$.

Solution   By Statement II above, this inequality is equivalent to
$$-5 < x - 3 \quad \text{and} \quad x - 3 < 5,$$
or
$$-5 < x - 3 < 5,$$
$$-2 < x < 8.$$
The solution set of the given inequality is the interval $(-2, 8)$.

Example   Solve $|x - 3| > 5$.

1.5 Equations and Inequalities Involving Absolute Values

Solution   By Statement III above, this inequality is equivalent to

$$x - 3 < -5 \quad \text{or} \quad x - 3 > 5.$$

The first inequality reduces to $x < -2$, so the solution set of that inequality is the interval $(-\infty, -2)$. The second reduces to $x > 8$ and has solution set $(8, \infty)$. The solution of the given inequality is therefore the union $(-\infty, -2) \cup (8, \infty)$.

Inequalities involving absolute values and $\leq$ or $\geq$ are treated similarly by using the following statements.

IV. $|x| \leq a$ is equivalent to $-a \leq x$ and $x \leq a$; i.e., $-a \leq x \leq a$;
V. $|x| \geq a$ is equivalent to $x \leq -a$ or $x \geq a$.

Alternative method of solution   We can also solve absolute-value inequalities by using a method similar to the method of critical numbers used in Section 1.4.

Example   Solve $|x^2 - 1| < 8$.

Solution   The critical numbers are obtained by solving the equation

$$|x^2 - 1| = 8$$

or, equivalently, the two equations:

$$x^2 - 1 = 8, \qquad x^2 - 1 = -8,$$
$$x^2 = 9, \qquad x^2 = -7.$$
$$x = \pm 3. \qquad \text{No real solutions.}$$

Thus, the critical numbers are the real numbers -3 and 3. These values separate the set of real numbers into three intervals as shown in the figure below.

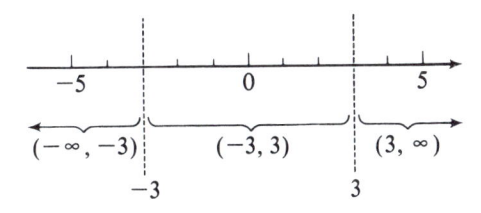

$$(-\infty, -3) \qquad (-3, 3) \qquad (3, \infty)$$

Substituting arbitrary values from each of the three intervals, say -4, 0, and 4, for the variable in the original inequality, we have

$$|(-4)^2 - 1| \overset{?}{<} 8 \qquad |(0)^2 - 1| \overset{?}{<} 8 \qquad |(4)^2 - 1| \overset{?}{<} 8$$

No. Yes. No.

Hence, the solution set is

$$\{x \mid -3 < x < 3\} = (-3, 3).$$

Exercise 1.5

A Solve. Write the solution set of the inequality in interval notation.

Examples a. $|x - 2| \leq 5$ b. $|x + 3| \geq 5$

Solutions a. From Statement IV on page 39: b. From Statement V on page 39:

$$-5 \leq x - 2 \leq 5$$
$$-3 \leq x \leq 7$$

The solution set is $[-3, 7]$.

$$x + 3 \geq 5 \quad \text{or} \quad x + 3 \leq -5$$
$$x \geq 2 \quad \text{or} \quad x \leq -8$$

The solution set is $(-\infty, -8] \cup [2, \infty)$.

1. $|x + 2| = 3$
2. $|4 - x| = 1$
3. $|3 - x| = 4$
4. $|x + 5| = 1$
5. $|2x + 1| = 3$
6. $|3x - 2| = 4$
7. $|x - 2| < 5$
8. $|x + 3| < 2$
9. $\left|x - \dfrac{1}{2}\right| < \dfrac{2}{3}$
10. $\left|x + \dfrac{3}{4}\right| < \dfrac{1}{2}$
11. $|x - 4| > 1$
12. $|x + 2| > 2$
13. $|3 + x| > 7$
14. $|2x + 1| > 3$
15. $|x - 4| \leq 1$
16. $|x + 2| \leq 5$
17. $|2x - 3| \leq 3$
18. $|3x + 4| \leq \dfrac{1}{2}$
19. $\left|x - \dfrac{1}{2}\right| \geq 2$
20. $|x + 1| \geq \dfrac{1}{2}$
21. $|3x - 1| \geq 3$
22. $\left|\dfrac{1}{2}x - \dfrac{1}{4}\right| \geq \dfrac{1}{4}$
23. $|x| \leq 3$
24. $|x| \geq 1$

Rewrite using absolute-value notation.

Examples a. $-1 < x < 1$ b. $-3 < x < 1$

Solutions a. $|x| < 1$ b. $-2 < x + 1 < 2$
$|x + 1| < 2$

25. $-3 \leq x \leq 3$
26. $-4 \leq x \leq 4$
27. $-2 < x < 6$
28. $0 \leq x \leq 8$
29. $-8 \leq x \leq -2$
30. $-10 < x < 0$

Review Exercises

B *Solve. Write the solution set in interval notation.*

31. $|4x^2 - 2| < 2$
32. $|2x^2 - 6| < 4$
33. $|x^2 - 4x| \geq 4$
34. $|x^2 + 5x| \geq 6$
35. $\left|\dfrac{x}{2x + 1}\right| \leq 3$
36. $\left|\dfrac{2x}{x - 1}\right| \leq 4$

Chapter Review

[1.1] *Solve the equation.*

1. $3 + \dfrac{x}{5} = \dfrac{7}{10}$
2. $\dfrac{3x}{4} - \dfrac{5x - 1}{8} = \dfrac{1}{2}$
3. $\dfrac{x}{x + 1} + \dfrac{1}{3} = 2$
4. $1 - \dfrac{y + 1}{y - 1} = \dfrac{3}{y}$

5. Solve $\dfrac{x + y}{5} = \dfrac{x - y}{3}$ for y in terms of x.

6. Solve $\dfrac{x + y}{5} = \dfrac{x - y}{3}$ for x in terms of y.

[1.2] *Solve by factoring.*

7. $(x - 2)(x + 1) = 4$
8. $x(x + 4) = 21$

Solve by extraction of roots.

9. $6x^2 = 30$
10. $(x - 5)^2 = 2$

Solve for x by using the quadratic formula.

11. $x^2 + 5x - 2 = 0$
12. $2x^2 + x - 1 = 0$
13. $8x^2 - 4x - 7 = 0$
14. $x^2 = 4 - 2x$

Solve for x in terms of k by using the quadratic formula.

15. $kx^2 - 3x + 1 = 0$
16. $x^2 + kx - 4 = 0$

[1.3] *Solve by using Theorem 1.4.*

17. $x - 5\sqrt{x} + 6 = 0$
18. $\sqrt{x+1} + \sqrt{x+8} = 7$

Solve by the method of substitution.

19. $y - 3y^{1/2} + 2 = 0$
20. $y^{-2} - y^{-1} - 42 = 0$

[1.4] *Solve the inequality.*

21. $\dfrac{x-3}{5} \geq 7$
22. $2(x+1) < \dfrac{1}{3}x$
23. $x^2 + 3x - 10 < 0$
24. $\dfrac{2}{1-x} > 3$

[1.5] *Solve the equation.*

25. $|3x + 1| = 5$
26. $\left|x - \dfrac{2}{3}\right| = \dfrac{5}{3}$

Solve the inequality and write the solution set in interval notation.

27. $|2x + 1| \leq 1$
28. $|-2x + 3| \geq 4$
29. $|x^2 - 2| < 1$
30. $\left|\dfrac{2}{x+1}\right| < 3$

Write the given inequality as a single inequality involving an absolute-value symbol.

31. $-5 \leq x \leq 3$
32. $-6 < x < 2$

2 Relations and Functions; Graphs

In Chapter 1 we were primarily concerned with the solutions of equations and inequalities involving one variable. In this chapter we shall consider the solutions of equations and inequalities involving two variables. In addition we shall consider the important mathematical concepts of relations and functions.

2.1 Pairings of Real Numbers

When we work with expressions involving one variable, the replacement sets are sets of numbers. When we deal with an expression involving two variables, however, each replacement for the variables must be a pair of numbers, one number replacing each variable.

Ordered pairs When the order for considering the numbers in a number pair is specified, the pair is called an **ordered pair**, and the pair is denoted by a symbol such as (3, 2), (2, 3), (−1, 5), or (0, 3). Each of the two numbers in an ordered pair is called a **component** of the ordered pair; in the ordered pair (x, y), x is called the **first component**, and y is called the **second component**.

R × R The sets of ordered pairs with which we shall be concerned in this chapter are subsets of the set of all ordered pairs of real numbers. This set is called the **Cartesian product** of R and R and is denoted by $R \times R$ (read "R cross R") or R^2. The facts that each member of R^2 corresponds to a point in the geometric plane and that the coordinates of each point in the geometric plane are the components of a member of R^2 are the bases for all plane graphing. As you probably recall from your earlier

43

study of algebra, the correspondence between points in the plane and ordered pairs of real numbers is usually established through a **Cartesian** (or **rectangular**) **coordinate system**, as shown in Figure 2.1. For this correspondence, the first component of an ordered pair is sometimes called the **abscissa** and the second component the **ordinate**.

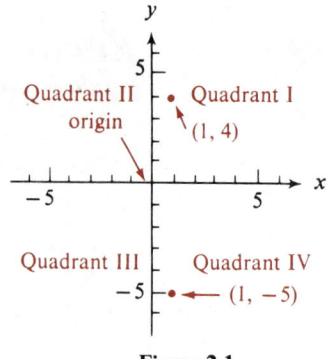

Figure 2.1

Solutions of an equation in two variables

Equations in two variables, such as

$$3x + 2y = 12, \qquad x^2y + 3x = y^5, \qquad \text{and} \qquad xy = y^2 - 5,$$

where x and y represent real numbers, have ordered pairs of numbers as solutions. For example, if the components of $(2,3)$ are substituted for the variables x and y, in that order, in the equation

$$3x + 2y = 12, \tag{1}$$

the result is

$$3(2) + 2(3) = 12,$$

which is true. Hence, $(2, 3)$ is a solution of Equation (1).

Since many equations (and inequalities) in two variables have an infinite number of solutions, we shall sometimes use the set-builder notation

$$\{(x, y) | \text{condition on } x \text{ and } y\}$$

to represent the set of all solutions. Thus the solution set of Equation (1) above can be denoted by

$$\{(x, y) | 3x + 2y = 12, \quad x, y \in R\}$$

[read "the set of all ordered pairs (x, y) such that $3x + 2y = 12$ and x and y are elements in the set of real numbers"].

Relations

In mathematics and its applications we are often concerned with relationships that involve the pairing of numbers. Such relationships as Fahrenheit and Celsius temperatures, standard height and weight tables, and square-root tables are common examples. The following definition gives us a basis for considering such relationships mathematically.

Definition 2.1   A *relation* is a set of ordered pairs.

2.1 Pairings of Real Numbers

It follows from Definition 2.1 that any subset of $R \times R$ is a relation. The set of all first components of the pairs in a relation is called the **domain** of the relation, and the set of all second components is called the **range** of the relation. For example,

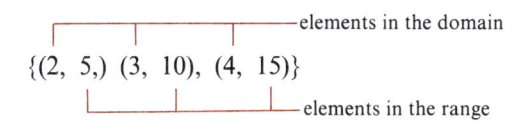

{(2, 5,) (3, 10), (4, 15)}

with first components being elements in the domain and second components being elements in the range, is a relation with domain {2, 3, 4} and range {5, 10, 15}.

Relations determined by rules

Relations are often defined by equations or inequalities. Such an equation or inequality can be thought of as a "rule" for finding the value or values in the range of a relation paired with a given value in the domain. For brevity, we shall simply refer to equations or inequalities such as

$$y = \frac{1}{x-2}, \quad y = \sqrt{x-2}, \quad \text{or} \quad y > 2x + 1$$

as relations.

If a relation is defined by an equation with variables x and y, and the domain (set of values for x) is not specified, we shall understand that *the domain is the set of all real numbers for which a real number exists in the range.*

Examples

Find the domain of each of the following relations.

a. $y = \dfrac{1}{x-2}$ **b.** $y = \sqrt{x-2}$ **c.** $y = \sqrt{4-x^2}$

Solutions

a. Since $1/(x-2)$ is a real number for every value x except 2, the domain of the relation is $\{x \mid x \neq 2\}$.

b. Since $\sqrt{x-2}$ is a real number only when $x - 2 \geq 0$, the domain of the relation is $\{x \mid x \geq 2\}$.

c. Since $\sqrt{4-x^2}$ is a real number only when $4 - x^2 \geq 0$, the domain of the relation is $\{x \mid -2 \leq x \leq 2\}$.

Functions

From Definition 2.1, a relation may be a pairing in which one or more elements in the domain are associated with *more than one element in the range*, as shown in Figure 2.2.

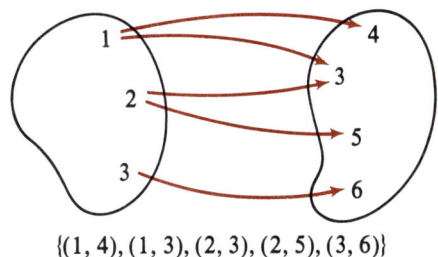

{(1, 4), (1, 3), (2, 3), (2, 5), (3, 6)}

Figure 2.2

{(1, 3), (2, 4), (3, 5)}
a

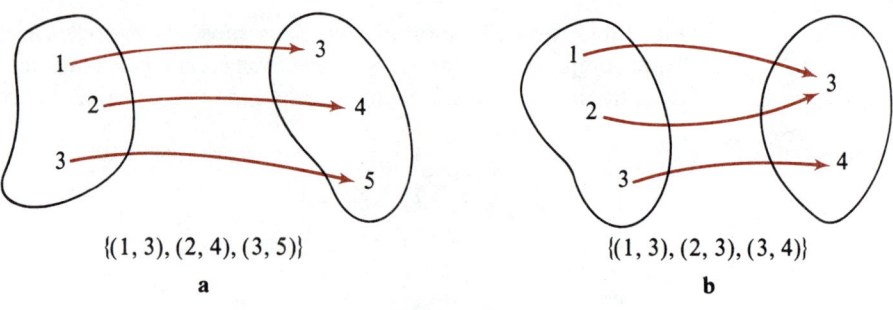

{(1, 3), (2, 3), (3, 4)}
b

Figure 2.3

A relation may also be a pairing in which each element in the domain is associated with *only one element in the range*, as illustrated in Figures 2.3-**a** and 2,3-**b**. Such a relation is called a *function*.

Definition 2.2   *A **function** is a relation in which each element of the domain is paired with one and only one element of the range.*

It follows from Definition 2.2 that no two ordered pairs belonging to a function can have the same first components and different second components. For example, both

$$\{(3, 4), (4, 5), (6, 8)\} \qquad (2)$$

and

$$\{(3, 4), (3, 5), (6, 8)\} \qquad (3)$$

are relations. However, only set (2) is a function. Set (3) is not a function because two elements 4 and 5 in the range are paired with the same element, 3, in the domain.

Graphical characterization of a function

The fact that a function associates each element in its domain with one and only one element in its range implies that no two of the ordered pairs in a function graph into points on the same vertical line. Imagine a vertical line moving across each of the graphs in Figure 2.4 from left to right. If the line at any position meets the graph of the relation in more than one point, then the relation is not a function.

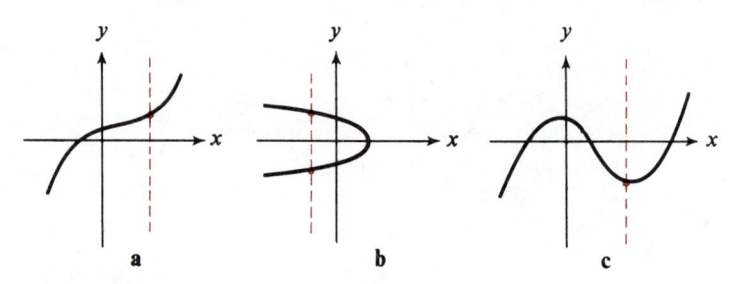

Figure 2.4

2.1 Pairings of Real Numbers

Thus, Figures 2.4-a and 2.4-c show the graphs of relations that are functions, but Figure 2.4-b shows the graph of a relation that is not a function.

Exercise 2.1

A *Supply the missing components so that the following ordered pairs are solutions of the given equations:* $(0, \ ), (1, \ ), (2, \ ), (-3, \ ), (2/3, \ )$.

1. $2x + y = 6$
2. $y = 9 - x^2$
3. $y = \dfrac{3x}{x^2 - 2}$
4. $y = 0$
5. $y = \sqrt{3x + 11}$
6. $y = |x - 1|$

Specify the domain of each relation. State whether or not each relation is a function.

Example $\{(3, 5), (4, 8), (4, 9), (5, 10)\}$

Solution Domain (the set of first components): $\{3, 4, 5\}$. The relation is not a function, because two ordered pairs, $(4, 8)$ and $(4, 9)$, have the same first components.

7. $\{(2, 3), (5, 7), (7, 8)\}$
8. $\{(-1, 6), (0, 2), (3, 3)\}$
9. $\{(2, -1), (3, 4), (3, 6)\}$
10. $\{(-4, 7), (-4, 8), (3, 2)\}$
11. $\{(5, 5), (6, 6), (7, 7)\}$
12. $\{(0, 0), (2, 4), (4, 2)\}$

Specify the domain of the relation.

Examples **a.** $y = \sqrt{16 - x^2}$ **b.** $y = \dfrac{1}{x(x + 2)}$

Solutions
a. For what values of x is $16 - x^2 \geq 0$? The domain is $\{x \mid -4 \leq x \leq 4\}$.
b. For what values of x is $x(x + 2) \neq 0$? The domain is $\{x \mid x \neq 0, -2\}$.

13. $y = x + 7$
14. $y = 2x - 3$
15. $y = x^2$
16. $y = \dfrac{1}{x}$
17. $y = \dfrac{1}{x - 2}$
18. $y = \dfrac{1}{x^2 + 1}$
19. $y = \sqrt{x}$
20. $y = \sqrt{4 - x}$
21. $y = \sqrt{4 - x^2}$
22. $y = \sqrt{x^2 - 9}$
23. $y = \dfrac{4}{x(x - 1)}$
24. $y = \dfrac{x}{(x - 1)(x + 2)}$

State whether or not the given equation defines a function.

Examples a. $x^2y = 3$ b. $x^2 + y^2 = 36$

Solutions Solve explicitly for y.

a. $y = \dfrac{3}{x^2}$

Yes. There is only one value of y associated with each value of x $(x \neq 0)$.

b. $y^2 = 36 - x^2$ or
$y = \pm\sqrt{36 - x^2}$

No. There are two values of y associated with values of x satisfying $|x| < 6$.

25. $x + y = 3$ 26. $y = -x^2$ 27. $y = \sqrt{x^2 - 5}$
28. $y = \sqrt{16 - x^2}$ 29. $x^2 + y^2 = 16$ 30. $y = \pm\sqrt{x^2}$
31. $y^2 = x^3$ 32. $y^4 = x^2$

2.2 Function Notation; Composition

Function notation In general, functions are denoted by single symbols; for example, f, g, h, and F might designate functions. Furthermore, a symbol such as $f(x)$, read "f of x" or "the value of f at x," represents the value in the range of a function f that is associated with the value x in the domain.

Examples If $f(x) = x^2 + 3$, find the member of the range associated with the given member of the domain.

a. 2 b. -3 c. $a + 2$ d. a^2

Solutions In each case the range value is obtained by replacing x with the given domain value.

a. $f(2) = (2)^2 + 3$
$ = 7$

b. $f(-3) = (-3)^2 + 3$
$ = 12$

c. $f(a + 2) = (a + 2)^2 + 3$
$ = a^2 + 4a + 7$

d. $f(a^2) = (a^2)^2 + 3$
$ = a^4 + 3$

Note that $f(x)$ represents a member of the range of a function and hence plays the same role as y in equations involving the variables x and y. For example,

$$y = x^2 + 3 \quad \text{and} \quad f(x) = x^2 + 3$$

define the same function.

2.2 Function Notation; Composition

Composition

In the example above, $f(x) = x^2 + 3$. If we name the expression $a + 2$ in part **c** by $g(a)$, then

$$f(g(a)) = f(a + 2)$$
$$= (a + 2)^2 + 3 = a^2 + 4a + 7.$$

The value $f(g(a))$ is called *the composite of f with g evaluated at a*.

Definition 2.3 *If f and g are functions, the function f composed with g is defined by*

$$(f \circ g)(x) = f(g(x)).$$

Examples If $f(x) = x^2 + 1$ and $g(x) = \dfrac{2}{x}$, find each of the indicated values.

a. $(f \circ g)(1)$ 　　　　　　　　　　**b.** $(g \circ f)(1)$

Solutions

a. Since $g(1) = 2/1 = 2$, we have

$(f \circ g)(1) = f(g(1)),$
$= f(2),$
$= (2)^2 + 1,$
$= 5.$

b. Since $f(1) = 1^2 + 1 = 2$, we have

$(g \circ f)(1) = g(f(1)),$
$= g(2),$
$= \dfrac{2}{(2)} = 1.$

In some applications in the calculus, it is useful to know the difference $f(b) - f(a)$ between two specific range values. We introduce the notation "$f(x)|_a^b$" for that difference. Thus,

$$f(x)|_a^b = f(b) - f(a).$$

Examples For the given function f and values a, and b, find $f(x)|_a^b$.

a. $f(x) = x^2$; $a = 1$, $b = 2$ 　　　**b.** $f(x) = 2x - 3$; $a = 0$, $b = 1$

Solutions

a. $x^2|_1^2 = (2)^2 - (1)^2$
$= 4 - 1$
$= 3$

b. $(2x - 3)|_0^1 = [2(1) - 3] - [2(0) - 3]$
$= -1 - (-3)$
$= 2$

Change in f

In particular, for a given function f we might want to know what change in the range values is produced by a given change h in the domain value. As the domain value changes from a given value a to the value $a + h$, the range value changes from $f(a)$ to $f(a + h)$. The change in the range value (sometimes referred to as "the change in y" or as "the change in f") is

$$f(a + h) - f(a) \quad \text{or} \quad f(x)|_a^{a+h}.$$

Example Find $f(a + h) - f(a)$ for $f(x) = 2x + 1$.

Solution We first compute

$$f(a + h) = 2(a + h) + 1$$
$$= 2a + 2h + 1.$$

We then subtract $f(a)$ to obtain

$$f(a + h) - f(a) = (2a + 2h + 1) - (2a + 1)$$
$$= 2h.$$

Exercise 2.2

A Find the indicated range value for the given function.

Examples a. $f(x) = 3x + 4$, $f(-1)$ b. $g(x) = x^2 - 1$, $g(a - 1)$

Solutions a. Substitute -1 for x.   b. Substitute $a - 1$ for x.
$f(-1) = 3(-1) + 4 = 1$ $g(a - 1) = (a - 1)^2 - 1 = a^2 - 2a$

1. $f(x) = x + 3$, $f(2)$
2. $f(x) = x - 2$, $f(-1)$
3. $g(x) = x^2 + 2x$, $g(-2)$
4. $g(x) = -3x^2 + x + 1$, $g(1)$
5. $s(x) = x^3 - 2x^2 + x - 1$, $s(1)$
6. $s(x) = 3x^3 - 2x^2 - 2x + 4$, $s(-2)$
7. $h(x) = \dfrac{1}{x^2 - 4}$, $h(-1)$
8. $h(x) = \dfrac{2x}{x - 3}$, $h(2)$
9. $h(x) = \sqrt{4 - x^2}$, $h(-1)$
10. $h(x) = \dfrac{1}{\sqrt{x^2 - 1}}$, $h(2)$
11. $f(x) = -2x + 4$, $f(a + 1)$
12. $f(x) = -\dfrac{3}{2}x + 1$, $f(a - 2)$
13. $g(x) = x^2 + 1$, $g(1/a)$
14. $g(x) = -2x^2 + x - 4$, $g(a + 1)$
15. $h(t) = t^4 - 2t^2 + 1$, $h(a - 2)$
16. $h(t) = -3t^4 + 2t^3 + t$, $h(1/a)$

For each pair of functions, find the indicated range values.

Examples $f(x) = 2x + 3$, $g(x) = x + 1$

a. $(f \circ g)(2)$ b. $(g \circ f)(2)$

2.2 Function Notation; Composition

Solutions

a. First compute $g(2) = 2 + 1 = 3$; then compute
$$(f \circ g)(2) = f(g(2)) = f(3),$$
$$= 2(3) + 3 = 9.$$

b. First compute $f(2) = 2(2) + 3 = 7$; then compute
$$(g \circ f)(2) = g(f(2)) = g(7),$$
$$= 7 + 1 = 8.$$

17. $f(x) = x + 4$, $g(x) = x^2$
    a. $(f \circ g)(3)$
    b. $(g \circ f)(3)$

18. $f(x) = x^2 + 1$, $g(x) = 1/x$
    a. $(f \circ g)(1)$
    b. $(g \circ f)(1)$

19. $f(x) = x$, $g(x) = 1/x$
    a. $(f \circ g)(a)$
    b. $(g \circ f)(a)$

20. $f(x) = 1/x$, $g(x) = 1/x^2$
    a. $(f \circ g)(a + 4)$
    b. $(g \circ f)(a + 4)$

Compute $f(x)|_a^b$ for the given function and values of a and b.

Examples

a. $f(x) = x^3$; $a = 0$, $b = 1$
b. $f(x) = \sqrt{x}$; $a = 1$, $b = 4$

Solutions

a. $x^3|_0^1 = (1)^3 - (0)^3 = 1$
b. $\sqrt{x}|_1^4 = \sqrt{4} - \sqrt{1} = 1$

21. $f(x) = x^2 - 4$; $a = -1$, $b = 2$
22. $f(x) = x^4$; $a = -1$, $b = 1$
23. $f(x) = \sqrt{x}$; $a = 4$, $b = 9$
24. $f(x) = \sqrt{x + 1}$; $a = 15$, $b = 24$

Find $f(a + h) - f(a)$ for the given function.

Example

$f(x) = x^2 + 1$

Solution

First compute
$$f(a + h) = (a + h)^2 + 1$$
$$= a^2 + 2ah + h^2 + 1.$$

Next subtract $f(a)$ and simplify to obtain
$$f(a + h) - f(a) = (a^2 + 2ah + h^2 + 1) - (a^2 + 1)$$
$$= 2ah + h^2.$$

25. $f(x) = 2x - 1$
26. $f(x) = 4x - 2$
27. $f(x) = -3x + 2$
28. $f(x) = -5x + 4$
29. $f(x) = 2 - x^2$
30. $f(x) = x^2 + 2x$

B Find and simplify $\dfrac{f(x + h) - f(x)}{h}$, $h \neq 0$, for each function.

31. $f(x) = x^2 - x + 1$
32. $f(x) = x^3 + x - 1$
33. $f(x) = \dfrac{1}{x}$
34. $f(x) = \dfrac{1}{x^2}$
35. $f(x) = \sqrt{x}$
36. $f(x) = \dfrac{1}{\sqrt{x}}$

2.3 Linear Equations and Inequalities in Two Variables

A **first-degree equation**, or **linear equation**, in x and y is an equation that can be written equivalently in the form

$$Ax + By + C = 0 \quad (A \text{ and } B \text{ not both } 0). \tag{1}$$

Graphs of first-degree equations

We shall call (1) the **standard form** for a linear equation. The graph of any such equation (technically, of its solution set) in R^2 is a straight line, although we do not prove this here.

Since two distinct points determine a straight line, it is evident that we need find only two solutions of such an equation to determine its graph. In practice, the two solutions easiest to find are usually those with first and second components, respectively, equal to zero—that is, the solutions $(0, y_1)$ and $(x_1, 0)$.

The numbers x_1 and y_1 are called the **x-** and **y-intercepts** of the graph. As an example, consider the function defined by the equation

$$3x + 4y = 12. \tag{2}$$

If $y = 0$, we have $x = 4$, and the x-intercept is 4. If $x = 0$, then $y = 3$, and the y-intercept is 3. Thus the graph of (2) appears as in Figure 2.5.

If the graph intersects both axes at or near the origin, then either the intercepts do not represent two separate points or the points are too close together to be of much use in drawing the graph. It is then necessary to plot at least one other point at a distance far enough removed from the origin to establish the line with pictorial accuracy.

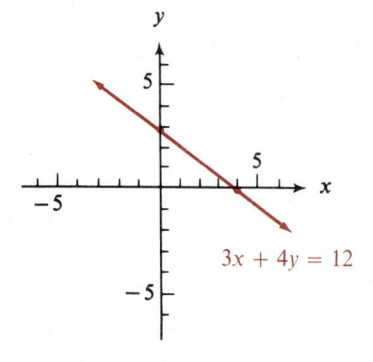

Figure 2.5

2.3 Linear Equations and Inequalities in Two Variables

Horizontal lines

Two special cases of linear equations are worth noting. First, an equation such as

$$y = 4$$

may be considered an equation in two variables,

$$0x + y = 4.$$

For each x, this equation assigns $y = 4$; that is, any ordered pair of the form $(x, 4)$ is a solution of the equation. For instance,

$$(1, 4), \quad (2, 4), \quad (3, 4), \quad \text{and} \quad (4, 4)$$

are all solutions of the equation. If we graph these points and connect them with a straight line, we have the graph shown in Figure 2.6-a.

Since the equation

$$y = 4$$

assigns to each x the same value for y, the function defined by this equation is called a **constant function**.

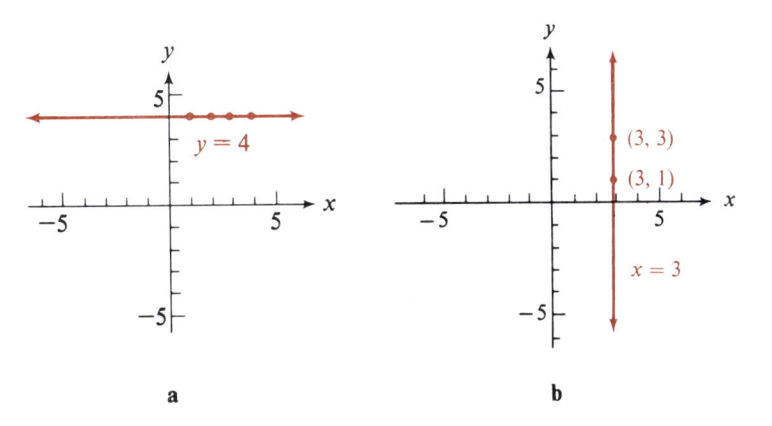

Figure 2.6

Vertical lines

The other special case of the linear equation is of the type

$$x = 3,$$

which may be looked upon as

$$x + 0y = 3.$$

Here, only one value is permissible for x, namely 3, whereas any value may be assigned to y; that is, any ordered pair of the form $(3, y)$ is a solution of this equation. If we choose two solutions, say $(3, 1)$ and $(3, 3)$, we can draw the graph shown in Figure 2.6-b. Note that in this case $B = 0$ and $A \neq 0$ in Equation (1) and this equation does not define a function with x as its first component.

Linear functions

If $B \neq 0$ in Equation (1), then this equation defines a function. Such a function is called a **linear function**.

Any two distinct points in a plane can be looked upon as the endpoints of a line segment. Two fundamental properties of a line segment are its **length** and its **inclination** with respect to the x-axis.

Distance between two points

The Pythagorean theorem can be used to find the length, d, of the line segment joining P_1 and P_2 in Figure 2.7. This theorem asserts that the square on the hypotenuse of any right triangle is equal to the sum of the squares on the legs.

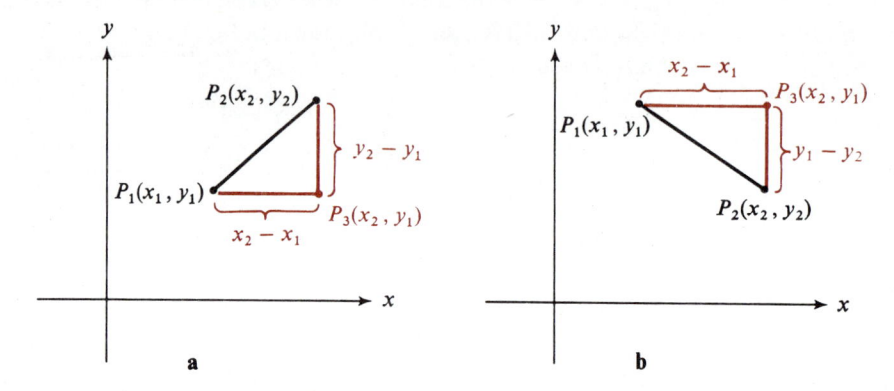

Figure 2.7

Thus, we have

$$d^2 = (x_2 - x_1)^2 + (y_2 - y_1)^2,$$

and, by considering only the positive (or nonnegative) square root of the right-hand member, we obtain the following theorem.

Theorem 2.1   The distance between the points (x_1, y_1) and (x_2, y_2) is

$$d = \sqrt{(x_2 - x_1)^2 + (y_2 - y_1)^2}.$$

Note from Theorem 2.1 that the distance is always positive—or 0 if the points coincide. If the points P_1 and P_2 lie on the same horizontal line (Figure 2.8-a), the distance between them is

$$\sqrt{(x_2 - x_1)^2 + (y_1 - y_1)^2} = \sqrt{(x_2 - x_1)^2}$$
$$= |x_2 - x_1|,$$

and if they lie on the same vertical line (Figure 2.8-b), then the distance is

$$\sqrt{(x_1 - x_1)^2 + (y_2 - y_1)^2} = \sqrt{(y_2 - y_1)^2}$$
$$= |y_2 - y_1|.$$

2.3 Linear Equations and Inequalities in Two Variables

Figure 2.8

Examples Find the distance between the given points.

  a. $P(2, 3)$ and $Q(2, -4)$ **b.** $P(3, 1)$ and $Q(-1, 1)$

Solutions
a. $d = \sqrt{(2 - 2)^2 + (-4 - 3)^2}$
$= \sqrt{(-7)^2} = |-7|$
$= 7$

b. $d = \sqrt{(-1 - 3)^2 + (1 - 1)^2}$
$= \sqrt{(-4)^2} = |-4|$
$= 4$

Notice that, for the points P_1 and P_2 in Figure 2.8-a, $x_2 - x_1$ is positive or negative as P_2 is to the right or left of P_1, respectively, and, for the points P_1 and P_2 in Figure 2.8-b, $y_2 - y_1$ is positive or negative as P_2 is above or below P_1, respectively. It is convenient to call the values $x_2 - x_1$ and $y_2 - y_1$ the **directed distance** from P_1 to P_2 in Figures 2.8-a and 2.8-b, respectively. We denote this distance by $\overrightarrow{P_1 P_2}$.

Examples Find the directed distances from P_1 to P_2.

  a. $P_1(3, 1), P_2(-1, 1)$ **b.** $P_1(2, 3), P_2(2, 4)$

Solutions
a. $\overrightarrow{P_1 P_2} = x_2 - x_1$
$= -1 - 3$
$= -4$

b. $\overrightarrow{P_1 P_2} = y_2 - y_1$
$= 4 - 3$
$= 1$

The second useful property of the line segment joining two points, its inclination, can be measured by comparing the *rise* of the segment with a given *run*, that is, by comparing the directed distances $\overrightarrow{P_1 P_3}$ and $\overrightarrow{P_3 P_2}$ (see Figure 2.9).

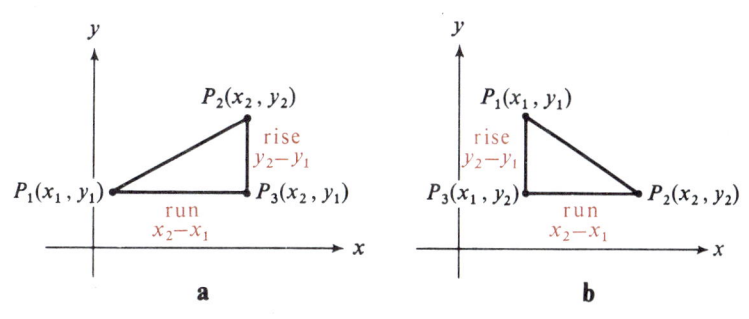

Figure 2.9

Slope of a line segment

The ratio of *rise* to *run* is called the **slope** of the line segment and is designated by the letter m. Since the rise is simply $y_2 - y_1$ and the run is $x_2 - x_1$, the slope of the line segment joining P_1 and P_2 is given by

$$m = \frac{y_2 - y_1}{x_2 - x_1}, \quad x_2 \neq x_1.$$

If P_2 is to the right of P_1, $x_2 - x_1$ will necessarily be positive, and the slope will be positive or negative as $y_2 - y_1$ is positive or negative. Thus positive slope indicates that a line rises to the right; negative slope indicates that it falls to the right. Since

$$\frac{y_2 - y_1}{x_2 - x_1} = \frac{-(y_1 - y_2)}{-(x_1 - x_2)} = \frac{y_1 - y_2}{x_1 - x_2},$$

the restriction that P_2 be to the right of P_1 is not necessary, and the order in which the points are considered is immaterial in determining slope.

Examples

Find the slope of the line segment joining the given points.

a. $P(2, -1)$ and $Q(3, 2)$ **b.** $P(-1, 1)$ and $Q(2, 2)$

Solutions

a. $m = \dfrac{2 - (-1)}{3 - 2}$

$= \dfrac{3}{1} = 3$

b. $m = \dfrac{2 - 1}{2 - (-1)}$

$= \dfrac{1}{3}$

If a line segment is parallel to the x-axis, then $y_2 - y_1 = 0$, and the line has slope 0; but if it is parallel to the y-axis, then $x_2 - x_1 = 0$, and its slope is not defined. These two special cases are shown in Figure 2.10.

a.

$P_1(x_1, y_1)$
$P_2(x_2, y_2)$
$y_2 - y_1 = 0$
$\therefore m = 0$

b.

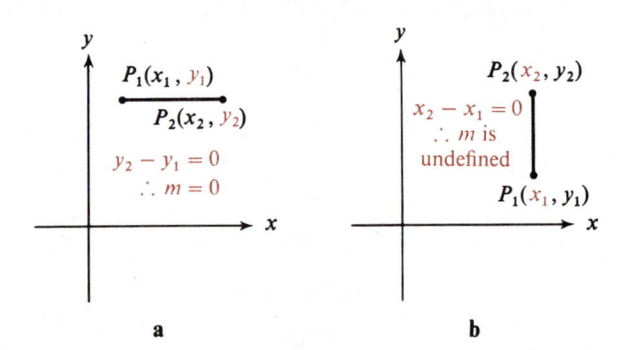

$P_2(x_2, y_2)$
$x_2 - x_1 = 0$
$\therefore m$ is undefined
$P_1(x_1, y_1)$

Figure 2.10

2.3 Linear Equations and Inequalities in Two Variables

Graphing inequalities

Observe that the graph of the equation

$$Ax + By + C = 0 \qquad (3)$$

is a straight line and thus divides the rest of the plane into exactly two regions, one on each side of the line. One of these regions is the graph of the solution set of the inequality

$$Ax + By + C < 0, \qquad (4)$$

and the other region is the graph of the solution set of

$$Ax + By + C > 0. \qquad (5)$$

One way to find the graph of an inequality such as (4) or (5) is to choose a point in one of the regions and determine if its coordinates satisfy the inequality. If the coordinates satisfy the inequality, then the corresponding region is the graph of the solution set of the inequality. If the coordinates of the chosen point do not satisfy the inequality, then the other region is the graph of the solution set of the inequality.

Example Graph $2x + y < 0$.

Solution First sketch the graph of

$$2x + y = 0.$$

Next choose a point that is not on the line, say $(1, 1)$, and check to see if the coordinates of this point satisfy the inequality. Since $2(1) + 1 < 0$ is false, the region that does *not* contain the point $(1, 1)$ is the graph of $2x + y < 0$. The appropriate region is then shaded.

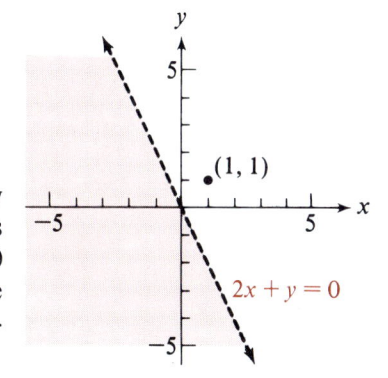

Note that a dashed line is used to indicate that the line $2x + y = 0$ is *not* part of the graph. If the inequality had been of the form $2x + y \leq 0$, then a solid line would have been used.

Exercise 2.3

A *Graph.*

Example $3x + 4y = 24$

Solution Determine the intercepts.

If $x = 0$, then $y = 6$;

if $y = 0$, then $x = 8$.

Sketch the line through $(0, 6)$ and $(8, 0)$, as shown.

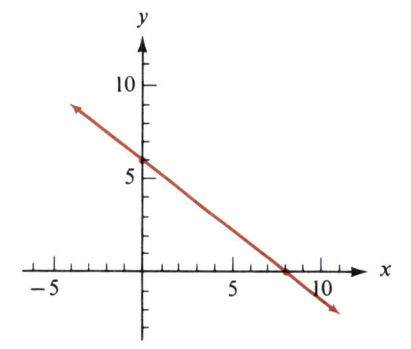

1. $y = 3x + 1$
2. $y = x - 5$
3. $3x - y = -2$
4. $2x + y = 3$
5. $y = -2x$
6. $2y = 3x$
7. $2x + 3y = 6$
8. $3x - 2y = 8$
9. $y = -3$
10. $y = 5$
11. $x = 2$
12. $x = -3$

Example

Graph $f(x) = x - 1$. Represent $f(5)$ and $f(3)$ by drawing line segments from $(5, 0)$ to $(5, f(5))$ and from $(3, 0)$ to $(3, f(3))$.

Solution

$f(5) = 5 - 1 = 4$

The ordinate at $x = 5$ is 4.

$f(3) = 3 - 1 = 2$

The ordinate at $x = 3$ is 2.

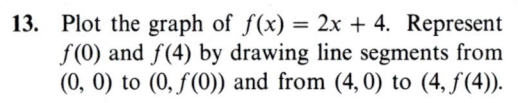

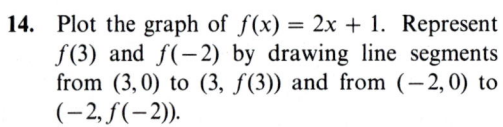

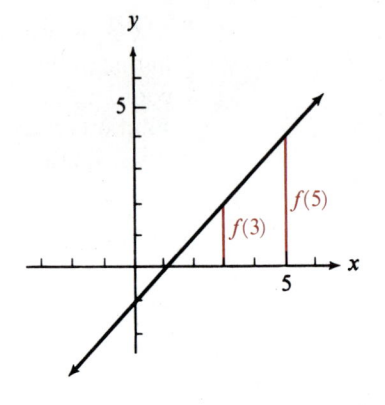

13. Plot the graph of $f(x) = 2x + 4$. Represent $f(0)$ and $f(4)$ by drawing line segments from $(0, 0)$ to $(0, f(0))$ and from $(4, 0)$ to $(4, f(4))$.

14. Plot the graph of $f(x) = 2x + 1$. Represent $f(3)$ and $f(-2)$ by drawing line segments from $(3, 0)$ to $(3, f(3))$ and from $(-2, 0)$ to $(-2, f(-2))$.

Find the distance between each of the given pairs of points, and find the slope of the line segment joining them.

Example

$(3, -5), (2, 4)$

Solution

Consider $(3, -5)$ as P_1 and $(2, 4)$ as P_2.

$$d = \sqrt{(x_2 - x_1)^2 + (y_2 - y_1)^2} \qquad m = \frac{y_2 - y_1}{x_2 - x_1}$$

$$= \sqrt{[2 - 3]^2 + [4 - (-5)]^2} \qquad = \frac{4 - (-5)}{2 - 3}$$

$$= \sqrt{1 + 81} \qquad = \frac{9}{-1}$$

Distance, $\sqrt{82}$; slope, -9.

15. $(1, 1), (4, 5)$
16. $(-1, 1), (5, 9)$
17. $(-3, 2), (2, 14)$
18. $(-4, -3), (1, 9)$
19. $(2, 1), (1, 0)$
20. $(-3, 2), (0, 0)$
21. $(5, 4), (-1, 1)$
22. $(2, -3), (-2, -1)$
23. $(3, 5), (-2, 5)$
24. $(2, 0), (-2, 0)$
25. $(0, 5), (0, -5)$
26. $(-2, -5), (-2, 3)$

2.3 Linear Equations and Inequalities in Two Variables

Find the lengths of the sides of the triangle having vertices as given.

27. (10, 1), (3, 1), (5, 9)
28. (0, 6), (9, −6), (−3, 0)
29. (5, 6), (11, −2), (−10, −2)
30. (−1, 5), (8, −7), (4, 1)

Graph each inequality.

Example $y \geq 4 - 2x$

Solution Graph the equation $y = 4 - 2x$. Observe that the origin is not part of the graph of the inequality, because $(0,0)$ is not a solution of the given inequality; that is, $0 \not\geq 4 - 2(0)$. Hence the region above the graph of the equation $y = 4 - 2x$ is shaded. The line is included in the graph.

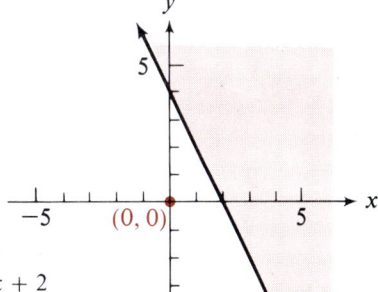

31. $y < x$
32. $y > x$
33. $y \leq x + 2$
34. $y \geq x - 2$
35. $x + y < 5$
36. $2x + y < 2$

B 37. Show that the triangle described in Exercise 28 is a right triangle. *Hint*: Use the converse of the Pythagorean theorem.

38. The two line segments with endpoints at $(0, -7)$, $(8, -5)$ and $(5, 7)$, $(8, -5)$ are perpendicular. Find the slope of each line segment. Compare the slopes. Do the same for the perpendicular line segments with endpoints at $(8, 0)$, $(6, 6)$ and $(-3, 3)$, $(6, 6)$. Can you make a conjecture about the slopes of perpendicular line segments?

39. The graph of a linear function contains the points $(2, -3)$ and $(6, -1)$. Find an equation that defines the function.

40. Determine algebraically whether or not the points lie on the same line.

   a. (2, 7), (−2, −5), (0, 1)
   b. (9, 5), (−3, −1), (0, 1)

Compute $\dfrac{f(x) - f(a)}{x - a}$, $(x \neq a)$ *for the given function. Interpret the result geometrically.*

41. $f(x) = 2x + 4$
42. $f(x) = -\dfrac{1}{3}x + 2$

Graph each inequality. Hint: Consider the graphs in each quadrant separately: $x, y \geq 0$; $x \leq 0$, $y \geq 0$; $x, y \leq 0$; and $x \geq 0$, $y \leq 0$.

43. $|x| + |y| \leq 1$
44. $|x| + |y| \geq 1$
45. $|x| - |y| \leq 1$
46. $|x| - |y| \geq 1$
47. $|x| - |y| \geq 0$
48. $|x| - |y| \leq 0$

2.4 Forms of Linear Equations

In the preceding section we observed that the graph of a relation of the standard form $Ax + By + C = 0$ (A and B not both 0) is a straight line. It is sometimes convenient to use the linear equation in a form other than the standard form. In this section we shall discuss some of the different forms and their uses.

Point-slope form

Assuming that the slope of the line segment joining any two points on a line does not depend on the points (see Exercise 34), consider a line in the plane with given slope m that passes through a given point (x_1, y_1), as shown in Figure 2.11. If we choose any other point on the line and assign to it the coordinates (x, y), it is evident that the slope of the line is given by

$$\frac{y - y_1}{x - x_1} = m,$$

from which

$$y - y_1 = m(x - x_1). \qquad (1)$$

Figure 2.11

Note that (1) is satisfied also by $(x, y) = (x_1, y_1)$. Since now x and y are the coordinates of *any* point on the line, (1) is an equation of the line passing through (x_1, y_1) with slope m. This is called the **point-slope form** for a linear equation. Any linear equation in point-slope form can be written equivalently in standard form.

Example

Find an equation in standard form of the line with slope 3/4 passing through the point $(-4, 1)$.

Solution

Substituting $3/4$, -4, and 1 for m, x_1, and y_1, respectively, in Equation (1), we have

$$y - 1 = \frac{3}{4}(x - (-4)),$$

$$y - 1 = \frac{3}{4}x + 3,$$

from which

$$3x - 4y + 16 = 0.$$

Slope-intercept form

Now consider the equation of the line with slope m passing through a given point on the y-axis having coordinates $(0, b)$ as shown in Figure 2.12. Substituting the components of $(0, b)$ in the point-slope form of a linear equation,

$$y - y_1 = m(x - x_1),$$

we obtain

$$y - b = m(x - 0),$$

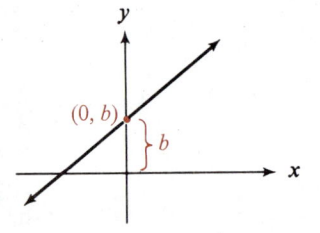

Figure 2.12

2.4 Forms of Linear Equations

from which
$$y = mx + b. \qquad (2)$$

Equation (2) is called the **slope-intercept form** for a linear equation. Any linear equation in standard form can be written equivalently in the slope-intercept form by solving for y in terms of x if $B \neq 0$ in Equation (1) of Section 2.3.

Example Find the slope and y-intercept of the line with equation $2x + 3y - 6 = 0$.

Solution We can write $2x + 3y - 6 = 0$ equivalently as
$$y = -\frac{2}{3}x + 2.$$

We can now obtain the slope of the line, $-2/3$, and the y-intercept, 2, directly from the last equation.

Intercept form If the x- and y-intercepts of the graph of
$$y = mx + b \qquad (3)$$
are a and b ($a, b \neq 0$), respectively, as shown in Figure 2.13, then the slope m is clearly equal to $-b/a$. Replacing m in Equation (3) with $-b/a$, we have
$$y = -\frac{b}{a}x + b,$$
$$ay = -bx + ab,$$
$$bx + ay = ab,$$
and multiplying each member by $1/ab$ produces
$$\frac{x}{a} + \frac{y}{b} = 1. \qquad (4)$$

Equation (4) is called the **intercept form** for a linear equation.

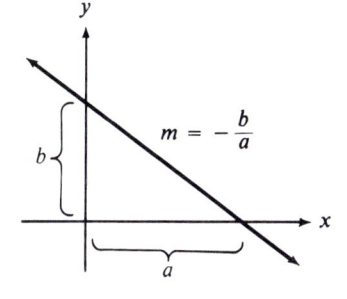

Figure 2.13

Example Find an equation in standard form of the line with x-intercept 3 and y-intercept -2.

Solution We substitute 3 and -2 for a and b, respectively, in $\frac{x}{a} + \frac{y}{b} = 1$, to obtain
$$\frac{x}{3} + \frac{y}{-2} = 1,$$
from which
$$-2x + 3y + 6 = 0 \quad \text{or} \quad 2x - 3y - 6 = 0.$$

Any of the four forms of linear equations discussed in this section may be used in working with linear functions and their graphs.

Exercise 2.4

A Find the equation, in standard form, of the line passing through each of the given points and having the given slope.

Example $(3, -5)$, $m = -2$

Solution Substitute the given values in the point-slope form of the linear equation and rewrite the equation in standard form.

$$y - y_1 = m(x - x_1)$$
$$y - (-5) = -2(x - 3)$$
$$y + 5 = -2x + 6$$
$$2x + y - 1 = 0$$

1. $(2, 1)$, $m = 4$
2. $(-2, 3)$, $m = 5$
3. $(5, 5)$, $m = -1$
4. $(-3, -2)$, $m = 1/2$
5. $(0, 0)$, $m = 3$
6. $(-1, 0)$, $m = 1$
7. $(0, -1)$, $m = -1/2$
8. $(2, -1)$, $m = 3/4$
9. $(-2, -2)$, $m = -3/4$
10. $(2, -3)$, $m = 0$
11. $(-4, 2)$, $m = 0$
12. $(-1, -2)$, parallel to y-axis

Write each equation in slope-intercept form; specify the slope of the line and the y-intercept.

Example $2x - 3y = 5$

Solution Solve explicitly for y.

$$-3y = 5 - 2x$$
$$3y = 2x - 5$$
$$y = \frac{2}{3}x - \frac{5}{3}$$

Compare with the general slope-intercept form $y = mx + b$. Slope, 2/3; y-intercept, $-5/3$.

13. $x + y = 3$
14. $2x + y = -1$
15. $3x + 2y = 1$
16. $3x - y = 7$
17. $x - 3y = 2$
18. $2x - 3y = 0$
19. $4x - 3y = 7$
20. $-2x + 6y = 8$
21. $y = 6$
22. $3y - 4 = 6$
23. $x = 5$
24. $2x - 5 = 0$

2.5 Parallel and Perpendicular Lines

Find the equation, in standard form, of the line with the given intercepts. The x-intercept is given first.

Example 3; −1/2

Solution Substitute 3 and $-\frac{1}{2}$ for a and b, respectively, in the intercept form $\frac{x}{a} + \frac{y}{b} = 1$.

$$\frac{x}{3} + \frac{y}{-1/2} = 1$$

from which

$$\frac{x}{3} - 2y = 1 \quad \text{or} \quad x - 6y - 3 = 0.$$

25. 2; 3   26. 4; −1   27. −2; −5

28. −1; 7   29. $-\frac{1}{2}; \frac{3}{2}$ 30. $\frac{2}{3}; -\frac{3}{4}$

B 31. Show that, for $x_2 \neq x_1$,

$$y - y_1 = \left(\frac{y_2 - y_1}{x_2 - x_1}\right)(x - x_1)$$

is an equation of the line joining the points (x_1, y_1) and (x_2, y_2). This is the **two-point form** of the linear equation.

32. Use the form given in Exercise 31 to find the equation of the line through (2, 1) and (−1, 3).

33. Consider the linear function $y = F(x)$. Assuming that (2, 3) and (−1, 4) are known to be in F, find $F(x)$ in terms of x.

34. Show that the slope of the line segment joining any two points on a line does not depend on the points.

2.5 Parallel and Perpendicular Lines

In Section 2.3 we obtained formulas for the distance between two points in a plane and for the slope of a line (or line segment) joining two points in a plane. In this section we shall consider some additional geometric properties of lines in a plane.

Parallel lines Similar triangles can be used to show that line segments (and lines) not perpendicular to the x-axis are parallel if and only if they have equal slopes. The proof of the following theorem depends on familiar geometric properties and is left as an exercise.

Theorem 2.2 *Two lines with slopes m_1 and m_2 are parallel if and only if $m_1 = m_2$. Two lines perpendicular to the x-axis are parallel to each other.*

Example Find the equation in standard form of the line through $(-1, 2)$ that is parallel to $3x - 2y = 6$.

Solution Make a sketch. The given equation can be written equivalently in slope-intercept form as

$$y = \frac{3}{2}x - 3,$$

and, by inspection, we see that the slope of its graph is $3/2$. Then, using the point-slope form for a linear equation, with $m = 3/2$, $x_1 = -1$, and $y_1 = 2$, we have

$$(y - 2) = \frac{3}{2}(x + 1),$$

which can be written in the standard form

$$3x - 2y + 7 = 0.$$

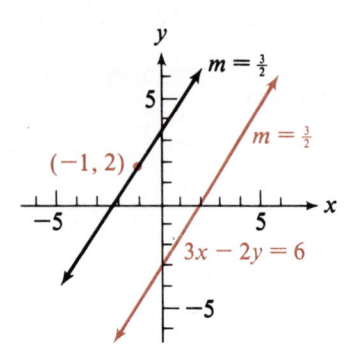

Perpendicular lines The relationship between the slopes of two perpendicular lines is given in the following theorem. The proof of this theorem also follows from familiar geometric properties and is left as an exercise.

Theorem 2.3 *Two lines with slopes m_1 and m_2 are perpendicular if and only if $m_1 \cdot m_2 = -1$. The x- and y-axes or two lines that are parallel to the x- and y-axes, respectively, are also perpendicular to each other.*

Example Find the equation in standard form of the line passing through $(3, -2)$ that is perpendicular to the graph of $2x + 5y = 10$.

Solution Make a sketch. The given equation can be written equivalently in slope-intercept form as

$$y = -\frac{2}{5}x + 2,$$

and, by inspection, the slope of its graph is $-2/5$. Hence, from Theorem 2.3, the slope of any line perpendicular to this graph must be $5/2$. Using the point-slope form for a linear equation with $m = 5/2$, $x_1 = 3$, and $y_1 = -2$, we have

$$[y - (-2)] = \frac{5}{2}(x - 3),$$

which can be written in the standard form

$$5x - 2y - 19 = 0.$$

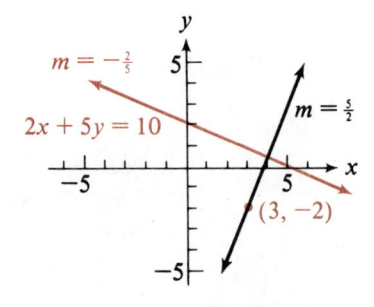

2.5 Parallel and Perpendicular Lines

Midpoint formula

The coordinates of the midpoint of the line segment joining the points (x_1, y_1) and (x_2, y_2) are given in the following theorem. The proof of the theorem is left as an exercise.

Theorem 2.4 *The coordinates of the midpoint of the line segment joining the points (x_1, y_1) and (x_2, y_2) are*

$$x = \frac{x_1 + x_2}{2} \quad \text{and} \quad y = \frac{y_1 + y_2}{2}.$$

Example Find the midpoint of the line segment joining the points $(2, 3)$ and $(-1, -4)$.

Solution From Theorem 2.4, we have

$$x = \frac{2 + (-1)}{2} = \frac{1}{2} \quad \text{and} \quad y = \frac{3 + (-4)}{2} = -\frac{1}{2}.$$

Thus the desired coordinates are $(1/2, -1/2)$.

Exercise 2.5

A *Find the equations in standard form of (a) the line that is parallel and (b) the line that is perpendicular to the graph of the given equation, both of which pass through the given point.*

Example $2x + 3y = 6; \quad (4, -2)$

Solution The given equation can be written equivalently in the slope-intercept form as

$$y = -\frac{2}{3}x + 2.$$

Hence, the slope of the line is $-2/3$. An equation of the line parallel to the given line and containing the point $(4, -2)$ is given by

$$y - (-2) = -\frac{2}{3}(x - 4),$$

which is equivalent to

$$3y + 6 = -2x + 8 \quad \text{or} \quad 2x + 3y - 2 = 0.$$

Any line perpendicular to the given line has slope

$$m = -\frac{1}{-2/3} = \frac{3}{2}.$$

An equation of the line with slope $3/2$ and containing the point $(4, -2)$ is

$$y - (-2) = \frac{3}{2}(x - 4),$$

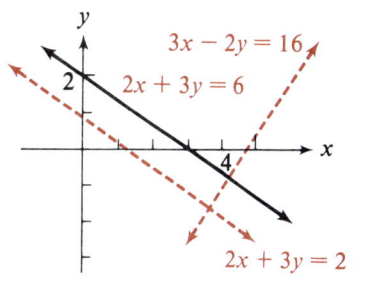

which is equivalent to

$$2y + 4 = 3x - 12 \quad \text{or} \quad 3x - 2y - 16 = 0.$$

Solution continued overleaf

1. $3x + y = 6$; $(5, 1)$
2. $2x - y = -3$; $(-2, 1)$
3. $3x - 2y = 5$; $(0, 0)$
4. $4x - 2y = -5$; $(0, 0)$
5. $4x + 6y = 3$; $(-2, -1)$
6. $5x - 3y = 2$; $(-4, -1)$
7. $2x + 4y = 9$; $(-3, 4)$
8. $3x - 9y = 4$; $(5, -7)$

Find an equation of the line that is the perpendicular bisector of the segment whose endpoints are given.

Example $(7, -4)$ and $(-5, -9)$

Solution The slope of the segment is

$$m = \frac{y_2 - y_1}{x_2 - x_1} = \frac{-9 - (-4)}{-5 - 7} = \frac{-5}{-12} = \frac{5}{12}.$$

From Theorem 2.3, it follows that the slope of the desired bisector is $-12/5$. From Theorem 2.4, the coordinates of the midpoint of the given segment are

$$x = \frac{x_1 + x_2}{2} = \frac{7 + (-5)}{2} = \frac{2}{2} = 1,$$

$$y = \frac{y_1 + y_2}{2} = \frac{-4 + (-9)}{2} = -\frac{13}{2}.$$

Using the point-slope form for a linear equation with $m = -12/5$ and given point $(1, -13/2)$, we have the desired equation

$$y - \left(-\frac{13}{2}\right) = -\frac{12}{5}(x - 1),$$

which simplifies to the standard form

$$24x + 10y + 41 = 0.$$

9. $(5, 3)$ and $(9, 7)$
10. $(9, 1)$ and $(-3, -5)$
11. $(5, -5)$ and $(-9, 1)$
12. $(9, 4)$ and $(-3, 4)$
13. $(a, 0)$ and $(0, b)$
14. (a, b) and $(a + k_1, b + k_2)$

15. Use slopes to show that the triangle with vertices at $A(0, 8)$, $B(6, 2)$, and $C(-4, 4)$ is a right triangle.

16. Use slopes to show that the triangle with vertices $D(2, 5)$, $E(5, 2)$, and $F(10, 7)$ is a right triangle.

17. Use slopes to show that the quadrilateral with vertices at $P(-1, 2)$, $Q(5, 4)$, $R(8, 2)$, and $S(2, 0)$ is a parallelogram.

18. Show that the quadrilateral with vertices at $A(-5, -1)$, $B(0, 0)$, $C(1, -5)$, and $D(-4, -6)$ is a square.

2.5 Parallel and Perpendicular Lines

B 19. Recall from geometry that if two parallel lines are cut by a transversal, then corresponding angles are congruent. Use this fact to show, in the figure at the right, where L_1 is parallel to L_2, that $\Delta P_1 Q_1 P_3$ is similar to $\Delta P_2 Q_2 P_4$ and, hence, that the slopes m_1 and m_2 of the lines are equal.

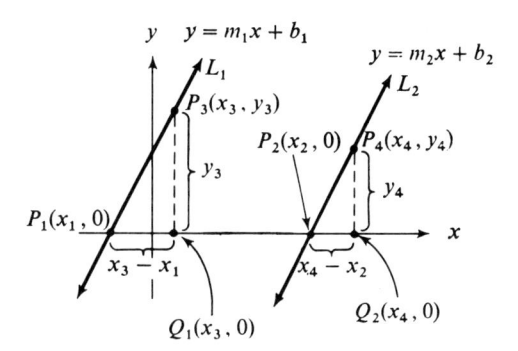

20. In the figure for Exercise 19, assume that $m_1 = m_2$. Deduce that $\Delta P_1 Q_1 P_3$ is similar to $\Delta P_2 Q_2 P_4$ and, hence, that L_1 is parallel to L_2.

21. In the adjoining figure, where L_1 is perpendicular to L_2, use the fact that $\Delta P_1 Q P_2$ is similar to $\Delta P_3 Q P_1$, so that

$$\frac{y_1 - y_2}{x_2 - x_1} = \frac{x_2 - x_1}{y_3 - y_1},$$

to deduce that $m_1 m_2 = -1$.

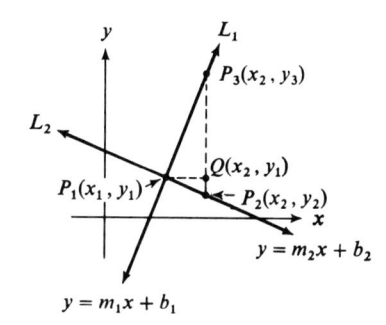

22. Use the figure in Exercise 21 to argue that, because

$$(m_1 - m_2)^2 = m_1^2 - 2m_1 m_2 + m_2^2$$

and because the converse of the Pythagorean theorem is true, the relation $m_1 m_2 = -1$ implies that L_1 is perpendicular to L_2.

23. Use the similar triangles shown in the figure to prove that

$$x = \frac{x_1 + x_2}{2} \quad \text{and} \quad y = \frac{y_1 + y_2}{2}.$$

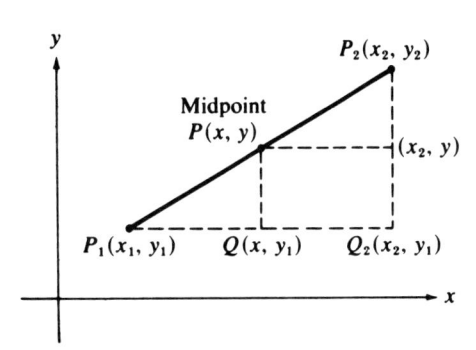

24. Use similar triangles to show that if the point $P(x, y)$ divides the segment from $P_1(x_1, y_1)$ to $P_2(x_2, y_2)$ in such a way that $P_1 P / P P_2 = m_2 / m_1$, then

$$x = \frac{m_1}{m_1 + m_2} x_1 + \frac{m_2}{m_1 + m_2} x_2$$

and

$$y = \frac{m_1}{m_1 + m_2} y_1 + \frac{m_2}{m_1 + m_2} y_2.$$

2.6 Quadratic Equations and Inequalities in Two Variables

In Sections 2.3 and 2.4 we considered linear or first-degree equations and inequalities. In this section we shall consider certain quadratic or second-degree equations and inequalities.

Graph of a quadratic function

Consider the quadratic equation in two variables,

$$y = x^2 - 4. \tag{1}$$

As with linear equations in two variables, solutions of this equation must be ordered pairs (x, y). We need replacements for both x and y in order to obtain a statement we may adjudge to be true or false. As before, such ordered pairs can be found by arbitrarily assigning values to x and computing related values for y. For instance, assigning the value -3 to x in Equation (1), we obtain

$$y = (-3)^2 - 4,$$
$$y = 5,$$

and $(-3, 5)$ is a solution. Similarly, we find that

$$(-2, 0), \quad (-1, -3), \quad (0, -4), \quad (1, -3), \quad (2, 0), \quad \text{and} \quad (3, 5)$$

are also solutions of (1). Locating the corresponding points on the plane, we have the graph in Figure 2.14-**a**. Clearly, these points do not lie on a straight line, and we might reasonably inquire whether the graph of the solution set of

$$y = x^2 - 4$$

forms any kind of a meaningful pattern on the plane. By graphing additional solutions of (1)—solutions with x-components between those already found—we may be able to obtain a clearer picture. Accordingly, we find the solutions

$$\left(-\frac{5}{2}, \frac{9}{4}\right), \left(-\frac{3}{2}, -\frac{7}{4}\right), \left(-\frac{1}{2}, -\frac{15}{4}\right), \left(\frac{1}{2}, -\frac{15}{4}\right), \left(\frac{3}{2}, -\frac{7}{4}\right), \left(\frac{5}{2}, \frac{9}{4}\right),$$

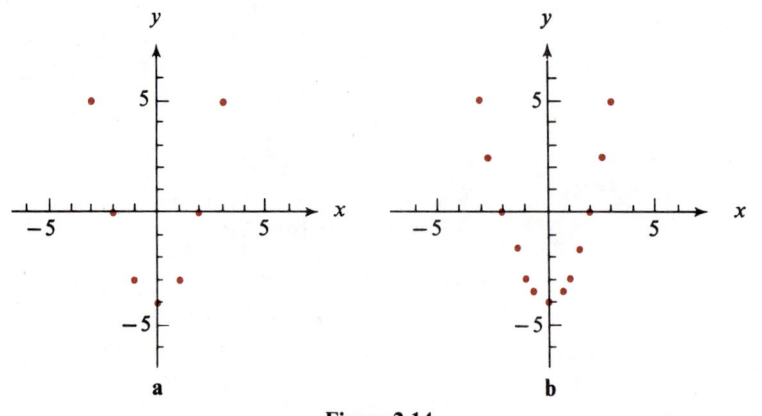

Figure 2.14

2.6 Quadratic Equations and Inequalities in Two Variables

and, by graphing these points in addition to those found earlier, we have the graph in Figure 2.14-b. It now appears reasonable to connect these points in sequence, say from left to right, by a smooth curve as in Figure 2.15, and to assume that the resulting curve is a good approximation to the graph of (1). This curve is an example of a **parabola**.

More generally, the graph of the solution set of any second-degree equation of the form

$$y = ax^2 + bx + c, \qquad (2)$$

where a, b, and c are real and $a \neq 0$, is a parabola. Since for each x an equation of the form (2) will determine only one y, such an equation defines a function, called a **quadratic function**, having as domain the entire set of real numbers and as range some subset of the reals. For example, we observe from the graph in Figure 2.15 that the range of the function defined by (1) is the set of real numbers

$$\{y \mid y \geq -4\}.$$

Figure 2.15

Axis of symmetry

The parabola that is the graph of an equation of the form (2) will have a lowest (minimum) point (will open upward) or a highest (maximum) point (will open downward), depending on whether $a > 0$ or $a < 0$, respectively. Such a point is called the **vertex** of the parabola. The line through the vertex and parallel to the y-axis is called the **axis of symmetry**, or simply the **axis**, of the parabola; it separates the parabola into two parts, each the mirror image of the other. If we observe that the graphs of

$$y = ax^2 + bx + c \qquad (3)$$

and

$$y = ax^2 + bx \qquad (4)$$

have the same axis (Figure 2.16), we can find an equation for the axis of (3) by inspecting Equation (4). Factoring the right-hand member of $y = ax^2 + bx$ yields

$$y = x(ax + b),$$

and thus we can see that 0 and $-b/a$ are the x-intercepts of the graph. Since the axis of symmetry bisects the segment joining the intercepts, an equation for the axis of symmetry is

$$x = -\frac{b}{2a}.$$

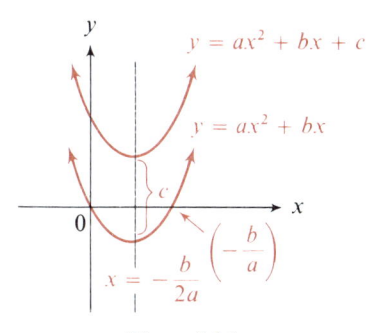

Figure 2.16

Example

Find an equation for the axis of symmetry of the graph of
$$y = 2x^2 - 5x + 7.$$

Solution

By comparing the given equation with $y = ax^2 + bx + c$, we see that $a = 2$ and $b = -5$. Hence, we have
$$x = -\frac{b}{2a} = -\frac{-5}{2(2)} = \frac{5}{4},$$

and $x = 5/4$ is an equation for the axis.

Since the vertex of a parabola lies on its axis, obtaining an equation for this axis will give us the x-coordinate of the vertex. The y-coordinate can then be easily obtained by substitution in the equation for the parabola.

Graphing quadratic functions

When graphing a quadratic function, it is desirable first to select components for ordered pairs that ensure that the more significant parts of the graph are displayed. For a parabola, these parts include the intercepts, if they exist, and the maximum or minimum point on the curve. Thus when graphing a quadratic function we perform the following five steps.

1. Find and plot the intercepts if possible.
2. Note whether the parabola opens upward $(a > 0)$ or downward $(a < 0)$.
3. Find and plot the highest (or lowest) point and the axis of symmetry.
4. If there are too few points obtained in steps 1 and 3 to find an accurate picture of the graph, then plot a few more selected points that will be helpful.
5. Sketch the curve through the points obtained above.

Example

Graph $y = x^2 - 3x - 4$.

Solution

Setting $x = 0$, we find that the y-intercept is -4. Setting $y = 0$, we have
$$0 = x^2 - 3x - 4 = (x - 4)(x + 1),$$

and the x-intercepts are 4 and -1. Since $a > 0$, the curve opens upward and the x-coordinate of the minimum point is
$$-\frac{b}{2a} = -\frac{-3}{2(1)} = \frac{3}{2},$$

and the equation of the axis of symmetry is
$$x = \frac{-b}{2a} = \frac{3}{2}.$$

By substituting $3/2$ for x in $y = x^2 - 3x - 4$, we obtain
$$y = \left(\frac{3}{2}\right)^2 - 3\left(\frac{3}{2}\right) - 4 = \frac{9}{4} - \frac{9}{2} - 4 = -\frac{25}{4}.$$

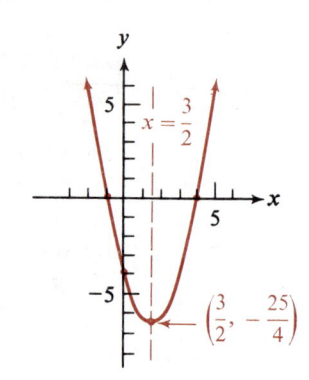

2.6 Quadratic Equations and Inequalities in Two Variables

Graphing the intercepts and the coordinates of the minimum point and then sketching the curve produce the graph shown.

In the above example we plotted four distinct points in steps 1 and 3; thus step 4 was unnecessary. The following example illustrates the use of step 4.

Example   Graph $y = x^2 + x + 3$.

Solution   Setting $x = 0$, we find that the y-intercept is 3, and by using the quadratic formula we find that $x^2 + x + 3 = 0$ has no real solutions; thus, there are no x-intercepts. Since a (equal to 1) is greater than 0, the curve opens upward and the x-coordinate of the minimum point is

$$-\frac{b}{2a} = -\frac{1}{2(1)} = -\frac{1}{2},$$

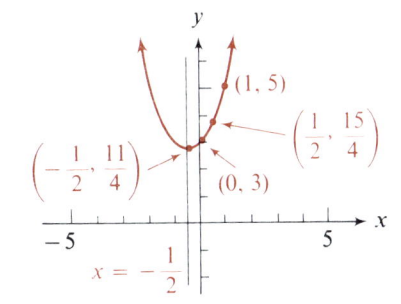

and the equation of the axis of symmetry is

$$x = \frac{-b}{2a} = -\frac{1}{2}.$$

By substituting $-1/2$ for x in $y = x^2 + x + 3$, we obtain $y = 11/4$. Thus the lowest point is at $(-1/2, 11/4)$. To obtain a more accurate picture of the graph, we can plot two more points. It is useful to plot these points on the same side of the axis of symmetry and then use the symmetry to sketch the curve. So we let $x = 1/2$ and find $y = 15/4$, and let $x = 1$ and find $y = 5$. Now we sketch the curve to obtain the graph shown.

Solutions, zeros, and x-intercepts

Consider the graph of the function

$$f(x) = ax^2 + bx + c, \qquad (5)$$

for $a \neq 0$, and the solution set of the equation

$$ax^2 + bx + c = 0. \qquad (6)$$

Any value of x for which $f(x) = 0$ in (5) will be a solution of (6). Since any point on the x-axis has y-coordinate zero [that is, $f(x) = 0$], the x-intercepts of the graph of (5) are the real solutions of (6). Values of x for which $f(x) = 0$ are called the **zeros of the function**. Thus we have three different names for a single idea:

1. The *real solutions* of the equation
$$ax^2 + bx + c = 0.$$

2. The *zeros of the function*
$$f(x) = ax^2 + bx + c.$$

3. The *x-intercepts* of the graph of
$$y = ax^2 + bx + c.$$

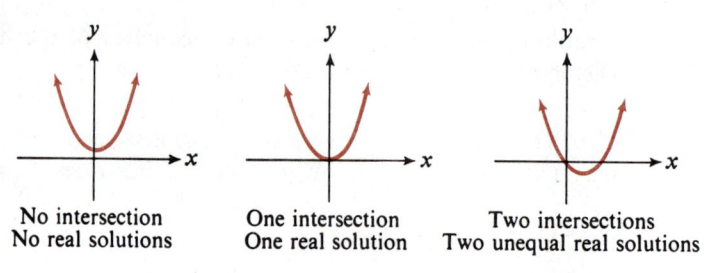

No intersection   One intersection   Two intersections
No real solutions   One real solution   Two unequal real solutions

Figure 2.17

A quadratic equation of the form $ax^2 + bx + c = 0$ may have no real solution, one real solution, or two real solutions. If the equation has no real solution, we find that the graph of the related quadratic equation in two variables does not touch the x-axis; if there is one real solution, the graph is tangent to the x-axis; if there are two real solutions, the graph crosses the x-axis in two distinct points. These cases are illustrated in Figure 2.17.

Quadratic inequalities

Relations of the form

$$y < ax^2 + bx + c \quad \text{and} \quad y > ax^2 + bx + c$$

can be graphed in the same manner in which we graphed relations defined by linear inequalities in two variables in Section 2.3. We first graph the relation defined by the equation having the same members as the defining inequality and then shade an appropriate region as required.

Example

Graph $y < x^2 + 2$.

Solution

We first graph the associated equation $y = x^2 + 2$. The coordinates of a selected point not on the graph of the associated equation can be used to determine which of the two resulting regions of the plane is the graph of the inequality. Upon substitution of 0 for x and 0 for y in $y < x^2 + 2$, we have $(0) < (0)^2 + 2$, a true statement. Hence the region containing the origin is shaded as shown in the figure. Since the graph of the equation is not part of the graph of the inequality, a broken curve is used.

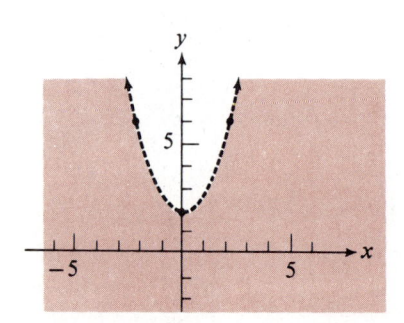

Exercise 2.6

A Graph. (Obtain the intercepts and the maximum or minimum point algebraically, and then sketch the curve.)

Example

$y = -x^2 + 7x - 6$

2.6 Quadratic Equations and Inequalities in Two Variables

Solution

Set $x = 0$ to find that the y-intercept is -6. Since the solutions of

$$-x^2 + 7x - 6 = -(x - 1)(x - 6) = 0$$

are 1 and 6, these are the x-intercepts. The axis of symmetry has equation

$$x = -\frac{b}{2a} = \frac{7}{2}.$$

Set $x = 7/2$ in $y = -x^2 + 7x - 6$ to obtain

$$y = \frac{25}{4}.$$

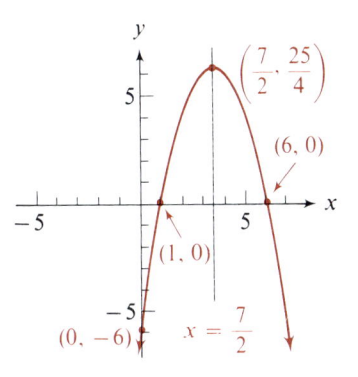

Since the curve opens downward ($a < 0$), the maximum point is $(7/2, 25/4)$. Use these points to sketch the graph as shown.

1. $y = x^2 - 5x + 4$
2. $y = x^2 - 5x - 6$
3. $y = x^2 - 6x - 7$
4. $y = x^2 - 3x + 2$
5. $f(x) = -x^2 + 5x - 4$
6. $f(x) = -x^2 - 8x + 9$
7. $g(x) = \frac{1}{2}x^2 + 2$
8. $g(x) = -\frac{1}{2}x^2 - 2$

9. Graph $f(x) = x^2 + 1$. Represent $f(0)$ and $f(4)$ by drawing line segments from $(0, 0)$ to $(0, f(0))$ and from $(4, 0)$ to $(4, f(4))$.

10. Graph $g(x) = x^2 + 1$. Represent $g(-3)$ and $g(2)$ by drawing line segments from $(-3, 0)$ to $(-3, g(-3))$ and from $(2, 0)$ to $(2, g(2))$.

11. Graph $f(x) = x^2 + 1$. Draw the line segment joining the points $(4, f(4))$ and $(0, f(0))$. Compute $\dfrac{f(4) - f(0)}{4 - 0}$ and interpret this quantity geometrically.

12. Graph $g(x) = x^2 + 1$. Draw the line segment joining the points $(2, g(2))$ and $(-3, g(-3))$. Compute $\dfrac{g(2) - g(-3)}{2 - (-3)}$ and interpret this quantity geometrically.

Example

Graph.

Solution

$y \geq x^2 + 2x$

First graph $y = x^2 + 2x$. Use $(-1, 0)$ as a test point. Since $0 \geq (-1)^2 + 2(-1)$, the point with coordinates $(-1, 0)$ is in the graph. Shade the portion of the plane above the curve, and show the curve as a continuous parabola since the points on it satisfy the inequality.

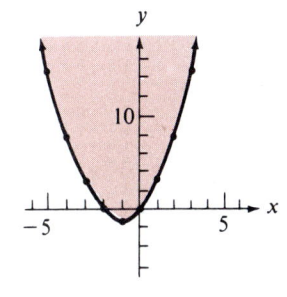

13. $y > x^2$
14. $y < x^2$
15. $y \geq x^2 + 3$
16. $y \leq x^2 + 3$
17. $y < 3x^2 + 2x$
18. $y > 3x^2 + 2x$
19. $y \leq x^2 + 3x + 2$
20. $y \geq x^2 + 3x + 2$

B 21. Find two numbers having sum 8 and product as great as possible.

22. Find the maximum possible area of a rectangle with perimeter 100 inches.

23. On a single set of axes, sketch the family of four curves that are the graphs of
$$y = x^2 + k \quad (k = -2, 0, 2, 4).$$
What effect does varying k have on the graph?

24. On a single set of axes, sketch the family of six curves that are the graphs of
$$y = kx^2 \quad \left(k = \frac{1}{2}, 1, 2, -\frac{1}{2}, -1, -2\right).$$
What effect does varying k have on the graph?

25. On a single set of axes, sketch the family of three curves that are the graphs of
$$y = x^2 + kx \quad (k = -1, 0, 1).$$
What effect does varying k have on the graph?

26. On a single set of axes, sketch the family of three curves that are the graphs of
$$y = -x^2 + kx \quad (k = -1, 0, 1).$$
What effect does varying k have on the graph?

Graph each of the following relations. The graphs are parabolas. Hint: The axis of symmetry of each parabola is parallel to the x-axis.

27. $x = y^2 - 4$
28. $x = y^2 + 4$
29. $x = y^2 - 4y + 4$
30. $x = 2y^2 + 3y - 2$

31. Graph the set of points whose coordinates satisfy both
$$y \leq 4 - x^2 \quad \text{and} \quad y \geq x^2 - 4.$$

32. Graph the set of points whose coordinates satisfy both
$$x \geq y^2 - 4 \quad \text{and} \quad y \geq x^2 - 4.$$

33. For $f(x) = 2x^2 + x + 1$, compute $\dfrac{f(x) - f(a)}{x - a}$ and interpret this quantity geometrically.

34. For $f(x) = -x^2 + 3x + 2$, compute $\dfrac{f(x) - f(a)}{x - a}$ and interpret this quantity geometrically.

2.7 Inverse Functions

In Section 2.1 we considered rules that can be used to find the element in the range of a function that is paired with a given element of the domain. We now consider how to find a rule that gives us the element in the domain of a given function that is

2.7 Inverse Functions

paired with a given element in the range. This rule will determine the *inverse* of the given function.

By Definition 2.2, a function is a set of ordered pairs (x, y) such that no two have the same first components and different second components. When the components of every ordered pair in a function f are interchanged, the resulting relation may or may not be a function. For example, if

$$(1, 5), \quad (2, 5), \quad \text{and} \quad (3, 6)$$

are ordered pairs in f, then

$$(5, 1), \quad (5, 2), \quad \text{and} \quad (6, 3)$$

are members of the relation formed by interchanging the components of these ordered pairs. Clearly these latter pairs cannot be members of a function, since two of them, $(5, 1)$ and $(5, 2)$, have the same first components and different second components. If, however, no two ordered pairs of a function f have the same second component, the function is called **one-to-one** and the function obtained by interchanging the first and second components of every pair in f is also a function. This new function is called the *inverse function* of f. The notation f^{-1} (read "f inverse") is frequently used to denote the inverse function of f.

Definition 2.4   If f denotes a one-to-one function, then the set of ordered pairs obtained by interchanging the first and second components of each ordered pair in f is called the **inverse function of f** and is denoted f^{-1}.

Example   If $f = \{(2, 1), (3, 2), (4, 3)\}$, find f^{-1}.

Solution   First note that f is a one-to-one function. Interchange the components of each ordered pair of f:

$$f^{-1} = \{(1, 2), (2, 3), (3, 4)\}.$$

It is evident from Definition 2.4 and the example above that the domain and range of f^{-1} are just the range and domain, respectively, of f. Thus, if $y = f(x)$ defines a function f, and if f is one-to-one, then the inverse function f^{-1} is obtained as follows:

1. Interchange x and y in $y = f(x)$ to write $x = f(y)$.
2. Solve $x = f(y)$ for y to obtain $y = f^{-1}(x)$.

For example, the inverse of the function defined by

$$y = 4x - 3 \tag{1}$$

is defined by

$$x = 4y - 3, \tag{2}$$

where the variables have been interchanged. When y is expressed in terms of x, we have

$$y = \frac{1}{4}(x + 3). \qquad (2')$$

Equations (2) and (2') are equivalent. Each of these equations determines the inverse of the function defined by Equation (1).

Graphs of inverse functions

The graphs of inverse functions are related in an interesting way. To see this, we first observe in Figure 2.18 that the graphs of the ordered pair, (a, b) and (b, a) are always located symmetrically with respect to the graph of $y = x$. Therefore, because for every ordered pair (a, b) in f the ordered pair (b, a) is in f^{-1}, the graphs of $y = f^{-1}(x)$ and $y = f(x)$ are reflections of each other about the graph of $y = x$ (see Exercise 33).

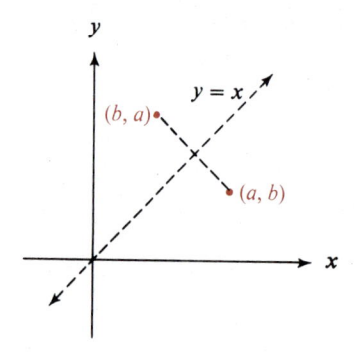

Figure 2.18

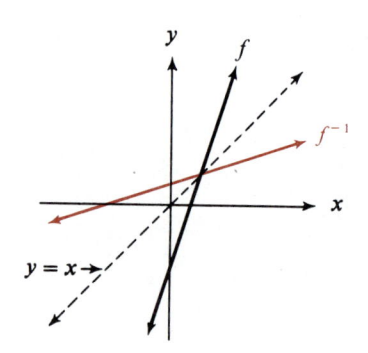

Figure 2.19

Figure 2.19 shows the graphs of the linear function

$$y = f(x) = 4x - 3 \qquad (3)$$

and its inverse

$$y = f^{-1}(x) = \frac{1}{4}(x + 3),$$

together with the graph of $y = x$.

Restricting domains

Note the graphs of

$$y = x^2$$

and

$$x = y^2 \quad \text{or} \quad y = \pm\sqrt{x},$$

2.7 Special Functions

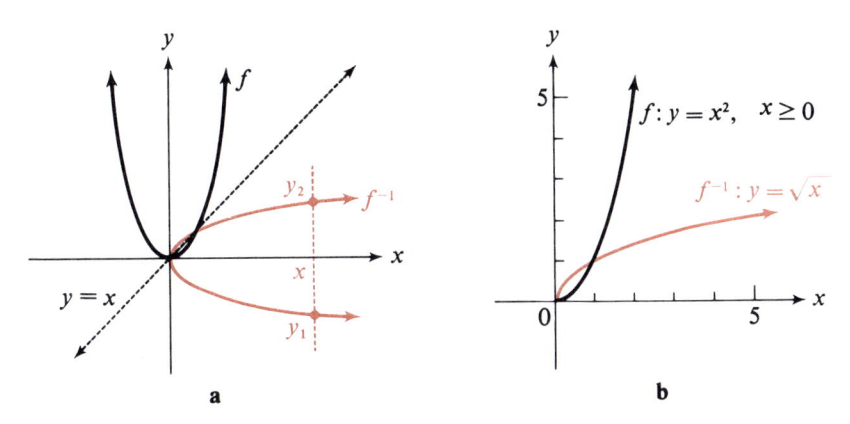

Figure 2.20

shown in Figure 2.20-a. We observe in the latter that two different y's are associated with each x for all but one value in its domain. The equation $x = y^2$ or $y = \pm\sqrt{x}$ does not define a function. However, if the domain of f is restricted so that $x \geq 0$, the function has an inverse f^{-1} with range $y \geq 0$, as shown in Figure 2.20-b.

$f^{-1}[f(x)] = x$
and
$f[f^{-1}(x)] = x$

Since an element in the domain of the inverse f^{-1} of a function f is the range element in the corresponding ordered pair of f and vice versa, it follows that for every x in the domain of f,

$$f^{-1}(f(x)) = x$$

(read "f inverse of f of x is equal to x"); and for every x in the domain of f^{-1},

$$f(f^{-1}(x)) = x$$

(read "f of f inverse of x is equal to x"). Using Equation (3) above as an example, we note that if f is the linear function defined by

$$f(x) = 4x - 3,$$

then f is a one-to-one function. Now f^{-1} is defined by

$$f^{-1}(x) = \frac{1}{4}(x + 3),$$

and we see that

$$f^{-1}(f(x)) = \frac{1}{4}[(4x - 3) + 3] = x$$

and

$$f(f^{-1}(x)) = 4\left[\frac{1}{4}(x + 3)\right] - 3 = x.$$

Exercise 2.7

A Graph each given function f and its inverse function f^{-1}, using the same set of axes.

Example $f: \{(4, 1), (-2, 3), (1, 2)\}$

Solution   Interchange the components of each ordered pair in f to obtain

$$f^{-1}: \{(1, 4), (3, -2), (2, 1)\}.$$

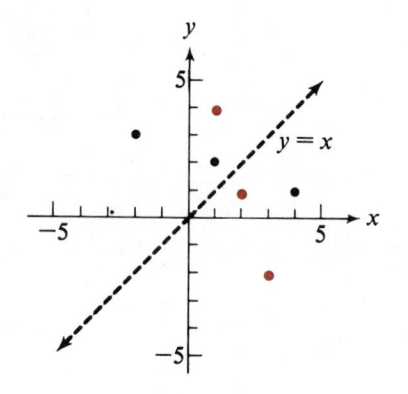

1. $f: \{(-2, 3), (4, 7), (5, 9)\}$
2. $f: \{(-3, 1), (2, -1), (3, 4)\}$
3. $f: \{(-1, -1), (2, 2), (3, 3)\}$
4. $f: \{(-4, 4), (0, 0), (4, -4)\}$

Example $f: x + 3y = 6$

Solution   Interchange the variables in the defining equation to obtain f^{-1}. Graph f and f^{-1}.

$$f^{-1}: y + 3x = 6$$

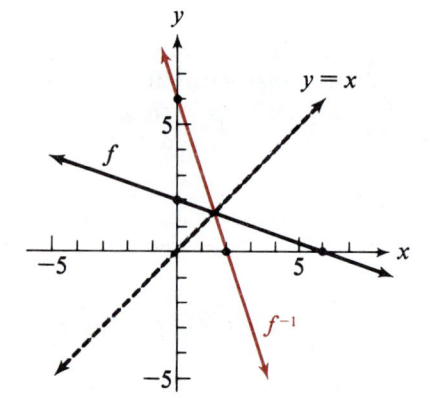

5. $f: y = 2x + 6$
6. $f: y = 3x - 6$
7. $f: y = 4 - 2x$
8. $f: y = 6 + 3x$
9. $f: 3x - 4y = 12$
10. $f: x - 6y = 6$
11. $f: 4x + y = 4$
12. $f: 2x - 3y = 12$
13. $f: y = x^2 + 1, \quad x \geq 0$
14. $f: y = x^2 + 1, \quad x \leq 0$
15. $f: y = x^2 - 1, \quad x \geq 0$
16. $f: y = x^2 - 1, \quad x \leq 0$
17. $f: y = x^2 - 4, \quad x \geq 0$
18. $f: y = x^2 + 4, \quad x \leq 0$
19. $f: y = x^2 - 4x + 4, \quad x \geq 2$
20. $f: y = x^2 - 2x - 3, \quad x \leq 1$

B In Exercises 21–28, each equation defines a one-to-one function f. Find an equation defining f^{-1} and show that $f[f^{-1}(x)] = f^{-1}[f(x)] = x$. *Hint:* Solve explicitly for y and let $y = f(x)$.

21. $y = x$ 22. $y = -x$ 23. $2x + y = 4$ 24. $x - 2y = 4$
25. $3x - 4y = 12$ 26. $3x + 4y = 12$ 27. $yx = 1$ 28. $x^3 - y = 1$
29. If $f(2) = 4$, find $f^{-1}(4)$.
30. If $f^{-1}(0) = 1$, find $f(1)$.
31. If $f(3) = -1$, find $f^{-1}(-1)$.
32. If $f^{-1}(-2) = -3$, find $f(-3)$.
33. Show that the line $y = x$ is the perpendicular bisector of the line segment between the points (a, b) and (b, a) for every value of a and b. Notice that this proves that the graphs of $y = f(x)$ and $y = f^{-1}(x)$ are reflections of one another in the line $y = x$.
34. Show that the graph of the inverse of a linear function is a straight line.

2.8 Special Functions

Functions defined piecewise

It is possible to define a function using different rules for different parts of the domain. To determine which rule to use to find the range value paired with a given domain value, we first determine the part of the domain in which the given domain value lies.

Examples

Let $f(x) = \begin{cases} x^2 & \text{if } x < 0 \\ -x + 1 & \text{if } x \geq 0 \end{cases}$. Find the range value.

a. $f(-1)$ **b.** $f(0)$ **c.** $f(1)$

Solutions

a. Since $-1 < 0$,
$f(-1) = (-1)^2$
$= 1.$

b. Since $0 \geq 0$,
$f(0) = -0 + 1$
$= 1.$

c. Since $1 \geq 0$,
$f(1) = -1 + 1$
$= 0.$

The graph of a function that is defined piecewise can be obtained by graphing it separately over each subset of the domain determined by the definition of the function.

Example

Graph $f(x) = \begin{cases} x & \text{if } x \leq 1 \\ -x + 2 & \text{if } x > 1 \end{cases}$.

Solution

We first graph $y = f(x) = x$ for $x \leq 1$ to obtain Figure **a**. We then graph $y = f(x) = -x + 2$ over $x > 1$ to obtain Figure **b**. We then sketch both graphs on the same set of coordinate axes to obtain Figure **c**, which shows the completed graph of $y = f(x)$. The figures are shown on page 80.

Solution continued overleaf

a

$y = x, \quad x \le 1$

b

$y = -x + 2, \quad x > 1$

c

$$y = \begin{cases} x & \text{if } x \le 1 \\ -x + 2 & \text{if } x > 1 \end{cases}$$

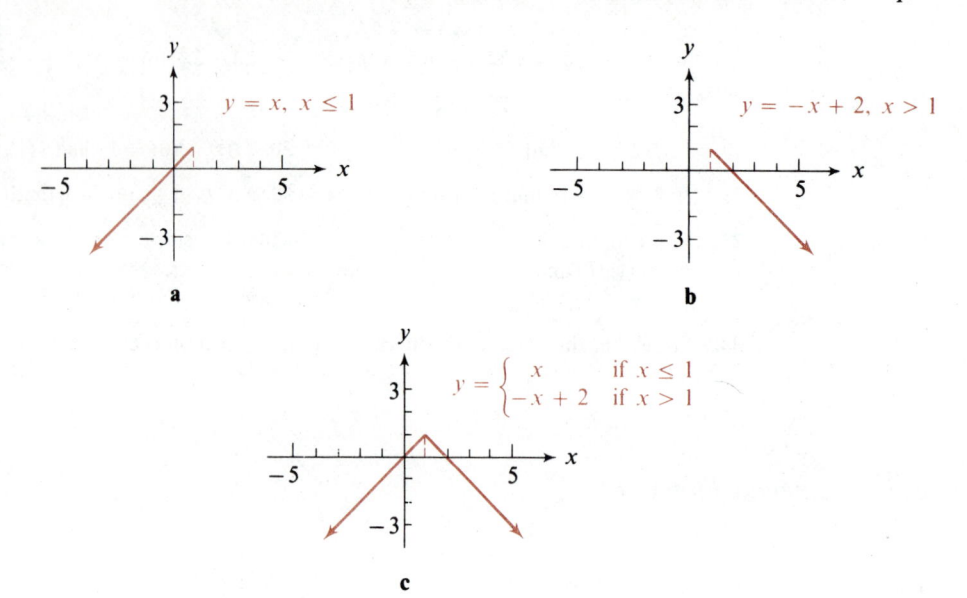

Absolute-value functions

Functions involving the absolute value of the variable provide some examples of piecewise-defined functions. Consider the function defined by

$$y = |x|. \tag{1}$$

From the definition of $|x|$, we have

$$y = x, \quad \text{for} \quad x \ge 0, \tag{2}$$

and

$$y = -x, \quad \text{for} \quad x < 0. \tag{3}$$

If we graph (2) and (3) on the same set of axes, we have the graph of $y = |x|$, shown in Figure 2.21.

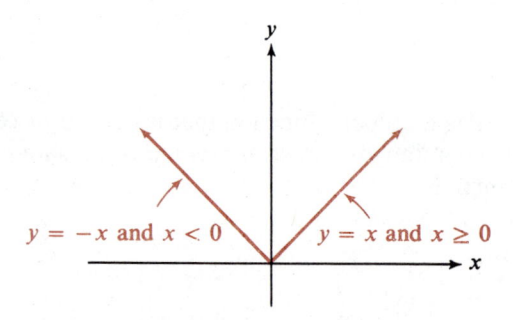

$y = -x$ and $x < 0$ $y = x$ and $x \ge 0$

Figure 2.21

We can graph any equation involving $|x|$ or $|f(x)|$ by first writing the equation in piecewise-defined form and then using the method described above.

Example   Graph $y = |x| + 1$.

2.8 Special Functions

Solution The definition of absolute value implies that $y = |x| + 1$ is equivalent to

$$y = \begin{cases} x + 1 & \text{if } x \geq 0 \\ -x + 1 & \text{if } x < 0. \end{cases}$$

Thus we graph $y = x + 1$, $x \geq 0$ and $y = -x + 1$, $x \leq 0$ on the same set of coordinate axes to obtain the figure.

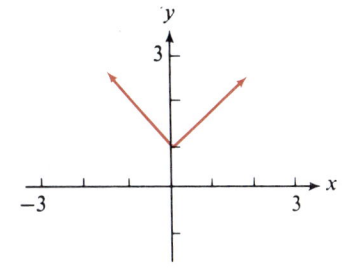

Reflection It is sometimes convenient to use an alternative method to graph functions involving the absolute value of a variable. Observe that if $y < 0$, then the x-axis is the perpendicular bisector of the line segment joining the points (x, y) and $(x, |y|)$. Intuitively, this means that the points (x, y) and $(x, |y|)$ are reflections of one another in the x-axis (see Figure 2.22-**a**). Therefore, we can obtain the graph of $y = |f(x)|$ by graphing $y = f(x)$ and then reflecting the portion of the graph with $y < 0$ in the x-axis (see Figure 2.22-**b**).

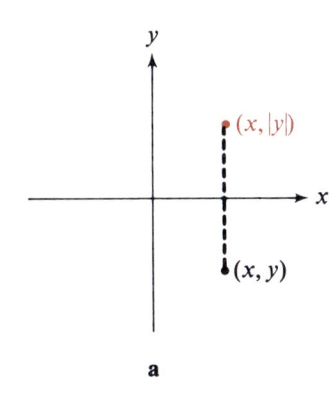

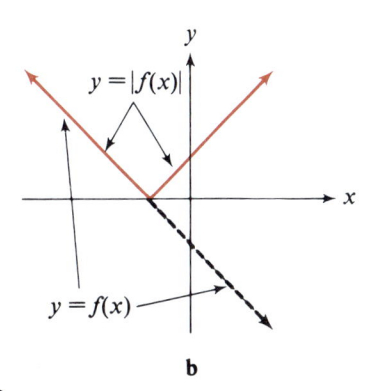

Figure 2.22

Example Graph $y = |2x + 1|$.

Solution We first sketch $y = 2x + 1$ as shown in Figure **a**. Next we reflect in the x-axis that portion of the graph below the x-axis to obtain the graph of $y = |2x + 1|$, which is shown in Figure **b**.

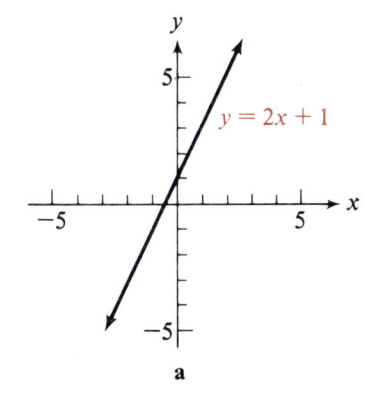

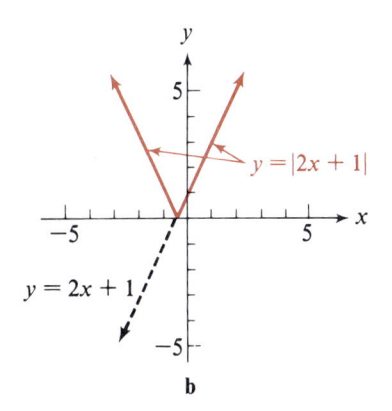

Vertical shifts

Equations of the form

$$y = |f(x)| + c \qquad (4)$$

can be graphed by observing that for each x the ordinate in (4) is c units different from the ordinate of $y = |f(x)|$. Thus the graph of (4) is simply the graph of $y = |f(x)|$ with each ordinate increased by c units if $c > 0$ or decreased by $|c|$ units if $c < 0$.

Example

Graph $y = |x| - 2$.

Solution

We first sketch $y = |x|$ as shown in Figure **a**. We then shift the graph 2 units down, as shown in Figure **b**, to obtain the graph of $y = |x| - 2$.

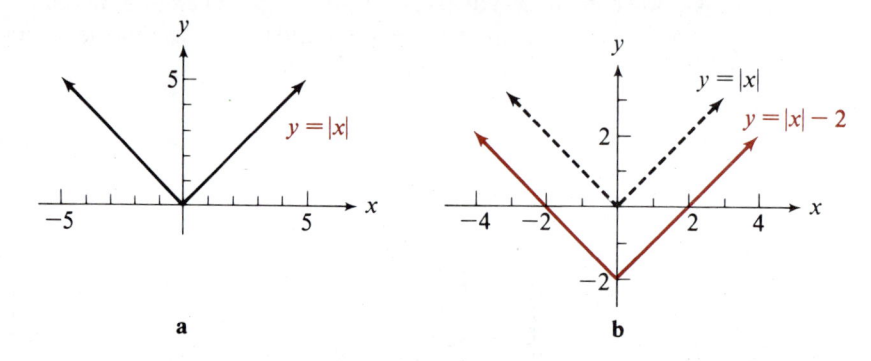

Greatest integer function

Another interesting function (sometimes called the **greatest integer function**) is defined by the equation

$$f(x) = j \quad \text{if} \quad j \leq x < j + 1, \quad j \in J. \qquad (5)$$

Note that for a given value x, $f(x)$ is the greatest integer less than or equal to x.

The greatest integer function is sometimes denoted by

$$f(x) = [x],$$

and hence this function is sometimes called the *bracket* function. Thus,

$$[-2] = -2, \quad \left[-\frac{1}{2}\right] = -1, \quad \left[\frac{1}{2}\right] = 0$$

$$\left[\frac{7}{4}\right] = 1, \quad \text{and} \quad [2] = 2.$$

To graph (5), we consider unit intervals along the x-axis. If $0 \leq x < 1$, $[x]$ is 0, since the greatest integer contained in any number between 0 and 1 is 0. Similarly, if $1 \leq x < 2$, $[x]$ is 1; for $-2 \leq x < -1$, $[x]$ is -2; etc., as shown in Table 2.1. The graph of (5) is therefore as shown in Figure 2.23. The heavy dot on the left-hand endpoint of each line segment indicates that the endpoint is a part of the graph. The function defined by (5) is sometimes called a *step* function, for an obvious reason.

2.8 Special Functions

Table 2.1

x	$[x]$
$[-2, -1)$	-2
$[-1, 0)$	-1
$[0, 1)$	0
$[1, 2)$	1
$[2, 3)$	2

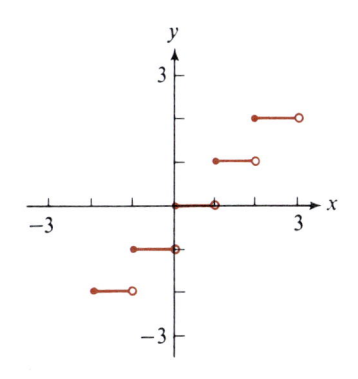

Figure 2.23

Exercise 2.8

A *Find the indicated range value for the given function.*

Example $f(x) = \begin{cases} 2x & \text{if } x < 1 \\ x - 1 & \text{if } x \geq 1 \end{cases}$; $f(3)$

Solution Since $3 \geq 1$ and for any $x \geq 1$, $f(x) = x - 1$,

$$f(3) = 3 - 1 = 2.$$

1. $f(x) = \begin{cases} x - 2 & \text{if } x \geq 2 \\ 2 - x & \text{if } x < 2 \end{cases}$; $f(3)$

2. $f(x) = \begin{cases} -3x - 1 & \text{if } x < -1 \\ 2x + 1 & \text{if } x \geq -1 \end{cases}$; $f(1)$

3. $f(x) = \begin{cases} x & \text{if } x < 0 \\ 2x + 1 & \text{if } 0 \leq x < 2 \\ 3x + 2 & \text{if } x \geq 2 \end{cases}$; $f(2)$

4. $f(x) = \begin{cases} -2x & \text{if } x \leq -1 \\ x & \text{if } -1 < x < 1 \\ 2x & \text{if } x \geq 1 \end{cases}$; $f(-1)$

Graph each function. For Exercises 19–22 and 35–38, graph the function over the interval $-5 \leq x \leq 5$.

5. $f(x) = \begin{cases} -x & \text{if } x \leq -1 \\ 3x + 4 & \text{if } x > -1 \end{cases}$

6. $f(x) = \begin{cases} -2x + 1 & \text{if } x \leq 2 \\ -3 & \text{if } x > 2 \end{cases}$

7. $f(x) = \begin{cases} x^2 - 2 & \text{if } x \le 1 \\ -x^2 & \text{if } x > 1 \end{cases}$

8. $f(x) = \begin{cases} -x^2 + 1 & \text{if } x \le 0 \\ x^2 & \text{if } x > 0 \end{cases}$

9. $f(x) = \begin{cases} x^2 & \text{if } x < 0 \\ x^2 + 1 & \text{if } x \ge 0 \end{cases}$

10. $f(x) = \begin{cases} -x^2 & \text{if } x < -1 \\ x^2 + 1 & \text{if } x \ge -1 \end{cases}$

11. $y = |x| + 2$

12. $y = -|x| + 3$

13. $f(x) = |x + 1|$

14. $f(x) = |x - 2|$

15. $y = -|2x - 1|$

16. $y = |3x + 2|$

17. $g(x) = |2x| - 3$

18. $y = |3x| + 2$

19. $y = 2[x]$

20. $y = [2x]$

21. $y = [x + 1]$

22. $f(x) = [x] + 1$

B 23. $f(x) = \begin{cases} -1 & \text{if } x < -1 \\ 1 & \text{if } -1 \le x \le 1 \\ \dfrac{1}{2} & \text{if } 1 < x \end{cases}$

24. $f(x) = \begin{cases} x + 1 & \text{if } x \le 0 \\ 1 - x & \text{if } 0 < x \le 1 \\ x - 1 & \text{if } 1 < x \end{cases}$

25. $f(x) = \begin{cases} -x^2 & \text{if } x < -1 \\ -1 & \text{if } -1 \le x < 1 \\ x^2 - 2 & \text{if } 1 \le x \end{cases}$

26. $f(x) = \begin{cases} x & \text{if } x < 0 \\ x^2 & \text{if } 0 \le x < 2 \\ -x + 6 & \text{if } 2 \le x \end{cases}$

27. $f(x) = |2x| + |x|$

28. $f(x) = |3x| - |x|$

29. $y = -2|x| + x$

30. $y = 3|x| - x$

31. $y = |x|^2$

32. $y = |x^2|$

33. $g(x) = |x + 1| - x$

34. $g(x) = x + 1 - |x|$

35. $y = [x] + x$

36. $y = \left[\dfrac{1}{2}x\right] + x$

37. $y = [x] - x$

38. $y = |[x]|$

39. The postage on a letter sent by first-class mail is c cents per ounce or fraction thereof. Write an equation relating the cost (C) of mailing a letter and the weight of the letter in ounces (x).

40. A travel agency is offering a charter flight from Los Angeles to Paris. The round trip fare is $1000 per person if more than 9 but fewer than 26 take the charter. The fare per person decreases $2.50 for every person after the 25th and up to the 75th, and it decreases $3.00 for every person after the 75th. If the capacity of the plane is 150 persons, write the fare per person as a function of the number of persons taking the charter flight.

Chapter Review

[2.1] *Specify the domain of each relation.*

1. $\{(4, -1), (2, -4), (3, -5)\}$

2. $\{(1, 2), (1, 3), (1, 6)\}$

3. $y = \dfrac{1}{x + 4}$

4. $y = \sqrt{x - 6}$

[2.2] Let $f(x) = x - 3$ and $g(x) = x^2 + 4$. Find each of the following.

5. $g(3)$ **6.** $f(-2)$ **7.** $f(a - h)$ **8.** $g(a + h)$

Evaluate $(f \circ g)(x)$ and $(g \circ f)(x)$ for each of the following.

9. $f(x) = x^2 - 2x$
$g(x) = \sqrt{x}$

10. $f(x) = 7x + 1$
$g(x) = \dfrac{1}{7}(x - 1)$

For each function, find $f(a + h) - f(a)$.

11. $f(x) = 2x + 3$ **12.** $f(x) = x^2 + 2$

Evaluate $f(x)|_2^4$ for each of the following.

13. $f(x) = x^2 + 2$ **14.** $f(x) = \dfrac{1}{\sqrt{x - 1}}$

[2.3] Graph each equation.

15. $3x - 4y = 6$ **16.** $x + 2y = \dfrac{1}{2}$ **17.** $4x - y = 1$ **18.** $3x - 2y = 4$

Find the distance between each of the given pairs of points, and find the slope of the line segment joining them.

19. $(2, 0)$ and $(-3, 4)$ **20.** $(-6, 1)$ and $(-8, 2)$

Graph each inequality.

21. $2x - y < 6$ **22.** $2y + x > 0$
23. $-2 \leq y < 3$ **24.** $3 < x \leq 5$

[2.4] Find the equation in standard form of the line passing through each of the given points and having the given slope.

25. $(2, -7)$, $m = 4$ **26.** $(-6, 3)$, $m = \dfrac{1}{2}$

Write each equation in slope-intercept form, and specify the slope and the y-intercept of the line.

27. $4x + y = 6$

28. $3x - 2y = 16$

Find the equation in standard form of the line with the given intercepts.

29. $x = -3;\ y = 2$

30. $x = \dfrac{1}{3};\ y = -4$

[2.5] Find an equation of the line that passes through the given point and (a) is parallel and (b) is perpendicular to the graph of the given equation.

31. $(1, 1),\ -3x + 2y = 7$

32. $(-1, 1),\ x - y = 0$

Find the midpoint of the line segment joining the given pair of points.

33. $(1, 1)$ and $(-1, 2)$

34. $(0, 0)$ and $\left(\dfrac{1}{2}, \dfrac{1}{3}\right)$

[2.6] Find the x-intercepts, the axis of symmetry, and the maximum or minimum point of the graph of each function by algebraic methods.

35. $f(x) = x^2 - x - 6$

36. $f(x) = -x^2 + 7x - 10$

37. Graph the function of Exercise 35.

38. Graph the function of Exercise 36.

Graph each inequality.

39. $y < x^2 - 9$

40. $y \geq x^2 + 6x + 5$

[2.7] Graph each function f and its inverse f^{-1} using the same set of axes.

41. $f: \{(-3, 1), (-1, 3), (2, 4)\}$

42. $f: \{(1, 1), (2, 4), (3, 6)\}$

43. $f: 2x - y = 6$

44. $f: 3x - 4y = 8$

45. $f: y = x^2 - 9,\ x \geq 0$

46. $f: y = x^2 - 4x - 5,\ x \leq 2$

[2.8] Find the indicated range value for each function.

47. $f(x) = \begin{cases} -3x + 1 & \text{if } x < 2 \\ x - 2 & \text{if } x \geq 2 \end{cases};\ f(-1)$

48. $f(x) = \begin{cases} x^2 & \text{if } x \leq 1 \\ -2x^2 + 1 & \text{if } x > 1 \end{cases};\ f(2)$

Graph each function.

49. $f(x) = \begin{cases} x - 1 & \text{if } x \leq 2 \\ \dfrac{1}{2}x & \text{if } x > 2 \end{cases}$

50. $f(x) = \begin{cases} x + 4 & \text{if } x \leq -1 \\ -3x & \text{if } x > -1 \end{cases}$

51. $y = \left| \dfrac{x}{5} \right|$

52. $y = -\left| x - \dfrac{1}{4} \right|$

53. $y = \left[\dfrac{x+1}{4} \right]; \quad -5 \leq x \leq 5$

54. $y = [x] - 1; \quad -5 \leq x \leq 5$

3 Polynomial and Rational Functions

In Chapter 2 we studied functions that are defined by linear and quadratic polynomials. In this chapter we shall study functions defined by polynomials of degree higher than two and functions defined by quotients of polynomials.

3.1 Quotients of Polynomials

Some results obtained by working with quotients of polynomials will be useful in our further work.

Long division   You may recall that if the divisor of a quotient contains more than one term, the familiar **long-division algorithm**, involving successive subtractions, can be used to rewrite the quotient. For example, the computation

$$\begin{array}{r} x - 3 \\ x^2 + 2x - 1 \overline{\smash{\big)}\, x^3 - x^2 - 7x + 3} \\ \underline{x^3 + 2x^2 - x } \\ -3x^2 - 6x + 3 \\ \underline{-3x^2 - 6x + 3} \\ 0 \end{array}$$

shows that, for $x^2 + 2x - 1 \neq 0$,

$$\frac{x^3 - x^2 - 7x + 3}{x^2 + 2x - 1} = x - 3.$$

It is most convenient to arrange the dividend and the divisor in descending powers of the variable before using the division algorithm and to leave an appropriate space for any missing terms (terms with coefficient 0) in the dividend.

3.1 Quotients of Polynomials

When the divisor is not a factor of the dividend, the division process will produce a nonzero remainder. For example, from

$$\begin{array}{r}x^3 - 3x^2 + 10x - 28\\x+3\overline{\smash{\big)}\,x^4+x^2+2x-1}\\\underline{x^4+3x^3}\\-3x^3+x^2\\\underline{-3x^3-9x^2}\\10x^2+2x\\\underline{10x^2+30x}\\-28x-1\\\underline{-28x-84}\\83\text{(remainder)},\end{array}$$

we see that

$$\frac{x^4+x^2+2x-1}{x+3} = x^3 - 3x^2 + 10x - 28 + \frac{83}{x+3} \quad (x \ne -3).$$

Observe in the foregoing example that a space is left in the dividend for a term involving x^3, even though the dividend contains no such term.

Synthetic division

If the divisor is of the form $x + c$, the process of dividing one polynomial by another can be simplified by a process called **synthetic division**. Consider the foregoing example. If we omit writing the variables and write only the coefficients of the terms, and use zero for the coefficient of any missing power, we have

$$\begin{array}{r}1-3+10-28\\1+3\overline{\smash{\big)}\,1+0+1+2-1}\\\underline{1+3}\\-3+(1)\\\underline{-3-9}\\10+(2)\\\underline{10+30}\\-28-(1)\\\underline{-28-84}\\83\text{(remainder)}.\end{array}$$

Now, observe that the numerals shown in color are repetitions of the numerals written immediately above and are also repetitions of the coefficients of the associated variable in the quotient; the numbers in parentheses are repetitions of the coefficients of the dividend. Therefore, the whole process can be written in compact form as

$$\begin{array}{rl}(1) & \underline{3\,|}1012-1\\(2) & 3-930-84\\(3) & \overline{1-310-2883}\text{(remainder: 83)},\end{array}$$

where the repetitions are omitted and where 1, the coefficient of x in the divisor, has also been omitted.

The entries in line (3), which are the coefficients of the variables in the quotient and the remainder, have been obtained by *subtracting* the **detached coefficients** in line (2) from the detached coefficients of terms of the same degree in line (1). We could obtain the same result by replacing 3 with -3 in the divisor and *adding* instead of subtracting at each step, and this is what is done in the *synthetic-division* process. The final form then appears as

$$
\begin{array}{rrrrrrr}
(1) & -3\,| & 1 & 0 & 1 & 2 & -1 \\
(2) & & & -3 & 9 & -30 & 84 \\
(3) & & 1 & -3 & 10 & -28 & 83 \quad \text{(remainder: 83).}
\end{array}
$$

Comparing the results using synthetic division with those obtained by the same process using long division, we observe that the entries in line (3) are the coefficients of the polynomial $x^3 - 3x^2 + 10x - 28$ and that there is a remainder of 83.

Example   Write $\dfrac{3x^3 - 4x - 1}{x - 2}$ in the form $Q(x) + \dfrac{r}{D(x)}$, where r is a constant.

Solution   Using synthetic division we first write

$$2\,|\quad 3 \quad 0 \quad -4 \quad -1,$$

where 0 has been inserted in the position that would be occupied by the coefficient of a second-degree term if such a term were present in the dividend. The divisor, $x - 2$, is indicated by the negative of -2, or 2. Continuing the synthetic division, we write

$$
\begin{array}{rrrrrr}
(1) & 2\,| & 3 & 0 & -4 & -1 \\
(2) & & & 6 & 12 & 16 \\
(3) & & 3 & 6 & 8 & 15 \quad \text{(remainder: 15).}
\end{array}
$$

This process employs these steps:

1. 3 is "brought down" from line (1) to line (3).
2. 6, the product of 2 and 3, is written in the next position on line (2).
3. 6, the sum of 0 and 6, is written on line (3).
4. 12, the product of 2 and 6, is written in the next position on line (2).
5. 8, the sum of -4 and 12, is written on line (3).
6. 16, the product of 2 and 8, is written in the next position on line (2).
7. 15, the sum of -1 and 16, is written on line (3).

We can use the first three entries in line (3) as coefficients to write a polynomial of degree one less than the degree of the dividend. This polynomial is the quotient lacking the remainder. The last number is the remainder. Thus, for $x - 2 \neq 0$, the quotient of $3x^3 - 4x - 1$ divided by $x - 2$ is

$$3x^2 + 6x + 8,$$

3.1 Quotients of Polynomials

with a remainder of 15; that is,

$$\frac{3x^3 - 4x - 1}{x - 2} = 3x^2 + 6x + 8 + \frac{15}{x - 2} \quad (x \neq 2).$$

The foregoing example illustrates a theorem that we shall state without proof.

Theorem 3.1 *If $P(x)$ is a real polynomial of degree $n \geq 1$ and c is any real number, then there exists a unique real polynomial $Q(x)$ of degree $n - 1$ and a unique real number r, such that*

$$P(x) = (x - c)Q(x) + r.$$

Although this theorem does not directly involve the quotient $\dfrac{P(x)}{x - c}$, it does assure us that, if $x \neq c$,

$$\frac{P(x)}{x - c} = Q(x) + \frac{r}{x - c}.$$

Remainder theorem

From Theorem 3.1, we know that, for every real number c, there exists a real polynomial $Q(x)$ and a real number r, such that

$$P(x) = (x - c)Q(x) + r.$$

Since this is true for all $x \in R$, it must be true for $x = c$. Thus we have

$$P(c) = (c - c)Q(c) + r$$
$$= 0 \cdot Q(c) + r = r.$$

This proves the following important result.

Theorem 3.2 *If $P(x)$ is a real polynomial, then for every real number c there exists a unique polynomial $Q(x)$ such that*

$$P(x) = (x - c)Q(x) + P(c).$$

This theorem is called the **remainder theorem** because it asserts that the remainder, when $P(x)$ is divided by $x - c$, is the value of P at c, that is, $P(c)$. Since synthetic division offers a quick means of obtaining this remainder, we can sometimes find values $P(c)$ more rapidly by synthetic division than by direct substitution.

Example Given $P(x) = x^3 - x^2 + 3$, find $P(3)$ by the remainder theorem.

Solution Synthetically dividing $x^3 - x^2 + 3$ by $x - 3$, we have

$$\begin{array}{r|rrrr} 3 & 1 & -1 & 0 & 3 \\ & & 3 & 6 & 18 \\ \hline & 1 & 2 & 6 & 21 \end{array}$$

and, by inspection, $r = P(3) = 21$.

Factor theorem

Observe that if $P(r) = 0$, then, by the remainder theorem,

$$P(x) = (x - r)Q(x) + P(r)$$
$$= (x - r)Q(x) + 0,$$

and $(x - r)$ is a factor of $P(x)$. This proves the following result, called the **factor theorem**.

Theorem 3.3 *If $P(x)$ is a polynomial with real-number coefficients, and $P(r) = 0$, then $(x - r)$ is a factor of $P(x)$.*

Converse of factor theorem

It might be noted, as the converse of the factor theorem, that if $(x - r)$ is a factor of $P(x)$, so that $P(x) = (x - r)Q(x)$, then $P(r) = 0$, since

$$P(r) = (r - r)Q(r)$$
$$= 0 \cdot Q(r) = 0.$$

Exercise 3.1

Write the given quotient $\dfrac{P(x)}{D(x)}$ in the form $Q(x) + \dfrac{R(x)}{D(x)}$, where the degree of $R(x)$ is less than the degree of $D(x)$. In Exercises 5–16 use synthetic divisions.

Example

$$\frac{y^3 + y^2 - 5y + 2}{y^2 - 2y}$$

Solution

$$\begin{array}{r} y + 3 \\ y^2 - 2y \overline{\smash{\big)} y^3 + y^2 - 5y + 2} \\ \underline{y^3 - 2y^2} \\ 3y^2 - 5y \\ \underline{3y^2 - 6y} \\ y + 2 \end{array}$$

$$\frac{y^3 + y^2 - 5y + 2}{y^2 - 2y} = y + 3 + \frac{y + 2}{y^2 - 2y} \quad (y \neq 0, 2)$$

1. $\dfrac{4y^3 + 12y + 5}{2y + 1}$

2. $\dfrac{2x^4 + 13x^3 - 7}{2x - 1}$

3. $\dfrac{4y^5 - 4y^2 - 5y + 1}{2y^2 + y + 1}$

4. $\dfrac{2x^3 - 3x^2 - 15x - 1}{x^2 + 5}$

Examples

a. $\dfrac{2x^4 + x^3 - 1}{x + 2}$

b. $\dfrac{x^3 - 1}{x - 1}$

3.1 Quotients of Polynomials

Solutions

a.
$$\begin{array}{r|rrrrr} -2 & 2 & 1 & 0 & 0 & -1 \\ & & -4 & 6 & -12 & 24 \\ \hline & 2 & -3 & 6 & -12 & 23 \end{array}$$

$$2x^3 - 3x^2 + 6x - 12 + \frac{23}{x+2} \quad (x \neq -2)$$

b.
$$\begin{array}{r|rrrrr} 1 & 1 & 0 & 0 & -1 \\ & & 1 & 1 & 1 \\ \hline & 1 & 1 & 1 & 0 \end{array}$$

$$x^2 + x + 1 \quad (x \neq 1)$$

5. $\dfrac{x^4 - 3x^3 + 2x^2 - 1}{x - 2}$
6. $\dfrac{x^4 + 2x^2 - 3x + 5}{x - 3}$
7. $\dfrac{2x^3 + x - 5}{x + 1}$
8. $\dfrac{3x^3 + x^2 - 7}{x + 2}$
9. $\dfrac{2x^4 - x + 6}{x - 5}$
10. $\dfrac{3x^4 - x^2 + 1}{x - 4}$
11. $\dfrac{x^3 + 4x^2 + x - 2}{x + 2}$
12. $\dfrac{x^3 - 7x^2 - x + 3}{x + 3}$
13. $\dfrac{x^6 + x^4 - x}{x - 1}$
14. $\dfrac{x^6 + 3x^3 - 2x - 1}{x - 2}$
15. $\dfrac{x^5 - 1}{x - 1}$
16. $\dfrac{x^5 + 1}{x + 1}$

Use synthetic division and Theorem 3.2 in Exercises 17–22.

17. $P(x) = x^3 + 2x^2 + x - 1$; find $P(1)$, $P(2)$, and $P(3)$.
18. $P(x) = x^3 - 3x^2 - x + 3$; find $P(1)$, $P(2)$, and $P(3)$.
19. $P(x) = 2x^4 - 3x^3 + x + 2$; find $P(-2)$, $P(2)$, and $P(4)$.
20. $P(x) = 3x^4 + 3x^2 - x + 3$; find $P(-2)$, $P(2)$, and $P(4)$.
21. $P(x) = 3x^5 - x^3 + 2x^2 - 1$; find $P(-3)$, $P(2)$, and $P(3)$.
22. $P(x) = 2x^6 - x^4 + 3x^3 + 1$; find $P(-3)$, $P(2)$, and $P(3)$.

Assume the given number is a solution of $P(x) = 0$ and factor $P(x)$ completely.

Example   3; $P(x) = x^3 - 3x^2 - x + 3$

Solution   Since 3 is a solution of $P(x) = 0$, we have $P(3) = 0$. Thus, by Theorem 3.3, $x - 3$ is a factor of $P(x)$. Dividing synthetically, we have

$$\begin{array}{r|rrrr} 3 & 1 & -3 & -1 & 3 \\ & & 3 & 0 & -3 \\ \hline & 1 & 0 & -1 & 0 \end{array}$$

and the quotient is $x^2 - 1$. Thus,

$$P(x) = (x - 3)(x^2 - 1) = (x - 3)(x + 1)(x - 1).$$

23. 2; $P(x) = x^3 - 3x - 2$
24. 3; $P(x) = x^3 - 5x^2 + 7x - 3$
25. -1; $P(x) = x^3 + 2x^2 - 5x - 6$
26. -1; $P(x) = x^3 - 7x - 6$

27. $\frac{1}{2}$; $P(x) = 2x^3 - x^2 - 8x + 4$
28. $\frac{2}{3}$; $P(x) = 3x^3 - 2x^2 - 3x + 2$
29. -1; $P(x) = x^4 - 2x^3 - x^2 + 2x$
30. 2; $P(x) = x^4 - x^3 - 4x^2 + 4x$
31. 1; $P(x) = 2x^4 + 3x^3 - 3x^2 - 2x$
32. 2; $P(x) = 3x^4 - 11x^3 + 12x^2 - 4x$

Find a value for k so that the second polynomial is a factor of the first.

33. $x^3 - 5x^2 - 16x + k$; $x - 5$
34. $x^3 - 9x^2 + 14x + k$; $x + 1$
35. $x^4 + 2x^3 - 21x^2 + kx + 40$; $x - 4$
36. $3x^4 - 40x^3 + 130x^2 + kx + 27$; $x - 9$

3.2 Zeros of a Polynomial Function

In our previous work we obtained x-intercepts to help us graph first- and second-degree equations in two variables. These intercepts are also helpful in graphing polynomial equations of degree greater than two. Since the x-intercepts of the graph of a polynomial function are the zeros of the function, we shall first consider some facts that will aid us in finding such zeros.

Continuity

For real-number replacements for x, we know that

$$y = P(x) = a_n x^n + a_{n-1} x^{n-1} + \cdots + a_0 \quad (a_n \neq 0)$$

can be represented by a graph in the coordinate plane. Moreover, although we shall not prove it here, we have the following theorem on continuity.

Theorem 3.4 If $P(x) = a_n x^n + a_{n-1} x^{n-1} + \cdots + a_0$ $(a_n \neq 0)$, and if k is between $P(x_1)$ and $P(x_2)$, then there exists at least one c between x_1 and x_2 such that $P(c) = k$.

More intuitively, there are no breaks or jumps in the graph of $P(x)$, so we know that $P(x)$ must assume all values between any two of its values. Thus, the graph of P must be a continuous, unbroken curve (see Figure 3.1).

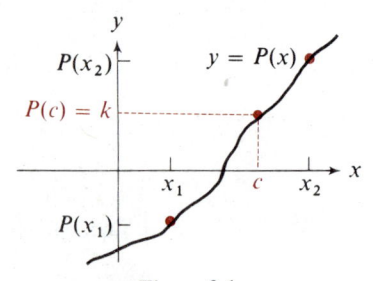

Figure 3.1

To assist us in the search for real zeros of a polynomial function, we have the following theorem that is a consequence of Theorem 3.4.

3.2 Zeros of a Polynomial Function

Theorem 3.5 Let $P(x)$ be a polynomial with real coefficients. If $P(x_1)$ and $P(x_2)$ are opposite in sign, then there is at least one value c between x_1 and x_2 such that $P(c) = 0$.

This theorem expresses the fact that if the points with coordinates $(x_1, P(x_1))$ and $(x_2, P(x_2))$ are on opposite sides of the x-axis, then the graph of $y = P(x)$ must cross the x-axis at some (at least one) point c between x_1 and x_2 (see Figure 3.2).

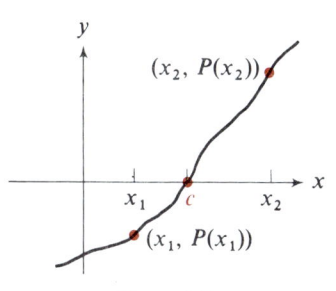

Figure 3.2

Example Show that $P(x) = x^3 + x - 1 = 0$ has a solution between 0 and 1.

Solution We observe that $P(0) = -1$ while $P(1) = (1)^3 + (1) - 1 = 1$. Since $P(0)$ and $P(1)$ have opposite signs, Theorem 3.5 guarantees that there is a solution to the given equation between 0 and 1.

Bounds on zeros

The following theorem is sometimes helpful in isolating real zeros of a polynomial function with real coefficients.

Theorem 3.6 Let $P(x)$ be a polynomial with real-number coefficients.

I. If $r_1 \geq 0$, and the coefficients of the terms in $Q(x)$ and the term $P(r_1)$ are all of the same sign in the right-hand member of

$$P(x) = (x - r_1)Q(x) + P(r_1),$$

then $P(x) = 0$ can have no solution greater than r_1.

II. If $r_2 \leq 0$, and the coefficients of the terms in $Q(x)$ and the term $P(r_2)$ alternate in sign (zero suitably denoted by $+0$ or -0) in the right-hand member of

$$P(x) = (x - r_2)Q(x) + P(r_2),$$

then $P(x) = 0$ can have no solution less than r_2.

We shall verify only the first part of the theorem here.
For all $x > r_1$, $(x - r_1) > 0$. Moreover, if all the coefficients in $Q(x)$ are of the same sign, say positive, then, since $x > r_1 \geq 0$, we have

$$Q(x) > 0 \quad \text{and} \quad (x - r_1)Q(x) > 0.$$

Since $P(r_1) \geq 0$ by hypothesis, we have

$$P(x) = (x - r_1)Q(x) + P(r_1) > 0,$$

and the first part is proved. The second part follows from the first by considering
$$P(-x) = (-x - r_2)Q(-x) + P(r_2).$$
This theorem permits us to place upper and lower bounds on the set of real zeros of the polynomial function
$$P(x) = a_n x^n + \cdots + a_0,$$
and consequently on the real members of the solution set of $P(x) = 0$.

Example   Show that 2 and -2 are upper and lower bounds, respectively, for the location of the zeros of
$$P(x) = 18x^3 - 12x^2 - 11x + 10.$$

Solution   Using synthetic division to divide $P(x)$ by $x - 2$, we have

$$\underline{2\,|}\ \ \begin{array}{cccc} 18 & -12 & -11 & 10 \\ & 36 & 48 & 74 \\ \hline 18 & 24 & 37 & 84. \end{array}$$

Thus, $Q(x) = 18x^2 + 24x + 37$ and $P(2) = 84 > 0$. Hence, by Theorem 3.6-I, 2 is an upper bound for the zeros of P. Next, dividing $P(x)$ by $x + 2$, we have

$$\underline{-2\,|}\ \ \begin{array}{cccc} 18 & -12 & -11 & 10 \\ & -36 & 96 & -170 \\ \hline 18 & -48 & 85 & -160. \end{array}$$

Here $Q(x) = 18x^2 - 48x + 85$ and $P(-2) = -160$. Therefore, by Theorem 3.6-II, -2 is a lower bound for the zeros of P.

Example   Find the least nonnegative integer and the greatest nonpositive integer that are, by Theorem 3.6, upper and lower bounds, respectively, for the real zeros of
$$P(x) = x^4 - x^3 - 10x^2 - 2x + 12.$$

Solution   We shall first seek an upper bound by dividing $P(x)$ successively by $x - 1$, $x - 2$, and so on. Each row after the first in the following array is the bottom row in the respective synthetic division involved.

$$\begin{array}{c|ccccc} & 1 & -1 & -10 & -2 & 12 \\ \hline 1 & 1 & 0 & -10 & -12 & 0 \\ 2 & 1 & 1 & -8 & -18 & -24 \\ 3 & 1 & 2 & -4 & -14 & -30 \\ 4 & 1 & 3 & 2 & 6 & 36 \end{array}$$

3.2 Zeros of a Polynomial Function

Since the numbers in the last row are all positive, 4 is an upper bound. Next, we divide by $x + 1$, $x + 2$, and so on, in search of a lower bound.

	1	−1	−10	−2	12
−1	1	−2	−8	6	6
−2	1	−3	−4	6	0
−3	1	−4	2	−8	36

Since the numbers in the row following −3 alternate from positive to negative, etc., −3 is a lower bound. (Had the numbers in the row been 1, 0, 2, −8, 36, then the sign "−" could arbitrarily have been assigned to 0 to give the desired pattern of alternating signs.)

Variations in sign

A **variation in sign** occurs in a polynomial with real coefficients if, as the polynomial is viewed from left to right, successive coefficients are opposite in sign. For example, in the polynomial

$$P(x) = 3x^5 - 2x^4 - 2x^2 + x - 1,$$

there are three variations in sign, and, in

$$P(-x) = -3x^5 - 2x^4 - 2x^2 - x - 1,$$

there are no variations in sign.

Descartes' Rule of Signs

Another useful theorem, which again we shall not prove here, is that which establishes **Descartes' Rule of Signs**.

Theorem 3.7 If $P(x)$ is a polynomial over the field R of real numbers, then the number of positive real solutions of $P(x) = 0$ either is equal to the number of variations in sign occurring in the coefficients of $P(x)$, or else is less than this number by an even natural number. Moreover, the number of negative real solutions of $P(x) = 0$ either is equal to the number of variations in sign occurring in $P(-x)$, or else is less than this number by an even natural number.

Example

Find the possible number of positive real solutions and the possible number of negative real solutions for the equation

$$3x^4 + 3x^3 - 2x^2 + x + 1 = 0. \tag{1}$$

Solution

Since $P(x) = 3x^4 + 3x^3 - 2x^2 + x + 1$ has but two variations in sign, the equation has either zero or two positive real solutions. Since

$$P(-x) = 3x^4 - 3x^3 - 2x^2 - x + 1$$

has two variations in sign, the number of negative real solutions of (1) is either zero or two.

Exercise 3.2

A In Exercises 1–6, use Theorem 3.5 to verify the given statement.

Example $f(x) = x^3 - 2x + 2$ has a zero between -2 and -1.

Solution Compute $f(-2)$ and $f(-1)$.

$$f(-2) = (-2)^3 - 2(-2) + 2 \quad \text{and} \quad f(-1) = (-1)^3 - 2(-1) + 2$$
$$= -2 \qquad\qquad\qquad\qquad\qquad = 3$$

Since $f(-2) < 0$ and $f(-1) > 0$, Theorem 3.5 guarantees that f has a zero between -2 and -1.

1. $f(x) = x^3 - 3x + 1$ has a zero between 0 and 1.
2. $f(x) = 2x^3 + 7x^2 + 2x - 6$ has a zero between -2 and -1.
3. $g(x) = x^4 - 2x^2 + 12x - 17$ has a zero between -3 and -2.
4. $g(x) = 2x^4 + 3x^3 - 14x^2 - 15x + 9$ has a zero between -2 and -1.
5. $P(x) = 2x^2 + 4x - 4$ has zeros between -3 and -2 and between 0 and 1.
6. $P(x) = x^3 - x^2 - 2x + 1$ has zeros between -2 and -1, between 0 and 1, and between 1 and 2.

Use Theorem 3.7 to find the possible number of positive real solutions and the possible number of negative real solutions of each polynomial equation.

7. $x^4 - 2x^3 + 2x + 1 = 0$
8. $3x^3 + 3x^2 + 2x^2 - x + 1 = 0$
9. $2x^5 + 3x^3 + 2x + 1 = 0$
10. $4x^5 - 2x^3 - 3x - 2 = 0$
11. $3x^4 + 1 = 0$
12. $2x^5 - 1 = 0$

In Exercises 13–16, use an application of Theorem 3.5 to approximate the given zero of the indicated polynomial to within one-half unit.

Example $P(x) = x^2 - 2$; the zero is between 1 and 2.

Solution It is given that there is a zero between 1 and 2. To obtain a closer approximation, subdivide the interval $[1, 2]$ into the two subintervals $[1, 3/2]$ and $[3/2, 2]$. The zero lies in one of these subintervals. Use Theorem 3.5 to check which one.

$$P(1) = -1, \quad P\left(\frac{3}{2}\right) = \frac{1}{4}$$

3.2 Zeros of a Polynomial Function

Since $P(1)$ and $P(3/2)$ are of opposite signs, there is a zero between 1 and 3/2. Thus the zero is within one-half unit of both 1 and 3/2.

13. $P(x) = x^2 - 3$; the zero between 1 and 2.
14. $P(x) = x^2 - 5$; the zero between 2 and 3.
15. $P(x) = x^2 - 7$; the zero between -3 and -2.
16. $P(x) = x^2 - 10$; the zero between -4 and -3.

Find an upper bound and a lower bound for the real zeros of each polynomial function.

Example $\quad P(x) = 12x^3 - 4x^2 - 3x + 1$

Solution $\quad$ Use synthetic division to divide $P(x)$ by $x - 1$.

$$
\begin{array}{r|rrrr}
1 & 12 & -4 & -3 & 1 \\
  &   & 12 & 8 & 5 \\
\hline
  & 12 & 8 & 5 & 6
\end{array}
$$

The quotient is $Q(x) = 12x^2 + 8x + 5$ and $P(1) > 0$. By Theorem 3.6-I, 1 is an upper bound for the zeros of P. Use synthetic division to divide $P(x)$ by $x + 1$.

$$
\begin{array}{r|rrrr}
-1 & 12 & -4 & -3 & 1 \\
   &   & -12 & 16 & -13 \\
\hline
   & 12 & -16 & 13 & -12
\end{array}
$$

The quotient is $Q(x) = 12x^2 - 16x + 13$ and $P(-1) < 0$. By Theorem 3.6-II, -1 is a lower bound for the zeros of P.

17. $P(x) = x^3 + 2x^2 - 7x - 8$
18. $P(x) = x^3 - 8x + 5$
19. $P(x) = x^4 - 2x^3 - 7x^2 + 10x + 10$
20. $P(x) = x^3 - 4x^2 - 4x + 12$
21. $P(x) = x^5 - 3x^3 + 24$
22. $P(x) = x^5 - 3x^4 - 1$
23. $P(x) = 2x^5 + x^4 - 2x - 1$
24. $P(x) = 2x^5 - 2x^2 + x - 2$
25. $P(x) = x^5 - 2x^4 - 19x^3 + 20x^2$
26. $P(x) = 4x^5 - 8x^4 - x^3 + 2x^2$

B *Find the given zero to the indicated polynomial to within one-eighth of a unit.*

27. $P(x) = x^3 - 4x^2 - 2x + 8$; the zero between -2 and -1.
28. $P(x) = 2x^3 + 2x^2 - x - 1$; the zero between 0 and 1.
29. $P(x) = 2x^4 - x^3 - 2$; the zero between -1 and 0.
30. $P(x) = 3x^4 - 2x - 1$; the zero between 0 and 1.

3.3 Rational Zeros of Polynomial Functions

In this section we shall develop a method for finding all the rational zeros for a polynomial function with integer coefficients. First we shall consider the notion of a prime number on which the results of this section depend.

Prime and composite numbers

If a is a natural number, and $a \neq 1$, then a is a **prime number** if and only if a has no factor in N other than itself and 1; otherwise, a is a **composite number**. For example, 2 and 3 are prime numbers, and 6 (equal to $2 \cdot 3$) is composite; but 1 is considered to be neither prime nor composite.

The **unique-factorization theorem**, or **fundamental theorem of arithmetic** (which we shall not prove), follows.

Theorem 3.8   *If a is a composite number, then a is the product of only one set of prime factors; that is, there are unique prime numbers $p_1, \ldots, p_n$ and natural numbers $k_1, \ldots, k_n$ so that*

$$a = p_1^{k_1} \cdot p_2^{k_2} \cdot \cdots \cdot p_n^{k_n}.$$

Examples

a. $24 = 2 \cdot 2 \cdot 2 \cdot 3 = 2^3 \cdot 3$
b. $36 = 2 \cdot 2 \cdot 3 \cdot 3 = 2^2 \cdot 3^2$
c. $30 = 2 \cdot 3 \cdot 5$
d. $90 = 2 \cdot 3 \cdot 3 \cdot 5 = 2 \cdot 3^2 \cdot 5$

Two integers a and b are said to be **relatively prime** if and only if they have no prime factors in common. For example, -6 and 35 are relatively prime, since $-6 = -2 \cdot 3$ and $35 = 5 \cdot 7$; but 6 and 8 are not, since they have the prime factor 2 in common. The fraction a/b is said to express a rational number in **lowest terms** if and only if a and b are relatively prime.

Identifying possible rational zeros

If all the coefficients of the polynomial

$$P(x) = a_n x^n + a_{n-1} x^{n-1} + \cdots + a_0$$

are integers, then we can identify all possible rational zeros of P by means of the following theorem.

Theorem 3.9   *If the rational number p/q, in lowest terms, is a solution of*

$$P(x) = a_n x^n + a_{n-1} x^{n-1} + \cdots + a_0 = 0,$$

where $a_j \in J$, then p is an integral factor of a_0 and q is an integral factor of a_n.

To see that this is true, let p/q be a solution of $P(x) = 0$, which is reduced to lowest terms. Then we have

$$a_n \left(\frac{p}{q}\right)^n + a_{n-1} \left(\frac{p}{q}\right)^{n-1} + \cdots + a_0 = 0,$$

3.3 Rational Zeros of Polynomial Functions

and we can multiply each member here by q^n to obtain

$$a_n p^n + a_{n-1} p^{n-1} q + \cdots + a_0 q^n = 0.$$

Adding $-a_0 q^n$ to each member and factoring p from each term in the left-hand member of the resulting equation, we have

$$p(a_n p^{n-1} + a_{n-1} p^{n-2} q + \cdots + a_1 q^{n-1}) = -a_0 q^n.$$

Since the set J of integers is closed with respect to addition and multiplication, the expression in parentheses in the left-hand member represents an integer, say r, so that we have

$$pr = -a_0 q^n,$$

where pr is an integer having p as a factor. Hence, p is a factor of $-a_0 q^n$. But, by Theorem 3.8, p and q^n have no factor in common, because, by hypothesis, p/q is in lowest terms; hence p must be a factor of a_0. In a similar manner, by writing the equation

$$a_n p^n + a_{n-1} p^{n-1} q + \cdots + a_0 q^n = 0$$

in the form

$$-a_n p^n = a_{n-1} p^{n-1} q + \cdots + a_0 q^n,$$

we can factor q from each term in the right-hand member and show that q must be a factor of a_n.

Example List all possible rational zeros of

$$P(x) = 2x^3 - 4x^2 + 3x + 9.$$

Solution Rational zeros, p/q, must, by Theorem 3.9, be such that p is an integral factor of 9 and q is an integral factor of 2. Hence

$$p \in \{-9, -3, -1, 1, 3, 9\}, \quad q \in \{-2, -1, 1, 2\},$$

and the possible rational zeros of P are

$$-9, -\frac{9}{2}, -3, -\frac{3}{2}, -1, -\frac{1}{2}, \frac{1}{2}, 1, \frac{3}{2}, 3, \frac{9}{2}, 9.$$

Test for rational zeros It is important to observe that Theorem 3.9 does not assure us that a polynomial function with integral coefficients indeed has a rational zero; it simply enables us to identify possibilities for rational zeros. These can then be checked by synthetic division.

As a special case of Theorem 3.9, it is evident that if a function P is defined by the equation

$$P(x) = x^n + a_{n-1} x^{n-1} + \cdots + a_0,$$

in which $a_i \in J$ and $a_n = 1$, then any rational zero of P must be an integer and, moreover, must be an integral factor of a_0.

Example Find all rational zeros of $P(x) = x^3 - 4x^2 + x + 6$.

Solution The only possible rational zeros of P are $-6, -3, -2, -1, 1, 2, 3,$ and 6. Using synthetic division, we set up the following array:

$$
\begin{array}{r|rrrr}
 & 1 & -4 & 1 & 6 \\
\hline
1 & 1 & -3 & -2 & 4 \\
\hline
-1 & 1 & -5 & 6 & 0
\end{array}
$$

The third row in this array is the bottom row in division of $x^3 - 4x^2 + x + 6$ by $x + 1$. Thus we can cease our trials with -1, because, by the remainder theorem, -1 is a zero of P, and the remaining zeros can be obtained from the equation $x^2 - 5x + 6 = 0$, whose coefficients are taken from the last line in the array. We can write the equation as $(x - 2)(x - 3) = 0$ and observe that 2 and 3 are also zeros.

Exercise 3.3

A Find all integral zeros of each function.

Example $P(x) = 9x^3 - 9x^2 - x + 1$

Solution By Theorem 3.9 the only possible rational zeros are

$$-1, -\frac{1}{3}, -\frac{1}{9}, \frac{1}{9}, \frac{1}{3}, 1.$$

Of these possibilities only 1 and -1 are integers. Use synthetic division to see which is a zero of the polynomial.

$$
\begin{array}{r|rrrr}
1 & 9 & -9 & -1 & 1 \\
 & & 9 & 0 & -1 \\
\hline
 & 9 & 0 & -1 & 0
\end{array}
\qquad
\begin{array}{r|rrrr}
-1 & 9 & -9 & -1 & 1 \\
 & & -9 & 18 & -17 \\
\hline
 & 9 & -18 & 17 & -16
\end{array}
$$

Thus $x = 1$ is the only integral zero of $P(x) = 9x^3 - 9x^2 - x + 1$.

1. $f(x) = 3x^3 - 13x^2 + 6x - 8$
2. $f(x) = x^4 - x^2 - 4x + 4$
3. $f(x) = x^4 + x^3 + 2x - 4$
4. $f(x) = 5x^3 + 11x^2 - 2x - 8$
5. $P(x) = 2x^4 - 3x^3 - 8x^2 - 5x - 3$
6. $P(x) = 3x^4 - 40x^3 + 130x^2 - 120x + 27$

Find all rational zeros of each function.

Example $P(x) = 2x^3 + x^2 - 4x - 2$

3.3 Rational Zeros of Polynomial Functions

Solution By Theorem 3.9 the only possible rational zeros are

$$-2, -1, -\frac{1}{2}, \frac{1}{2}, 1, 2.$$

Use synthetic division to determine which is a root.

$$\begin{array}{r|rrrr} & 2 & 1 & -4 & -2 \\ \frac{1}{2} & & 1 & 1 & -\frac{3}{2} \\ \hline & 2 & 2 & -3 & -\frac{7}{2} \\ -\frac{1}{2} & & -1 & & \\ \hline & 2 & 0 & -4 & 0 \end{array}$$

The third row in this array is the bottom row in the synthetic division of $2x^3 + x^2 - 4x - 2$ by $x + \frac{1}{2}$. Thus

$$2x^3 + x^2 - 4x - 2 = \left(x + \frac{1}{2}\right)(2x^2 - 4).$$

Since the solutions of $2x^2 - 4 = 0$ are $\pm\sqrt{2}$, which are irrational, the only rational zero of $P(x) = 2x^3 + x^2 - 4x - 2$ is $x = -1/2$.

7. $f(x) = 2x^3 + 3x^2 - 14x - 21$
8. $f(x) = 3x^4 - 11x^3 + 9x^2 + 13x - 10$
9. $P(x) = 4x^4 - 13x^3 - 7x^2 + 41x - 14$
10. $P(x) = 2x^3 - 4x^2 + 3x + 9$
11. $Q(x) = 2x^3 - 7x^2 + 10x - 6$
12. $Q(x) = x^3 + 3x^2 - 4x - 12$

Find all zeros of each function. Hint: First find all rational zeros.

Example $P(x) = x^3 - 2x^2 - 4x + 3$

Solution Use Theorem 3.9 to find that the only rational zero is $x = 3$. Divide $x^3 - 2x^2 - 4x + 3$ by $x - 3$.

$$\begin{array}{r|rrrr} 3 & 1 & -2 & -4 & 3 \\ & & 3 & 3 & -3 \\ \hline & 1 & 1 & -1 & 0 \end{array}$$

The quotient is $x^2 + x - 1$. Use the quadratic formula to find that the zeros of $x^2 + x - 1$ are

$$\frac{-1 + \sqrt{5}}{2} \text{ and } \frac{-1 - \sqrt{5}}{2}.$$

Thus the zeros of $P(x) = x^3 - 2x^2 - 4x + 3$ are $3, \frac{-1 + \sqrt{5}}{2}$, and $\frac{-1 - \sqrt{5}}{2}$.

13. $P(x) = 3x^3 - 5x^2 - 14x - 4$
14. $P(x) = x^3 + 2x^2 - 7x - 14$
15. $P(x) = 8x^4 + 12x^3 - 4x^2 - 6x + 2$
16. $P(x) = 8x^4 - 22x^3 - 19x^2 + 66x - 15$
17. $P(x) = 12x^4 + 7x^3 - 36x^2 - 14x + 24$
18. $P(x) = 2x^4 + 5x^3 - x^2 - 5x + 2$

B 19. Factor the polynomial $2x^3 + 3x^2 - 2x - 3$.

20. Factor the polynomial $2x^4 - 3x^3 - 6x^2 + 6x + 4$.

21. Show that $\sqrt{3}$ is irrational. *Hint*: Consider the equation $x^2 - 3 = 0$.

22. Show that $\sqrt{2}$ is irrational.

23. Show that if p is a prime number, then $\sqrt{p}$ is irrational.

24. Show that if n is any natural number *not* of the form $n = k^2$, $k \in N$, then $\sqrt{n}$ is irrational.

3.4 *Graphing Polynomial Functions*

In Section 2.3 we graphed linear functions
$$f(x) = a_1 x + a_0 \quad (a_1 \neq 0),$$
and in Section 2.5 we graphed quadratic functions
$$f(x) = a_2 x^2 + a_1 x + a_0 \quad (a_2 \neq 0).$$
We can graph any real polynomial function
$$P(x) = a_n x^n + a_{n-1} x^{n-1} + \cdots + a_0 \quad (a_n \neq 0)$$
by similar methods—that is, by combining the plotting of points with a consideration of certain general properties of the defining equations. The set of zeros of a polynomial function considered in Sections 3.2 and 3.3 is one such property.

General form of a polynomial graph

To determine whether the graph of a polynomial ultimately goes up or down to the right, that is, for large values of x, we can examine the leading coefficient a_n. If $a_n > 0$, then the graph ultimately goes up to the right, as in Figure 3.3-**a** and **c**. If $a_n < 0$, then the graph ultimately goes down to the right, as in Figure 3.3-**b** and **d**. To determine whether the graph of a polynomial ultimately goes up or down to the left, that is, for negative values of x with large magnitude, we must examine both the leading coefficients a_n and the degree n.

a   **b**

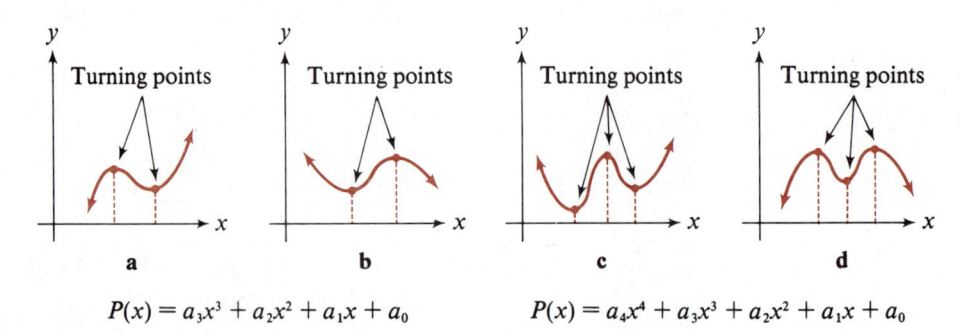

$P(x) = a_3 x^3 + a_2 x^2 + a_1 x + a_0$

c   **d**

$P(x) = a_4 x^4 + a_3 x^3 + a_2 x^2 + a_1 x + a_0$

Figure 3.3

3.4 Graphing Polynomial Functions

If $a_n > 0$ and n is even, then the graph ultimately goes up to the left, as in Figure 3.3-c. If $a_n > 0$ and n is odd, then the graph ultimately goes down to the left, as in Figure 3.3-a.

If $a_n < 0$, we obtain just the opposite results: If n is even, the graph ultimately goes down to the left; if n is odd, the graph ultimately goes up to the left. In each case, then, the leading term $a_n x^n$ governs the behavior of the graph of the polynomial function for large values of $|x|$.

Notice that for each graph in Figure 3.3 there are intervals on which the function values increase as x increases and intervals on which the function values decrease as x increases. A function is said to be **increasing** on an interval if the function values increase as x increases on the given interval. Similarly, a function is said to be **decreasing** on an interval if the function values decrease as x increases on the given interval.

Finding turning points

Another fact about polynomial functions involves *turning points*, or local maximum and minimum values of y. Thus, for example, in Figure 3.3-a there is one local maximum as well as one local minimum, or a total of two turning points. In Figure 3.3-c there are two local minima and one local maximum, or a total of three turning points. In general, we have the following result.

Theorem 3.10 If $P(x) = a_n x^n + a_{n-1} x^{n-1} + \cdots + a_0$ $(a_n \neq 0)$ *as a real polynomial of degree n, then its graph is a smooth curve that has at most $n - 1$ turning points.*

Since the proof of this theorem involves ideas from the calculus, it is omitted. As a consequence of this theorem, the graphs of third- and fourth-degree polynomial functions might appear as in Figure 3.3-**a** and **d**, respectively.

To determine where the turning points of the graph of a polynomial lie, we use the following result. Its proof also involves ideas from the calculus, so it is omitted.

Theorem 3.11 *The first components of the turning points of the graph of*

$$P(x) = a_n x^n + a_{n-1} x^{n-1} + \cdots + a_1 x + a_0$$

are solutions of the equation

$$P'(x) = n a_n x^{n-1} + (n-1) a_{n-1} x^{n-2} + \cdots + 2 a_2 x + a_1 = 0.$$

To find the turning points of the graph of a polynomial function $y = P(x)$, we first solve $P'(x) = 0$, where $P'(x)$ is as defined in Theorem 3.11. We then substitute these values for x that are real in $P(x)$ to obtain the second coordinates of the points. By plotting these points we can often determine by inspection whether each point is a local maximum or a local minimum. If inspection is not sufficient, then plotting a few more points on either side of the point will enable us to determine its nature.

Example Find the turning points of
$$P(x) = x^3 - 3x^2 - 9x + 15.$$

Solution We first note from Theorem 3.10 that there are at most two turning points. By Theorem 3.11 the x-components of these turning points are solutions of $P'(x) = 0$ where
$$P'(x) = 3(1)x^{3-1} + 2(-3)x^{2-1} + (-9).$$
Then we solve
$$\begin{aligned} P'(x) &= 3x^2 - 6x - 9 \\ &= 3(x-3)(x+1) = 0, \end{aligned}$$
obtaining $x = 3$ and $x = -1$. Next we use the remainder theorem and synthetic division to compute $P(3)$ and $P(-1)$. We have

$$\begin{array}{r|rrrr} -1 & 1 & -3 & -9 & 15 \\ & & -1 & 4 & 5 \\ \hline & 1 & -4 & -5 & 20 \end{array} \qquad \begin{array}{r|rrrr} 3 & 1 & -3 & -9 & 15 \\ & & 3 & 0 & -27 \\ \hline & 1 & 0 & -9 & -12 \end{array}$$

and the possible turning points are $(-1, 20)$ and $(3, -12)$. Now we plot $(3, -12)$ and $(-1, 20)$. Since the curve ultimately goes down to the left and up to the right, we see by inspection of the figure that $(-1, 20)$ is a local maximum and $(3, -12)$ is a local minimum.

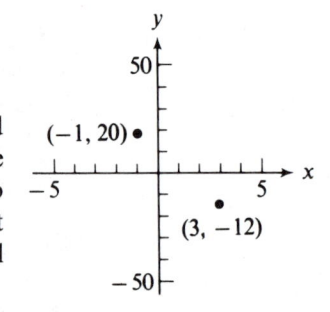

Graphing polynomial functions When graphing a polynomial function, we determine the general form of the graph, locate and graph the turning points and intercepts, if necessary plot a few more points $(x, P(x))$, and then sketch the curve through these points.

Example Graph $P(x) = 4x^3 + 3x^2 - 6x$.

Solution Since P is defined by a cubic polynomial with positive leading coefficient, we expect a graph of form similar to that in Figure 3.3-a. To find the possible turning points of the graph, we solve
$$\begin{aligned} P'(x) &= (3)4x^{3-1} + (2)3x^{2-1} + (-6) \\ &= 12x^2 + 6x - 6 \\ &= 6(2x-1)(x+1) = 0, \end{aligned}$$

3.4 Graphing Polynomial Functions

obtaining $x = 1/2$ and $x = -1$. Using the remainder theorem and synthetic division to compute $P(-1)$ and $P(1/2)$, we have

$$\begin{array}{r|rrrr} -1 & 4 & 3 & -6 & 0 \\ & & -4 & 1 & 5 \\ \hline & 4 & -1 & -5 & 5 \end{array} \qquad \begin{array}{r|rrrr} 1/2 & 4 & 3 & -6 & 0 \\ & & 2 & \frac{5}{2} & -\frac{7}{4} \\ \hline & 4 & 5 & -\frac{7}{2} & -\frac{7}{4}. \end{array}$$

Thus, $(-1, 5)$ and $(1/2, -7/4)$ are on the graph.

By inspection of the figure we see that $(-1, 5)$ is a local maximum and $(1/2, -7/4)$ is a local minimum. The y-intercept is $P(0) = 0$. Thus $(0, 0)$ is on the graph. To obtain the values of the x-intercepts, we solve $P(x) = 4x^3 + 3x^2 - 6x = 0$. Factoring, we obtain

$$4x^3 + 3x^2 - 6x = x(4x^2 + 3x - 6) = 0.$$

By inspection, one solution is $x = 0$; by using the quadratic formula to solve $4x^2 + 3x - 6 = 0$, we see that

$$x = \frac{-3 + \sqrt{105}}{8} \approx 0.91$$

and

$$x = \frac{-3 - \sqrt{105}}{8} \approx -1.66.$$

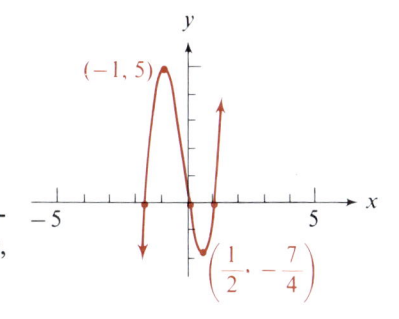

We approximate the points with these coordinates on the x-axis and sketch the curve, as shown in the accompanying figure.

In the foregoing examples, all the solutions of the equation $P'(x) = 0$ yielded turning points. This is *not* always the case, as the following example illustrates.

Example Find the turning points of $P(x) = x^3 - 3x^2 + 3x - 1$.

Solution We first obtain

$$P'(x) = 3(1)x^{3-1} + 2(-3)x^{2-1} + 3.$$

Solving

$$\begin{aligned} P'(x) &= 3x^2 - 6x + 3, \\ &= 3(x - 1)^2 = 0, \end{aligned}$$

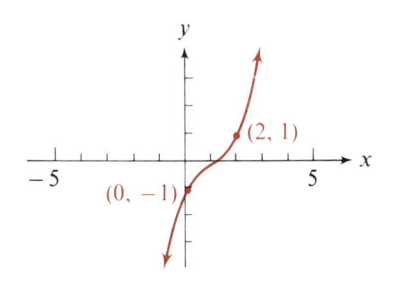

we obtain $x = 1$. Hence, the *only possible* turning point [since 1 is the only solution to $P'(x) = 0$] is $(1, 0)$. The degree of $P(x)$ is 3 and the leading coefficient is positive; thus the graph ultimately goes down to the left and up to the right. If the point $(1, 0)$ were a turning point, then there would have to be at least one more turning point for the graph to have this form. Hence, the graph has no turning points. By obtaining several additional ordered pairs we obtain the graph shown.

Exercise 3.4

A Find all the possible turning points, and state which are local maxima, which are local minima, and which are neither.

Example $P(x) = x^3 - 3x - 1$

Solution Use Theorem 3.11 to find the x-coordinates of the possible turning points.

$$P'(x) = 3(1)x^{3-1} - 3$$
$$= 3x^2 - 3$$

Solve $P'(x) = 0$.

$$P'(x) = 3x^2 - 3$$
$$= 3(x - 1)(x + 1) = 0$$
$$x = 1 \quad \text{or} \quad x = -1$$

Evaluate P at the solutions to $P'(x) = 0$.

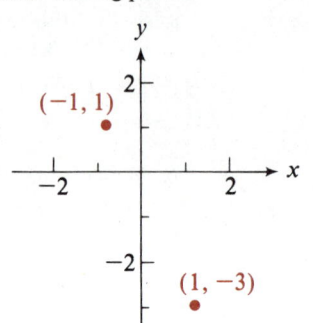

$$P(-1) = 1 \quad \text{and} \quad P(1) = -3$$

The possible turning points are $(-1, 1)$ and $(1, -3)$. Plot these points and note that since the degree of P is 3 and the leading coefficient is positive, the curve ultimately goes down to the left and up to the right. Thus $(-1, 1)$ is a local maximum and $(1, -3)$ is a local minimum.

1. $P(x) = 2x^3 - 9x^2 + 12x + 10$
2. $P(x) = 2x^3 + 3x^2 - 72x + 50$
3. $P(x) = x^4 - 2x^2 + 1$
4. $P(x) = 3x^4 + 4x^3 - 12x^2 + 10$
5. $P(x) = 6x^4 - 8x^3 + 10$
6. $P(x) = 6x^5 - 15x^4 + 10x^3 + 20$

Graph. Include enough of the domain to show all turning points.

7. $P(x) = x^3 - 3x$
8. $P(x) = 2x^3 - 3x^2 - 12x + 18$
9. $P(x) = 4x^3 - 12x^2 + 9x$
10. $P(x) = 2x^3 - 5x^2 + 4x + 1$
11. $P(x) = \dfrac{x^3}{3} - \dfrac{x^2}{2} - 6x$
12. $P(x) = \dfrac{2x^3}{3} + \dfrac{5x^2}{2} - 3x$
13. $P(x) = x^3 - 4x^2 - 3x$
14. $P(x) = 3x^3 + 2x^2 - \dfrac{4}{3}x + 1$
15. $P(x) = -x^3 + 2x^2 - 1$
16. $P(x) = -10x^3 + 9x^2 + 12x$
17. $P(x) = -3x^3 + 9x^2 + 7x + 6$
18. $P(x) = \dfrac{-x^3}{3} + x$

B Graph. Include enough of the domain to show all turning points.

19. $P(x) = -x^4 + x^2$
20. $P(x) = -x^4 - 4x$
21. $P(x) = x^4 - 10x^2 + 9$
22. $P(x) = x^4 - 5x^2 + 4$

23. $P(x) = x^4 - 4x^3 + 4x^2 + 12$ 24. $P(x) = x^4 - x^3 - 2x^2 + 3x - 3$

25. $P(x) = x^3$ 26. $P(x) = -x^3 + 8$

27. Use Theorem 3.11 to find the coordinates of the vertex of the graph of the function $P(x) = ax^2 + bx + c$.

28. Find conditions on the real numbers a, b, and c that guarantee that the graph of the function $P(x) = ax^3 + bx^2 + cx + d$ has no turning points. Has one turning point. Has two turning points. If no such conditions exist, then so state.

3.5 Rational Functions

A function defined by an equation of the form

$$y = \frac{P(x)}{Q(x)}, \qquad (1)$$

where $P(x)$ and $Q(x)$ are polynomials in x and $Q(x)$ is not the zero polynomial, is called a **rational function**. Notice, in particular, that $Q(x)$ might be a nonzero constant, say $Q(x) = 1$, so that polynomial functions are special cases of rational functions. Rational functions with real coefficients are of importance in the calculus and provide some interesting problems with respect to their graphs.

Vertical asymptotes

Since $P(x)/Q(x)$ is not defined for values of x for which $Q(x) = 0$, it is evident that we shall not be able to find points on the graph of (1) having such x-coordinates. We can, however, consider the graph for values of x as close as we please to a value, say x_0, for which $Q(x_0) = 0$, but still with $x \neq x_0$. This consideration is usually described by saying that x "approaches" x_0, "grows close" to x_0, etc., and correspondingly that $Q(x)$ approaches 0. Thus, the closer $Q(x)$ approaches 0, if $P(x)$ does not approach 0 at the same time, then the larger $|y|$ becomes in (1). For example, Figure 3.4-**a** shows the behavior of

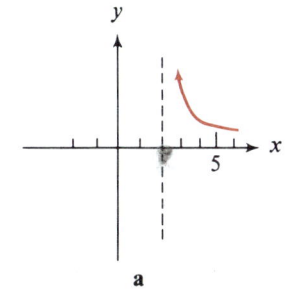

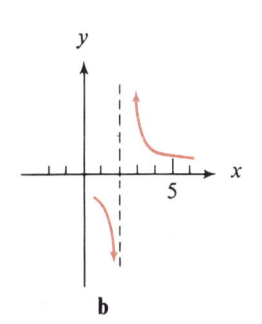

$$y = \frac{2}{x-2}, \qquad (2)$$

Figure 3.4

as x approaches 2 from the right. We are equally interested in its behavior as x approaches 2 from the left. Figure 3.4-b illustrates this. As long as $x > 2$, we have $x - 2 > 0$ and $2/(x - 2) > 0$; but if $x < 2$, then we have $x - 2 < 0$ and $2/(x - 2) < 0$.

In situations such as this, the vertical line that the curve approaches is called a **vertical asymptote**.

Theorem 3.12 The graph of the function defined by $y = P(x)/Q(x)$ has a vertical asymptote $x = a$ for each value a at which $Q(x)$ vanishes and $P(x)$ does not vanish.

Example Find all vertical asymptotes of the graph of

$$y = \frac{1}{x^2 - 3x + 2}.$$

Solution We first find the values at which the denominator vanishes. Solving the equation

$$x^2 - 3x + 2 = (x - 1)(x - 2) = 0,$$

we have $x = 1$ and $x = 2$. Next we note that the numerator never vanishes. Thus by Theorem 3.12, $x = 1$ and $x = 2$ are both vertical asymptotes.

Horizontal asymptotes The graphs of some rational functions have **horizontal asymptotes**, which can, in general, be identified by using the following theorem.

Theorem 3.13 The graph of the rational function defined by

$$y = R(x) = \frac{a_n x^n + a_{n-1} x^{n-1} + \cdots + a_0}{b_m x^m + b_{m-1} x^{m-1} + \cdots + b_0}, \tag{3}$$

where $a_n, b_m \neq 0$, and n, m are nonnegative integers, has

    I. a horizontal asymptote at $y = 0$ if $n < m$;
    II. a horizontal asymptote at $y = a_n/b_m$ if $n = m$;
    III. no horizontal asymptotes if $n > m$.

Though we shall not give a rigorous proof of this theorem, we can certainly make the results plausible. If $n < m$, we can divide the numerator and denominator of the right-hand member of (3) by x^m to obtain, for $x \neq 0$,

$$y = R(x) = \frac{\dfrac{a_n}{x^{m-n}} + \dfrac{a_{n-1}}{x^{m-n+1}} + \cdots + \dfrac{a_0}{x^m}}{b_m + \dfrac{b_{m-1}}{x} + \cdots + \dfrac{b_0}{x^m}}.$$

Now, as $|x|$ grows greater and greater, each term containing an x in its denominator grows closer and closer to 0, and we find the expression on the right approaching

3.5 Rational Functions

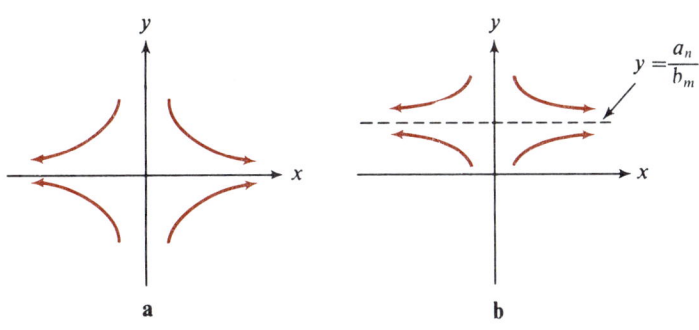

Figure 3.5

$0/b_0$, so that y approaches 0. But if, as $|x|$ increases without bound, y grows close to 0, then the graph of $y = R(x)$ approaches the line $y = 0$ asymptotically and actually must approach 0 from one of the directions shown in Figure 3.5-**a**. A similar argument shows that if $n = m$, then, as $|x|$ increases without bound, the graph of $y = R(x)$ approaches the line $y = a_n/b_m$ from one of the directions shown in Figure 3.5-**b**. If $n > m$, then as $|x|$ becomes greater and greater, so does $|y|$. Thus the curve has no horizontal asymptote.

Example Find all horizontal and vertical asymptotes of the graph of the function

$$y = \frac{x^2 - 4}{x - 1}.$$

Solution We observe by Theorem 3.12 that $x = 1$ is a vertical asymptote and by Theorem 3.13 that there are no horizontal asymptotes.

Oblique asymptotes In Theorem 3.13 if $n = m + 1$, that is, if the degree of the numerator is one greater than that of the denominator, we can argue that though the graph has no horizontal asymptote, it does have an **oblique asymptote**. We shall illustrate with a special case only, but the technique involved is quite general.

Example Find all asymptotes for the graph of the function

$$y = \frac{x^2 - 4}{x - 1}.$$

Solution This is the same function we investigated in the preceding example, where we found the vertical asymptote $x = 1$ and no horizontal asymptote. If we now rewrite $y = (x^2 - 4)/(x - 1)$ by dividing $x^2 - 4$ by $x - 1$, we obtain

$$y = x + 1 - \frac{3}{x - 1}.$$

Now, as $|x|$ grows greater and greater, $3/(x - 1)$ becomes closer and closer to 0 and the graph of $y = (x^2 - 4)/(x - 1)$ approaches the graph of $y = x + 1$. Hence, the graph of $y = x + 1$, which is an oblique line, is an asymptote to the curve.

112   3   Polynomial and Rational Functions

Aids for graphing

Identifying asymptotes is one aid to graphing a rational function. Other helpful items are the following:

1. The zeros of the function, because these give us the x-intercepts;
2. The domain and range, because these let us know where we can expect to find parts of the graph and where we cannot;
3. Some specific points on the graph, because these give us guidelines in sketching.

Example

Graph $y = \dfrac{x-1}{x-2}$.

Solution

We can begin by observing that the numerator of the right-hand member will be equal to 0 when x is equal to 1. Therefore, when $x = 1$, we have $y = 0$, and 1 is an x-intercept. Also, when $x = 0$, we have $y = 1/2$, so that $1/2$ is a y-intercept. Thus we can begin our graph as shown in Figure **a**. By inspection, $y = (x-1)/(x-2)$

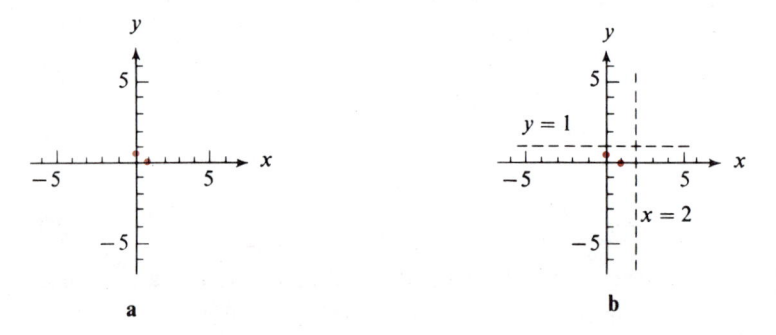

is defined for all x except $x = 2$, so that $\{x \mid x \neq 2\}$ is the domain. Similarly, if we solve the defining equation for x in terms of y, we have

$$x = \frac{2y-1}{y-1},$$

which is defined for all values of y except 1. Hence $\{y \mid y \neq 1\}$ is the range of the function. From the defining equation and Theorems 3.12 and 3.13, we see that there is a vertical asymptote at $x = 2$ and a horizontal asymptote at $y = 1$. We can then add this information to our graph, as indicated in Figure **b**. To determine the behavior of the function near the vertical asymptote, we consider one side of the asymptote at a time. On the left side x is just less than 2, say $2 - p$, $0 < p < 1/10$, so the denominator $x - 2$ in the expression for y is

$$2 - p - 2 = -p,$$

which is barely negative, whereas the numerator

$$x - 1 = 2 - p - 1 = 1 - p$$

is definitely positive; hence y is negative and $|y|$ large. Thus the graph goes down on the left side of the vertical asymptote. The graph appears in Figure **c**. To the right

3.5 Rational Functions

of the vertical asymptote, we consider $x = 2 + p$, $0 < p < 1/10$, to find that the graph goes up on the right of the vertical asymptote. This along with the knowledge that $y = 1$ is a horizontal asymptote leads us to the complete graph of the function $y = (x - 1)/(x - 2)$, as shown in Figure **d**.

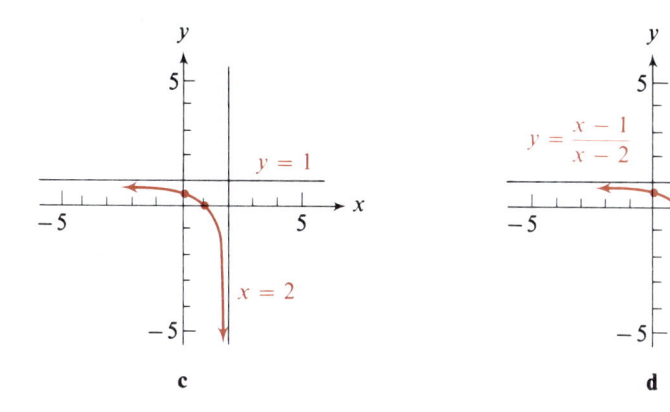

c

d

In the next example the rational function has an oblique asymptote.

Example Graph $y = \dfrac{x^2 - 4}{x - 1}$.

Solution This is the same function we investigated in the example on page 111, for which we found the vertical asymptote $x = 1$ and the oblique asymptote $y = x + 1$. If $x = 0$, then $y = 4$, and if $y = 0$, then $x = 2$ or $x = -2$, so that the y-intercept is 4 and the x-intercepts are 2 and -2. We therefore have the situation shown in Figure **a**. Without further information, we have reason to suspect that the graph will

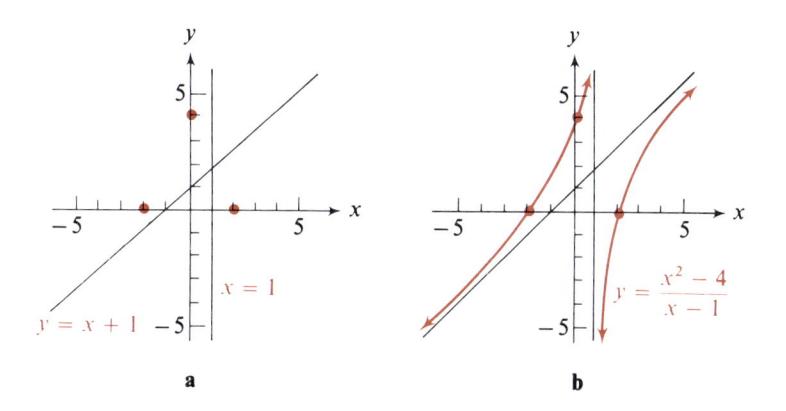

a

b

appear as shown in Figure **b**. Plotting a few check points, say $(-1, 3/2)$, $(1/2, 15/2)$, $(3/2, -7/2)$, and $(3, 5/2)$, would tend to confirm our conjecture.

Exercise 3.5

A Determine the vertical asymptotes of the graph of each function.

Example $x^2y - 4y = 1$

Solution Express y explicitly in terms of x.

$$y(x^2 - 4) = 1$$

$$y = \frac{1}{x^2 - 4} = \frac{1}{(x - 2)(x + 2)}$$

By Theorem 3.12 there are vertical asymptotes at $x = 2$ and $x = -2$.

1. $y = \dfrac{1}{x - 3}$
2. $y = \dfrac{1}{x + 4}$
3. $y = \dfrac{4}{(x + 2)(x - 3)}$
4. $y = \dfrac{8}{(x - 1)(x + 3)}$
5. $y = \dfrac{2x - 1}{x^2 + 5x + 4}$
6. $y = \dfrac{x + 3}{2x^2 - 5x - 3}$
7. $xy + y = 4$
8. $x^2y + xy = 3$

Graph.

9. $y = \dfrac{1}{x}$
10. $y = \dfrac{1}{x + 4}$
11. $y = \dfrac{1}{x - 3}$
12. $y = \dfrac{1}{x - 6}$
13. $y = \dfrac{4}{(x + 2)(x - 3)}$
14. $y = \dfrac{8}{(x - 1)(x + 3)}$
15. $y = \dfrac{2}{(x - 3)^2}$
16. $y = \dfrac{1}{(x + 4)^2}$

Determine any vertical, horizontal, or oblique asymptotes of the graphs of each of the following functions.

Example $y = \dfrac{6x^2 + 1}{2x^2 + 5x - 3}$

3.5 Rational Functions

Solution   The equation can be written equivalently as

$$y = \frac{6x^2 + 1}{(2x - 1)(x + 3)}.$$

By Theorem 3.12 there are vertical asymptotes at $x = 1/2$ and $x = -3$. By Theorem 3.13 there is a horizontal asymptote at $y = 6/2 = 3$.

17. $y = \dfrac{x}{x^2 - 4}$ 18. $y = \dfrac{3x - 6}{x^2 + 3x + 2}$ 19. $y = \dfrac{x^2 - 9}{x - 4}$

20. $y = \dfrac{x^3 - 27}{x^2 - 1}$ 21. $y = \dfrac{x^2 - 3x + 2}{x^2 - 3x - 4}$ 22. $y = \dfrac{x^2}{x^2 - x - 6}$

Graph. Use information concerning the zeros of the function and concerning vertical, horizontal, and oblique asymptotes.

23. $y = \dfrac{x}{x - 2}$ 24. $y = \dfrac{x - 1}{x + 3}$ 25. $y = \dfrac{2x - 4}{x^2 - 9}$

26. $y = \dfrac{3x}{x^2 - 5x + 4}$ 27. $y = \dfrac{x^2 - 4}{x^3}$ 28. $y = \dfrac{x - 2}{x^2}$

29. $y = \dfrac{x + 1}{x(x^2 - 4)}$ 30. $y = \dfrac{x^2 + x - 2}{x(x^2 - 9)}$ 31. $y = \dfrac{x^2 - 4x + 4}{x - 1}$

32. $y = \dfrac{x^2 + 4}{x - 2}$ 33. $y = \dfrac{x^3 + 1}{x^2}$ 34. $\dfrac{x^3 - x^2 - 2x}{x^2 - 1}$

B   *Graph. Use information concerning the zeros of the function and vertical, horizontal, and oblique asymptotes if they exist. Hint: Reduce the rational expression to lowest terms, and keep in mind that the function is undefined for **all** values at which the **original** denominator vanishes.†*

35. $y = \dfrac{x^2 - 9}{x + 3}$ 36. $y = \dfrac{x^2 - 1}{x + 1}$ 37. $y = \dfrac{x^3 - 3x^2 + 2x}{x - 1}$

38. $y = \dfrac{-x^3 + 2x^2 - x}{x}$ 39. $y = \dfrac{x - 1}{x^2 - x}$ 40. $y = \dfrac{x}{x^2 - 4x}$

41. Describe the asymptotic behavior of $y = \dfrac{x^3 + x^2 + 2}{x + 1}$.

42. Sketch the graph of $y = \dfrac{x^3 + x^2 + 2}{x + 1}$.

---

† Often in the definition of rational function the numerator and denominator are restricted from having common zeros.

43. Is it possible for the graph of a rational function to cross a vertical asymptote? Why or why not?

44. Find the point where the graph of $y = \dfrac{x^2 + 1}{x^2 - 2x - 1}$ crosses its horizontal asymptote.

45. Find the point where the graph of $y = \dfrac{2x^3 + x^2 + 3x}{x^2 + 1}$ crosses its oblique asymptote.

46. Give an example of a rational function with the property that its graph crosses its oblique asymptote more than once.

Chapter Review

[3.1] *Write the given quotient as a polynomial.*

1. $\dfrac{2n^2 - 5n - 3}{n - 3}$
2. $\dfrac{2r^2 + 3r + 1}{2r + 1}$

Use synthetic division to write the given quotient $P(x)/D(x)$ either in the form $Q(x)$ or in the form $Q(x) + \dfrac{r}{D(x)}$, where $Q(x)$ is a polynomial and r is a constant.

3. $\dfrac{2x^3 - 11x^2 + 13x - 4}{x - 4}$
4. $\dfrac{3x^3 + 7x^2 - 3x + 10}{x + 3}$

Assume the given number is a solution of $P(x) = 0$ and factor $P(x)$ completely.

5. $1;\ P(x) = x^3 - 2x^2 - x + 2$
6. $-2;\ P(x) = 2x^3 - x^2 - 13x - 6$

[3.2]
7. Use Theorem 3.5 to show that $P(x) = 2x^3 + x^2 - 4x - 2$ has a zero between 1 and 2.
8. Use Theorem 3.5 to show that $P(x) = x^4 - x$ has a zero between $1/2$ and $3/2$.

Find an upper bound and a lower bound for the real zeros of the given polynomial function.

9. $P(x) = 2x^3 + x^2 - 4x - 2$
10. $P(x) = x^4 + 2x^3 - 7x^2 - 10x + 10$

Use Descartes' Rule of Signs to find the possible number of positive real solutions and the possible number of negative real solutions of each polynomial equation.

11. $2x^3 + x^2 - 4x - 2 = 0$
12. $x^4 + 2x^3 - 7x^2 - 10x + 10 = 0$

Review Exercises

[3.3] 13. Find all integral zeros of $Q(x) = x^4 - 3x^3 - x^2 - 11x - 4$.

14. Find all rational zeros of $P(x) = 2x^3 - 11x^2 + 12x + 9$.

15. Find all zeros of $P(x) = x^3 + 2x^2 + 2x + 4$.

16. Find all zeros of $P(x) = 6x^4 - x^3 - 13x^2 + 2x - 2$.

[3.4] *Find all possible turning points and state which are local maxima, which are local minima, and which are neither.*

17. $P(x) = 2x^3 - 9x^2 + 12x$ 18. $P(x) = x^3 - x^2$

Graph the function.

19. $P(x) = 2x^3 - 9x^2 + 12x$ 20. $P(x) = 2x^3 + 3x^2$

21. $P(x) = x^3 - x^2$ 22. $P(x) = x^4 - 2x^2 - 3$

[3.5] *Determine all asymptotes, and graph each function.*

23. $y = \dfrac{3}{x+2}$ 24. $y = \dfrac{2x}{(x+3)(x-4)}$

25. $y = \dfrac{x+2}{x-3}$ 26. $y = \dfrac{2x}{x^2-3x+2}$

4 Exponential and Logarithmic Functions

The functions studied in Chapters 2 and 3 are rational functions as defined in Section 3.5. In this chapter we consider two functions that are not rational.

4.1 Exponential Functions

Properties of powers of the form b^x, where $b \in R$, $b > 0$, and x denotes a *rational number*, are noted in "Preliminary Concepts." These powers can be used to define functions. Notice that the base b is restricted here to positive real numbers to ensure that b^x be real for all rational numbers x.

Since we now want to define a function over R using b^x, we must be able to interpret powers with *irrational exponents*, such as

$$b^{\sqrt{2}}, \quad b^{-\sqrt{3}}, \quad \text{and} \quad b^{\pi},$$

to be real numbers. The following theorem, which is presented without proof, will be useful in doing this.

Theorem 4.1   Let $x, y \in Q$, and let $x > y$. Then

$$b^x > b^y \text{ if } b > 1, \quad b^x = b^y \text{ if } b = 1, \quad \text{and} \quad b^x < b^y \text{ if } 0 < b < 1.$$

Powers with real-number exponents

You should recall that irrational numbers can be approximated by rational numbers to as great a degree of accuracy as desired. For example, $\sqrt{2} \approx 1.4$, or $\sqrt{2} \approx 1.414$, etc. Since 2^x is defined for rational x, by Theorem 4.1 we can write

4.1 Exponential Functions

the sequence of inequalities

$$2^1 < 2^2,$$
$$2^{1.4} < 2^{1.5},$$
$$2^{1.41} < 2^{1.42},$$
$$2^{1.414} < 2^{1.415},$$
$$\vdots$$

It seems plausible and is actually true, though we shall not prove it, that the difference between the number on the left and that on the right can be made as small as we please and that there is just one number, which we denote by $2^{\sqrt{2}}$, that lies between the number on the left and the number on the right, no matter how long this process of approximation is continued. Since we can produce the same type of argument for any irrational exponent x and any positive base b, we shall assume that b^x ($b > 0$) is defined in this way for all real values of x and that the laws of exponents, appropriately reworded for real-number exponents, are valid for such powers.

Since for any given $b > 0$ and for each $x \in R$, there is only one value for b^x, the equation

$$f(x) = b^x \quad (b > 0) \tag{1}$$

defines a function. Because $1^x = 1$ for all $x \in R$, (1) defines a constant function if $b = 1$. If $b \neq 1$, we say that (1) defines an **exponential function**. Exponential functions can perhaps be visualized most clearly by considering their graphs. We illustrate two typical examples, in which $0 < b < 1$ and $b > 1$, respectively. Assigning values to x in the equations

$$f(x) = \left(\frac{1}{2}\right)^x \quad \text{and} \quad f(x) = 2^x,$$

we find some ordered pairs in each solution set and sketch the graphs shown in Figure 4.1-a and b, respectively.

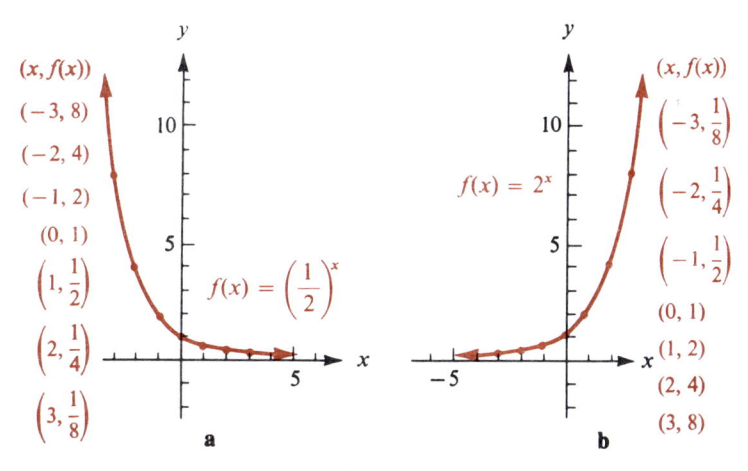

Figure 4.1

Notice that, in accordance with Theorem 4.1, the graph of the function determined by $f(x) = (1/2)^x$ goes *down* to the right, and the graph of the function determined by $f(x) = 2^x$ goes *up* to the right. For this reason, we say that the former function is a *decreasing* function and the latter is an *increasing* function. In either case, the domain is the set of real numbers, and the range is the set of positive real numbers.

Frequently used bases

There are two exponential functions that are of special interest because they are frequently used in mathematics and other sciences. The base for one of these functions is 10 and the base for the other is an irrational number that we shall denote by e. The value of e to eight decimal places is 2.71828183. Table I on page 481 shows values of e^x and e^{-x} for selected values of x; alternatively, a calculator can be used to find such powers. We will discuss methods for finding values of 10^x, where x is not an integer, in Section 4.3.

Exercise 4.1

A Find the second component of each of the ordered pairs that makes the pair a solution of the corresponding equation.

1. $y = 3^x$; (0,), (1,), (2,)
2. $y = 4^x$; (0,), (1,), (2,)
3. $y = -5^x$; (0,), (1,), (2,)
4. $y = -2^x$; (0,), (1,), (2,)
5. $y = \left(\frac{1}{2}\right)^x$; (−3,), (0,), (3,)
6. $y = \left(\frac{1}{3}\right)^x$; (−3,), (0,), (3,)
7. $y = 10^x$; (−2,), (1,), (0,)
8. $y = 10^{-x}$; (0,), (1,), (2,)

Graph the function.

9. $y = 4^x$
10. $y = 5^x$
11. $y = 10^x$
12. $y = 10^{-x}$
13. $y = 2^{-x}$
14. $y = 3^{-x}$
15. $y = \left(\frac{1}{3}\right)^x$
16. $y = \left(\frac{1}{4}\right)^x$
17. $y = \left(\frac{2}{3}\right)^x$
18. $y = \left(\frac{2}{3}\right)^{-x}$
19. $y = \left(\frac{3}{2}\right)^{x+1}$
20. $y = \left(\frac{3}{2}\right)^{-x+1}$

21. Graph $f: y = 10^x$ and $f^{-1}: x = 10^y$ on the same set of axes.
22. Graph $f: y = 2^x$ and $f^{-1}: x = 2^y$ on the same set of axes.

Solve for x by inspection.

23. $10^x = \dfrac{1}{100}$
24. $\left(\dfrac{1}{2}\right)^x = 16$
25. $16^x = 8$
26. $8^x = 16$

4.2 Logarithmic Functions

B *Determine an integer n such that $n < x < n + 1$.*

27. $3^x = 16.2$ 28. $4^x = 87.1$ 29. $10^x = 0.016$ 30. $2^x = 6$

31. Graph $y = b^x$, $b = 2, 3,$ and $4,$ on the same set of coordinate axes. What effect does increasing b have on the curves if $b > 1$?

32. Graph $y = b^x$, $b = \frac{1}{4}, \frac{1}{3},$ and $\frac{1}{2},$ on the same set of coordinate axes. What effect does increasing b have on the curves if $0 < b < 1$?

4.2 Logarithmic Functions

Inverse of an exponential function

In the exponential function

$$y = b^x \quad (b > 0, b \neq 1), \tag{1}$$

illustrated in Figure 4.1 for $b = 1/2$ and $b = 2,$ there is only one x associated with each y. Thus, by Definition 2.4 we have the inverse function

$$x = b^y \quad (b > 0, b \neq 1). \tag{2}$$

Observe that Equation (2) implies $x > 0,$ because there is no real number y for which b^y is not positive.

The graphs of functions of this form can be illustrated by the example

$$x = 10^y.$$

We consider x the variable denoting an element in the domain and, in the defining equation, we assign specific values for x, say

$$(0.01, \quad), (0.1, \quad), (1, \quad), (10, \quad), (100, \quad),$$

to obtain the ordered pairs

$$(0.01, -2), (0.1, -1), (1, 0), (10, 1), (100, 2).$$

These can be graphed and connected with a smooth curve as shown in Figure 4.2.

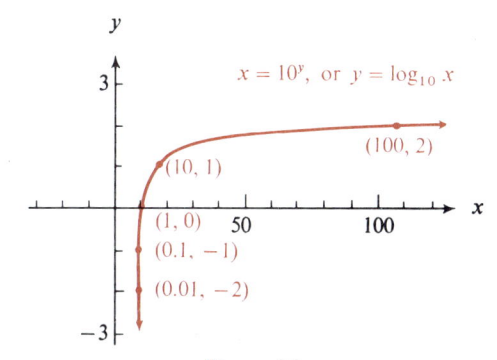

Figure 4.2

It is always useful to be able to express the variable denoting an element in the range explicitly in terms of the variable denoting an element in the domain. To do this in the equation defining function (2), we use the notation

$$y = \log_b x \quad (x > 0, b > 0, b \neq 1). \tag{3}$$

Thus for $x > 0$, $b > 0$, $b \neq 1$, we have

$$y = \log_b x \quad \text{if and only if} \quad x = b^y.$$

Here, $\log_b x$ is read "the logarithm to the base b of x," or "the logarithm of x to the base b." The function defined by Equation (3) is called a **logarithmic function**.

Observe that $\log_b x$ is the *exponent* y such that the power b^y is equal to x; that is,

$$b^{\log_b x} = x.$$

Properties of logarithmic functions

From the graph in Figure 4.2, we generalize from $\log_{10} x$ to $\log_b x$ and observe that, for $b > 1$, a logarithmic function has the following properties:

1. The domain is the set of positive real numbers, and the range is the set of all real numbers.
2. $\log_b x < 0$ for $0 < x < 1$, $\log_b x = 0$ for $x = 1$, and $\log_b x > 0$ for $x > 1$.

Logarithmic and exponential statements

It should be recognized that Equations (2) and (3) are two equations determining the same function, in the same way that $x = y + 4$ and $y = x - 4$ determine the same function, and we may use whichever equation suits our purpose. Thus, exponential statements may be written in logarithmic form, and logarithmic statements may be written in exponential form.

Examples

a. $5^2 = 25$ can be written as $\log_5 25 = 2$.

b. $8^{1/3} = 2$ can be written as $\log_8 2 = \dfrac{1}{3}$.

c. $3^{-3} = \dfrac{1}{27}$ can be written as $\log_3 \dfrac{1}{27} = -3$.

Examples

a. $\log_{10} 100 = 2$ can be written as $10^2 = 100$.

b. $\log_3 81 = 4$ can be written as $3^4 = 81$.

c. $\log_2 \dfrac{1}{2} = -1$ can be written as $2^{-1} = \dfrac{1}{2}$.

Notice in each of the examples above that the logarithm and the exponent are equal.

Since a logarithm is an exponent, the following theorem follows directly from the properties of powers with real-number exponents.

4.2 Logarithmic Functions

Theorem 4.2 If $x_1, x_2 \in R$, $x_1, x_2, b > 0$, $b \neq 1$, $m \in R$, then

I. $\log_b (x_1 x_2) = \log_b x_1 + \log_b x_2$;

II. $\log_b \dfrac{x_2}{x_1} = \log_b x_2 - \log_b x_1$;

III. $\log_b (x_1)^m = m \log_b x_1$.

The validity of I can be shown as follows: Since

$$x_1 = b^{\log_b x_1} \quad \text{and} \quad x_2 = b^{\log_b x_2},$$

it follows from the properties of powers that

$$x_1 x_2 = b^{\log_b x_1} \cdot b^{\log_b x_2} = b^{\log_b x_1 + \log_b x_2},$$

and then, by the definition of a logarithm, that

$$\log_b (x_1 x_2) = \log_b x_1 + \log_b x_2.$$

The validity of II and III can be established in a similar way and are left as exercises.

Examples

a. $\log_{10} 100 = \log_{10} (10 \cdot 10)$
$= \log_{10} 10 + \log_{10} 10$
$= 1 + 1 = 2$

b. $\log_2 \dfrac{1}{4} = \log_2 1 - \log_2 4$
$= 0 - 2 = -2$

c. $\log_3 81 = \log_3 3^4$
$= 4 \log_3 3 = 4 \cdot 1 = 4$

d. $\log_2 4^{-8} = -8 \log_2 4$
$= -8 \cdot 2 = -16$

Logarithmic equations

An equation that involves logarithms can sometimes be solved by using Theorem 4.2 to rewrite the equation equivalently as an equation involving a single logarithm with coefficient 1 and then writing the equation in exponential notation.

Example Solve $\log_{10} (x + 6) = 1 - \log_{10} (x - 3)$.

Solution We first rewrite the given equation as

$$\log_{10} (x + 6) + \log_{10} (x - 3) = 1.$$

We then use Theorem 4.2-I to obtain

$$\log_{10} [(x + 6)(x - 3)] = 1.$$

Next we rewrite this equation in exponential form to get

$$(x + 6)(x - 3) = 10^1.$$

Solving this equation for x, we obtain $x = -7$ and $x = 4$. Neither $\log_{10}(-7 + 6)$ nor $\log_{10}(-7 - 3)$ is defined. Thus -7 does not satisfy the original equation. Since 4 satisfies the equation, the solution set is $\{4\}$.

Exercise 4.2

A *Graph the function.*

Example $y = \log_4 x$

Solution   Some solutions to this equation are $(1, 0)$, $(4, 1)$, and $(16, 2)$. Plot these points and join them with a smooth curve to obtain the figure shown.

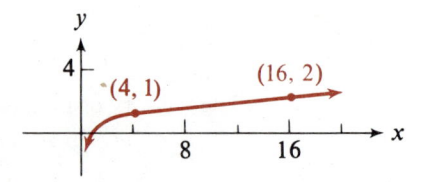

1. $y = \log_2 x$
2. $y = \log_3 x$
3. $y = \log_2 4x$
4. $y = \log_3 9x$
5. $y = \log_2 (x + 3)$
6. $y = \log_3 (x + 4)$

Express in logarithmic notation.

Examples   **a.** $2^3 = 8$ **b.** $\left(\dfrac{1}{2}\right)^{-1} = 2$ **c.** $4^{1/2} = 2$

Solutions   **a.** $\log_2 8 = 3$ **b.** $\log_{1/2} 2 = -1$ **c.** $\log_4 2 = \dfrac{1}{2}$

7. $4^2 = 16$
8. $5^3 = 125$
9. $3^3 = 27$
10. $8^2 = 64$
11. $\left(\dfrac{1}{2}\right)^2 = \dfrac{1}{4}$
12. $\left(\dfrac{1}{3}\right)^2 = \dfrac{1}{9}$
13. $8^{-1/3} = \dfrac{1}{2}$
14. $64^{-1/6} = \dfrac{1}{2}$
15. $10^2 = 100$
16. $10^0 = 1$
17. $10^{-1} = 0.1$
18. $10^{-2} = 0.01$

Express in exponential notation.

Examples   **a.** $\log_4 16 = 2$ **b.** $\log_{1/2} 8 = -3$ **c.** $\log_{64} 8 = \dfrac{1}{2}$

Solutions   **a.** $4^2 = 16$ **b.** $\left(\dfrac{1}{2}\right)^{-3} = 8$ **c.** $64^{1/2} = 8$

19. $\log_2 64 = 6$
20. $\log_5 25 = 2$
21. $\log_3 9 = 2$
22. $\log_{16} 256 = 2$
23. $\log_{1/3} 9 = -2$
24. $\log_{1/2} 8 = -3$
25. $\log_{10} 1000 = 3$
26. $\log_{10} 1 = 0$

4.2 Logarithmic Functions

Find the value of each expression.

27. $\log_7 49$
28. $\log_2 32$
29. $\log_4 64$
30. $\log_3 \frac{1}{3}$

31. $\log_5 \frac{1}{5}$
32. $\log_3 3$
33. $\log_2 2$
34. $\log_{10} 10$

35. $\log_{10} 100$
36. $\log_{10} 1$
37. $\log_{10} 0.1$
38. $\log_{10} 0.01$

Solve for x, y, or b.

Examples

a. $\log_2 x = 3$
b. $\log_b 2 = \frac{1}{2}$
c. $\log_{1/4} 16 = y$

Determine the solution by inspection or by first writing the equation equivalently in exponential form.

Solutions

a. $2^3 = x$

$x = 8$

b. $b^{1/2} = 2$

$(b^{1/2})^2 = (2)^2$

$b = 4$

c. $\left(\frac{1}{4}\right)^y = 16$

$y = -2$

39. $\log_3 9 = y$
40. $\log_5 125 = y$
41. $\log_b 8 = 3$
42. $\log_b 625 = 4$
43. $\log_4 x = 3$
44. $\log_{1/2} x = -5$
45. $\log_2 \frac{1}{8} = y$
46. $\log_5 5 = y$
47. $\log_b 10 = \frac{1}{2}$
48. $\log_b 0.1 = -1$
49. $\log_2 x = 2$
50. $\log_{10} x = -3$

Express as the sum or difference of simpler logarithmic quantities.

Example

$\log_b \left(\frac{xy}{z}\right)^{1/2}$

Solution

By Theorem 4.2-III,

$$\log_b \left(\frac{xy}{z}\right)^{1/2} = \frac{1}{2} \log_b \left(\frac{xy}{z}\right).$$

By Theorem 4.2-I and II,

$$\frac{1}{2} \log_b \left(\frac{xy}{z}\right) = \frac{1}{2} (\log_b x + \log_b y - \log_b z).$$

51. $\log_b (xy)$
52. $\log_b (xyz)$
53. $\log_b \left(\dfrac{x}{y}\right)$
54. $\log_b \left(\dfrac{xy}{z}\right)$
55. $\log_b x^5$
56. $\log_b x^{1/2}$
57. $\log_b \sqrt[3]{x}$
58. $\log_b \sqrt[3]{x^2}$
59. $\log_b \sqrt{\dfrac{x}{z}}$
60. $\log_b \sqrt{xy}$
61. $\log_{10} \sqrt[3]{\dfrac{xy^2}{z}}$
62. $\log_{10} \sqrt[5]{\dfrac{x^2 y}{z^3}}$

Express as a single logarithm with coefficient 1.

Example

$\dfrac{1}{2}(\log_b x - \log_b y)$

Solution

By Theorem 4.2-II and III,
$$\dfrac{1}{2}(\log_b x - \log_b y) = \dfrac{1}{2}\log_b \left(\dfrac{x}{y}\right) = \log_b \left(\dfrac{x}{y}\right)^{1/2}.$$

63. $\log_b 2x + 3 \log_b y$
64. $3 \log_b x - \log_b 2y$
65. $\dfrac{1}{2} \log_b x + \dfrac{2}{3} \log_b y$
66. $\dfrac{1}{4} \log_b x - \dfrac{3}{4} \log_b y$
67. $3 \log_b x + \log_b y - 2 \log_b z$
68. $\dfrac{1}{3}(\log_b x + \log_b y - 2 \log_b z)$
69. $\log_{10} (x - 2) + \log_{10} x - 2 \log_{10} z$
70. $\dfrac{1}{2}(\log_{10} x - 3 \log_{10} y - 5 \log_{10} z)$

Solve for x.

Example

$\log_{10} (x + 9) + \log_{10} x = 1$

Solution

By Theorem 4.2-I, the given equation is equivalent to
$$\log_{10} x(x + 9) = 1$$
for $x + 9 > 0$ and $x > 0$. Write this equation in exponential form to obtain
$$x(x + 9) = 10^1,$$
$$x^2 + 9x - 10 = 0.$$
Solve for x to obtain $x = -10$ and $x = 1$. Since neither $\log_{10}(-10 + 9)$ nor $\log_{10}(-10)$ is defined, -10 does not satisfy the original equation. Since 1 satisfies the original equation, the solution set is {1}.

71. $\log_{10} x + \log_{10} 2 = 3$
72. $\log_{10} (x - 1) - \log_{10} 4 = 2$
73. $\log_{10} x + \log_{10} (x + 21) = 2$
74. $\log_{10} (x + 3) + \log_{10} x = 1$
75. $\log_{10} (x + 2) + \log_{10} (x - 1) = 1$
76. $\log_{10} (x - 3) - \log_{10} (x + 1) = 1$

4.3 Special Logarithms and Powers

B 77. Show that $\log_b 1 = 0$ for $b > 0$, $b \neq 1$.

78. Show that $\log_b b = 1$ for $b > 0$, $b \neq 1$.

79. Prove Theorem 4.2-II.

80. Prove Theorem 4.2-III.

81. Graph $y = \log_b x$ for $b = 2, 4, 8$ on the same set of axes. What effect does increasing b have on the curves if $b > 1$?

82. Graph $y = \log_b x$ for $b = 1/2, 1/4, 1/8$ on the same set of axes. For $0 < b < 1$, where is $\log_b x < 0$? Where is $\log_b x > 0$?

4.3 Special Logarithms and Powers

The two logarithmic functions

$$y = \log_{10} x \quad \text{and} \quad y = \log_e x,$$

where e is the irrational number introduced in Section 4.1, are of special interest. We shall first consider the logarithm and exponential functions with the base 10.

Determination of $\log_{10} x$

Values for $\log_{10} x$ are called **logarithms to the base 10**, or **common logarithms**. From the definition of $\log_{10} x$,

$$10^{\log_{10} x} = x \quad (x > 0); \tag{1}$$

that is, $\log_{10} x$ is the exponent that must be placed on 10 so that the resulting power is x. Thus, $\log_{10} x$ can easily be determined for all values of x that are *integral* powers of 10 by inspection, directly from the definition of a logarithm:

$$\log_{10} 10 = \log_{10} 10^1 = 1,$$
$$\log_{10} 100 = \log_{10} 10^2 = 2,$$

and so forth; similarly,

$$\log_{10} 1 = \log_{10} 10^0 = 0,$$
$$\log_{10} 0.1 = \log_{10} 10^{-1} = -1,$$
$$\log_{10} 0.01 = \log_{10} 10^{-2} = -2,$$
$$\log_{10} 0.001 = \log_{10} 10^{-3} = -3.$$

Using tables

Values for $\log_{10} x$ in case x is not an integral power of 10 can be obtained by using a table of logarithms (Table II on pages 482 and 483) and Theorem 4.2-I or by using a calculator. We shall first consider how these values are obtained by using the table.

For $1 < x < 10$, the table on pages 482 and 483 provides values of $\log_{10} x$ directly. Consider the excerpt from this table shown in Figure 4.3. Each number in

x	0	1	2	3	4	5	6	7	8	9
3.8	0.5798	0.5809	0.5821	0.5832	0.5843	0.5855	0.5866	0.5877	0.5888	0.5899
3.9	0.5911	0.5922	0.5933	0.5944	0.5955	0.5966	0.5977	0.5988	0.5999	0.6010
4.0	0.6021	0.6031	0.6042	0.6053	0.6064	0.6075	0.6085	0.6096	0.6107	0.6117
4.1	0.6128	0.6138	0.6149	0.6160	0.6170	0.6180	0.6191	0.6201	0.6212	0.6222
4.2	0.6232	0.6243	0.6253	0.6263	0.6274	0.6284	0.6294	0.6304	0.6314	0.6325
4.3	0.6335	0.6345	0.6355	0.6365	0.6375	0.6385	0.6395	0.6405	0.6415	0.6425
4.4	0.6435	0.6444	0.6454	0.6464	0.6474	0.6484	0.6493	0.6503	0.6513	0.6522
4.5	0.6532	0.6542	0.6551	0.6561	0.6571	0.6580	0.6590	0.6599	0.6609	0.6618
4.6	0.6628	0.6637	0.6646	0.6656	0.6665	0.6675	0.6684	0.6693	0.6702	0.6712

Figure 4.3

the column headed by x represents the first two digits of x, while each number in the row opposite x contains the third digit of x. The digits located at the intersection of a row and a column form the logarithm of x. For example, $\log_{10} 4.25$, is at the intersection of the row opposite 4.2 under x and the column headed by 5. Thus,

$$\log_{10} 4.25 = 0.6284.$$

The equality sign is used here in an inexact sense; $\log_{10} 4.25 \approx 0.6284$ is more proper, because $\log_{10} 4.25$ is irrational and cannot be precisely represented by a rational number. We shall follow customary usage, however, writing $=$ instead of $\approx$, and leave the intent to the context.

Now suppose we wish to find $\log_{10}$ for values of x outside the range of the tables—that is, for $0 < x < 1$ or $x > 10$ as shown in color in Figure 4.4. This

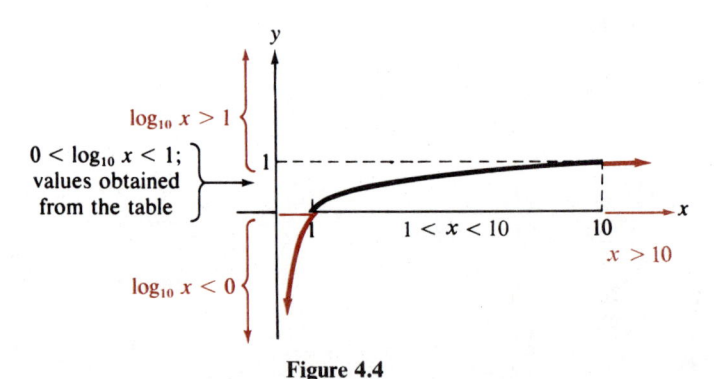

Figure 4.4

can be done quite readily by first representing the number in scientific notation and then applying Theorem 4.2-I.

Examples a. $\log_{10} 42.5 = \log_{10} (4.25 \times 10^1)$
$= \log_{10} 4.25 + \log_{10} 10^1$
$= 0.6284 + 1 = 1.6284$

b. $\log_{10} 425 = \log_{10} (4.25 \times 10^2)$
$= \log_{10} 4.25 + \log_{10} 10^2$
$= 0.6284 + 2 = 2.6284$

4.3 Special Logarithms and Powers

Characteristic and mantissa

Observe that in the preceding examples the decimal portion of the logarithm is always 0.6284 and the integral portion is just the exponent on 10 when the number is written in scientific notation. This process can be reduced to a mechanical one by considering $\log_{10} x$ to consist of two parts, an integral part (called the **characteristic**) and a nonnegative decimal fraction part (called the **mantissa**). Thus the table of values for $\log_{10} x$ for $1 < x < 10$ can be looked upon as a table of mantissas for $\log_{10} x$ for all $x > 0$.

Example

Find $\log_{10} 43700$.

Solution

We first write
$$\log_{10} 43700 = \log_{10} (4.37 \times 10^4).$$
From the table of logarithms on page 482, we find that $\log_{10} 4.37 = 0.6405$, so that
$$\log_{10} 43700 = 0.6405 + 4 = 4.6405.$$

Now consider an example of the form $\log_{10} x$, for $0 < x < 1$.

Example

Find $\log_{10} 0.00402$.

Solution

We first write
$$\log_{10} 0.00402 = \log_{10} (4.02 \times 10^{-3}).$$
Examining the table on page 482, we find that $\log_{10} 4.02 = 0.6042$. Upon adding 0.6042 to the characteristic -3, we obtain
$$\log_{10} 0.00402 = 0.6042 - 3 = -2.3958.$$

In the preceding example, the decimal portion -0.3958 of -2.3958 is negative. Because the table of common logarithms has only positive entries, if we plan to locate this number in the table in any further work, it is more convenient to leave the logarithm in the form $0.6042 - 3$, in which the decimal part is positive. Alternatively, we can write
$$\log_{10} 0.00402 = 0.6042 - 3$$
$$= 0.6042 + (7 - 10)$$
$$= 7.6042 - 10,$$
and again the decimal part is positive. There are many other representations such as $6.6042 - 9$, $5.6042 - 8$, etc. which are equally valid. In any case, *if we wish to use Table II, then the decimal part of the logarithm must be positive.* The characteristic may be positive, negative, or zero.

In case x is between 1 and 10 but is not in the table, we can approximate $\log_{10} x$ with the value in Table II corresponding to the value closest to x, or with a value between the two function values corresponding to the two values closest to x.†

Determination of 10^x

Since $y = 10^x$ and $x = \log_{10} y$ are equivalent equations, if, for a given value of x, we want to find the value of y or 10^x, we reverse the process outlined above.

Example

Find $10^{1.6395}$.

Solution

We begin by locating the mantissa, 0.6395, of the exponent 1.6395 in the body of Table II and observing that it is paired with 4.36. Thus,

$$10^{1.6395} = 10^{0.6395} \times 10^1$$
$$= 4.36 \times 10^1 = 43.6.$$

If the decimal portion of the given value for x is negative, then in order to use the table to determine 10^x, we must first rewrite x as a positive decimal plus a negative integer.

Example

Find $10^{-2.3605}$.

Solution

We first add $(+3 - 3)$ to -2.3605 to obtain a positive decimal part. Thus,

$$10^{-2.3605} = 10^{(-2.3605+3)-3}$$
$$= 10^{0.6395-3} = 10^{0.6395} \times 10^{-3}.$$

We then use Table II and note that $10^{0.6395} = 4.36$. Hence,

$$10^{-2.3605} = 10^{0.6395} \times 10^{-3}$$
$$= 4.36 \times 10^{-3} = 0.00436.$$

In case the value of x cannot be found in the body of Table II, we can approximate 10^x with the power of 10 corresponding to the value in the table closest to x or with a value between the two powers of 10 corresponding to the two values in the body of the table closest to x.

Antilog$_{10}$ x

Because the process for finding values of $\log_{10} x$ is reversed to find values of 10^x, we sometimes use the notation **antilog$_{10}$ x** for 10^x.

Determination of $\log_e x$ or $\ln x$

Values for $\log_e x$ are called **logarithms to the base e**, or **natural logarithms**. Frequently $\log_e x$ is denoted by **ln x**; we shall adopt that convention. Thus,

$$\ln x = \log_e x.$$

---

† There are interpolation procedures which yield more accurate values; however, in this section we will use the nearest value in the table.

4.3 Special Logarithms and Powers

From the definition of $\log_b x$,

$$e^{\ln x} = x \quad (x > 0);$$

that is, ln x is the exponent that must be placed on e so that the resulting power is x.

Table III on page 484 includes some values of ln x, $0.1 \leq x \leq 190$. Values for ln x outside this interval can be obtained in the same way that we obtained values for $\log_{10} x$, except that we no longer have integer characteristics.

Example   Find ln 270.

Solution   We first observe that $270 = 2.70 \times 10^2$. Using Table III, we have

$$\begin{aligned} \ln 270 = \ln (2.70 \times 10^2) &= \ln 2.70 + 2 \ln 10 \\ &= 0.9933 + 2(2.3026) \\ &= 0.9933 + 4.6052 \\ &= 5.5985. \end{aligned}$$

Example   Find ln 0.27.

Solution   This time, let us observe that $0.27 = 2.70 \times 10^{-1}$. Hence,

$$\begin{aligned} \ln 0.27 = \ln (2.70 \times 10^{-1}) &= \ln 2.70 + \ln 10^{-1} \\ &= \ln 2.70 - \ln 10 \\ &= 0.9933 - 2.3026 \\ &= -1.3093. \end{aligned}$$

As one would expect, the logarithm is negative. In this case we may leave the decimal part negative since we will be using a different table to compute powers of e.

Determination of e^x   The process used in the preceding examples can be reversed to find e^x. However, if a table for the exponential function $y = e^x$ is available (see Table I on page 481), e^x can be read directly from this table.

Examples   **a.** Find $e^{3.2}$.   **b.** Find $e^{-3.2}$.

Solutions   Reading directly from Table I, we obtain

     a. $e^{3.2} = 24.533$.   **b.** $e^{-3.2} = 0.0408$.

As in the case with base 10, we sometimes use **antilog$_e$** x to denote e^x.

Determination of b^x ($b \neq 10, b \neq e$) Since tables of values of b^x for $b \neq 10$ and $b \neq e$ are not readily available, to compute the value of b^x for bases other than 10 and e we must use Theorem 4.2-III as the following example illustrates.

Example Find $(1.01)^{14}$.

Solution We first let $y = (1.01)^{14}$. Since $\log_{10}$ is a function, we have
$$\log_{10} y = \log_{10}(1.01)^{14}.$$
We now use Theorem 4.2-III to write
$$\log_{10} y = 14 \log_{10} 1.01$$
$$= 14(0.0043)$$
$$= 0.0602.$$
Since the equations $\log_{10} y = 0.0602$ and $y = 10^{0.0602}$ are equivalent, we have
$$(1.01)^{14} = y = 10^{0.0602} = 1.15.$$

Using a calculator It is a simple process to find values for $\log_{10} x$, $\ln x$, 10^x, e^x, and b^x by using a calculator that has such function keys. Most calculators display these values in standard decimal form with more than four-decimal-place accuracy. To obtain values consistent with those obtained in the tables, we shall round off all results to an appropriate number of decimal places for the table involved.

Examples
a. $\log_{10} 0.234 = -0.6307814$
$\phantom{\log_{10} 0.234} = -0.6308$

b. $\ln 234 = 5.45532112$
$ = 5.4553$

c. $10^{0.7796} = 6.02004864$
$\phantom{10^{0.7796}} = 6.02$

d. $e^{0.24} = 1.27124915$
$\phantom{e^{0.24}} = 1.2712$

e. $10^{-0.6308} = 0.23399146$
$\phantom{10^{-0.6308}} = 0.234$

f. $e^{-2.4} = 0.09071795$
$\phantom{e^{-2.4}} = 0.0907$

Note that in Example **a** the decimal part of the logarithm is negative. Since we can obtain values for 10^x, $x < 0$, directly by using a calculator, as shown in Example **e**, there is no need to change -0.6308 to a number with a positive decimal part.

Accuracy It should be noted that, for values of x that do not appear in the tables, the values for $\log_{10} x$, 10^x, $\ln x$, and e^x that are obtained on a calculator are more accurate than those obtained by using an approximation from the tables.

4.3 Special Logarithms and Powers

Exercise 4.3

A *Find the characteristic of the given logarithm.*

Examples

    a. $\log_{10} 248$ **b.** $\log_{10} 0.0057$

Solutions

Represent the number in scientific notation.

  a. $\log_{10}(2.48 \times 10^2)$ **b.** $\log_{10}(5.7 \times 10^{-3})$

The exponent on the base 10 is the characteristic.

    2   -3, or $7 - 10$

1. $\log_{10} 312$ **2.** $\log_{10} 0.02$ **3.** $\log_{10} 0.00851$

4. $\log_{10} 8.012$ **5.** $\log_{10} 0.00031$ **6.** $\log_{10} 0.0004$

7. $\log_{10}(15 \times 10^3)$ **8.** $\log_{10}(820 \times 10^4)$

Find the logarithm directly from Table II or by using a hand calculator.

Examples

  a. $\log_{10} 16.8$ **b.** $\log_{10} 0.043$

Solutions using the tables

Represent the number in scientific notation.

  a. $\log_{10}(1.68 \times 10^1)$ **b.** $\log_{10}(4.3 \times 10^{-2})$

Determine the mantissa from the table.

    0.2253   0.6335

Add the characteristic as determined by the exponent on the base 10.

    1.2253 $0.6335 - 2$

Calculator solutions

  a. $\log_{10} 16.8 = 1.2253093$ **b.** $\log_{10} 0.043 = -1.36653154$

               $= 1.2253$ $= -1.3665$

In Example **b**, note that $0.6335 - 2$ is equal to -1.3665.

9. $\log_{10} 6.73$ **10.** $\log_{10} 891$ **11.** $\log_{10} 0.813$

12. $\log_{10} 0.00214$ **13.** $\log_{10} 0.08$ **14.** $\log_{10} 0.000413$

15. $\log_{10}(2.48 \times 10^2)$ **16.** $\log_{10}(5.39 \times 10^3)$ **17.** $\log_{10}(3.41 \times 10^{-4})$

18. $\log_{10}(6.42 \times 10^{-6})$ **19.** $\log_{10}(2.71 \times 10^2)$ **20.** $\log_{10}(5.42 \times 10^4)$

Find the power directly from Table II or by using a hand calculator.

Examples

a. $10^{2.7364}$
b. $10^{-1.1221}$

[$10^{2.7364}$ and $10^{-1.1221}$ are sometimes written as antilog$_{10}$ 2.7364 and antilog$_{10}$ (−1.2211), respectively.]

Solutions using the table

a. Locate the mantissa, 0.7364, in the body of Table II, and determine the associated power.

$$10^{2.7364} = 10^{0.7364+2} = 10^{0.7364} \times 10^2$$
$$= 5.45 \times 10^2 = 545$$

b. Add (+2 − 2) to −1.1221 to obtain 0.8779 − 2. Use Table II to find the associated power.

$$10^{-1.1221} = 10^{0.8779-2} = 10^{0.8779} \times 10^{-2}$$
$$= 7.55 \times 10^{-2} = 0.0755$$

Calculator solutions

a. $10^{2.7364} = 545.0043890$
$= 545$

b. $10^{-1.1221} = 0.07549184$
$= 0.0755$

21. $10^{0.6128}$ 22. $10^{0.2504}$ 23. $10^{0.5647}$
24. $10^{0.9258}$ 25. $10^{2.8470}$ 26. $10^{2.3874}$
27. $10^{3.3139}$ 28. $10^{2.8000}$ 29. $10^{-2.1851}$
30. $10^{-1.0991}$ 31. $10^{-2.3893}$ 32. $10^{-3.1543}$

Find the logarithm directly from Table III or by using a calculator.

Examples

a. ln 120
b. ln 0.075

Solutions using the table

a. ln 120 = ln (1.2 × 10²)
= ln 1.2 + ln 10²
= ln 1.2 + ln 100
= 0.1823 + 4.6052
= 4.7875

b. ln 0.075 = ln (7.5 × 10⁻²)
= ln 7.5 + ln 10⁻²
= ln 7.5 + (−2) ln 10
= 2.0149 + (−4.6052)
= −2.5903

Calculator solutions

a. ln 120 = 4.78749174
= 4.7875

b. ln 0.075 = −2.59026717
= −2.5903

33. ln 3   34. ln 8   35. ln 17   36. ln 98
37. ln 450   38. ln 650   39. ln 0.065   40. ln 0.035

4.3 Special Logarithms and Powers

Find the power directly from Table I or by using a calculator.

Examples
  a. $e^{0.85}$
  b. $e^{-3.4}$

[$e^{0.85}$ and $e^{-3.4}$ are sometimes written antilog$_e$ 0.85 and antilog$_e$ (−3.4), respectively.]

Solutions using the table

Obtain both values from Table I.

  a. $e^{0.85} = 2.3396$
  b. $e^{-3.4} = 0.0334$

Calculator solutions

  a. $e^{0.85} = 2.33964685$
     $= 2.3396$
  b. $e^{-3.4} = 0.03337327$
     $= 0.0334$

41. $e^{0.50}$ 42. $e^{1.5}$ 43. $e^{3.4}$ 44. $e^{4.5}$

45. $e^{0.23}$ 46. $e^{1.4}$ 47. $e^{-0.15}$ 48. $e^{-0.95}$

49. $e^{-2.5}$ 50. $e^{-4.2}$ 51. $e^{-0.25}$ 52. $e^{-3.6}$

Compute the following powers.

Example $(1.02)^8$

Solution using the table

Write $y = (1.02)^8$. Use the fact that $\log_{10}$ is a function and Theorem 4.2-III to write

$$\log_{10} y = \log_{10} (1.02)^8 = 8 \log_{10} (1.02)$$
$$= 8(0.0086) = 0.0688.$$

Use the fact that $\log_{10} y = 0.0688$ is equivalent to $y = 10^{0.0688}$ and use Table II to determine $10^{0.0688}$. Then

$$(1.02)^8 = y = 10^{0.0688} = 1.17.$$

Calculator solution $(1.02)^8 = 1.17165938 = 1.17$

53. $(1.01)^{12}$ 54. $(1.01)^7$ 55. $(0.99)^8$

56. $(0.98)^7$ 57. $(1.03)^8$ 58. $(1.02)^{10}$

B 59. By definition, $\log_e x = N$ can be written in exponential form as $e^N = x$. Show that $N = \log_{10} x / \log_{10} e$ and hence that

$$\log_e x = \frac{\log_{10} x}{\log_{10} e}.$$

60. Use the results of Exercise 59 to show that

$$\log_e x = 2.3026 \log_{10} x.$$

The following exercises require the use of a calculator. Evaluate each of the following functions at $x = 0.1, 0.01, 0.001,$ and 0.0001. Show four-decimal-place accuracy.

61. $f(x) = 1 - x^x$
62. $f(x) = (1 + x)^{1/x} - e$
63. $f(x) = x \ln x$
64. $f(x) = x (\ln x)^2$

4.4 Solution of Exponential Equations; Applications

In this section, calculator values of powers and logarithms are used in the examples and exercises, with answers rounded off to four decimal places. If you use the tables, slightly different values may result.

Solving exponential equations

When solving an exponential equation, $y = b^{f(x)}$, we are interested in finding the value for y, given a value of x, or in finding the value(s) for x, given a value of y.

Finding powers

The first case in which we want to find y or $b^{f(x)}$ for a given value of x is a direct application of the methods of Section 4.3.

Example

Solve $y = ce^{kt}$ for y, given that $c = 25$, $k = -5$, and $t = 0.5$.

Solution

Substituting 25, -5, and 0.5 for c, k, and t, respectively, we have

$$y = 25e^{-5(0.5)} = 25e^{-2.5}$$
$$= 25(0.0821) = 2.0521.$$

Example

The amount in a sample of a radioactive element is initially 25 grams. If the amount y (measured in grams) remaining at any time t is given by $y = 25e^{-0.5t}$, where t is in seconds, then how much of the element is remaining after 3 seconds?

Solution

Substituting 3 for t, we have

$$y = 25e^{-0.5(3)} = 25e^{-1.5}$$
$$= 25(0.2231) = 5.5783.$$

Thus, there is approximately 5.5783 grams of material remaining after 3 seconds.

Example

Solve $A = \left(1 + \dfrac{r}{n}\right)^{nt}$ for A, given that $r = 0.07$, $n = 4$, and $t = 3$.

4.4 Solution of Exponential Equations; Applications

Solution Substituting 0.07, 4, and 3 for r, n, and t, respectively, we have

$$A = \left(1 + \frac{0.07}{4}\right)^{4(3)} = (1.0175)^{12}.$$

Using the techniques of Section 4.3 or the y^x key on a calculator, we find that $A = 1.2314$.

Finding exponents For a given value of y, in order to solve an exponential equation $y = b^{f(x)}$ for x, it is necessary first to rewrite the equation so that the variable no longer appears in the exponent. In the cases when the base is either 10 or e, we simply rewrite the equation using logarithmic notation with the appropriate base.

Example Find the solution of $10^{2x} = 6$.

Solution Rewriting the equation in logarithmic notation with base 10, we have

$$2x = \log_{10} 6 = 0.7782.$$

Thus,

$$x = 0.3891.$$

Example The amount A that results from a deposit of P dollars after t years when interest is compounded continuously at an annual rate of $r\%$ is given by the formula

$$A = Pe^{rt/100}.$$

If a savings and loan company compounds interest continuously, and if the annual rate is $5\frac{3}{4}\%$, how long will it take for a deposit of $5000 to grow to $6000?

Solution Substituting appropriate values in the given formula, we obtain the equation

$$6000 = 5000e^{0.0575t}.$$

This is equivalent to

$$\frac{6}{5} = e^{0.0575t}.$$

Rewriting the equation in logarithmic notation with base e, we have

$$\ln \frac{6}{5} = 0.0575t,$$

from which

$$t = \frac{1}{0.0575} \ln \frac{6}{5} = 3.1708.$$

Hence, it takes slightly more than 3 years for the value of the account to reach $6000.

Example

If a particular component of a certain manufactured system shows no aging effect (in other words, it does not wear out with use), then the probability that the component will not fail within t years is

$$P = e^{-kt},$$

where $k > 0$ is a constant that depends on the component. If $k = 2$ for a given component, find t so that the probability that the component will not fail within t years is 1/2.

Solution

Since we want $P = 1/2$ and we know $k = 2$, we have

$$\frac{1}{2} = e^{-2t}.$$

Rewriting this equation in logarithmic notation with base e, we have

$$\ln\left(\frac{1}{2}\right) = -2t,$$

$$-0.6931 = -2t,$$

$$t = \frac{-0.6931}{-2} = 0.3466.$$

Hence the probability is 1/2 that the component will not fail within 0.3466 year.

The foregoing examples involved powers of 10 and e. When the base b is not 10 or e, logarithm tables for the base b are not readily available and most calculators do not have $\log_b$ keys; thus we shall use $\log_{10}$ and Theorem 4.2-III to rewrite the equation so that the variable does not appear in the exponent.

Example

Find the solution of $5^x = 7$.

Solution

We have

$$\log_{10} 5^x = \log_{10} 7;$$

from Theorem 4.2-III,

$$x \log_{10} 5 = \log_{10} 7.$$

Dividing each member by $\log_{10} 5$, we obtain

$$x = \frac{\log_{10} 7}{\log_{10} 5} = \frac{0.8451}{0.6990} \approx 1.2091.$$

Note that in seeking the decimal approximation, the logarithms are *divided*, not subtracted. It should be noted that the technique used in the preceding example is applicable in solving an exponential equation that involves any positive base b.

4.4 Solution of Exponential Equations; Applications

Exponential equations involving more than one variable can be solved for one of the variables in terms of the others.

Example Solve $y = Ce^{kt}$ for k in terms of y, C, and t.

Solution We first rewrite the equation as

$$\frac{y}{C} = e^{kt}.$$

We then rewrite this equation in logarithmic notation with the base e to obtain

$$\ln\left(\frac{y}{C}\right) = kt.$$

Solving for k, we have

$$k = \frac{1}{t} \ln\left(\frac{y}{C}\right).$$

Example Solve $y = (1 + r)^n$ for n in terms of y and r using logarithms to the base 10.

Solution By equating $\log_{10}$ of both sides we have

$$\log_{10} y = \log_{10} (1 + r)^n;$$

from Theorem 4.2-III,

$$\log_{10} y = n \log_{10} (1 + r).$$

Thus,

$$n = \frac{\log_{10} y}{\log_{10} (1 + r)}.$$

Exercise 4.4

A *Solve for the indicated variable. Show the solution to two decimal places.*

1. $y = k10^t$, for y given $k = 5$, $t = 0.2455$.
2. $y = k10^t$, for y given $k = 5$, $t = 2.2455$.
3. $y = ke^{2t}$, for y given $k = 0.01$, $t = 5$.
4. $y = ke^{-4t}$, for y given $k = 10$, $t = 0.5$.
5. $A = (1 + r)^n$, for A given $r = 0.01$, $n = 8$.
6. $A = (1 + r)^n$, for A given $r = 0.0025$, $n = 20$.

Solve. Give the solution in logarithmic form using the base 10 and also as an approximation in decimal form. Show four decimal place accuracy.

Example $10^{2x+1} = 4$

Solution Rewrite the equation in logarithmic notation with base 10; then solve for x.

$$2x + 1 = \log_{10} 4$$
$$2x = -1 + \log_{10} 4$$
$$x = -\frac{1}{2} + \frac{1}{2} \log_{10} 4$$

The solution is $-\frac{1}{2} + \frac{1}{2} \log_{10} 4 \approx -0.1990$.

7. $10^{3x} = 6$
8. $10^{-2x} = 15$
9. $10^{3x+1} = 9$
10. $10^{-2x+3} = 20$
11. $10^{x^2} = 150$
12. $10^{2x^2+1} = 200$

Solve. Give the solution in logarithmic form with base e and also as an approximation in decimal form. Show four-decimal-place accuracy.

Example $e^{3x+2} = 5$

Solution Rewrite the equation in logarithmic notation with base e; then solve for x.

$$3x + 2 = \ln 5$$
$$3x = -2 + \ln 5$$
$$x = -\frac{2}{3} + \frac{1}{3} \ln 5$$

The solution is $-\frac{2}{3} + \frac{1}{3} \ln 5 \approx -0.1302$.

13. $e^{3x} = 5$
14. $e^{-2x} = 10$
15. $e^{2x+1} = 25$
16. $e^{-x+2} = 8$
17. $e^{x^2/2} = 15$
18. $e^{x^2+1} = 4$

Solve. Give the solution in logarithmic form with base 10 and also as an approximation in decimal form. Show four-decimal-place accuracy.

Example $3^{x-2} = 16$

Solution Equate the $\log_{10}$ of both sides.

$$\log_{10} 3^{x-2} = \log_{10} 16$$

4.4 Solution of Exponential Equations; Applications

By Theorem 4.2-III,

$$(x - 2) \log_{10} 3 = \log_{10} 16,$$

from which

$$x - 2 = \frac{\log_{10} 16}{\log_{10} 3},$$

$$x = \frac{\log_{10} 16}{\log_{10} 3} + 2.$$

The solution is $\dfrac{\log_{10} 16}{\log_{10} 3} + 2 \approx 4.5237$.

19. $2^x = 7$
20. $3^x = 4$
21. $7^{2x-1} = 3$
22. $3^{x+2} = 10$
23. $4^{x^2} = 16$
24. $8^{x^2+1} = 64$

Solve. Leave the result in the form of an equation equivalent to the given equation.

25. $A = 10^{kt}$, for t
26. $A = B10^{-kt}$, for t
27. $y = e^{kt}$, for t
28. $y = Ce^{-kt}$, for t
29. $y = x^n$, for n
30. $y = Cx^{-n}$, for n

31. The atmospheric pressure p, in inches of mercury, is given approximately by the formula $p = 30.0(10^{-0.09a})$, where a is the altitude in miles above sea level. What is the atmospheric pressure at sea level? At 3 miles above sea level?

32. Using the information in Exercise 31, find the atmospheric pressure 6 miles above sea level.

33. The amount of a radioactive element remaining at any time t is given by $y = y_0 e^{-0.4t}$, where t is in seconds and y_0 is the amount present initially. How much of the element would remain after 3 seconds if 40 grams were present initially?

34. The number of bacteria present in a culture is given by the formula $N = N_0 e^{0.04t}$, where N_0 is the amount of bacteria present at time $t = 0$, and t is time in hours. If 10,000 bacteria are present 10 hours after the beginning of an experiment, how many were present when $t = 0$?

The pH (hydrogen potential) of a solution is given by

$$\text{pH} = -\log_{10} [\text{H}^+],$$

where $[\text{H}^+]$ is a numerical value for the concentration of hydrogen ions in aqueous solution in moles per liter.

Example Calculate the pH of a solution whose hydrogen ion concentration is 3.7×10^{-6}.

Solution overleaf

Solution Substitute 3.7×10^{-6} for $[H^+]$ in the relationship $pH = -\log_{10}[H^+]$

$$pH = -\log_{10}(3.7 \times 10^{-6})$$
$$= -(\log_{10} 3.7 + \log_{10} 10^{-6})$$
$$= -0.5682 + 6 \approx 5.4$$

35. Calculate the pH of a solution whose hydrogen ion concentration $[H^+]$ is 2.0×10^{-8}.
36. Calculate the pH of a solution whose hydrogen ion concentration is 6.3×10^{-7}.
37. Calculate the hydrogen ion concentration of a solution whose pH is 5.6.
38. Calculate the hydrogen ion concentration of a solution whose pH is 7.2.

P dollars invested at an annual interest rate r compounded yearly yields an amount A after n years, given by $A = P(1 + r)^n$. If the interest is compounded t times yearly, the amount is given by

$$A = P\left(1 + \frac{r}{t}\right)^{tn}.$$

Example One dollar compounded annually for 12 years yields \$1.90. What is the annual rate of interest to the nearest $\frac{1}{2}\%$?

Solution Use the given equation to obtain $(1 + r)^{12} = 1.90$. Equate $\log_{10}$ of each member and apply Theorem 4.2-III.

$$12 \log_{10}(1 + r) = \log_{10} 1.90 = 0.2788$$

Multiply each member by 1/12.

$$\log_{10}(1 + r) = \frac{1}{12}(0.2788) = 0.0232$$

Rewrite this equation in exponential notation.

$$1 + r = 10^{0.0232}$$

Determine $10^{0.0232}$ and solve for r.

$$1 + r = 10^{0.0232} = 1.0549$$

$$r = 0.0549, \quad \text{or} \quad r = 5\frac{1}{2}\%$$

The annual rate of interest is $5\frac{1}{2}\%$.

39. One dollar compounded annually for 10 years yields \$1.95. What is the annual rate of interest to the nearest $\frac{1}{2}\%$?
40. How many years (nearest year) would it take for \$1.00 to yield \$2.19 if compounded annually at 9%?

4.4 Solution of Exponential Equations; Applications

41. Find the compounded amount of $5000 invested at an annual rate of $9\frac{1}{2}\%$ for 10 years when compounded annually; when compounded semiannually.

42. Two men, A and B, each invested $10,000 at an annual rate of $9\frac{1}{2}\%$ for 20 years with a bank that computed interest quarterly. A withdrew his interest at the end of each 3-month period, but B let his investment be compounded. How much more than A did B earn over the period of 20 years?

If S dollars is borrowed at an annual interest rate r (simple interest) and is to be repaid in equal monthly payments over a period of n months, then each monthly payment p is given by

$$p = S\left[\frac{r/12}{1 - (1 + r/12)^{-n}}\right].$$

Example A person borrows $64,000 at 12% annual simple interest for a period of 30 years. What are the monthly payments?

Solution Using the given equation, we have

$$p = 64{,}000\left[\frac{0.01}{1 - (1.01)^{-360}}\right] = 658.31.$$

The monthly payments are $658.31.

In Exercises 43–46, assume simple interest.

43. A family buying a house costing $95,000 has $19,000 to use as a down payment. If they borrow the remaining amount at an annual rate of $11\frac{3}{4}\%$ for a 30-year period, what is their monthly payment?

44. A person wishes to borrow $100,000. One loan company will lend it at $11\frac{1}{2}\%$ annual interest for a period of 30 years. A second loan company will lend it at 11% annual interest but only for a period of 25 years. Which loan results in the lower monthly payment, and how much difference is there in the monthly payments?

45. A family wishes to buy a house costing $125,000. They can obtain a loan at 12% annual interest. They want to pay the loan off in 30 years. If they want to keep their monthly house payment at $1000, how much of the $125,000 do they have to have as a down payment?

46. Suppose in Exercise 45 the family had $50,000 to use as a down payment. How long would the term of the loan have to be (to the nearest month) to keep the monthly payments at $1000?

The present value of A dollars in t years when compounded continuously at $r\%$ annually is the amount that must be deposited in an account at the present so that it is worth A dollars in t years. The formula for present value (PV) is

$$PV = Ae^{-rt/100}$$

47. What is the annual rate of interest of an account if when compounded continuously the present value of $10,000 in 10 years is $5000?

48. At an annual interest rate of 7.25% when compounded continuously, how long must $1000 be kept in an account to attain a value of $2000?

49. If a component in a system shows no aging effect and the probability of failure after 10 units of time is 1/10, using the formula on page 138, find the value k associated with this component. The value $1/k$ is the average time until a failure occurs.

50. If a component in a system shows no aging effect and the probability of failure after 3 units of time is 0.001, find the average time until a failure occurs (see Exercise 49).

Chapter Review

[4.1] 1. Graph $y = 3^x$. 2. Graph $y = 3^{5x}$.

[4.2] *Express in logarithmic notation.*

3. $16^{-1/2} = \dfrac{1}{4}$ 4. $7^3 = 343$

Express in exponential notation.

5. $\log_2 8 = 3$ 6. $\log_{10} 0.0001 = -4$

Solve.

7. $\log_2 16 = y$ 8. $\log_{10} x = 3$

Express as the sum or difference of simpler logarithmic quantities.

9. $\log_{10} \sqrt[3]{xy^2}$ 10. $\log_{10} \dfrac{2R^3}{\sqrt{PQ}}$

Express as a single logarithm with coefficient 1.

11. $2 \log_b x - \dfrac{1}{3} \log_b y$ 12. $\dfrac{1}{3}(2 \log_{10} x + \log_{10} y) - 3 \log_{10} z$

[4.3] *Find the logarithm. Show four-decimal-place accuracy.*

13. $\log_{10} 42$
14. $\log_{10} 0.00314$
15. $\log_{10} 682$
16. $\log_{10} 0.0414$

Find the power. Show three-decimal-place accuracy.

17. $10^{1.8287}$
18. $10^{0.4240}$
19. $10^{-1.3316}$
20. $10^{-0.1776}$

Find the logarithm. Show four-decimal-place accuracy.

21. $\ln 7$
22. $\ln 23$
23. $\ln 241$
24. $\ln 510$

Find the power. Show four-decimal-place accuracy.

25. $e^{6.0}$
26. $e^{1.8}$
27. $e^{-3.5}$
28. $e^{-0.42}$

Find the power. Show three-decimal-place accuracy.

29. $(1.01)^{36}$
30. $(0.99)^{36}$

[4.4] *Solve. Give the solution in logarithmic form using the base 10 and also as an approximation in decimal form. Show four-decimal-place accuracy.*

31. $10^{-2x} = 8$
32. $10^{-x-2} = 25$

Solve. Give the solution in logarithmic form with the base e and also as an approximation in decimal form. Show four-decimal-place accuracy.

33. $e^{4x} = 8$
34. $e^{-2x+2} = 16$

Solve. Give the solution in logarithmic form with the base 10 and also as an approximation in decimal form. Show four-decimal-place accuracy.

35. $3^x = 2$
36. $3^{x+1} = 80$

Solve. Leave the result in the form of an equation equivalent to the given equation.

37. $y = A + ke^{-t}$, for t
38. $y = A + ke^{-(t+c)}$, for t

Solve. Use the appropriate formula from pages 138 and 141–143.

39. One dollar compounded quarterly for 5 years yields $1.50. What is the annual rate of interest to the nearest 1/4%?

40. Calculate the hydrogen ion concentration of a solution if its pH is 6.2.

41. Find the interest rate at which $10,000 has a present value of $7000 in 5 years with continuously compounded interest.

42. Find the average time until failure of a component that shows no aging effect if the probability of failure after 3 units of time is 0.01.

43. A person borrows $25,000 at 18% annual simple interest. If the loan is to be repaid in 5 years, what are the monthly payments?

44. A person borrows $30,000 at 15% annual simple interest. If the monthly payments are $400, how long is the term of the loan?

Supplemental Exercises for Chapters 1–4

The problems in this supplement are significantly more difficult than those in the exercise sets at the end of each section and the chapter reviews. They are designed to challenge the student, and a solution may draw on a combination of ideas and techniques from any of the preceding chapters. Many of these problems contain results that are used in more advanced classes in mathematics.

Find the real solutions of each equation.

1. $(x - 2)^3(4)(x + 1)^3 + (x + 1)^4(3)(x - 2)^2 = 0$

2. $2x(x + 1)^3(x + 2) + 3x^2(x + 1)^2(x + 2) + x^2(x + 1)^3 = 0$

3. $(6x - 7)^3(2)(8x^2 + 9)(16x) + (8x^2 + 9)^2(3)(6x - 7)^2(6) = 0$

4. $\dfrac{m^2}{(3m + 1)\sqrt{m^2 + 1}} + \dfrac{-\sqrt{m^2 + 1}}{(3m + 1)^2} = 0$

5. Solve $\pi R^3 \left[\dfrac{2y(R - y)(R + y) - (R + y)^2(R - 2y)}{(R - y)^2 y^2} \right] = 0$ for y in terms of R.

6. Solve $\dfrac{-2kA}{x^3} + \dfrac{2kB}{(L - x)^3} = 0$ for x in terms of k, L, A, and B. Assume $A \neq -B$.

7. Solve $\dfrac{2(V_0 - Cx)Cx - C(V_0 - Cx)^2}{C^2 x^2} = 0$ for x in terms of V_0 and C. Assume $C \neq 0$, $C \neq -2$.

8. Solve $\dfrac{2(Ax - B)A(Bx - A)^2 - 2(Bx - A)B(Ax - B)^2}{(Bx - A)^4} = 0$ for x in terms of A and B. Assume $A \neq 0$, $B \neq 0$, and $B \neq A$.

Solve. Represent the solution set in interval notation and also graphically.

9. $\dfrac{4}{3}(x^{1/3} + x^{-2/3}) > 0$

10. $\dfrac{x^{3/2} - (\frac{3}{2})(x - 4)x^{1/2}}{x^3} > 0$

11. $\dfrac{5x - 15\sqrt{x^2 + 4}}{75\sqrt{x^2 + 4}} < 0$

12. $x\left(\dfrac{-2x}{2\sqrt{4 - x^2}}\right) + \sqrt{4 - x^2} > 0$

13. $-\dfrac{2}{9}(x^2 + 4)^{-4/3}(4x^2) + \left(\dfrac{4}{3}\right)(x^2 + 4)^{-1/3} > 0$

14. $2(x^2 + 1)^{2/3} + (\frac{2}{3})(x^2 + 1)^{-2/3}(4x^2) > 0$

15. $e^x - e^{-x} \geq 0$

16. $e^x - e^{-2x} \geq 0$

Graph each of the following functions.

17. $f(x) = 2^{-x^2}$

18. $f(x) = 2^{x^2+1}$

19. $f(x) = \begin{cases} x^2 & \text{if } x \leq 1 \\ 2^x & \text{if } x > 1 \end{cases}$

20. $f(x) = \begin{cases} e^{-x} & \text{if } x \leq 0 \\ -x + 1 & \text{if } 0 < x \leq 1 \\ \ln x & \text{if } x > 1 \end{cases}$

21. $f(x) = \log_{10} x^2$

22. $f(x) = \log_{10}\left(\dfrac{1}{x^2}\right)$

23. $f(x) = \ln(1 - x^2)$

24. $f(x) = \ln(x^2 - 1)$

25. $P(x) = x^4 - \dfrac{8}{3}x^3 - 10x^2 + 24x + 20$

26. $P(x) = 3x^4 - 8x^3 + 6x^2$

27. $P(x) = \dfrac{x^5}{5} + \dfrac{x^4}{2} + \dfrac{x^3}{3}$

28. $f(x) = \dfrac{x^4}{x^3 - x}$

29. $f(x) = \dfrac{x^3 + x}{x^2 + x - 2}$

30. $R(x) = \dfrac{x^4 - x^3 + 1}{x - 1}$

31. $f(x) = \left|\,||x| - 1| - \dfrac{1}{2}\right|$

32. $f(x) = \left|\,||2x| - 4| - 2\right|$

Find the range of each of the following functions. Hint: Solve for x in terms of y and determine allowable y values, or graph the given function and determine the range from the graph.

33. $y = \dfrac{1 - x^2}{x^2}$

34. $y = \dfrac{1}{\sqrt{1 - x}}$

35. $y = 2^{-x^2}$

36. $y = \ln(1 - x^2)$

Find $f^{-1}(x)$ for the given function f.

37. $f(x) = e^x - e^{-x}$

38. $f(x) = e^x + e^{-x}, \quad x \geq 0$.

Supplemental Exercises for Chapters 1–4 149

39. Use similar triangles to show that if the point $P(x, y)$ divides the segment from $P_1(x_1, y_1)$ to $P_2(x_2, y_2)$ in the ratio m_1/m_2; that is, if $P_1P/PP_2 = m_2/m_1$, then

$$x = \frac{m_1}{m_1 + m_2} x_1 + \frac{m_2}{m_1 + m_2} x_2$$

and

$$y = \frac{m_1}{m_1 + m_2} y_1 + \frac{m_2}{m_1 + m_2} y_2.$$

40. Assume a and b are known positive constants with $a \neq 1$ and $b \neq 1$. Find $\log_a x$ in terms of $\log_b x$.

5   Circular Functions

In Chapter 4 we discussed the exponential and logarithmic functions, which are useful in describing growth. Neither these functions, however, nor in fact the polynomial or rational functions that we have also considered, would be useful in describing cyclic phenomena. In this chapter we shall discuss the *circular functions*; because of their periodic nature, these are particularly appropriate in studying such phenomena.

5.1   The Functions Sine and Cosine

Arc on a unit circle

Consider, intuitively, a point moving steadily in a counterclockwise direction around the **unit circle**, that is, the circle with equation $x^2 + y^2 = 1$. At any given time, the moving point occupies a position on the circle; the point, in turn, is associated with an ordered pair (x, y). If the distance along the circle from the point $(1, 0)$ to the point (x, y) is designated by s, then we can associate the real number s with the ordered pair (x, y) (Figure 5.1-**a**). We shall assume that there is

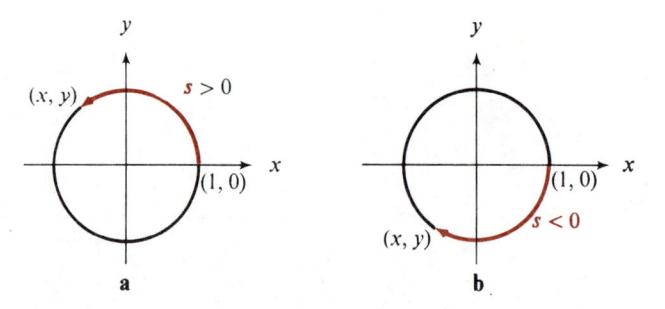

Figure 5.1

5.1 The Functions Sine and Cosine

a one-to-one correspondence between the set of nonnegative real numbers and the lengths of all arcs of the unit circle measured in the counterclockwise direction from the point (1, 0) to points (x, y) on the circle. We shall also assume that there is a one-to-one correspondence between the sets of negative real numbers and the lengths of all arcs of the circle measured in the clockwise direction from the point (1, 0) to points (x, y) on the circle (Figure 5.1-**b**).

Since the circumference of a circle is $2\pi r$ and the radius of the unit circle is 1, the distance once around the unit circle is 2π, twice around is 4π, and so on. Furthermore, the distance halfway around is π, one-fourth of the way around is $\pi/2$, and so forth. The endpoints of selected arcs from the point (1, 0) and their corresponding arc lengths are shown in Figure 5.2. The arcs in **a** are in the *counterclockwise direction*, and the arcs in **b** are in the *clockwise direction*. Especially

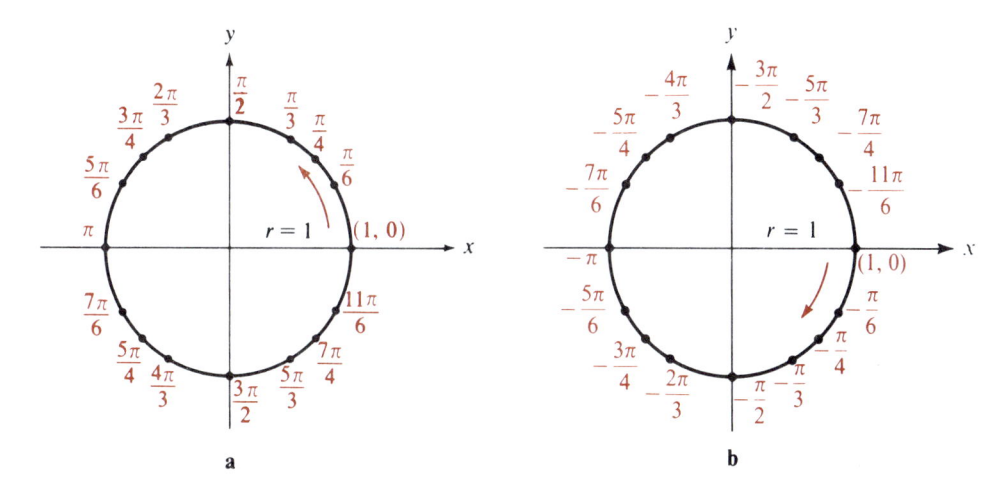

Figure 5.2

useful are the two functions that associate a real number with the components of the ordered pairs (x, y) corresponding to points on the unit circle. These functions are called the **cosine** and the **sine**.

Definition 5.1 If (x, y) is the point at arc length s from (1, 0) on the unit circle, then the **cosine of** s is x and the **sine of** s is y. This is denoted by writing

$$\cos s = x \quad \text{and} \quad \sin s = y.$$

In other words, the first component of the point (x, y) located at arc length s from (1, 0) on the unit circle is called the cosine of s, and the second component is called the sine of s; thus, $(x, y) = (\cos s, \sin s)$.

Domain and range of cosine and sine

The domain of each of these functions is the set R. The range of the cosine function is the set of all first components of the ordered pairs corresponding to points on the unit circle, and hence is the set $\{x | -1 \le x \le 1\}$; the range of the sine function is $\{y | -1 \le y \le 1\}$, the set of all second components.

Signs of cosine and sine Because cos s and sin s are simply the coordinates of points on the unit circle, they are positive or negative in accord with the values of x and y in the various quadrants. Table 5.1 summarizes in a convenient way the sign associated with cos s and sin s in each quadrant.

Table 5.1

Quadrant II	Quadrant I
x or cos s < 0	x or cos s > 0
y or sin s > 0	y or sin s > 0
Quadrant III	Quadrant IV
x or cos s < 0	x or cos s > 0
y or sin s < 0	y or sin s < 0

cos s and sin s in terms of each other Because cos $s = x$ and sin $s = y$ are subject to the condition that $x^2 + y^2 = 1$, we have the following basic identity relating cos s and sin s.

Theorem 5.1   For every $s \in R$,

$$\cos^2 s + \sin^2 s = 1. \tag{1}$$

Note that for convenience we write $\cos^2 s$ for $(\cos s)^2$ and $\sin^2 s$ for $(\sin s)^2$. Now Theorem 5.1 and Table 5.1 can be used to write

$$\sin s = \begin{cases} \sqrt{1 - \cos^2 s} & \text{in Quadrants I and II,} \quad (2a) \\ -\sqrt{1 - \cos^2 s} & \text{in Quadrants III and IV,} \quad (2b) \end{cases}$$

and

$$\cos s = \begin{cases} \sqrt{1 - \sin^2 s} & \text{in Quadrants I and IV,} \quad (3a) \\ -\sqrt{1 - \sin^2 s} & \text{in Quadrants II and III.} \quad (3b) \end{cases}$$

Therefore, if either sin s or cos s is known and the quadrant in which the terminal point of the arc of length s lies can be determined, we can find the value for the other function.

Example Given that cos $s = -3/5$ and $\pi < s < 3\pi/2$, find sin s.

Solution Using (2b) above, we have

$$\sin s = -\sqrt{1 - \cos^2 s}$$
$$= -\sqrt{1 - \left(-\frac{3}{5}\right)^2} = -\sqrt{\frac{16}{25}} = -\frac{4}{5}.$$

5.1 The Functions Sine and Cosine

Exercise 5.1

Determine the point on the figure that corresponds to the given arc length when measured from the point (1, 0). Do not refer to Figure 5.2.

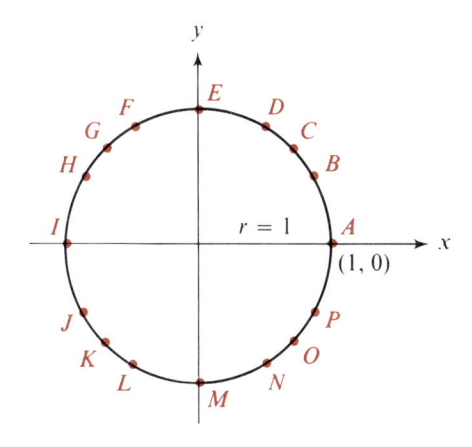

Examples

a. $\dfrac{3\pi}{2}$

b. $-\dfrac{\pi}{4}$

Solutions

a. Since the circumference of the circle is 2π, and $3\pi/2$ is three-fourths of 2π, the point corresponding to the length $3\pi/2$ is three-fourths of the way around the circle in the counter-clockwise direction from point A. The point is M.

b. Since the circumference of the circle is 2π, and $\pi/4$ is one-eighth of 2π, the point corresponding to the length $-\pi/4$ is one-eighth of the way around the circle in the clockwise direction from point A. The point is O.

1. π
2. $\dfrac{\pi}{2}$
3. $\dfrac{\pi}{4}$
4. $\dfrac{\pi}{3}$
5. $\dfrac{5\pi}{6}$
6. $\dfrac{2\pi}{3}$
7. $\dfrac{5\pi}{4}$
8. $\dfrac{11\pi}{6}$
9. 3π
10. $\dfrac{7\pi}{2}$
11. $\dfrac{13\pi}{4}$
12. $\dfrac{13\pi}{6}$
13. $-\dfrac{\pi}{2}$
14. $-\dfrac{3\pi}{2}$
15. $-\dfrac{3\pi}{4}$
16. -3π

Use Table 5.1 to state whether the given function value is positive or negative.

Examples

a. $\sin \dfrac{11\pi}{4}$

b. $\cos \dfrac{17\pi}{6}$

Solution overleaf

Solutions

a. Since $11\pi/4$ terminates in the second quadrant, we find from Table 5.1 that $\sin 11\pi/4$ is positive.

b. Since $17\pi/6$ terminates in the second quadrant, we find from Table 5.1 that $\cos 17\pi/6$ is negative.

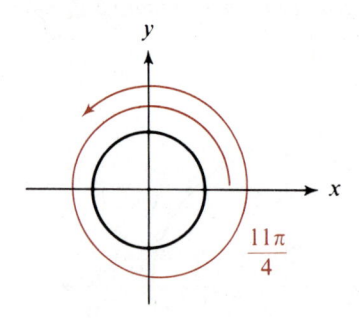

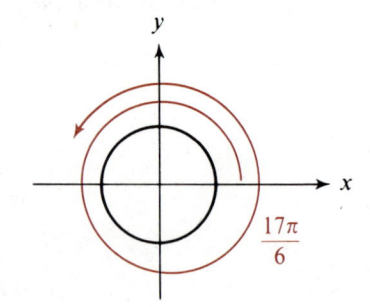

17. $\cos \dfrac{13\pi}{3}$

18. $\cos \dfrac{27\pi}{5}$

19. $\sin \dfrac{17\pi}{7}$

20. $\sin \dfrac{29\pi}{4}$

21. $\cos \dfrac{41\pi}{5}$

22. $\sin \dfrac{36\pi}{7}$

23. $\sin \left(-\dfrac{\pi}{6}\right)$

24. $\cos \left(-\dfrac{\pi}{3}\right)$

25. $\sin \left(-\dfrac{4\pi}{3}\right)$

26. $\cos \left(-\dfrac{7\pi}{5}\right)$

27. $\cos \left(-\dfrac{31\pi}{4}\right)$

28. $\sin \left(-\dfrac{17\pi}{3}\right)$

In Exercises 29–38, find the required function value and state the quadrant in which s terminates.

Example $\sin s$, given that $\cos s = \sqrt{7}/4$ and $\sin s < 0$.

Solution Since $\cos s = \sqrt{7}/4$ and $\sin s < 0$, s terminates in Quadrant IV; and we see from (2b) on page 152 that

$$\sin s = -\sqrt{1 - \cos^2 s} = -\sqrt{1 - \left(\dfrac{\sqrt{7}}{4}\right)^2}$$

$$= -\sqrt{1 - \dfrac{7}{16}} = -\sqrt{\dfrac{9}{16}} = -\dfrac{3}{4}.$$

29. $\cos s$, given that $\sin s = 3/5$ and $\cos s > 0$.
30. $\sin s$, given that $\cos s = 4/5$ and $\sin s < 0$.
31. $\sin s$, given that $\cos s = 5/13$ and $\sin s < 0$.
32. $\cos s$, given that $\sin s = 12/13$ and $\cos s < 0$.
33. $\cos s$, given that $\sin s = -2/3$ and $\cos s > 0$.

5.2 Special Function Values of Cosine and Sine

34. sin s, given that cos s = −1/4 and sin s > 0.

35. sin s, given that cos s = −$\sqrt{3}/2$ and sin s < 0.

36. cos s, given that sin s = −$\sqrt{3}/2$ and cos s > 0.

37. cos s, given that sin s = −3/5 and cos s > 0.

38. sin s, given that cos s = −4/5 and sin s > 0.

5.2 Special Function Values of Cosine and Sine

Quadrantal function values

In general, finding cos s and sin s for a given real number s is a difficult matter. However, some values of cos s and sin s, corresponding to special values of s, are readily available.

Figure 5.3 shows the coordinates of four points on the unit circle with which we can associate specific values of s. For s = 0,

$$\cos 0 = 1 \quad \text{and} \quad \sin 0 = 0. \qquad (1)$$

Furthermore, since the arc of a circle included in a quadrant is one-fourth of the circumference of the circle, that is, $2\pi/4 = \pi/2$, we know immediately that

$$\cos \frac{\pi}{2} = 0 \quad \text{and} \quad \sin \frac{\pi}{2} = 1, \qquad (2)$$

$$\cos \pi = -1 \quad \text{and} \quad \sin \pi = 0, \qquad (3)$$

$$\cos \frac{3\pi}{2} = 0 \quad \text{and} \quad \sin \frac{3\pi}{2} = -1. \qquad (4)$$

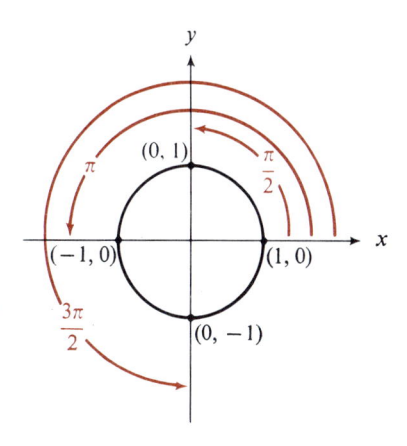

Figure 5.3

Notice that in finding values for cos s and sin s we are finding *coordinates of points* on the unit circle. The function values obtained in (1), (2), (3), and (4) are sometimes called **quadrantal values**, because the terminal point of each arc lies on an axis.

We can find other special values for cos s and sin s by using the geometry of the unit circle and the distance formula,

$$d^2 = (x_2 - x_1)^2 + (y_2 - y_1)^2.$$

Values for
s = π/4

First consider $s = \pi/4$. Figure 5.4 shows the unit circle and the designated value for s. Since (x, y) bisects the arc from $(1, 0)$ to $(0, 1)$, it follows from geometric considerations that $x = y$. Because $x^2 + y^2 = 1$, we have

$$x^2 + x^2 = 1,$$
$$2x^2 = 1,$$
$$x^2 = \frac{1}{2},$$

and

$$x = \frac{1}{\sqrt{2}} \quad \text{or} \quad x = -\frac{1}{\sqrt{2}}.$$

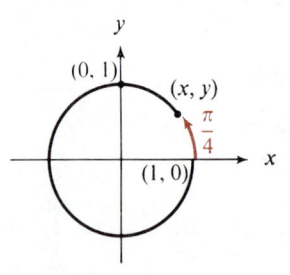

Figure 5.4

Now, both x and y are positive in the first quadrant, so the desired value for x is $1/\sqrt{2}$. Since $x = y$, y is also equal to $1/\sqrt{2}$, and hence we have

$$\cos\frac{\pi}{4} = \frac{1}{\sqrt{2}} \quad \text{and} \quad \sin\frac{\pi}{4} = \frac{1}{\sqrt{2}}.$$

Values for
s = π/6

Next, consider $s = \pi/6$, as pictured in Figure 5.5. If the ordered pair corresponding to $s = \pi/6$ is (x, y), then the ordered pair corresponding to $s = -\pi/6$ is $(x, -y)$. Now the arc from $(x, -y)$ to (x, y) is of length

$$\frac{\pi}{6} + \frac{\pi}{6} = \frac{\pi}{3}$$

and so is the length of the arc from (x, y) to $(0, 1)$. Because equal arcs of a circle subtend equal chords, the distance from $(0, 1)$ to (x, y) is the same as the distance from (x, y) to $(x, -y)$. Using the distance formula, we therefore have

$$(x - 0)^2 + (y - 1)^2 = (x - x)^2 + (-y - y)^2$$

or

$$x^2 + y^2 - 2y + 1 = 4y^2.$$

Since $x^2 + y^2 = 1$, we substitute 1 for $x^2 + y^2$ in this equation and obtain

$$1 - 2y + 1 = 4y^2,$$
$$4y^2 + 2y - 2 = 0,$$
$$2(2y - 1)(y + 1) = 0,$$

so that

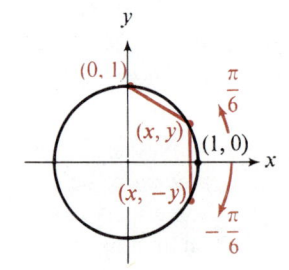

Figure 5.5

$$y = \frac{1}{2} \quad \text{or} \quad y = -1.$$

5.2 Special Function Values of Cosine and Sine

Because (x, y) is in the first quadrant, we must select the value $1/2$ as the y-coordinate. Next, we use $x^2 + y^2 = 1$ to find a value for x. Thus,

$$x^2 + \left(\frac{1}{2}\right)^2 = 1,$$

from which

$$x = \pm\frac{\sqrt{3}}{2}.$$

Because (x, y) is in the first quadrant, we have $x = \sqrt{3}/2$, and it follows that

$$\cos\frac{\pi}{6} = \frac{\sqrt{3}}{2} \quad \text{and} \quad \sin\frac{\pi}{6} = \frac{1}{2}. \tag{5}$$

Values for $s = \pi/3$

From Equation (5), we can quickly find values for $\cos(\pi/3)$ and $\sin(\pi/3)$ by symmetry. Figure 5.6 shows the point (x, y) associated with s equal to $\pi/3$; the abscissa of (x, y) is the ordinate of $(\sqrt{3}/2, 1/2)$, and the ordinate of (x, y) is the abscissa of $(\sqrt{3}/2, 1/2)$. Hence

$$\cos\frac{\pi}{3} = \frac{1}{2} \quad \text{and} \quad \sin\frac{\pi}{3} = \frac{\sqrt{3}}{2}.$$

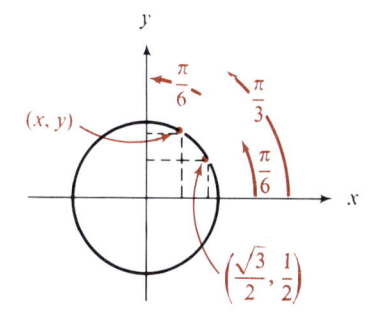

Figure 5.6

For ease of reference, the values of $\cos s$ and $\sin s$ for the special values of s discussed above are given in Table 5.2.

Table 5.2

s	0	$\dfrac{\pi}{6}$	$\dfrac{\pi}{4}$	$\dfrac{\pi}{3}$	$\dfrac{\pi}{2}$	π	$\dfrac{3\pi}{2}$
$\cos s$	1	$\dfrac{\sqrt{3}}{2}$	$\dfrac{1}{\sqrt{2}}$	$\dfrac{1}{2}$	0	-1	0
$\sin s$	0	$\dfrac{1}{2}$	$\dfrac{1}{\sqrt{2}}$	$\dfrac{\sqrt{3}}{2}$	1	0	-1

Values for $0 < s < \dfrac{\pi}{2}$ using a table

The values in Table 5.2 were obtained using geometric considerations. These techniques are not adequate for most values of s, but all values can be approximated using more sophisticated methods.

In Table IV on pages 485–488, decimal approximations are given for $\sin s$ and $\cos s$ for selected values of s between 0 and $\pi/2 (\approx 1.57)$. The table is graduated in

Examples   **a.** $\cos 0.73 \approx 0.7452$ **b.** $\sin 1.29 \approx 0.9608$

Notice that in Table IV the letter *x* is used to represent an element in the domain. This is customary in many tables. In this usage, of course, *x* can be thought of as representing an arc length along the unit circle, just as *s* did. In fact, because *x* is ordinarily the variable used to represent an element in the domain of a function, we shall hereafter use x, t, or other symbols where we heretofore used s.

Using a calculator   A scientific calculator can be used to obtain the function values listed in Table IV. Furthermore, it enables us to find function values for numbers with more than three digits. Because the operating procedures for different types of calculators vary, you should refer to the instruction booklet to obtain this information for your particular calculator.

The circular-function values are obtained on a calculator by using the radian (RAD or "R") mode. (Radians will be discussed in detail in Chapter 6.) The number of digits that are shown in the display of a calculator vary. To be consistent with values obtained from Table IV, all values obtained using a calculator will be rounded off to four places.

Examples   Find:

a. $\cos 0.73$ **b.** $\sin 1.43$

Solutions   Set the calculator in radian mode.

a. $\cos 0.73 \approx 0.74517 = 0.7452$ **b.** $\sin 1.43 \approx 0.99010 = 0.9901$

Notice that whether we use the tables or a hand calculator, the values that we obtain for cos *x* and sin *x* are, in most cases, approximations. However, for convenience, we shall generally use the symbol "=" instead of "≈" in relationships that involve such approximations.

Exercise 5.2

A *Use Table 5.2 to find the value of the given expression.*

Examples   **a.** $\cos \dfrac{\pi}{2} + \sin \pi$ **b.** $\left(\sin \dfrac{\pi}{2}\right)(\cos \pi)$

5.2 Special Function Values of Cosine and Sine

Solutions

a. From Table 5.2,
$$\cos\frac{\pi}{2} = 0 \quad \text{and} \quad \sin \pi = 0.$$
Thus,
$$\cos\frac{\pi}{2} + \sin \pi = 0.$$

b. From Table 5.2,
$$\sin\frac{\pi}{2} = 1 \quad \text{and} \quad \cos \pi = -1.$$
Thus,
$$\left(\sin\frac{\pi}{2}\right)(\cos \pi) = -1.$$

1. $\cos\dfrac{\pi}{3}$
2. $\cos\dfrac{\pi}{6}$
3. $\sin\dfrac{\pi}{4}$
4. $\sin\dfrac{\pi}{6}$
5. $\cos 0$
6. $\cos\dfrac{3\pi}{2}$
7. $\sin\dfrac{\pi}{2}$
8. $\sin \pi$
9. $\cos\dfrac{\pi}{4} + \sin\dfrac{\pi}{2}$
10. $\sin\dfrac{\pi}{3} + \cos\dfrac{\pi}{6}$
11. $\left(\sin\dfrac{\pi}{4}\right)\left(\cos\dfrac{\pi}{4}\right)$
12. $(\cos 0)\left(\sin\dfrac{3\pi}{2}\right)$

Use a hand calculator or Table IV to find the approximate value of the given expression.

Example $\cos 1.43 + \sin 0.02$

Solution   From Table IV or a calculator, we find that
$$\cos 1.43 = 0.1403 \quad \text{and} \quad \sin 0.02 = 0.0200.$$
Thus,
$$\cos 1.43 + \sin 0.02 = 0.1403 + 0.0200 = 0.1603.$$

13. $\cos 0.33$
14. $\cos 0.67$
15. $\sin 0.75$
16. $\sin 0.25$
17. $\cos 1.41$
18. $\cos 1.17$
19. $\sin 1.34$
20. $\sin 1.05$
21. $\cos 0.11$
22. $\cos 0.64$
23. $\sin 1.25$
24. $\sin 1.48$
25. $\cos 1.17 + \sin 0.11$
26. $\cos 1.08 + \cos 1.00$
27. $(\sin 0.01)(\cos 1.10)$
28. $(\sin 1.50)(\cos 1.50)$
29. $\dfrac{\sin 0.15}{\cos 1.51}$
30. $\dfrac{\cos 1.11}{\sin 1.11}$
31. $\dfrac{\sin 1.15 + 2}{\cos 1.01 + 1}$
32. $\dfrac{\sin 1.21 + \cos 0.47}{\sin 0.47 + \cos 1.21}$

33. $\dfrac{(\cos^2 0.01)(\sin^2 1.00)}{\cos 1.00}$

34. $\dfrac{(\sin^2 1.27)(\cos^2 1.25)}{\cos 0.01}$

35. $\dfrac{3\cos^2 0.45 + \sin 0.43}{2 \sin 1.52}$

36. $\dfrac{2\cos^2 1.51 - 3\sin 0.51}{\sin 1.50}$

Assume that $d = K \sin 2\pi t$. Find the indicated value under the given conditions.

Example K, given $d = 15$ and $t = 1/8$.

Solution We have
$$15 = K \sin 2\pi \left(\dfrac{1}{8}\right) = K \sin \dfrac{\pi}{4}.$$

Thus, from Table 5.2, we obtain
$$15 = \dfrac{1}{\sqrt{2}} K.$$

Therefore, $K = 15\sqrt{2}$.

37. K, given $d = 25$ and $t = 1/4$.
38. K, given $d = 10$ and $t = 3/4$.
39. d, given $K = 100$ and $t = 0.5$.
40. d, given $K = 75$ and $t = 0.25$.

Assume that $E = 1.2 \cos 10\pi t$. Find the value of E for the given value of t.

41. $t = 0$
42. $t = \dfrac{1}{10}$
43. $t = \dfrac{1}{30}$
44. $t = \dfrac{1}{60}$
45. $t = 0.05$
46. $t = 0.15$

B Exercises 47–54 require the use of a calculator. Evaluate each of the following functions to four decimal places at $x = 0.1, 0.01, 0.001,$ and 0.0001. What value is $f(x)$ approaching as the value of x approaches 0 but remains positive?

47. $f(x) = \dfrac{\sin x}{x}$
48. $f(x) = \dfrac{\sin 3x}{x}$
49. $f(x) = \dfrac{\sin^2 x}{x}$
50. $f(x) = \dfrac{\sin^2 4x}{x}$
51. $f(x) = \dfrac{\cos x - 1}{x}$
52. $f(x) = \dfrac{\cos 2x - 1}{2x}$
53. $f(x) = (1 - x) \sin x$
54. $f(x) = (1 - x) \cos x$.

5.3 Function Values of cos x and sin x, for x ∈ R

In Section 5.2 we obtained function values of the sine and the cosine for certain arc lengths between 0 and $\pi/2$ on a unit circle. In this section we shall see how to find function values of the sine and cosine for any arc length on the unit circle.

Reference arcs

In order to find the values sin x and cos x where x is not between 0 and $\pi/2$, we shall use the notion of a reference arc.

Definition 5.2 *For any arc A on the unit circle, with initial point (1, 0) and with measure x, the **reference arc** is the arc on the unit circle with least nonnegative measure between the terminal point of A and the horizontal axis. The length of the reference arc is denoted $\bar{x}$.*

The reference arcs for positive arcs terminating in each of the four quadrants are illustrated in Figure 5.7.

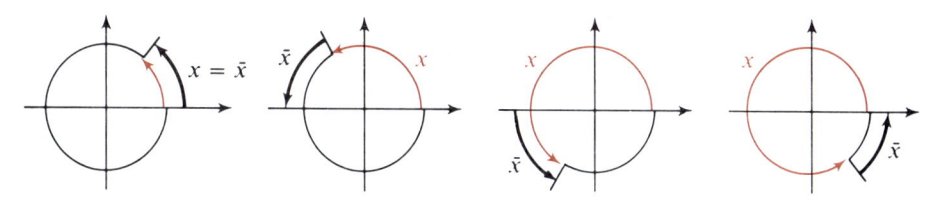

Figure 5.7

Examples Find the length of the reference arc for the arc with the given length.

a. $\dfrac{8\pi}{9}$ b. $\dfrac{7\pi}{6}$

Solutions

a. Observe that the arc with length $x = 8\pi/9$ terminates in the second quadrant, as shown in the figure. Thus,

$$\bar{x} = \pi - \frac{8\pi}{9} = \frac{\pi}{9}.$$

b. Observe that the arc with length $x = 7\pi/6$ terminates in the third quadrant, as shown in the figure. Thus,

$$\bar{x} = \frac{7\pi}{6} - \pi = \frac{\pi}{6}.$$

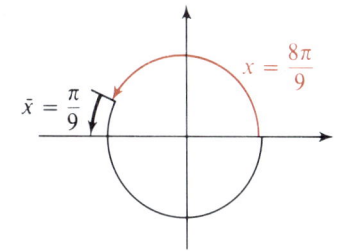

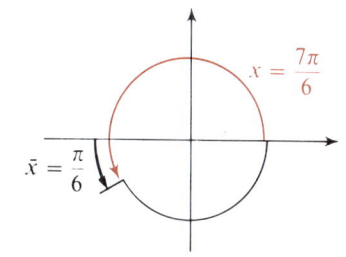

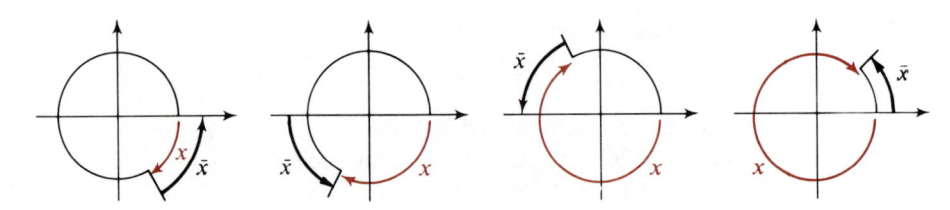

Figure 5.8

Negative arcs also have reference arcs, as illustrated in Figure 5.8.

Example Find the length of the reference arc for the arc with length $x = -2.30$.

Solution Note that the arc terminates in Quadrant III, as shown in the figure. We use $\pi \approx 3.14$ to obtain

$$\bar{x} = \pi - 2.30$$
$$\approx 3.14 - 2.30$$
$$= 0.84.$$

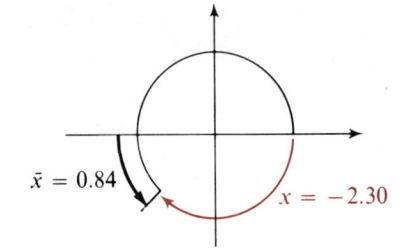

$|\cos x|$
$= \cos \bar{x}$,
$|\sin x|$
$= \sin \bar{x}$

Consider an arc of length x, starting from the point $(1, 0)$, which terminates in the second quadrant (see Figure 5.9-**a**).

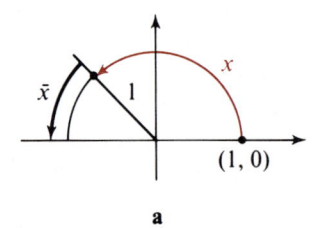

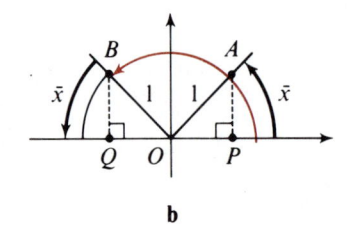

Figure 5.9

Construct an arc of length $\bar{x}$ (the reference arc length) from the point $(1, 0)$ (see Figure 5.9-**b**). It can be shown using elementary geometry that triangles POA and QOB are congruent. This implies that $PA = QB$ and $OP = OQ$. But note that $\cos x$ and $\sin x$ are the abscissa and ordinate, respectively, of the point B and that $\cos \bar{x}$ and $\sin \bar{x}$ are the abscissa and ordinate, respectively, of the point A. Thus, from Figure 5.9-**b**, we have

$$|\cos x| = QB = PA = \cos \bar{x}$$

and

$$|\sin x| = OQ = OP = \sin \bar{x}.$$

5.3 Function Values of cos x and sin x, for $x \in R$

Although we assumed that x terminates in the second quadrant, the argument we used is quite general and can be used to prove the following theorem.

Theorem 5.2   For $x \in R$,

$$|\cos x| = \cos \bar{x} \quad \text{and} \quad |\sin x| = \sin \bar{x},$$

where $\bar{x}$ is the length of the reference arc corresponding to x.

Calculating cos x, sin x

The fact that every reference arc length is between 0 and $\pi/2$, along with Theorem 5.2, enables us to compute $\cos x$ and $\sin x$ for any arc length x. We simply perform the following steps.

1. Find the reference arc length $\bar{x}$.
2. Use Theorem 5.2 to find $|\cos x|$ and $|\sin x|$.
3. Use Table 5.1 to determine the correct algebraic sign for $\cos x$ and $\sin x$.

Example

Compute $\cos(3\pi/4)$ and $\sin(3\pi/4)$ using Table 5.2.

Solution

The length of the reference arc is

$$\bar{x} = \pi - \frac{3\pi}{4} = \frac{\pi}{4},$$

as shown in the figure. From Theorem 5.2 and Table 5.2,

$$\left|\cos \frac{3\pi}{4}\right| = \cos \frac{\pi}{4} = \frac{1}{\sqrt{2}} \quad \text{and} \quad \left|\sin \frac{3\pi}{4}\right| = \sin \frac{\pi}{4} = \frac{1}{\sqrt{2}}.$$

Since the arc of length $3\pi/4$ terminates in the second quadrant, we know from Table 5.1 that $\cos(3\pi/4) < 0$ and $\sin(3\pi/4) > 0$. Thus, we have

$$\cos \frac{3\pi}{4} = -\frac{1}{\sqrt{2}} \quad \text{and} \quad \sin \frac{3\pi}{4} = \frac{1}{\sqrt{2}}.$$

In case x is not one of the special arc lengths, the procedure is the same except that Table IV or a calculator is used to determine the values of $|\cos x|$ and $|\sin x|$.

Using a calculator

Most calculators will determine values for $\cos x$ and $\sin x$ directly for all real numbers x. Furthermore, because of the fact that calculators use a more accurate approximation to π than is used to obtain reference arcs ($\pi \approx 3.14$), the values obtained on a calculator will be more accurate than those that are obtained by using reference arcs. To be consistent with answers obtained by using reference arcs with the tables, *we shall use reference arcs in conjunction with a calculator to compute all values of the circular functions in all examples and exercises in this chapter.*

Example   Compute cos 5.31 and sin 5.31 using a hand calculator or Table V.

Solution   The length of the reference arc is

$$\bar{x} = 2\pi - 5.31$$
$$\approx 6.28 - 5.31$$
$$= 0.97,$$

as shown in the figure. From Theorem 5.2, and a calculator or Table IV,

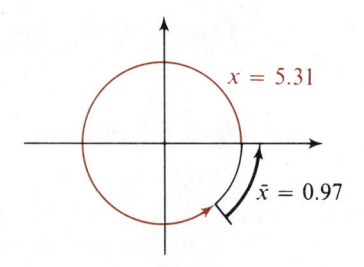

$$|\cos 5.31| = \cos 0.97 = 0.5653 \quad \text{and} \quad |\sin 5.31| = \sin 0.97 = 0.8249.$$

Since the arc of length 5.31 terminates in the fourth quadrant, we know from Table 5.1 that cos 5.31 > 0 and sin 5.31 < 0. Thus, we have

$$\cos 5.31 = 0.5653 \quad \text{and} \quad \sin 5.31 = -0.8249.$$

Example   Compute cos (−2.31) and sin (−2.31) using a hand calculator or Table V.

Solution   The length of the reference arc is

$$\bar{x} = \pi - 2.31$$
$$\approx 3.14 - 2.31$$
$$= 0.83,$$

as shown in the figure. From a calculator or Table IV, and Theorem 5.2, we have

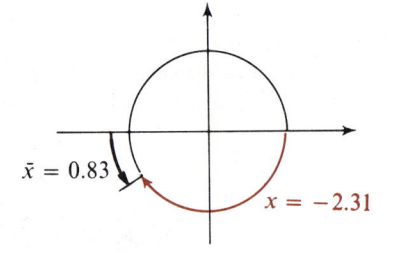

$$|\cos(-2.31)| = \cos 0.83 = 0.6749 \quad \text{and} \quad |\sin(-2.31)| = \sin 0.83 = 0.7379.$$

Since the arc of length −2.31 terminates in the third quadrant, we know from Table 5.1 that cos x < 0 and sin x < 0. Thus, we have

$$\cos(-2.31) = -0.6749 \quad \text{and} \quad \sin(-2.31) = -0.7379.$$

Exercise 5.3

A   *Use Table 5.2, to exactly determine the value.*

Example $\sin \dfrac{11\pi}{6}$

5.3 Function Values of cos x and sin x, for x ∈ R

Solution

The length of the reference arc is

$$\bar{x} = 2\pi - \frac{11\pi}{6} = \frac{\pi}{6}.$$

From Table 5.2 and Theorem 5.2,

$$\left|\sin\frac{11\pi}{6}\right| = \sin\frac{\pi}{6} = \frac{1}{2}.$$

Since $x = 11\pi/6$ terminates in the fourth quadrant,

$$\sin\frac{11\pi}{6} = -\frac{1}{2}.$$

1. $\cos\dfrac{5\pi}{6}$
2. $\sin\dfrac{2\pi}{3}$
3. $\cos\dfrac{4\pi}{3}$
4. $\sin\dfrac{7\pi}{6}$
5. $\cos\dfrac{7\pi}{4}$
6. $\sin\dfrac{5\pi}{3}$
7. $\sin\dfrac{4\pi}{3}$
8. $\cos\dfrac{2\pi}{3}$
9. $\sin\dfrac{7\pi}{4}$
10. $\cos\dfrac{7\pi}{6}$
11. $\sin\dfrac{3\pi}{4}$
12. $\cos\dfrac{5\pi}{3}$

Use Table IV or a calculator to find an approximation for the function value.

Example cos 1.73

Solution

The length of the reference arc is

$$\bar{x} = \pi - 1.73$$
$$\approx 3.14 - 1.73$$
$$= 1.41.$$

Now, from a calculator or Table IV, by Theorem 5.2,

$$|\cos 1.73| = \cos 1.41 = 0.1601.$$

Since $x = 1.73$ terminates in the second quadrant, we find from Table 5.1 that

$$\cos 1.73 = -0.1601.$$

13. cos 2.11
14. sin 1.69
15. cos 3.51
16. sin 4.00
17. cos 6.15
18. sin 5.15
19. sin 1.58
20. cos 1.59
21. sin 3.16
22. cos 3.18
23. sin 4.75
24. cos 4.72

Example $\cos(-1.68)$

Solution   The length of the reference arc is

$$\bar{x} = \pi - 1.68$$
$$\approx 3.14 - 1.68$$
$$= 1.46.$$

From a calculator of Table IV, by Theorem 5.2,

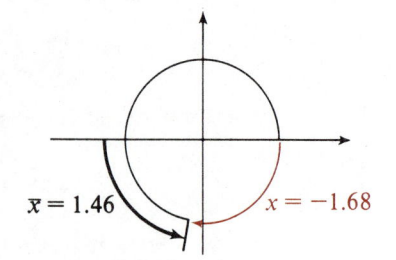

$$|\cos(-1.68)| = \cos 1.46 = 0.1106.$$

Since $x = -1.68$ terminates in the third quadrant,

$$\cos(-1.68) = -0.1106.$$

25. $\cos\left(-\dfrac{\pi}{3}\right)$ 26. $\sin\left(-\dfrac{\pi}{6}\right)$ 27. $\cos\left(-\dfrac{3\pi}{4}\right)$

28. $\sin\left(-\dfrac{5\pi}{6}\right)$ 29. $\cos\left(-\dfrac{4\pi}{3}\right)$ 30. $\sin\left(-\dfrac{7\pi}{6}\right)$

31. $\sin(-1.85)$ 32. $\cos(-3.00)$ 33. $\sin(-3.57)$

34. $\cos(-4.11)$ 35. $\sin(-5.11)$ 36. $\cos(-6.20)$

Use Table 5.2 to exactly determine the indicated value.

Example $\cos\dfrac{7\pi}{3}$

Solution   Notice that the arc with length $x = 7\pi/3$ makes more than one full revolution. The length of the reference arc is

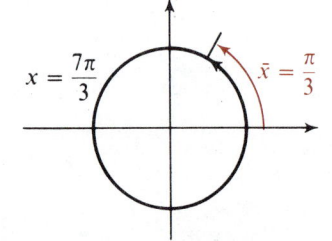

$$\bar{x} = \dfrac{7\pi}{3} - 2\pi = \dfrac{\pi}{3}.$$

Now, from Table 5.2 and Theorem 5.2,

$$\left|\cos\dfrac{7\pi}{3}\right| = \cos\dfrac{\pi}{3} = \dfrac{1}{2}.$$

Since $x = 7\pi/3$ terminates in the first quadrant,

$$\cos\dfrac{7\pi}{3} = \dfrac{1}{2}.$$

37. $\cos\dfrac{13\pi}{6}$ 38. $\sin\dfrac{9\pi}{4}$ 39. $\cos\dfrac{17\pi}{4}$

40. $\sin\dfrac{14\pi}{3}$ 41. $\cos\left(-\dfrac{7\pi}{3}\right)$ 42. $\sin\left(-\dfrac{11\pi}{4}\right)$

5.4 Graphs of $y = \cos x$ and $y = \sin x$

Use Table 5.2, Table IV, or a hand calculator to determine the indicated function value.

43. sin 6.40
44. cos 7.04
45. sin 14.25
46. cos 16.00
47. sin (−6.85)
48. cos (−15.00)
49. $\sin \dfrac{2\pi}{3} + \cos \dfrac{7\pi}{4}$
50. $\sin \dfrac{7\pi}{4} \cos \dfrac{3\pi}{4}$
51. $2 \sin^2 6.01 - 3 \cos 2.36$
52. $-2 \sin (-2.14) + 6 \cos (-3.27)$
53. $\dfrac{\sin^2 12.47}{\cos 6.27}$
54. $\dfrac{\cos^2 8.65}{\sin 4.26}$

B Compute each of the following values: (a) using Table IV and a reference $arc(\pi \approx 3.14)$, (b) to four decimal places directly using a calculator. Which of the answers in (a) and (b) is more accurate? Why?

55. sin 150.00
56. sin (−201.00)
57. cos 210.00
58. cos (−105.00)

5.4 Graphs of $y = \cos x$ and $y = \sin x$

In Sections 5.1–5.3 we discussed the definition and evaluation of the cosine function and the sine function. In this section we shall define the concept of a periodic function and use it to graph the periodic functions $y = \cos x$ and $y = \sin x$.

Periodic functions

The graphs of some functions display interesting cyclical characteristics. For example, the x-axis shown in Figure 5.10 can clearly be divided into equal successive subintervals in such a way that the graph over each subinterval is a repetition of the graph over every other equal subinterval. Functions that display this property are

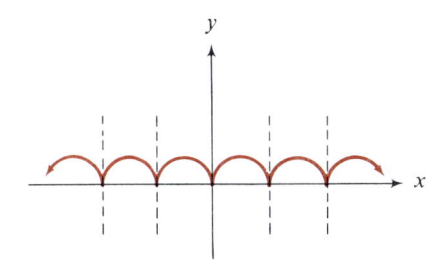

Figure 5.10

said to be *periodic* and the length of each of the equal subintervals is called a **period** of the function.

Definition 5.3   If f is a function, such that for some $p \in R$, $p \neq 0$,
$$f(x + p) = f(x - p) = f(x),$$
then f is **periodic** with period p.

If there is a least positive number p for which the function is periodic, then p is called the **fundamental period** of the function. Thus, for example, Figure 5.11 shows part of the graph of a periodic function with fundamental period 5.

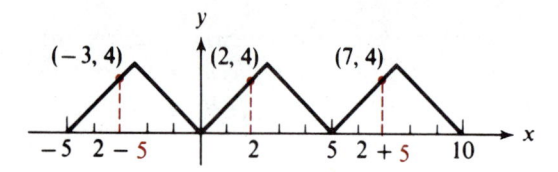

Figure 5.11

Period of the sine and cosine functions

Both sine and cosine are periodic functions. In particular, for any real number x we have (see Exercises 47 and 48 on page 176)
$$\sin(x + 2\pi) = \sin x$$
and
$$\cos(x + 2\pi) = \cos x.$$
Hence, 2π is a period for both functions. It can be shown (although we shall not do so here) that 2π is the fundamental period of both.

Graph of $y = \sin x$

The graph of $y = \sin x$ on the interval $0 \leq x \leq 2\pi$ can be obtained by plotting several ordered pairs over this interval, as shown in Figure 5.12. The graphs of the

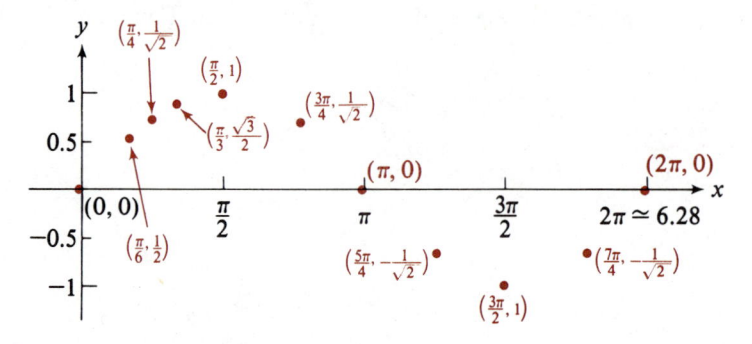

Figure 5.12

5.4 Graphs of $y = \cos x$ and $y = \sin x$

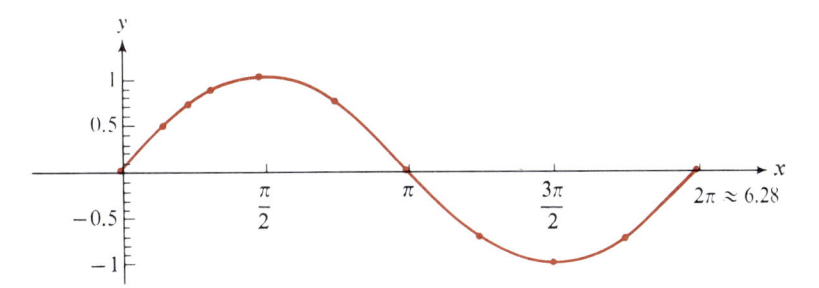

Figure 5.13

ordered pairs are then joined with a smooth curve to obtain the graph shown in Figure 5.13.

Since the sine function has fundamental period 2π, we can obtain the graph of $y = \sin x$ over an extended interval by repeating the pattern in Figure 5.13 in both directions along the x-axis. For example, the graph of $y = \sin x$ over the interval $-\pi \leq x \leq 4\pi$ is shown in Figure 5.14.

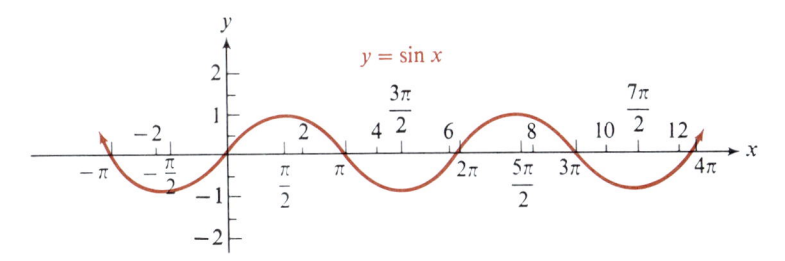

Figure 5.14

The zeros of the function—namely the values of x associated with the points at which the curve crosses the x-axis—appear on the graph. They are $k\pi$, where $k \in J$. Graphs with this characteristic form are called **sine waves** or **sinusoids**. The portion of the graph over any fundamental period of the function is called a **cycle** of the sine wave. Half the difference of the maximum and minimum ordinates on such a curve is called the **amplitude** of the wave. Thus, for the graph of $y = \sin x$, the amplitude of the wave is

$$\frac{1}{2}[1 - (-1)] = 1.$$

Graph of
$y = \cos x$

The graph of the cosine function can be obtained in the same manner as the graph of the sine function. We first obtain the coordinates of several points on the graph; then by connecting these points with a smooth curve, as shown in Figure 5.15, we obtain one cycle of the graph of $y = \cos x$. Duplicating this pattern in both directions along the x-axis, we have a representative portion of the entire graph of the cosine function, as shown in Figure 5.16.

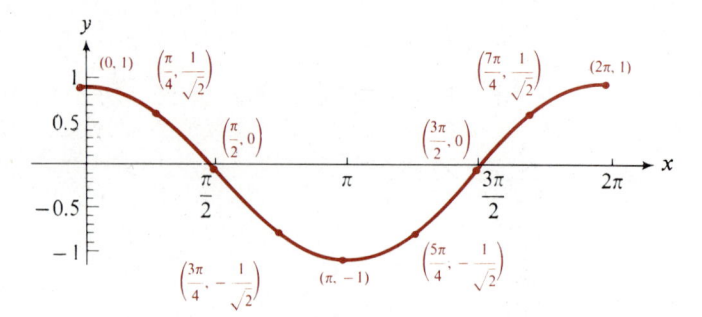

Figure 5.15

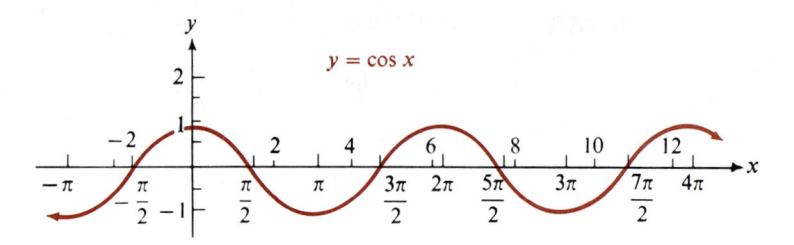

Figure 5.16

Notice that the graph of the cosine function is also a sinusoid, with fundamental period 2π and amplitude 1. Furthermore, the zeros of the function, $\pi/2 + k\pi$, where $k \in J$, are evident.

Functions defined by equations of the form

$$y = A \sin (Bx + C) \quad \text{and} \quad y = A \cos (Bx + C),$$

where A, B, and C are constants and A and B are not 0, always have sine waves for graphs. With variations in the numbers A, B, and C, the graphs are variously situated with respect to the origin and have a variety of amplitudes and periods. To analyze such graphs, let us first consider the effect of A on the graphs of the functions defined by $y = A \sin x$ and $y = A \cos x$.

Graphs of $y = A \sin x$ **and** $y = A \cos x$

For each value of x, each ordinate to the graph of $y = A \sin x$ is A times the ordinate to the graph of $y = \sin x$. Therefore, the amplitude of the graph of $y = A \sin x$ is $|A|$ times the amplitude of the graph of $y = \sin x$; that is, the amplitude of $y = A \sin x$ is $|A|$. Of course, the graph of $y = A \cos x$ is a similar modification of the graph of $y = \cos x$.

Example

Graph $y = 3 \sin x$, $-\pi \leq x \leq 4\pi$.

Solution

It may be helpful first to sketch $y = \sin x$, $0 \leq x \leq 2\pi$, as a reference. Then, since the amplitude of $y = 3 \sin x$ is 3, we can sketch the desired graph on the same

5.4 Graphs of $y = \cos x$ and $y = \sin x$

coordinate system over the interval $0 \leq x \leq 2\pi$ by making each ordinate three times the corresponding ordinate of the graph of $y = \sin x$.

For $y = 3 \sin x$, the fundamental period p is the same as for $y = \sin x$, namely $p = 2\pi$. Thus we can then extend this cycle to include the entire interval $-\pi \leq x \leq 4\pi$, as shown in the figure. In this figure, one cycle is sketched with a colored line for emphasis. Note also that, in this and some of the succeeding figures, different unit lengths are used on the x- and y-axes.

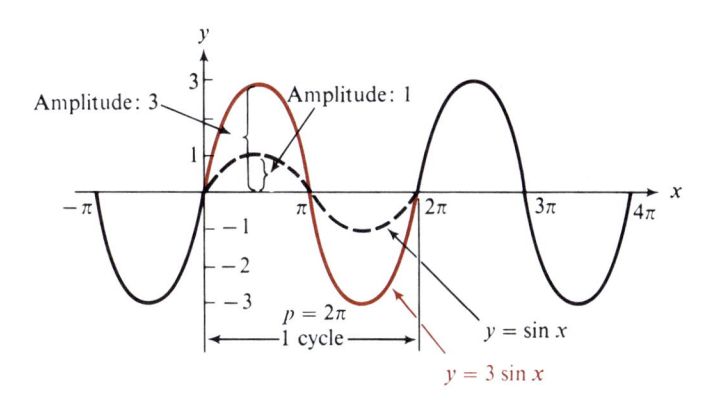

Graphs of $y = \sin Bx$ and $y = \cos Bx$

Next let us examine the way the graph of

$$y = \sin Bx, \quad B \neq 0,$$

differs from the graph of $y = \sin x$. Note first that $\sin Bx$ has values between -1 and 1 inclusive, as has $\sin x$. Also,

$$\sin(Bx + 2\pi) = \sin Bx,$$

just as $\sin(x + 2\pi) = \sin x$. If we factor B from $Bx + 2\pi$, however, we have $B(x + 2\pi/B)$, and hence the function defined by $y = \sin Bx$ has period

$$p = \frac{2\pi}{|B|}.$$

We use $|B|$ instead of B to ensure a positive number for the period. It can be shown, although it is not done here, that $2\pi/|B|$ is the fundamental period of the function. Hence the graph of $y = \sin Bx$ is a sine wave with amplitude 1; it completes one cycle over the interval $0 \leq x \leq 2\pi/|B|$.

Example Graph $y = \cos 2x$, $-3\pi/2 \leq x \leq 2\pi$.

Solution Let us first sketch a cycle of $y = \cos x$, $0 \leq x \leq 2\pi$, as a reference. Since

$$p = \frac{2\pi}{|B|} = \frac{2\pi}{2} = \pi,$$

Solution continued overleaf

we next sketch a cycle of the graph of $y = \cos 2x$ over the interval $0 \le x \le \pi$ (shown in color) on the same coordinate system and extend the cycle obtained over the interval $-3\pi/2 \le x \le 2\pi$.

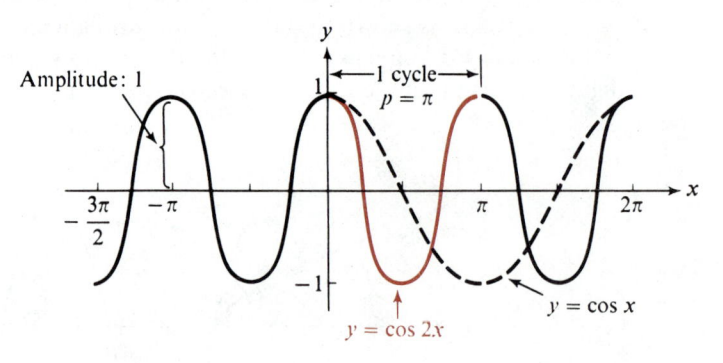

Example Graph $y = -4 \sin \dfrac{1}{2} x$, $-2\pi \le x \le 4\pi$.

Solution We first sketch the graph of $y = \sin x$ as a reference. Since the coefficient of $\sin(x/2)$ is -4, each ordinate of the graph of $y = -4 \sin(x/2)$ is the negative of the ordinate of the graph of $y = 4 \sin(x/2)$. Since

$$p = \frac{2\pi}{|B|} = \frac{2\pi}{\frac{1}{2}} = 4\pi,$$

there is one cycle in the interval $0 \le x \le 4\pi$. With this information, we sketch one cycle (shown in color) and then extend the graph over the interval $-2\pi \le x \le 4\pi$.

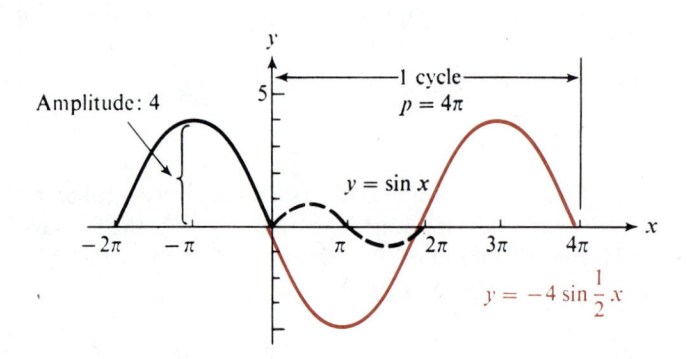

Graphs of $y = \sin(x + C)$ and $y = \cos(x + C)$

Finally, let us compare the graphs of

$$y = \sin(x + C) \quad \text{and} \quad y = \sin x,$$

where $C > 0$. For $x = -C$, we have

$$\sin(x + C) = \sin(-C + C) = \sin 0 = 0.$$

5.4 Graphs of $y = \cos x$ and $y = \sin x$

Similarly, for any real number x_1, the value of $\sin(x + C)$ at $x_1 - C$ will be the same as the value of $\sin x$ at x_1. Thus if x is thought of as measuring time, each value y occurs C units of time earlier on $y = \sin(x + C)$ than on $y = \sin x$. Hence the graph of $y = \sin(x + C)$ is said to **lead** the graph of $y = \sin x$ by C. The number C, itself, is called the **phase shift** of the wave. If $C < 0$, then the graph of $y = \sin(x + C)$ is obtained by shifting the graph of $y = \sin x$ $|C|$ units to the right and is said to **lag** the graph of $y = \sin x$.

We use all the foregoing information about the effect of A, B, and C on the graph of $y = A \sin(Bx + C)$, or of $y = A \cos(Bx + C)$, to help sketch the graph of such an equation.

Example Sketch the graph of $y = 3 \sin\left(2x + \dfrac{\pi}{3}\right)$.

Solution First, we rewrite the equation by factoring 2 from the expression in parentheses:

$$y = 3 \sin 2\left(x + \dfrac{\pi}{6}\right).$$

By inspecting this equation, we note the following things about the graph:

1. It is a sine wave.
2. It has amplitude 3.
3. It has period $2\pi/2 = \pi$.
4. It leads the graph of $y = 3 \sin 2x$ by $\pi/6$.

With these facts, we can quickly sketch the graph shown.

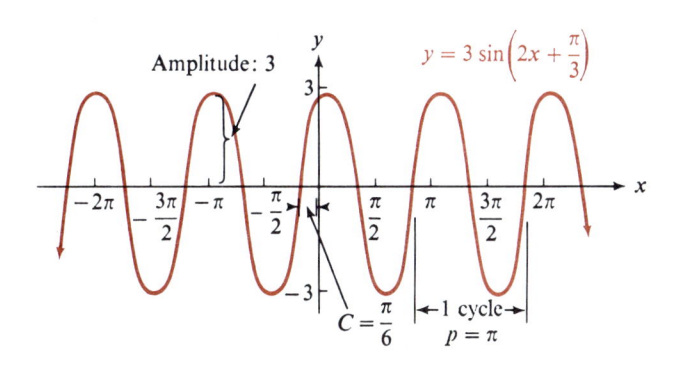

For some sine waves the period is not a rational multiple of π. In this case it is convenient to scale the x-axis in integral units.

Example Sketch the graph of $y = \frac{1}{2} \sin \frac{\pi x}{2}$.

Solution By inspection, we note the following things concerning the graph:

1. It is a sine wave.
2. It has amplitude 1/2.
3. It has period $2\pi/(\pi/2) = 4$.

Since the period is 4, we scale the x-axis in integral units to facilitate sketching the graph as shown.

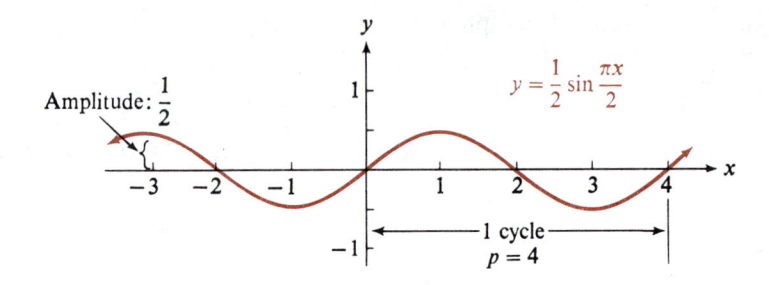

The ordinates of the graph of an equation of the form $y = A \sin Bx + C \cos Dx$ can be obtained by adding the ordinates for the graphs of $y = A \sin Bx$ and $y = C \cos Dx$ at each point x on the x-axis. An example is shown in the exercise set.

Exercise 5.4

A Sketch the graph of the given equation over the interval $-2\pi \leq x \leq 2\pi$. Give the amplitude, period, and phase shift.

1. $y = 2 \sin x$
2. $y = 3 \cos x$
3. $y = \frac{1}{2} \cos x$
4. $y = \frac{1}{3} \cos x$
5. $y = -4 \sin x$
6. $y = -\frac{1}{2} \cos x$
7. $y = \sin 2x$
8. $y = \cos 3x$
9. $y = \cos \frac{1}{3} x$
10. $y = \cos \frac{1}{2} x$
11. $y = -3 \sin 2x$
12. $y = -\frac{1}{2} \sin 3x$
13. $y = \sin (x + \pi)$
14. $y = \cos \left(x - \frac{\pi}{2} \right)$

5.4 Graphs of $y = \cos x$ and $y = \sin x$

15. $y = 2 \cos \left(x - \dfrac{\pi}{4} \right)$
16. $y = 3 \sin \left(x + \dfrac{\pi}{6} \right)$
17. $y = 3 \sin 2 \left(x - \dfrac{\pi}{3} \right)$
18. $y = 2 \cos 3 \left(x + \dfrac{\pi}{4} \right)$
19. $y = 2 \sin \pi x$
20. $y = -3 \cos \dfrac{\pi}{2} x$
21. $y = -\dfrac{1}{2} \cos \dfrac{\pi}{3} x$
22. $y = \dfrac{1}{4} \sin \dfrac{\pi}{4} x$

From the respective graph, determine the zeros (over the specified domain) of the function defined by the equation in the given exercise.

23. Exercise 7
24. Exercise 8
25. Exercise 9
26. Exercise 10
27. Exercise 19
28. Exercise 20

B *Sketch the graph of the given equation over the interval* $-2\pi \le x \le 2\pi$.

Example $y = \sin x + 2 \cos x$

Solution First, sketch the graphs of $y = \sin x$ and $y = 2 \cos x$ on the same coordinate system over the given interval. The ordinate of the graph of $y = \sin x + 2 \cos x$ at each point x on the x-axis is the sum of the corresponding ordinates of $y = \sin x$ and $y = 2 \cos x$. Thus for $x = \pi/6$,

$$\sin x = 0.5, \quad 2 \cos x = \dfrac{2\sqrt{3}}{2} \approx 1.7, \quad \text{and} \quad \sin x + 2 \cos x \approx 2.2.$$

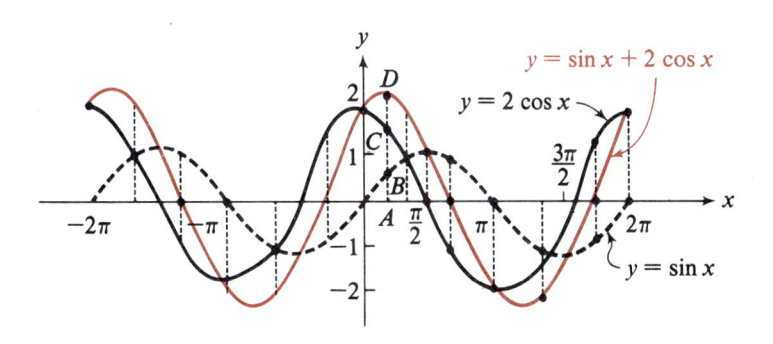

Now the ordinate can be approximated graphically by adding the directed line segments from the x-axis to the curves at $x = \pi/6$:

$$\overline{AB} + \overline{AC} = \overline{AB} + \overline{BD} = \overline{AD}.$$

If this is done for a few selected values of x, we obtain a good approximation for the curve.

176 5 Circular Functions

29. $y = \sin x + \cos x$
30. $y = 3 \sin x + \cos x$
31. $y = \sin 2x + \frac{1}{2} \cos x$
32. $y = \sin 3x + 2 \cos \frac{1}{2} x$
33. $y = \sin x - 2 \cos x$
34. $y = 3 \cos x - \sin 2x$
35. $y = 3 \sin x + 1$
36. $y = 4 \sin 2x - 3$
37. $y = x + \cos x$
38. $y = 2x - \cos x$

Hint: For Exercises 39–46, use the methods described in Section 2.8.

39. $y = |\sin x|$
40. $y = |\cos x|$
41. $y = |\sin 2x|$
42. $y = |\cos 2x|$
43. $y = |2 \sin 2x|$
44. $y = |2 \cos 2x|$
45. $y = \left|2 \sin \left(2x + \frac{\pi}{2}\right)\right|$
46. $y = \left|2 \cos \left(2x + \frac{\pi}{2}\right)\right|$

47. Show for any real number x, $\sin(x + 2\pi) = \sin x$.
48. Show for any real number x, $\cos(x + 2\pi) = \cos x$.

5.5 Tangent Function

The functions sine and cosine can be used to define other periodic functions. One of these is the tangent function.

Definition 5.4 For $x \in R$, $\cos x \neq 0$, the **tangent of** x is denoted by **tan** x, and

$$\tan x = \frac{\sin x}{\cos x}.$$

Domain and range of tangent

Because $\tan x = \sin x / \cos x$, the domain of the tangent function is R, with the exception of the real numbers x for which $\cos x = 0$. In other words, we except real numbers of the form $\pi/2 + k\pi$, $k \in J$. The range of the tangent function is R. We can use Definition 5.4 together with Table 5.2, page 157, to find $\tan x$ for all values of x included in the table. For example,

$$\tan 0 = \frac{\sin 0}{\cos 0} = \frac{0}{1} = 0,$$

$$\tan \frac{\pi}{6} = \frac{\sin(\pi/6)}{\cos(\pi/6)}$$

$$= \frac{1/2}{\sqrt{3}/2} = \frac{1}{\sqrt{3}},$$

and so forth. The values of $\tan x$ obtained in this way are displayed in Table 5.3.

5.5 Tangent Function

Table 5.3

x	0	$\dfrac{\pi}{6}$	$\dfrac{\pi}{4}$	$\dfrac{\pi}{3}$	$\dfrac{\pi}{2}$	π	$\dfrac{3\pi}{2}$
$\tan x$	0	$\dfrac{1}{\sqrt{3}}$	1	$\sqrt{3}$	Not def.	0	Not def.

Computing tan x

By Definition 5.4 and Theorem 5.2

$$|\tan x| = \left|\frac{\sin x}{\cos x}\right|$$

$$= \frac{|\sin x|}{|\cos x|}$$

$$= \frac{\sin \bar{x}}{\cos \bar{x}} = \tan \bar{x}$$

where $\bar{x}$ denotes the length of the reference arc corresponding to x. Hence the value of $\tan x$ can be computed using a reference arc in the same way that $\sin x$ and $\cos x$ are computed using a reference arc.

Example Find $\tan(4\pi/3)$.

Solution The length of the reference arc is

$$\bar{x} = \frac{4\pi}{3} - \pi = \frac{\pi}{3}.$$

Thus from Table 5.3 we have

$$\left|\tan \frac{4\pi}{3}\right| = \tan \frac{\pi}{3} = \sqrt{3}.$$

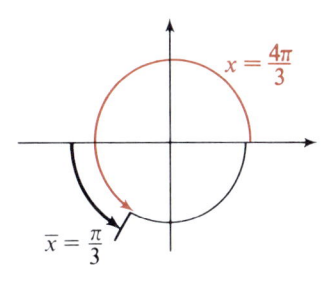

Since $4\pi/3$ terminates in the third quadrant, we have $\sin(4\pi/3) < 0$ and $\cos(4\pi/3) < 0$. Thus $\tan(4\pi/3) > 0$ and we have

$$\tan \frac{4\pi}{3} = \sqrt{3}.$$

We can also use the identity $\cos^2 x + \sin^2 x = 1$ to compute $\tan x$ if we are given the value of either $\sin x$ or $\cos x$ and the quadrant in which x lies.

Example If $\sin x = 3/5$ and $\pi/2 \leq x \leq \pi$, find $\cos x$ and $\tan x$.

Solution By Equation (3b) on page 152, for $\pi/2 \leq x \leq \pi$,

$$\cos x = -\sqrt{1 - \sin^2 x}.$$

Solution continued overleaf

Then, since $\sin x = 3/5$,

$$\cos x = -\sqrt{1 - \left(\frac{3}{5}\right)^2} = -\frac{4}{5}.$$

By Definition 5.4,

$$\tan x = \frac{\sin x}{\cos x} = \frac{3/5}{-4/5} = -\frac{3}{4}.$$

Period of $y = \tan x$

Observe that since the sine and cosine functions have period 2π, so has the tangent function. Note, however, that the points at arc length x and $x + \pi$ from $(1, 0)$ on the unit circle lie on the same line through the origin (see Figure 5.17). Thus, if

$$x \neq \frac{\pi}{2} + k\pi, \quad k \in J,$$

the slope of segment OA is the same as the slope of segment OB, and we have

$$\frac{\sin(x + \pi) - 0}{\cos(x + \pi) - 0} = \text{slope of } OB = \text{slope of } OA = \frac{\sin x - 0}{\cos x - 0}.$$

Thus,

$$\tan(x + \pi) = \frac{\sin(x + \pi)}{\cos(x + \pi)} = \frac{\sin x}{\cos x} = \tan x,$$

and π is also a period of the tangent function. It can be shown (although we shall not do so) that π is the fundamental period of the tangent function.

Graph of $y = \tan x$

We can obtain the graph of $y = \tan x$ by first plotting several points over an interval of length π (the fundamental period of the tangent function). Then we join these points with a smooth curve as shown in Figure 5.18. Observe that the line

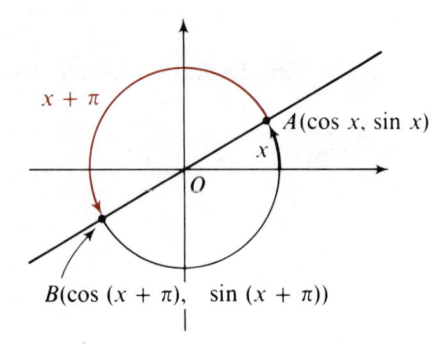

Figure 5.17

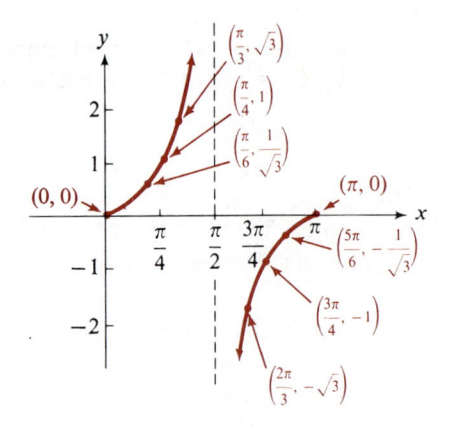

Figure 5.18

5.5 Tangent Function

$x = \pi/2$ is a vertical asymptote of the graph of the function $y = \tan x$ since $\tan x = \sin x/\cos x$ and $\sin \pi/2 \neq 0$ but $\cos \pi/2 = 0$.

We can now duplicate the pattern in Figure 5.18 in both directions along the x-axis to obtain the graph of $y = \tan x$ over an extended interval. Several cycles of the graph are shown in Figure 5.19.

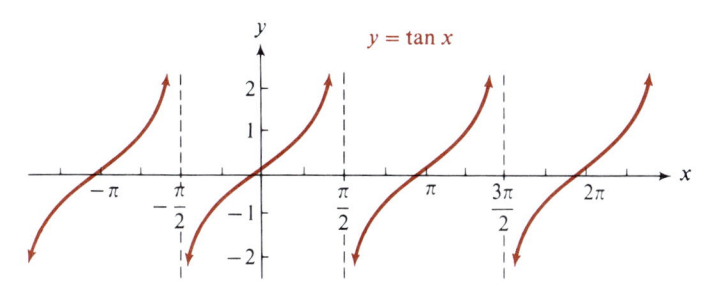

Figure 5.19

It is evident from the graph that the range of the tangent function is the set of all real numbers. Furthermore, it is evident that any integral multiple of π is a zero of $y = \tan x$ and that the curve has vertical asymptotes at $x = \pi/2 + k\pi$ for every integer k.

Graph of
$y = \tan Bx$

Because, for any real number x,

$$\tan (Bx + \pi) = \tan Bx, \quad B \neq 0,$$

and $Bx + \pi = B(x + \pi/B)$, we have

$$\tan B\left(x + \frac{\pi}{B}\right) = \tan Bx.$$

Thus, $\pi/|B|$ is a period of $y = \tan Bx$ and is in fact the fundamental period since π is the fundamental period of $\tan x$. Thus, the graph of $y = \tan Bx$ has the same basic shape as the graph of $y = \tan x$, but it repeats over intervals of length $\pi/|B|$ instead of π.

Example

Graph $y = \tan 2x$ over the interval $-\pi \leq x \leq \pi$.

Solution

It is helpful first to sketch the graph of $y = \tan x$ over the interval $-\pi/2 \leq x \leq \pi/2$ as a reference (see the dashed curve in the figure). Since the fundamental period of $y = \tan 2x$ is

$$p = \frac{\pi}{|B|} = \frac{\pi}{2},$$

we next sketch a cycle of $y = \tan 2x$ over the interval $0 \leq x \leq \pi/2$ on the same coordinate system and extend the cycle obtained over the interval $-\pi \leq x \leq \pi$.

Solution continued overleaf

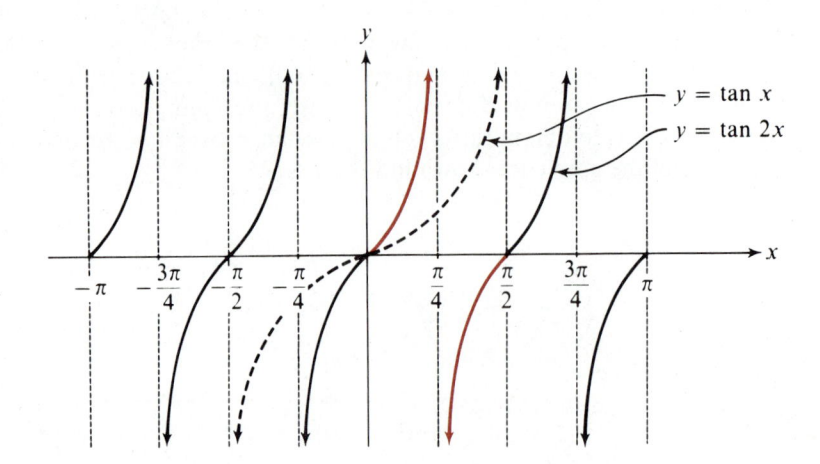

We can sketch the graph of the function $y = A \tan (Bx + C)$ using the same procedure that was used in Section 5.4 to graph

$$y = A \sin (Bx + C) \quad \text{and} \quad y = A \cos (Bx + C).$$

Note, however, that since the graph of $y = A \tan (Bx + C)$ is not a sinusoid, we cannot interpret A as the amplitude.

Exercise 5.5

A Use Table 5.2 and Definition 5.4 to compute the indicated value if it exists; if not, then so indicate.

1. $\tan 0$
2. $\tan \dfrac{\pi}{6}$
3. $\tan \dfrac{\pi}{4}$
4. $\tan \dfrac{\pi}{3}$
5. $\tan \dfrac{\pi}{2}$
6. $\tan \pi$
7. $\tan \dfrac{3\pi}{2}$
8. $\tan 2\pi$

Use a reference arc and Table 5.3 to find the exact value if it exists; if not, then so indicate.

Example $\tan \dfrac{5\pi}{3}$

Solution The length of the reference arc is

$$\bar{x} = 2\pi - \dfrac{5\pi}{3} = \dfrac{\pi}{3}.$$

By Table 5.3,

$$\left| \tan \dfrac{5\pi}{3} \right| = \tan \dfrac{\pi}{3} = \sqrt{3}.$$

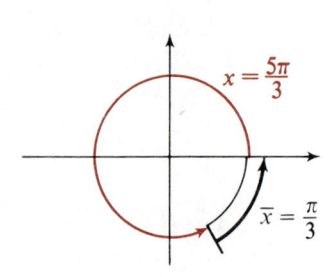

5.5 Tangent Function

Since $x = 5\pi/3$ terminates in the fourth quadrant,

$$\tan \frac{5\pi}{3} = -\sqrt{3}.$$

9. $\tan \dfrac{2\pi}{3}$ **10.** $\tan \dfrac{3\pi}{4}$ **11.** $\tan \dfrac{7\pi}{6}$ **12.** $\tan \dfrac{7\pi}{4}$

13. $\tan \left(-\dfrac{\pi}{6}\right)$ **14.** $\tan \left(-\dfrac{3\pi}{4}\right)$ **15.** $\tan \dfrac{5\pi}{2}$ **16.** $\tan \dfrac{13\pi}{6}$

Use a reference arc, together with Table IV or a calculator, to obtain an approximation for the given function value.

Example $\tan 3.27$

Solution   The length of the reference arc is

$$\bar{x} = 3.27 - \pi$$
$$\approx 3.27 - 3.14$$
$$= 0.13.$$

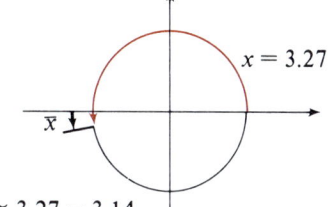

By Table IV or a calculator,

$$|\tan 3.27| = \tan 0.13 = 0.1291.$$

$\bar{x} \approx 3.27 - 3.14 = 0.13$

Since $x = 3.27$ terminates in the third quadrant,

$$\tan 3.27 = 0.1291.$$

17. $\tan 4.52$ **18.** $\tan 5.61$ **19.** $\tan 7.45$ **20.** $\tan 8.30$

21. $\tan 12.50$ **22.** $\tan 27.30$ **23.** $\tan (-9.32)$ **24.** $\tan (-6.41)$

Use the given information to compute $\tan x$.

Example $\cos x = \dfrac{2}{5}; \quad \dfrac{3\pi}{2} \leq x \leq 2\pi$

Solution   By Equation (3a) on page 152,

$$\sin x = -\sqrt{1 - \cos^2 x}$$
$$= -\sqrt{1 - \left(\frac{2}{5}\right)^2}$$
$$= \frac{-\sqrt{21}}{5}.$$

Thus by Definition 5.4,

$$\tan x = \frac{\sin x}{\cos x} = -\frac{\sqrt{21}/5}{2/5} = -\frac{\sqrt{21}}{2}.$$

25. $\sin x = \frac{2}{3}$; $\frac{\pi}{2} \leq x \leq \pi$

26. $\sin x = -\frac{7}{8}$; $\frac{3\pi}{2} \leq x \leq 2\pi$

27. $\cos x = \frac{1}{8}$; $\frac{3\pi}{2} \leq x \leq 2\pi$

28. $\cos x = -\frac{1}{3}$; $\pi \leq x \leq \frac{3\pi}{2}$

Sketch the graph of each function over the interval $-2\pi \leq x \leq 2\pi$, *and list the zeros and vertical asymptotes of each.*

29. $y = \tan \frac{x}{2}$

30. $y = \tan \frac{x}{3}$

31. $y = \tan\left(x + \frac{\pi}{2}\right)$

32. $y = \tan\left(2x + \frac{\pi}{2}\right)$

B Exercises 33–36 require the use of a calculator.

Evaluate each of the following functions to four decimal places at $x = 0.1, 0.01, 0.001$, *and* 0.0001. *What value is* $f(x)$ *approaching as the value of* x *approaches 0 but remains positive?*

33. $f(x) = \dfrac{\tan x}{x}$

34. $f(x) = \dfrac{\tan 2x}{x}$

35. $f(x) = x \tan x$

36. $f(x) = x \tan^2 x$

5.6 Other Circular Functions

In Section 5.5 we used the sine and cosine functions to define the tangent function. In this section we shall consider three other periodic functions that are defined in terms of the sine and cosine.

Cotangent

The first of these periodic functions is the cotangent function.

Definition 5.5 For $x \in R$, $\sin x \neq 0$, the **cotangent of** x is denoted by **cot** x, and

$$\cot x = \frac{\cos x}{\sin x}.$$

Since $\cot x = \cos x/\sin x$, the domain of the cotangent function is R, excepting those real numbers for which $\sin x = 0$, that is, all real numbers except those of the form $k\pi$, where $k \in J$. The range of cotangent is R, and, like the tangent function,

5.6 Other Circular Functions

cotangent has fundamental period π. Note that because $\cot x = \cos x / \sin x$ and $\tan x = \sin x / \cos x$,

$$\cot x = \frac{1}{\tan x} \quad \text{and} \quad \tan x = \frac{1}{\cot x} \tag{1}$$

for those values of x for which each expression is defined.

Since $\cot x$ is the reciprocal of $\tan x$, $x \neq k\pi$, $k \in J$, we can find the function values $\cot x$ by first finding $\tan x$ and then computing its reciprocal. Since most calculators will not compute $\cot x$ with a single keystroke, this method is ordinarily used to compute $\cot x$.

Examples Compute each of the following values.

a. $\cot \dfrac{\pi}{4}$
b. $\cot 1.32$

Solutions **a.** By using Equation (1) and Table 5.3, we have

$$\cot \frac{\pi}{4} = \frac{1}{\tan (\pi/4)} = \frac{1}{1} = 1.$$

b. By reading directly from Table IV, we have

$$\cot 1.32 = 0.2562.$$

Alternatively, by using Equation (1) and a calculator, we obtain

$$\cot 1.32 = \frac{1}{\tan 1.32} = 0.2562.$$

Graph of
$y = \cot x$

We can sketch the graph of the cotangent function by plotting some points over $0 < x < \pi$ and joining the points with a curve. The graph is shown in Figure 5.20. Note that the graph has vertical asymptotes at $x = k\pi$, $k \in J$ and x-intercepts at $x = (\pi/2) + k\pi$, $k \in J$.

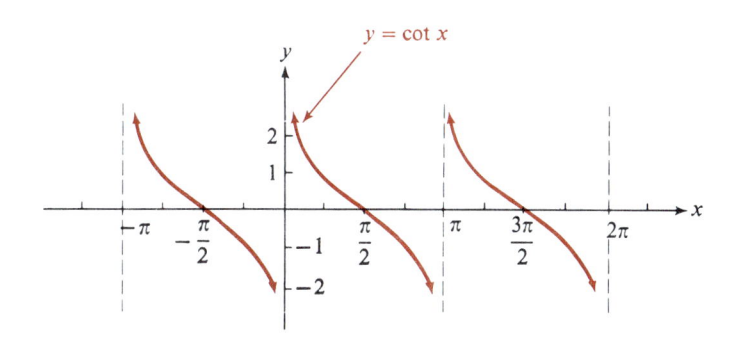

Figure 5.20

Secant

Another periodic function is the secant.

Definition 5.6 For $x \in R$, $\cos x \neq 0$, the **secant of** x is denoted by **sec** x, and

$$\sec x = \frac{1}{\cos x}.$$

Since sec x is the reciprocal of cos x, the domain of the secant function is R, except for those values of x for which $\cos x = 0$, namely, $x = \pi/2 + k\pi$, $k \in J$. Because $|\cos x| \leq 1$ for all $x \in R$, sec x is never less than 1 in absolute value. Thus the range of the secant function is $\{y \mid |y| \geq 1\}$. Since the cosine function has fundamental period 2π, the secant function also has fundamental period 2π.

As in the case of the cotangent, the values sec x usually cannot be obtained with a single keystroke on a calculator. The fact that sec x is the reciprocal of cos x, $x \neq \pi/2 + k\pi$, $k \in J$, is used to compute sec x.

Examples Compute each of the following values.

a. $\sec \dfrac{\pi}{6}$ **b.** $\sec 1.41$

Solutions **a.** By using Definition 5.6 and Table 5.2, we have

$$\sec \frac{\pi}{6} = \frac{1}{\cos (\pi/6)} = \frac{1}{\sqrt{3}/2} = \frac{2}{\sqrt{3}}.$$

b. By reading directly from Table IV, we have

$$\sec 1.41 = 6.246.$$

Alternatively, by using a calculator, we obtain

$$\sec 1.41 = \frac{1}{\cos 1.41} = 6.2459.$$

Graph of We can sketch the graph of the secant function by plotting some points over
$y = \sec x$ $0 < x < 2\pi$ and joining the points with a curve. The graph is shown in Figure 5.21. Note that the graph has vertical asymptotes at $x = \pi/2 + k\pi$, $k \in J$.

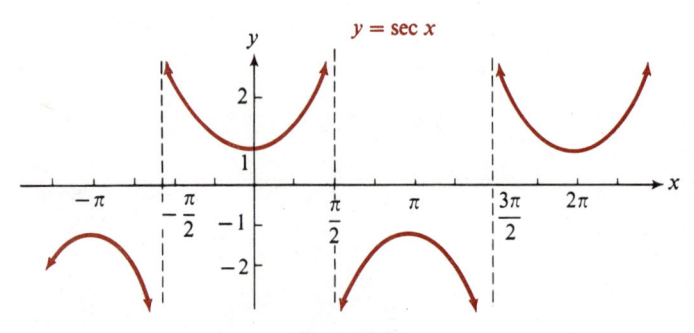

Figure 5.21

5.6 Other Circular Functions

Cosecant

The last function we shall discuss in this section is the cosecant.

Definition 5.7 For $x \in R$, $\sin x \neq 0$, the **cosecant of x** is denoted by **csc x**, and

$$\csc x = \frac{1}{\sin x}.$$

Because csc x is the reciprocal of sin x, the domain of the cosecant function contains all real numbers except those for which $\sin x = 0$; that is, the domain is all R except the numbers $k\pi$, $k \in J$. The cosecant function has range $\{y \mid |y| \geq 1\}$ and fundamental period 2π.

As in the case of the secant function, the values csc x usually cannot be obtained with a single keystroke on a calculator, and thus the fact that csc x is the reciprocal of sin x, $x \neq k\pi$, $k \in J$, is used to compute csc x.

Examples

Compute each of the following values.

a. $\csc \dfrac{\pi}{6}$

b. csc 1.32

Solutions

a. By using Definition 5.7 and Table 5.2, we have

$$\csc \frac{\pi}{6} = \frac{1}{\sin (\pi/6)} = \frac{1}{1/2} = 2.$$

b. By reading directly from Table IV, we have

$$\csc 1.32 = 1.032.$$

Alternatively, by using a calculator, we obtain

$$\csc 1.32 = \frac{1}{\sin 1.32} = 1.0323.$$

Graph of $y = \csc x$

We can sketch the graph of the cosecant function by plotting some points over $0 < x < 2\pi$ and joining the points with a curve. The graph is shown in Figure 5.22. Note that the graph has vertical asymptotes at $x = k\pi$, $k \in J$.

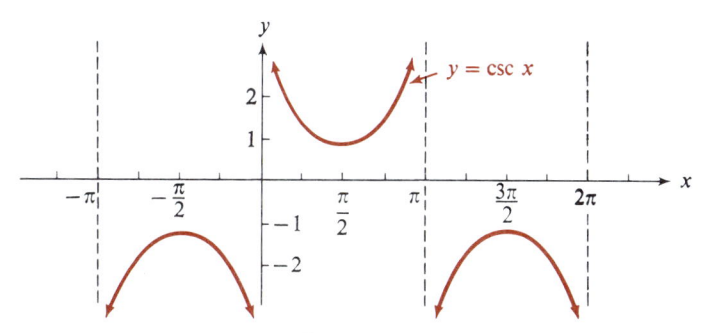

Figure 5.22

Functions of the form $y = A \cot (Bx + C)$, $y = A \sec (Bx + C)$, and $y = A \csc (Bx + C)$ can be graphed using the same methods that were used for $y = A \sin (Bx + C)$, and $y = A \cos (Bx + C)$ in Section 5.4. Since these curves are not sinusoids, we do not interpret A as an amplitude.

Exercise 5.6

A *Use the appropriate definition and Table 5.2 or 5.3 to find the exact function value if it exists; if not, then so indicate.*

Example $\cot \pi$

Solution Using Table 5.2 and Definition 5.5, we have
$$\cot \pi = \frac{\cos \pi}{\sin \pi} = \frac{-1}{0}.$$
Thus, $\cot \pi$ is undefined.

1. $\cot \dfrac{\pi}{2}$
2. $\cot \dfrac{\pi}{3}$
3. $\sec \dfrac{\pi}{4}$
4. $\sec \dfrac{\pi}{3}$
5. $\csc \dfrac{\pi}{3}$
6. $\csc \dfrac{\pi}{2}$
7. $\csc \pi$
8. $\sec \dfrac{\pi}{2}$

Example $\sec \dfrac{4\pi}{3}$

Solution Using a reference arc, we obtain
$$\cos \frac{4\pi}{3} = -\frac{1}{2}.$$
Then, from Definition 5.6,
$$\sec \frac{4\pi}{3} = \frac{1}{\cos (4\pi/3)} = \frac{1}{-1/2} = -2.$$

9. $\cot \dfrac{7\pi}{6}$
10. $\cot \dfrac{5\pi}{4}$
11. $\sec \dfrac{7\pi}{4}$
12. $\sec \dfrac{3\pi}{4}$
13. $\csc \dfrac{5\pi}{6}$
14. $\csc \dfrac{5\pi}{3}$

5.6 Other Circular Functions

15. $\cot(-\pi)$
16. $\cot\left(\dfrac{13\pi}{4}\right)$
17. $\sec\left(-\dfrac{\pi}{6}\right)$
18. $\sec\dfrac{13\pi}{4}$
19. $\csc\left(-\dfrac{5\pi}{2}\right)$
20. $\csc\dfrac{11\pi}{4}$

Use the approximate definition and a hand calculator or Table IV to approximate the indicated value.

21. $\cot 4.31$
22. $\cot 3.51$
23. $\sec 1.05$
24. $\sec 2.41$
25. $\csc 1.65$
26. $\csc 5.32$
27. $\cot(-2.10)$
28. $\cot 8.31$
29. $\sec 3.64$
30. $\sec 12.51$
31. $\csc(-3.00)$
32. $\csc 8.65$

Sketch the graph of each function over the indicated interval, and list the zeros and vertical asymptotes of each.

Example $y = \sec 2x$; $[0, 2\pi]$

Solution Since $\sec 2x = 1/\cos 2x$ and the period of $y = \cos 2x$ is

$$p = \dfrac{2\pi}{2} = \pi,$$

the period of $y = \sec 2x$ is also π. We first plot a number of points over the interval $0 \le x \le \pi$ and then join these points with a smooth curve. We can then repeat the pattern obtained in both directions along the x-axis to obtain the graph shown. There are no zeros. The vertical asymptotes are $x = \pi/4, 3\pi/4, 5\pi/4, 7\pi/4$.

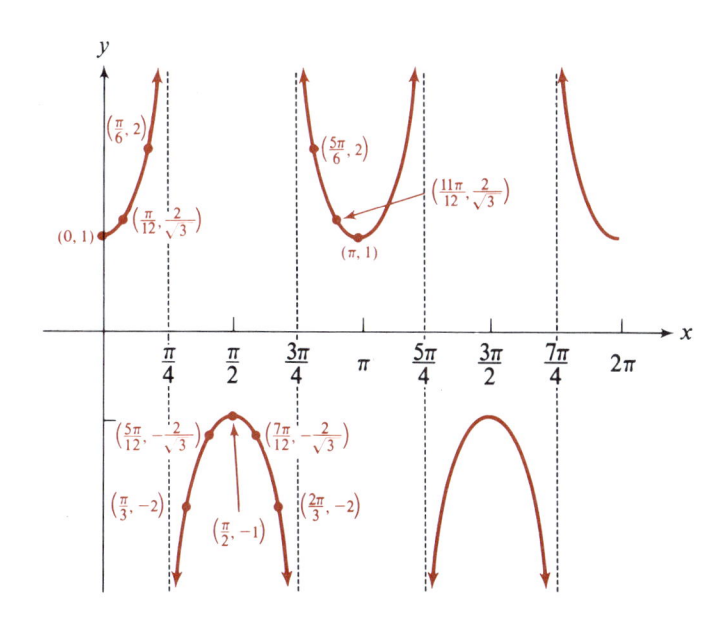

33. $y = \cot 3x$; $[0, \pi]$

34. $y = \cot \dfrac{x}{2}$; $[-2\pi, 2\pi]$

35. $y = \sec 3x$; $[0, \pi]$

36. $y = \sec \dfrac{x}{2}$; $[-2\pi, 2\pi]$

37. $y = \csc 3x$; $[0, \pi]$

38. $y = \csc \dfrac{x}{2}$; $[-2\pi, 2\pi]$

5.7 Inverse Circular Functions

In many applications it is necessary to find a value x that yields a given value of y for a function $y = f(x)$. For example, we may want to find a number x such that $\sin x = 0.5$. We encountered a similar problem with the exponential function in Section 4.2 and solved it by considering its inverse, the logarithm function. In this section we shall consider the inverses of the circular functions.

Inverse relations

Recall from Section 2.7 that the inverse of a function is the relation obtained by interchanging the components x and y of each ordered pair in the function and that the graph of the inverse of a function can be obtained by reflecting the graph of the function in the line with equation $y = x$. Furthermore, the resulting relation is a function if and only if the original function is one-to-one. Now consider the sine function,

$$y = \sin x, \tag{1}$$

and the relation

$$x = \sin y. \tag{2}$$

Both graphs are shown in Figure 5.23 where it is evident that each is the reflection of the other in the line $y = x$. Equation (2) does not define a function, because for

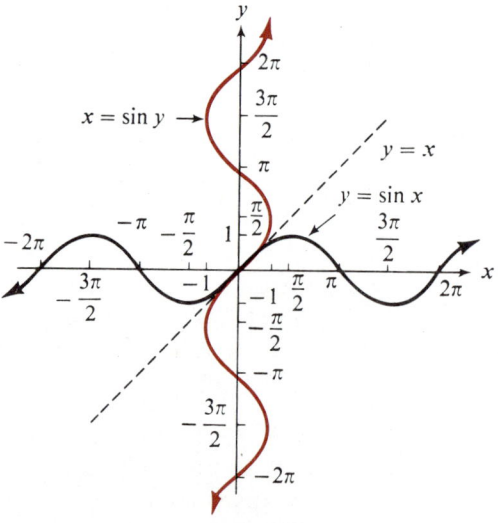

Figure 5.23

5.7 Inverse Circular Functions

each element x in its domain, $\{x \mid -1 \leq x \leq 1\}$, there are an unlimited number of elements in its range.

The graphs of the relations

$$x = \cos y$$
$$x = \tan y$$

are shown in Figure 5.24. Note that these equations also do not define functions.

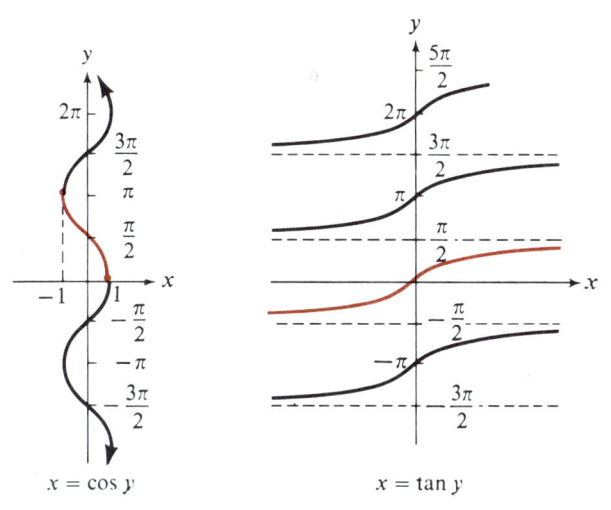

Figure 5.24

Inverse functions

By suitably restricting the domains of the circular functions—that is, by suitably restricting the ranges of the relations obtained by interchanging the variables—we can define an inverse function for each circular function. For the sine function, we restrict the range to the interval $-\pi/2 \leq y \leq \pi/2$, and we have the inverse function

$$x = \sin y, \quad -\frac{\pi}{2} \leq y \leq \frac{\pi}{2}. \tag{3}$$

The graph of this inverse is shown in Figure 5.25.

To write (3) in a form in which y is expressed explicitly in terms of x, we use the special notation

$$y = \mathrm{Sin}^{-1} x$$

(read "y is equal to the inverse sine of x"), with the understanding, of course, that $\mathrm{Sin}^{-1} x$ does not denote $1/\mathrm{Sin}\, x$. Note that we use a capital S with this inverse notation. Similar notation is used for the other inverse circular functions, where the ranges have been suitably restricted in accordance with the following definition.

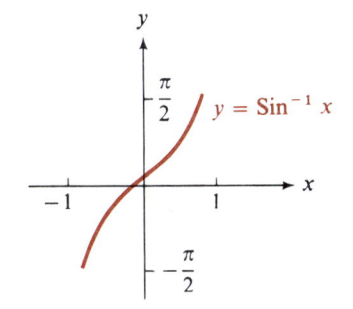

Figure 5.25

Definition 5.8 *The six inverse circular functions are*

$$y = \text{Sin}^{-1} x \quad \leftrightarrow \quad x = \sin y, \quad -\frac{\pi}{2} \leq y \leq \frac{\pi}{2},^\dagger$$

$$y = \text{Cos}^{-1} x \quad \leftrightarrow \quad x = \cos y, \quad 0 \leq y \leq \pi,$$

$$y = \text{Tan}^{-1} x \quad \leftrightarrow \quad x = \tan y, \quad -\frac{\pi}{2} < y < \frac{\pi}{2},$$

$$y = \text{Csc}^{-1} x \quad \leftrightarrow \quad x = \csc y, \quad -\frac{\pi}{2} \leq y \leq \frac{\pi}{2} \quad (y \neq 0)$$

$$y = \text{Sec}^{-1} x \quad \leftrightarrow \quad x = \sec y, \quad 0 \leq y \leq \pi \quad \left(y \neq \frac{\pi}{2}\right),$$

$$y = \text{Cot}^{-1} x \quad \leftrightarrow \quad x = \cot y, \quad -\frac{\pi}{2} \leq y \leq \frac{\pi}{2} \quad (y \neq 0).$$

Although the particular ranges in Definition 5.8 are the ones customarily chosen, the choices actually are quite arbitrary. They have generally been selected to involve small values of y, to have relatively simple graphs, and of course to yield a one-to-one correspondence between domain and range. They are also values that can be obtained with most scientific calculators. The graphs of $y = \text{Cos}^{-1} x$ and $y = \text{Tan}^{-1} x$ are shown in Figure 5.26-**a** and **b**, respectively. The remaining inverse circular functions are less frequently used, and their graphs are not shown.

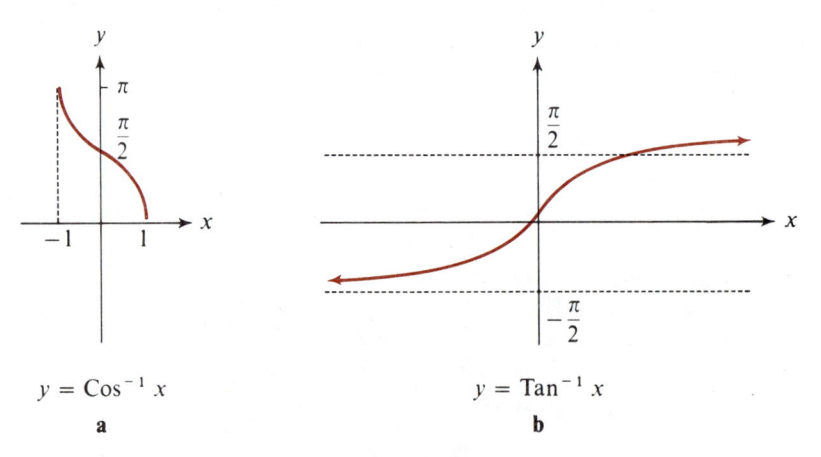

$y = \text{Cos}^{-1} x$
a

$y = \text{Tan}^{-1} x$
b

Figure 5.26

The inverse-circular-function values are sometimes named by using the prefix "Arc" with the function name. For example, $\text{Sin}^{-1} x$ is sometimes represented by Arcsin x (read "arcsin x").

† The symbol "$\leftrightarrow$" should be read "is equivalent to."

5.7 Inverse Circular Functions

It is sometimes helpful to interpret an expression involving inverse notation by first rewriting the expression equivalently without using this notation.

Examples Find the value of each of the following in terms of rational multiples of π.

    a. $\operatorname{Sin}^{-1} \dfrac{1}{2}$ **a.** $\operatorname{Arcsec} \dfrac{2}{\sqrt{3}}$

Solutions   **a.** Let $y = \operatorname{Sin}^{-1}(1/2)$. Then, by Definition 5.8,

$$y = \operatorname{Sin}^{-1}\frac{1}{2} \quad \leftrightarrow \quad \sin y = \frac{1}{2}, \quad -\frac{\pi}{2} \leq y \leq \frac{\pi}{2}.$$

From Table 5.2, $y = \pi/6$. Hence,

$$y = \operatorname{Sin}^{-1}\frac{1}{2} = \frac{\pi}{6}.$$

b. Let $y = \operatorname{Arcsec}(2/\sqrt{3})$. Then, by Definition 5.8,

$$y = \operatorname{Arcsec}\frac{2}{\sqrt{3}} \quad \leftrightarrow \quad \sec y = \frac{2}{\sqrt{3}}, \quad 0 \leq y \leq \pi,$$

which by Definition 5.6 is equivalent to

$$\cos y = \frac{\sqrt{3}}{2}, \quad 0 \leq y \leq \pi.$$

From Table 5.2, $y = \pi/6$. Hence,

$$y = \operatorname{Arcsec}\frac{2}{\sqrt{3}} = \frac{\pi}{6}.$$

With a little practice, you will be able to find an inverse function value directly. The following examples show the process.

Examples Find the value of each of the following in terms of rational multiples of π.

    a. $\operatorname{Arctan} \dfrac{1}{\sqrt{3}}$ **b.** $\operatorname{Cos}^{-1} \dfrac{1}{\sqrt{2}}$

Solutions   **a.** The number y between 0 and $\pi/2$ such that $\tan y = 1/\sqrt{3}$ is $\pi/6$. Hence,

$$\operatorname{Arctan}\frac{1}{\sqrt{3}} = \frac{\pi}{6}.$$

b. The number between 0 and π such that $\cos y = 1/\sqrt{2}$ is $\pi/4$. Hence,

$$\operatorname{Cos}^{-1}\frac{1}{\sqrt{2}} = \frac{\pi}{4}.$$

We must give special attention to the ranges specified in Definition 5.8 when finding values of functions that have negative values in their domains.

Examples Find the value of the following expressions in terms of rational multiples of π.

a. $\text{Arctan}\left(-\dfrac{1}{\sqrt{3}}\right)$ **b.** $\text{Cos}^{-1}\left(-\dfrac{1}{\sqrt{2}}\right)$

Solutions **a.** The number y between $-\pi/2$ and $\pi/2$ such that $\tan y = -1/\sqrt{3}$ is $-\pi/6$. Hence,

$$\text{Arctan}\left(-\dfrac{1}{\sqrt{3}}\right) = -\dfrac{\pi}{6}.$$

b. The number y between 0 and π such that $\cos y = -1/\sqrt{2}$ is $3\pi/4$. Hence,

$$\text{Cos}^{-1}\left(-\dfrac{1}{\sqrt{2}}\right) = \dfrac{3\pi}{4}.$$

Using a table or a calculator Table IV or a hand calculator, *in radian mode*, with inverse-circular-function capability can be used to find approximations for function values that are not listed in Table 5.2. It sometimes happens that the exact value being sought is not in Table IV. If this occurs, simply choose the closest value in the table. A calculator will be used to compute inverse-circular-function values in all examples and exercises. Answers obtained in this way may vary slightly from those obtained using Table IV if the exact value sought is not in the table.

Some calculators have a key marked INV or ARC that is used to calculate the values of the inverse circular function.

Example Compute $\text{Tan}^{-1}\, 0.2236$.

Solution It may be helpful first to let $y = \text{Tan}^{-1}\, 0.2236$ and then rewrite the equation equivalently as

$$\tan y = 0.2236, \quad -\dfrac{\pi}{2} < y < \dfrac{\pi}{2}.$$

By using a calculator in radian mode or from Table IV, we obtain $y = 0.22$. Since $-\pi/2 < 0.22 < \pi/2$, we have

$$\text{Tan}^{-1}\, 0.2236 = 0.22.$$

Exercise 5.7

A Find the value of the expression. Give the result in terms of rational multiples of π.

Examples **a.** $\text{Arccos}\,\dfrac{1}{2}$ **b.** $\text{Tan}^{-1}(-1)$

5.7 Inverse Circular Functions

Solutions

a. We seek a number y such that
$$\cos y = \frac{1}{2}, \quad 0 \le y \le \pi.$$
The number is $\pi/3$.

b. We seek a number y such that
$$\tan y = -1, \quad -\frac{\pi}{2} < y < \frac{\pi}{2}.$$
The number is $-\pi/4$.

1. $\operatorname{Arcsin} \dfrac{1}{2}$
2. $\operatorname{Arctan} \sqrt{3}$
3. $\operatorname{Cot}^{-1} 1$
4. $\operatorname{Cos}^{-1} \dfrac{1}{\sqrt{2}}$
5. $\operatorname{Cos}^{-1} \dfrac{1}{2}$
6. $\operatorname{Arcsin} \dfrac{\sqrt{3}}{2}$
7. $\operatorname{Arctan}\left(-\dfrac{1}{\sqrt{3}}\right)$
8. $\operatorname{Cot}^{-1}(-1)$

Use a calculator or Table IV in the Appendix to find an approximation of the expression.

Example $\operatorname{Cos}^{-1} 0.9940$

Solution It may be helpful to let $y = \operatorname{Cos}^{-1} 0.9940$ and then rewrite this equation equivalently as
$$\cos y = 0.9940, \quad 0 \le y \le \pi.$$
By using a calculator in radian mode or by using Table IV, we obtain $y = 0.11$. Since $0 \le 0.11 \le \pi$, we have
$$\operatorname{Cos}^{-1} 0.9940 = 0.11.$$

9. $\operatorname{Tan}^{-1} 0.1003$
10. $\operatorname{Arctan} 3.467$
11. $\operatorname{Sin}^{-1} 0.3802$
12. $\operatorname{Cos}^{-1} 0.6675$
13. $\operatorname{Arcsin}(-0.8624)$
14. $\operatorname{Arccos}(-0.3902)$

Example $\operatorname{Arcsec}(2.571)$

Solution Let $y = \operatorname{Arcsec} 2.571$ and rewrite this equivalently as
$$\sec y = 2.571, \quad 0 \le y \le \pi \quad \left(y \ne \frac{\pi}{2}\right).$$
This is equivalent to
$$\cos y = \frac{1}{2.571} = 0.3890, \quad 0 \le y \le \pi \quad \left(y \ne \frac{\pi}{2}\right).$$
By using a calculator in radian mode or from Table IV, $y = 1.17$. Since $0 \le 1.17 < \pi$,
$$\operatorname{Arcsec} 2.571 = 1.17.$$

15. $\operatorname{Arccsc} 2.697$
16. $\operatorname{Csc}^{-1} 1.422$
17. $\operatorname{Arcsec} 1.053$
18. $\operatorname{Sec}^{-1} 2.807$
19. $\operatorname{Arccot} 1.688$
20. $\operatorname{Cot}^{-1} 1.895$

Find the exact value for the given expression if the value exists.

Examples

a. $\text{Cos}^{-1}(\tan \pi)$

b. $\sin\left(\text{Cos}^{-1}\dfrac{1}{2}\right)$

Solutions

a. Since $\tan \pi = 0$,

$$\text{Cos}^{-1}(\tan \pi) = \text{Cos}^{-1} 0 = \dfrac{\pi}{2}.$$

b. Since $\text{Cos}^{-1}(1/2) = \pi/3$,

$$\sin\left(\text{Cos}^{-1}\dfrac{1}{2}\right) = \sin\left(\dfrac{\pi}{3}\right) = \dfrac{\sqrt{3}}{2}.$$

21. $\text{Sin}^{-1}\left(\cos\dfrac{\pi}{4}\right)$

22. $\text{Cos}^{-1}\left(\sin\dfrac{\pi}{2}\right)$

23. $\text{Tan}^{-1}\left(\tan\dfrac{\pi}{3}\right)$

24. $\text{Sin}^{-1}\left(\sin\dfrac{3\pi}{2}\right)$

25. $\sin\left(\text{Arccos}\dfrac{1}{2}\right)$

26. $\tan\left(\text{Arcsin}\dfrac{\sqrt{3}}{2}\right)$

27. $\cos(\text{Cot}^{-1}(-\sqrt{3}))$

28. $\sin(\text{Tan}^{-1}(-1))$

29. $\sin\left(\text{Arcsin}\dfrac{1}{2}\right)$

30. $\cos(\text{Tan}^{-1} 0)$

31. $\text{Arccos}(\sin(\text{Arctan}(-1)))$

32. $\sin(\text{Cos}^{-1}(\tan 0))$

Solve the given equation for x in terms of y.

Example

$y = 2 \text{ Arcsin } 3x$

Solution

Divide both members by 2.

$$\dfrac{y}{2} = \text{Arcsin } 3x$$

From the definition of Arcsine, write the expression as

$$3x = \sin\dfrac{y}{2},$$

and then as

$$x = \dfrac{1}{3}\sin\dfrac{y}{2},$$

with

$$-\dfrac{\pi}{2} \leq \dfrac{y}{2} \leq \dfrac{\pi}{2} \quad \text{or} \quad -\pi \leq y \leq \pi.$$

33. $y = 3 \text{ Arccos } 2x$

34. $y = 2 \text{ Sin}^{-1} \dfrac{x}{5}$

35. $y = \dfrac{1}{2} \text{Tan}^{-1}(x + \pi)$

36. $y = \dfrac{2}{3} \text{Tan}^{-1}(\pi x)$

37. $y = \dfrac{1}{2} \text{Cos}^{-1}(2x + \pi)$

38. $y = \dfrac{1}{2} \text{Arcsin } (\pi - 2x)$

39. $y = -\dfrac{1}{3} \text{Tan}^{-1}\left(\pi x + \dfrac{\pi}{2}\right)$

40. $y = -\dfrac{2}{3} \text{Arctan}\left(3x + \dfrac{\pi}{2}\right)$

Chapter Review

[5.1] *State whether the given function value is positive or negative.*

1. $\cos \dfrac{2\pi}{3}$
2. $\cos \dfrac{16\pi}{3}$
3. $\sin\left(\dfrac{-8\pi}{7}\right)$
4. $\sin\left(\dfrac{-\pi}{6}\right)$

Find the required function value and state the quadrant in which s terminates.

5. $\sin s$, given that $\cos s = 2/3$ and $\sin s < 0$.
6. $\cos s$, given that $\sin s = 3/8$ and $\cos s > 0$.

[5.2] *Use Table 5.2 to find the exact value of the given expression.*

7. $\cos \dfrac{\pi}{6} + \sin \pi$

8. $\sin \dfrac{\pi}{4} - \cos \dfrac{\pi}{4}$

9. $(\sin \pi)\left(\cos \dfrac{\pi}{3}\right)$

10. $\dfrac{\cos \pi/3}{\sin \pi/3}$

Use Table IV or a hand calculator to find the approximate value of the given expression.

11. $\cos 1.48 + \sin 1.05$

12. $\cos 1.35 - \sin 0.45$

13. $(\sin 0.31)(\cos 1.25)$

14. $\dfrac{\sin 1.01}{\cos 1.31}$

Assume that $d = K \cos 2\pi t$. Find the indicated value under the given conditions. Use $\pi \approx 3.14$.

15. K, given $d = 25$ and $t = 1/2$.
16. d, given $K = 75$ and $t = 0.05$.

[5.3] Use Table 5.2 or a hand calculator or Table IV to compute the indicated value.

17. $\cos \dfrac{3\pi}{4}$ 18. $\cos \dfrac{11\pi}{6}$ 19. $\sin \dfrac{5\pi}{4}$

20. $\sin \dfrac{5\pi}{6}$ 21. $\cos 3.28$ 22. $\cos 6.11$

23. $\sin 2.51$ 24. $\sin 3.35$ 25. $\cos(-1.53)$

26. $\cos 6.56$ 27. $\sin(-1.65)$ 28. $\sin 7.21$

[5.4] Find the amplitude, period, and phase shift for the indicated function.

29. $y = 2 \sin 3x$ 30. $y = \dfrac{3}{2} \cos\left(x - \dfrac{\pi}{4}\right)$ 31. $y = 4 \sin(2x - \pi)$

Sketch the graph of the indicated function over the interval $-\pi \le x \le 2\pi$.

32. $y = \dfrac{1}{3} \cos 2x$ 33. $y = \sin\left(x + \dfrac{\pi}{3}\right)$ 34. $y = \cos\left(3x + \dfrac{\pi}{2}\right)$

[5.5] Use Definition 5.4 and Table 5.2, Table IV, or a hand calculator to compute the indicated value.

35. $\tan \dfrac{5\pi}{4}$ 36. $\tan \dfrac{4\pi}{3}$ 37. $\tan 1.95$

38. $\tan 3.65$ 39. $\tan(-2.18)$ 40. $\tan 7.25$

Sketch the graph of each function over the interval $-\pi \le x \le 2\pi$.

41. $y = \tan \dfrac{x}{4}$ 42. $y = 2 \tan 2x$

[5.6] Use the appropriate definition and Table 5.2 to find the exact value of each expression if it exists; if not, then so indicate.

43. $\cot \dfrac{\pi}{6}$ 44. $\sec \dfrac{\pi}{6}$ 45. $\csc \dfrac{\pi}{2}$ 46. $\sec \dfrac{\pi}{2}$

Use the appropriate definition to find an approximation of the given expression if it exists; if not, then so indicate.

47. $\cot \dfrac{3\pi}{4}$ 48. $\sec \dfrac{9\pi}{4}$ 49. $\csc\left(-\dfrac{\pi}{6}\right)$

50. $\cot 3.85$ 51. $\sec 9.56$ 52. $\csc(-1.65)$

[5.7] *Find the value of the given expression if the value exists. Express the result in terms of rational multiples of π.*

53. $\text{Sin}^{-1} \dfrac{1}{\sqrt{2}}$
54. $\text{Cos}^{-1} 1$
55. $\text{Tan}^{-1} \dfrac{1}{\sqrt{3}}$
56. $\text{Arccot } \sqrt{3}$
57. $\text{Arcsec}(-2)$
58. $\text{Csc}^{-1}(-2)$

Use Table IV or a hand calculator to find an approximation of the given expression.

59. $\text{Arctan } 1.398$
60. $\text{Arcsin } 0.9320$
61. $\text{Cos}^{-1} 0.9550$
62. $\text{Sec}^{-1} 1.609$
63. $\text{Arccsc } 1.771$
64. $\text{Cot}^{-1} 0.9712$

Find an approximation of the given expression if the value exists.

65. $\sin(\text{Sin}^{-1} 0.25)$
66. $\text{Cos}^{-1}(\cos 4\pi)$
67. $\tan\left(\text{Sin}^{-1} \dfrac{\sqrt{2}}{2}\right)$
68. $\sin(\text{Tan}^{-1} \sqrt{3})$

6 Trigonometric Functions; Vectors

6.1 Angles and Their Measure

In studying geometry, we learn that an angle is the union of two rays with a common endpoint (Figure 6.1-a) and that an angle can be designated by naming a point on each ray together with the common endpoint of the rays or else by simply assigning a single symbol, say α, to the angle. The common endpoint of the rays is called the **vertex** of the angle, and the rays are called the **sides** of the angle.

Standard position

Now each angle in the plane is congruent ($\cong$) to an angle with one side along the positive x-axis and vertex at the origin (Figure 6.1-b). Such an angle is said to be in **standard position**. The side $\overline{OC'}$ of $\angle A'OC'$ in Figure 6.1-b is called the **initial side** of the angle; the side $\overline{OA'}$ is called the **terminal side**; and the angle can be visualized as being formed by a rotation from the initial side $\overline{OC'}$ into the terminal side $\overline{OA'}$. In trigonometry, the *amount of rotation* is considered along with the angle itself.

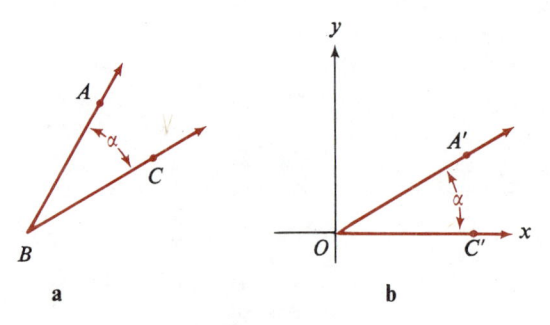

Figure 6.1

6.1 Angles and Their Measure

If the terminal side of an angle in standard position lies in a given quadrant, we say that the angle is *in* that quadrant.

Angle measurement

A measure is assigned to an angle by means of a circle with center at the vertex of the angle. For convenience, we shall restrict this discussion to the set of angles in standard position, although the process described is perfectly general. If the circumference of a circle of radius $r > 0$ is divided into p arcs of equal length, then each of the arcs will have length $2\pi r/p$. Starting at the point $(r, 0)$, we can scale the circumference of the circle in both the counterclockwise and clockwise directions from this point, in terms of this arc length as a unit. In doing this, it is customary to assign *negative* numbers in the *clockwise* direction and *positive* numbers in the *counterclockwise* direction. The terminal side of any angle α in standard position intercepts the circumference at a point, and one of the scale numbers (arc lengths) assigned to that point becomes the measure of the angle, depending on the amount of rotation involved.

The two most commonly encountered units of angle measure are the **degree** and the **radian**. In degree measure, the circumference of a circle is divided into 360 arcs of equal length, and hence the unit arc is of length

$$\frac{2\pi r}{360} = \frac{\pi}{180} r,$$

where r is the radius of the circle (see Figure 6.2-a). For this measure, we use the notation $°$. Thus, the angle shown in Figure 6.2-a has measure approximately $1°$.

In radian measure, the circumference is divided into 2π arcs of equal length, and hence the length of the unit arc (see Figure 6.2-b) is

$$\frac{2\pi r}{2\pi} = r.$$

For this measure we use the notation ^{R}. Thus, in radian measure, the length of the unit arc is the length of the radius of the circle, and the circumference contains just 2π of these units. The angle shown in Figure 6.2-b has measure approximately 1^R.

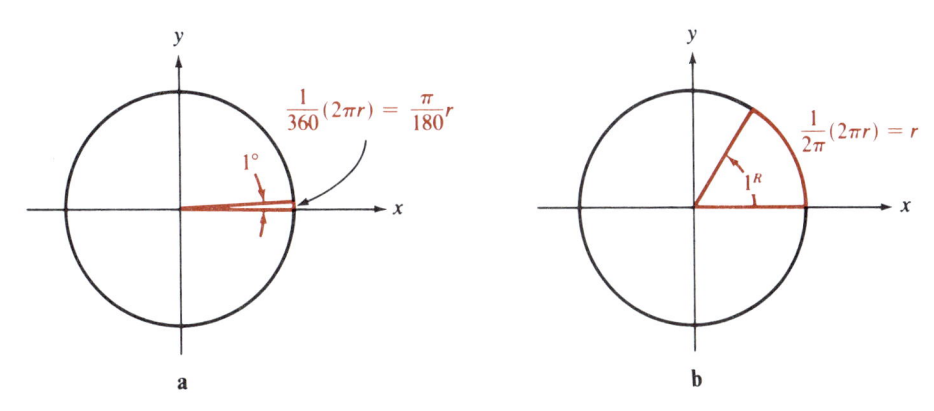

Figure 6.2

Recall from geometry that the lengths of arcs intercepted by a given central angle on concentric circles are proportional to the circumferences of the circles. *It follows that the measure assigned to an angle in terms of a given unit is independent of the radius of the measuring circle.* In particular, if the circle that is used as the basis for radian measure has radius 1, then the length of the unit arc is 1. For this reason, the unit circle is customarily used as a reference for the measure of an angle in radians.

Conversion formulas

Since one full rotation of the terminal side of an angle in standard position yields an angle with measures of $360°$ or $2\pi^R$, we write

$$360° = 2\pi^R.$$

Thus,

$$1^R = \left(\frac{180}{\pi}\right)°, \qquad (1)$$

and

$$1° = \left(\frac{\pi}{180}\right)^R. \qquad (2)$$

Equations (1) and (2) are **conversion formulas** for expressing the relationships between degree and radians.

Examples

Find the degree measure, to the nearest tenth of a degree, of the angle whose radian measure is given.

a. $\dfrac{\pi^R}{3}$ **b.** 0.62^R

Solutions

a. $\dfrac{\pi^R}{3} = \left(\dfrac{180}{\pi} \cdot \dfrac{\pi}{3}\right)° = 60°$

b. $0.62^R = \left(\dfrac{180}{\pi} \cdot 0.62\right)° = \dfrac{111.6°}{\pi} \approx 35.5°$

In Example **b** above, we used the calculator value for π, 3.1415926. We shall use this value for π in our computations in examples and in the answers for the exercises. Slightly different values may be expected if a different approximation for π is used.

Examples

Find the radian measure, to the nearest hundredth of a radian, of the angle whose degree measure is given.

a. $30°$ **b.** $300°$

Solutions

a. $30° = \left(\dfrac{\pi}{180} \cdot 30\right)^R = \dfrac{\pi^R}{6} \approx 0.52^R$

b. $300° = \left(\dfrac{\pi}{180} \cdot 300\right)^R = \dfrac{5\pi^R}{3} \approx 5.24^R$

6.1 Angles and Their Measure

Note that, in these examples, the number of units in one measure does not equal the number of units in another measure; for example, $30 \neq \pi/6$. But an angle whose measure is $30°$ is congruent to an angle whose measure is $(\pi/6)^R$, and we express this fact by writing $30° = (\pi/6)^R$. This is analogous to writing 36 inches = 3 feet, 6 feet = 2 yards, and so forth, when indicating lengths of line segments in different units.

If the radian measure of an angle is α^R, then the length s of the intercepted arc of a circle with given radius r can be found directly by noting that since arc lengths are proportional to corresponding central angle measure, we have

$$\frac{s}{2\pi r} = \frac{\alpha^R}{2\pi^R}.$$

Thus

$$s = r \cdot \alpha. \qquad (3)$$

If the measure of an angle is given in degrees, it can first be changed to radian measure and then the length of the intercepted arc can be found directly from (3).

Example Find the length of the arc intercepted by a central angle of 135° in a circle with a radius of 5 centimeters.

Solution We first change 135° to radian measure.

$$135° = \left(\frac{\pi}{180} \cdot 135\right)^R = \frac{3\pi^R}{4}.$$

Then

$$s = r \cdot \alpha$$
$$= 5 \cdot \frac{3\pi}{4} = \frac{15\pi}{4} \approx 11.8.$$

Therefore, the arc length to the nearest tenth is 11.8 centimeters.

Note that if the intercepted arc is part of a unit circle, then Equation (3) guarantees that the length of the arc is precisely the radian measure of the angle subtending the arc.

Coterminal angles Note that although angles such as α_1, α_2, and α_3 in Figure 6.3 have the same initial side and the same terminal side, their measures are different because the amounts of rotation involved are different. Such angles are called **coterminal**. The measures of α_1 and α_3 are positive, and the measure of α_2 is negative. An angle in standard position is often specified by giving its measure.

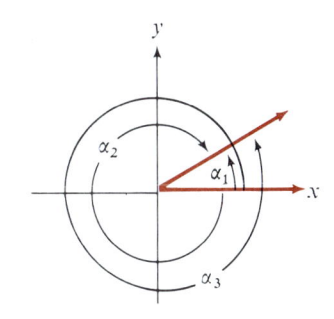

Figure 6.3

Examples Find all angles α satisfying $-360° \leq \alpha \leq 720°$ and coterminal with the given angle.

    a. 85°   **b.** $-15°$ **c.** 390°

Solutions
a. $85° + 360° = 445°$ **b.** $-15° + 360° = 345°$ **c.** $390° - 360° = 30°$
$85° - 360° = -275°$ $-15° + 720° = 705°$ $390° - 720° = -330°$

Exercise 6.1

Find the degree measure of the angle whose radian measure is as given.

1. **a.** 0^R **b.** $\dfrac{\pi^R}{2}$ **c.** π^R **d.** $\dfrac{3\pi^R}{2}$ **e.** $2\pi^R$

2. By filling in the blank spaces, complete the following table, which compares the radian measure and degree measure of angles that are of frequent occurrence.

°	30°	45°		120°	135°	
R			$\dfrac{\pi^R}{3}$			$\dfrac{5\pi^R}{6}$

°	210°			315°	330°
R		$\dfrac{5\pi^R}{4}$	$\dfrac{4\pi^R}{3}$	$\dfrac{5\pi^R}{3}$	

Find the degree measure, to the nearest tenth of a degree, of the angle whose radian measure is as given.

Examples   **a.** $\dfrac{2\pi^R}{5}$ **b.** 1.65^R

Solutions
a. By Equation (1), $1^R = \left(\dfrac{180}{\pi}\right)°$.   **b.** By Equation (1), $1^R = \left(\dfrac{180}{\pi}\right)°$.

Thus,   Thus,

$\dfrac{2\pi^R}{5} = \dfrac{2\pi}{5}\left(\dfrac{180}{\pi}\right)° = 72°.$ $1.65^R = 1.65\left(\dfrac{180}{\pi}\right)° \approx 94.5°.$

6.1 Angles and Their Measure

3. $\dfrac{2\pi^R}{9}$ 4. $\dfrac{3\pi^R}{5}$ 5. $\dfrac{7\pi^R}{5}$ 6. $\dfrac{5\pi^R}{8}$

7. 0.30^R 8. 1.25^R 9. 3.62^R 10. 9.14^R

Find the radian measure, to the nearest hundredth of a radian, of the angle whose degree measure is as given. Use $\pi \approx 3.14$.

Examples

a. $85°$ b. $640°$

Solutions

a. By Equation (2), $1° = \left(\dfrac{\pi}{180}\right)^R$.

Thus,
$$85° = 85\left(\dfrac{\pi}{180}\right)^R \approx 1.48^R.$$

b. By Equation (2), $1° = \left(\dfrac{\pi}{180}\right)^R$.

Thus,
$$640° = 640\left(\dfrac{\pi}{180}\right)^R \approx 11.16^R.$$

11. $20°$ 12. $50°$ 13. $130°$ 14. $310°$

15. $420°$ 16. $580°$ 17. $750°$ 18. $800°$

19. What is the degree measure, to the nearest hundredth of a degree, of an angle whose radian measure is 1^R?

20. What is the radian measure, to the nearest thousandth of a radian, of an angle whose degree measure is $1°$?

On a circle with given radius, find the length of the arc intercepted by the angle in standard position whose measure is as given.

Example

$r = 3$ inches; $120°$

Solution

The measure of the angle is first changed to radian measure. Thus
$$120° = \left(\dfrac{\pi}{180} \cdot 120\right)^R = \dfrac{2\pi^R}{3}.$$

Then, by Equation (3) on page 201,
$$s = r \cdot \alpha$$
$$= 3 \cdot \dfrac{2\pi}{3} = 2\pi \approx 6.28.$$

Therefore, the desired arc length is approximately 6.28 inches.

21. $r = 4$ centimeters; $(\pi/6)^R$ 22. $r = 5.2$ inches; π^R

23. $r = 1.2$ meters; 0.60^R 24. $r = 3.6$ feet; 1^R

25. $r = 2$ meters; $135°$ 26. $r = 4.3$ yards; $180°$

204   6 Trigonometric Functions; Vectors

Find the degree measure of the angle subtending an arc of the given length on a unit circle.

Example

$$\frac{\pi}{6}$$

Solution   Since the arc in question lies on a unit circle, its length is precisely the radian measure α^R of the angle subtending the arc. Thus, we need to convert $(\pi/6)^R$ to degrees. By Equation (1) on page 200,

$$1^R = \left(\frac{180}{\pi}\right)^\circ,$$

so

$$\frac{\pi^R}{6} = \left(\frac{\pi}{6} \cdot \frac{180}{\pi}\right)^\circ = 30°.$$

27. $\dfrac{\pi}{2}$ **28.** $\dfrac{3\pi}{2}$ **29.** 0.57

30. 2.30   **31.** 8.64   **32.** 11.56

Find all angles α satisfying $-360° \leq \alpha \leq 720°$ that are coterminal with the angle whose measure is given. Write a representation for the measures of all angles coterminal with the angle.

Examples

   a. 24°   **b.** $-184°$

Solutions

   a. $24° + 360° = 384°$ **b.** $-184° + 360° = 176°$
      $24° - 360° = -336°$ $-184° + 720° = 536°$
      $24° + 360°k, \quad k \in J$ $-184° + 360°k, \quad k \in J$

33. 30°   **34.** $-150°$ **35.** $-240°$ **36.** 190°

37. 420°   **38.** 683°   **39.** $-330°$ **40.** $-271°$

6.2   Functions of Angles

Historically, interest in the circular functions arose from the study of angles and triangles. In defining the circular functions, we used the notion of arc length on a unit circle to associate a real number [the arc length from the point (1, 0) to a point

6.2 Functions of Angles

on the circle] with another real number. Thus we defined functions with real numbers for *both* domain and range.

Geometric ratios in the coordinate plane

Now, consider any angle α in standard position, let (x_1, y_1) and (x_2, y_2) be any two distinct points (except the origin) in its terminal side or ray, and then let $r_1 = \sqrt{x_1^2 + y_1^2}$ and $r_2 = \sqrt{x_2^2 + y_2^2}$ (Figure 6.4). We can show that the following equalities of ratios hold:

$$\frac{x_1}{r_1} = \frac{x_2}{r_2},$$

$$\frac{y_1}{r_1} = \frac{y_2}{r_2},$$

and

$$\frac{y_1}{x_1} = \frac{y_2}{x_2} \quad (x_1, x_2 \neq 0).$$

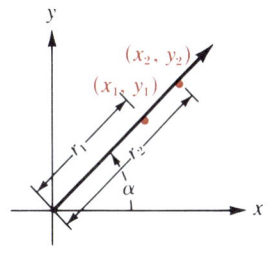

Figure 6.4

The proofs of these are left as exercises.

Although the figure shows the terminal side of α in the first quadrant, the ratios are equal for coordinates of points on the terminal side of an angle in any quadrant. Because these ratios do not depend on the choice of a point in the terminal side of α, we can use them to define some new functions, each with *the set of angles in standard position* as domain, and with sets of real numbers as ranges. Traditionally, six such assignments are made, and the names given the resulting functions are those we have used to name the circular functions.

Trigonometric ratios

Definition 6.1   Let α be an angle in standard position, let $(x, y) \neq (0, 0)$ be any point on the terminal side of α, and let $r = \sqrt{x^2 + y^2}$. Then

$$\cos \alpha = \frac{x}{r}, \qquad \sec \alpha = \frac{r}{x} \quad (x \neq 0),$$

$$\sin \alpha = \frac{y}{r}, \qquad \csc \alpha = \frac{r}{y} \quad (y \neq 0),$$

$$\tan \alpha = \frac{y}{x} \quad (x \neq 0), \quad \cot \alpha = \frac{x}{y} \quad (y \neq 0).$$

These ratios define the six **trigonometric functions**.

Example

Find the value of each of the six trigonometric functions of α, if the terminal side of α contains the point $(-3, 5)$.

Solution

First we compute r as the *positive* square root of $x^2 + y^2$.

$$r = \sqrt{x^2 + y^2} = \sqrt{(-3)^2 + 5^2} = \sqrt{9 + 25} = \sqrt{34}.$$

Solution continued overleaf

Then, by Definition 6.1 we have

$$\cos \alpha = \frac{x}{r} = -\frac{3}{\sqrt{34}}, \quad \sec \alpha = \frac{r}{x} = -\frac{\sqrt{34}}{3},$$

$$\sin \alpha = \frac{y}{r} = \frac{5}{\sqrt{34}}, \quad \csc \alpha = \frac{r}{y} = \frac{\sqrt{34}}{5},$$

$$\tan \alpha = \frac{y}{x} = -\frac{5}{3}, \quad \cot \alpha = \frac{x}{y} = -\frac{3}{5}.$$

Because every angle in the plane is congruent to an angle in standard position, the definitions of the trigonometric functions can be extended to assign the same numbers to every angle congruent to a given angle α. Thus, while these functions are defined in terms of angles in standard position, they can be viewed as applying to the set of all angles in the plane. Moreover, since congruent angles have the same measure, we can identify angles in the domain of each function with a particular unit of measure. Thus, we write

$$\sin 30° \quad \text{and} \quad \sin \frac{\pi^R}{6}$$

as abbreviations for "the sine of an angle whose measure is 30 degrees" and "the sine of an angle whose measure is $\pi/6$ radians," respectively.

Relationship between circular and trigonometric functions

The fact that the circular functions are defined using the unit circle, together with the fact that the unit circle can be used to assign measures to angles, makes it reasonable to expect a very close relationship to exist between the circular and trigonometric functions. Such is indeed the case.

Since the trigonometric functions of an angle α have been defined in terms of the coordinates of *any* point other than the origin on the terminal side of α, we can arbitrarily choose the point where the terminal side of the angle α intersects the unit circle. This is the point $P(\cos x, \sin x)$, where x is the radian measure of α (Figure 6.5). Thus, the trigonometric functions cosine and sine are related to the circular functions cosine and sine by

$$\cos \alpha = \frac{\cos x}{1} = \cos x, \quad (1)$$

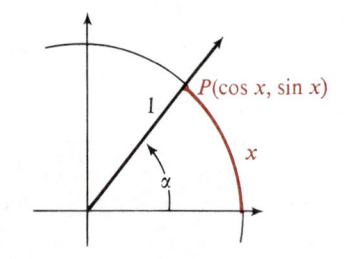

Figure 6.5

and, similarly,

$$\sin \alpha = \frac{\sin x}{1} = \sin x. \quad (2)$$

6.2 Functions of Angles

Furthermore the remaining four trigonometric functions are related to the corresponding circular functions in a similar way.

In Section 6.1 we agreed to name an angle by its measure. In particular, if the radian measure of α is x, then we write $\alpha = x^R$; and if the degree measure of α is t, then we write $\alpha = t°$. Thus, by replacing α by x^R and $t°$ in (1), we obtain

$$\cos x^R = \cos x \quad \text{and} \quad \cos t° = \cos x.$$

For example,

$$\cos \frac{\pi^R}{3} = \cos \frac{\pi}{3} = \frac{1}{2} \quad \text{and} \quad \cos 60° = \cos \frac{\pi}{3} = \frac{1}{2}. \tag{3}$$

The values of the trigonometric functions for some special angles are given in Table 6.1 below. These values can be obtained by the same method used to obtain Equation (3). The entries are left for you to verify.

Table 6.1

°	R	sin α	csc α	cos α	sec α	tan α	cot α
0°	0^R	0	Not defined	1	1	0	Not defined
30°	$\frac{\pi^R}{6}$	$\frac{1}{2}$	2	$\frac{\sqrt{3}}{2}$	$\frac{2}{\sqrt{3}}$	$\frac{1}{\sqrt{3}}$	$\sqrt{3}$
45°	$\frac{\pi^R}{4}$	$\frac{1}{\sqrt{2}}$	$\sqrt{2}$	$\frac{1}{\sqrt{2}}$	$\sqrt{2}$	1	1
60°	$\frac{\pi^R}{3}$	$\frac{\sqrt{3}}{2}$	$\frac{2}{\sqrt{3}}$	$\frac{1}{2}$	2	$\sqrt{3}$	$\frac{1}{\sqrt{3}}$
90°	$\frac{\pi^R}{2}$	1	1	0	Not defined	Not defined	0
180°	π^R	0	Not defined	−1	−1	0	Not defined
270°	$\frac{3\pi^R}{2}$	−1	−1	0	Not defined	Not defined	0

Examples   a. $\cos 45° = \frac{1}{\sqrt{2}}$ b. $\tan \pi^R = 0$

Signs of the trigonometric functions   The quadrant in which an angle lies determines whether a given trigonometric function of this angle is positive or negative. Figure 6.6 on page 208 presents this in convenient tabular form. The algebraic sign of a trigonometric function is the same as the algebraic sign of the corresponding circular function given in Table 5.1.

QUADRANT	I. $x > 0, y > 0$	II. $x < 0, y > 0$	III. $x < 0, y < 0$	IV. $x > 0, y < 0$
$\sin \alpha = y/r$ $\csc \alpha = r/y$	+	+	−	−
$\cos \alpha = x/r$ $\sec \alpha = r/x$	+	−	−	+
$\tan \alpha = y/x$ $\cot \alpha = x/y$	+	−	+	−

Figure 6.6

Tables for trigonometric functions

Trigonometric function values that are not listed in Table 6.1 can be obtained from Table IV in the appendix if the radian measure of α is known. Table V in the appendix gives trigonometric function values of angles with given degree measure. Table V is graduated in intervals of 6 minutes (6′), one minute being equal to one-sixtieth of a degree. Observe that the table reads from top to bottom for $0° \leq \alpha \leq 45°$, where the function values are identified at the top of the page, and from bottom to top for $45° \leq \alpha \leq 90°$, where the function values are identified at the bottom of the page.

Examples

a. $\tan 1.36^R$ **b.** $\csc 54°$

Solutions

a. From Table IV, $\tan 1.36^R = 4.673$.   **b.** From Table V, $\csc 54° = 1.236$.

Reference angles

Note that Table IV contains only trigonometric function values of angles with measure between 0^R and 1.57^R. Similarly, Table V contains only trigonometric function values for angles with measures between $0°$ and $90°$. For angles whose measures are not in these ranges, we use the concept of a reference angle in the same way that we used a reference arc with the circular functions.

Definition 6.2   *For any angle α in standard position, the **reference angle** $\bar{\alpha}$ is the least positive angle that the terminal ray of α forms with the x-axis.*

Figure 6.7 shows the reference angle for angles lying in each of the four quadrants.

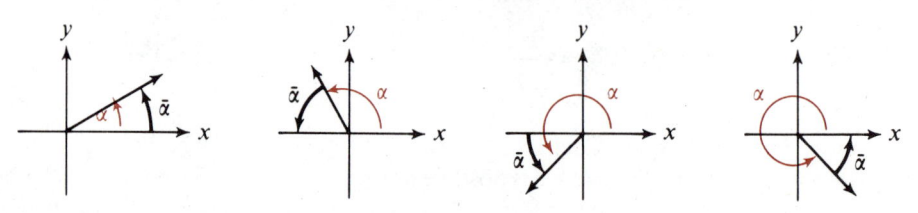

Figure 6.7

6.2 Functions of Angles

Examples Find the measure of the reference angle for each angle. Use $\pi \approx 3.14$.

  a. $\alpha = 135°$ **b.** $\alpha = 3.50^R$

Solutions

a. Since α terminates in the second quadrant, we have

$$\bar{\alpha} = 180° - 135°$$
$$= 45°.$$

$\bar{\alpha} = 180° - 135°$
$= 45°$ $\alpha = 135°$

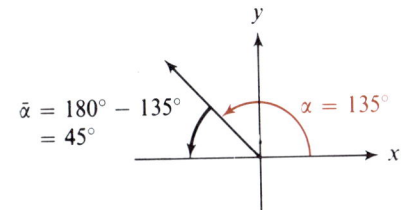

b. Since α terminates in the third quadrant, we have

$$\bar{\alpha} = 3.50^R - \pi^R$$
$$\approx 3.50^R - 3.14^R$$
$$= 0.36^R.$$

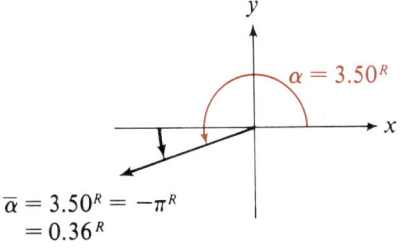

$\alpha = 3.50^R$

$\bar{\alpha} = 3.50^R = -\pi^R$
$= 0.36^R$

The following theorem is analogous to Theorem 5.2 on page 163. The proof is similar, and is omitted.

Theorem 6.1   If T is a trigonometric function and $\bar{\alpha}$ is the reference angle for the angle α, then

$$|T(\alpha)| = T(\bar{\alpha}).$$

We can now use reference angles to compute values of the trigonometric functions in the same way that reference arcs were used to compute the values of circular functions.

Example Find $\tan 225°$.

Solution The reference angle for $225°$ is

$$\bar{\alpha} = 225° - 180°$$
$$= 45°.$$

Thus, from Table 6.1 and Theorem 6.1, we have

$$|\tan 225°| = \tan 45° = 1.$$

$\alpha = 225°$

$\bar{\alpha} = 45°$

Hence, either $\tan 225° = 1$ or $\tan 225° = -1$. Since the terminal side of $\alpha = 225°$ lies in the third quadrant, we must have

$$\tan 225° = 1.$$

Hand calculators

Hand calculators can be used to obtain values for the trigonometric functions in the same way they were used to compute values for the circular functions when the measure of the angle is given in either radians or degrees. Care must be taken, however, to ensure that the calculator is in degree mode when the measure of the angle is given in degrees and in radian mode when the measure of the angle is given in radians. Although the values for the trigonometric functions obtained using a calculator directly are at least as accurate as those obtained using a reference angle and the tables or a calculator, in order to be consistent with the values obtained using the tables we shall use a reference angle in the computations in the exercises and examples in this section as we did in Chapter 5 when we found function values for the circular functions.

Example Find tan 149°.

Solution The reference angle for 149° is

$$\bar{\alpha} = 180° - 149°$$
$$= 31°.$$

Thus, we have

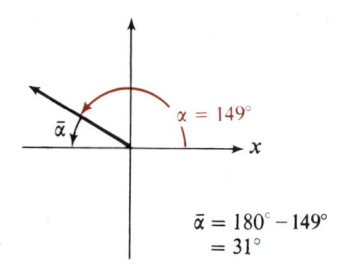

$$|\tan 149°| = \tan 31°$$
$$= 0.6009.$$

Hence, either $\tan 149° = 0.6009$ or $\tan 149° = -0.6009$. Since the terminal side of $\alpha = 149°$ lies in the second quadrant, we must have

$$\tan 149° = -0.6009.$$

Example Find sec 27°12'.

Solution Reading directly from Table V, we have

$$\sec 27°12' = 1.1243.$$

Alternatively, using a calculator to determine the value of cos 27°12', we have

$$\sec 27°12' = \frac{1}{\cos 27°12'} = \frac{1}{\cos 27.2°}$$
$$= \frac{1}{0.8894} = 1.1243.$$

Inverse trigonometric functions

Inverse functions for each of the six trigonometric functions can be defined by considering the domains of these functions to contain angles in standard position with measures suitably restricted. The restrictions ordinarily used are the same as those used in Definition 5.8 (page 190) with y representing the radian measure of an angle. The same notation is used for both the inverse trigonometric functions and the inverse circular functions.

6.2 Functions of Angles

Note that the elements of the ranges of the inverse trigonometric functions are angles in standard position. However, we are frequently interested in the *measure* of the angle rather than the angle itself.

In some cases, exact values for the inverse trigonometric functions can be found by using Table 6.1.

Example   Compute $\sin^{-1}(1/\sqrt{2})$. Express your answer in

    a. radians   **b.** degrees

Solution   We want an angle y with

$$\sin y = \frac{1}{\sqrt{2}} \quad \text{and} \quad -\frac{\pi^R}{2} \le y \le \frac{\pi^R}{2}.$$

From Table 6.1, we find $y = \dfrac{\pi^R}{4}$, and $y = 45°$.

If the angle in which we are interested is not one of these special angles and if the radian measure of the angle is desired, a hand calculator in radian mode or Table IV should be used; if the degree measure of the angle is desired, a hand calculator in degree mode or Table V should be used. As in the case of the inverse circular functions, it sometimes happens that the exact value desired is not in the table. If this occurs, then we choose the value in the table closest to the desired value. In such a case the value obtained from the table may vary slightly from the value obtained on a hand calculator. In all examples and exercises a hand calculator has been used to compute values of the inverse trigonometric functions.

Example   Compute Arccos 0.8253. Express your answer in:

    a. radians   **b.** degrees

Solution   We want an angle y with

$$\cos y = 0.8253, \quad \text{and} \quad 0^R \le y \le \pi^R, \quad \text{or} \quad 0° \le y \le 180°.$$

    a. From Table IV or a hand calculator, we find $y = 0.60^R$.

    b. From Table V or a hand calculator, we find $y = 34.4°$.

Example   Compute $\cot^{-1} 1.6577$. Express your answer in:

    a. radians   **b.** degrees

Solution   We want an angle y with

$$\cot y = 1.6577 \quad \text{and} \quad -\frac{\pi^R}{2} \le y \le \frac{\pi^R}{2} \quad (y \ne 0^R),$$

$$\text{or} \quad -90° \le y \le 90° \quad (y \ne 0°).$$

Solution continued overleaf

Requiring $\cot y = 1.6577$ is equivalent to requiring

$$\tan y = \frac{1}{1.6577} = 0.6032.$$

a. From Table IV or a hand calculator, we find $y = 0.54^R$.

b. From Table V or a hand calculator, we find $y = 31.1°$.

Exercise 6.2

A Find each of the six trigonometric function values of α where the terminal side of α contains the given point.

1. $(3, 4)$
2. $(5, -12)$
3. $(\sqrt{2}, -\sqrt{2})$
4. $(-1, \sqrt{3})$
5. $(-6, -8)$
6. $(-4, -3)$
7. $(0, 3)$
8. $(-7, 0)$

Determine in which quadrant the terminal side of the angle α lies.

Example $\sin \alpha > 0$, $\cos \alpha < 0$

Solution Since $\sin \alpha > 0$, the terminal side of α lies in Quadrant I or II; since $\cos \alpha < 0$, the terminal side of α lies in Quadrant II or III. Therefore, the terminal side of α lies in Quadrant II.

9. $\sin \alpha < 0$, $\cos \alpha > 0$
10. $\cos \alpha > 0$, $\tan \alpha < 0$
11. $\sec \alpha > 0$, $\sin \alpha > 0$
12. $\tan \alpha < 0$, $\sin \alpha > 0$
13. $\sec \alpha > 0$, $\csc \alpha < 0$
14. $\cot \alpha > 0$, $\sin \alpha < 0$

Use Table 6.1 and reference angles to find the value of the given expression.

Example $\cos 150°$

Solution The reference angle for $150°$ is

$$\bar{\alpha} = 180° - 150°$$
$$= 30°.$$

Thus, we have

$$|\cos 150°| = \cos 30° = \frac{\sqrt{3}}{2}.$$

$\bar{\alpha} = 180° - 150°$
$= 30°$

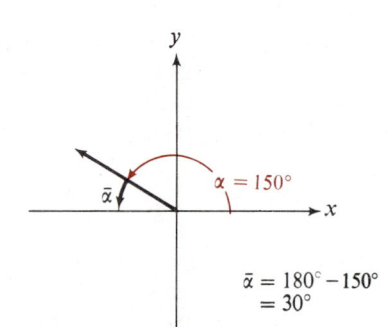

6.2 Functions of Angles

Hence, either $\cos 150° = \sqrt{3}/2$ or $\cos 150° = -\sqrt{3}/2$. Since the terminal side of the angle lies in the second quadrant, we must have

$$\cos 150° = -\frac{\sqrt{3}}{2}.$$

15. $\tan 330°$
16. $\cot 120°$
17. $\cos 225°$
18. $\tan 150°$
19. $\sin 330°$
20. $\cos 240°$
21. $\sin(-30°)$
22. $\tan(-60°)$
23. $\sin(-240°)$
24. $\cos(-135°)$
25. $\tan(-300°)$
26. $\tan(-750°)$

Use a hand calculator or Table IV or Table V, as appropriate, to find a decimal approximation for each value.

Example $\cot 75°$

Solution   Reading directly from Table V, we have

$$\cot 75° = 0.2679.$$

Alternatively, using a calculator to determine the value of $\tan 75°$, we have

$$\cot 75° = \frac{1}{\tan 75°}$$

$$= \frac{1}{3.7321} = 0.2679.$$

27. $\sin 32°$
28. $\cos 49°$
29. $\tan 33°$
30. $\cot 51°$
31. $\sec 17°$
32. $\csc 84°$
33. $\cos 0.63^R$
34. $\sin 0.42^R$
35. $\cot 1.42^R$
36. $\tan 1.03^R$
37. $\csc 1.74^R$
38. $\sec 1.83^R$
39. $\sin 132°$
40. $\cos 153°$
41. $\tan 320°$
42. $\sin 312°$
43. $\cos(-130°)$
44. $\tan(-605°)$
45. $\sin 2.07^R$
46. $\cos 3.51^R$
47. $\tan 4.21^R$
48. $\tan 6.00^R$
49. $\cos(-1.63^R)$
50. $\sin(-12.32^R)$

Use a hand calculator or Table IV and Table V to approximate (a) the radian measure and (b) the degree measure of the indicated value.

Example $\text{Cos}^{-1} 0.7071$

Solution overleaf

Solution We are seeking the measure of an angle α such that

$$\cos \alpha = 0.7071 \quad \text{and} \quad 0^R \leq \alpha \leq \pi^R, \quad \text{or} \quad 0° \leq \alpha \leq 180°.$$

Using Tables IV and V or a calculator, we have

a. 0.79^R **b.** $45°$

51. $\sin^{-1} 1.000$
52. $\cos^{-1} 0.5000$
53. $\tan^{-1} 1.000$
54. Arcsin 0.5000
55. Arccos 0.9816
56. Arctan 0.3249
57. $\sin^{-1} 0.6131$
58. $\tan^{-1} 1.116$
59. $\csc^{-1} 1.6353$
60. $\sec^{-1} 1.1547$
61. Arccot 0.8878
62. Arccsc 1.4020

B 63. Show (geometrically) that if (x_1, y_1) and (x_2, y_2) are the coordinates of any two points (except the origin) on the terminal side (ray) of an angle in the first quadrant, and $r_1 = \sqrt{x_1^2 + y_1^2}$, $r_2 = \sqrt{x_2^2 + y_2^2}$, then the following equalities of ratios hold:

$$\frac{x_1}{r} = \frac{x_2}{r}, \quad \text{and} \quad \frac{y_1}{r} = \frac{y_2}{r},$$

$$\frac{y_1}{x_1} = \frac{y_2}{x_2} \quad (x_1, x_2 \neq 0).$$

64. Under the same conditions as in Exercise 63, explain how the equalities of these ratios hold for the coordinates of two points on the terminal side of an angle whose terminal side is in the second, third, or fourth quadrant.

6.3 Right Triangles

An examination of Figure 6.8-a makes it evident that, in the first quadrant, any point (x, y) on the terminal side of an angle α determines a right triangle with sides measuring x and y and with hypotenuse of length $r = \sqrt{x^2 + y^2}$. Therefore, as

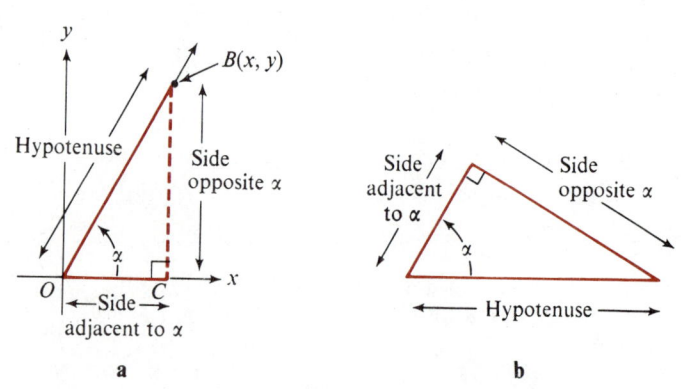

Figure 6.8

6.3 Right Triangles

special cases of trigonometric function values, we can consider the following **trigonometric ratios** of the sides of a right triangle:

$$\cos \alpha = \frac{\text{length of side adjacent to } \alpha}{\text{length of hypotenuse}},$$

$$\sin \alpha = \frac{\text{length of side opposite } \alpha}{\text{length of hypotenuse}},$$

$$\tan \alpha = \frac{\text{length of side opposite } \alpha}{\text{length of side adjacent to } \alpha}.$$

Similar statements can be made about csc α, sec α, and cot α.

It is not necessary that the right triangle be oriented in such a way that α is in standard position. The foregoing relations involving ratios of the lengths of the sides of a right triangle are equally applicable to a right triangle in any position, as suggested in Figure 6.8-b.

Solution of a triangle

If we are given some parts of a triangle, that is, the measures of some angles and the lengths of some sides, and asked to find the remaining measures, we are asked to **solve** the triangle. In general, we shall give solutions to the nearest tenth of a unit for lengths and to the nearest 6' (tenth of a degree) for the degree measures of angles.

Example

If one angle of a right triangle measures 47° and the side adjacent to this angle has length 24, solve the triangle.

Solution

We first make a sketch illustrating the situation. Generally the hypotenuse is labeled c, the legs a and b, and the angles opposite a, b, and c are labeled α, β, and γ, respectively, or A, B, and C, respectively. In this case, we note first that

$$\beta = 90° - \alpha = 90° - 47° = 43°.$$

Next we have

$$\tan 47° = \frac{a}{b} = \frac{a}{24}, \quad \text{or} \quad a = 24 \tan 47°,$$

and

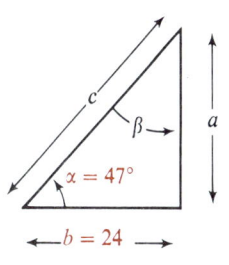

$$\cos 47° = \frac{b}{c} = \frac{24}{c}, \quad \text{or} \quad c = \frac{24}{\cos 47°}.$$

From Table V or by using a calculator, we obtain

$$\tan 47° = 1.072 \quad \text{and} \quad \cos 47° = 0.6820,$$

so that

$$a = 24(1.072) \approx 25.7,$$

and

$$c = \frac{24}{0.6820} \approx 35.2.$$

We selected the trigonometric ratios for tan 47° and cos 47° in the above example instead of cot 47° and sec 47°, because the values for tangent and cosine could more easily be obtained on a calculator.

Right triangles play a useful role in finding one trigonometric function value, given another.

Example If tan α = 2/3 and α is in Quadrant I, find sin α.

Solution Sketch a right triangle and label the sides so that tan α is as given. By the Pythagorean theorem, the hypotenuse of the triangle has length
$$\sqrt{2^2 + 3^2} = \sqrt{13},$$
so that
$$\sin \alpha = \frac{2}{\sqrt{13}}.$$

Note in this example that once the length of the hypotenuse is determined, the values for *all* trigonometric functions of α are available by inspection.

Right triangles are helpful also in quadrants other than the first if care is taken to assign *directed* (*positive or negative*) *distances* as lengths of the sides, while viewing the length of the hypotenuse as always positive.

Example If cot α = 3/7 and sin α < 0, find cos α.

Solution Since cot α > 0 while sin α < 0, α can lie only in Quadrant III. Sketch a right triangle in Quadrant III and label the sides with directed lengths so that cot α = 3/7. By the Pythagorean theorem,
$$OA = \sqrt{(-3)^2 + (-7)^2}$$
$$= \sqrt{9 + 49} = \sqrt{58}.$$
By inspection, then, cos α = $-3/\sqrt{58}$.

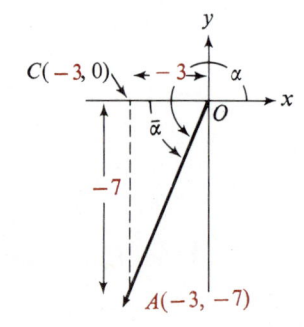

Triangle interpretation of inverse notation

It is sometimes useful to interpret an expression including inverse notation by using right triangles.

6.3 Right Triangles

Example Evaluate cos [Arcsin($\sqrt{3}/4$)].

Solution We construct a right triangle as shown, with hypotenuse of length 4 and one leg of length $\sqrt{3}$, such that

$$\alpha = \text{Arcsin} \frac{\sqrt{3}}{4} \quad \left(\sin \alpha = \frac{\sqrt{3}}{4}\right).$$

From the triangle, we see that

$$\cos\left(\text{Arcsin} \frac{\sqrt{3}}{4}\right) = \cos \alpha = \frac{\sqrt{13}}{4}.$$

Right triangles can also be used if the element in the domain is positive but not specified.

Example Write $\sin(\text{Cos}^{-1} x)$ as an equivalent expression without inverse notation.

Solution We construct a right triangle as shown, with hypotenuse of length 1 and one leg of length x, such that

$$\alpha = \text{Cos}^{-1} x \quad (\cos \alpha = x).$$

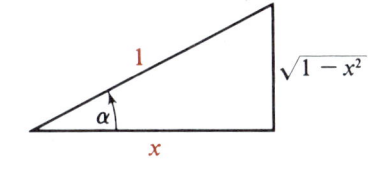

From the triangle, we see that

$$\sin(\text{Cos}^{-1} x) = \sin \alpha = \sqrt{1 - x^2}.$$

Solutions of triangles in general, and of right triangles in particular, occur in many applications in engineering and the physical sciences. Some of these applications occur in the exercises and in later sections.

Exercise 6.3

Solve the right triangle ABC. Consider $C = 90°$. Give lengths to the nearest tenth of a unit and angle measures to the nearest tenth of a degree.

1. $a = 6$, $b = 8$
2. $a = 5.7$, $c = 6.2$
3. $b = 120$, $B = 54°$
4. $a = 3.5$, $B = 48°$
5. $c = 16$, $A = 22°42'$
6. $b = 12.5$, $A = 14°24'$

7. In the right triangle ABC, find a if $\sin A = 4/5$ and $c = 20$.
8. In the right triangle ABC, find b if $\tan A = 3/4$ and $a = 12$.

9. Find the other trigonometric function values of θ if $\sin \theta = -1/2$ and the terminal side of θ is in the third quadrant.

10. Find the other trigonometric function values of θ if $\tan \theta = -7/24$ and the terminal side of θ is in the fourth quadrant.

11. Find the other trigonometric function values of θ if $\cos \theta = -3/7$ and the terminal side of θ is in the second quadrant.

12. Find the other trigonometric function values of θ if $\sin \theta = 2/3$ and the terminal side of θ is in the second quadrant.

Use right triangles to compute the indicated value.

Example $\quad \sin\left(\operatorname{Tan}^{-1} \dfrac{4}{3}\right)$

Solution We draw a right triangle as shown, with legs of length 4 and 3, such that

$$\alpha = \operatorname{Tan}^{-1} \dfrac{4}{3} \quad \left(\tan \alpha = \dfrac{4}{3}\right).$$

From the triangle, we have

$$\sin\left(\operatorname{Tan}^{-1} \dfrac{4}{3}\right) = \sin \alpha = \dfrac{4}{5}.$$

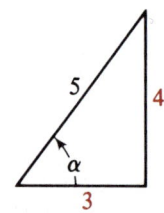

13. $\tan\left(\operatorname{Arcsin} \dfrac{1}{3}\right)$ 14. $\sin\left(\operatorname{Cos}^{-1} \dfrac{2}{3}\right)$

15. $\cos\left(\operatorname{Tan}^{-1} \dfrac{2}{3}\right)$ 16. $\cot(\operatorname{Tan}^{-1} 5)$

Write the given expression without inverse notation. Assume x and y are positive.

17. $\cos(\operatorname{Arcsin} x)$ 18. $\cot(\operatorname{Arctan} x)$

19. $\tan(\operatorname{Sin}^{-1} y)$ 20. $\sin(\operatorname{Tan}^{-1} y)$

21. $\sec(\operatorname{Arccos} x)$ 22. $\csc(\operatorname{Sin}^{-1} y)$

23. Show that $\operatorname{Arcsin}(2/5) = \operatorname{Arctan}(2/\sqrt{21})$. *Hint:* Sketch a triangle.

24. Show that $\operatorname{Arccos}(1/3) = \operatorname{Arccot}(1/2\sqrt{2})$.

In Exercises 25 and 26, given the information pertaining to the figure, find the length of the line segment denoted by x.

6.4 *Law of Sines*

25. Given: $BA = 25$,

    $\alpha = 16°$,

    $\beta = 12°$.

    Line BC is a straight line.

26. Given: $AB = 350$,

    $\alpha = 21°$,

    $\beta = 8°$.

    Line BD is a straight line.

27. A rectangle is 80 meters long and 64 meters wide. Find the degree measures of the angles between a diagonal and the sides.

28. The sides of an isosceles triangle have lengths 6, 6, and 8 meters. Find the degree measure of each angle in the triangle.

29. The angle of elevation (the angle between the line of sight and the horizontal) from a point 58.3 meters from the base of a cliff to the base of a fortress wall on top of the cliff is 60°. The angle of elevation from the same spot to the top of the wall is 65°. How tall is the wall?

30. A pilot is preparing to fly from city A to city B. City A is 100 miles due south of city C, which is 60 miles due west of city B. Determine the heading the pilot should fly (assume that there is no wind and that due North is a 0° heading and due East is a 90° heading).

31. Government surveyors are measuring the height of a mountain. They measure the angle of elevation from point A to the top of the mountain as 68° and from point B to the top of the mountain as 34°. Assume point B is at the same elevation as point A and is 1 kilometer (1000 meters) farther from the mountain than point A. How high above A and B is the top of the mountain in meters? (See Exercise 25.)

32. A man standing on the top of a building 35 meters high measures the angle of elevation of the top of the building across the street to be 32°. He measures the angle of depression of the base of the same building to be 55°. How far away is the building across the street and how tall is it?

33. A radar antenna is atop a 30-meter building and is elevated to an angle of 28° with the horizontal. Assume an airplane is flying toward the radar station at an altitude of 5000 meters AGL (above ground level). How far away from the radar station will the airplane be when it first appears on the air traffic controller's radar screen?

34. If the radar antenna in Exercise 33 were on the ground, how would the answer be affected? How much is the answer actually changed?

6.4 *Law of Sines*

In Section 6.3 we used trigonometric ratios to solve right triangles. In this section we shall discuss one technique that can sometimes be used in solving triangles that are not right triangles.

The fact that the area $\mathscr{A}$ of a triangle is equal to one-half the product of the length of its base and the length of its altitude gives us an immediate expression for its area in terms of the lengths of two sides of the triangle and the measure of the included angle. Thus, from Figure 6.9, we have

$$\mathscr{A} = \frac{1}{2} c(b \sin \alpha) = \frac{1}{2} bc \sin \alpha$$

and

$$\mathscr{A} = \frac{1}{2} c(a \sin \beta) = \frac{1}{2} ac \sin \beta.$$

It can also be shown that

$$\mathscr{A} = \frac{1}{2} ab \sin \gamma.$$

Figure 6.9

Because each triangle has only one area, we can equate the right-hand members of these equations to obtain

$$\frac{1}{2} bc \sin \alpha = \frac{1}{2} ac \sin \beta = \frac{1}{2} ab \sin \gamma.$$

Multiplying each member here by $2/(abc)$, we obtain the following result, called the **law of sines**.

Theorem 6.2 *If α, β, and γ are the angles of a triangle and a is the length of the side opposite α, b is the length of the side opposite β, and c is the length of the side opposite γ, then*

$$\frac{\sin \alpha}{a} = \frac{\sin \beta}{b} = \frac{\sin \gamma}{c}.$$

Solving triangles

The law of sines can be used to solve certain triangles. Let us first consider the case in which two angles and one side of a triangle are given.

Example

Solve the triangle for which

$$\alpha = 45°, \quad \beta = 60°, \quad \text{and} \quad a = 10.$$

Solutions

It is helpful first to make a sketch. Then, from Theorem 6.2, we have

$$\frac{\sin 45°}{10} = \frac{\sin 60°}{b},$$

from which

$$b = \frac{10 \sin 60°}{\sin 45°} = 10 \left(\frac{0.8660}{0.7071} \right)$$

$$\approx 12.2.$$

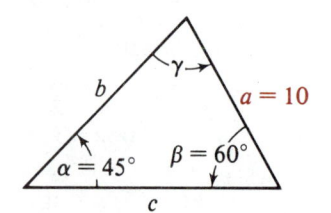

6.4 Law of Sines

Next, we observe that

$$\gamma = 180° - \alpha - \beta = 180° - 45° - 60° = 75°.$$

From Theorem 6.2, we have

$$\frac{\sin 45°}{10} = \frac{\sin 75°}{c}.$$

Then, from Table V or a calculator,

$$c = \frac{10 \sin 75°}{\sin 45°} = 10\left(\frac{0.9659}{0.7071}\right) \approx 13.7.$$

Thus, we have

$$b \approx 12.2, \quad c \approx 13.7, \quad \text{and} \quad \gamma = 75°.$$

Ambiguous case

If the lengths of two sides of a triangle, say a and b, and the measure of an angle opposite one of them, say α, are given, we may encounter ambiguity, depending on the value of a in relation to those of b and α. In Figure 6.10, where α is acute, we hold b and α constant and observe the possible situations as a assumes different values.

First we determine the length h of the altitude of a triangle with the given measures for the acute angle α and the adjacent side b. Since $\sin \alpha = h/b$,

$$h = b \sin \alpha.$$

Now consider the four cases illustrated in Figure 6.10.

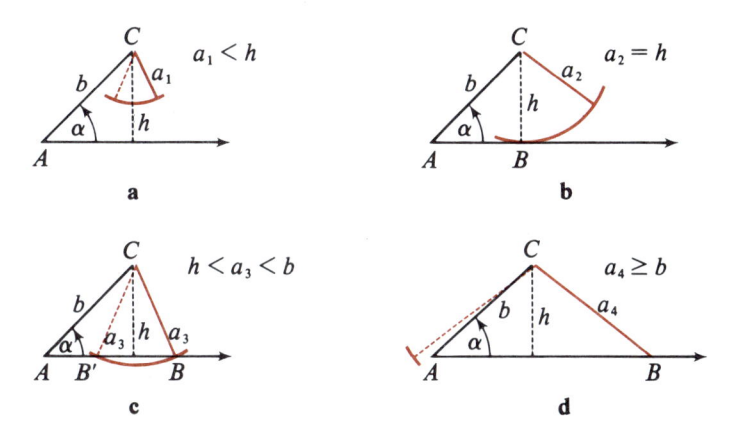

Figure 6.10

Case I (Figure 6.10-a). The length a_1 of the given side satisfies the inequality $a_1 < b \sin \alpha$. Since a_1 is less than h, the given values are such that no triangle is possible.

Case II (Figure 6.10-b). The length a_2 of the given side satisfies the equation

$$a_2 = b \sin \alpha.$$

We have a unique solution—a right triangle.

Case III (Figure 6.10-c). The length a_3 of the given side is such that
$$b \sin \alpha < a_3 < b$$
We have two triangles possible, $\triangle ABC$ and $\triangle AB'C$.

Case IV (Figure 6.10-d). The length a_4 of the given side is such that
$$a_4 \geq b.$$
Here we also have a unique solution.

Of course, if the given angle α is obtuse (Figure 6.11), then any specified lengths of the sides and measures of the angles permit only two possibilities:

1. $a \leq b$, no triangle (as shown in Figure 6.11-a);
2. $a > b$, one oblique triangle (as shown in Figure 6.11-b).

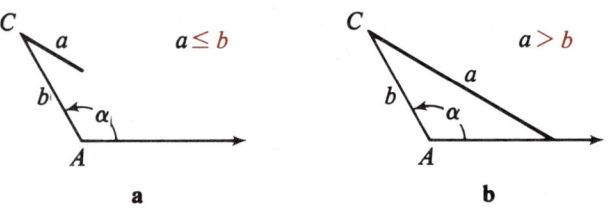

Figure 6.11

Example

Solve the triangle for which $a = 4$, $b = 3$, and $\beta = 45°$.

Solution

We first sketch a figure and observe that this gives rise to the ambiguous case in which we are given the lengths of two sides and the measure of an angle opposite one of them. We then check for the number of possible solutions and observe that $a \sin \beta < b < a$. Therefore, there are two triangles with the given measurements. Using Theorem 6.2, we next determine a value for $\sin \alpha$. We have

$$\frac{\sin \alpha}{4} = \frac{\sin 45°}{3},$$

$$\sin \alpha = \frac{4}{3}(0.7071) = 0.9428.$$

Because $\sin \alpha > 0$ in Quadrants I and II, from Table V or by using a calculator, we have either

$$\alpha \approx 70.5° \quad \text{or} \quad \alpha' \approx 180° - 70.5° = 109.5°.$$

If $\alpha \approx 70.5°$, then $\gamma \approx 180° - 45° - 70.5° = 64.5°$, and we therefore have

$$\frac{\sin 45°}{3} \approx \frac{\sin 64.5°}{c},$$

6.4 Law of Sines

so that

$$c \approx 3\left(\frac{0.9026}{0.7071}\right) \approx 3.8.$$

If $\alpha' \approx 109.5°$, then $\gamma' \approx 180° - 45° - 109.5° = 25.5°$, and we have

$$\frac{\sin 45°}{3} \approx \frac{\sin 25.5°}{c'},$$

so that

$$c' \approx 3\left(\frac{0.4305}{0.7071}\right) \approx 1.8.$$

Thus, the two solutions for the triangle are

$$\alpha \approx 70.5°, \quad \gamma \approx 64.5°, \quad c \approx 3.8$$

and

$$\alpha' \approx 109.5°, \quad \gamma' \approx 25.5°, \quad c' \approx 1.8.$$

Exercise 6.4

A *In all exercises, give lengths to the nearest tenth of a unit and angle measures to the nearest tenth of a degree.*

In Exercises 1–6, solve the triangle.

1. $b = 10$, $\beta = 30°$, $\alpha = 80°$
2. $a = 64$, $\beta = 36°$, $\gamma = 82°$
3. $c = 78.1$, $\alpha = 58.1°$, $\gamma = 63.2°$
4. $b = 1.02$, $B = 41.6°$, $C = 80.6°$
5. $a = 84.2$, $A = 110°$, $C = 22°6'$
6. $c = 0.94$, $A = 41°18'$, $B = 96°54'$

Determine the number of triangles that satisfy the given conditions.

Example $b = 34$, $c = 12$, $C = 30°$

Solution   Sketch a figure and determine

$h = 34 \sin 30° = 34(0.5000) = 17.$

Since $12 < 17$, we have $c < b \sin C$, and no triangle exists for the given data.

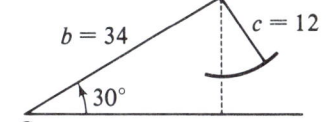

7. $a = 5.4$, $b = 7.0$, $B = 30°$
8. $b = 4.9$, $c = 3.2$, $C = 30°$
9. $a = 31.1$, $c = 41.3$, $C = 30°$
10. $a = 42.3$, $b = 20.7$, $B = 30°$

11. $b = 16.2$, $c = 14.3$, $C = 20°$
12. $a = 141$, $b = 182$, $B = 20°$
13. $a = 4.6$, $b = 2.3$, $B = 30°$
14. $b = 68.1$, $c = 41.3$, $C = 30°$

Solve the triangle.

15. $a = 4.8$, $c = 3.9$, $\alpha = 113°$
16. $a = 3.2$, $b = 2.6$, $\beta = 54°$
17. $b = 6.21$, $c = 4.39$, $\beta = 42°42'$
18. $a = 179$, $b = 212$, $\beta = 114°12'$
19. $b = 1.8$, $c = 1.5$, $\gamma = 32.5°$
20. $a = 9.4$, $b = 8.6$, $\beta = 54.4°$
21. $a = 8.4$, $b = 6.9$, $B = 62°12'$
22. $b = 0.42$, $c = 0.21$, $B = 31°54'$
23. $a = 13.84$, $b = 6.92$, $A = 60°$
24. $a = 4.72$, $c = 9.44$, $A = 30°$
25. $b = 420$, $c = 610$, $B = 33°24'$
26. $a = 5.42$, $b = 6.82$, $A = 43°30'$

27. From a window in a tower, 85 feet above the ground, the angle of elevation to the top of a nearby building measures $34°30'$. From a point on the ground directly below the window, the angle of elevation to the top of the same building measures $50°24'$. Find the height of the building.

28. A pilot is flying away from city A at a heading of $138°$ ($0°$ is due north and $90°$ is due east). When he is 10 kilometers away from A, he reports to the weather station in city A that there is a large storm centered due east of the city and $15°$ north of east from his position. How far is the storm centered from the city?

29. Engineers are planning to tunnel straight through a mountain. They take three measurements. Point A is at the base of the western side of the mountain, and the angle of elevation from point A to the top of the mountain is $48°$. Point B is 1 kilometer west of point A, and the angle of elevation to the top of the mountain is $25°$. Point C is at the base of the mountain due east of point A, and the angle of elevation from point C to the top of the mountain is $28°$. How far must they tunnel; that is, how wide is the mountain at the base?

30. A woman standing on a river bank is trying to decide if she can swim across the river. On the opposite bank there are two towers which she knows to be about 1 kilometer from one another. She approximates the angles formed by her lines of sight and the river bank to be $30°$ for tower A and $45°$ for tower B. Approximately how wide is the river?

31. If an equilateral triangle is inscribed in a circle of radius 4, what is the length of each side of the triangle?

32. A regular pentagon is inscribed in a circle of radius 25. Find the area and the circumference of the pentagon. *Hint*: Divide the pentagon into five congruent triangles, compute the area of each, and multiply by 5.

6.5 Law of Cosines

The law of sines introduced in Section 6.4 cannot be used to solve a triangle when we are given only the length of each side of the triangle or when we are given two sides and the included angle. In this section we shall consider a relationship that will help us solve such triangles.

6.5 Law of Cosines

Let us denote the angles of a triangle by α, β, and γ and the sides opposite these angles by a, b, and c, respectively (Figure 6.12-**a**). Assume all three angles are acute (the case in which the triangle contains an obtuse angle is left as an exercise). Assume that the line segment through the vertex of γ and perpendicular to side c is of length h and that side c of the triangle lies along the positive x-axis of a coordinate system, as shown in Figure 6.12-**b**.

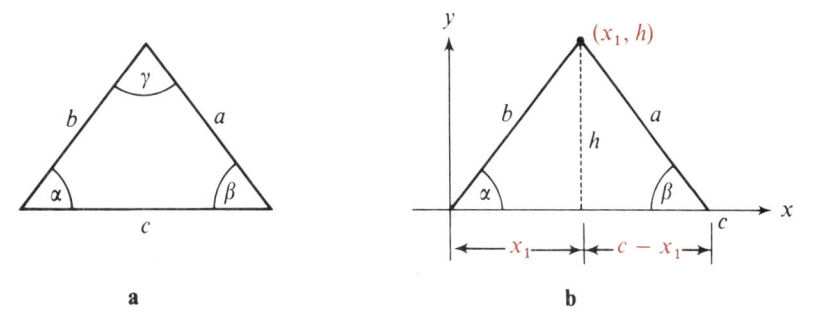

Figure 6.12

We observe from this figure that

$$x_1^2 = b^2 - h^2 \tag{1}$$

and

$$x_1 = b \cos \alpha. \tag{2}$$

Using the distance formula and Equations (1) and (2), we have

$$\begin{aligned} a^2 &= (c - x_1)^2 + h^2 \\ &= c^2 - 2cx_1 + x_1^2 + h^2 \\ &= c^2 - 2cb \cos \alpha + b^2 - h^2 + h^2 \\ &= b^2 + c^2 - 2bc \cos \alpha. \end{aligned}$$

This argument can be repeated using the other sides of the triangle to establish the following theorem, which is known as the **law of cosines**.

Theorem 6.3 *If α, β, and γ are the angles of a triangle, and a, b, and c are the lengths of the sides opposite α, β, and γ, respectively, then*

$$c^2 = a^2 + b^2 - 2ab \cos \gamma, \tag{3}$$

$$b^2 = a^2 + c^2 - 2ac \cos \beta, \tag{4}$$

$$a^2 = b^2 + c^2 - 2bc \cos \alpha. \tag{5}$$

Solving triangles

This theorem has many applications, among them the solution of certain triangles. If we are given the measure of an angle and the lengths of the adjacent sides, we can use the law of cosines to help us find the remaining parts of the triangle.

Example Solve the triangle for which

$$\alpha = 100°, \quad b = 10, \quad \text{and} \quad c = 12.$$

Solution Make a sketch of the triangle and label the sides and angles. We can first find a by using the law of cosines in the form

$$a^2 = b^2 + c^2 - 2bc \cos \alpha.$$

Thus,

$$a^2 = 10^2 + 12^2 - 2(10)(12) \cos 100°$$
$$= 100 + 144 - 240(-0.1736)$$
$$= 244 + 41.68 \approx 285.7.$$

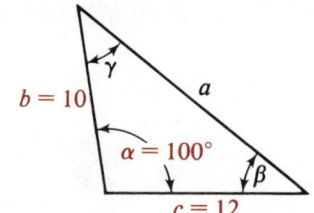

Using a calculator, or Table VI, we find $\sqrt{285.7} \approx 16.9$. Next, to find β, we again use the law of cosines (we could also use the law of sines because now we know the length of a side opposite a given angle) in the form

$$b^2 = a^2 + c^2 - 2ac \cos \beta,$$

or

$$\cos \beta = \frac{b^2 - a^2 - c^2}{-2ac}.$$

With $a^2 = 285.7$, $b^2 = 100$, and $c^2 = 144$, we obtain

$$\cos \beta = \frac{100 - 285.7 - 144}{-2(16.9)(12)} \approx 0.8129.$$

Hence, $\beta \approx 35.6°$. Since the sum of the angles of a triangle is 180°, we can find an approximation for γ by noting that

$$\gamma = 180° - \alpha - \beta$$
$$\approx 180° - 100° - 35.6° = 44.4°.$$

For the remaining parts of the triangle, we accordingly have

$$a \approx 16.9, \quad \beta \approx 35.6°, \quad \text{and} \quad \gamma \approx 44.4°.$$

Since Equations (3), (4), and (5) in Theorem 6.3 are relationships between three sides and one angle of a triangle, we can, as we did in the preceding example, use any of these forms to solve for the measure of an angle, given the lengths of the three sides.

Exercise 6.5

A *Find the remaining parts of the triangle. In all exercises in this set, give lengths to the nearest tenth of a unit and angle measurements to the nearest tenth of a degree.*

1. $a = 10$, $b = 4$, $\gamma = 30.6°$
2. $b = 14.2$, $c = 7.9$, $\alpha = 64.2°$
3. $a = 4.9$, $c = 6.8$, $\beta = 122° \, 24'$
4. $a = 241$, $c = 104$, $\beta = 148° \, 12'$

5. $b = 14.6$, $c = 6.21$, $A = 80° 42'$
6. $b = 9.4$, $c = 10.2$, $A = 100° 54'$
7. $a = 5$, $b = 8$, $c = 7$
8. $a = 4.5$, $b = 5.3$, $c = 2.8$

9. Find the greatest angle of the triangle whose sides are 5.1, 4.2, and 4.5.

10. Find the least angle of the triangle whose sides are 29.5, 33.2, and 41.4.

11. Two points A and B are on opposite sides of a pond. A surveyor at a point C determines that A is 92 feet from C, B is 128 feet from C, and the measure of the angle between CA and CB is 37°. Determine the distance between A and B.

12. A ship navigator determines his position C to be 48 miles from Port A and 80 miles from Port B. The distance between Ports A and B is 57 miles. Find the degree measure of the angle between CA and CB.

13. A professional golfer has hit his drive to the right. He estimates the angle formed by the line from the tee to the hole and the line from the tee to his ball is 30°. He paces the distance from the tee to his ball and finds that distance to be 240 yards. Assume the distance from the tee to the hole is 460 yards. How far is his ball from the hole?

14. A weather balloon is hovering between cities A and B at an altitude of 7000 meters AGL (above ground level). The angle of elevation from city B to the balloon is 38°. Assume that the distance from city A to city B is 20 kilometers. Find the distance between the balloon and city A, and the angle of elevation from city A to the balloon.

15. A hiker starts out from ranger station A on a heading of 20° east of north. He walks to a campground and stops to rest. His pedometer reads 5.75 miles. Assume ranger station B is 10 miles due north of station A. The hiker now wants to hike from his resting place to station B. What compass heading should he take?

16. The distance between two U.S. Forest Service observation platforms is 4 miles. A fire is spotted by observers on both platforms, and the angles formed by the lines joining the fire and the stations and the line joining the stations are 12° and 20°. How far is each platform from the fire?

17. Show that $a^2 = b^2 + c^2 - 2bc \cos \alpha$, where a, b, c, and α are as shown in the figure.

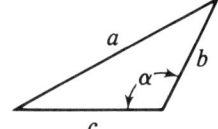

18. Use the law of cosines to show that if a, b, and c are the lengths of the sides of a triangle and α is the angle opposite side a, then

   a. $a^2 < b^2 + c^2$ if α is acute,
   b. $a^2 = b^2 + c^2$ if α is a right angle,
   c. $a^2 > b^2 + c^2$ if α is obtuse.

6.6 Geometric Vectors

A single real number is adequate to measure a physical quantity such as length, mass, or temperature that can be described by a magnitude only. Furthermore, the real number can be associated with a point or a line segment on the real-number line. However, a complete description of a physical quantity such as velocity, acceleration, and force requires that direction as well as magnitude be specified. Such quantities can be represented by a pair of real numbers (a *vector*) and associated with a line segment in the plane pointing in an assigned direction (a *geometric vector*).

Two-dimensional geometric vectors

In this section we shall only consider the notion of a geometric vector in a plane and several applications in which these vectors can be used to set up models in applied problems.

Definition 6.3 *A geometric vector in a plane is a line segment in that plane with a specified direction.*

Throughout this section, when we refer to geometric vectors, we shall mean *geometric vectors in a plane.*

A geometric vector can be denoted by a letter such as **v**. It can also be named by using the names of the endpoints. For example, in Figure 6.13 the vector can be denoted by **AB**, where A is the initial point and B is the terminal point.

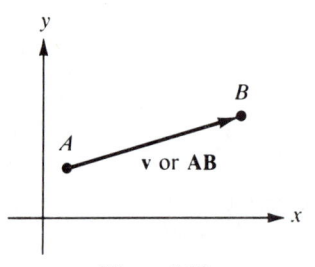

Figure 6.13

The following concepts are useful when working with vectors (see Figure 6.14).

$\alpha_1 = \alpha_2 = \alpha_3$
$\|\mathbf{v}_1\| = \|\mathbf{v}_2\| = \|\mathbf{v}_3\|$

a   **b**   **c**

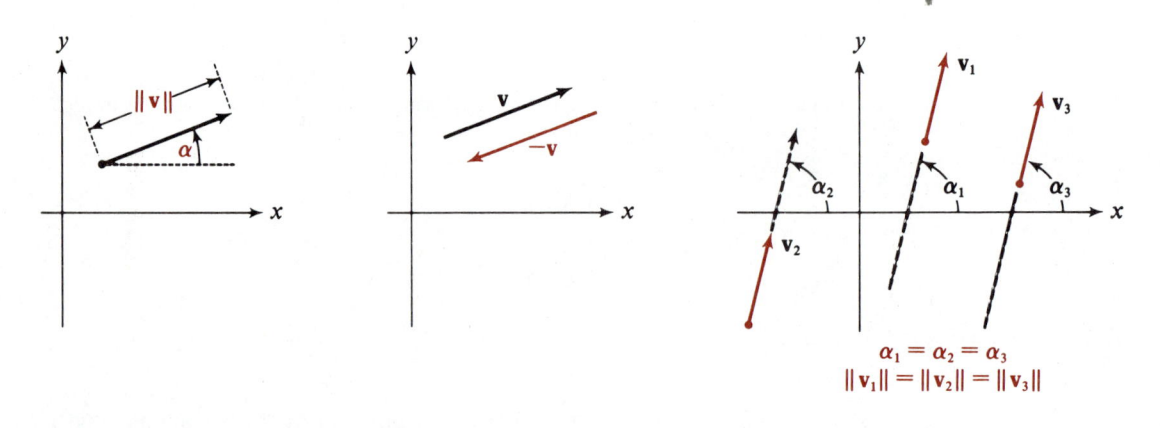

Figure 6.14

6.6 Geometric Vectors

Definition 6.4   *For each geometric vector* **v**, *the* **norm** *or* **magnitude** *of* **v** *is the length of* **v** *and is denoted by* $\|\mathbf{v}\|$.

Definition 6.5   *The direction angle of a geometric vector* **v** *is the angle* α, *where* $-180° < \alpha \leq 180°$, *such that the initial side of* α *is the ray from the initial point of* **v** *parallel to the x-axis and directed in the positive x-direction, and the terminal side of* α *is the ray from the initial point of* **v** *and containing* **v**.

Definition 6.6   *For each geometric vector* **v**, *the geometric vector* $-\mathbf{v}$ *has the same norm as* **v** *and has a direction angle whose measure differs from the direction angle of* **v** *by* $180°$.

Definition 6.7   *Geometric vectors* $\mathbf{v}_1$ *and* $\mathbf{v}_2$ *are equivalent if they have the same norm and same direction angle.*

The symbol = is used to denote the relationship of equivalent geometric vectors. Note that we can imagine "sliding" a geometric vector onto any vector equivalent to it, without at any time altering the direction angle of the vector (see Figure 6.14-c).

Operations on geometric vectors

The following two operations pair two geometric vectors to form a third geometric vector.

Definition 6.8   *For all geometric vectors* $\mathbf{v}_1$ *and* $\mathbf{v}_2$, *their* **sum** *or* **resultant** $\mathbf{v}_1 + \mathbf{v}_2$ *is a geometric vector having as its initial point the initial point of* $\mathbf{v}_1$ *(or an equivalent geometric vector) and as its terminal point the terminal point of* $\mathbf{v}_2$ *(or an equivalent geometric vector), where the terminal point of* $\mathbf{v}_1$ *is the initial point of* $\mathbf{v}_2$.

From Definitions 6.7 and 6.8, the sum $\mathbf{v}_1 + \mathbf{v}_2$ in Figure 6.15-a equals $\mathbf{v}_1 + \mathbf{v}_2'$, where $\mathbf{v}_2'$ is equivalent to $\mathbf{v}_2$. Note that a triangle is formed by $\mathbf{v}_1, \mathbf{v}_2$, and $\mathbf{v}_3$. Hence, we say that geometric vectors can be added according to the "the triangle law."

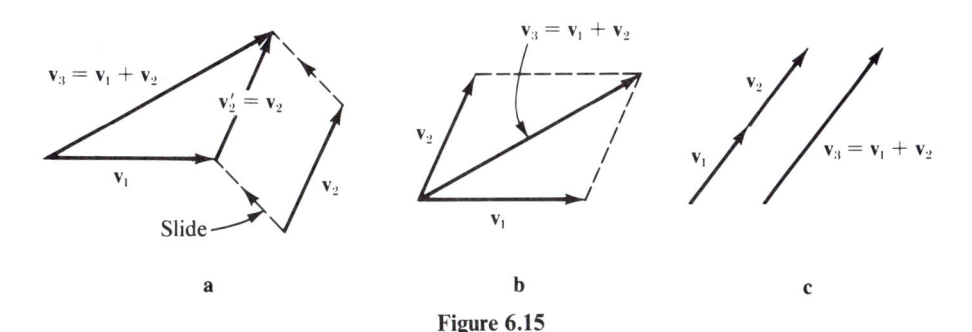

Figure 6.15

Also note in Figure 6.15-b that if v_1 and v_2 have the same initial point, the sum $v_1 + v_2$ is a diagonal of the parallelogram with adjacent sides v_1 and v_2. Hence, we also say that geometric vectors can be added according to "the parallelogram law." Figure 6.15-c illustrates $v_1 + v_2$ in the case where v_1 and v_2 are collinear and have the same direction.

From a vector viewpoint, real numbers are called **scalars**.

Definition 6.9 *For any geometric vector* **v** *and any scalar c, the **product of the vector by a scalar**, c**v**, is a geometric vector with magnitude* $\|c\mathbf{v}\| = |c| \cdot \|\mathbf{v}\|$, *having the same direction angle as* **v** *if* $c > 0$ *and the same direction angle as* $-\mathbf{v}$ *if* $c < 0$.

In Figure 6.16, $2\mathbf{v}$ and $\frac{1}{2}\mathbf{v}$ are in the same direction as **v**, and $\|2\mathbf{v}\| = 2\|\mathbf{v}\|$ and $\|\frac{1}{2}\mathbf{v}\| = \frac{1}{2}\|\mathbf{v}\|$. Also, $-2\mathbf{v}$ is in the opposite direction to **v**, and $\|-2\mathbf{v}\| = 2\|\mathbf{v}\|$.

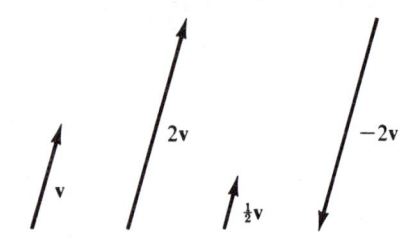

Figure 6.16

In Definition 6.9, if $c = 0$, then $0\mathbf{v}$ is the **zero geometric vector** with magnitude 0. It is considered to have any direction angle.

Applications As we noted, many physical quantities are specified by magnitude and direction. Therefore, geometric vectors are sometimes helpful in setting up models that can be solved by trigonometric methods. In doing this, we sometimes make use of *projections* of vectors on the coordinate axes, as illustrated in the following examples.

Example Find the magnitudes of the vector projections on the *x*- and *y*-axes of the vector **v** with magnitude $\|\mathbf{v}\| = 8$ and direction angle $\alpha = 52°$.

Solution Sketching the vector **v**, with initial point at the origin, we want to find $\|\mathbf{v}_x\|$ and $\|\mathbf{v}_y\|$. The projections $\mathbf{v}_x$ and $\mathbf{v}_y$ are the geometric vectors that are obtained by "constructing" a perpendicular line from the terminal point of **v** to the *x*- and *y*-axis, respectively. Hence,

$$\cos 52° = \frac{\|\mathbf{v}_x\|}{8} \quad \text{and} \quad \sin 52° = \frac{\|\mathbf{v}_y\|}{8},$$

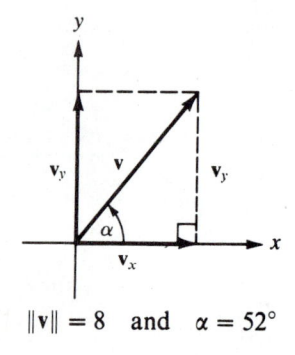

$\|\mathbf{v}\| = 8$ and $\alpha = 52°$

6.6 Geometric Vectors

from which

$$\|v_x\| = 8 \cos 52° \quad \text{and} \quad \|v_y\| = 8 \sin 52°$$
$$\approx 8(0.6157) \qquad\qquad \approx 8(0.7880)$$
$$= 4.93 \qquad\qquad\qquad = 6.30$$

Hence, the magnitude of the x-projection is 4.93 and the magnitude of the y-projection is 6.30.

Example   Two forces of 3 and 7 pounds act on an object at an angle of 30° with respect to each other. Find the magnitude of the resultant force and the angle it forms with the 3-pound force.

Solution   The conditions can be described by geometric vectors as suggested in Figure **a**. Since v'_2 is parallel to v_2, we have from geometry that $\alpha_2 = 150°$ (the interior angles on the same side of a transversal are supplementary).

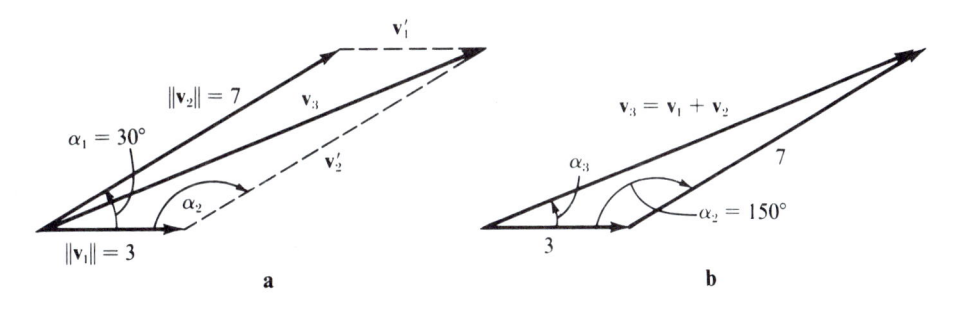

Now, using the law of cosines for the triangle in Figure **b**, we obtain

$$\|v_3\| = \sqrt{3^2 + 7^2 - 2(3)(7)\cos 150°}$$
$$\approx 9.71.$$

Hence, the magnitude of the resultant force is about 9.71. Then, using the law of sines,

$$\frac{7}{\sin \alpha_3} = \frac{9.71}{\sin 150°}$$

$$\sin \alpha_3 = \frac{7 \sin 150°}{9.71}$$

$$\approx \frac{7(0.5000)}{9.71} \approx 0.3603.$$

From Table V or a calculator, to the nearest tenth of a degree, $\alpha_3 = 21.1°$.

Example   An airplane flies 340 miles per hour in a direction 30° counterclockwise from the east for 2 hours and then flies due north at the same speed for 1 hour. How far from the starting point is the airplane at the end of this time?

Solution overleaf

Solution The conditions can be described by geometric vectors as suggested in Figure **a**, where we wish to find $\|\mathbf{v}_3\| = \|\mathbf{v}_1 + \mathbf{v}_2\|$. Since α_2 in Figure **b** is 60° and since $\mathbf{v}_2$

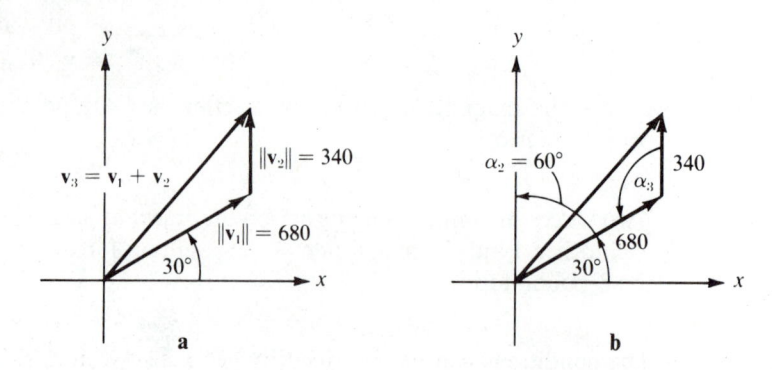

is parallel to the y-axis (due north), we have from geometry that $\alpha_3 = 120°$. Now, using the law of cosines, we obtain

$$\|\mathbf{v}_3\| = \sqrt{(680)^2 + (340)^2 - 2(680)(340)\cos 120°}$$
$$\approx \sqrt{809{,}200}.$$

Hence, the airplane is approximately $\sqrt{809{,}200}$, or about 900, miles from its starting point.

Exercise 6.6

A *In Exercises 1–10, use the following geometric vectors.*

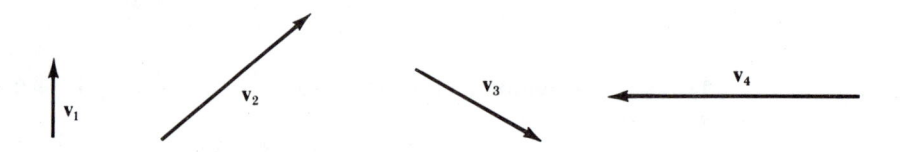

If $\|\mathbf{v}_1\| = 1$ and $\mathbf{v}_1$ is in the direction of the positive y-axis, estimate the magnitude and the measure of the direction angle for each geometric vector in Exercises 1–4.

1. $\mathbf{v}_1$ 2. $\mathbf{v}_2$ 3. $\mathbf{v}_3$ 4. $\mathbf{v}_4$

Using freehand methods, represent each of the following by a single geometric vector.

5. $\mathbf{v}_1 + \mathbf{v}_2$ 6. $\mathbf{v}_2 + \mathbf{v}_3$ 7. $2\mathbf{v}_3$

8. $(1/2)\mathbf{v}_4$ 9. $2\mathbf{v}_1 + \mathbf{v}_4$ 10. $\mathbf{v}_2 + 2\mathbf{v}_3$

Review Exercises

Find the magnitudes of the vector projections on the x- and y-axis of the vector **v** *with the given magnitude and direction angle.*

11. $\|\mathbf{v}\| = 3\sqrt{2}$ and $\alpha = 45°$
12. $\|\mathbf{v}\| = 4$ and $\alpha = 60°$
13. $\|\mathbf{v}\| = 7$ and $\alpha = 150°$
14. $\|\mathbf{v}\| = 2$ and $\alpha = 90°$

In Exercises 15–18, two forces act on a point in a plane at an angle with the given measures. Find the magnitude of the resultant force and the angle this force makes with the smaller given force.

15. Forces of 5 and 7 pounds, acting at an angle of 90° with respect to each other.
16. Forces of 6 and 12 pounds, acting at an angle of 45° with respect to each other.
17. Forces of 2 and 8 pounds, acting at an angle of 120° with respect to each other.
18. Forces of 3 and 9 pounds, acting at an angle of 148° with respect to each other.
19. An airplane flies 800 miles due south and then flies 600 miles 60° clockwise from north. How far from the starting point is the airplane at the end of this time?
20. An airplane is flying at 20,000 feet on a heading of 300° clockwise from north at 450 miles per hour. The wind velocity at this level is 90 miles per hour from 240° clockwise from north. Find the ground speed and the direction of the flight over the ground.
21. A weight of 160 pounds is placed on a smooth plane inclined at an angle of 30° with the horizontal (see figure). What is the minimum force that must be exerted against the weight and parallel to the plane to prevent the weight from slipping? *Hint:* Resolve the vertical force into components in the direction of the plane and perpendicular to it.
22. A force of 40 pounds is directed to the right. What vertical force must be applied so that the magnitude of the resultant force is 80 pounds? What angle does the resultant force make with the horizontal?

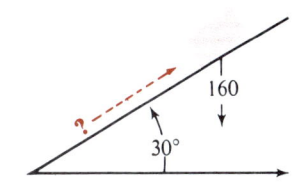

Chapter Review

[6.1] *Find the degree measure, to the nearest tenth of a degree, of the angle whose radian measure is as given.*

1. $\dfrac{3\pi^R}{8}$
2. 8.72^R

Find the radian measure, to the nearest hundredth of a radian, of the angle whose degree measure is as given.

3. 40°
4. 610°

On a circle with given radius, find the length of the arc intercepted by the angle whose measure is as given.

5. $r = 4.8$ inches; $\dfrac{\pi^R}{3}$
6. $r = 1.6$ feet; 150°

Find all angles α, where $-360° \leq \alpha \leq 720°$, that are coterminal with the angle whose measure is given.

7. $-330°$
8. 518°

[6.2] *Determine in which quadrant the terminal side of the angle α lies.*

9. $\tan \alpha > 0$ and $\cos \alpha < 0$
10. $\cos \alpha < 0$ and $\sin \alpha < 0$

Find the exact value of each expression.

11. $\cos 210°$
12. $\sin 330°$
13. $\tan 225°$
14. $\cos 120°$
15. $\sin(-210°)$
16. $\cos(-315°)$

Find approximations for the given expression. Use an appropriate table or a hand calculator.

17. $\cos 23°$
18. $\tan 1.24^R$
19. $\sin 120°$
20. $\cos(-135°)$
21. $\sin 192°$
22. $\tan 422°$
23. $\cos 3.68^R$
24. $\sin(-5.12^R)$

Use the appropriate table or a hand calculator to compute (a) the radian measure and (b) the degree measure of each of the following.

25. Arcsin 0.5000
26. $\text{Cos}^{-1}\, 0.9963$

In the following exercises in this review, give lengths to the nearest tenth of a unit and angle measure to the nearest tenth of a degree.

Review Exercises

[6.3] *Solve the right triangle ABC. Assume* $C = 90°$.

27. $a = 9$, $b = 12$
28. $c = 11$, $\alpha = 41°$
29. $a = 5$, $A = 12°$
30. $a = 5$, $c = 11$

Find the other trigonometric function values of θ, where θ is one acute angle of a right triangle.

31. $\sin \theta = 3/5$
32. $\tan \theta = 5/12$

33. Each of two surveyors is located 62.7 meters from a flagpole. If the angle formed by the lines of sight from the first surveyor to the second and from the first to the flagpole is 36°, how far apart are the two surveyors?

34. A TV antenna installer needs to run a guy wire from the top of an antenna 2 meters tall to the edge of a building. If the angle of depression (the angle between the line of sight and the horizontal) from the top of the antenna to the edge of the building is 28°, how much wire is needed? Assume the roof of the building is horizontal.

[6.4] *Solve the triangle.*

35. $b = 20$, $\beta = 25°$, $\alpha = 75°$
36. $a = 12$, $c = 15$, $\alpha = 30°$
37. $b = 11.4$, $B = 35°$, $C = 95°$
38. $c = 12.5$, $A = 100°$, $C = 20°$

39. Two men, 500 feet apart, observe a balloon between them that is in the same vertical plane with them. The respective angles of elevation of the balloon are observed by the men to measure 80.2° and 52.9°. Find the height of the balloon above the ground.

40. An airplane flies straight and level between two cities. A person in city A measures the angle of elevation to the airplane as 58°. At the same time a person in city B measures the angle of elevation to the airplane as 30°. If the distance between the two cities is 10 kilometers and if both cities are at sea level, what is the altitude of the airplane?

[6.5] *Solve the triangle.*

41. $a = 12$, $b = 3$, $\gamma = 25°$
42. $b = 12.8$, $c = 8.35$, $\alpha = 80.7°$
43. $a = 8$, $b = 10$, $c = 9$
44. $a = 5$, $b = 4$, $c = 3$

45. The straight-line distance between city A and city B is approximately 1566 miles, between city A and city C is approximately 608 miles, and between city B and city C is approximately 1169 miles. Find the measures of the angles of the triangle formed by the three cities.

46. The captain of a rescue ship receives messages in the form of light signals from two ships that have collided. The captain estimates that his ship is 1800 meters from one of the damaged ships and 2100 meters from the other. From the light signals he finds that the angle formed by the lines joining his ship with the two damaged ships is 4°. How far apart have the damaged ships drifted?

[6.6] 47. Find the magnitude of the projections on the x- and y-axis of the vector **v** with a direction angle $\alpha = 30°$ and $\|\mathbf{v}\| = 12$.

48. Find the magnitude of the projections on the x- and y-axis of the vector **v** with a direction angle $\alpha = 135°$ and $\|\mathbf{v}\| = 16.8$.

49. Two forces with magnitudes of 6 and 10 pounds act at an angle of 72° with respect to each other. Find the magnitude of the resultant force and the angle this force makes with the larger of the given forces.

50. A boat has a maximum speed of 18.2 miles per hour in still water. If the boat crossed a river at right angles to the current and has a "drift angle" of 8°, find the speed of the current.

7 Identities and Conditional Equations

In this chapter we shall consider identities and conditional equations that involve trigonometric function values. We shall use lowercase Greek letters such as α, β, γ, and θ to represent angles only; and we shall use lowercase English letters such as x, y, and z to represent either angles or real numbers, so that the functions involved might be considered as being either circular or trigonometric.

7.1 Basic Identities

The fact that $\sin x$ and $\cos x$ are related by the equation

$$\sin^2 x + \cos^2 x = 1, \tag{1}$$

and the facts that

$$\tan x = \frac{\sin x}{\cos x}, \quad \cot x = \frac{\cos x}{\sin x},$$

$$\sec x = \frac{1}{\cos x}, \quad \text{and} \quad \csc x = \frac{1}{\sin x},$$

suggest that the trigonometric function values are related in a variety of ways. We shall first consider several other basic relationships involving trigonometric functions and then use these relationships to rewrite expressions involving trigonometric function values in equivalent forms.

Because $\cos^2 x \neq 0$ for any x for which $\tan x$ and $\sec x$ are defined, we can multiply each member of Equation (1) by $1/\cos^2 x$ to obtain the equation

$$\frac{\sin^2 x}{\cos^2 x} + \frac{\cos^2 x}{\cos^2 x} = \frac{1}{\cos^2 x},$$

7 Identities and Conditional Equations

$1 + \tan^2 x$

from which we obtain
$$\tan^2 x + 1 = \sec^2 x. \tag{2}$$

Equation (2) and the other relationships above are called **identities** because they are true for all values of x for which all the function values involved are defined. Hereafter, in the absence of specific statements of restrictions on variables in identities, we shall assume that the replacement sets of the variables are restricted to values for which the expressions are defined.

$1 + \cot^2 x$

We can obtain another useful identity by starting with Equation (1). Since $\sin^2 x \neq 0$ for any x for which $\cot x$ and $\csc x$ are defined, we can multiply each member of Equation (1) by $1/\sin^2 x$ to obtain the equation

$$\frac{\sin^2 x}{\sin^2 x} + \frac{\cos^2 x}{\sin^2 x} = \frac{1}{\sin^2 x},$$

from which we obtain
$$1 + \cot^2 x = \csc^2 x. \tag{3}$$

$\sin(-x)$
$\cos(-x)$
$\tan(-x)$

The relationships
$$\sin(-x) = -\sin x, \tag{4}$$
$$\cos(-x) = \cos x, \tag{5}$$
and
$$\tan(-x) = -\tan x \tag{6}$$

are also important identities. To see why they are valid, note that for the arcs measuring x and $-x$ in Figure 7.1 triangles AOP and BOP are congruent. Thus, $OA = OB$ and $AP = BP$. Since the ordinate of the point B is negative, we have

$$\sin(-x) = -\frac{BP}{OB} = -\frac{AP}{OA}$$
$$= -\sin x$$

and
$$\cos(-x) = \frac{OP}{OB} = \frac{OP}{OA}$$
$$= \cos x,$$

from which we obtain
$$\tan(-x) = \frac{\sin(-x)}{\cos(-x)} = \frac{-\sin x}{\cos x}$$
$$= -\tan x.$$

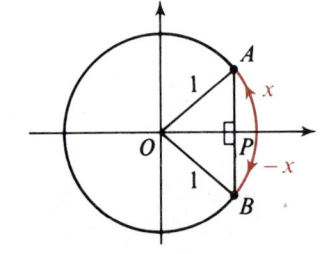

Figure 7.1

The geometric arguments to demonstrate the validity of identities (4), (5), and (6) for arcs terminating in quadrants other than the first quadrant are similar.

Methods to prove identities

There is no general method for verifying identities. However, it is sometimes helpful first to change one member of an equation to expressions involving only sines and/or cosines and then to use algebraic procedures to rewrite that member as an expression identical to the other member.

7.1 Basic Identities

Example Verify that $\dfrac{\sin \alpha}{1 - \cos \alpha} - \cot \alpha = \dfrac{1}{\sin \alpha}$ is an identity.

Solution In this case, it is helpful to rewrite the left-hand member in terms of $\sin \alpha$ and $\cos \alpha$ only. Writing $\cos \alpha/\sin \alpha$ for $\cot \alpha$, we have

$$\frac{\sin \alpha}{1 - \cos \alpha} - \frac{\cos \alpha}{\sin \alpha} = \frac{1}{\sin \alpha}.$$

Writing the left-hand member as a single fraction, we obtain

$$\frac{\sin \alpha}{1 - \cos \alpha} - \frac{\cos \alpha}{\sin \alpha} = \frac{\sin^2 \alpha - (1 - \cos \alpha) \cos \alpha}{(1 - \cos \alpha) \sin \alpha}$$

$$= \frac{\sin^2 \alpha - \cos \alpha + \cos^2 \alpha}{(1 - \cos \alpha) \sin \alpha}$$

$$= \frac{(\sin^2 \alpha + \cos^2 \alpha) - \cos \alpha}{(1 - \cos \alpha) \sin \alpha}$$

$$= \frac{1 - \cos \alpha}{(1 - \cos \alpha) \sin \alpha}.$$

Since $1 - \cos \alpha$ is restricted from 0 in the original equation, we can rewrite the last expression as $1/\sin \alpha$. Thus the equation

$$\frac{\sin \alpha}{1 - \cos \alpha} - \cot \alpha = \frac{1}{\sin \alpha}$$

is an identity.

Example Verify that $\dfrac{\sec \alpha}{\sin \alpha} = \tan \alpha + \cot \alpha$ is an identity.

Solution In this case, we shall first rewrite the right-hand member in terms of $\sin \alpha$ and $\cos \alpha$. Writing $\sin \alpha/\cos \alpha$ and $\cos \alpha/\sin \alpha$ for $\tan \alpha$ and $\cot \alpha$, respectively, we have

$$\tan \alpha + \cot \alpha = \frac{\sin \alpha}{\cos \alpha} + \frac{\cos \alpha}{\sin \alpha}$$

$$= \frac{\sin^2 \alpha + \cos^2 \alpha}{\cos \alpha \sin \alpha}$$

$$= \frac{1}{\cos \alpha \sin \alpha}$$

$$= \frac{1/\cos \alpha}{\sin \alpha}$$

$$= \frac{\sec \alpha}{\sin \alpha}.$$

Solution continued overleaf

Thus,
$$\frac{\sec \alpha}{\sin \alpha} = \tan \alpha + \cot \alpha$$
is an identity.

Example Verify that $\dfrac{1 - \sin x}{\cos x} = \dfrac{\cos x}{1 + \sin x}$ is an identity.

Solution In this case, we shall show that the left-hand member is equivalent to the right-hand member. Since $\cos x \neq 0$, we can multiply the left-hand member by $\cos x/\cos x$ to obtain

$$\frac{1 - \sin x}{\cos x} = \frac{\cos x(1 - \sin x)}{\cos^2 x}. \qquad (7)$$

Because $\sin^2 x + \cos^2 x = 1$ can be written equivalently as

$$\cos^2 x = 1 - \sin^2 x = (1 - \sin x)(1 + \sin x),$$

we can rewrite the right-hand member of (7) to obtain

$$\frac{1 - \sin x}{\cos x} = \frac{\cos x(1 - \sin x)}{(1 + \sin x)(1 - \sin x)}.$$

Hence
$$\frac{1 - \sin x}{\cos x} = \frac{\cos x}{1 + \sin x}$$
is an identity.

Proving equations are not identities   If an equation is not an identity, this fact can be established by simply finding a *counterexample*. For example, substituting $\pi/6$ for x in the equation

$$\sin x \cdot \cos x = \tan x \qquad (8)$$

yields

$$\frac{1}{2} \cdot \frac{\sqrt{3}}{2} = \frac{1}{\sqrt{3}},$$

which is a false statement. Thus, by finding a counterexample, we have shown that Equation (8) is not an identity.

In using a counterexample, you must, of course, select a value for the variable for which each member of the equation is defined.

7.1 Basic Identities

Exercise 7.1

A *Transform the first expression so that it is identical to the second expression. Assume suitable restrictions on the variable.*

1. $\cos x \tan x$; $\sin x$
2. $\sin x \sec x$; $\tan x$
3. $\sin^2 x \cot^2 x$; $\cos^2 x$
4. $\dfrac{\cos^2 x}{\cot^2 x}$; $\sin^2 x$
5. $\cos^2 \alpha (1 + \tan^2 \alpha)$; 1
6. $(\csc^2 \alpha - 1)$; $\cot^2 \alpha$
7. $\sec \alpha \csc \alpha - \cot \alpha$; $\tan \alpha$
8. $(1 - \cos^2 \alpha)(1 + \cot^2 \alpha)$; 1
9. $\dfrac{\sin \theta \sec \theta}{\tan \theta}$; 1
10. $\cos \theta \tan \theta \csc \theta$; 1
11. $(\sec^2 \theta - 1)(\csc^2 \theta - 1)$; 1
12. $\dfrac{\cos \theta - \sin \theta}{\cos \theta}$; $1 - \tan \theta$
13. $\dfrac{1}{1 + \sin \theta} + \dfrac{1}{1 - \sin \theta}$; $2 \sec^2 \theta$
14. $\dfrac{1 + \tan^2 \theta}{\csc^2 \theta}$; $\tan^2 \theta$

Verify that the equation is an identity.

15. $\sin x \cot x = \cos x$
16. $\tan x \csc x = \sec x$
17. $\dfrac{\sin \theta}{\tan \theta} = \cos(-\theta)$
18. $\dfrac{\sin(-\theta) \sec \theta}{\tan(-\theta)} = 1$
19. $\tan^2 x \cos^2 x = \sin^2 x$
20. $(\csc^2 x - 1)(\sin^2 x) = \cos^2 x$
21. $\dfrac{(1 + \sin \theta)(1 - \sin \theta)}{\cos \theta} = \cos \theta$
22. $\dfrac{(1 + \cos \theta)(1 - \cos \theta)}{\sin \theta} = \sin \theta$
23. $\sec x - \cos x = \sin x \tan x$
24. $\sec x - \sin x \tan x = \cos x$
25. $\dfrac{1 + \tan^2 x}{\tan^2 x} = \csc^2 x$
26. $\dfrac{\sin^2 x}{1 - \cos x} = \dfrac{1 + \sec x}{\sec x}$
27. $\tan^2 \alpha - \sin^2 \alpha = \sin^2 \alpha \tan^2 \alpha$
28. $\cot^2 \alpha + \sec^2 \alpha = \tan^2 \alpha + \csc^2 \alpha$
29. $\tan \alpha + \sec \alpha = \dfrac{1}{\sec \alpha - \tan \alpha}$
30. $\dfrac{\sec \alpha + \csc \alpha}{1 + \tan \alpha} = \csc \alpha$
31. $\dfrac{\cos x + \sin x}{\sin x} = 1 + \cot x$
32. $\dfrac{1 - \tan^2 x}{\tan x} = \cot x - \tan x$
33. $\dfrac{1}{1 + \sin x} + \dfrac{1}{1 - \sin x} = 2 \sec^2 x$
34. $\tan x + \cot x = \sec x \csc x$
35. $\dfrac{1 + \sec \theta}{\sec \theta} = \dfrac{\sin^2 \theta}{1 - \cos \theta}$
36. $\dfrac{1 - \cos \alpha}{\sin \alpha} = \dfrac{\sin \alpha}{1 + \cos \alpha}$

37. $\dfrac{1 + \sin \alpha}{\cos \alpha} = \dfrac{\cos \alpha}{1 - \sin \alpha}$

38. $\dfrac{1}{\sec x - \tan x} = \sec x + \tan x$

39. $\dfrac{1 - \cos \alpha}{1 + \cos \alpha} = \dfrac{\sec \alpha - 1}{\sec \alpha + 1}$

40. $\dfrac{\cos x}{\sec x - \tan x} = \dfrac{\cos^2 x}{1 - \sin x}$

41. $\tan^4 \theta + \tan^2 \theta = \sec^4 \theta - \sec^2 \theta$

42. $\cos^4 \theta - \sin^4 \theta = \cos^2 \theta - \sin^2 \theta$

Use a counterexample to show that the equation is not an identity.

43. $\tan^2 \theta + \sec^2 \theta = 1$

44. $\sin^2 \theta - \cos^2 \theta = 1$

45. $\sin \theta + \tan \theta = \cos \theta$

46. $\tan^2 \theta = 2 \tan \theta$

B *Verify that the equation is an identity.*

47. $(a \sin \alpha + b \cos \alpha)^2 + (a \cos \alpha - b \sin \alpha)^2 = a^2 + b^2$

48. $\dfrac{\cot \alpha - 1}{\cot \alpha + 1} = \dfrac{1 - \tan \alpha}{1 + \tan \alpha}$

49. $\left(\dfrac{\sec \theta - 1}{\sec \theta + 1}\right) \sin^2 \theta + \sin^2 \theta - 2 = 2 \cos \theta$

50. $\dfrac{\tan^2 \theta + 2 \tan \theta + 1}{\cot^2 \theta + 2 \cot \theta + 1} + 1 = \dfrac{1}{1 - \sin^2 \theta}$

7.2 Special Formulas for the Cosine Function

In this section we shall derive several important formulas involving the cosine that are used in calculus and other advanced mathematics courses.

$\cos(x_1 + x_2)$

Figure 7.2 shows selected points on the unit circle with their coordinates. Since the length of arc from $(1, 0)$ to $(\cos(x_1 + x_2), \sin(x_1 + x_2))$ is $x_1 + x_2$, and from $(\cos x_1, -\sin x_1)$ to $(\cos x_2, \sin x_2)$ the arc length is also $x_1 + x_2$, the respective chords have equal lengths. Using the distance formula to express this fact, we have

$$\sqrt{[\cos(x_1 + x_2) - 1]^2 + [\sin(x_1 + x_2) - 0]^2}$$
$$= \sqrt{(\cos x_2 - \cos x_1)^2 + [\sin x_2 - (-\sin x_1)]^2},$$

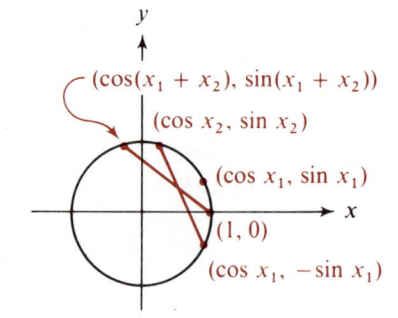

Figure 7.2

7.2 Special Formulas for the Cosine Function

or, squaring both members,

$$[\cos(x_1 + x_2) - 1]^2 + [\sin(x_1 + x_2)]^2$$
$$= (\cos x_2 - \cos x_1)^2 + (\sin x_2 + \sin x_1)^2.$$

If we now perform the operations indicated, we have

$$\cos^2(x_1 + x_2) - 2\cos(x_1 + x_2) + 1 + \sin^2(x_1 + x_2)$$
$$= \cos^2 x_2 - 2\cos x_1 \cos x_2 + \cos^2 x_1 + \sin^2 x_2 + 2\sin x_1 \sin x_2 + \sin^2 x_1.$$

Regrouping terms, we obtain

$$[\cos^2(x_1 + x_2) + \sin^2(x_1 + x_2)] - 2\cos(x_1 + x_2) + 1$$
$$= (\cos^2 x_2 + \sin^2 x_2) + (\cos^2 x_1 + \sin^2 x_1) - 2\cos x_1 \cos x_2 + 2\sin x_1 \sin x_2,$$

and since $\cos^2 x + \sin^2 x = 1$, it follows that

$$1 - 2\cos(x_1 + x_2) + 1 = 1 + 1 - 2\cos x_1 \cos x_2 + 2\sin x_1 \sin x_2.$$

This simplifies to the formula

$$\cos(x_1 + x_2) = \cos x_1 \cos x_2 - \sin x_1 \sin x_2. \qquad (1)$$

Example Find an exact expression for cos 105°.

Solution Write 105° as 60° + 45°. By Equation (1),

$$\cos 105° = \cos(60° + 45°) = \cos 60° \cos 45° - \sin 60° \sin 45°$$
$$= \left(\frac{1}{2}\right)\left(\frac{\sqrt{2}}{2}\right) - \left(\frac{\sqrt{3}}{2}\right)\left(\frac{\sqrt{2}}{2}\right)$$
$$= \frac{\sqrt{2}(1 - \sqrt{3})}{4}.$$

$\cos(x_1 - x_2)$ If we replace x_2 in Equation (1) by $-x_2$, we obtain

$$\cos(x_1 - x_2) = \cos x_1 \cos(-x_2) - \sin x_1 \sin(-x_2),$$

and since $\cos(-x_2) = \cos x_2$ and $\sin(-x_2) = -\sin x_2$, we have the formula

$$\cos(x_1 - x_2) = \cos x_1 \cos x_2 + \sin x_1 \sin x_2. \qquad (2)$$

Example Find an exact expression for cos 15°.

Solution Write 15° as 60° − 45°. By Equation (2),

$$\cos 15° = \cos(60° - 45°)$$
$$= \cos 60° \cos 45° + \sin 60° \sin 45°$$
$$= \left(\frac{1}{2}\right)\left(\frac{\sqrt{2}}{2}\right) + \left(\frac{\sqrt{3}}{2}\right)\left(\frac{\sqrt{2}}{2}\right)$$
$$= \frac{\sqrt{2}(1 + \sqrt{3})}{4}.$$

Equations (1) and (2) are called the **sum formula** and **difference formula**, respectively, for the cosine function.

We can make use of Equations (1) and (2) to obtain formulas for cos 2x and cos (x/2). These formulas are known, respectively, as the **double-angle** and **half-angle formulas**, although they are equally valid when x is a real number.

cos 2x

By letting $x_1 = x_2 = x$ in Equation (1), we obtain

$$\cos(x + x) = \cos x \cos x - \sin x \sin x,$$

or

$$\mathbf{\cos 2x = \cos^2 x - \sin^2 x.} \tag{3}$$

We can obtain alternative forms of the equation for cos 2x by using the substitutions

$$\cos^2 x = 1 - \sin^2 x \quad \text{and} \quad \sin^2 x = 1 - \cos^2 x$$

in the right-hand member of (3). The results are the formulas

$$\mathbf{\cos 2x = 1 - 2\sin^2 x,} \tag{4}$$

$$\mathbf{\cos 2x = 2\cos^2 x - 1.} \tag{5}$$

Example

Given that $\cos x = -3/5$, find cos 2x.

Solution

Using Equation (5), we have

$$\cos 2x = 2\left(-\frac{3}{5}\right)^2 - 1 = 2\left(\frac{9}{25}\right) - 1,$$

$$= \frac{18}{25} - 1 = -\frac{7}{25}.$$

$\cos\left(\dfrac{x}{2}\right)$

By replacing x with $x/2$ in Equation (5), we find

$$\cos 2\left(\frac{x}{2}\right) = 2\cos^2\frac{x}{2} - 1,$$

or

$$\cos x = 2\cos^2\frac{x}{2} - 1.$$

Solving for cos (x/2) we obtain

$$\cos\frac{x}{2} = \pm\sqrt{\frac{1 + \cos x}{2}}.$$

7.2 Special Formulas for the Cosine Function

Now, because $\cos(x/2) > 0$ for $x/2$ terminating in Quadrants I and IV, we take the positive square root for $x/2$ terminating in these quadrants. Because $\cos(x/2) < 0$ for $x/2$ terminating in Quadrants II and III, we take the negative square root in these quadrants. Thus, we have the identity

$$\cos\frac{x}{2} = \begin{cases} \sqrt{\dfrac{1 + \cos x}{2}} & \text{when } \dfrac{x}{2} \text{ terminates in Quadrant I or IV;} \quad (6a) \\[2ex] -\sqrt{\dfrac{1 + \cos x}{2}} & \text{when } \dfrac{x}{2} \text{ terminates in Quadrant II or III.} \quad (6b) \end{cases}$$

Example   Given that $\cos x = 3/7$ and $0 \leq x/2 \leq \pi/2$, find $\cos(x/2)$.

Solution   Using Equation (6a), we have

$$\cos\frac{x}{2} = \sqrt{\frac{1 + \cos x}{2}} = \sqrt{\frac{1 + 3/7}{2}} = \sqrt{\frac{5}{7}}.$$

Identities   The methods used in the preceding section can be applied to show that some equations involving $\cos 2x$ and/or $\cos(x/2)$ are identities.

Example   Verify that $\sec^2 x \cos 2x = 1 - \tan^2 x$ is an identity.

Solution   We first use the definition of $\sec x$ and Equation (3) to write the left-hand member in terms of $\sin x$ and $\cos x$. We then simplify to obtain

$$\sec^2 x \cos 2x = \frac{1}{\cos^2 x}(\cos^2 x - \sin^2 x)$$

$$= \frac{\cos^2 x}{\cos^2 x} - \frac{\sin^2 x}{\cos^2 x}$$

$$= 1 - \tan^2 x.$$

Thus $\sec^2 x \cos 2x = 1 - \tan^2 x$ is an identity.

Exercise 7.2

A   *Find an exact value for each expression.*

Example $\cos 285°$

Solution overleaf

Solution Write $285° = 45° + 240°$. Then

$$\cos 285° = \cos 240° \cos 45° - \sin 240° \sin 45°$$

$$= \frac{-1}{2} \cdot \frac{1}{\sqrt{2}} - \frac{-\sqrt{3}}{2} \cdot \frac{1}{\sqrt{2}}$$

$$= \frac{1 + \sqrt{3}}{2\sqrt{2}}.$$

1. $\cos 195°$
2. $\cos 345°$
3. $\cos 255°$
4. $\cos 165°$
5. $\cos 75°$
6. $\cos 22.5°$

Example Given that $\cos x = -4/5$, find $\cos 2x$ and $\cos (x/2)$, $\pi/2 < x < \pi$.

Solution By Equation (5) on page 244,

$$\cos 2x = 2 \cos^2 x - 1 = 2\left(-\frac{4}{5}\right)^2 - 1$$

$$= \frac{32}{25} - 1 = \frac{7}{25}.$$

Since $\pi/2 < x < \pi$, $x/2$ terminates in Quadrant I. By Equation (6a),

$$\cos \frac{x}{2} = \sqrt{\frac{1 + \cos x}{2}} = \sqrt{\frac{1 + (-4/5)}{2}}$$

$$= \sqrt{\frac{1}{10}}.$$

7. Given that $\cos x = -3/5$, find

   a. $\cos 2x$, $\dfrac{\pi}{2} \le x \le \pi$
   b. $\cos \dfrac{x}{2}$, $\dfrac{\pi}{2} \le x \le \pi$

8. Given that $\cos x = 3/4$, find

   a. $\cos 2x$, $\dfrac{3\pi}{2} \le x \le 2\pi$
   b. $\cos \dfrac{x}{2}$, $\dfrac{3\pi}{2} \le x \le \pi$

Prove the identity.

9. $\cos (\pi + x) = -\cos x$
10. $\cos (\pi - x) = -\cos x$
11. $\cos \left(\dfrac{\pi}{2} + x\right) = -\sin x$
12. $\cos \left(\dfrac{\pi}{2} - x\right) = \sin x$
13. $\cos \left(\dfrac{3\pi}{2} + x\right) = \sin x$
14. $\cos \left(\dfrac{3\pi}{2} - x\right) = -\sin x$

15. $\sin^2 x = \dfrac{1 - \cos 2x}{2}$ **16.** $\cos^2 x = \dfrac{1 + \cos 2x}{2}$

17. $\cos 2\theta = \dfrac{1 - \tan^2 \theta}{1 + \tan^2 \theta}$ **18.** $\dfrac{2}{1 + \cos \theta} = \sec^2 \theta$

19. $\cos 3x = 4\cos^3 x - 3\cos x$, $0 \leq x \leq \dfrac{\pi}{2}$ *Hint:* $\cos 3x = \cos(2x + x)$

20. $\cos 4x = 1 - 8\sin^2 x + 8\sin^4 x$ *Hint:* $\cos 4x = \cos 2(2x)$

21. Show that, for any real numbers x and $h \neq 0$,
$$\dfrac{\cos(x + h) - \cos x}{h} = \cos x \left(\dfrac{\cos h - 1}{h}\right) - \sin x \left(\dfrac{\sin h}{h}\right).$$

7.3 Special Formulas for the Sine Function

In Section 7.2 we derived the sum, difference, double-angle, and half-angle formulas for the cosine function. In this section we obtain similar formulas for the sine function.

$\sin(x_1 + x_2)$

Note that from Equation (2) on page 243 we have
$$\cos\left(\dfrac{\pi}{2} - x\right) = \cos\dfrac{\pi}{2}\cos x + \sin\dfrac{\pi}{2}\sin x$$
$$= 0 \cdot \cos x + 1 \cdot \sin x,$$

from which
$$\cos\left(\dfrac{\pi}{2} - x\right) = \sin x. \tag{1}$$

If we now set $x = x_1 + x_2$, we obtain
$$\cos\left(\dfrac{\pi}{2} - (x_1 + x_2)\right) = \sin(x_1 + x_2).$$

This can be rewritten as
$$\cos\left[\left(\dfrac{\pi}{2} - x_1\right) - x_2\right] = \sin(x_1 + x_2),$$

and if the left-hand member is expanded by means of Equation (2) on page 243, we find that
$$\cos\left(\dfrac{\pi}{2} - x_1\right)\cos x_2 + \sin\left(\dfrac{\pi}{2} - x_1\right)\sin x_2 = \sin(x_1 + x_2). \tag{2}$$

248   7   *Identities and Conditional Equations*

From Equation (1) above with $x = \pi/2 - x_1$, we have

$$\sin\left(\frac{\pi}{2} - x_1\right) = \cos\left[\frac{\pi}{2} - \left(\frac{\pi}{2} - x_1\right)\right]$$

$$= \cos x_1.$$

Thus, we obtain

$$\cos\left(\frac{\pi}{2} - x_1\right)\cos x_2 + \sin\left(\frac{\pi}{2} - x_1\right)\sin x_2 = \sin x_1 \cos x_2 + \cos x_1 \sin x_2,$$

which in conjunction with Equation (2) above establishes the identity

$$\sin(x_1 + x_2) = \sin x_1 \cos x_2 + \cos x_1 \sin x_2. \tag{3}$$

$\sin(x_1 - x_2)$ By replacing x_2 in Equation (3) with $-x_2$, we get

$$\sin(x_1 - x_2) = \sin x_1 \cos(-x_2) + \cos x_1 \sin(-x_2).$$

Substituting $\cos x_2$ for $\cos(-x_2)$ and $-\sin x_2$ for $\sin(-x_2)$ in the right-hand member, we obtain the identity

$$\sin(x_1 - x_2) = \sin x_1 \cos x_2 - \cos x_1 \sin x_2. \tag{4}$$

Relationships (3) and (4), respectively, are called the **sum formula** and the **difference formula** for the sine function.

Example   Find an exact value for $\sin 75°$.

Solution   Write $75°$ as $30° + 45°$. Then, by Equation (3),

$$\sin 75° = \sin(30° + 45°)$$

$$= \sin 30° \cos 45° + \cos 30° \sin 45°$$

$$= \left(\frac{1}{2}\right)\left(\frac{\sqrt{2}}{2}\right) + \left(\frac{\sqrt{3}}{2}\right)\left(\frac{\sqrt{2}}{2}\right)$$

$$= \frac{\sqrt{2}(1 + \sqrt{3})}{4}.$$

$\sin 2x$ We can obtain the double-angle formula for the sine function by letting $x_1 = x_2 = x$ in Equation (3):

$$\sin(x + x) = \sin x \cos x + \cos x \sin x,$$

or

$$\sin 2x = 2 \sin x \cos x. \tag{5}$$

Example   Given that $\sin x = 2/3$, find $\sin 2x$, $\pi/2 \leq x \leq \pi$.

7.3 Special Formulas for the Sine Function

Solution Since $\sin x = 2/3$ and x terminates in Quadrant II, we have

$$\cos x = -\sqrt{1 - \sin^2 x} = -\sqrt{1 - (2/3)^2}$$
$$= -\sqrt{\frac{5}{9}} = \frac{-\sqrt{5}}{3}.$$

From Equation (5) we obtain

$$\sin 2x = 2 \sin x \cos x$$
$$= 2\left(\frac{2}{3}\right)\left(-\frac{\sqrt{5}}{3}\right) = \frac{-4\sqrt{5}}{9}.$$

$\sin\left(\dfrac{x}{2}\right)$ The half-angle formula for the sine function can be obtained from the identity

$$\cos 2x = 1 - 2 \sin^2 x.$$

By replacing x with $x/2$, we have

$$\cos 2\left(\frac{x}{2}\right) = 1 - 2 \sin^2 \frac{x}{2},$$

from which

$$\cos x = 1 - 2 \sin^2 \frac{x}{2},$$

$$\sin^2 \frac{x}{2} = \frac{1 - \cos x}{2},$$

and

$$\sin \frac{x}{2} = \pm\sqrt{\frac{1 - \cos x}{2}}.$$

Now, because $\sin(x/2) > 0$ for $x/2$ terminating in Quadrants I and II, we take the positive square root for $x/2$ terminating in these quadrants and because $\sin(x/2) < 0$ for $x/2$ terminating in Quadrants III and IV, we take the negative square root for $x/2$ terminating in these quadrants. Thus, we have the identity

$$\sin \frac{x}{2} = \begin{cases} \sqrt{\dfrac{1 - \cos x}{2}} & \text{if } \dfrac{x}{2} \text{ terminates in Quadrant I or II;} \quad (6a) \\ -\sqrt{\dfrac{1 - \cos x}{2}} & \text{if } \dfrac{x}{2} \text{ terminates in Quadrant III or IV.} \quad (6b) \end{cases}$$

Example Given that $\cos x = 3/7$ and $0 \leq x/2 \leq \pi/2$, find $\sin(x/2)$.

Solution Using Equation (6a), we have

$$\sin \frac{x}{2} = \sqrt{\frac{1 - \cos x}{2}} = \sqrt{\frac{1 - (3/7)}{2}} = \sqrt{\frac{2}{7}}.$$

Identities The methods used in Section 7.1 can be applied to show that some equations involving sin 2x and/or sin (x/2) are identities.

Example Verify that $\left(\frac{1}{2}\right) \sec x \csc x \sin 2x = 1$ is an identity.

Solution We first use the definitions of sec x and csc x and Equation (5) to write the left-hand member in terms of sin x and cos x. We then simplify to obtain

$$\frac{1}{2} \sec x \csc x \sin 2x = \frac{1}{2} \cdot \frac{1}{\cos x} \cdot \frac{1}{\sin x} \cdot 2 \sin x \cos x$$

$$= 1.$$

Thus $\left(\frac{1}{2}\right) \sec x \csc x \sin 2x = 1$ is an identity.

Exercise 7.3

Find an exact value for each expression.

Example sin 345°

Solution Write 345° as $345° = 45° + 300°$. Then

$$\sin 345° = \sin 45° \cos 300° + \cos 45° \sin 300°$$

$$= \frac{1}{\sqrt{2}} \cdot \frac{1}{2} + \frac{1}{\sqrt{2}} \cdot \frac{-\sqrt{3}}{2}$$

$$= \frac{1 - \sqrt{3}}{2\sqrt{2}}.$$

1. sin 15°
2. sin 105°
3. sin 255°
4. sin 165°
5. sin 195°
6. sin 22.5°

Example Given that $\sin x = 4/5$, find sin 2x, and sin (x/2), $\pi/2 \leq x \leq \pi$.

Solution Since $\sin x = 4/5$ and x terminates in Quadrant II,

$$\cos x = -\sqrt{1 - \sin^2 x} = -\sqrt{1 - \left(\frac{4}{5}\right)^2}$$

$$= -\sqrt{\frac{9}{25}} = -\frac{3}{5}.$$

From Equation (5) on page 248, we have

$$\sin 2x = 2 \sin x \cos x$$

$$= 2\left(\frac{4}{5}\right)\left(-\frac{3}{5}\right) = -\frac{24}{25}.$$

Since $\pi/2 \le x \le \pi$, $x/2$ terminates in Quadrant I. By Equation (6a),

$$\sin \frac{x}{2} = \sqrt{\frac{1 - \cos x}{2}} = \sqrt{\frac{1 - (-3/5)}{2}}$$

$$= \sqrt{\frac{4}{5}} = \frac{2}{\sqrt{5}}.$$

7. Given that $\cos x = -3/5$, find

   a. $\sin 2x$, $\pi \le x \le \frac{3\pi}{2}$

   b. $\sin \frac{x}{2}$, $3\pi \le x \le \frac{7\pi}{2}$

8. Given that $\sin x = \sqrt{5}/3$, find

   a. $\sin 2x$, $\frac{\pi}{2} \le x \le \pi$

   b. $\sin \frac{x}{2}$, $0 \le x \le \frac{\pi}{2}$

Prove the identity.

9. $\sin (\pi + x) = -\sin x$

10. $\sin (\pi - x) = \sin x$

11. $\sin \left(\frac{\pi}{2} + x\right) = \cos x$

12. $\sin \left(\frac{\pi}{2} - x\right) = \cos x$

13. $\sin \left(\frac{3\pi}{2} + x\right) = -\cos x$

14. $\sin \left(\frac{3\pi}{2} - x\right) = -\cos x$

15. $\sin 2\alpha = \dfrac{2 \tan \alpha}{1 + \tan^2 \alpha}$

16. $\dfrac{2}{\sin 2\alpha} = \tan \alpha + \cot \alpha$

17. $\dfrac{1 + \cos 2x}{\sin 2x} = \cot x$

18. $\sin 2x \sec x = 2 \sin x$

19. $\sin 3x = 3 \sin x - 4 \sin^3 x$ *Hint:* $\sin 3x = \sin (2x + x)$.

20. $\sin 4x = 4 \sin x \cos x - 8 \sin^3 x \cos x$ *Hint:* $\sin 4x = \sin 2(2x)$.

21. Show that, for any real numbers x and $h \ne 0$,

$$\frac{\sin (x + h) - \sin x}{h} = \sin x \left(\frac{\cos h - 1}{h}\right) + \cos x \left(\frac{\sin h}{h}\right).$$

7.4 Special Formulas for the Tangent Function

In this section we derive the sum, difference, double-angle, and half-angle formulas for the tangent function.

tan $(x_1 + x_2)$ By Definition 5.4,

$$\tan (x_1 + x_2) = \frac{\sin (x_1 + x_2)}{\cos (x_1 + x_2)}.$$

By Equation (1) on page 243 and Equation (3) on page 248, we have

$$\tan(x_1 + x_2) = \frac{\sin x_1 \cos x_2 + \cos x_1 \sin x_2}{\cos x_1 \cos x_2 - \sin x_1 \sin x_2}.$$

Dividing numerator and denominator of the right-hand member by $\cos x_1 \cos x_2$, we find

$$\tan(x_1 + x_2) = \frac{\dfrac{\sin x_1 \cos x_2}{\cos x_1 \cos x_2} + \dfrac{\cos x_1 \sin x_2}{\cos x_1 \cos x_2}}{\dfrac{\cos x_1 \cos x_2}{\cos x_1 \cos x_2} - \dfrac{\sin x_1 \sin x_2}{\cos x_1 \cos x_2}}$$

After simplifying each term in the right-hand member, we obtain

$$\tan(x_1 + x_2) = \frac{\tan x_1 + \tan x_2}{1 - \tan x_1 \tan x_2}. \qquad (1)$$

$\tan(x_1 - x_2)$ By first replacing x_2 in Equation (1) with $-x_2$, we have

$$\tan(x_1 - x_2) = \frac{\tan x_1 + \tan(-x_2)}{1 - \tan x_1 \tan(-x_2)}.$$

Substituting $-\tan x_2$ for $\tan(-x_2)$ in the right-hand member, we obtain

$$\tan(x_1 - x_2) = \frac{\tan x_1 - \tan x_2}{1 + \tan x_1 \tan x_2}. \qquad (2)$$

Example Find an exact value for $\tan 15°$.

Solution Write $15°$ as $60° - 45°$. Then by Equation (2) we have

$$\tan 15° = \tan(60° - 45°)$$

$$= \frac{\tan 60° - \tan 45°}{1 + \tan 60° \tan 45°} = \frac{\sqrt{3} - 1}{1 + \sqrt{3}}.$$

$\tan 2x$ By replacing both x_1 and x_2 with x in Equation (1), we have

$$\tan(x + x) = \frac{\tan x + \tan x}{1 - \tan x \tan x},$$

from which we obtain the identity

$$\tan 2x = \frac{2 \tan x}{1 - \tan^2 x}. \qquad (3)$$

$\tan\left(\dfrac{x}{2}\right)$ By Definition 5.4, we have

$$\tan \frac{x}{2} = \frac{\sin(x/2)}{\cos(x/2)}.$$

7.4 Special Formulas for the Tangent Function

Then, by multiplying numerator and denominator by $2 \sin(x/2)$, we have

$$\tan \frac{x}{2} = \frac{2 \sin^2(x/2)}{2 \sin(x/2) \cos(x/2)}.$$

Substituting $1 - \cos x$ for $2 \sin^2(x/2)$ and $\sin x$ for $2 \sin(x/2) \cos(x/2)$, we obtain

$$\tan \frac{x}{2} = \frac{1 - \cos x}{\sin x}. \tag{4}$$

Example Given that $\cos x = -3/5$, $\pi/2 < x < \pi$, find

a. $\tan 2x$ **b.** $\tan \frac{x}{2}$

Solution Since $\cos x = -3/5$ and x terminates in Quadrant II,

$$\sin x = \sqrt{1 - \cos^2 x}$$
$$= \sqrt{1 - \left(-\frac{3}{5}\right)^2}$$
$$= \sqrt{\frac{16}{25}} = \frac{4}{5}.$$

Hence,

$$\tan x = \frac{\sin x}{\cos x} = \frac{4/5}{-3/5} = -\frac{4}{3}.$$

a. From Equation (3),

$$\tan 2x = \frac{2 \tan x}{1 - \tan^2 x} = \frac{2(-4/3)}{1 - (-4/3)^2}$$
$$= \frac{-8/3}{1 - 16/9} = \frac{-8/3}{-7/9} = \frac{24}{7}.$$

b. From Equation (4),

$$\tan \frac{x}{2} = \frac{1 - \cos x}{\sin x} = \frac{1 - (-3/5)}{4/5}$$
$$= \frac{8/5}{4/5} = 2.$$

Identities The methods used in Section 7.1 can be applied to show that some equations involving $\tan 2x$ and/or $\tan(x/2)$ are identities.

Example Verify that $\tan \dfrac{x}{2} = \dfrac{\sin x}{1 + \cos x}$ is an identity.

Solution overleaf

Solution We first use Equation (4) to write $\tan(x/2)$ in terms of $\sin x$ and $\cos x$. We then multiply the right-hand member of Equation (4) by $(1 + \cos x)/(1 + \cos x)$. We then simplify to obtain

$$\tan \frac{x}{2} = \frac{(1 - \cos x)(1 + \cos x)}{\sin x(1 + \cos x)}$$

$$= \frac{1 - \cos^2 x}{\sin x(1 + \cos x)}$$

$$= \frac{\sin^2 x}{\sin x(1 + \cos x)}$$

$$= \frac{\sin x}{1 + \cos x}.$$

Thus $\tan \dfrac{x}{2} = \dfrac{\sin x}{1 + \cos x}$ is an identity.

Summary of Identities

1. $\sin^2 x + \cos^2 x = 1$ 2. $\cos(-x) = \cos x$ 3. $\sin(-x) = -\sin x$
4. $\cos(x_1 + x_2) = \cos x_1 \cos x_2 - \sin x_1 \sin x_2$
5. $\cos(x_1 - x_2) = \cos x_1 \cos x_2 + \sin x_1 \sin x_2$
6. $\sin(x_1 + x_2) = \sin x_1 \cos x_2 + \cos x_1 \sin x_2$
7. $\sin(x_1 - x_2) = \sin x_1 \cos x_2 - \cos x_1 \sin x_2$
8. **a.** $\cos 2x = \cos^2 x - \sin^2 x$ 9. $\sin 2x = 2 \sin x \cos x$
   b. $\cos 2x = 2 \cos^2 x - 1$
   c. $\cos 2x = 1 - 2 \sin^2 x$
10. $\cos \dfrac{x}{2} = \pm \sqrt{\dfrac{1 + \cos x}{2}}$ 11. $\sin \dfrac{x}{2} = \pm \sqrt{\dfrac{1 - \cos x}{2}}$
12. $\tan x = \dfrac{\sin x}{\cos x}$ 13. $\tan(-x) = -\tan x$
14. $\tan^2 x + 1 = \sec^2 x$ 15. $\cot^2 x + 1 = \csc^2 x$
16. $\sec x = \dfrac{1}{\cos x}$ 17. $\csc x = \dfrac{1}{\sin x}$
18. $\cot x = \dfrac{\cos x}{\sin x}$ 19. $\cot x = \dfrac{1}{\tan x}$
20. $\tan(x_1 + x_2) = \dfrac{\tan x_1 + \tan x_2}{1 - \tan x_1 \tan x_2}$ 21. $\tan(x_1 - x_2) = \dfrac{\tan x_1 - \tan x_2}{1 + \tan x_1 \tan x_2}$
22. $\tan 2x = \dfrac{2 \tan x}{1 - \tan^2 x}$ 23. $\tan \dfrac{x}{2} = \dfrac{1 - \cos x}{\sin x}$

7.4 Special Formulas for the Tangent Function

Relationships similar to those developed in this and the preceding sections can be derived for cotangent, secant, and cosecant. For practical purposes, however, the formulas for cosine, sine, and tangent are the only ones necessary in view of the fact that

$$\sec x = \frac{1}{\cos x}, \quad \csc x = \frac{1}{\sin x}, \quad \text{and} \quad \cot x = \frac{1}{\tan x}.$$

For convenience, a single list of some of the more important trigonometric identities is given on page 254. While no restrictions are given for the variables, it is necessary to keep such restrictions in mind when using any identity. Furthermore, the variable x can be viewed as representing either an element in the set of real numbers or an element in the set of all angles.

Exercise 7.4

Find the exact value of each expression.

Example $\tan 75°$

Solution   Write $75°$ as $75° = 30° + 45°$. Then

$$\tan 75° = \tan(30° + 45°) = \frac{\tan 30° + \tan 45°}{1 - \tan 30° \tan 45°}$$

$$= \frac{(1/\sqrt{3}) + 1}{1 - (1/\sqrt{3})(1)} = \frac{\sqrt{3} + 3}{3 - \sqrt{3}}.$$

1. $\tan 105°$
2. $\tan 255°$
3. $\tan 22.5°$
4. $\tan 112.5°$
5. $\tan 67.5°$
6. $\tan 195°$

Example   Given that $\cos x = 1/3$, $3\pi/2 < x < 2\pi$, find $\tan 2x$ and $\tan x/2$.

Solution   Since $\cos x = 1/3$, and x terminates in Quadrant IV,

$$\sin x = -\sqrt{1 - \cos^2 x} = -\sqrt{1 - (1/3)^2}$$

$$= -\sqrt{\frac{8}{9}} = \frac{-2\sqrt{2}}{3},$$

and

$$\tan x = \frac{\sin x}{\cos x} = -2\sqrt{2}.$$

Solution continued overleaf

By Equation (3) on page 252,

$$\tan 2x = \frac{2 \tan x}{1 - \tan^2 x} = \frac{2(-2\sqrt{2})}{1 - (-2\sqrt{2})^2}$$

$$= \frac{-4\sqrt{2}}{-7} = \frac{4\sqrt{2}}{7}.$$

By Equation (4) on page 253,

$$\tan \frac{x}{2} = \frac{1 - \cos x}{\sin x} = \frac{1 - (1/3)}{(-2\sqrt{2}/3)}$$

$$= -\frac{1}{\sqrt{2}}.$$

7. Given that $\sin x = 3/5$, $0 < x < \pi/2$, find

   a. $\tan 2x$ **b.** $\tan \dfrac{x}{2}$

8. Given that $\cos x = -4/5$, $\pi/2 < x < \pi$, find

   a. $\tan 2x$ **b.** $\tan \dfrac{x}{2}$

Prove each identity

9. $\tan(\pi - x) = -\tan x$ **10.** $\tan(\pi + x) = \tan x$

11. $\tan(2\pi - x) = -\tan x$ **12.** $\tan(2\pi + x) = \tan x$

13. $\cot \theta - \cot 2\theta = \dfrac{\sec^2 \theta}{2 \tan \theta}$ **14.** $\tan \dfrac{x}{2} = \dfrac{\sin x}{1 + \cos x}$

15. $\tan 3x = \dfrac{3 \tan x - \tan^3 x}{1 - 3 \tan^2 x}$ *Hint*: $\tan 3x = \tan(2x + x)$

16. $\tan 4x = \dfrac{4 \tan x - 4 \tan^3 x}{1 - 6 \tan^2 x + \tan^4 x}$ *Hint*: $\tan 4x = \tan(2x + 2x)$.

17. $2 \tan 4x = \dfrac{4 \tan 2x}{1 - \tan^2 2x}$ **18.** $\tan 2x = \dfrac{1 - \cos 4x}{\sin 4x}$

19. $\tan\left(\dfrac{3x}{2}\right) = \dfrac{1 + 3 \cos x - 4 \cos^3 x}{3 \sin x - 4 \sin^3 x}$

Hint: See Exercise 19 in Sections 7.2 and 7.3.

20. $\tan\left(\dfrac{3x}{2}\right) = \dfrac{\sin x + \sin 2x}{\cos x + \cos 2x}$ *Hint:* $\tan\left(\dfrac{3x}{2}\right) = \tan\left(x + \dfrac{x}{2}\right)$.

21. $\tan\left(\dfrac{\pi}{2} - x\right) = \cot x$

22. $\tan\left(\dfrac{\pi}{2} + x\right) = -\cot x$

23. $\tan\left(\dfrac{3\pi}{2} - x\right) = \cot x$

24. $\tan\left(\dfrac{3\pi}{2} + x\right) = -\cot x$

7.5 Conditional Equations

In Section 7.1 we observed that certain equations involving trigonometric function values are *identities*; that is, they are true for all permissible values of any variables involved. In this section we shall be concerned with *conditional equations* involving these function values, namely equations that are *not* satisfied by all permissible values of the variables involved.

Various procedures, including the methods applicable to algebraic equations, exist for solving conditional trigonometric equations. We shall use Tables 5.2, 5.3, and 6.1 when the solutions can be expressed in terms of the special values in those tables, and Table IV or V or a calculator otherwise. In the examples and in the exercises where Table IV or V or a calculator are required, the solutions are expressed to the nearest tenth of a unit.

Because the trigonometric functions are periodic, we should expect conditional equations involving trigonometric function values to have infinite solution sets.

Example   Solve $\sin x = -1$ for

a. x a real number,

b. x an angle with measure given in radians,

c. x an angle with measure given in degrees.

Solution   From the entries in Table 5.2 (page 157), Table 6.1 (page 207), and the fact that the sine is periodic, we observe that the solutions are

a. $\left\{x \mid x = \dfrac{3\pi}{2} + k \cdot 2\pi,\ k \in J\right\}$,

b. $\left\{x \mid x = \left(\dfrac{3\pi}{2} + k \cdot 2\pi\right)^R,\ k \in J\right\}$,

c. $\{x \mid x = (270 + k \cdot 360)°,\ k \in J\}$.

where J is the set of integers.

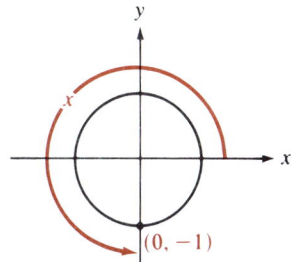

Next let us consider an example in which a reference angle proves helpful.

Example Solve $\sin \theta = \sqrt{3}/2$ for θ an angle measured in degrees.

Solution From the entries in Table 5.2 we observe that

$$\sin 60° = \frac{\sqrt{3}}{2}.$$

Hence, $\theta_1 = 60°$ is the positive acute angle that satisfies the equation. Using $60°$ as the measure of the reference angle and the fact that $\sin \theta$ is positive only in the first and second quadrants, we see that $\theta_2 = 120°$ is also a solution. Since the sine function has period $360°$, members of the solution set are members of either

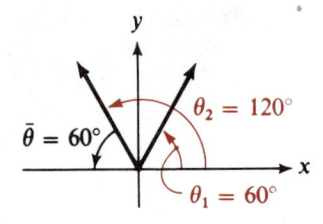

$$\{\theta | \theta = (60 + k \cdot 360)°, \quad k \in J\} \quad \text{or} \quad \{\theta | \theta = (120 + k \cdot 360)°, \quad k \in J\}.$$

Using the symbol $\cup$ to denote the union of sets, we can write the solution set as

$$\{\theta | \theta = (60 + k \cdot 360)°, \quad k \in J\} \cup \{\theta | \theta = (120 + k \cdot 360)°, \quad k \in J\}.$$

To solve some conditional equations it may be necessary to use identities to rewrite equations involving more than one function value as equivalent equations involving values of only one function.

Example Solve $\sin \alpha = \cos \alpha$ over the interval $0^R \leq \alpha \leq (\pi/2)^R$.

Solution Since $\sin \alpha \neq 0$ when $\cos \alpha = 0$, the equation is not satisfied if $\cos \alpha = 0$. We can accordingly assume that $\cos \alpha \neq 0$ and hence can multiply each member by $1/\cos \alpha$ to obtain

$$\frac{\sin \alpha}{\cos \alpha} = 1$$

or, equivalently,

$$\tan \alpha = 1.$$

Therefore, by Table 6.1 on page 207, the solution set in $0^R \leq \alpha \leq \pi^R/2$ is $\{\pi^R/4\}$.

Example Solve $\tan^2 \alpha + \sec^2 \alpha = 3$ over the interval $0° \leq \alpha \leq 90°$.

7.5 Conditional Equations

Solution Because $\tan^2 \alpha = \sec^2 \alpha - 1$, we can rewrite the given equations as

$$2 \sec^2 \alpha - 1 = 3,$$

or, equivalently,

$$\sec^2 \alpha = 2.$$

Thus $\sec \alpha = \sqrt{2}$ or $\sec \alpha = -\sqrt{2}$. Since $\sec \alpha > 0$ in the first quadrant, the equation $\sec \alpha = -\sqrt{2}$ has no solutions over $0° \leq \alpha \leq 90°$. Therefore by Table 6.1 the solution set in $0° \leq \alpha \leq 90°$ is $\{45°\}$.

Recall that a product ab equals 0, if and only if $a = 0$ or $b = 0$ (or both). This property is sometimes useful in solving trigonometric equations in which one member is 0 and the other member is in factored form.

Example Solve $(2 \cos \theta - 1)(\sin \theta - 1) = 0$ over $0^R \leq \theta < 2\pi^R$.

Solution Setting each factor of the left-hand member equal to zero and solving, we have

$$2 \cos \theta - 1 = 0 \qquad \text{or} \qquad \sin \theta - 1 = 0$$

$$\cos \theta = \frac{1}{2}; \qquad\qquad \sin \theta = 1.$$

Hence over the interval $0^R \leq \theta < 2\pi^R$, the solution set is

$$\left\{\frac{\pi^R}{3}, \frac{5\pi^R}{3}\right\} \cup \left\{\frac{\pi^R}{2}\right\} = \left\{\frac{\pi^R}{3}, \frac{5\pi^R}{3}, \frac{\pi^R}{2}\right\}.$$

Note that, in the foregoing example, if the equation had been presented in the form

$$2 \sin \theta \cos \theta - \sin \theta - 2 \cos \theta + 1 = 0,$$

it would first have been necessary to factor the equation into the given form before proceeding with the solution.

In case we have an equation that is quadratic in form, we can either factor or use the method of substitution (see Section 1.3) and the quadratic formula.

Example Solve $2 \sin^2 \theta - 3 \sin \theta + 1 = 0$ over the interval $0° \leq \theta \leq 180°$.

Solution Factoring the left-hand member, we obtain the equivalent equation

$$(2 \sin \theta - 1)(\sin \theta - 1) = 0.$$

Setting each factor of the left-hand member equal to zero and solving as in the preceding example, we obtain

$$\sin \theta = \frac{1}{2}; \qquad \text{or} \qquad \sin \theta = 1.$$

Solution continued overleaf

Thus over the interval $0° \leq \theta \leq 180°$, the solution set is

$$\{30°, 150°\} \cup \{90°\} = \{30°, 90°, 150°\}.$$

Alternatively, we can set $p = \sin \theta$. Then substituting in the original equation, we have

$$2p^2 - 3p + 1 = 0.$$

Solving for p yields

$$p = \frac{1}{2} \quad \text{or} \quad p = 1;$$

hence,

$$\sin \theta = \frac{1}{2} \quad \text{or} \quad \sin \theta = 1.$$

These yield the same solutions as those obtained above.

Exercise 7.5

In Exercises 1–12, find the solution set of the given equation in (a) degree measure and (b) radian measure.

1. $\cos \theta = \frac{1}{2}$
2. $\sin \theta = \frac{\sqrt{2}}{2}$
3. $\tan \theta = \sqrt{3}$
4. $\cot \theta = -1$
5. $\sec \theta - \sqrt{2} = 0$
6. $\csc \theta + 1 = 0$
7. $4 \sin \theta - 1 = 0$
8. $2 \tan \theta + 3 = 0$
9. $3 \cot \theta - 1 = 0$
10. $3 \cos \theta + 1 = 0$
11. $2 \sec \theta - 5 = 0$
12. $3 \csc \theta + 8 = 0$

In Exercises 13–18, find the solution set of the given equation, where x denotes a real number.

Example $\sqrt{3} \sin x + \cos x = 0$

Solution First note that $\cos x \neq 0$ in any solution. Divide each member by $\cos x$ to produce

$$\sqrt{3} \frac{\sin x}{\cos x} + 1 = 0,$$

from which

$$\sqrt{3} \tan x + 1 = 0, \quad \text{or} \quad \tan x = -\frac{1}{\sqrt{3}},$$

an equation that involves only one function value. By inspection, the solution set is

$$\left\{ x \mid x = \frac{5\pi}{6} + k\pi, \quad k \in J \right\}.$$

7.5 Conditional Equations

13. $\sin x - \sqrt{3} \cos x = 0$
14. $3 \sin x + \cos x = 0$
15. $\tan^2 x - 1 = 0$
16. $2 \sin^2 x - 1 = 0$
17. $\sin^2 x - \cos^2 x = 1$
18. $\sin x + \cos x \tan x = 3$

In Exercises 19–22, find the solutions of the given equation over the interval $0° \leq \theta < 360°$.

Example

$2 \cos^2 \theta \tan \theta - \tan \theta = 0$

Solution

Factor $\tan \theta$ from the terms in the left-hand member.

$$\tan \theta \, (2 \cos^2 \theta - 1) = 0$$

From this, either

$$\tan \theta = 0 \quad \text{or} \quad 2 \cos^2 \theta - 1 = 0$$

$$\cos \theta = \pm \frac{1}{\sqrt{2}}.$$

Therefore the solution set over $0° \leq \theta < 360°$ is

$$\{0°, 180°\} \cup \{45°, 135°, 225°, 315°\} = \{0°, 45°, 135°, 180°, 225°, 315°\}.$$

19. $(2 \sin \theta - 1)(2 \sin^2 \theta - 1) = 0$
20. $(\tan \theta - 1)(2 \cos \theta + 1) = 0$
21. $2 \sin \theta \cos \theta + \sin \theta = 0$
22. $\tan \theta \sin \theta - \tan \theta = 0$

In Exercises 23–28, find the solutions of the given equation over the interval $0^R \leq \alpha < 2\pi^R$.

Example

$\tan^2 \alpha + \sec \alpha - 1 = 0$

Solution

Since $\tan^2 \alpha = \sec^2 \alpha - 1$, the given equation can be written as

$$\sec^2 \alpha - 1 + \sec \alpha - 1 = 0,$$
$$\sec^2 \alpha + \sec \alpha - 2 = 0,$$
$$(\sec \alpha + 2)(\sec \alpha - 1) = 0,$$

from which either

$$\sec \alpha + 2 = 0 \quad \text{or} \quad \sec \alpha - 1 = 0$$
$$\sec \alpha = -2 \qquad\qquad \sec \alpha = 1.$$

Therefore the solution set over $0^R \leq \alpha < 2\pi^R$ is

$$\left\{\frac{2\pi^R}{3}, \frac{4\pi^R}{3}\right\} \cup \{0^R\} = \left\{0^R, \frac{2\pi^R}{3}, \frac{4\pi^R}{3}\right\}.$$

23. $\tan^2 \alpha - 2 \tan \alpha + 1 = 0$
24. $4 \sin^2 \alpha - 4 \sin \alpha + 1 = 0$
25. $\cos^2 \alpha + \cos \alpha = 2$
26. $\cot^2 \alpha = 5 \cot \alpha - 4$
27. $\sec^2 \alpha + 3 \tan \alpha - 11 = 0$
28. $\tan^2 \alpha + 4 = 2 \sec^2 \alpha$
29. Solve $\sin^2 \alpha + \sin \alpha - 1 = 0$. *Hint*: Use the quadratic formula.
30. Solve $\tan^2 \alpha = \tan \alpha + 3$. *Hint*: Use the quadratic formula.
31. Approximate a solution of $(x/2) - \sin x = 0$, where x denotes a real number, $x \neq 0$, by graphical methods. *Hint*: Graph $y_1 = x/2$ and $y_2 = \sin x$ on the same coordinate system, and determine a value of x for which $y_1 = y_2$.
32. Approximate a positive solution of $\cos x = x^2$, where x denotes a real number, by graphical methods.

7.6 *Conditional Equations for Multiples*

Equations that contain trigonometric function values such as $\sin 2\alpha$ and $\cos 3\alpha$ need further consideration.

Example Solve $\sqrt{2} \cos 3\alpha = 1$ over the interval $0° \leq \alpha \leq 360°$.

Solution The equation can be written equivalently as $\cos 3\alpha = 1/\sqrt{2}$. Therefore, we have
$$3\alpha = 45° + k \cdot 360° \quad \text{or} \quad 3\alpha = 315° + k \cdot 360°, \quad k \in J,$$
from which, by dividing each member by 3, we obtain
$$\alpha = 15° + k \cdot 120° \quad \text{or} \quad \alpha = 105° + k \cdot 120°, \quad k \in J.$$
Notice that 0, 1, and 2 are the only replacements for k that will yield values of α in the interval $0° \leq \alpha \leq 360°$. Therefore, the solution set over this interval is
$$\{15°, 135°, 255°, 105°, 225°, 345°\}.$$

A judicious selection of one or more of the identities encountered earlier can frequently help you solve certain kinds of equations. The following examples illustrate two such selections.

Example Solve $\sin \alpha \cos \alpha = 1/4$ over the interval $0^R \leq \alpha \leq 2\pi^R$.

Solution Since $2 \sin \alpha \cos \alpha = \sin 2\alpha$ for every value α, we multiply each member of the given equation by 2 to obtain the equivalent equation
$$2 \sin \alpha \cos \alpha = \frac{1}{2}$$

7.6 Conditional Equations for Multiples

or, in terms of a single function value,

$$\sin 2\alpha = \frac{1}{2}.$$

Therefore, we have

$$2\alpha = \left(\frac{\pi}{6} + k \cdot 2\pi\right)^R \quad \text{or} \quad 2\alpha = \left(\frac{5\pi}{6} + k \cdot 2\pi\right)^R, \quad k \in J,$$

so that

$$\alpha = \left(\frac{\pi}{12} + k\pi\right)^R \quad \text{or} \quad \alpha = \left(\frac{5\pi}{12} + k\pi\right)^R, \quad k \in J.$$

Because we want the solutions over the interval $0^R \leq \alpha < 2\pi^R$, we consider 0 and 1 as replacements for k and obtain the solution set

$$\left\{\frac{\pi^R}{12}, \frac{5\pi^R}{12}, \frac{13\pi^R}{12}, \frac{17\pi^R}{12}\right\}.$$

Example Solve $\cos 2x = \sin x$ for $x \in R$.

Solution Since $\cos 2x = 1 - 2 \sin^2 x$, the equation given can be written equivalently as

$$1 - 2 \sin^2 x = \sin x,$$
$$2 \sin^2 x + \sin x - 1 = 0,$$
$$(2 \sin x - 1)(\sin x + 1) = 0,$$

from which

$$\sin x = \frac{1}{2} \quad \text{or} \quad \sin x = -1.$$

Then as solution set we have

$$\left\{x \mid x = \frac{\pi}{6} + k \cdot 2\pi\right\} \cup \left\{x \mid x = \frac{5\pi}{6} + k \cdot 2\pi\right\} \cup \left\{x \mid x = \frac{3\pi}{2} + k \cdot 2\pi\right\}, \quad k \in J.$$

Exercise 7.6

In Exercises 1–12, solve the given equation for θ in the interval $0° \leq \theta < 360°$.

1. $\cos 2\theta = \dfrac{\sqrt{2}}{2}$
2. $\tan 2\theta = \sqrt{3}$
3. $\sin \dfrac{1}{2}\theta = \dfrac{1}{2}$
4. $\cot \dfrac{1}{3}\theta = -1$
5. $\tan 3\theta = 0$
6. $\sin 4\theta = 1$

7. $\sin\theta\cos\theta = \dfrac{1}{2}$
8. $\cos^2\theta - \sin^2\theta = -1$
9. $\cos 2\theta + \sin 2\theta = 0$
10. $\sin\theta\cos\theta = \dfrac{\cos 2\theta}{2}$
11. $2\cos^2 2\theta + \cos 2\theta - 1 = 0$
12. $\tan^2 2\theta + 2\tan 2\theta + 1 = 0$

In Exercises 13–20, solve the given equation for real numbers x.

13. $\sin 2x - \cos x = 0$
14. $\cos 2x = \cos^2 x - 1$
15. $\cos 2x = \cos x - 1$
16. $\sin x = \sin 2x$
17. $\cos 2x \sin x + \sin x = 0$
18. $\sin 2x \cos x - \sin x = 0$
19. $\sin 4x - 2\sin 2x = 0$ *Hint:* $\sin 4x = \sin 2(2x)$.
20. $\sin 3x + 4\sin^2 x = 0$ *Hint:* $\sin 3x = \sin(x + 2x)$.

Chapter Review

[7.1] *Verify that the given equation is an identity.*

1. $\tan x = \sin x \cdot \sec x$
2. $\cos x - \sin x = (1 - \tan x)\cos x$
3. $\dfrac{1 - \tan^2\alpha}{\tan\alpha} = \cot\alpha - \tan\alpha$
4. $\dfrac{\cos^2\alpha}{1 - \sin\alpha} = \dfrac{\cos\alpha}{\sec\alpha - \tan\alpha}$
5. $\dfrac{1}{\cot^2\theta\csc^2\theta} = \dfrac{1}{\cot^2\theta} - \dfrac{1}{\csc^2\theta}$
6. $\dfrac{4}{\cos^2\theta} = \dfrac{2}{1 - \sin\theta} + \dfrac{2}{1 + \sin\theta}$
7. $\dfrac{1 + \sin x}{\cos x} = \dfrac{1}{\sec x - \tan x}$
8. $\dfrac{1}{\cos^2 x} + \dfrac{1}{\sin^2 x} = \dfrac{1}{\cos^2 x \sin^2 x}$

[7.2] *Given that $\cos\theta = -2/3$ and $90° \leq \theta \leq 180°$, find the value of the given expression.*

9. $\cos(30° + \theta)$
10. $\cos(60° + \theta)$
11. $\cos(30° - \theta)$
12. $\cos(60° - \theta)$
13. $\cos 2\theta$
14. $\cos(\theta/2)$

[7.3] *Given that $\sin\theta = 2/3$ and $90° \leq \theta \leq 180°$, find the value of the given expression.*

15. $\sin(30° + \theta)$
16. $\sin(60° + \theta)$
17. $\sin(30° - \theta)$
18. $\sin(60° - \theta)$
19. $\sin 2\theta$
20. $\sin(\theta/2)$

[7.4] *Given that* $\sin \theta = 2/3$ *and* $90° \leq \theta \leq 180°$, *find the value of the given expression.*

21. $\tan(30° + \theta)$
22. $\tan(60° + \theta)$
23. $\tan(30° - \theta)$
24. $\tan(60° - \theta)$
25. $\tan 2\theta$
26. $\tan(\theta/2)$

[7.5] *Solve the given equation over the specified replacement set.*

27. $\cos^2 \theta - 1 = 0; \quad 0° \leq \theta \leq 360°$
28. $2 \cos \theta \sin \theta - \cos \theta = 0; \quad 0° \leq \theta \leq 360°$
29. $\sin^2 \alpha - 3 \sin \alpha + 2 = 0; \quad 0^R \leq \alpha \leq 2\pi^R$
30. $\sqrt{3} \tan x - 1 = 0,$ where x is a real number.

[7.6] *Solve the given equation over the interval* $0° \leq \theta \leq 360°$.

31. $\sin 3\theta = \dfrac{\sqrt{3}}{2}$
32. $\cos^2 2\theta + 2 \cos 2\theta + 1 = 0$

Solve the given equation for x a real number.

33. $\sin 2x - \cos 2x = 0$
34. $\tan^2 2x - 3 = 0$

8 Complex Numbers

In Chapter 2 we considered real-number solutions of polynomial equations. Some polynomial equations, such as $x^2 + 1 = 0$, have no solutions in the set of real numbers. In this chapter we shall define a set of numbers, called the *complex numbers*, that contains a subset that can be identified with the set of real numbers, and that contains solutions to all polynomial equations with real coefficients.

8.1 Definitions; Basic Operations

The imaginary unit

Observe that for $k > 0$ the equation $x^2 + k = 0$ can be satisfied only by a number whose square is $-k$. But there are no real numbers whose squares are negative. We shall now define a new set of numbers, based on the real numbers together with an imaginary unit i satisfying $i^2 = -1$, that contains solutions of such equations.

Complex numbers

The complex numbers are defined in terms of the real numbers and the imaginary unit i, as follows.

Definition 8.1   The set C of complex numbers is the set
$$C = \{a + bi \,|\, a, b \in R\}.$$

Equality

Definition 8.2   The complex numbers $z_1 = a + bi$ and $z_2 = c + di$ are equal; $z_1 = z_2$, if and only if
$$a = c \quad \text{and} \quad b = d.$$

8.1 Definitions; Basic Operations

Examples **a.** $\sqrt{4} + (-7)i = \frac{6}{3} + (1-8)i$ **b.** $3 + 2i \neq 2 + 3i$

Sums and products

We wish to define addition and multiplication of complex numbers in such a way that all the field axioms (see "Preliminary Concepts") hold. The following definitions accomplish that purpose.

Definition 8.3   The sum of $z_1 = a + bi$ and $z_2 = c + di$ is

$$z_1 + z_2 = (a + c) + (b + d)i.$$

Examples   Express each sum in the form $a + bi$.

a. $(2 + 3i) + (3 + 0i)$ **b.** $(3 + 2i) + (-2 + 4i)$

Solutions
a. $(2 + 3i) + (3 + 0i)$
$= (2 + 3) + (3 + 0)i$
$= 5 + 3i$

b. $(3 + 2i) + (-2 + 4i)$
$= (3 - 2) + (2 + 4)i$
$= 1 + 6i$

Definition 8.4   The product of $z_1 = a + bi$ and $z_2 = c + di$ is

$$z_1 z_2 = (ac - bd) + (ad + bc)i.$$

Example   Express $(2 + 3i)(6 + 1i)$ in the form $a + bi$.

Solution $(2 + 3i)(6 + 1i) = ((2)(6) - (3)(1)) + ((2)(1) + (3)(6))i$
$= (12 - 3) + (2 + 18)i = 9 + 20i$

For convenience, we write

a for $a + 0i$, bi for $0 + bi$, $a - bi$ for $a + (-b)i$, i for $0 + 1i$.

With a identified with $a + 0i$, every real number is contained in C. Moreover, addition and multiplication of real numbers in C are no different than they are in R. For instance,

$$3 + 5 = (3 + 0i) + (5 + 0i)$$
$$= (3 + 5) + (0 + 0)i$$
$$= 8 + 0i = 8,$$

and

$$3 \cdot 5 = (3 + 0i) \cdot (5 + 0i)$$
$$= (3 \cdot 5 - 0 \cdot 0) + (3 \cdot 0 + 0 \cdot 5)i$$
$$= 15 + 0i = 15.$$

Special role of i

With $0 + 1i$ written as i, the special role of this specific complex number becomes apparent. We have

$$i^2 = (0 + 1i) \cdot (0 + 1i)$$
$$= (0 \cdot 0 - 1 \cdot 1) + (0 \cdot 1 + 1 \cdot 0)i$$
$$= -1 + 0i,$$

from which,

$$i^2 = -1.$$

Making use of the fact that $i^2 = -1$, we can multiply complex numbers just as we multiply real polynomial expressions.

Examples

a. $(2 + 3i)(6 + i)$.

b. $(4 - 2i)(3i)$.

Solutions

a. $(2 + 3i)(6 + i) = 12 + 18i + 2i + 3i^2$
$= 12 + 20i - 3$
$= 9 + 20i$

b. $(4 - 2i)(3i) = 12i - 6i^2$
$= 12i - 6(-1)$
$= 6 + 12i$

The complex numbers $a + bi$ with $b \neq 0$ are called **imaginary numbers**; the imaginary numbers bi with $a = 0$ are called **pure imaginary numbers.** Thus, our number systems are related as shown in Figure 8.1.

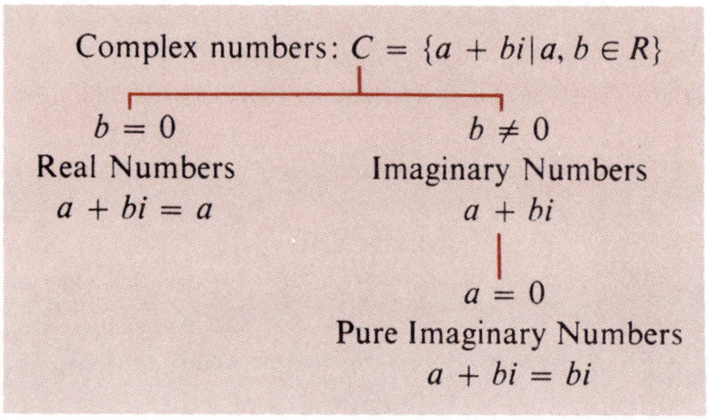

Complex numbers: $C = \{a + bi \,|\, a, b \in R\}$

$b = 0$
Real Numbers
$a + bi = a$

$b \neq 0$
Imaginary Numbers
$a + bi$

$a = 0$
Pure Imaginary Numbers
$a + bi = bi$

Figure 8.1

Additive identity element, 0

Note that, for any complex number $z = a + bi$,

$$z + 0 = (a + bi) + (0 + 0i)$$
$$= (a + 0) + (b + 0)i$$
$$= a + bi = z.$$

Similarly, $0 + z = z$. We therefore have the following result.

8.1 Definitions; Basic Operations

Theorem 8.1   For every $z \in C$,
$$z + 0 = 0 + z = z.$$

Thus the number $0 = 0 + 0i$ is the **additive identity element** in C.

Additive inverse element, $-z$

Note that, for any complex number $z = a + bi$,
$$(a + bi) + (-a - bi) = (a - a) + (b - b)i$$
$$= 0 + 0i = 0.$$

Similarly $(-a - bi) + (a + bi) = 0$. If we denote $-a - bi$ by $-z$, then we have the following theorem.

Theorem 8.2   For every $z \in C$,
$$z + (-z) = (-z) + z = 0.$$

Thus $-z$ is the **additive inverse** of z in C.

Examples

a. $-(3 + 0i) = -3 - 0i$
$ = -3$

b. $-(1 - 2i) = -1 - (-2)i$
$ = -1 + 2i$

Differences

We now use additive inverses to define the difference of two complex numbers.

Definition 8.5   For $z_1, z_2 \in C$, the difference $z_1 - z_2$ is
$$z_1 - z_2 = z_1 + (-z_2).$$

Examples

a. $(2 + i) - (3 - i)$
$= (2 + i) + (-3 + i)$
$= -1 + 2i$

b. $(3 - 2i) - i$
$= (3 - 2i) + (0 - i)$
$= 3 - 3i$

Multiplicative identity element, 1

Now, for any complex number $z = a + bi$,
$$z \cdot 1 = (a + bi)(1 + 0i)$$
$$= (a)(1) + (a)(0)i + (b)(1)i + (b)(0)i^2$$
$$= a + bi = z.$$

Similarly, $1 \cdot z = z$. We therefore have the following theorem.

Theorem 8.3   For every $z \in C$,
$$z \cdot 1 = 1 \cdot z = z.$$

Thus the number $1 = 1 + 0i$ is the **multiplicative identity** element in C.

Conjugate of $a + bi$

In order to investigate whether or not there is a multiplicative inverse of $z \in C$, $z \neq 0$, it is useful first to define the conjugate of a complex number.

Definition 8.6   For any complex number $z = a + bi$, the conjugate of z is

$$\bar{z} = a - bi.$$

Examples

a. $\overline{2 + 3i} = 2 - 3i$

b. $\overline{2 - i} = 2 - (-i)$
$= 2 + i$

c. $\overline{-6} = \overline{-6 + 0i}$
$= -6 - 0i = -6$

d. $\overline{2i} = \overline{0 + 2i}$
$= 0 - 2i = -2i$

Multiplicative inverse of $a + bi$

Note that, for any complex number $z = a + bi$, we have

$$z\bar{z} = (a + bi)(a - bi)$$
$$= a^2 + b^2.$$

Thus, $z\bar{z} \in R$, and therefore, for $z \neq 0$,

$$\frac{1}{z\bar{z}} = \frac{1}{a^2 + b^2},$$

which is a real number.

Now, for any nonzero complex number $a + bi$, we have

$$\left(\frac{1}{z\bar{z}} \cdot \bar{z}\right)z = \left(\frac{a}{a^2 + b^2} - \frac{b}{a^2 + b^2}i\right)(a + bi)$$

$$= \frac{1}{a^2 + b^2}(a^2 + b^2) = 1$$

Similarly, $z\left(\frac{1}{z\bar{z}} \cdot \bar{z}\right) = 1$. We denote $\frac{1}{z\bar{z}} \cdot \bar{z}$ by $\frac{1}{z}$.

The argument above establishes the following result.

Theorem 8.4   For every $z \in C$, $z \neq 0$,

$$\left(\frac{1}{z}\right)z = z\left(\frac{1}{z}\right) = 1.$$

Thus, for $z \neq 0$, $z \in C$, $\frac{1}{z} = \frac{1}{z\bar{z}} \cdot \bar{z}$ is the **multiplicative inverse** of z.

8.1 Definitions; Basic Operations

Examples

Write $1/z$ in the form $a + bi$ (or a or bi).

a. $z = i$
b. $z = 1 + 2i$

Solutions

a. $\dfrac{1}{0 + i} = \dfrac{1}{(0 + i)(0 - i)} \cdot (0 - i)$

$= \dfrac{-i}{1}$

$= -i$

Note that $i(-i) = 1$.

b. $\dfrac{1}{1 + 2i} = \dfrac{1}{(1 + 2i)(1 - 2i)} \cdot (1 - 2i)$

$= \dfrac{1}{5}(1 - 2i)$

$= \dfrac{1}{5} - \dfrac{2}{5}i$

Note that $(1 + 2i)\left(\dfrac{1}{5} - \dfrac{2}{5}i\right) = 1$.

Quotients

We use multiplicative inverses of complex numbers to define the quotient of two complex numbers.

Definition 8.7 For $z_1, z_2 \in C$, $z_2 \neq 0$, the quotient z_1/z_2 is

$$\dfrac{z_1}{z_2} = z_1\left(\dfrac{1}{z_2}\right).$$

Note that

$$\dfrac{1}{z_2} = \dfrac{1}{z_2 \overline{z_2}} \cdot \overline{z_2},$$

and thus

$$\dfrac{z_1}{z_2} = \dfrac{z_1 \overline{z_2}}{z_2 \overline{z_2}}. \qquad (1)$$

Equation (1) can be used to write quotients of complex numbers in the form $a + bi$.

Examples

Express the indicated quotients in the form $a + bi$.

a. $\dfrac{1 + 2i}{2i}$
b. $\dfrac{2 + 3i}{1 - 2i}$

Solutions

a. $\dfrac{1 + 2i}{2i}$

$= \dfrac{(1 + 2i)(0 - 2i)}{(0 + 2i)(0 - 2i)}$

$= \dfrac{4 - 2i}{4}$

$= 1 - \dfrac{1}{2}i$

b. $\dfrac{2 + 3i}{1 - 2i}$

$= \dfrac{(2 + 3i)(1 + 2i)}{(1 - 2i)(1 + 2i)}$

$= \dfrac{-4 + 7i}{5}$

$= -\dfrac{4}{5} + \dfrac{7}{5}i$

Field axioms

Theorems 8.1–8.4 show that four of the field axioms (see "Preliminary Concepts," page 6) are valid for the complex numbers. In fact, all the field axioms can be shown to be valid for these numbers. The proofs are omitted.

The complex numbers cannot be ordered, however, in such a way as to satisfy the order properties on page 5.

Radical notation

Recall that if $b > 0$, then b has two square roots: one positive, denoted by $\sqrt{b}$, and the other negative, denoted by $-\sqrt{b}$.

Similarly, since

$$(\sqrt{b}\,i)^2 = (\sqrt{b})^2 i^2 = -b \quad \text{and} \quad (-\sqrt{b}\,i)^2 = (-\sqrt{b})^2 i^2 = -b,$$

$-b$ also has two square roots, $\sqrt{b}\,i$ and $-\sqrt{b}\,i$.

Just as we use two symbols to distinguish between the two square roots of the positive number b, namely

$$\sqrt{b} \quad \text{for the positive square root,}$$

and

$$-\sqrt{b} \quad \text{for the negative square root,}$$

so do we use two symbols to distinguish between the two square roots of $-b$, namely

$$\sqrt{-b} \quad \text{for the square root } \sqrt{b}\,i,$$

and

$$-\sqrt{-b} \quad \text{for the square root } -\sqrt{b}\,i:$$

$$\sqrt{-b} = \sqrt{b}\,i, \quad -\sqrt{-b} = -\sqrt{b}\,i.$$

In particular, we have

$$\sqrt{-1} = \sqrt{1}\,i = 1i = i \quad \text{and} \quad -\sqrt{-1} = -\sqrt{1}\,i = -1i = -i.$$

Since the product and quotient of two negative numbers are both positive, care must be exercised in choosing the correct symbol for the product or quotient of square roots of negative numbers. Thus if $b_1 > 0$ and $b_2 > 0$, then

$$\frac{\sqrt{-b_1}}{\sqrt{-b_2}} = \frac{\sqrt{b_1}\,i}{\sqrt{b_2}\,i} = \frac{\sqrt{b_1}}{\sqrt{b_2}} = \sqrt{\frac{b_1}{b_2}}$$

but

$$\sqrt{-b_1}\sqrt{-b_2} = (\sqrt{b_1}\,i)(\sqrt{b_2}\,i) = -\sqrt{b_1}\sqrt{b_2} = -\sqrt{b_1 b_2} \neq \sqrt{b_1 b_2}.$$

Examples

a. $\dfrac{\sqrt{-8}}{\sqrt{-2}} = \dfrac{\sqrt{8}\,i}{\sqrt{2}\,i}$

$= \dfrac{\sqrt{8}}{\sqrt{2}} = \sqrt{\dfrac{8}{2}}$

$= \sqrt{4} = 2$

b. $\sqrt{-8}\sqrt{-2} = (\sqrt{8}\,i)(\sqrt{2}\,i)$

$= -\sqrt{8}\sqrt{2} = -\sqrt{16}$

$= -4 \neq \sqrt{16}$

8.1 Definitions; Basic Operations

To help ensure correct calculations, it is best to write $\sqrt{-b}$ as $\sqrt{b}\,i$ before proceeding, as illustrated in the examples just given.

Exercise 8.1

A *Write each expression in the form $a + bi$.*

Examples
    a. $(2 + i) + (3 - 2i)$ b. $3i + (4 + i)$

Solutions
    a. $(2 + i) + (3 - 2i)$ b. $3i + (4 + i)$
         $= (2 + 3) + (1 - 2)i$ $= (0 + 4) + (3 + 1)i$
         $= 5 - i$ $= 4 + 4i$

1. $(5 + i) + (-4 - 2i)$
2. $(1 + i) + (10 - 2i)$
3. $(3 - i) + 6$
4. $13 + (3 - 10i)$
5. $(4 - 6i) + 3i$
6. $i + (1 + 6i)$
7. $i + (6 - i)$
8. $-2i + (10 + 2i)$

Examples
    a. $(1 + 3i)(2 - i)$ b. $(2 - i)(2 + i)$

Solutions
    a. $(1 + 3i)(2 - i)$ b. $(2 - i)(2 + i)$
         $= 2 + 5i - 3i^2$ $= 4 + 0i - i^2$
         $= 2 + 5i + 3 = 5 + 5i$ $= 4 + 0i - (-1) = 5$

9. $(2 + i)(3 - i)$
10. $(1 - i)(5 + i)$
11. $2(3 + i)$
12. $-4(8 - i)$
13. $i(1 + i)$
14. $-2i(1 - 2i)$
15. $(4 - i)(1 + 2i)$
16. $(3 + 4i)(1 + i)$
17. $(1 + i)^2$
18. $(2 - i)^2$
19. $(1 - 2i)(1 + 2i)$
20. $(2 + 3i)(2 - 3i)$

Examples
    a. $(2 + i) - (1 + i)$ b. $(3 - 2i) - (2 - i)$

Solutions
    a. $(2 + i) - (1 + i)$ b. $(3 - 2i) - (2 - i)$
         $= (2 + i) + (-1 - i)$ $= (3 - 2i) + (-2 + i)$
         $= (2 - 1) + (1 - 1)i$ $= (3 - 2) + (-2 + 1)i$
         $= 1 + 0i = 1$ $= 1 - i$

21. $(5 - i) - (5 + i)$
22. $(6 + i) - (1 + 6i)$
23. $(1 + 2i) - (1 - i)$
24. $(1 - i) - (2 + i)$
25. $(3 + 4i) - (5 - 2i)$
26. $(3 - i) - (7 - 2i)$

Examples

a. $\dfrac{1}{2 + i}$

b. $\dfrac{1}{2i}$

Solutions

a. $\dfrac{1}{2 + i} = \dfrac{2 - i}{(2 + i)(2 - i)}$
$= \dfrac{2 - i}{5} = \dfrac{2}{5} - \dfrac{1}{5}i$

b. $\dfrac{1}{2i} = \dfrac{1 \cdot (-2i)}{(2i)(-2i)}$
$= \dfrac{-2i}{4} = -\dfrac{1}{2}i$

27. $\dfrac{1}{1 + 2i}$
28. $\dfrac{1}{2 + 3i}$
29. $\dfrac{1}{1 - i}$
30. $\dfrac{1}{3 - 2i}$
31. $\dfrac{1}{4i}$
32. $\dfrac{1}{-6i}$

Examples

a. $\dfrac{1 + 4i}{i}$

b. $\dfrac{1 + 3i}{2 - i}$

Solutions

a. $\dfrac{1 + 4i}{i} = \dfrac{(1 + 4i)(-i)}{i \cdot (-i)}$
$= \dfrac{-i - 4i^2}{-i^2}$
$= 4 - i$

b. $\dfrac{1 + 3i}{2 - i} = \dfrac{(1 + 3i)(2 + i)}{(2 - i)(2 + i)}$
$= \dfrac{2 + 7i + 3i^2}{4 - i^2}$
$= \dfrac{-1 + 7i}{5} = -\dfrac{1}{5} + \dfrac{7}{5}i$

33. $\dfrac{2 + i}{3 - i}$
34. $\dfrac{1 - i}{5 + i}$
35. $\dfrac{4 - i}{1 + 2i}$
36. $\dfrac{3 + 4i}{1 + i}$
37. $\dfrac{i}{3 + i}$
38. $\dfrac{3i}{2 + i}$
39. $\dfrac{3}{1 - i}$
40. $\dfrac{-2}{2 + i}$
41. $\dfrac{2 - i}{2i}$
42. $\dfrac{3 - 2i}{-4i}$
43. $\dfrac{1 - i}{i}$
44. $\dfrac{2 + 3i}{-i}$

Write each expression in the form $a + bi$.

Examples

a. $1 + \sqrt{-2} + \sqrt{-4}$

b. $(1 + \sqrt{-2})(2 + \sqrt{-3})$

8.1 Definitions; Basic Operations 275

Solutions

a. $1 + \sqrt{-2} + \sqrt{-4}$
$= 1 + \sqrt{2}i + \sqrt{4}i$
$= 1 + (\sqrt{2} + 2)i$

b. $(1 + \sqrt{-2})(2 + \sqrt{-3})$
$= (1 + \sqrt{2}i)(2 + \sqrt{3}i)$
$= 2 + (2\sqrt{2} + \sqrt{3})i + \sqrt{6}i^2$
$= (2 - \sqrt{6}) + (2\sqrt{2} + \sqrt{3})i$

45. $(1 + \sqrt{-9}) + (3 - \sqrt{-2})$
46. $(2 - \sqrt{-4}) + (1 + \sqrt{-3})$
47. $(1 + \sqrt{-9})(3 - \sqrt{-2})$
48. $(2 - \sqrt{-4})(1 + \sqrt{-3})$
49. $\dfrac{1 + \sqrt{-9}}{3 - \sqrt{-2}}$
50. $\dfrac{2 - \sqrt{-4}}{1 + \sqrt{-3}}$
51. $\sqrt{-16} + \sqrt{-12}$
52. $\sqrt{-18} + \sqrt{-25}$

Write each expression in the form $a + bi$, a, or bi.

Examples

a. i^3

b. $\dfrac{i^2}{i^4}$

Solutions

a. $i^3 = i^2 \cdot i = -1 \cdot i = -i$

b. $\dfrac{i^2}{i^4} = \dfrac{1}{i^2} = \dfrac{1}{-1} = -1$

53. i^4
54. i^5
55. i^7
56. i^6
57. $\dfrac{i}{i^5}$
58. $\dfrac{i}{i^4}$
59. $\dfrac{i}{i^6}$
60. $\dfrac{i}{i^7}$
61. $(1 + i)^3$
62. $(2 - i)^3$
63. $\dfrac{(1 + i)^2}{1 - i}$
64. $\dfrac{(-1 - 2i)^2}{2 - i}$
65. $\dfrac{1 - 2i}{(2 + i)^3}$
66. $\dfrac{3 - i}{(2 - 3i)^3}$

Use the quadratic formula to write the solutions of each quadratic equation in the form $a + bi$.

Example $x^2 + 2x + 4 = 0$

Solution

$$x = \dfrac{-2 \pm \sqrt{2^2 - 4 \cdot 4}}{2} = \dfrac{-2 \pm \sqrt{-12}}{2}$$

$$= \dfrac{-2 \pm \sqrt{4(-3)}}{2} = -1 \pm \sqrt{-3}$$

$$= -1 \pm \sqrt{3}\,i$$

67. $x^2 + 2x + 2 = 0$
68. $x^2 + x + 1 = 0$
69. $2x^2 + x + 1 = 0$
70. $3x^2 + x + 2 = 0$
71. $-x^2 + 2x - 4 = 0$
72. $-2x^2 - 2x - 5 = 0$
73. $x^2 + 2ix - 1 = 0$
74. $x^2 - 4ix - 1 = 3$
75. $x^2 + 2ix + 8 = 0$
76. $x^2 + 3ix + 10 = 0$

B Prove each of the following statements.

Example If $z \in R$, then $\bar{z} = z$.

Solution Since $z \in R$, $z = a + 0i$ where $a \in R$. Thus, $\bar{z} = a - 0i = a + 0i = z$.

77. If $z_1, z_2 \in C$, then $\overline{z_1 \pm z_2} = \bar{z}_1 \pm \bar{z}_2$.
78. If $z_1, z_2 \in C$, then $\overline{z_1 z_2} = \bar{z}_1 \bar{z}_2$.
79. If $z \in C$, then $\overline{z^n} = \bar{z}^n$.
80. If $z \in C$, then $\bar{\bar{z}} = z$.
81. If $z = a + bi$, then $z + \bar{z} = 2a$.
82. If $z = a + bi$, then $z - \bar{z} = 2bi$.

8.2 Complex Zeros of Polynomial Functions

In Sections 3.2 and 3.3 we studied several methods for locating real roots of polynomials with real-number coefficients. In this section we shall consider the replacement set for the variable to be the set of complex numbers. This will enable us to discuss some very useful theorems concerning zeros of polynomial functions.

Remainder and factor theorems

In Section 3.1 we proved both the remainder theorem and the factor theorem for polynomials with real-number coefficients and replacement sets of real numbers. Since these proofs depend only on the field axioms and since C is a field, both theorems apply to polynomials with complex-number coefficients and replacement sets of complex numbers.

Conjugate complex roots

Recalling from Section 8.1 that the conjugate of the complex number $z = a + bi$ is $\bar{z} = a - bi$, where $a, b \in R$, we state an important property of a polynomial. The proof is left as an exercise.

Theorem 8.5 If $P(z)$ is a polynomial with real-number coefficients, and $P(z) = 0$ for some $z \in C$, then $P(\bar{z}) = 0$.

8.2 Complex Zeros of Polynomial Functions

This theorem guarantees that the *complex zeros* of a polynomial function with *real coefficients* always occur in *conjugate pairs*.

Example Given that $2 - i$ is a zero of

$$P(x) = x^3 - 6x^2 + 13x - 10,$$

find all zeros of P.

Solution By Theorem 8.5, $\overline{2 - i}$, or $2 + i$, is a zero of P. Thus, by the factor theorem,

$$\begin{aligned} P(x) &= [x - (2 - i)][x - (2 + i)]Q(x) \\ &= (x^2 - 4x + 5)Q(x). \end{aligned}$$

Hence,

$$Q(x) = \frac{x^3 - 6x^2 + 13x - 10}{x^2 - 4x + 5} = x - 2.$$

Thus,

$$P(x) = [x - (2 - i)][x - (2 + i)](x - 2) = 0,$$

and, by the factor theorem, the solutions of this equation—and hence the zeros of P—are $2 - i$, $2 + i$, and 2.

Fundamental theorem of algebra

When Theorem 8.5 is coupled with the following theorem, which is called the **fundamental theorem of algebra**, a great deal of information relative to the zeros of polynomial functions becomes readily available.

Theorem 8.6 *Every polynomial function of degree $n \geq 1$ over the complex numbers has at least one real or complex zero.*

The proof of this theorem involves concepts beyond those available to us and is omitted.

An nth-degree polynomial function has n zeros

Repeated applications of the fundamental theorem of algebra and the factor theorem can be used to prove the following theorem. We leave the details as an exercise.

Theorem 8.7 *Every polynomial of degree $n \geq 1$ over the complex numbers can be expressed as a product of a constant and n linear factors of the form $x - x_j$, where $x_j \in C$.*

If a factor $(x - x_i)$ occurs k times in such a linear factorization of $P(x)$, then x_i is said to be a **zero of multiplicity** k. With this agreement, Theorem 8.7 shows that every polynomial function defined by a polynomial $P(x)$ of degree n with complex coefficients has exactly n zeros.

Note that any theorem stated in terms of zeros of polynomial functions applies to solutions of polynomial equations and vice versa; a *zero* of

$$P(x) = a_n x^n + a_{n-1} x^{n-1} + \cdots + a_0$$

is a *solution* of $P(x) = 0$.

Examples **a.** $x^3 + 2x - 1$ has three complex zeros. **b.** $x^5 - 2x^3 - x + 1 = 0$ has five complex solutions.

Exercise 8.2

A In Exercises 1–12, one or more zeros are given for each of the polynomial functions; find the other zeros.

Example $P(x) = x^3 - 2x^2 + x - 2$; i is one zero.

Solution By Theorem 8.5, $\bar{i} = -i$ is a zero of P. By the factor theorem,

$$P(x) = (x - i)[x - (-i)]Q(x)$$
$$= (x^2 + 1)Q(x).$$

Hence,

$$Q(x) = \frac{x^3 - 2x^2 + x - 2}{x^2 + 1} = x - 2.$$

Thus,

$$P(x) = (x - i)(x + i)(x - 2).$$

By the factor theorem, the zeros of P are i, $-i$, and 2.

1. $P(x) = x^2 + 4$; $2i$ is a zero.
2. $P(x) = 3x^2 + 27$; $-3i$ is a zero.
3. $Q(x) = x^3 - 3x^2 + x - 3$; 3 and i are zeros.
4. $Q(x) = x^3 - 5x^2 + 7x + 13$; -1 and $3 - 2i$ are zeros.
5. $P(x) = 2x^3 - x^2 + 2x - 1$; $-i$ is a zero.
6. $P(x) = 3x^3 - 10x^2 + 7x + 10$; $2 - i$ is a zero.
7. $Q(x) = x^4 + 5x^2 + 4$; $-i$ and $2i$ are zeros.
8. $Q(x) = x^4 + 11x^2 + 18$; $3i$ and $i\sqrt{2}$ are zeros.
9. $P(x) = x^4 + 3x^3 + 4x^2 + 27x - 45$; $-3i$ is a zero.

10. $P(x) = x^4 - 7x^3 + 18x^2 - 22x + 12$; $1 - i$ is a zero.

11. $Q(x) = x^5 - 2x^4 + 8x^3 - 16x^2 + 16x - 32$; $2i$ is a zero of multiplicity 2.

12. $Q(x) = 2x^5 - 9x^4 + 20x^3 - 24x^2 + 16x - 4$; $1 + i$ is a zero of multiplicity 2.

13. One zero of $P(x) = 2x^3 - 11x^2 + 28x - 24$ is $2 - 2i$. Factor $P(x)$ over the complex numbers.

14. One zero of $Q(x) = 3x^3 - 10x^2 + 7x + 10$ is $2 + i$. Factor $Q(x)$ over the complex numbers.

15. One solution of $x^4 - 10x^3 + 35x^2 - 50x + 34 = 0$ is $4 - i$. Find the remaining solutions.

16. One solution of $5x^4 + 34x^3 + 40x^2 - 78x + 51 = 0$ is $-4 - i$. Find the remaining solutions.

17. A cubic equation with real coefficients has solutions -2 and $1 + i$. What is the third solution? Write the equation in the form $P(x) = 0$, given that the leading coefficient (the coefficient of the highest power of x) is 1.

18. A cubic equation with real coefficients has solutions 4 and $2 - i$. What is the third solution? Write the equation in the form $P(x) = 0$, given that the leading coefficient is 1.

19. Argue that every polynomial equation with real coefficients and of odd degree has at least one real solution.

20. One zero of a polynomial P is i. Is $-i$ necessarily a zero of P? Why or why not? The number i is a zero of $P(x) = x^3 + i$. Determine whether or not $-i$ is a zero.

21. Determine the three cube roots of -1. *Hint*: Consider the solutions of the equation $x^3 + 1 = 0$.

22. Determine the three cube roots of 1. *Hint*: Consider the solutions of the equation $x^3 - 1 = 0$.

B 23. Prove Theorem 8.5. *Hint*: Show $\overline{P(z)} = P(\bar{z})$.

24. Fill in the details of the proof of Theorem 8.7.

25. Show that if P and Q are two polynomials of degree n and $P(x) = Q(x)$ for more than n values of x, then $P(x) = Q(x)$ for every x.

8.3 Trigonometric Form of Complex Numbers

Graphs of complex numbers

In Chapter 2, we used Cartesian coordinates to establish a one-to-one correspondence between the set of ordered pairs (a, b) in R^2 and the set of points P in the geometric plane. By pairing each complex number $a + bi$ with the ordered pair (a, b) (we call a the **real part** and b the **imaginary part** of the complex number), we can establish a one-to-one correspondence between the set of all complex numbers and R^2. Hence each point in the plane can be viewed as the graph of a complex number (see Figure 8.2 on page 280). Since the real part of $a + bi$ is taken as the x-coordinate of P, in this context the x-axis is called the **real axis**. Similarly,

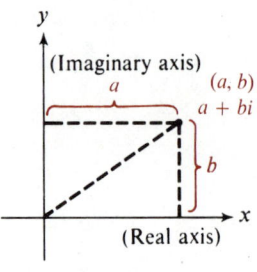

Figure 8.2

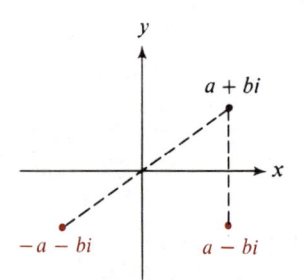

Figure 8.3

since the imaginary part is taken as the ordinate or y-coordinate of P, the y-axis is called the **imaginary axis**.

A plane on which complex numbers are thus represented is often called a **complex plane**. It is also sometimes called an **Argand plane**, after the French mathematician Jean Robert Argand (1768–1822), who systematically used it, or a **Gauss plane**, after the great German mathematician Karl Friedrich Gauss (1777–1855).

A complex number $z = a + bi$, its conjugate $\bar{z} = a - bi$, and also its negative $-z = -a - bi$ are represented in Figure 8.3. It is evident that $\bar{z}$ is the reflection of z in the real axis and $-z$ is the reflection of z in the origin as well as the reflection of $\bar{z}$ in the imaginary axis.

Modulus and argument

Since each nonzero complex number $z = a + bi$ lies on a ray with the origin as endpoint (Figure 8.4), we can associate with each z two useful concepts.

Definition 8.8 *The **absolute value**, or **modulus**, of the complex number $z = a + bi$ is denoted by r, $|z|$, or $|a + bi|$ and is given by*

$$r = |z| = |a + bi| = \sqrt{a^2 + b^2}.$$

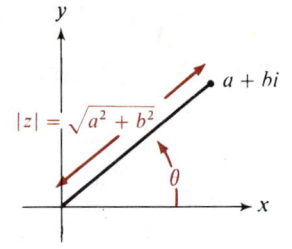

Figure 8.4

Thus the modulus $|a + bi|$ is just the distance from the origin to the point $a + bi$.

Examples

a. $|1 + 0i| = \sqrt{1^2 + 0^2}$
    $= 1$

b. $|0 + 2i| = \sqrt{0^2 + 2^2}$
    $= 2$

c. $|1 + 3i| = \sqrt{1^2 + 3^2}$
    $= \sqrt{10}$

d. $|-1 + i| = \sqrt{(-1)^2 + 1^2}$
    $= \sqrt{2}$

Definition 8.9 *An **argument**, or an **amplitude**, of the complex number $z = a + bi$ is an angle θ with initial side the positive x-axis and terminal side the ray from the origin containing $a + bi$.*

Note that if θ is an argument of $a + bi$, then so is $\theta + k \cdot 2\pi^R$, or $\theta + k \cdot 360°$, for each $k \in J$. For $a + bi = 0$, that is, for $a^2 + b^2 = 0$, any angle θ might be used as

8.3 Trigonometric Form of Complex Numbers

an argument of $a + bi$. Also if θ is an argument of $a + bi$, then $b/a = \tan \theta$ when $a \neq 0$.

Example Find all arguments of the complex number $-\sqrt{3} + i$.

Solution We note that the graph of the complex number $-\sqrt{3} + i$ is in Quadrant II and that

$$\tan \theta = -\frac{1}{\sqrt{3}}.$$

Thus $\theta = 2\pi^R/3 = 120°$ is one argument of $-\sqrt{3} + i$. Hence the arguments of $-\sqrt{3} + i$ are the angles of measure

$$\frac{2\pi^R}{3} + k \cdot 2\pi^R = (120° + k \cdot 360°), \quad k \in J.$$

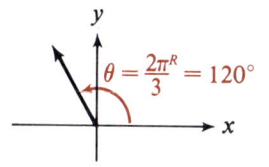

Trigonometric form

As seen in Figure 8.5, where $r = \sqrt{a^2 + b^2}$ is the modulus of $a + bi$ and θ is an argument of $a + bi$, any complex number $a + bi$ can be written in the form

$$r \cos \theta + ir \sin \theta, \quad \text{or} \quad r(\cos \theta + i \sin \theta),$$

which is called the **trigonometric form**, or **polar form**, for a complex number. More generally, using degree measure for angles, we have

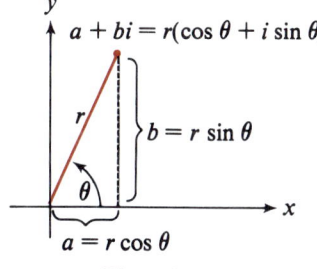

$$a + bi = r[\cos(\theta + k \cdot 360°) + i \sin(\theta + k \cdot 360°)], \quad k \in J.$$

A convenient abbreviation that is used for the expression $\cos \theta + i \sin \theta$ is **cis** θ (read "cosine θ plus i sine θ"), so that we can write

$$a + bi = r \operatorname{cis}(\theta + k \cdot 360°), \quad k \in J.$$

We ordinarily use for θ the angle of least nonnegative measure that is a solution of $a + bi = r \operatorname{cis} \theta$.

Example Represent $1 - \sqrt{3}i$ in trigonometric form.

Solution We first compute

$$r = |1 - \sqrt{3}i| = \sqrt{1 + 3} = 2$$

Then, noting that the graph of the complex number is in Quadrant IV and that

$$\tan \theta = -\frac{\sqrt{3}}{1} = -\sqrt{3},$$

we find that $\theta = 300°$. Hence,

$$1 + \sqrt{3}i = 2(\cos 300° + i \sin 300°) = 2 \operatorname{cis} 300°$$

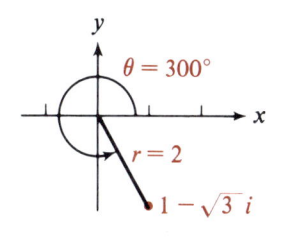

Example Represent 4 cis 225° graphically, and write the number in rectangular form.

Solution
$$a = r \cos \theta = 4\left(-\frac{\sqrt{2}}{2}\right) = -2\sqrt{2},$$

$$b = r \sin \theta = 4\left(-\frac{\sqrt{2}}{2}\right) = -2\sqrt{2},$$

and
$$a + bi = -2\sqrt{2} - 2\sqrt{2}i.$$

Products and quotients

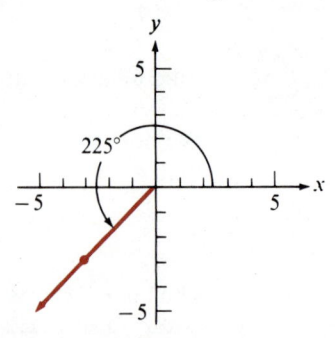

Products and quotients of complex numbers can be found quite easily when the complex numbers are in trigonometric form. Observe that

$$z_1 = r_1(\cos \theta_1 + i \sin \theta_1) \quad \text{and} \quad z_2 = r_2(\cos \theta_2 + i \sin \theta_2),$$

from which

$$z_1 \cdot z_2 = r_1(\cos \theta_1 + i \sin \theta_1) \cdot r_2(\cos \theta_2 + i \sin \theta_2)$$
$$= r_1 \cdot r_2 \cdot (\cos \theta_1 \cos \theta_2 + i \cos \theta_1 \sin \theta_2 + i \sin \theta_1 \cos \theta_2 + i^2 \sin \theta_1 \sin \theta_2)$$
$$= r_1 \cdot r_2 \cdot [(\cos \theta_1 \cos \theta_2 - \sin \theta_1 \sin \theta_2) + i(\cos \theta_1 \sin \theta_2 + \sin \theta_1 \cos \theta_2)].$$

By using the sum formula derived in Sections 7.2 and 7.3, the right-hand member can be written

$$r_1 r_2 [\cos(\theta_1 + \theta_2) + i \sin(\theta_1 + \theta_2)],$$

so that

$$z_1 \cdot z_2 = r_1 r_2 \operatorname{cis}(\theta_1 + \theta_2).$$

This proves part I of the following theorem. The proof of part II is left as an exercise.

Theorem 8.8 If $z_1, z_2 \in C$, with $z_1 = r_1 \operatorname{cis} \theta_1$ and $z_2 = r_2 \operatorname{cis} \theta_2$, then

I. $z_1 \cdot z_2 = r_1 r_2 \operatorname{cis}(\theta_1 + \theta_2)$;

II. $\dfrac{z_1}{z_2} = \dfrac{r_1}{r_2} \operatorname{cis}(\theta_1 - \theta_2) \quad (z_2 \neq 0 + 0i).$

Example Write the product $3 \operatorname{cis} 80° \cdot 5 \operatorname{cis} 40°$ in the form $a + bi$.

Solution By Theorem 8.8-I,

$$3 \operatorname{cis} 80° \cdot 5 \operatorname{cis} 40° = 15 \operatorname{cis} 120° = 15(\cos 120° + i \sin 120°).$$

Since $\cos 120° = -1/2$ and $\sin 120° = \sqrt{3}/2$, we have

$$15(\cos 120° + i \sin 120°) = 15\left(-\frac{1}{2} + \frac{\sqrt{3}}{2}i\right) = -\frac{15}{2} + \frac{15\sqrt{3}}{2}i.$$

8.3 Trigonometric Form of Complex Numbers

Example Write the quotient $\dfrac{8 \text{ cis } 540°}{2 \text{ cis } 225°}$ as a complex number in the form $a + bi$.

Solution By Theorem 8.8-II,

$$\frac{8 \text{ cis } 540°}{2 \text{ cis } 225°} = \frac{8}{2} \text{ cis } (540° - 225°) = 4 \text{ cis } 315° = 4(\cos 315° + i \sin 315°).$$

Since $\cos 315° = 1/\sqrt{2} = \sqrt{2}/2$ and $\sin 315° = -1/\sqrt{2} = -\sqrt{2}/2$, we have

$$4(\cos 315° + i \sin 315°) = 4\left(\frac{\sqrt{2}}{2} - \frac{\sqrt{2}}{2}i\right) = 2\sqrt{2} - 2\sqrt{2}i.$$

Exercise 8.3

A Graph the given complex number, its conjugate, its negative, and the negative of its conjugate. Draw line segments joining each pair of these four points.

1. $2 + 3i$
2. $-3 + 4i$
3. $4 - i$
4. $-2 - i$
5. $4i$
6. $-3i$

Write without absolute-value notation.

Examples a. $|-3|$ b. $|2 + 5i|$ c. $|-3 + 2i|$

Solutions By Definition 8.8,

a. $\sqrt{(-3)^2} = 3$ b. $\sqrt{2^2 + 5^2} = \sqrt{29}$ c. $\sqrt{(-3)^2 + (2)^2} = \sqrt{13}$

7. $|4|$
8. $|-2|$
9. $|3 + 2i|$
10. $|4 - i|$
11. $|2i|$
12. $|-5i|$
13. $|-2 - i|$
14. $|-7 - i|$

Write the complex number in the form $r \text{ cis } \theta$.

Example $3\sqrt{3} - 3i$

Solution $r = |3\sqrt{3} - 3i| = \sqrt{27 + 9} = 6$

Noting that the graph of the complex number is in Quadrant IV and also that $\tan \theta = -3/3\sqrt{3} = -1/\sqrt{3}$, we find that $\theta = 330°$. Hence,

$3\sqrt{3} - 3i = 6(\cos 330° + i \sin 330°)$
$= 6 \text{ cis } 330°.$

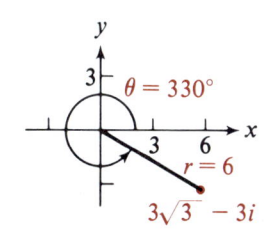

15. $3 + 3i$
16. $2 + 2i$
17. $-5 + 5i$
18. $-7 + 7i$
19. $2\sqrt{3} - 2i$
20. $-3\sqrt{3} - 3i$
21. $-2 - 2i$
22. $-\sqrt{3} - i$
23. 3
24. -2
25. $2i$
26. $-3i$

Write the complex number in the form $a + bi$.

Example   2 cis 120°

Solution   Since $r = 2$ and $\theta = 120°$, we have

$$a = r \cos \theta = 2 \cos 120°$$
$$= 2\left(-\frac{1}{2}\right) = -1,$$

and

$$b = r \sin \theta = 2 \sin 120°$$
$$= 2\left(\frac{\sqrt{3}}{2}\right) = \sqrt{3}.$$

Hence, $a + bi = -1 + \sqrt{3}i$.

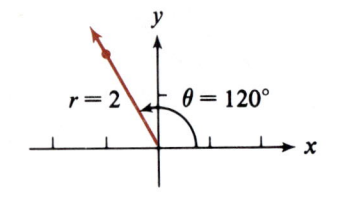

27. 4 cis 240°
28. 3 cis 300°
29. 6 cis (−30°)
30. 5 cis 180°
31. 12 cis 420°
32. 10 cis (−480°)
33. 6 cis 90°
34. 5 cis 270°
35. $\frac{3}{2}$ cis 40°
36. $\frac{2}{3}$ cis 75°
37. 2 cis (−50°)
38. 4 cis (−160°)

For the given pair of complex numbers z_1 and z_2, find (a) $z_1 \cdot z_2$, and (b) z_1/z_2. Express each result in the form $a + bi$. Use Table V as needed.

39. $z_1 = 3$ cis 90°   and $z_2 = \sqrt{2}$ cis 45°
40. $z_1 = 4$ cis 30°   and $z_2 = 2$ cis 60°
41. $z_1 = 6$ cis 150°   and $z_2 = 18$ cis 570°
42. $z_1 = 14$ cis 210°   and $z_2 = 2$ cis 120°
43. $z_1 = 2$ cis 15°   and $z_2 = 3$ cis 25°
44. $z_1 = 6$ cis (−10°)   and $z_2 = 4$ cis 25°

B 45. Prove that for any $z \in C$, $z\bar{z} = |z|^2$.

46. Show that if $a + bi = r \text{ cis } \theta$, then $(a + bi)^2 = r^2 \text{ cis } 2\theta$.

47. Use the result of Exercise 46 to show that if $a + bi = r \text{ cis } \theta$, then $(a + bi)^3 = r^3 \text{ cis } 3\theta$.

48. Show that if $z = r \text{ cis } \theta$, then $\bar{z} = r \text{ cis } (-\theta)$.

8.4 De Moivre's Theorem — Powers and Roots

Powers of complex numbers

Since $a + bi = r \text{ cis } \theta$, an application of Theorem 8.8-I to $(a + bi)^2$ results in

$$(a + bi)^2 = (r \text{ cis } \theta)(r \text{ cis } \theta) = r^2 \text{ cis } 2\theta. \quad (1)$$

Because

$$(a + bi)^3 = (a + bi)^2(a + bi),$$

from (1) we have

$$(a + bi)^3 = (r^2 \text{ cis } 2\theta)(r \text{ cis } \theta) = r^3 \text{ cis } 3\theta.$$

In a similar way, we can show that

$$(a + bi)^4 = (r^3 \text{ cis } 3\theta)(r \text{ cis } \theta) = r^4 \text{ cis } 4\theta,$$

and it seems plausible to make the following assertion, known as **De Moivre's theorem**. The proof is omitted.

Theorem 8.9 If $z \in C$, $z = r \text{ cis } \theta$ and $n \in N$, then

$$z^n = r^n \text{ cis } n\theta.$$

Example Write $(\sqrt{3} + i)^7$ in the form $a + bi$.

Solution For the modulus, we have

$$r = \sqrt{(\sqrt{3})^2 + 1^2} = 2.$$

If θ is an argument, then

$$\tan \theta = \frac{1}{\sqrt{3}}$$

and from the fact that the graph of $\sqrt{3} + i$ is in Quadrant I, we obtain $\theta = 30°$. Thus, $(\sqrt{3} + i)^7 = (2 \text{ cis } 30°)^7$. Then, by De Moivre's theorem,

$$(2 \text{ cis } 30°)^7 = 2^7 \text{ cis } (7 \cdot 30°) = 128 \text{ cis } 210°.$$

Solution continued overleaf

Converting to the form $a + bi$, we find that

$$128(\cos 210° + i \sin 210°) = 128\left(-\frac{\sqrt{3}}{2} - \frac{1}{2}i\right)$$
$$= -64\sqrt{3} - 64i,$$

so

$$(\sqrt{3} + i)^7 = -64\sqrt{3} - 64i.$$

By appropriately defining z^0 and z^{-n}, we can extend De Moivre's theorem to include as exponents all $n \in J$. The proof is left as an exercise.

Definition 8.10   If $z \neq 0 + 0i$, then

    I. $z^0 = 1 + 0i$;

    II. $z^{-n} = \dfrac{1}{z^n}$,   for $n \in J$.

Theorem 8.10   If $z \in C$, $z \neq 0 + 0i$, $z = r \operatorname{cis} \theta$ and $n \in J$, then

$$z^n = r^n \operatorname{cis} n\theta.$$

Example   Write $(1 + i)^{-6}$ in the form $a + bi$.

Solution   For the modulus we have

$$r = \sqrt{1^2 + 1^2} = \sqrt{2}$$

If θ is an argument then

$$\tan \theta = \frac{1}{1}.$$

Noting that the graph of $1 + i$ is in Quadrant I we have $\theta = 45°$. Hence,

$$(1 + i)^{-6} = (\sqrt{2} \operatorname{cis} 45°)^{-6}.$$

By Theorem 8.10,

$$(\sqrt{2} \operatorname{cis} 45°)^{-6} = (\sqrt{2})^{-6} \operatorname{cis}(-6 \cdot 45°)$$
$$= \frac{1}{8} \operatorname{cis}(-270°)$$
$$= \frac{1}{8}[\cos(-270°) + i \sin(-270°)].$$

Since $\cos(-270°) = 0$ and $\sin(-270°) = 1$, we obtain

$$(1 + i)^{-6} = \frac{1}{8}(0 + i) = \frac{1}{8}i.$$

8.4 De Moivre's Theorem—Powers and Roots

Roots of complex numbers

Yet another extension of De Moivre's theorem is possible if we make the following definition.

Definition 8.11   For $z \in C$, $n \in N$, w is an nth root of z provided
$$w^n = z.$$

Theorem 8.11   If $z \in C$, $z = r \operatorname{cis} \theta$ and $n \in N$, then
$$r^{1/n} \operatorname{cis}\left(\frac{\theta}{n}\right)$$
is an nth root of z.

This theorem follows directly from De Moivre's theorem. The proof is left as an exercise. The fact that
$$\operatorname{cis} \theta = \operatorname{cis}(\theta + k \cdot 360°),$$
for $k \in J$, enables us to find n distinct complex nth roots for each $z \in C, z \neq 0 + 0i$ as illustrated in the following example.

Example   Write each of the four fourth roots of $z = 2 + 2\sqrt{3}i$ in the form $r \operatorname{cis} \theta$.

Solution   For the modulus we have
$$r = \sqrt{2^2 + (2\sqrt{3})^2} = 4.$$
The graph of $2 + 2\sqrt{3}i$ is in Quadrant I and for any argument θ,
$$\tan \theta = \sqrt{3}.$$
Hence, $\theta = 60°$ and
$$z = 2 + 2\sqrt{3}i = 4 \operatorname{cis} 60°,$$
$$= 4 \operatorname{cis}(60° + k \cdot 360°). \quad k \in J.$$
From Theorem 8.11 each number
$$4^{1/4} \operatorname{cis}\left(\frac{60° + k \cdot 360°}{4}\right), \quad k \in J,$$
is a fourth root of z. Taking $k = 0, 1, 2,$ and 3, in turn, gives the roots
$$w_0 = \sqrt{2} \operatorname{cis} 15°, \quad w_1 = \sqrt{2} \operatorname{cis} 105°, \quad w_2 = \sqrt{2} \operatorname{cis} 195°, \quad w_3 = \sqrt{2} \operatorname{cis} 285°.$$
Substitution of any other integer for k will give one of these four complex numbers.

Exercise 8.4

A Use Theorem 8.8, and/or 8.10 to write the given expression as a complex number of the form $a + bi$. Use Table V as necessary.

1. $[2 \text{ cis } (-30°)]^7$
2. $(4 \text{ cis } 36°)^5$
3. $\left(-\dfrac{1}{2} + \dfrac{1}{2}\sqrt{3}i\right)^3$
4. $(1 + i)^{12}$
5. $(\sqrt{3} \text{ cis } 45°)^{12}$
6. $(\sqrt{2} \text{ cis } 30°)^7$
7. $(\sqrt{3} - i)^{-5}$
8. $(1 - i)^{-6}$
9. $\dfrac{(1 + i)^3}{(1 + \sqrt{3}i)^5(1 - i)^2}$
10. $\dfrac{4(\sqrt{3} + i)^3}{(1 - i)^3}$
11. $\dfrac{(1 + i)^{-6}}{(1 - i)^{-5}}$
12. $\dfrac{(1 + \sqrt{3}i)^{-4}}{(\sqrt{3} + i)^{-6}}$

Find the nth roots of z by applying Theorem 8.11. Leave the results in trigonometric form and list all n of the nth roots.

13. $z = 32 \text{ cis } 45°, \quad n = 5$
14. $z = 27 \text{ cis } 180°, \quad n = 3$
15. $z = -16\sqrt{3} + 16i, \quad n = 5$
16. $z = 1 - i, \quad n = 4$
17. $z = -i, \quad n = 6$
18. $z = 2 + 2\sqrt{3}i, \quad n = 3$

Solve the given equation over C.

19. $x^5 = 16 - 16\sqrt{3}i$
20. $x^3 + 4i = 4\sqrt{3}$
21. $x^7 + 1 = 0$
22. $x^7 - 1 = 0$

B
23. Factor $x^4 + 16$ into linear factors.
24. Factor $x^5 - 1$ into linear factors.
25. Show that the sum of the four fourth roots of 1 is $0 + 0i$.
26. Show that the sum of the fifth roots of 1 is $0 + 0i$.
27. Prove Theorem 8.10.
28. Prove Theorem 8.11.

Chapter Review

[8.1] *Write each expression in the form a + bi.*

1. $(3 + i) + (1 - 3i)$
2. $(3 + i) \cdot (1 - 3i)$
3. $(3 + i) - (1 - 3i)$
4. $(3 + i) \div (1 - 3i)$
5. $(1 + \sqrt{-1})^2$
6. $(1 - \sqrt{-16})\sqrt{-4}$
7. i^9
8. $\dfrac{1}{i^7}$

[8.2] *Find all the zeros of the given polynomial function.*

9. $P(x) = x^3 - 2x^2 + 4x - 8$; $2i$ is one zero.
10. $P(x) = 2x^3 - 11x^2 + 28x - 24$; $2 + 2i$ is one zero.
11. $Q(x) = x^4 + x^3 + 2x^2 + 4x - 8$; $-2i$ is one zero.
12. $Q(x) = x^4 - 3x^3 + 7x^2 + 21x - 26$; $2 - 3i$ is one zero.

[8.3] *Write the complex number in the form r cis θ.*

13. $2 + 2i$
14. $5 - 5\sqrt{3}i$

Write the complex number in the form a + bi.

15. $3 \text{ cis } 150°$
16. $4 \text{ cis } (-60°)$

For the pair of complex numbers, find (a) $z_1 \cdot z_2$ and (b) z_1/z_2. Express each result in the form a + bi.

17. $z_1 = 2 \text{ cis } 30°$ and $z_2 = 9 \text{ cis } 120°$
18. $z_1 = 3 \text{ cis } 15°$ and $z_2 = 2 \text{ cis } 45°$

[8.4] *Use Theorem 8.8 and/or 8.10 to write the given expression as a complex number of the form a + bi. Use Table V as necessary.*

19. $(3 \text{ cis } 45°)^5$
20. $\dfrac{2(1 - i\sqrt{3})^2}{(1 + i)^4}$

Find the nth roots of z by applying Theorem 8.11. Leave the results in trigonometric form and list all n of the nth roots.

21. $z = 1 + i, \quad n = 3$

22. $z = 3\sqrt{3} - 3i, \quad n = 4$

Solve the given equation over C.

23. $x^4 + 1 = 0$

24. $x^5 - 1 = 0$

Supplemental Exercises for Chapters 5–8

The problems in this supplement are significantly more difficult than those in the exercise sets at the end of each section and the chapter reviews. They are designed to challenge the student, and a solution may draw on a combination of ideas and techniques from any of the preceding chapters. Many of these problems contain results that are used in more advanced classes in mathematics.

Exercises 1–6 require the use of a calculator. Evaluate each of the following functions to six decimal places at $x = 0.01, 0.001, 0.0001, 0.00001,$ and 0.000001. Is the function value getting closer to some value as x gets closer to 0 but remains positive? If so, what is that value?

1. $f(x) = \sin(1/x)$
2. $f(x) = \cos(1/x)$
3. $f(x) = x \sin(1/x)$
4. $f(x) = x \cos(1/x)$
5. $f(x) = \sqrt{x} \sin(1/x)$
6. $f(x) = \sqrt{x} \cos(1/x)$

Graph each of the following functions.

7. $f(x) = \sin|x|$
8. $f(x) = \cos|x|$
9. $f(x) = \sin(\text{Arcsin } x)$
10. $f(x) = \cos(\text{Arccos } x)$
11. $f(x) = \text{Arcsin}(\sin x)$
12. $f(x) = \text{Arccos}(\cos x)$
13. $f(x) = x \sin x$
14. $f(x) = x \cos x$
15. $f(x) = e^{-x} \sin x$
16. $f(x) = e^{-x} \cos x$
17. $f(x) = e^{-x^2/10} \sin x$
18. $f(x) = e^{-x^2/10} \cos x$

Draw the graph of each function on the interval $[1/4\pi, \pi/4]$ and describe the graphs of each of the following functions in the interval $[-1, 1]$.

19. $f(x) = \sin(1/x)$ **20.** $f(x) = \cos(1/x)$

21. $f(x) = x \sin(1/x)$ **22.** $f(x) = x \cos(1/x)$

23. Show that $\cos x = \cos |x|$ is an identity

24. Show that $\sin^2 x = \sin^2 |x|$ is an identity.

25. Show that, for any two complex numbers z and w, $|z \pm w| \leq |z| + |w|$.

26. Show that, for any two complex numbers z and w, $||z| - |w|| \leq |z + w|$.

27. Explain why for all $n > 1$ the sum of the nth roots of any complex number is zero.

28. Show that the sum of the squares of the nth roots of 1 is 0, if n is even.

For Exercises 29–32, use the figure below.

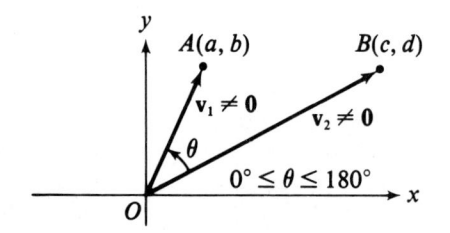

29. Show that $\cos \theta = \dfrac{ac + bd}{\|\mathbf{v}_1\| \|\mathbf{v}_2\|}$.

30. Show that $ac + bd \leq \|\mathbf{v}_1\| \|\mathbf{v}_2\|$.

31. Show that $\mathbf{v}_1$ is perpendicular to $\mathbf{v}_2$ if and only if $ac + bd = 0$.

32. Show that $\mathbf{v}_1$ is perpendicular to $\mathbf{v}_2$ if and only if

$$\|\mathbf{v}_1 + \mathbf{v}_2\|^2 = \|\mathbf{v}_1\|^2 + \|\mathbf{v}_2\|^2.$$

Solve each of the following inequalities in the interval $[-2\pi, 2\pi]$.

33. $\sin x \leq \dfrac{1}{2}$ **34.** $\cos x \leq \dfrac{1}{2}$

35. $\sin^2 x \leq \dfrac{3}{4}$ **36.** $\cos^2 x \leq \dfrac{3}{4}$

9 Systems of Equations and Inequalities

For a given collection of equations or inequalities in two variables, it is often necessary to determine the ordered pairs which satisfy all of the equations or inequalities in the collection. In such a case, the collection of equations or inequalities is called a **system in two variables**. Any ordered pair which is a solution of all of the equations or inequalities in a system is called a **solution of the system**, and the set of all solutions of a system is called the **solution set of the system**. Systems in more than two variables are defined similarly.

9.1 Systems of Linear Equations in Two Variables

We shall begin by considering the system

$$a_1 x + b_1 y + c_1 = 0 \quad (a_1, b_1 \text{ not both } 0) \tag{1}$$

$$a_2 x + b_2 y + c_2 = 0 \quad (a_2, b_2 \text{ not both } 0). \tag{2}$$

Linear dependence The left-hand members of Equations (1) and (2) are said to be **linearly dependent** if one of them can be obtained from the other through multiplication by a constant. If the left-hand members of Equations (1) and (2) are *not* linearly dependent, then they are said to be **linearly independent**.

Examples **a.** The expressions $2x + 4y - 8$ and $6x + 12y - 24$ are linearly dependent since

$$3(2x + 4y - 8) = 6x + 12y - 24.$$

b. The expressions $2x + 4y - 8$ and $6x + 12y - 23$ are linearly independent.

Consistency A system is said to be **consistent** if its solution set is nonempty. A system which is not consistent is **inconsistent**.

Examples

a. The system
$$x - 2 = 0$$
$$y + 3 = 0$$
is consistent. The solution set is $\{(2, -3)\}$.

b. The system
$$x + y = 0$$
$$x + y - 1 = 0$$
is inconsistent The solution set is empty.

Geometric interpretation

In a geometric sense, because the graphs of both Equations (1) and (2) above are straight lines, we are confronted with three possibilities. One possibility is that the graphs of the equations are the same line, as illustrated in Figure 9.1. In this case, the solution set of the system is the same as the solution set of each equation. The left-hand members of the two equations in standard form are linearly dependent and the system is consistent.

A second possibility is that the graphs of the equations are parallel lines, as illustrated in Figure 9.2. Since the lines have no point in common, there is no ordered pair which satisfies both equations. Hence the solution set of the system is empty. In this case, the left-hand members are linearly independent and the system is inconsistent.

The last possibility is that the graphs of the equations intersect in a single point, as illustrated in Figure 9.3. Since the lines have exactly one point in common, there is exactly one ordered pair which satisfies both equations. Hence, the solution set of the system contains one ordered pair. In this case the left-hand members are linearly independent and the system is consistent.

$$a_1x + b_1y + c_1 = 0$$
$$a_2x + b_2y + c_2 = 0$$

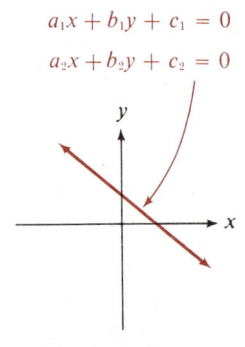

Figure 9.1

$$a_1x + b_1y + c_1 = 0$$
$$a_2x + b_2y + c_2 = 0$$

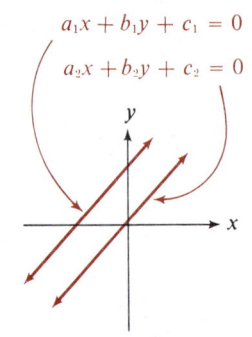

Figure 9.2

$$a_1x + b_1y + c_1 = 0$$
$$a_2x + b_2y + c_2 = 0$$

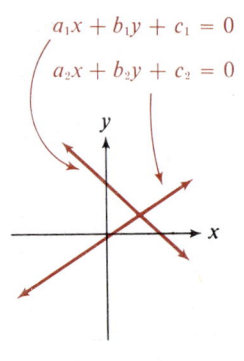

Figure 9.3

You will recall that equivalent equations have the same solution set. Systems may be said to be equivalent in a similar sense.

Equivalent systems

Definition 9.1 *If the solution set of one system is equal to (the same as) the solution set of another system, then the systems are* **equivalent**.

In seeking the solution set of a system of equations, our procedure will be to generate equivalent systems until we arrive at a system for which the solution set is obvious. One way to obtain an equivalent system is to replace one equation by a certain *linear combination* of the equations in the system. If $f(x, y)$ and $g(x, y)$ are

9.1 Systems of Linear Equations in Two Variables

polynomials, the polynomial $af(x, y) + bg(x, y)$, where a and b are not both zero, is said to be a linear combination of $f(x, y)$ and $g(x, y)$. We shall then refer to the equation

$$af(x, y) + bg(x, y) = 0$$

as a **linear combination** of the equations $f(x, y) = 0$ and $g(x, y) = 0$.

We now compare the two systems:

$$\begin{aligned} f(x, y) &= 0 \\ g(x, y) &= 0 \end{aligned} \qquad (3)$$

and

$$\begin{aligned} af(x, y) + bg(x, y) &= 0 \\ g(x, y) &= 0 \end{aligned} \qquad (4)$$

where $a \neq 0$. If (x_1, y_1) is a solution to (3), then $f(x_1, y_1) = 0$ and $g(x_1, y_1) = 0$, so

$$af(x_1, y_1) + bg(x_1, y_1) = 0,$$

and (x_1, y_1) is a solution to (4). On the other hand, suppose (x_2, y_2) is a solution to (4). Then $g(x_2, y_2) = 0$ and

$$af(x_2, y_2) + bg(x_2, y_2) = 0.$$

Thus, $af(x_2, y_2) = 0$, and since $a \neq 0$, $f(x_2\ y_2) = 0$, so that (x_2, y_2) is a solution to (3). Therefore, (3) and (4) have the same solution set. This proves the following theorem.

Theorem 9.1 *If either equation in the system*

$$\begin{aligned} f(x, y) &= 0 \\ g(x, y) &= 0 \end{aligned}$$

is replaced by a linear combination of the two equations with nonzero coefficient for the replaced equation, the result is an equivalent system.

Theorem 9.1 is useful in solving linear systems of the form

$$a_1 x + b_1 y + c_1 = 0 \quad (a_1, b_1 \text{ not both } 0) \qquad (5)$$

$$a_2 x + b_2 y + c_2 = 0 \quad (a_2, b_2 \text{ not both } 0). \qquad (6)$$

By appropriate choice of multipliers a and b, the linear combination

$$a(a_1 x + b_1 y + c_1) + b(a_2 x + b_2 y + c_2) = 0 \qquad (7)$$

will be free of one variable; that is, the coefficient of one variable will be 0. We can then form an equivalent system by substituting Equation (7) for either Equation (5) or Equation (6).

Example Solve

$$x - 3y + 5 = 0$$
$$2x + y - 4 = 0. \tag{8}$$

Solution We first form the linear combination

$$1(x - 3y + 5) + 3(2x + y - 4) = 0,$$

or

$$7x - 7 = 0,$$

from which

$$x - 1 = 0.$$

Replacing $x - 3y + 5 = 0$ with $x - 1 = 0$, we then have the equivalent system

$$x - 1 = 0$$
$$2x + y - 4 = 0. \tag{9}$$

We can next replace the second equation in (9) with the linear combination

$$-2(x - 1) + 1(2x + y - 4) = 0$$

or

$$y - 2 = 0$$

to obtain the equivalent system

$$x - 1 = 0$$
$$y - 2 = 0, \tag{10}$$

from which, by inspection, we can obtain the unique solution (1, 2). Hence the solution set of (8) is {(1, 2)}. Graphs of the systems (8), (9), and (10) appear in Figures **a**, **b**, and **c**. The point of intersection in each case has coordinates (1, 2).

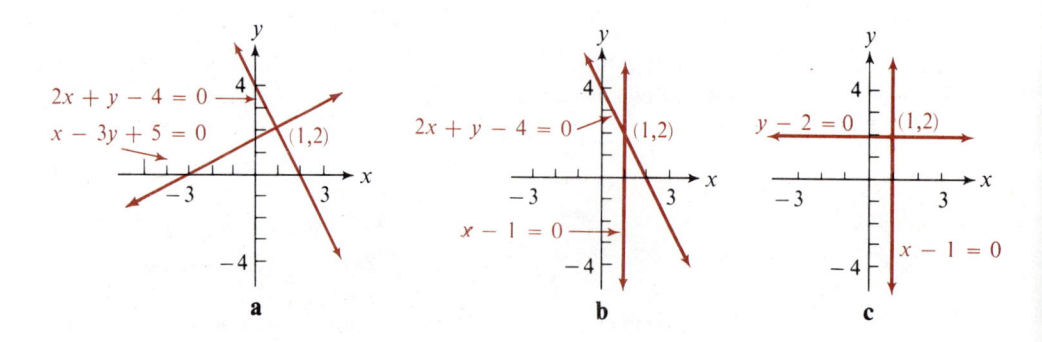

a

b

c

Solution of a system by substitution

Note that we could have proceeded somewhat differently from (9), as follows. A solution of system (9) must be of the form (1, y) (because any such ordered pair is a solution of $x - 1 = 0$); when x is replaced by 1 in $2x + y - 4 = 0$, we obtain $y = 2$, the only ordered pair that satisfies both of the equations

9.1 Systems of Linear Equations in Two Variables

in (9) is (1, 2). Hence, again the solution set of (8) is {(1, 2)}. In this latter way of proceeding from (9) to the solution of (8), we have used what is known as the **method of substitution**.

Choice of multipliers for linear combinations

Observe that the multipliers used in the foregoing example were first 1 and 3, and later -2 and 1. These were chosen because they produced coefficients that were additive inverses, first for the terms in y and later for the terms in x. In general, the linear combination

$$b_2(a_1 x + b_1 y + c_1) - b_1(a_2 x + b_2 y + c_2) = 0$$

will always be free of y, and

$$a_2(a_1 x + b_1 y + c_1) - a_1(a_2 x + b_2 y + c_2) = 0$$

will be free of x.

Examples

Form a linear combination of the equations in the system

$$2x + 6y + 7 = 0$$
$$5x + 4y + 3 = 0$$

to obtain an equation that is free of

a. the variable x.   **b.** the variable y.

Solutions

a. $5(2x + 6y + 7) - 2(5x + 4y + 3) = 0$
$22y + 29 = 0$

b. $4(2x + 6y + 7) - 6(5x + 4y + 3) = 0$
$-22x + 10 = 0$

Criteria for dependent or inconsistent equations

If the coefficients of the variables in one equation in a system are proportional to the corresponding coefficients in the other equation, then the equations might be either dependent or inconsistent. The equations in the system

$$a_1 x + b_1 y + c_1 = 0$$
$$a_2 x + b_2 y + c_2 = 0 \quad (a_2, b_2, c_2 \neq 0)$$

are dependent if

$$\frac{a_1}{a_2} = \frac{b_1}{b_2} = \frac{c_1}{c_2},$$

and inconsistent if

$$\frac{a_1}{a_2} = \frac{b_1}{b_2} \neq \frac{c_1}{c_2}.$$

(See Exercises 9.1-36 and 9.1-37.)

If
$$\frac{a_1}{a_2} \neq \frac{b_1}{b_2},$$
then the system has a unique solution.

Systems of linear equations are quite useful in expressing relationships in practical applications. By assigning separate variables to represent separate physical quantities, we can usually decrease the difficulty in symbolically representing these relationships.

Example A company offers split-rail fence for sale in two prepackaged options. One option consists of 4 posts and 6 rails for $31; the other consists of 3 posts and 4 rails for $22. What are the individual values of posts and rails?

Solution Let x represent the value of a post in dollars and y represent the value of a rail in dollars. Then
$$4x + 6y = 31$$
$$3x + 4y = 22.$$

This system is solved in the first example in the exercises.

Exercise 9.1

A Solve each system in Exercises 1–18.

Example
$$4x + 6y = 31 \qquad (1)$$
$$3x + 4y = 22 \qquad (2)$$

Solution Replace (1) by 3 times itself and (2) by 4 times itself to obtain
$$12x + 18y = 93 \qquad (1')$$
$$12x + 16y = 88. \qquad (2')$$

Replace (1') by itself plus -1 times (2').
$$2y = 5 \qquad (1'')$$
$$12x + 16y = 88 \qquad (2'')$$

From (1''), $y = 5/2$. Substituting this value into (2''), we obtain
$$12x + 40x = 88,$$
$$x = 4.$$

The solution set is $\{(4, 5/2)\}$.

9.1 Systems of Linear Equations in Two Variables

1. $x - y = 1$
   $x + y = 5$

2. $2x - 3y = 6$
   $x + 3y = 3$

3. $3x + y = 7$
   $2x - 5y = -1$

4. $2x - y = 7$
   $3x + 2y = 14$

5. $5x - y = -29$
   $2x + 3y = 2$

6. $6x + 4y = 12$
   $3x + 2y = 12$

7. $5x + 2y = 3$
   $x = 0$

8. $2x - y = 0$
   $x = -3$

9. $3x - 2y = 4$
   $y = -1$

10. $x + 2y = 6$
    $x = 2$

11. $\frac{1}{4}x - \frac{1}{3}y = -\frac{5}{12}$
    $\frac{1}{10}x + \frac{1}{5}y = \frac{1}{2}$

12. $\frac{2}{3}x - y = 4$
    $x - \frac{3}{4}y = 6$

13. $\frac{1}{7}x - \frac{3}{7}y = 1$
    $2x - y = -4$

14. $\frac{1}{3}x - \frac{2}{3}y = 2$
    $x - 2y = 6$

15. $6x + 4y = 12$
    $3x + 2y = 6$

16. $\frac{1}{3}x - \frac{2}{3}y = 2$
    $x - 2y = 6$

17. $3x - y = 4$
    $18x - 6y = -24$

18. $\frac{1}{2}x - \frac{3}{4}y = 1$
    $x - \frac{3}{2}y = 6$

19. Find a and b so that the graph of $ax + by + 3 = 0$ passes through the points $(-1, 2)$ and $(-3, 0)$.

20. Find a and b so that the graph of $ax + by - 4 = 0$ passes through the points $(0, 2)$ and $(2, -4)$.

21. Find a and b so that the solution set of the system
    $$ax + by = 2$$
    $$bx - ay = 2$$
    is $\{(1, 1)\}$.

22. Find a and b so that the solution set of the system
    $$ax + by = 4$$
    $$bx - ay = -3$$
    is $\{(1, 2)\}$.

23. Recall that the slope-intercept form of the equation of a straight line is $y = mx + b$. Find an equation of the line that passes through the points $(0, 2)$ and $(3, -8)$.

24. Find an equation of the line that passes through the points $(-6, 2)$ and $(4, 1)$.

25. Find a linear relationship between centigrade temperature, C, and Fahrenheit temperature, F, given that $F = 32°$ when $C = 0°$, and $F = 212°$ when $C = 100°$.

26. A man has $1.80 in nickels and dimes, with three more dimes than nickels. How many dimes and nickels does he have?

27. How many pounds of an alloy containing 45% silver must be melted with an alloy containing 60% silver to obtain 40 pounds of a 48% silver alloy?

28. A man has three times as much money invested in 9% bonds as he has in stocks paying 8%. How much does he have invested in each if his yearly income from the investments is $1680?

29. A man has $1000 more invested at 9% than he has invested at 11%. If his annual income from the two investments together is $698, how much does he have invested at each rate?

30. An airplane travels 1260 miles in the same time that an automobile travels 420 miles. If the rate of the airplane is 120 miles per hour greater than the rate of the automobile, find the rate of each.

31. Two cars start together and travel in the same direction, one going twice as fast as the other. At the end of 3 hours, they are 96 miles apart. How fast is each traveling?

B Solve each system. Hint: Using substitutions, change the given system to one of the form

$$a_1 u + b_1 v = c_1$$
$$a_2 u + b_2 v = c_2.$$

32. $\dfrac{1}{x} + \dfrac{1}{y} = 2$

$\dfrac{1}{x} - \dfrac{2}{y} = -1$

33. $\dfrac{4}{x} - \dfrac{3}{y} = -7$

$\dfrac{-1}{x} - \dfrac{2}{y} = -1$

34. $\dfrac{1}{x+2} - \dfrac{2}{y+3} = 0$

$\dfrac{2}{x+2} - \dfrac{1}{y+3} = \dfrac{3}{4}$

35. $\dfrac{3}{x+4} + \dfrac{1}{y-4} = \dfrac{1}{6}$

$\dfrac{2}{x+4} + \dfrac{1}{y-4} = 0$

36. Show that the linear polynomials

$$a_1 x + b_1 y + c_1 \quad (a_1, b_1, c_1 \neq 0)$$
$$a_2 x + b_2 y + c_2 \quad (a_2, b_2, c_2 \neq 0)$$

are dependent if and only if

$$\dfrac{a_1}{a_2} = \dfrac{b_1}{b_2} = \dfrac{c_1}{c_2}.$$

Hint: Write the equations in slope-intercept form.

37. Show that the equations in the system

$$a_1 x + b_1 y + c_1 = 0$$
$$a_2 x + b_2 y + c_2 = 0 \quad (a_2, b_2, c_2 \neq 0)$$

are inconsistent if and only if

$$\dfrac{a_1}{a_2} = \dfrac{b_1}{b_2} \neq \dfrac{c_1}{c_2}.$$

Hint: Write the equations in slope-intercept form.

38. Find a and b so that the system

$$ax + by = -1$$
$$bx - ay = 3$$

is inconsistent.

9.2 Systems of Linear Equations in Three Variables

Solutions of an equation in three variables

A solution of an equation in three variables, such as

$$x + 2y - 3z + 4 = 0, \qquad (1)$$

is an ordered triple of numbers (x, y, z), because all three of the variables must be replaced before we can decide whether or not the statement is true. Thus, $(0, -2, 0)$ and $(-1, 0, 1)$ are solutions of Equation (1), whereas $(1, 1, 1)$ is not. There are, of course, infinitely many members in the solution set of such an equation.

Solution of a system

The solution set of a system of linear (first-degree) equations in three variables is the intersection of the solution sets of the separate equations in the system. We are primarily interested in systems involving three equations, such as

$$x + 2y - 3z + 4 = 0$$
$$2x - y + z - 3 = 0$$
$$3x + 2y + z - 10 = 0.$$

We can find the members of this set by methods analogous to those used in the preceding section.

Theorem 9.2 *If any equation in the system*

$$f(x, y, z) = 0$$
$$g(x, y, z) = 0$$
$$h(x, y, z) = 0$$

is replaced by a linear combination, with nonzero coefficients, of itself and any one of the other equations in the system, then the result is an equivalent system.

The proof of this theorem is similar to that of Theorem 9.1 and is omitted here.

Example   Find the solution set of the system

$$x + 2y - 3z + 4 = 0 \qquad (1)$$
$$2x - y + z - 3 = 0 \qquad (2)$$
$$3x + 2y + z - 10 = 0. \qquad (3)$$

Solution overleaf

Solution We begin by replacing Equation (2) with the linear combination formed by multiplying Equation (1) by -2 and Equation (2) by 1, that is, with

$$-2(x + 2y - 3z + 4) + 1(2x - y + z - 3) = 0,$$

or

$$-5y + 7z - 11 = 0.$$

We obtain the equivalent system

$$x + 2y - 3z + 4 = 0 \tag{1}$$
$$-5y + 7z - 11 = 0 \tag{2'}$$
$$3x + 2y + z - 10 = 0, \tag{3}$$

where (2') is free of x. Next, if we replace (3) by the sum of -3 times (1) and 1 times (3), we have

$$x + 2y - 3z + 4 = 0 \tag{1}$$
$$-5y + 7z - 11 = 0 \tag{2'}$$
$$-4y + 10z - 22 = 0, \tag{3'}$$

where both (2') and (3') are free of x. If now (3') is replaced by the sum of 4 times (2') and -5 times (3'), we have

$$x + 2y - 3z + 4 = 0 \tag{1}$$
$$-5y + 7z - 11 = 0 \tag{2'}$$
$$-22z + 66 = 0, \tag{3''}$$

which is equivalent to the original (1), (2), and (3). But, from (3''), we see that for any solution of (1), (2'), and (3''), $z = 3$; that is, the solution will be of the form $(x, y, 3)$. If 3 is substituted for z in (2'), we have

$$-5y + 7(3) - 11 = 0, \quad \text{or} \quad y = 2,$$

and any solution of the system must be of the form $(x, 2, 3)$. Substituting 2 for y and 3 for z in (1) gives

$$x + 2(2) - 3(3) + 4 = 0, \quad \text{or} \quad x = 1,$$

so that the single member of the solution set of (1), (2'), and (3'') is (1, 2, 3). Therefore, the solution set of the original system is $\{(1, 2, 3)\}$.

The foregoing process of solving a system of linear equations can be reduced to a series of mechanical procedures as shown in the following example.

Example Solve.

$$x + 2y - z + 1 = 0 \tag{1}$$
$$x - 3y + z - 2 = 0 \tag{2}$$
$$2x + y + 2z - 6 = 0 \tag{3}$$

9.2 Systems of Linear Equations in Three Variables

Solution

We first multiply (1) by -1 and add the result to 1 times (2). We then multiply (1) by -2 and add the result to 1 times (3). In each linear combination the x-terms vanish.

$$-5y + 2z - 3 = 0 \tag{4}$$

$$-3y + 4z - 8 = 0 \tag{5}$$

Next we multiply (4) by -3 and add the result to 5 times (5). In this linear combination the y-terms vanish.

$$14z - 31 = 0 \tag{6}$$

Now (1), (4), and (6) constitute a set of equations equivalent to (1), (2), and (3).

$$x + 2y - z + 1 = 0 \tag{1}$$

$$-5y + 2z - 3 = 0 \tag{4}$$

$$14z - 31 = 0 \tag{6}$$

Solving for z in (6), we obtain

$$z = \frac{31}{14}.$$

We then substitute $31/14$ for z in (4) and solve for y to obtain

$$-5y + 2\left(\frac{31}{14}\right) - 3 = 0,$$

$$y = \frac{2}{7}.$$

Substituting $2/7$ for y and $31/14$ for z in either (1), (2), or (3), say (1), and solving for x, we obtain

$$x + 2\left(\frac{2}{7}\right) - \frac{31}{14} + 1 = 0,$$

$$x = \frac{9}{14}.$$

The solution set is $\left\{\left(\frac{9}{14}, \frac{2}{7}, \frac{31}{14}\right)\right\}$.

If at any step in the procedure used in the examples above, the resulting linear combination vanishes or yields a contradiction, the system contains linearly dependent left-hand members or else inconsistent equations, or both, and it either has an infinite number of solutions or else has no member in its solution set.

Graphs in three dimensions

By establishing a three-dimensional Cartesian coordinate system as shown in Figure 9.4, a one-to-one correspondence can be established between the points in a three-dimensional space and ordered triples of real numbers. If this is done, it can be shown that the graph of a linear equation in three variables is a plane. For example, the equation

$$x + y + z = 1$$

represents the plane through the points (1, 0, 0), (0, 1, 0), and (0, 0, 1), as illustrated in Figure 9.4. Hence, the solution set of a system of three linear equations in three variables consists of the coordinates of the common intersection of three planes. Figure 9.5 shows the possibilities for the relative positions of the plane graphs of three linear equations in three variables. In case (a), the common

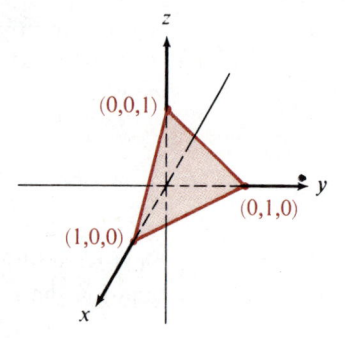

Figure 9.4

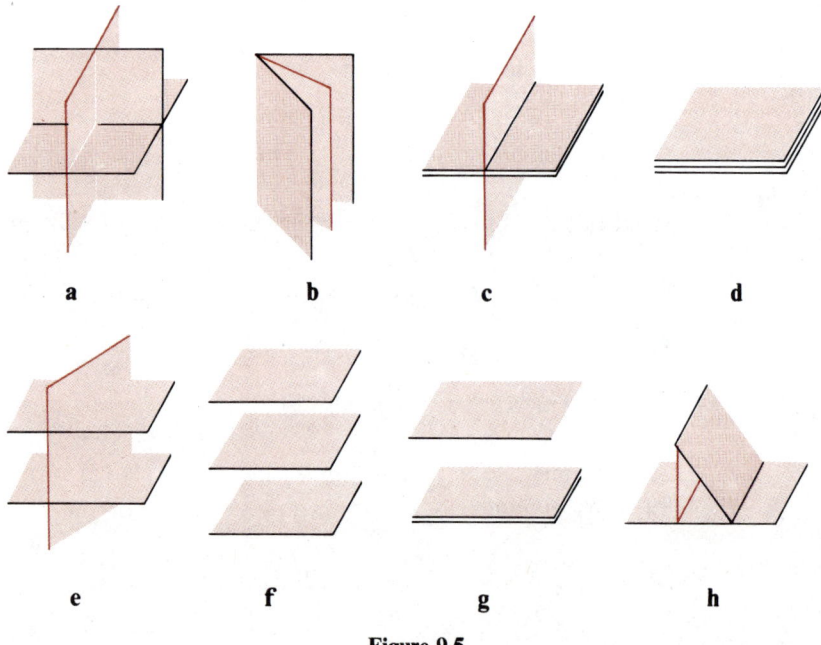

a   b   c   d

e   f   g   h

Figure 9.5

intersection consists of a single point, and hence the solution set of the corresponding system of three equations contains a single member. In cases (b), (c), and (d), the intersection is a line or a plane, and the solution of the corresponding system has infinitely many members. In cases (e), (f), (g) and (h), the three planes have no common intersection, and the solution set of the corresponding system is the null set. Linear equations corresponding to cases (a), (b), (c), and (d) are consistent, and the others inconsistent.

9.2 Systems of Linear Equations in Three Variables

The use of linear combinations to solve systems of linear equations can be extended to cover cases of n equations in n variables. Clearly, as n grows larger, the time and effort necessary to find the solution set of a system increase correspondingly. Fortunately, modern high-speed computers can handle such computations in stride for fairly large values of n. There are analytic methods other than those exhibited here which become preferable for very large systems.

Exercise 9.2

A Solve.

Example

$$x + 2y + z = 4 \quad (1)$$
$$2x + y - z = -1 \quad (2)$$
$$-x + y + z = 2 \quad (3)$$

Solution Replace (2) by itself plus -2 times (1), and (3) by itself plus (1).

$$x + 2y + z = 4 \quad (1')$$
$$-3y - 3z = -9 \quad (2')$$
$$3y + 2z = 6 \quad (3')$$

Replace (3') by itself plus (2').

$$x + 2y + z = 4 \quad (1'')$$
$$-3y - 3z = -9 \quad (2'')$$
$$-z = -3 \quad (3'')$$

From (3''), obtain $z = 3$. Substitute $z = 3$ into (2'') to obtain $y = 0$. Substitute $z = 3$ and $y = 0$ into (1'') to obtain $x = 1$. The solution set is therefore $\{(1, 0, 3)\}$.

1. $x + y + z = 2$
   $2x - y + z = -1$
   $x - y - z = 0$

2. $x + y + z = 1$
   $2x - y + 3z = 2$
   $2x - y - z = 2$

3. $x + y + 2z = 0$
   $2x - 2y + z = 8$
   $3x + 2y + z = 2$

4. $2x - 3y + z = 3$
   $x - y - 3z = -1$
   $-x + 2y - 3z = -4$

5. $x - 2y + z = -1$
   $2x + y - 3z = 3$
   $3x + 3y - 2z = 10$

6. $x - 2y + 4z = -3$
   $3x + y - 2z = 12$
   $2x + y - 3z = 11$

7. $4x - 2y + 3z = 4$
   $2x - y + z = 1$
   $3x - 3y + 4z = 5$

8. $x + 5y - z = 2$
   $3x - 9y + 3z = 6$
   $x - 3y + z = 4$

9. $x + z = 5$
   $y - z = -4$
   $x + y = 1$

10. $5y - 8z = -19$
    $5x - 8z = 6$
    $3x - 2y = 12$

11. $x - \frac{1}{2}y - \frac{1}{2}z = 4$
    $x - \frac{3}{2}y - 2z = 3$
    $\frac{1}{4}x + \frac{1}{4}y - \frac{1}{4}z = 0$

12. $x + 2y + \frac{1}{2}z = 0$
    $x + \frac{3}{5}y - \frac{2}{5}z = \frac{1}{5}$
    $4x - 7y - 7z = 6$

Use three variables in the solution of each of the following problems.

Example The parabola $y = ax^2 + bx + c$ passes through the points $(-1, 9), (1, 3)$, and $(3, 5)$. Find the coefficients $a, b,$ and c.

Solution Since $(-1, 9)$ is on the parabola,
$$9 = a(-1)^2 + b(-1) + c,$$
or
$$a - b + c = 9. \tag{1}$$

The other two points yield the equations
$$a + b + c = 3, \tag{2}$$
$$9a + 3b + c = 5. \tag{3}$$

Replace (2) by itself plus -1 times (1), and replace (3) by itself plus -9 times (1).
$$a - b + c = 9 \tag{1'}$$
$$2b = -6 \tag{2'}$$
$$12b - 8c = -76 \tag{3'}$$

From (2'), $b = -3$. Substitute $b = -3$ into (3') to obtain $c = 5$. Substitute $b = -3$ and $c = 5$ into (1') to obtain $a = 1$. Therefore the equation of the parabola is $y = x^2 - 3x + 5$.

13. The sum of three numbers is 15. The second equals two times the first and the third equals the second. Find the numbers.

14. The sum of three numbers is 2. The first number is equal to the sum of the other two, and the third number is the result of subtracting the first from the second. Find the numbers.

15. A box contains $6.25 in nickels, dimes, and quarters. There are 85 coins in all, with three times as many nickels as dimes. How many coins of each kind are there?

16. A man had $446 in ten-dollar, five-dollar, and one-dollar bills. There were 94 bills in all and 10 more five-dollar bills than ten-dollar bills. How many bills of each kind did he have?

17. The perimeter of a triangle is 155 centimeters. The side x is 20 centimeters shorter than the side y, and the side y is 5 centimeters longer than the side z. Find the lengths of the sides of the triangle.

18. The perimeter of a triangle is 120 centimeters. The side x is the same length as side y and 10 centimeters shorter than side z. Find the length of each side.

19. Find values for $a, b,$ and c so that the graph of $x^2 + y^2 + ax + by + c = 0$ will contain the points $(0, 0), (6, 0),$ and $(0, 8)$.

20. The equation for a circle can be written $x^2 + y^2 + ax + by + c = 0$. Find the equation of the circle with graph containing the points $(2, 3), (3, 2),$ and $(-4, -5)$.

21. Find values for a, b and c so that the graph of $y = ax^2 + bx + c$ contains the points $(-1, 0), (2, 12)$ and $(-2, 8)$.

9.3 Partial Fractions

22. Find values for a, b, and c so that the graph of $y = ax^2 + bx + c$ contains the points $(-1, 2)$, $(1, 6)$, and $(2, 11)$.

23. Three solutions of the equation $ax + by + cz = 1$ are $(0, 2, 1)$, $(6, -1, 2)$, and $(0, 2, 0)$. Find the coefficients a, b, and c.

24. Three solutions of the equation $ax + by + cz = 1$ are $(2, 1, 0)$, $(-1, 3, 2)$, and $(3, 0, 0)$. Find the coefficients a, b, and c.

B Solve each system. *Hint:* Using substitutions, change the given system to one of the form

$$a_1 u + b_1 v + c_1 w = d_1$$
$$a_2 u + b_2 v + c_2 w = d_2$$
$$a_3 u + b_3 v + c_3 w = d_3.$$

25. $\dfrac{1}{x} + \dfrac{1}{y} - \dfrac{1}{z} = 1$

$\dfrac{2}{x} - \dfrac{2}{y} + \dfrac{1}{z} = 1$

$\dfrac{-3}{x} + \dfrac{1}{y} - \dfrac{1}{z} = -3$

26. $\dfrac{4}{x} - \dfrac{2}{y} + \dfrac{1}{z} = 4$

$\dfrac{3}{x} - \dfrac{1}{y} + \dfrac{2}{z} = 0$

$\dfrac{-1}{x} + \dfrac{3}{y} - \dfrac{2}{z} = 0$

27. $\dfrac{2}{x-1} + \dfrac{1}{y+2} - \dfrac{1}{z-3} = \dfrac{-2}{3}$

$\dfrac{-3}{x-1} + \dfrac{1}{y+2} - \dfrac{2}{z-3} = \dfrac{16}{3}$

$\dfrac{2}{x-1} + \dfrac{3}{y+2} - \dfrac{1}{z-3} = 0$

28. $\dfrac{2}{x+1} - \dfrac{1}{y-3} + \dfrac{1}{z} = -3$

$\dfrac{-3}{x+1} + \dfrac{2}{y-3} - \dfrac{2}{z} = \dfrac{13}{2}$

$\dfrac{-2}{x+1} + \dfrac{4}{y-3} - \dfrac{1}{z} = 9$

29. Show that the system

$$x + y + 2z = 2$$
$$2x - y - z = 3$$

has an infinite number of members in its solution set. List two ordered triples that are solutions. *Hint:* Express x in terms of z alone, and express y in terms of z alone.

30. Give a geometric argument to show that any system of two consistent linear equations in three variables has an infinite number of solutions.

9.3 Partial Fractions

We can use the method of solving systems of equations that we considered in Sections 9.1 and 9.2 to help us rewrite certain kinds of fractions in what may be more useful forms. Recall that the sum $\dfrac{P(x)}{Q(x)} + \dfrac{R(x)}{S(x)}$ can be rewritten as the single fraction

$$\dfrac{P(x) \cdot S(x) + Q(x) \cdot R(x)}{Q(x) \cdot S(x)}.$$

It is frequently useful to be able to reverse this process in the case of certain kinds of fractions—in particular, for rational expressions in which the numerator is a polynomial of lesser degree than the denominator. Such rational expressions are customarily referred to as "proper" fractions.

Notice first that by the long-division algorithm, any rational expression can be written as a polynomial or the sum of a polynomial and a proper fraction. For example, by the long-division algorithm we find that

$$\dfrac{x^3 + 2x^2 + 2x + 3}{x^2 - 1} = x + 2 + \dfrac{3x + 5}{x^2 - 1}.$$

Distinct factors in the denominator

Let us now consider the problem of finding values c_1 and c_2 (if they exist) so that

$$\dfrac{ax + b}{(x - r_1)(x - r_2)} = \dfrac{c_1}{x - r_1} + \dfrac{c_2}{x - r_2}, \qquad (1)$$

where a, b, r_1, r_2 ($r_1 \neq r_2$) are known constants.

Suppose first that there are such constants $c_1, c_2 \in R$. Multiplying each member of Equation (1) by $(x - r_1)(x - r_2)$, we get

$$ax + b = c_1(x - r_2) + c_2(x - r_1).$$

Since we wish this equation to hold for *every value of* x other than r_1 and r_2, it must hold also for $x = r_1$ and $x = r_2$ (see Exercise 8.2-25). Substituting r_1 and r_2 for x in turn, we get the system of equations in c_1 and c_2,

$$ar_1 + b = c_1(r_1 - r_2) + c_2(r_1 - r_1)$$
$$ar_2 + b = c_1(r_2 - r_2) + c_2(r_2 - r_1),$$

from which, since $r_1 \neq r_2$,

$$c_1 = \dfrac{ar_1 + b}{r_1 - r_2} \quad \text{and} \quad c_2 = \dfrac{ar_2 + b}{r_2 - r_1}.$$

That these values satisfy the given fractional equation can be verified by direct substitution. Thus, not only do c_1 and c_2 exist, but they have the values shown. This proves the following theorem.

Theorem 9.3 *If $a, b, r_1, r_2 \in R$ are given, with $r_1 \neq r_2$, then there exist constants $c_1, c_2 \in R$ such that for $x \neq r_1, r_2$,*

$$\dfrac{ax + b}{(x - r_1)(x - r_2)} = \dfrac{c_1}{x - r_1} + \dfrac{c_2}{x - r_2}.$$

9.3 Partial Fractions

While proof of the foregoing theorem produces formulas for c_1 and c_2, it is preferable from the standpoint of efficiency to rely on the method used in the proof to obtain the numbers c_1 and c_2 in any particular problem. The technique is known as the method of **partial fractions**.

Example Apply the method of partial fractions to express $\dfrac{3x-2}{x^2-x}$ as a sum of fractions.

Solution We first observe that

$$\frac{3x-2}{x^2-x} = \frac{3x-2}{x(x-1)}. \tag{2}$$

Then by Theorem 9.3, we know there exist numbers $c_1, c_2 \in R$, such that

$$\frac{3x-2}{x(x-1)} = \frac{c_1}{x} + \frac{c_2}{x-1}. \tag{3}$$

Multiplying each member of the equation by $x(x-1)$, we obtain

$$3x - 2 = c_1(x-1) + c_2(x). \tag{4}$$

We wish the members of Equation (3) to be equal for all values of x except 0 and 1. However, if this is to be the fact, then the members of (4) must be equal for all values of x, *including* 0 and 1. Hence, while we can arbitrarily select *any* values for x and solve for c_1 and c_2 in (4), let us take 0 and 1 because these values of x yield the values of c_1 and c_2 by inspection. Thus, if $x = 0$, we have from Equation (4)

$$3(0) - 2 = c_1(0-1) + c_2(0),$$

from which

$$c_1 = 2.$$

If $x = 1$, Equation (4) becomes

$$3(1) - 2 = c_1(1-1) + c_2(1),$$

$$c_2 = 1.$$

Therefore,

$$\frac{3x-2}{x(x-1)} = \frac{2}{x} + \frac{1}{x-1}.$$

Theorem 9.3 generalizes to the following form, although the proof is omitted.

Theorem 9.4 *If $P(x)$ and $Q(x)$ are real polynomials, with $P(x)$ of degree less than $Q(x)$, and if $Q(x) = (x - r_1)(x - r_2) \cdots (x - r_n)$, where no two factors are identical, then there exist constants $c_1, c_2, \ldots, c_n \in R$ such that*

$$\frac{P(x)}{Q(x)} = \frac{c_1}{x - r_1} + \frac{c_2}{x - r_2} + \cdots + \frac{c_n}{x - r_n}.$$

Repeated factors in the denominator

If the denominator of a proper fraction can be factored into *equal* linear factors, then we need the following result, which is stated without proof.

Theorem 9.5 If $P(x)$ and $Q(x)$ are real polynomials with $P(x)$ of degree less than $Q(x)$, and if $Q(x) = (x - r_1)^n$, then there exist constants $c_1, c_2, \ldots, c_n \in R$ such that

$$\frac{P(x)}{Q(x)} = \frac{c_1}{(x - r_1)} + \frac{c_2}{(x - r_1)^2} + \cdots + \frac{c_n}{(x - r_1)^n}.$$

A further extension of Theorems 9.4 and 9.5 can be used to apply the method of partial fractions to rational expressions in which the denominator has one or more repeated factors. Each repeated factor contributes a sum such as the one in Theorem 9.5.

Example Express $\dfrac{5x^2 + 2}{x^3 - 2x^2 + x}$ as a sum of fractions.

Solution We first observe that $\dfrac{5x^2 + 2}{x^3 - 2x^2 + x} = \dfrac{5x^2 + 2}{x(x - 1)(x - 1)}$. Then we write

$$\frac{5x^2 + 2}{x(x - 1)(x - 1)} = \frac{c_1}{x} + \frac{c_2}{x - 1} + \frac{c_3}{(x - 1)^2}.$$

Multiplying each member of the equation by $x(x - 1)^2$, we obtain

$$5x^2 + 2 = c_1(x - 1)^2 + c_2 x(x - 1) + c_3 x.$$

Arbitrarily selecting any values of x, we can solve for $c_1, c_2,$ and c_3. Let us use 0, 1, and 2, because these numbers yield simple computations. Substituting 0 for x gives

$$5(0)^2 + 2 = c_1(-1)^2 + 0 + 0,$$

$$c_1 = 2.$$

Substituting 1 for x gives

$$5(1)^2 + 2 = 0 + 0 + c_3(1),$$

$$c_3 = 7.$$

Substituting 2 for x gives

$$5(2)^2 + 2 = c_1(1)^2 + c_2(2)(1) + c_3(2).$$

9.3 Partial Fractions

Now, since $c_1 = 2$ and $c_3 = 7$, we have

$$22 = 2 + 2c_2 + 14,$$

$$c_2 = 3.$$

Hence,

$$\frac{5x^2 + 2}{x^3 - 2x^2 + x} = \frac{2}{x} + \frac{3}{x - 1} + \frac{7}{(x - 1)^2}.$$

Alternative method

The constants c_i used to write a rational expression as a sum of fractions of the form $c_i/(x - r)^n$ can be obtained by equating the coefficients of two polynomials. This method is illustrated in the following example.

Example

Express $\dfrac{2x + 1}{(x - 1)^3}$ as a sum of fractions.

Solution

We first write

$$\frac{2x + 1}{(x - 1)^3} = \frac{c_1}{x - 1} + \frac{c_2}{(x - 1)^2} + \frac{c_3}{(x - 1)^3}.$$

Multiplying both members of this equation by $(x - 1)^3$, we obtain

$$2x + 1 = c_1(x - 1)^2 + c_2(x - 1) + c_3$$
$$= c_1 x^2 + (c_2 - 2c_1)x + (c_1 - c_2 + c_3).$$

Since both members of this equation are polynomials, and they are equal for all values of x except $x = 1$, their coefficients must be equal (Exercise 8.2-25). Thus, we have the system

$$c_1 = 0$$
$$-2c_1 + c_2 = 2$$
$$c_1 - c_2 + c_3 = 1.$$

Using the methods of Section 9.2, we solve this system to obtain

$$c_1 = 0, \quad c_2 = 2, \quad \text{and} \quad c_3 = 3.$$

Hence,

$$\frac{2x + 1}{(x - 1)^3} = \frac{2}{(x - 1)^2} + \frac{3}{(x - 1)^3}.$$

Note that Theorems 9.4 and 9.5 apply to fractions whose denominators can be factored into linear factors. Similar theorems apply to fractions whose denominators cannot be factored completely into linear factors. In particular, the numerator of each term of the partial-fraction expansion whose denominator is a nonfactorable quadratic expression will be a linear expression of the form $c_1 x + c_2$. Several such fractions are given in the exercises.

Exercise 9.3

A Use the method of partial fractions to express each fraction as a sum of fractions with powers of linear polynomials as denominators. The numerators may be rational numbers.

Example

$$\frac{2}{x^2 - x}$$

Solution

Factor $x^2 + x$ as $x(x + 1)$ and write

$$\frac{2}{x^2 + x} = \frac{c_1}{x} + \frac{c_2}{x + 1} = \frac{c_1(x + 1) + c_2 x}{x^2 + x}.$$

Thus, $2 = c_1(x + 1) + c_2 x$. Substitute $x = -1$ and $x = 0$ to obtain

$$2 = c_1(-1 + 1) + c_2(-1) = -c_2,$$

$$2 = c_1(0 + 1) + c_2 \cdot 0 = c_1.$$

Therefore, $c_1 = 2$, $c_2 = -2$, and

$$\frac{2}{x^2 + x} = \frac{2}{x} - \frac{2}{x + 1}.$$

1. $\dfrac{4}{x^2 + x}$
2. $\dfrac{x + 2}{x^2 + x}$
3. $\dfrac{x}{(x + 1)(x + 2)}$
4. $\dfrac{3}{x^2 + 4x + 3}$
5. $\dfrac{x - 2}{x^2 + 5x + 6}$
6. $\dfrac{2x - 1}{x^2 + 5x - 14}$

Example

$$\frac{x + 1}{x(x + 2)(x + 4)}$$

Solution

Write

$$\frac{x + 1}{x(x + 2)(x + 4)} = \frac{c_1}{x} + \frac{c_2}{x + 2} + \frac{c_3}{x + 4}$$

$$= \frac{c_1(x + 2)(x + 4) + c_2 x(x + 4) + c_3 x(x + 2)}{x(x + 2)(x + 4)}.$$

Thus, $x + 1 = c_1(x + 2)(x + 4) + c_2 x(x + 4) + c_3 x(x + 2)$. Substitute $x = 0, -2, -4$ to obtain

$$1 = c_1 \cdot 2 \cdot 4 + c_2 \cdot 0 + c_3 \cdot 0 = 8c_1$$

$$-1 = c_1 \cdot 0 \cdot 2 + c_2(-2) \cdot 2 + c_3(-2) \cdot 0 = -4c_2$$

$$-3 = c_1 \cdot (-2) \cdot 0 + c_2(-4) \cdot 0 + c_3(-4)(-2) = 8c_3.$$

9.3 Partial Fractions

Therefore, $c_1 = \frac{1}{8}$, $c_2 = \frac{1}{4}$, $c_3 = \frac{-3}{8}$, so that

$$\frac{x+1}{x(x+2)(x+4)} = \frac{\frac{1}{8}}{x} + \frac{\frac{1}{4}}{x+2} - \frac{\frac{3}{8}}{x+4}.$$

7. $\dfrac{2x+1}{x(x+1)(x+2)}$

8. $\dfrac{x}{(x-1)(x+1)(x+3)}$

9. $\dfrac{x^2-x+1}{x(x+1)(x+2)}$

10. $\dfrac{x^2+1}{x(x-1)(x+1)}$

11. $\dfrac{1}{x^3+3x^2+2x}$

12. $\dfrac{x-1}{(x+1)(x^2-2x)}$

Example $\dfrac{x+2}{x(x+1)^2}$

Solution Write $\dfrac{x+2}{x(x+1)^2} = \dfrac{c_1}{x} + \dfrac{c_2}{x+1} + \dfrac{c_3}{(x+1)^2}$

$$= \frac{c_1(x+1)^2 + c_2 x(x+1) + c_3 x}{x(x+1)^2}.$$

Thus, $x + 2 = c_1(x+1)^2 + c_2 x(x+1) + c_3 x$. Substitute $x = 0, -1$ to obtain

$$2 = c_1 \cdot 1 + c_2 \cdot 0 + c_3 \cdot 0 = c_1$$

$$1 = c_1 \cdot 0 + c_2 \cdot 0 + c_3(-1) = -c_3.$$

Substitute any other value for x to find c_2; here we will use $x = 1$ to obtain

$$3 = c_1 \cdot 4 + c_2 \cdot 2 + c_3 = 2 \cdot 4 + 2 \cdot c_2 - 1,$$

so that $3 = 7 + 2c_2$, or $c_2 = -2$. Therefore,

$$\frac{x+2}{x(x+1)^2} = \frac{2}{x} - \frac{2}{x+1} - \frac{1}{(x+1)^2}.$$

13. $\dfrac{x+1}{x^2(x-2)}$

14. $\dfrac{x+3}{(x+1)(x+2)^2}$

15. $\dfrac{x^3}{(x-1)^2(x+1)^2}$

16. $\dfrac{x^2+1}{x^4-2x^2+1}$

17. $\dfrac{1}{x^3-x^2}$

18. $\dfrac{1}{x^4+x^2}$

Use long division and the method of partial fractions to express each fraction as the sum of a polynomial and fractions with powers of linear polynomials for denominators.

Example $\dfrac{x^3+2x^2+x+2}{x^2+x}$

Solution overleaf

Solution Long division yields

$$\frac{x^3 + 2x^2 + x + 2}{x^2 + x} = x + 1 + \frac{2}{x^2 + x}.$$

The method of partial fractions yields

$$\frac{2}{x^2 + x} = \frac{2}{x} - \frac{2}{x + 1}.$$

Thus,

$$\frac{x^3 + 2x^2 + x + 2}{x^2 + x} = x + 1 + \frac{2}{x} - \frac{2}{x + 1}.$$

19. $\dfrac{x^4 + x^3 + x + 2}{x^2 + x}$ 20. $\dfrac{2x^2 + 9x + 10}{x^2 + 5x + 6}$

21. $\dfrac{x^3}{(x - 1)^2}$ 22. $\dfrac{x^3}{x^2 - 1}$

B 23. Find constants a, b, and c so that

$$\frac{x + 1}{x^3 + x^2 + x} = \frac{a}{x} + \frac{bx + c}{x^2 + x + 1}.$$

(*Hint*: Multiply each member of the equation by $x^3 + x^2 + x = x(x^2 + x + 1)$. Then simplify the right-hand member and equate the coefficients of like powers of x. Solve the resulting system of equations for a, b, and c).

24. Find constants a, b, and c so that

$$\frac{3x + 1}{x^3 - 8} = \frac{a}{x - 2} + \frac{bx + c}{x^2 + 2x + 4}.$$

25. Find constants a, b, and c so that

$$\frac{2x - 3}{x^3 - 2x^2 + 4x} = \frac{a}{x} + \frac{bx + c}{x^2 - 2x + 4}.$$

26. Find constants a, b, and c so that

$$\frac{3x - 2}{(x - 2)(x^2 - 3x + 3)} = \frac{a}{x - 2} + \frac{bx + c}{x^2 - 3x + 3}.$$

9.4 Systems of Nonlinear Equations

In Sections 9.1 and 9.2 we discussed systems of *linear* equations. In this section we introduce methods which can sometimes be used to solve systems containing non-linear equations.

9.4 Systems of Nonlinear Equations

Use of graphs In solving a system containing nonlinear equations, it is frequently helpful to sketch the graphs of all of the equations in the system. By studying the graphs, we can approximate the intersection points of the curves. This usually tells us how many solutions with real components the system has, and it gives us approximate values for the solutions, which we can use as a rough check on any solutions we find algebraically.

Example Determine the number of solutions of the system

$$x^2 + y^2 = 25$$

$$x + y = 1$$

and approximate them.

Solution We sketch the graphs of $x^2 + y^2 = 25$ and $x + y = 1$ on the same set of coordinate axes. Observe from the figure that there are two points of intersection with coordinates approximately $(-3, 4)$ and $(4, -3)$. Thus, the system has two solutions with real components and they are approximately $(-3, 4)$ and $(4, -3)$.

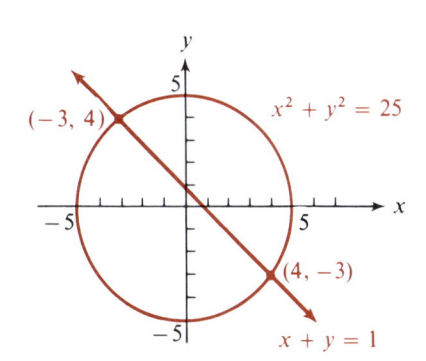

Substitution The substitution method is sometimes a convenient means of finding solution sets of systems in which nonlinear equations are present.

Example Solve.

$$x^2 + y^2 = 25 \tag{1}$$

$$x + y = 1$$

Solution In the foregoing example, we found that the system has two solutions with real components. We now use the method of substitution to determine the solutions algebraically. Equation (2) can be written equivalently in the form

$$y = 1 - x, \tag{3}$$

and we can replace y in (1) by $(1 - x)$ from (3). This produces

$$x^2 + (1 - x)^2 = 25, \tag{4}$$

Solution continued overleaf

which has as a solution set those values of x for which the ordered pair (x, y) is a common solution of (1) and (2). We can now find the solution set of (4):

$$x^2 + 1 - 2x + x^2 = 25,$$
$$2x^2 - 2x - 24 = 0,$$
$$2(x + 3)(x - 4) = 0,$$

which is satisfied if x is either -3 or 4. Now, by replacing x in the *first-degree equation* (3) by each of these numbers in turn, we obtain

$$y = 1 - (-3) = 4$$

and

$$y = 1 - 4 = -3,$$

respectively, so that the solution set of the system (1) and (2) is $\{(-3, 4), (4, -3)\}$.

If, in the preceding example, we had substituted the values -3 and 4 that we obtained for x in the *second-degree equation* (1), we would have obtained some extraneous solutions that do not satisfy Equation (2).

Linear combination If both of the equations in a system of two equations in two variables are of the second degree in both variables, the use of linear combinations of the equations often provides a simpler means of solution than does substitution.

Example Solve.

$$3x^2 - 7y^2 + 15 = 0 \qquad (1)$$
$$3x^2 - 4y^2 - 12 = 0 \qquad (2)$$

Solution We graph the two equations on the same set of coordinate axes and observe that there are four solutions with real components. By forming the linear combination

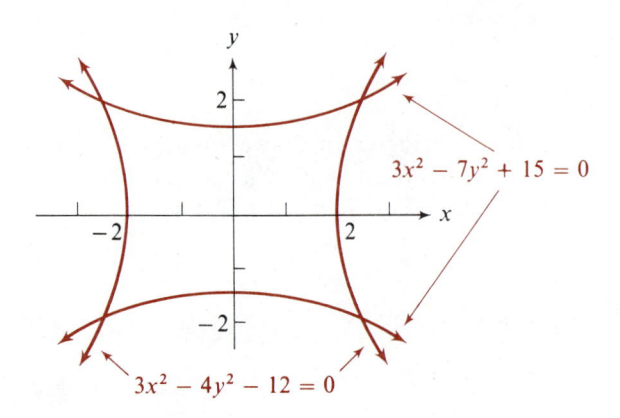

9.4 Systems of Nonlinear Equations

of -1 times (1) and 1 times (2), we obtain

$$3y^2 - 27 = 0$$
$$y^2 = 9,$$

from which

$$y = 3 \quad \text{or} \quad y = -3,$$

and we have the y-components of the members of the solution set of Equations (1) and (2). Substituting 3 for y in either (1) or (2), say (1), we have

$$3x^2 - 7(3)^2 + 15 = 0,$$
$$x^2 = 16,$$

from which

$$x = 4 \quad \text{or} \quad x = -4.$$

Thus, the ordered pairs (4, 3) and $(-4, 3)$ are solutions of the system. Substituting for y in (1) or (2) (this time we shall use (2)) gives us

$$3x^2 - 4(-3)^2 - 12 = 0,$$
$$x^2 = 16,$$

so that

$$x = 4 \quad \text{or} \quad x = -4.$$

Thus the ordered pairs $(4, -3)$ and $(-4, -3)$ are solutions of the system, and accordingly the complete solution set is $\{(4, 3), (4, -3), (-4, 3), (-4, -3)\}$.

Linear combination and substitution

The solution of some systems requires the application of both linear combinations *and* substitution.

Example Solve.

$$x^2 + y^2 = 5 \qquad (1)$$
$$x^2 - 2xy + y^2 = 1 \qquad (2)$$

Solution We graph the two equations on the same set of coordinate axes. Note that $x^2 - 2xy + y^2 = 1$ is equivalent to $(x - y)^2 = 1$, or $x - y = \pm 1$. Thus the graph of this equation consists of the two parallel lines $x - y = 1$ and $x - y = -1$. Observe that there are four solutions with real components.

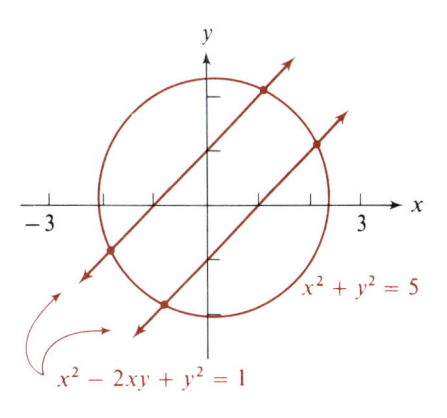

Solution continued overleaf

By forming the linear combination of 1 times (1) and -1 times (2), we obtain
$$2xy = 4,$$
$$xy = 2. \tag{3}$$

The system (1) and (2) is equivalent to the system (1) and (3). This latter system can be solved by substitution. From (3), we have
$$y = \frac{2}{x}.$$

Replacing y in (1) by $2/x$, we obtain
$$x^2 + \left(\frac{2}{x}\right)^2 = 5,$$
$$x^2 + \frac{4}{x^2} = 5, \tag{4}$$
$$x^4 + 4 = 5x^2,$$
$$x^4 - 5x^2 + 4 = 0, \tag{5}$$

which is a quadratic in x^2. The left-hand member of (5), when factored, yields
$$(x^2 - 1)(x^2 - 4) = 0,$$

from which we obtain
$$x^2 - 1 = 0 \quad \text{or} \quad x^2 - 4 = 0,$$

so that
$$x = 1, \quad x = -1, \quad x = 2, \quad x = -2.$$

Since the step from (4) to (5) was not an elementary transformation, we are careful to note that these values of x all satisfy (4). Substituting 1, -1, 2, and -2 in turn for x in (3), we obtain the corresponding values for y, and thus the solution set of the system (1) and (3) or, equivalently, the solution set of the system (1) and (2), is $\{(1, 2), (-1, -2), (2, 1), (-2, -1)\}$.

There are other techniques involving substitution in conjunction with linear combinations that are useful in handling systems of higher-degree equations, but they all bear similarity to those illustrated.

Imaginary solutions In the foregoing examples, each solution is a member of $R \times R$ and their graphs are the points of intersection of the graphs of each equation. If one or more of the components of the solutions are complex numbers, we find these solutions in
$$C \times C = \{(x, y) | x \in C \text{ and } y \in C\}.$$

However, the graphs in the real plane of the equations do *not* have points of intersection corresponding to these solutions.

9.4 Systems of Nonlinear Equations

Example Solve.

$$x^2 + y^2 = 26 \qquad (1)$$

$$x + y = 8 \qquad (2)$$

Solution We graph the two equations on the same set of coordinate axes and observe that there are no solutions with real components. We use substitution to find the solutions.

Equation (2) can be written equivalently as

$$y = 8 - x. \qquad (3)$$

Substituting $8 - x$ for y in (1) and simplifying yields

$$x^2 + (8 - x)^2 = 26$$

$$x^2 + 64 - 16x + x^2 = 26$$

$$2x^2 - 16x + 38 = 0$$

$$x^2 - 8x + 19 = 0.$$

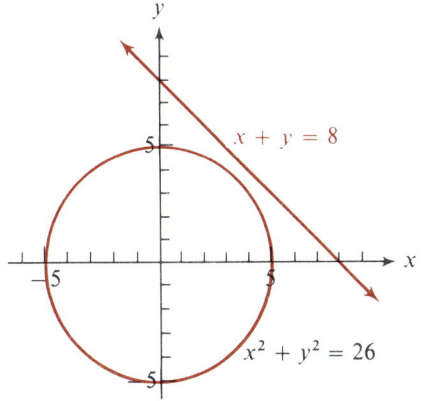

Using the quadratic formula to solve for x, we obtain

$$x = \frac{8 \pm \sqrt{64 - 76}}{2(1)} = \frac{8 \pm \sqrt{-12}}{2}$$

$$= \frac{8 \pm 2i\sqrt{3}}{2} = \frac{2(4 \pm i\sqrt{3})}{2} = 4 \pm i\sqrt{3}.$$

Then, substituting $4 + i\sqrt{3}$ for x in (3) gives $y = 4 - i\sqrt{3}$, and substituting $4 - i\sqrt{3}$ for x in (3) gives $y = 4 + i\sqrt{3}$. Hence, the solution set is

$$\{(4 + i\sqrt{3}, 4 - i\sqrt{3}), (4 - i\sqrt{3}, 4 + i\sqrt{3})\}.$$

Approximations to real solutions When the degree of any equation in a set of equations is greater than two, or when the left-hand member of one of the equations of the form $f(x, y) = 0$ is not a polynomial, it may be very difficult or impossible to find common solutions analytically. In this case, we can at least obtain approximations to any real solutions by graphical methods.

Example Approximate the solutions to the system

$$y = 2^{-x}$$

$$y = \sqrt{0.01x}.$$

Solution overleaf

Solution   We first graph these equations.

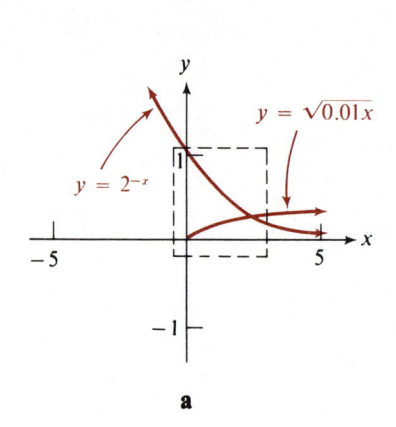

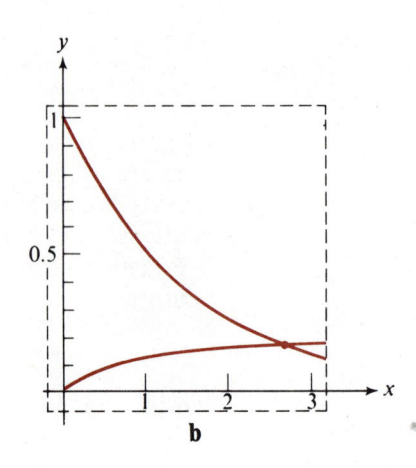

a

b

Examining Figure **b**, an enlargement of part of Figure **a**, we observe that the curves intersect at approximately (2.6, 0.16), and from geometric considerations we conclude that this ordered pair approximates the only member in the solution set of the system.

The preceding example shows, in particular, that the equation

$$2^{-x} - \sqrt{0.01x} = 0$$

has just one solution, approximately 2.6, and suggests a method of approximating solutions of similar equations.

Exercise 9.4

A   *Solve by the method of substitution. Check the solutions by sketching the graphs of the equations and estimating the coordinates of any points of intersection.*

Example

$$y = x^2 + 2x + 1 \qquad (1)$$
$$y - x = 3 \qquad (2)$$

Solution   Solve Equation (2) explicitly for y.

$$y = x + 3 \qquad (2')$$

Substitute $x + 3$ for y in (1).

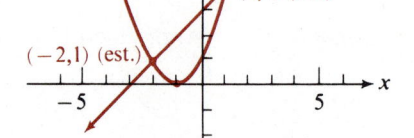

$$x + 3 = x^2 + 2x + 1 \qquad (3)$$

$(1, 4)$ (est.)

$(-2, 1)$ (est.)

9.4 Systems of Nonlinear Equations

Solve for x.
$$x^2 + x - 2 = 0$$
$$(x+2)(x-1) = 0$$
$$x = -2, \quad x = 1$$

Substitute each of these values in turn in (2′) to determine values for y.

If $x = -2$, then $y = 1$.

If $x = 1$, then $y = 4$.

The solution set is $\{(-2, 1), (1, 4)\}$.

1. $y = x^2 - 5$
   $y = 4x$

2. $y = x^2 - 2x + 1$
   $y + x = 3$

3. $x^2 + y^2 = 13$
   $x + y = 5$

4. $x^2 + 2y^2 = 12$
   $2x - y = 2$

5. $x + y = 1$
   $xy = -12$

6. $2x - y = 9$
   $xy = -4$

Solve using linear combinations.

Example

$$2x^2 + y^2 = 7 \qquad (1)$$
$$x^2 - y^2 = 2 \qquad (2)$$

Solution

Replace (1) by itself minus 2 times (2).
$$3y^2 = 3 \qquad (1')$$
$$x^2 - y^2 = 2 \qquad (2)$$

From (1′) $y^2 = 1$; substitution in (2) yields $x^2 = 3$. Therefore
$$x = \pm\sqrt{3} \text{ and } y = \pm 1,$$
so the solution set is $\{(\sqrt{3}, 1), (\sqrt{3}, -1), (-\sqrt{3}, 1), (-\sqrt{3}, -1)\}$.

7. $x^2 + y^2 = 10$
   $9x^2 + y^2 = 18$

8. $x^2 + 4y^2 = 52$
   $x^2 + y^2 = 25$

9. $x^2 - y^2 = 7$
   $2x^2 + 3y^2 = 24$

10. $x^2 + 4y^2 = 25$
    $4x^2 + y^2 = 25$

11. $4x^2 - 9y^2 + 132 = 0$
    $x^2 + 4y^2 - 67 = 0$

12. $16y^2 + 5x^2 - 26 = 0$
    $25y^2 - 4x^2 - 17 = 0$

Solve.

Example

$$x^2 - 3xy + 2y^2 = 0 \qquad (1)$$
$$2x^2 - xy - 3y^2 = 12 \qquad (2)$$

Solution overleaf

Solution Factor the left-hand member of (1).

$$(x - y)(x - 2y) = 0$$

Since the product is 0, we must have

$$x - y = 0 \quad \text{or} \quad x - 2y = 0.$$

Thus,

$$x = y, \qquad (1')$$

or

$$x = 2y. \qquad (1'')$$

Substitute y for x in (2) to obtain

$$2y^2 - y^2 - 3y^2 = 12,$$
$$-2y^2 = 12,$$
$$y = i\sqrt{6} \quad \text{or} \quad y = -i\sqrt{6}.$$

Substitute each of these values in turn in (1') to determine values for x.

If $y = i\sqrt{6}$, then $x = i\sqrt{6}$.

If $y = -i\sqrt{6}$, then $x = -i\sqrt{6}$.

Thus $(i\sqrt{6}, i\sqrt{6})$ and $(-i\sqrt{6}, -i\sqrt{6})$ are solutions of the system.

We now substitute $2y$ for x in (2) and solve for y.

$$8y^2 - 2y^2 - 3y^2 = 12$$
$$3y^2 = 12$$
$$y = 2 \quad \text{or} \quad y = -2.$$

Substitute each of these values in turn in (1") to determine values for x.

If $y = 2$, then $x = 2(2) = 4$.

If $y = -2$, then $x = 2(-2) = -4$.

The solution set is $\{(4, 2), (-4, -2)(i\sqrt{6}, i\sqrt{6}), (-i\sqrt{6}, -i\sqrt{6})\}$.

13. $x^2 - xy + y^2 = 7$
    $x^2 + y^2 = 5$

14. $3x^2 - 2xy + 3y^2 = 34$
    $x^2 + y^2 = 17$

15. $3x^2 + 3xy - y^2 = 35$
    $x^2 - xy - 6y^2 = 0$

16. $x^2 - xy + y^2 = 21$
    $x^2 + 2xy - 8y^2 = 0$

17. $2x^2 - xy - 6y^2 = 0$
    $x^2 + 3xy + 2y^2 = 4$

18. $2x^2 + xy - y^2 = 0$
    $6x^2 + xy - y^2 = 1$

19. $x^2 + y^2 = 9$
    $y = 4$

20. $x^2 + 2y^2 = 6$
    $x + y = 10$

21. $2x^2 + xy - 4y^2 = 12$
    $x^2 - 2y^2 = 4$

22. $x^2 + 3xy - y^2 = -3$
    $x^2 - xy - y^2 = 1$

9.4 Systems of Nonlinear Equations

23. $y - 10^x = 0$
    $\log_{10} y - 2x = 1$

24. $y - 10^x = 0$
    $\log_{10} y + 3x = 4$

25. $10^y = \dfrac{1}{x}$
    $\log_{10} x = y^2$

26. $10^y = \sqrt{x}$
    $\log_{10} x^2 = y^2$

Solve by graphing. Approximate components of solutions to the nearest half unit.

27. $y = 10^x$
    $x + y = 2$

28. $y = 2^x$
    $y - x = 2$

29. $y = \log_{10} x$
    $y = x^2 - 2x + 1$

30. $y = 10^x$
    $y = x^2$

31. The sum of the squares of two positive numbers is 13. If twice the first number is added to the second, the sum is 7. Find the numbers.

32. The sum of two numbers is 6 and their product is 35/4. Find the numbers.

33. The annual income from an investment is $32. If the amount invested were $200 more and the rate 1/2% less, the annual income would be $35. What are the amount and rate of the investment?

34. At a constant temperature, the pressure P and volume V of a gas are related by the equation $PV = K$. The product of the pressure (in pounds per square inch) and the volume (in cubic inches) of a certain gas is 30 inch-pounds. If the temperature remains constant as the pressure is increased 4 pounds per square inch, the volume is decreased by 2 cubic inches. Find the original pressure and volume of the gas.

B 35. How many real solutions are possible for a simultaneous system of linearly independent equations that consists of:

   a. two linear equations in two variables?
   b. one linear equation and one quadratic equation in two variables?
   c. two quadratic equations in two variables?

   Support each of your answers with sketches.

36. What relationships must exist between the numbers a and b so that the solution set of the system

$$x^2 + y^2 = 25$$

$$y = ax + b$$

has two ordered pairs of real numbers? One ordered pair of real numbers? No ordered pairs of real numbers? *Hint:* Use substitution and consider the nature of the roots of the resulting quadratic equation.

37. Consider the system

$$x^2 + y^2 = 8 \qquad (1)$$

$$xy = 4. \qquad (2)$$

We can solve this system by substituting $4/x$ for y in (1) to obtain

$$x^2 + \frac{16}{x^2} = 8,$$

from which we obtain $x = 2$ or $x = -2$. Now if we obtain the y-components of the solution from (2), we find that for $x = 2$ we have $y = 2$, and that for $x = -2$ we have $y = -2$. But if we seek y-components from (1), for $x = 2$ we have $y = \pm 2$, and for $x = -2$, $y = \pm 2$. Discuss the fact that we seem to obtain two more solutions from (1) than from (2). What is the solution set of the system?

9.5 Systems of Inequalities

Graphs of systems of inequalities

In Section 2.3, we observed that the graph of the solution set of an inequality in two variables might be a region in the plane. The graph of the solution set of a system of inequalities in two variables, which consists of the intersection of the graphs of the inequalities in the system, might also be a plane region.

Example

Graph the solution set of the system

$$x + 2y \leq 6$$

$$2x - 3y \geq 12.$$

Solution

Graphing each inequality by the method discussed in Section 2.3, we obtain the figure shown, where the doubly shaded region is the graph of the solution set of the system.

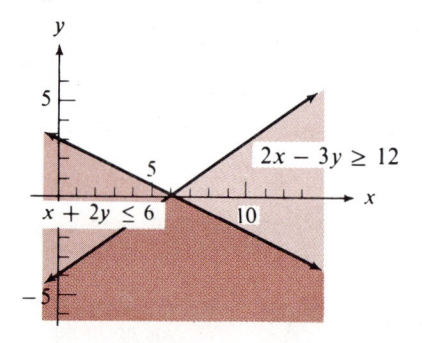

Example

Graph the solution set of the system

$$y \leq x + 2$$

$$y \geq x^2.$$

Solution

The graph of each inequality is shaded as shown in the figure. The doubly shaded region constitutes the graph of the solution set of the system.

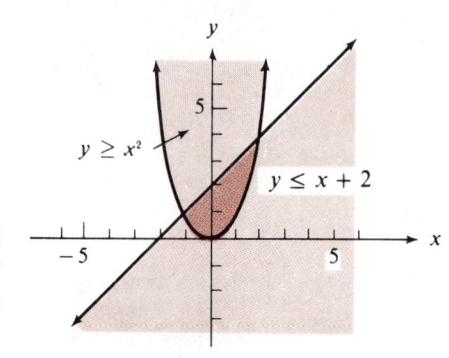

9.5 Systems of Inequalities

Example Graph the solution set of the system

$$y > 2$$
$$x > -2$$
$$x + y > 1.$$

Solution The triply shaded region in the figure constitutes the graph of the solution set of the system.

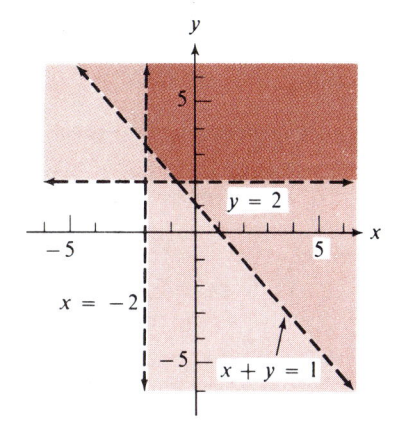

Exercise 9.5

A By double or triple shading, indicate the region in the plane representing the solution set of each system.

1. $y \geq x + 1$
   $y \geq 5 - x$

2. $y \geq 4$
   $x \geq 2$

3. $y - x \geq 0$
   $y + x \geq 0$

4. $2y - x \geq 1$
   $x < -3$

5. $y + 3x < 6$
   $y > 2$

6. $x > 3$
   $x + y \geq 5$

7. $x = 3$
   $x + y < 4$

8. $y = 2$
   $2y + x < 3$

9. $x - 3y < 6$
   $y + x = 1$

10. $2x + y \geq 4$
    $x - y = -2$

11. $y > x^2 + 1$
    $x + y > 4$

12. $y < x^2 + 4$
    $x - y \leq 4$

13. $x^2 + y^2 < 25$
    $y > 3$

14. $x^2 + y^2 \geq 25$
    $y > x^2$

15. $9x^2 + 16y^2 \geq 144$
    $y < x^2$

16. $y \geq 2$
    $x \geq 2$
    $y \geq x$

17. $y \leq 3$
    $x \leq 2$
    $y < x$

18. $x^2 + y^2 \leq 36$
    $x \geq 3$
    $y \geq 3$

B 19. $x^2 + y^2 \leq 25$
    $y \geq x^2 - 4$
    $y \leq -x^2 + 4$

20. $y \leq \log_{10} x$
    $y \geq x - 1$

21. $y \leq 10^x$
    $y \leq 2 - x^2$

22. $9 \leq x^2 + y^2 \leq 16$
    $-1 \leq x - y \leq 1$

23. $9 \leq x^2 + y^2 \leq 16$
    $x^2 + 1 \leq y \leq x^2 + 3$

24. $x^2 - 2 \leq y \leq 2 - x^2$
    $|x| \geq 1$

9.6 Convex Sets—Polygonal Regions

From the examples in Section 9.5 it is apparent that the graph of the solution set of a system of linear inequalities in two variables is simply the common intersection of a number of half-planes. Any such intersection is an example of a *convex set*.

Definition 9.2 A set $\mathscr{S}$ of points is a **convex set** if and only if, for each two points P and Q in $\mathscr{S}$, the line segment $\overline{PQ}$ lies entirely in $\mathscr{S}$.

Figure 9.6 shows three sets of points in a plane. Figures 9.6-a and 9.6-b show examples of convex sets, while the set in Figure 9.6-c is not convex because part of

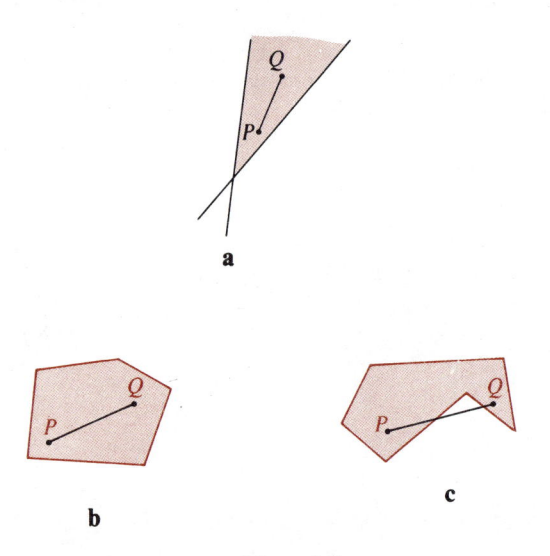

Figure 9.6

the line segment PQ does not lie in the set. If the boundary of a convex set is a (closed) polygon, then the boundary is called a **convex polygon**, and the region enclosed by the polygon (including the boundary) is called a **closed convex polygonal set**.

We have two immediate results from Definition 9.2 which are intuitively true and whose proofs are omitted.

Theorem 9.6   Any half-plane is a convex set.

Theorem 9.7   The intersection of two convex sets is a convex set.

9.6 Convex Sets—Polygonal Regions

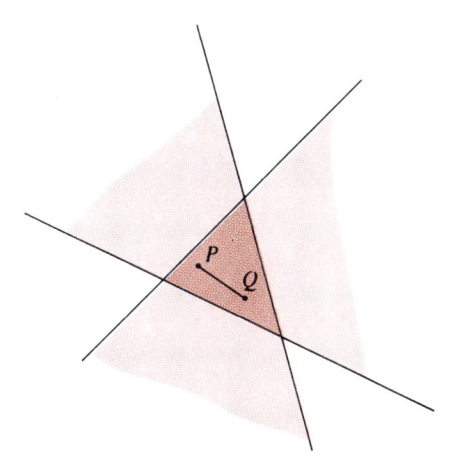

Figure 9.7

As suggested by Figure 9.7, this latter result extends to any number of convex sets. If the intersection of the graphs of a system of linear inequalities constitutes a (closed) polygonal set $\mathscr{P}$, then we can locate the set $\mathscr{P}$ in the plane by graphing the equations associated with the given inequalities and identifying the vertices of $\mathscr{P}$.

Example Shade the polygonal set specified by the system

$$x + y \leq 5 \qquad x \geq 0$$
$$x - y \leq 2 \qquad y \geq 0.$$
$$x - y \geq -2$$

Solution Graph the associated equations

$$x + y = 5$$
$$x - y = 2$$
$$x - y = -2$$
$$x = 0$$

and

$$y = 0,$$

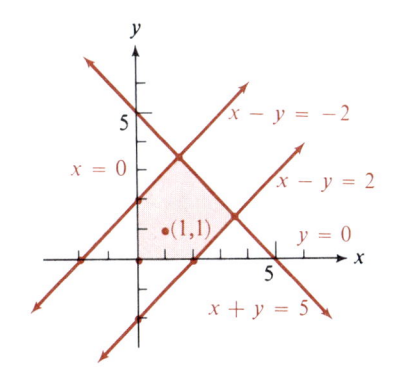

as shown. Note which half-plane is determined by each of the five inequalities, and shade the intersection of the five half-planes. The intersection is the pentagonal region shown. To check the result, note that (1, 1) is in the region and its coordinates satisfy each given inequality.

We should not conclude from the foregoing example that, simply because the five equations associated with the set of inequalities determine a convex pentagon, the pentagonal region is necessarily the graph of the system. If the inequality $y \geq 0$ in the example is replaced with $y \leq 0$, then the graph of the equations remains unchanged, but the graph of the system of inequalities is the triangular region shown in Figure 9.8-a. On the other hand, if the inequality $x + y \leq 5$ is replaced with $x + y \geq 5$, then the intersection of the graphs becomes the (unbounded) rectangular region shown in Figure 9.8-b. Finally we observe that if both of the foregoing changes are made, that is, if $x + y \geq 5$ replaces $x + y \leq 5$ and $y \leq 0$ replaced $y \geq 0$, then the intersection of the graphs of the inequalities in the system is $\emptyset$.

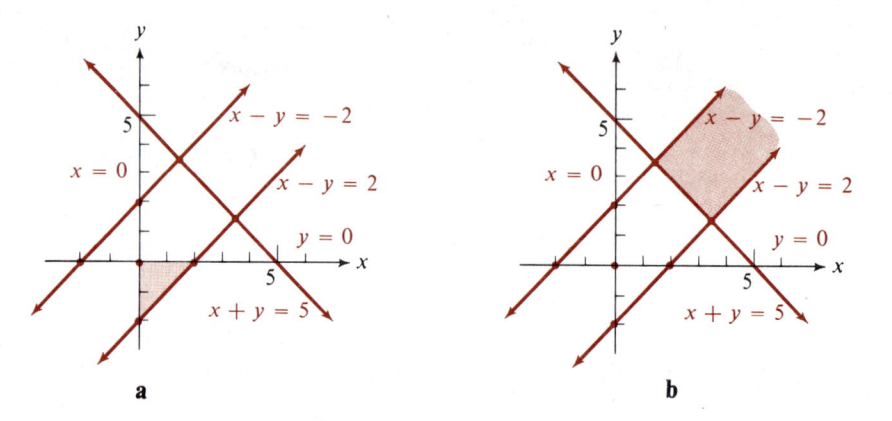

Figure 9.8

Exercise 9.6

Graph the convex polygonal set defined by the given system of inequalities.

1. $x - y \leq 2$
   $x - y \geq -2$
   $x + y \geq 2$
   $x + y \leq 6$

2. $0 \leq x \leq 5$
   $0 \leq y \leq 5$
   $x + y \leq 6$

3. $0 \leq x \leq 5$
   $0 \leq y \leq 4$
   $x + 2y \leq 10$
   $2x + y \leq 10$

4. $0 \leq x$
   $0 \leq y \leq 4$
   $x - 2y \leq 6$
   $2x + y \leq 12$
   $2x - y \leq 10$

5. $0 \leq x \leq 5$
   $y \geq x - 3$
   $x + y \leq 9$
   $3y \leq 2x + 12$

6. $0 \leq x$
   $0 \leq y \leq 6$
   $x + 2y \leq 13$
   $2x + y \leq 11$
   $3x + y \leq 15$

9.7 Linear Programming

7. Repeat Exercise 1 with $x - y \leq 2$ replaced with $x - y \geq 2$. Is the graph a closed polygonal set?

8. Repeat Exercise 3 with $x + 2y \leq 10$ replaced with $x + 2y \geq 10$. Is the graph a closed polygonal set?

9. Repeat Exercise 4 with $0 \leq y \leq 4$ replaced with $y \leq 4$. Is the graph a closed polygonal set?

10. Repeat Exercise 6 without the condition $x \geq 0$. Is the graph a closed polygonal set?

9.7 Linear Programming

For each ordered pair $(x, y) \in R^2$, if a, b are real numbers then the linear expression $ax + by$ has a value. In particular, if $\mathscr{P} \subset R^2$, and $\mathscr{P}$ is a closed polygonal subset of R^2, then $ax + by$ has a value for each $(x, y) \in \mathscr{P}$. It can be shown that the following result holds for the values of $ax + by$ over $\mathscr{P}$.

Theorem 9.8 *If $\mathscr{P} \subset R^2$ is a closed convex polygonal set, then the expression $ax + by$ takes on its maximum and minimum values over $\mathscr{P}$ at one of the vertices of $\mathscr{P}$.*

Thus, if $\mathscr{P}$ is the closed convex polygonal set with vertices $(1, 2), (3, 1), (5, 3), (4, 5)$, and $(3, 6)$, as pictured in Figure 9.9, then the linear expression $3x - 2y$ has a value for every ordered pair (x, y) in, or on the boundary of, $\mathscr{P}$. In particular, at the vertices we have the values shown in the table below.

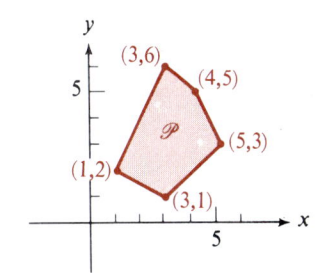

Figure 9.9

Vertex	(1, 2)	(3, 1)	(5, 3)	(4, 5)	(3, 6)
Value of $3x - 2y$	-1	7	9	2	-3

The maximum of these values is 9 and the minimum is -3. By Theorem 9.8, then, 9 is the greatest value $3x - 2y$ has over the entire set $\mathscr{P}$, and its least value over $\mathscr{P}$ is -3.

Theorem 9.8 has many important applications.

Example

A company manufactures two kinds of electric shavers, one using a cord and the other a cordless model. The company can make up to 500 cord models and 400 cordless models per day, but it can make a total of only 600 shavers per day. It takes 2 man-hours to manufacture the cord model and 3 man-hours to manufacture the cordless model, and the company has available at most 1400 man-hours per day. If there is a profit of $2.50 on each cord shaver and $3.50 on each cordless shaver, how many of each kind should the company make each day in order to realize the greatest profit?

Solution

Let x = number of cord shavers made in one day;
y = number of cordless shavers made in one day.

We have the following constraints on x and y:

$$0 \leq x \leq 500, \quad 0 \leq y \leq 400,$$
$$x + y \leq 600, \quad 2x + 3y \leq 1400.$$

Solving the associated equations for these inequalities in pairs yields the vertices of the polygonal region shown in the figure. We wish to maximize

$$2.5x + 3.5y.$$

We find the values for this expression at the vertices to be:

Vertex	Profit
(0, 0)	$ 0
(0, 400)	$1400
(100, 400)	$1650
(400, 200)	$1700
(500, 100)	$1601
(500, 0)	$1250

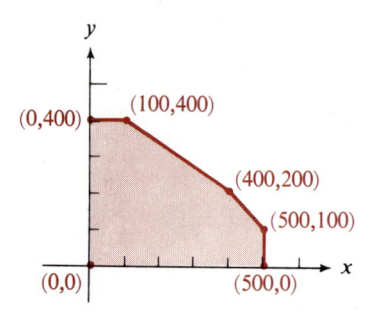

Clearly, the most profitable numbers of shavers are 400 cord models and 200 cordless models.

Exercise 9.7

Find the maximum and minimum values of the given expression over the given closed convex polygonal set of the specified set of Exercise 9.6, page 328.

1. $x + 3y$; the set in Exercise 1.
2. $3x + 5y$; the set in Exercise 2.
3. $3x + 5y$; the set in Exercise 3.
4. $x + 2y$; the set in Exercise 4.
5. $5x - 4y$; the set in Exercise 5.
6. $10x - 2y$; the set in Exercise 6.

9.7 Linear Programming

For each Exercise 7–10, use Tables A and B below, which show the number of units of labor, machinery, materials, and overhead necessary for a manufacturer to produce one unit of each of two products x and y. The tables also show the total available units of each of these items.

Table A

	Units needed for 1 unit of product x	Units needed for 1 unit of product y	Total units available
Labor	1	6	120
Machinery	1	3	66
Material	3	2	86
Overhead	10	3	250

Table B

	Units needed for 1 unit of product x	Units needed for 1 unit of product y	Total units available
Labor	2	7	63
Machinery	5	1	50
Material	1	2	21
Overhead	1	1	14

7. Using Table A, find the maximum profit if each unit of product x earns a profit of $50 and each unit of product y earns a profit of $60.

8. Using Table A, find the maximum profit if each unit of product x earns a profit of $60 and each unit of product y earns a profit of $50.

9. Using Table B, find the maximum profit if each unit of product x earns a profit of $50 and each unit of product y earns a profit of $60.

10. Using Table B, find the maximum profit if each unit of product x earns a profit of $60 and each unit of product y earns a profit of $50.

11. An electronics company makes two kinds of electronic ranges, a standard model that earns a $50 profit and a deluxe model that earns a $60 profit. The company has machinery capable of producing any number of deluxe ranges up to 400 per month and any number of standard ranges up to 500 per month, but it has enough man-hours available to produce

only 600 ranges of both kinds in a month. How many of each kind of range should the company produce to realize a maximum profit?

12. A pharmacy has 300 ounces of a drug it can use to make two different kinds of medicines. From each ounce of the drug, 15 bottles of medicine A or 25 bottles of medicine B can be produced. The pharmacy must keep all the bottles in its own storage room, which has a capacity of 6000 bottles in all. At least 600 bottles of medicine A and 1250 bottles of medicine B must be retained in inventory and the rest will be used. If medicine A sells for $2.75 per bottle and medicine B sells for $2.00 per bottle, how many ounces of the drug should be devoted to each medicine to maximize the total return?

Chapter Review

[9.1] Solve each system.

1. $x + 5y = 18$
   $x - y = -3$

2. $x + 5y = 11$
   $2x + 3y = 8$

3. $2x - 3y = 8$
   $3x + 2y = 7$

4. $3x - y = -7$
   $x - 2y = -4$

5. $2x - 3y = 1$
   $4x - 6y = 2$

6. $9x + 12y = 1$
   $6x + 8y = 0$

7. Find values for a and b so that the graph of $ax + by = 19$ passes through the points $(2, 3)$ and $(-3, 5)$.

8. Find a and b so that the solution set of

$$ax + by = 3$$
$$2ax - by = 1$$

is $\{(2, 1)\}$.

[9.2] Solve each system.

9. $x + 3y - z = 3$
   $2x - y + 3z = 1$
   $3x + 2y + z = 5$

10. $x + y + z = 2$
    $3x - y + z = 4$
    $2x + y + 2z = 3$

11. $2x + 3y - z = -2$
    $x - y + z = 6$
    $3x - y + z = 10$

12. Find value for a, b, and c so that the graph of $y = ax^2 + bx + c$ contains the points $(-1, 9), (0, 4),$ and $(1, 3)$.

[9.3] Use the method of partial fractions to express each fraction as a sum of fractions with powers of linear polynomials for denominators.

13. $\dfrac{5}{2x^2 - x - 3}$

14. $\dfrac{x^2 + 1}{x^3 - x}$

[9.4] Solve each system.

15. $x^2 + y = 3$
    $5x + y = 7$

16. $x^2 - xy + y^2 = 1$
    $2x^2 - xy + 2y^2 = 3$

17. $2x^2 + 5y^2 - 53 = 0$
    $4x^2 + 3y^2 - 43 = 0$

18. $x^2 + 3xy + 2y^2 = 0$
    $x^2 - 3xy + y^2 = 1$

19. $y - 10^{2x} = 0$
    $\log_{10} y^2 - x + 1 = 0$

20. $10^y - 2x = 0$
    $\log_{10} x + \log_{10} 2 = -1$

21. Approximate the solution of the system

$$y = 3^x$$
$$y = 3 - x$$

by graphical methods.

[9.5] By double shading, indicate the region representing the solution set of each system.

22. $y > x^2 - 4$
    $y < 2 - x$

23. $y + x^2 < 0$
    $y + 3 > 0$
    $y < x - 3$

24. $y - x^2 < 4$
    $y - 3 > 0$
    $y - x > 0$

[9.6] Graph the polygonal set defined by each system of inequalities.

25. $0 \leq x \leq 3$
    $0 \leq y \leq 4$
    $x + y \leq 5$

26. $0 \leq x \leq 3$
    $0 \leq y$
    $y \geq 2 + x$
    $x + y \leq 4$

[9.7]

27. Find the maximum and minimum values of $2x + y$ over the set in Exercise 25.

28. Find the maximum and minimum values of $3x + 2y$ over the set in Exercise 26.

29. Using Table A on page 331, find the maximum profit if each unit of product x earns a profit of $80 and each unit of product y earns a profit of $60.

30. Using Table B on page 331, find the maximum profit if each unit of product x earns a profit of $10 and each unit of product y earns a profit of $20.

10 Matrices and Determinants

Matrices, as introduced in this chapter, are today much used in mathematics and engineering, and also in the physical, social, and life sciences.

10.1 Definitions; Matrix Addition

A **matrix** is a rectangular array of numbers (or other suitable entities), which are called the **entries** or **elements** of the matrix. In this book, we shall consider only real numbers as entries. A matrix is customarily displayed in a pair of brackets or parentheses (we shall use brackets). Thus,

$$\begin{bmatrix} 1 & 2 & 3 \\ 4 & 5 & 6 \end{bmatrix} \text{ and } \begin{bmatrix} 2 \\ 1 \end{bmatrix}$$

are matrices. The **order**, or **dimension**, of a matrix is the ordered pair having as first component the number of (horizontal) **rows** and as second component the number of (vertical) **columns** in the matrix. Thus,

$$\begin{bmatrix} 1 & 2 & 3 \\ 4 & 5 & 6 \end{bmatrix}, \begin{bmatrix} 1 \\ 2 \\ 3 \end{bmatrix}, \text{ and } \begin{bmatrix} a_1 & a_2 & a_3 & a_4 \\ b_1 & b_2 & b_3 & b_4 \\ c_1 & c_2 & c_3 & c_4 \\ d_1 & d_2 & d_3 & d_4 \end{bmatrix}$$

are 2×3 (read "two by three"), 3×1 (read "three by one"), and 4×4 (read "four by four") matrices, respectively. Note that the number of *rows* is given first, and then the number of *columns*. A matrix consisting of a single row is called a **row matrix** or a **row vector**, whereas a matrix consisting of a single column is called a **column matrix** or a **column vector**.

10.1 Definitions; Matrix Addition

Matrices are frequently represented by capital letters. Thus, we might want to talk about the matrices A and B, where

$$A = \begin{bmatrix} a_1 & a_2 \\ b_1 & b_2 \end{bmatrix} \quad \text{and} \quad B = [b_1 \ b_2].$$

To indicate that A is a 2×2 matrix, we can write $A_{2 \times 2}$. Similarly, $B_{1 \times 2}$ is a matrix with one row and two columns.

To represent the entries of a matrix, double-subscript notation can be used. A single letter, say a, is used to denote an entry in a matrix, and then *two* subscripts are appended, the first subscript telling in which *row* the entry occurs, and the second telling in which *column*. Thus, we write

$$A = \begin{bmatrix} a_{11} & a_{12} & a_{13} \\ a_{21} & a_{22} & a_{23} \\ a_{31} & a_{32} & a_{33} \end{bmatrix},$$

where a_{21} is the element in the second *row* and first *column*, a_{33} is the element in the third *row* and third *column*, and, in general, a_{ij} is the element in the *i*th *row* and *j*th *column*.

Definition 10.1 Two matrices, A and B, are **equal** if and only if both matrices are of the same order and $a_{ij} = b_{ij}$ for each i, j.

Examples

a. $\begin{bmatrix} 2 & 1 \\ 3 & 0 \end{bmatrix} = \begin{bmatrix} \frac{4}{2} & 2-1 \\ \sqrt{9} & 0 \end{bmatrix}$
b. $\begin{bmatrix} 2 & 1 \\ 3 & 0 \end{bmatrix} \neq \begin{bmatrix} 2 & 3 \\ 1 & 0 \end{bmatrix}$
c. $\begin{bmatrix} 1 \\ 2 \end{bmatrix} \neq [1 \ 2]$

Definition 10.2 The **transpose** of a matrix A, denoted by A^t, is the matrix in which the rows are the columns of A and the columns are the rows of A.

Thus, if a_{ij} and b_{ij} represent the entries in the *i*th row and *j*th column of A and A^t, respectively, then we have

$$b_{ij} = a_{ji}$$

for each i and j.

Examples

a. $\begin{bmatrix} 2 & 1 \\ 3 & 0 \end{bmatrix}^t = \begin{bmatrix} 2 & 3 \\ 1 & 0 \end{bmatrix}$
b. $\begin{bmatrix} 1 & 2 & 3 \\ 4 & 5 & 6 \end{bmatrix}^t = \begin{bmatrix} 1 & 4 \\ 2 & 5 \\ 3 & 6 \end{bmatrix}$
c. $\begin{bmatrix} 1 \\ 2 \end{bmatrix}^t = [1 \ 2]$

Definition 10.3 The **sum** of two matrices of the same order, $A_{m \times n}$ and $B_{m \times n}$, is the matrix $(A + B)_{m \times n}$, in which the entry in the *i*th row and *j*th column is $a_{ij} + b_{ij}$, for $i = 1, 2, 3, \ldots, m$ and $j = 1, 2, 3, \ldots, n$.

Examples

a. $\begin{bmatrix} 3 & 1 & 2 \\ 2 & 1 & 4 \end{bmatrix} + \begin{bmatrix} 1 & 0 & 2 \\ -1 & 3 & 0 \end{bmatrix} = \begin{bmatrix} 3+1 & 1+0 & 2+2 \\ 2+(-1) & 1+3 & 4+0 \end{bmatrix} = \begin{bmatrix} 4 & 1 & 4 \\ 1 & 4 & 4 \end{bmatrix}$

b. $\begin{bmatrix} 1 & 2 \\ 1 & -3 \end{bmatrix} + \begin{bmatrix} 1 & -2 \\ 2 & 3 \end{bmatrix} = \begin{bmatrix} 1+1 & 2+(-2) \\ 1+2 & -3+3 \end{bmatrix} = \begin{bmatrix} 2 & 0 \\ 3 & 0 \end{bmatrix}$

Thus, the sum of two matrices of the same order is obtained by adding the corresponding entries; the sum of two matrices of different orders is not defined.

Definition 10.4 A matrix with each entry equal to 0 is a *zero matrix*.

Zero matrices are generally denoted by the symbol **0**. This distinguishes the zero matrix from the real number 0. For example,

$$\mathbf{0}_{2 \times 4} = \begin{bmatrix} 0 & 0 & 0 & 0 \\ 0 & 0 & 0 & 0 \end{bmatrix}$$

is the 2 × 4 zero matrix.

Definition 10.5 The *negative of a matrix* $A_{m \times n}$, denoted by $-A_{m \times n}$, is formed by replacing each entry a_{ij} in the matrix $A_{m \times n}$ with $-a_{ij}$.

For example, if

$$A_{3 \times 2} = \begin{bmatrix} 3 & -1 \\ 2 & -2 \\ -4 & 5 \end{bmatrix}, \quad \text{then} \quad -A_{3 \times 2} = \begin{bmatrix} -3 & 1 \\ -2 & 2 \\ 4 & -5 \end{bmatrix}.$$

The sum $B_{m \times n} + (-A_{m \times n})$ is called the **difference** of $B_{m \times n}$ and $A_{m \times n}$ and is denoted $B_{m \times n} - A_{m \times n}$.

Properties of sums

At this point, we are able to establish the following facts concerning sums of matrices with real-number entries. We shall prove Parts III and IV; the rest are left as exercises.

Theorem 10.1 If A, B, and C are $m \times n$ matrices with real-number entries, then;

I $(A + B)_{m \times n}$ is a matrix with real-number entries.   Closure law for addition.

II $(A + B) + C = A + (B + C)$ Associative law for addition.

III The matrix $\mathbf{0}_{m \times n}$ has the property that for every matrix $A_{m \times n}$,   Additive-identity law.

$A + \mathbf{0} = A$ and $\mathbf{0} + A = A$.

10.1 Definitions; Matrix Addition

IV For every matrix $A_{m \times n}$, the matrix $-A_{m \times n}$ has the property that
$$A + (-A) = \mathbf{0} \quad \text{and} \quad (-A) + A = \mathbf{0}.$$
Additive-inverse law.

V $A + B = B + A$ *Commutative law for addition.*

To prove 10.1-III, observe that since each entry of the zero matrix is 0, it follows that the entries of $A_{m \times n} + \mathbf{0}_{m \times n}$ are $a_{ij} + 0 = a_{ij}$ and the entries of $\mathbf{0}_{m \times n} + A_{m \times n}$ are $0 + a_{ij} = a_{ij}$. Thus, $A + \mathbf{0} = \mathbf{0} + A = A$.

Examples

a. $\begin{bmatrix} a_{11} & a_{12} \\ a_{21} & a_{22} \end{bmatrix} + \begin{bmatrix} 0 & 0 \\ 0 & 0 \end{bmatrix} = \begin{bmatrix} a_{11} & a_{12} \\ a_{21} & a_{22} \end{bmatrix}$ **b.** $\begin{bmatrix} a_1 \\ a_2 \\ a_3 \end{bmatrix} + \begin{bmatrix} 0 \\ 0 \\ 0 \end{bmatrix} = \begin{bmatrix} a_1 \\ a_2 \\ a_3 \end{bmatrix}$

To see that 10.1-IV is true, let the entries of $A_{m \times n}$ and $-A_{m \times n}$ be a_{ij} and $-a_{ij}$, respectively. Since each entry of $A + (-A)$ is $a_{ij} - a_{ij}$, or 0, we have $A + (-A) = \mathbf{0}$. Similarly, $(-A) + A = \mathbf{0}$.

Example If
$$A = \begin{bmatrix} 1 & -1 & 2 \\ 3 & -1 & 1 \end{bmatrix},$$
then
$$A + (-A) = \begin{bmatrix} 1 & -1 & 2 \\ 3 & -1 & 1 \end{bmatrix} + \begin{bmatrix} -1 & 1 & -2 \\ -3 & 1 & -1 \end{bmatrix} = \begin{bmatrix} 0 & 0 & 0 \\ 0 & 0 & 0 \end{bmatrix} = \mathbf{0}.$$

Example As an example of Theorem 10.1-II, consider the matrices
$$A = \begin{bmatrix} 1 & 3 \\ -1 & 2 \end{bmatrix}, \quad B = \begin{bmatrix} 2 & 1 \\ -3 & 2 \end{bmatrix}, \quad \text{and} \quad C = \begin{bmatrix} -1 & 1 \\ 0 & 2 \end{bmatrix}.$$
Then
$$(A + B) + C = \left(\begin{bmatrix} 1 & 3 \\ -1 & 2 \end{bmatrix} + \begin{bmatrix} 2 & 1 \\ -3 & 2 \end{bmatrix} \right) + \begin{bmatrix} -1 & 1 \\ 0 & 2 \end{bmatrix}$$
$$= \begin{bmatrix} 3 & 4 \\ -4 & 4 \end{bmatrix} + \begin{bmatrix} -1 & 1 \\ 0 & 2 \end{bmatrix} = \begin{bmatrix} 2 & 5 \\ -4 & 6 \end{bmatrix},$$
and
$$A + (B + C) = \begin{bmatrix} 1 & 3 \\ -1 & 2 \end{bmatrix} + \left(\begin{bmatrix} 2 & 1 \\ -3 & 2 \end{bmatrix} + \begin{bmatrix} -1 & 1 \\ 0 & 2 \end{bmatrix} \right)$$
$$= \begin{bmatrix} 1 & 3 \\ -1 & 2 \end{bmatrix} + \begin{bmatrix} 1 & 2 \\ -3 & 4 \end{bmatrix} = \begin{bmatrix} 2 & 5 \\ -4 & 6 \end{bmatrix}.$$

Example continued overleaf

Hence,
$$A + (B + C) = (A + B) + C.$$

Example As an example of Theorem 10.1-V, consider the matrices
$$A = \begin{bmatrix} 1 & -1 \\ 2 & 3 \end{bmatrix} \quad \text{and} \quad B = \begin{bmatrix} 3 & 4 \\ -2 & -4 \end{bmatrix}.$$
Then
$$A + B = \begin{bmatrix} 1 & -1 \\ 2 & 3 \end{bmatrix} + \begin{bmatrix} 3 & 4 \\ -2 & -4 \end{bmatrix} = \begin{bmatrix} 4 & 3 \\ 0 & -1 \end{bmatrix}$$
and
$$B + A = \begin{bmatrix} 3 & 4 \\ -2 & -4 \end{bmatrix} + \begin{bmatrix} 1 & -1 \\ 2 & 3 \end{bmatrix} = \begin{bmatrix} 4 & 3 \\ 0 & -1 \end{bmatrix}.$$
Hence,
$$A + B = B + A.$$

Solution of matrix equations Theorem 10.1 can be used to solve matrix equations of the form $X + A = B$, as follows:
$$(X + A) + (-A) = B + (-A)$$
$$X + (A + (-A)) = B + (-A) \quad \text{by Theorem 10.1-II}$$
$$X + 0 = B + (-A) \quad \text{by Theorem 10.1-IV}$$
$$X = B + (-A) \quad \text{by Theorem 10.1-III}$$

Example Solve $X + \begin{bmatrix} -3 & 1 \\ 2 & -1 \end{bmatrix} = \begin{bmatrix} 1 & -1 \\ 2 & -2 \end{bmatrix}.$

Solution
$$X = \begin{bmatrix} 1 & -1 \\ 2 & -2 \end{bmatrix} + \left(-\begin{bmatrix} -3 & 1 \\ 2 & -1 \end{bmatrix} \right)$$
$$= \begin{bmatrix} 1 & -1 \\ 2 & -2 \end{bmatrix} + \begin{bmatrix} 3 & -1 \\ -2 & 1 \end{bmatrix}$$
$$= \begin{bmatrix} 4 & -2 \\ 0 & -1 \end{bmatrix}$$

10.1 Definitions; Matrix Addition

Exercise 10.1

A State the order and find the transpose of each matrix.

Examples

a. $\begin{bmatrix} 2 & 4 \\ 1 & -3 \\ 6 & 0 \end{bmatrix}$

b. $\begin{bmatrix} 0 & 2 \\ -1 & 0 \end{bmatrix}$

Solutions

a. 3×2 matrix; $\begin{bmatrix} 2 & 4 \\ 1 & -3 \\ 6 & 0 \end{bmatrix}^t = \begin{bmatrix} 2 & 1 & 6 \\ 4 & -3 & 0 \end{bmatrix}$

b. 2×2 matrix; $\begin{bmatrix} 0 & 2 \\ -1 & 0 \end{bmatrix}^t = \begin{bmatrix} 0 & -1 \\ 2 & 0 \end{bmatrix}$

1. $\begin{bmatrix} 6 & -1 \\ 2 & 3 \end{bmatrix}$
2. $\begin{bmatrix} 4 & 1 \\ 0 & -2 \end{bmatrix}$
3. $\begin{bmatrix} 2 & -7 & 3 \\ 1 & 4 & 0 \end{bmatrix}$

4. $\begin{bmatrix} -3 & 1 \\ 6 & 0 \\ 0 & 2 \end{bmatrix}$
5. $\begin{bmatrix} 2 & 3 & -1 \\ 4 & 0 & 1 \\ -2 & 3 & 1 \end{bmatrix}$
6. $\begin{bmatrix} 4 & -1 & -2 \\ 3 & 0 & 0 \\ 2 & 1 & 1 \end{bmatrix}$

7. $\begin{bmatrix} 4 & -3 & -1 & 0 \\ 2 & 1 & 1 & 6 \end{bmatrix}$
8. $\begin{bmatrix} -2 & 1 & 3 & 2 \\ 4 & 0 & 0 & -2 \\ -1 & 3 & 2 & 4 \end{bmatrix}$

Write each sum or difference as a single matrix.

Example

$\begin{bmatrix} 2 & 1 & 4 \\ 3 & -1 & 0 \end{bmatrix} + \begin{bmatrix} 6 & 3 & 0 \\ -2 & 1 & 0 \end{bmatrix}$

Solution

$\begin{bmatrix} 2 & 1 & 4 \\ 3 & -1 & 0 \end{bmatrix} + \begin{bmatrix} 6 & 3 & 0 \\ -2 & 1 & 0 \end{bmatrix} = \begin{bmatrix} 2+6 & 1+3 & 4+0 \\ 3-2 & -1+1 & 0+0 \end{bmatrix} = \begin{bmatrix} 8 & 4 & 4 \\ 1 & 0 & 0 \end{bmatrix}$

9. $\begin{bmatrix} 2 & 3 \\ 1 & 6 \end{bmatrix} + \begin{bmatrix} 1 & -2 \\ 2 & 3 \end{bmatrix}$
10. $\begin{bmatrix} 4 & -1 & 3 \\ 2 & 1 & 0 \end{bmatrix} + \begin{bmatrix} 3 & -1 & 0 \\ 4 & 0 & -2 \end{bmatrix}$

11. $\begin{bmatrix} 3 & 0 & -1 \\ 2 & 1 & 2 \end{bmatrix} + \begin{bmatrix} 6 & -1 & 0 \\ 0 & 2 & 4 \end{bmatrix}$
12. $[1 \quad 3 \quad 5] + [0 \quad -2 \quad 1]$

13. $\begin{bmatrix} 4 & -3 \\ 2 & 1 \end{bmatrix} - \begin{bmatrix} 6 & 0 \\ -2 & 1 \end{bmatrix}$
14. $\begin{bmatrix} 4 & -1 & 2 \\ 3 & 1 & -4 \end{bmatrix} - \begin{bmatrix} -1 & -1 & 2 \\ 3 & 1 & 4 \end{bmatrix}$

15. $\begin{bmatrix} 10 & 3 & 2 \\ 5 & 1 & 7 \\ 6 & 1 & 9 \end{bmatrix} - \begin{bmatrix} 8 & 12 & 15 \\ -2 & 5 & 6 \\ -3 & 1 & 9 \end{bmatrix}$
16. $\begin{bmatrix} 3 & -1 & 2 \\ 4 & -2 & 1 \\ 6 & 3 & 2 \end{bmatrix} - \begin{bmatrix} 2 & -1 & 2 \\ 4 & -1 & 1 \\ 6 & 3 & 1 \end{bmatrix}$

17. $\begin{bmatrix} 4 \\ 3 \\ -1 \end{bmatrix} + \begin{bmatrix} 6 \\ 0 \\ -2 \end{bmatrix}$

18. $\begin{bmatrix} 2 & 3 \\ 1 & 0 \\ -1 & 2 \end{bmatrix} + \begin{bmatrix} -2 & 0 \\ -3 & 0 \\ 4 & -1 \end{bmatrix}$

19. $\begin{bmatrix} 2 & 3 & 4 \\ -1 & 6 & 2 \\ 1 & 0 & 3 \end{bmatrix} + \begin{bmatrix} 0 & 0 & 0 \\ 0 & 0 & 0 \\ 0 & 0 & 0 \end{bmatrix}$

20. $\begin{bmatrix} 2 & -3 \\ 4 & -1 \\ -2 & 1 \end{bmatrix} + \begin{bmatrix} -2 & 3 \\ -4 & 1 \\ 2 & -1 \end{bmatrix}$

Solve each of the following matrix equations.

Examples

a. $X + \begin{bmatrix} 2 & 3 \\ 1 & 7 \end{bmatrix} = \begin{bmatrix} 9 & -4 \\ 2 & 0 \end{bmatrix}$

b. $X - \begin{bmatrix} 2 & 1 \\ -3 & 0 \end{bmatrix} = \begin{bmatrix} -1 & 2 \\ 1 & 0 \end{bmatrix}$

Solutions

a. $X = \begin{bmatrix} 9 & -4 \\ 2 & 0 \end{bmatrix} - \begin{bmatrix} 2 & 3 \\ 1 & 7 \end{bmatrix}$
$= \begin{bmatrix} 7 & -7 \\ 1 & -7 \end{bmatrix}$

b. $X = \begin{bmatrix} -1 & 2 \\ 1 & 0 \end{bmatrix} + \begin{bmatrix} 2 & 1 \\ -3 & 0 \end{bmatrix}$
$= \begin{bmatrix} 1 & 3 \\ -2 & 0 \end{bmatrix}$

21. $X - \begin{bmatrix} -1 & 0 \\ 0 & 0 \end{bmatrix} = \begin{bmatrix} 3 & -1 \\ 2 & 1 \end{bmatrix}$

22. $X + \begin{bmatrix} 3 & -1 \\ 2 & 1 \end{bmatrix} = \begin{bmatrix} 5 & 1 \\ -3 & 5 \end{bmatrix}$

B 23. $\begin{bmatrix} 1 & 3 \\ -1 & 0 \end{bmatrix}^t = \begin{bmatrix} 2 & -2 \\ -1 & 3 \end{bmatrix}^t - X$

24. $X + \begin{bmatrix} 3 & 2 \\ -1 & 4 \end{bmatrix}^t = \begin{bmatrix} 2 & 6 \\ 1 & 5 \end{bmatrix}^t$

25. Show that $[A^t_{2 \times 2}]^t = A_{2 \times 2}$. Does an analogous result seem valid for $n \times n$ matrices? For $m \times n$ matrices?

26. Show that $[A_{2 \times 2} + B_{2 \times 2}]^t = A^t_{2 \times 2} + B^t_{2 \times 2}$. Does an analogous result seem valid for $n \times n$ matrices?

27. Prove Theorem 10.1-I.

28. Prove Theorem 10.1-II.

29. Prove Theorem 10.1-V.

10.2 Matrix Multiplication

We shall be interested in two kinds of products involving matrices: (1) the product of a matrix and a real number, and (2) the product of two matrices.

Definition 10.6 The **product** of a real number c and an $m \times n$ matrix A with entries a_{ij} is the matrix cA with corresponding entries ca_{ij}, where $i = 1, 2, 3, \ldots, m$ and $j = 1, 2, 3, \ldots, n$.

10.2 Matrix Multiplication

Examples

a. $3\begin{bmatrix} 2 & 1 \\ 0 & 5 \end{bmatrix} = \begin{bmatrix} 3(2) & 3(1) \\ 3(0) & 3(5) \end{bmatrix} = \begin{bmatrix} 6 & 3 \\ 0 & 15 \end{bmatrix}$ b. $-4\begin{bmatrix} 1 \\ 2 \\ -1 \end{bmatrix} = \begin{bmatrix} -4(1) \\ -4(2) \\ -4(-1) \end{bmatrix} = \begin{bmatrix} -4 \\ -8 \\ 4 \end{bmatrix}$

Properties of products of matrices and real numbers

The following theorem states some simple algebraic laws for the multiplication of matrices by real numbers.

Theorem 10.2   If A and B are $m \times n$ matrices, and $c, d \in R$, then

   I cA is an $m \times n$ matrix,   V $1A = A$,

   II $c(dA) = (cd)A$,   VI $(-1)A = -A$,

   III $(c + d)A = cA + dA$,   VII $0A = \mathbf{0}$,

   IV $c(A + B) = cA + cB$,   VIII $c\mathbf{0} = \mathbf{0}$.

We shall prove only Part IV, leaving the remaining parts as exercises. Since the elements of $A + B$ are of the form $a_{ij} + b_{ij}$, it follows, by definition, that the elements of $c(A + B)$ are of the form $c(a_{ij} + b_{ij})$. But, since a_{ij}, b_{ij}, and c denote real numbers,

$$c(a_{ij} + b_{ij}) = ca_{ij} + cb_{ij}.$$

Now, the elements of cA are of the form ca_{ij}, and those of cB are of the form cb_{ij}, so that the elements of $cA + cB$ are of the form $ca_{ij} + cb_{ij}$ and part IV is proved.

Products of matrices

Turning now to the *product of two matrices*, we first define the product of a row matrix and a column matrix.

Definition 10.7   For $A_{1 \times p} = [a_1 \ \cdots \ a_p]$ and $B_{p \times 1} = \begin{bmatrix} b_1 \\ \vdots \\ b_p \end{bmatrix}$, the product

$$A_{1 \times p} B_{p \times 1} = a_1 b_1 + a_2 b_2 + \cdots + a_p b_p.$$

Examples

a. $[1 \ -2]\begin{bmatrix} 3 \\ -4 \end{bmatrix} = 1(3) + (-2)(-4) = 11$

b. $[1 \ \ 0 \ -1 \ \ 2]\begin{bmatrix} 3 \\ 1 \\ -2 \\ 4 \end{bmatrix} = 1(3) + 0(1) + (-1)(-2) + 2(4) = 13$

Note from Definition 10.7 that the product $A_{1 \times p} B_{p \times 1}$ is always a real number. However, the product of matrices with more than one row or column is a matrix.

Definition 10.8 *The product of matrices $A_{m \times p}$ and $B_{p \times n}$ is the $m \times n$ matrix whose i, jth entry is the product of the ith row of A and the jth column of B. This product is denoted AB or $A \cdot B$.*

Example

Multiply: $\begin{bmatrix} 3 & 0 & 1 \\ 0 & 1 & 2 \end{bmatrix} \begin{bmatrix} 1 & -2 \\ -1 & 2 \\ 1 & 1 \end{bmatrix}$.

Solution

Since the matrix on the left is 2×3 and the one on the right is 3×2, the product will be 2×2.

The 1, 1 entry is $\begin{bmatrix} 3 & 0 & 1 \end{bmatrix} \begin{bmatrix} 1 \\ -1 \\ 1 \end{bmatrix} = 3 \cdot 1 + 0(-1) + 1 \cdot 1 = 4;$

the 1, 2 entry is $\begin{bmatrix} 3 & 0 & 1 \end{bmatrix} \begin{bmatrix} -2 \\ 2 \\ 1 \end{bmatrix} = 3(-2) + 0(2) + 1 \cdot 1 = -5;$

the 2, 1 entry is $\begin{bmatrix} 0 & 1 & 2 \end{bmatrix} \begin{bmatrix} 1 \\ -1 \\ 1 \end{bmatrix} = 0 \cdot 1 + 1(-1) + 2 \cdot 1 = 1;$

the 2, 2 entry is $\begin{bmatrix} 0 & 1 & 2 \end{bmatrix} \begin{bmatrix} -2 \\ 2 \\ 1 \end{bmatrix} = 0(-2) + 1 \cdot 2 + 2 \cdot 1 = 4.$

Therefore, $\begin{bmatrix} 3 & 0 & 1 \\ 0 & 1 & 2 \end{bmatrix} \begin{bmatrix} 1 & -2 \\ -1 & 2 \\ 1 & 1 \end{bmatrix} = \begin{bmatrix} 4 & -5 \\ 1 & 4 \end{bmatrix}.$

Example

If $A = \begin{bmatrix} 1 & 2 \\ -1 & 3 \end{bmatrix}$ and $B = \begin{bmatrix} 2 & 1 \\ 1 & 1 \end{bmatrix}$, find AB and BA.

Solution

$AB = \begin{bmatrix} 1 & 2 \\ -1 & 3 \end{bmatrix} \begin{bmatrix} 2 & 1 \\ 1 & 1 \end{bmatrix} = \begin{bmatrix} 2+2 & 1+2 \\ -2+3 & -1+3 \end{bmatrix} = \begin{bmatrix} 4 & 3 \\ 1 & 2 \end{bmatrix};$

$BA = \begin{bmatrix} 2 & 1 \\ 1 & 1 \end{bmatrix} \begin{bmatrix} 1 & 2 \\ -1 & 3 \end{bmatrix} = \begin{bmatrix} 2-1 & 4+3 \\ 1-1 & 2+3 \end{bmatrix} = \begin{bmatrix} 1 & 7 \\ 0 & 5 \end{bmatrix}.$

10.2 Matrix Multiplication

Properties of products

The preceding example shows very clearly that the multiplication of matrices, in general, is *not commutative*. Thus, when discussing products of matrices, we must specify the *order* in which the matrices are to be considered as factors. For the product AB, we say that A is *right-multiplied* by B, and that B is *left-multiplied* by A.

Note that the definition of the product of two matrices, A and B, requires that the matrix A have the same number of *columns* as B has *rows*; the result, AB, then has the same number of rows as A and the same number of columns as B. Such matrices A and B are said to be **conformable** for multiplication. The fact that two matrices are conformable in the order AB, however, does not mean that they necessarily are conformable in the order BA.

Example

If $A = \begin{bmatrix} 3 & 1 \\ 1 & 0 \\ 2 & 1 \end{bmatrix}$ and $B = \begin{bmatrix} 1 & -1 \\ 2 & 1 \end{bmatrix}$, find AB.

Solution

Since A is a 3×2 matrix, and B is a 2×2 matrix, they are conformable for multiplication in the order AB. We have

$$AB = \begin{bmatrix} 3 & 1 \\ 1 & 0 \\ 2 & 1 \end{bmatrix} \begin{bmatrix} 1 & -1 \\ 2 & 1 \end{bmatrix} = \begin{bmatrix} 3+2 & -3+1 \\ 1+0 & -1+0 \\ 2+2 & -2+1 \end{bmatrix} = \begin{bmatrix} 5 & -2 \\ 1 & -1 \\ 4 & -1 \end{bmatrix}.$$

Note that the matrices A and B in the example above are not conformable in the order BA.

In much of the matrix work in this book, we shall focus our attention on matrices having the same number of rows as columns. For brevity, a matrix of order $n \times n$ is often called a **square matrix** of order n. Although many of the ideas we shall discuss are applicable to conformable matrices of any order, we shall apply the notions only to square matrices.

Theorem 10.3   If A, B, and C are $n \times n$ square matrices, then

$$(AB)C = A(BC).$$

If A is a square matrix, then A^2, A^3, etc., denote AA, $(AA)A$, etc.

Theorem 10.4   If A, B, and C are $n \times n$ square matrices, then

$$A(B + C) = AB + AC$$

and

$$(B + C)A = BA + CA.$$

The proofs of these theorems involve some complicated symbolism and are omitted here, but you will be asked to show their validity for the case of 2 × 2 matrixes in the exercises. Observe that, because matrix multiplication is not, in general, commutative, we must establish both the left-hand and the right-hand distributive property.

Definition 10.9   The **principal diagonal** *of a square matrix is the ordered set of entries* a_{jj}, *extending from the upper left-hand corner to the lower right-hand corner of the matrix. Thus, the principal diagonal contains* a_{11}, a_{22}, a_{33}, *etc.*

For example, the principal diagonal of

$$\begin{bmatrix} 1 & 3 & -1 \\ 5 & 2 & 3 \\ 6 & 4 & 0 \end{bmatrix}$$

consists of 1, 2, and 0, in that order.

Definition 10.10   *A **diagonal matrix** is a square matrix in which all entries not in the principal diagonal are 0.*

Thus,

$$\begin{bmatrix} 4 & 0 \\ 0 & 2 \end{bmatrix} \quad \text{and} \quad \begin{bmatrix} 1 & 0 & 0 \\ 0 & 1 & 0 \\ 0 & 0 & 0 \end{bmatrix}$$

are diagonal matrices.

Definition 10.11 $I_{n \times n}$ *denotes the diagonal matrix having 1's for entries on the principal diagonal.*

For example,

$$I_{2 \times 2} = \begin{bmatrix} 1 & 0 \\ 0 & 1 \end{bmatrix} \quad \text{and} \quad I_{4 \times 4} = \begin{bmatrix} 1 & 0 & 0 & 0 \\ 0 & 1 & 0 & 0 \\ 0 & 0 & 1 & 0 \\ 0 & 0 & 0 & 1 \end{bmatrix}.$$

The following properties are consequences of the definitions we have adopted.

Theorem 10.5   *For each matrix* $A_{n \times n}$,

$$A_{n \times n} I_{n \times n} = I_{n \times n} A_{n \times n} = A_{n \times n}.$$

Furthermore, $I_{n \times n}$ *is the unique matrix having this property for all matrices* $A_{n \times n}$.

10.2 Matrix Multiplication

Accordingly, $I_{n \times n}$ is the **identity element for multiplication** in the set of $n \times n$ square matrices. The proof of this theorem, for the illustrative case $n = 2$, is left as an exercise.

Order of multiplication

The following result relates the order in which matrices can be multiplied by real numbers and by other matrices.

Theorem 10.6 If A and B are $n \times n$ square matrices, and a is a real number, then
$$a(AB) = (aA)B = A(aB).$$

The proof for the case $n = 2$ is left as an exercise.

Example Let
$$A = \begin{bmatrix} 1 & 3 \\ 2 & -4 \end{bmatrix} \quad \text{and} \quad B = \begin{bmatrix} 4 & 1 \\ -2 & -1 \end{bmatrix}.$$

Then

$$4(AB) = 4\left(\begin{bmatrix} 1 & 3 \\ 2 & -4 \end{bmatrix} \begin{bmatrix} 4 & 1 \\ -2 & -1 \end{bmatrix} \right) = 4 \begin{bmatrix} 4-6 & 1-3 \\ 8+8 & 2+4 \end{bmatrix}$$

$$= 4 \begin{bmatrix} -2 & -2 \\ 16 & 6 \end{bmatrix} = \begin{bmatrix} -8 & -8 \\ 64 & 24 \end{bmatrix},$$

$$(4A)B = \begin{bmatrix} 4 & 12 \\ 8 & -16 \end{bmatrix} \begin{bmatrix} 4 & 1 \\ -2 & -1 \end{bmatrix} = \begin{bmatrix} -8 & -8 \\ 64 & 24 \end{bmatrix},$$

and

$$A(4B) = \begin{bmatrix} 1 & 3 \\ 2 & -4 \end{bmatrix} \begin{bmatrix} 16 & 4 \\ -8 & -4 \end{bmatrix} = \begin{bmatrix} -8 & -8 \\ 64 & 24 \end{bmatrix}.$$

Thus, $4(AB) = (4A)B = A(4B)$.

Exercise 10.2

A *Write each product as a single matrix.*

Examples

a. $3 \begin{bmatrix} 2 & 1 \\ -1 & 3 \\ 2 & 0 \end{bmatrix}$

b. $\begin{bmatrix} 3 & 1 & -1 \\ 0 & -1 & 2 \end{bmatrix} \cdot \begin{bmatrix} 1 & -1 \\ 0 & 2 \\ 1 & 0 \end{bmatrix}$

Solutions

a. $\begin{bmatrix} 6 & 3 \\ -3 & 9 \\ 6 & 0 \end{bmatrix}$

b. $\begin{bmatrix} 3+0-1 & -3+2+0 \\ 0+0+2 & 0-2+0 \end{bmatrix} = \begin{bmatrix} 2 & -1 \\ 2 & -2 \end{bmatrix}$

1. $-5 \begin{bmatrix} 0 & 1 & -1 \\ 3 & -1 & 2 \end{bmatrix}$
2. $2 \begin{bmatrix} 2 & 1 & 3 & -2 \\ 4 & 2 & 0 & -1 \\ 0 & 0 & -1 & 2 \end{bmatrix}$

3. $\begin{bmatrix} 1 & -2 \end{bmatrix} \cdot \begin{bmatrix} 3 \\ 2 \end{bmatrix}$
4. $\begin{bmatrix} 3 & -2 & 2 \end{bmatrix} \cdot \begin{bmatrix} 1 \\ 0 \\ -2 \end{bmatrix}$

5. $\begin{bmatrix} 3 & -1 \\ 2 & 1 \end{bmatrix} \cdot \begin{bmatrix} 1 & -4 \\ 2 & 1 \end{bmatrix}$
6. $\begin{bmatrix} 1 & -5 \\ 0 & 2 \end{bmatrix} \cdot \begin{bmatrix} 3 & 1 \\ -1 & 2 \end{bmatrix}$

7. $\begin{bmatrix} 4 & -5 \\ 7 & 3 \end{bmatrix} \cdot \begin{bmatrix} 5 & -1 \\ -2 & 7 \end{bmatrix}$
8. $\begin{bmatrix} 1 & -2 \\ -3 & 1 \end{bmatrix} \cdot \begin{bmatrix} 5 & 1 \\ 0 & 2 \end{bmatrix}$

9. $\begin{bmatrix} -3 & 1 & 0 \\ 2 & 1 & 1 \end{bmatrix} \cdot \begin{bmatrix} 2 & 0 \\ 1 & -1 \\ 3 & 0 \end{bmatrix}$
10. $\begin{bmatrix} 1 & -1 & 0 \\ 2 & 1 & 3 \end{bmatrix} \cdot \begin{bmatrix} 4 & -1 \\ 2 & 0 \\ 1 & 1 \end{bmatrix}$

11. $\begin{bmatrix} -1 & 0 & 1 \\ 2 & 1 & 0 \\ 1 & 0 & 0 \end{bmatrix} \cdot \begin{bmatrix} 0 & 1 & 3 \\ 1 & 0 & 2 \\ -1 & 1 & 1 \end{bmatrix}$
12. $\begin{bmatrix} 2 & -3 & 1 \\ 0 & 1 & -1 \\ 2 & 0 & 0 \end{bmatrix} \cdot \begin{bmatrix} 1 & 0 & 0 \\ 0 & 1 & 0 \\ 0 & 0 & 1 \end{bmatrix}$

13. $\begin{bmatrix} 2 & -2 & -1 \\ 1 & 1 & -2 \\ 1 & 0 & -1 \end{bmatrix} \cdot \begin{bmatrix} -1 & -2 & 5 \\ -1 & -1 & 3 \\ -1 & -2 & 4 \end{bmatrix}$
14. $\begin{bmatrix} -1 & -2 & 5 \\ -1 & -1 & 3 \\ -1 & -2 & 4 \end{bmatrix} \cdot \begin{bmatrix} 2 & -2 & -1 \\ 1 & 1 & -2 \\ 1 & 0 & -1 \end{bmatrix}$

Let $A = \begin{bmatrix} 1 & -2 \\ 1 & 0 \end{bmatrix}$ and $B = \begin{bmatrix} -1 & 2 \\ -1 & 1 \end{bmatrix}$. Compute each of the following products.

15. AB
16. BA
17. $(AB)A$
18. $(BA)B$
19. A^tB
20. AB^t

Find a matrix X satisfying each matrix equation.

21. $3X + \begin{bmatrix} 1 & 0 \\ 2 & 1 \end{bmatrix} = \begin{bmatrix} -2 & 3 \\ -1 & -2 \end{bmatrix}$
22. $2X + 3 \begin{bmatrix} 1 & 1 \\ 0 & 1 \end{bmatrix} = \begin{bmatrix} 7 & -1 \\ 3 & -5 \end{bmatrix}$

23. $X + 2I = \begin{bmatrix} 3 & -1 \\ 1 & 2 \end{bmatrix}$
24. $3X - 2I = \begin{bmatrix} 7 & 3 \\ 6 & 4 \end{bmatrix}$

B 25. Show that if $A = \begin{bmatrix} -1 & 2 \\ 0 & 1 \end{bmatrix}$ and $B = \begin{bmatrix} 1 & 0 \\ -1 & 2 \end{bmatrix}$, then

    a. $(A + B)(A + B) \neq A^2 + 2AB + B^2$.
    b. $(A + B)(A - B) \neq A^2 - B^2$.

26. a. Show that for each matrix $A_{2 \times 2}$,

$$A_{2 \times 2} \cdot I_{2 \times 2} = I_{2 \times 2} \cdot A_{2 \times 2} = A_{2 \times 2}.$$

10.3 Solution of Linear Systems by Using Row-Equivalent Matrices

b. Show that if, for a given matrix $B_{2 \times 2}$ and for all $A_{2 \times 2}$,

$$A_{2 \times 2} \cdot B_{2 \times 2} = B_{2 \times 2} \cdot A_{2 \times 2} = A_{2 \times 2},$$

then $B_{2 \times 2} = I_{2 \times 2}$.

27. Show that

$$\left(\begin{bmatrix} a_{11} & a_{12} \\ a_{21} & a_{22} \end{bmatrix} \cdot \begin{bmatrix} b_{11} & b_{12} \\ b_{21} & b_{22} \end{bmatrix}\right) \cdot \begin{bmatrix} c_{11} & c_{12} \\ c_{21} & c_{22} \end{bmatrix} = \begin{bmatrix} a_{11} & a_{12} \\ a_{21} & a_{22} \end{bmatrix} \cdot \left(\begin{bmatrix} b_{11} & b_{12} \\ b_{21} & b_{22} \end{bmatrix} \cdot \begin{bmatrix} c_{11} & c_{12} \\ c_{21} & c_{22} \end{bmatrix}\right).$$

28. Show that

$$\begin{bmatrix} a_{11} & a_{12} \\ a_{21} & a_{22} \end{bmatrix} \cdot \left(\begin{bmatrix} b_{11} & b_{12} \\ b_{21} & b_{22} \end{bmatrix} + \begin{bmatrix} c_{11} & c_{12} \\ c_{21} & c_{22} \end{bmatrix}\right)$$

$$= \begin{bmatrix} a_{11} & a_{12} \\ a_{21} & a_{22} \end{bmatrix} \cdot \begin{bmatrix} b_{11} & b_{12} \\ b_{21} & b_{22} \end{bmatrix} + \begin{bmatrix} a_{11} & a_{12} \\ a_{21} & a_{22} \end{bmatrix} \cdot \begin{bmatrix} c_{11} & c_{12} \\ c_{21} & c_{22} \end{bmatrix}.$$

29. Show that

$$\left(\begin{bmatrix} b_{11} & b_{12} \\ b_{21} & b_{22} \end{bmatrix} + \begin{bmatrix} c_{11} & c_{12} \\ c_{21} & c_{22} \end{bmatrix}\right) \cdot \begin{bmatrix} a_{11} & a_{12} \\ a_{21} & a_{22} \end{bmatrix}$$

$$= \begin{bmatrix} b_{11} & b_{12} \\ b_{21} & b_{22} \end{bmatrix} \cdot \begin{bmatrix} a_{11} & a_{12} \\ a_{21} & a_{22} \end{bmatrix} + \begin{bmatrix} c_{11} & c_{12} \\ c_{21} & c_{22} \end{bmatrix} \cdot \begin{bmatrix} a_{11} & a_{12} \\ a_{21} & a_{22} \end{bmatrix}.$$

30. Show that $\begin{bmatrix} 0 & a \\ a & 0 \end{bmatrix}^2 = a^2 I.$

31. Show that $(A_{2 \times 2} \cdot B_{2 \times 2})^t = B^t_{2 \times 2} \cdot A^t_{2 \times 2}.$

32. Show that $A^2_{2 \times 2} = (-A_{2 \times 2})^2.$

In Exercises 33–39, prove the specified part of Theorem 10.2 for 2×2 matrices.

33. Part I **34.** Part II **35.** Part III **36.** Part V

37. Part VI **38.** Part VII **39.** Part VIII

40. Prove that if A and B are 2×2 matrices and a is a real number, then

$$a(AB) = (aA)B = A(aB).$$

10.3 Solution of Linear Systems by Using Row-Equivalent Matrices

Elementary transformations

An **elementary transformation** of a matrix $A_{n \times m}$ is one of the following three operations upon the rows of the matrix:

1. Multiply the entries of any row of $A_{n \times m}$ by k, where $k \in R$, $k \neq 0$.
2. Interchange any two rows of $A_{n \times m}$.
3. Multiply the entries of any row of $A_{n \times m}$ by k, where $k \in R$, and add to the corresponding entries of any other row.

Examples

Let $A = \begin{bmatrix} 1 & 3 \\ -2 & 4 \end{bmatrix}$.

a. The matrix $B = \begin{bmatrix} 1 & 3 \\ -1 & 2 \end{bmatrix}$ is obtained from A by multiplying each entry of Row 2 by 1/2.

b. The matrix $C = \begin{bmatrix} -2 & 4 \\ 1 & 3 \end{bmatrix}$ is obtained from A by interchanging Rows 1 and 2.

c. The matrix $D = \begin{bmatrix} 0 & 5 \\ -2 & 4 \end{bmatrix}$ is obtained from A by multiplying each entry of Row 2 by 1/2 and adding the result to the corresponding entry of Row 1.

Notice that the inverse of an elementary transformation is an elementary transformation. That is, you can undo an elementary transformation by means of an elementary transformation. For example, if A is transformed into B by interchanging two rows, then you can regain A from B by again interchanging the same two rows. It follows that if B results from performing a *succession* of elementary transformations on A, then A can similarly be obtained from B by performing the inverse operations in reverse order.

Row-equivalent matrices

Definition 10.12 If B is a matrix resulting from a succession of a finite number of elementary transformations on a matrix A, then A and B are **row-equivalent** matrices. This is expressed by writing $A \sim B$ or $B \sim A$.

Example

Show that $\begin{bmatrix} 1 & -2 & -1 \\ 1 & 0 & 2 \\ -4 & 3 & 1 \end{bmatrix} \sim \begin{bmatrix} 3 & -6 & -3 \\ 1 & 0 & 2 \\ -4 & 3 & 1 \end{bmatrix}$.

Solution

Multiplying each entry of the first row of the left-hand matrix by 3, we obtain the right-hand matrix.

Example

Show that $\begin{bmatrix} 1 & -2 & -1 \\ 1 & 0 & 2 \\ -4 & 3 & 1 \end{bmatrix} \sim \begin{bmatrix} -4 & 3 & 1 \\ 1 & 0 & 2 \\ 1 & -2 & -1 \end{bmatrix}$.

Solution

Interchanging the first and third row of the left-hand matrix, we obtain the right-hand matrix.

Sometimes it is convenient to make several elementary transformations on the same matrix.

10.3 Solution of Linear Systems by Using Row-Equivalent Matrices

Example

Show that $\begin{bmatrix} 1 & -2 & -1 \\ 1 & 0 & 2 \\ -4 & 3 & 1 \end{bmatrix} \sim \begin{bmatrix} 1 & -2 & -1 \\ 0 & 2 & 3 \\ 0 & -5 & -3 \end{bmatrix}$.

Solution

Multiplying the entries of the first row of the left-hand matrix by -1 and adding these products to the corresponding entries of the second row, and then multiplying the entries of the first row by 4 and adding these products to the corresponding entries of the third row, we obtain the right-hand matrix.

The $n \times n$ matrices that are row-equivalent to $I_{n \times n}$ are called **nonsingular** matrices. If an $n \times n$ matrix is not row-equivalent to $I_{n \times n}$, it is called **singular**.

Example

Show that $A = \begin{bmatrix} 1 & 2 & 1 \\ -1 & 1 & 0 \\ 1 & 0 & 1 \end{bmatrix}$ is a nonsingular matrix.

Solution

We first make appropriate transformations to obtain "0" elements in each entry (except for the principal diagonal) in columns 1, 2, and 3. Each reference to a row indicates the row of the preceding matrix.

$\begin{bmatrix} 1 & 2 & 1 \\ -1 & 1 & 0 \\ 1 & 0 & 1 \end{bmatrix} \sim \begin{bmatrix} 1 & 2 & 1 \\ 0 & 3 & 1 \\ 0 & -2 & 0 \end{bmatrix}$ Row 2 + Row 1
Row 3 + $[(-1) \times$ Row 1]

$\sim \begin{bmatrix} 1 & 0 & 1 \\ 0 & 3 & 1 \\ 0 & 0 & \frac{2}{3} \end{bmatrix}$ Row 1 + Row 3

Row 3 + $\left[\frac{2}{3} \times \text{Row 2}\right]$

$\sim \begin{bmatrix} 1 & 0 & 0 \\ 0 & 3 & 0 \\ 0 & 0 & \frac{2}{3} \end{bmatrix}$ Row 1 + $\left[-\frac{3}{2} \times \text{Row 3}\right]$

Row 2 + $\left[-\frac{3}{2} \times \text{Row 3}\right]$

$\sim \begin{bmatrix} 1 & 0 & 0 \\ 0 & 1 & 0 \\ 0 & 0 & 1 \end{bmatrix}$ $\frac{1}{3} \times$ Row 2

$\frac{3}{2} \times$ Row 3

Since the last operation yields the matrix $I_{3 \times 3}$, A is nonsingular.

Matrix solution of linear systems

Notice that in the preceding example each elementary transformation is noted next to the appropriate row. This provides a convenient record which is helpful when checking the work. It is advisable to keep such a record when doing problems.

In a linear system of the form

$$a_{11}x + a_{12}y + a_{13}z = c_1$$
$$a_{21}x + a_{22}y + a_{23}z = c_2$$
$$a_{31}x + a_{32}y + a_{33}z = c_3,$$

the matrices

$$\begin{bmatrix} a_{11} & a_{12} & a_{13} \\ a_{21} & a_{22} & a_{23} \\ a_{31} & a_{32} & a_{33} \end{bmatrix} \quad \text{and} \quad \begin{bmatrix} a_{11} & a_{12} & a_{13} & c_1 \\ a_{21} & a_{22} & a_{23} & c_2 \\ a_{31} & a_{32} & a_{33} & c_3 \end{bmatrix}$$

are called the **coefficient matrix** and the **augmented matrix**, respectively. Similar definitions hold for a system of n linear equations.

Starting with the augmented matrix of a linear system, and generating a sequence of row-equivalent matrices, we can obtain a matrix from which the solution set of the system is evident simply by inspection. The validity of the method, which is illustrated by example below, follows from the fact that performing elementary transformations on the augmented matrix of a system corresponds to performing the same sorts of operations on the equations of the system itself. Neither multiplying an equation by a nonzero constant nor interchanging two equations has any effect on the solution set of the system. Multiplying one equation by a constant and adding it to another equation in effect replaces the other by a linear combination of equations; a generalization of Theorems 9.1 and 9.2 ensures that the solution set remains the same.

Example

Solve the system

$$x + 2y - 3z = -4$$
$$2x - y + z = 3$$
$$3x + 2y + z = 10.$$

Solution

The augmented matrix of the system is

$$\begin{bmatrix} 1 & 2 & -3 & -4 \\ 2 & -1 & 1 & 3 \\ 3 & 2 & 1 & 10 \end{bmatrix}.$$

We perform elementary transformations on this matrix as follows.

$$\begin{bmatrix} 1 & 2 & -3 & -4 \\ 0 & -5 & 7 & 11 \\ 3 & 2 & 1 & 10 \end{bmatrix} \quad \text{Row } 2 + [-2 \times \text{Row } 1]$$

10.3 Solution of Linear Systems by Using Row-Equivalent Matrices

$$\begin{bmatrix} 1 & 2 & 3 & \vdots & -4 \\ 0 & -5 & 7 & \vdots & 11 \\ 0 & -4 & 10 & \vdots & 22 \end{bmatrix} \quad \text{Row 3} + [-3 \times \text{Row 1}]$$

$$\begin{bmatrix} 1 & 2 & -3 & \vdots & -4 \\ 0 & -5 & 7 & \vdots & 11 \\ 0 & 0 & -22 & \vdots & -66 \end{bmatrix} \quad -5 \times \text{Row 3, then Row 3} + [4 \times \text{Row 2}]$$

The resulting system is

$$x + 2y - 3z = -4$$
$$-5y + 7z = 11$$
$$-22z = -66$$

which may be solved by reverse substitution. This example was solved in Section 9.2 and the augmented matrices here correspond exactly to the equivalent systems obtained there. Matrix notation, however, so facilitates the operations on the equations (rows) that it is easy to obtain even simpler equivalent systems. We can rework the above example as follows; the system of equations corresponding to each successive matrix is shown on the right of the matrix in the following solution.

$$\begin{array}{c} \text{Row 2} + [-2 \times \text{Row 1}] \\ \text{Row 3} + [-3 \times \text{Row 1}] \end{array} \begin{bmatrix} 1 & 2 & -3 & \vdots & -4 \\ 0 & -5 & 7 & \vdots & 11 \\ 0 & -4 & 10 & \vdots & 22 \end{bmatrix} \begin{array}{l} x + 2y - 3z = -4 \\ 0x - 5y + 7z = 11 \\ 0x - 4y + 10z = 22 \end{array}$$

$$\text{Row 1} + \left[\frac{2}{5} \times \text{Row 2} \right] \begin{bmatrix} 1 & 0 & -\frac{1}{5} & \vdots & \frac{2}{5} \\ 0 & -5 & 7 & \vdots & 11 \\ \end{bmatrix} \begin{array}{l} x + 0y - \frac{1}{5}z = \frac{2}{5} \\ 0x - 5y + 7z = 11 \end{array}$$

$$\text{Row 3} + \left[-\frac{4}{5} \times \text{Row 2} \right] \begin{bmatrix} 0 & 0 & \frac{22}{5} & \vdots & \frac{66}{5} \end{bmatrix} \quad 0x + 0y + \frac{22}{5}z = \frac{66}{5}$$

$$\begin{array}{c} 5 \times \text{Row 1} \\ \\ \frac{5}{22} \times \text{Row 3} \end{array} \begin{bmatrix} 5 & 0 & -1 & \vdots & 2 \\ 0 & -5 & 7 & \vdots & 11 \\ 0 & 0 & 1 & \vdots & 3 \end{bmatrix} \begin{array}{l} 5x + 0y - z = 2 \\ 0x - 5y + 7z = 11 \\ 0x + 0y + z = 3 \end{array}$$

$$\begin{array}{c} \text{Row 1} + \text{Row 3} \\ \text{Row 2} + [-7 \times \text{Row 3}] \end{array} \begin{bmatrix} 5 & 0 & 0 & \vdots & 5 \\ 0 & -5 & 0 & \vdots & -10 \\ 0 & 0 & 1 & \vdots & 3 \end{bmatrix} \begin{array}{l} 5x + 0y + 0z = 5 \\ 0x - 5y + 0z = -10 \\ 0x + 0y + z = 3 \end{array}$$

$$\begin{array}{c} \frac{1}{5} \times \text{Row 1} \\ \\ -\frac{1}{5} \times \text{Row 2} \end{array} \begin{bmatrix} 1 & 0 & 0 & \vdots & 1 \\ 0 & 1 & 0 & \vdots & 2 \\ 0 & 0 & 1 & \vdots & 3 \end{bmatrix} \begin{array}{l} x + 0y + 0z = 1 \\ 0x + y + 0z = 2 \\ 0x + 0y + z = 3 \end{array}$$

Solution continued overleaf

The last system is equivalent to

$$x = 1$$
$$y = 2$$
$$z = 3.$$

From this, the solution set, $\{(1, 2, 3)\}$, for the given system is evident by inspection.

For any given $n \times n$ linear system with *nonsingular* coefficient matrix, there are many sequences of row operations which will transform the augmented matrix of a system equivalently to one of the form

$$\begin{bmatrix} 1 & 0 & 0 & \cdots & 0 & \vdots & x_1 \\ 0 & 1 & 0 & & 0 & \vdots & \cdot \\ 0 & 0 & 1 & & 0 & \vdots & \cdot \\ \vdots & & & & \vdots & \vdots & \vdots \\ 0 & 0 & 0 & \cdots & 1 & \vdots & x_n \end{bmatrix}, \tag{1}$$

from which the solution set, $\{(x_1, \ldots, x_n)\}$, of the original system is evident by inspection. Finding the most efficient sequence depends on experience and insight, but several good systematic procedures exist. The procedure used above was, briefly, the following one: obtain a 1 on the diagonal using type 1 or 2 transformations if necessary; then "clear out" the remainder of the column using type 3 transformations to obtain 0's; then proceed to the next column.

Systems with singular coefficient matrices

If the row-reduction process yields a row of the form

$$0 \quad 0 \quad \cdots \quad 0 \quad a,$$

then that row in the system corresponds to an equation of the form

$$0 \cdot x_1 + 0 \cdot x_2 + \cdots + 0 \cdot x_n = a.$$

This equation has a solution only if $a = 0$, and in that case has an infinite number of solutions. Thus, the system will have no solution when the row-reduction process yields a row of the form

$$0 \quad 0 \quad \cdots \quad 0 \quad a,$$

where $a \neq 0$. The system will have an infinite number of solutions when the process does not yield such a row but does yield a row of zeros.

Example

Solve the system

$$2x + y = 1$$
$$4x + 2y = 0.$$

10.3 Solution of Linear Systems by Using Row-Equivalent Matrices

Solutions The augmented matrix of the system is

$$\begin{bmatrix} 2 & 1 & \vdots & 1 \\ 4 & 2 & \vdots & 0 \end{bmatrix}.$$

Thus, we have

$$\begin{bmatrix} 2 & 1 & \vdots & 1 \\ 4 & 2 & \vdots & 0 \end{bmatrix} \sim \begin{bmatrix} 2 & 1 & \vdots & 1 \\ 0 & 0 & \vdots & -2 \end{bmatrix} \quad \text{Row 2} + [-2 \times \text{Row 1}]$$

Since the last row in the matrix on the right is of the form

$$0 \quad 0 \quad a$$

with $a = -2$, the system has no solution.

Example Solve the system

$$2x + y = 1$$
$$4x + 2y = 2.$$

Solution The augmented matrix of this system is

$$\begin{bmatrix} 2 & 1 & \vdots & 1 \\ 4 & 2 & \vdots & 2 \end{bmatrix}.$$

Thus, we have

$$\begin{bmatrix} 2 & 1 & \vdots & 1 \\ 4 & 2 & \vdots & 2 \end{bmatrix} \sim \begin{bmatrix} 2 & 1 & \vdots & 1 \\ 0 & 0 & \vdots & 0 \end{bmatrix} \quad \text{Row 2} + [-2 \times \text{Row 1}].$$

Since the last row in the matrix on the right is of the form

$$0 \quad 0 \quad a$$

with $a = 0$ and there are no rows of this form with $a \neq 0$, the system has an infinite number of solutions.

The system corresponding to the matrix

$$\begin{bmatrix} 2 & 1 & \vdots & 1 \\ 0 & 0 & \vdots & 0 \end{bmatrix}$$

is

$$2x + y = 1$$
$$0x + 0y = 0.$$

Since the second equation is satisfied by any ordered pair, the solution set of the system is the same as that of the first equation. When this occurs we usually write the general form of the solution by expressing one variable in terms of the other. In this case, we have $y = 1 - 2x$ and we write the solution to this system as $(x, 1 - 2x)$, $x \in R$.

Exercise 10.3

A Use row transformations of the augmented matrix to solve each system of equations. Note the elementary transformations that are used. (Each coefficient matrix is nonsingular.)

Example

$x + 2y + z = 4$
$2x + y - z = -1$
$-x + y + z = 2$

Solution

$$\begin{bmatrix} 1 & 2 & 1 & \vdots & 4 \\ 2 & 1 & -1 & \vdots & -1 \\ -1 & 1 & 1 & \vdots & 2 \end{bmatrix} \text{ (augmented matrix)}$$

$$\begin{bmatrix} 1 & 2 & 1 & \vdots & 4 \\ 0 & -3 & -3 & \vdots & -9 \\ 0 & 3 & 2 & \vdots & 6 \end{bmatrix} \begin{array}{l} \text{Row 2} + [-2 \times \text{Row 1}] \\ \text{Row 1} + \text{Row 3} \end{array}$$

$$\begin{bmatrix} 1 & 2 & 1 & \vdots & 4 \\ 0 & 1 & 1 & \vdots & 3 \\ 0 & 3 & 2 & \vdots & 6 \end{bmatrix} -\frac{1}{3} \times \text{Row 2}$$

$$\begin{bmatrix} 1 & 0 & -1 & \vdots & -2 \\ 0 & 1 & 1 & \vdots & 3 \\ 0 & 0 & -1 & \vdots & -3 \end{bmatrix} \begin{array}{l} \text{Row 1} + [-2 \times \text{Row 2}] \\ \text{Row 3} + [-3 \times \text{Row 2}] \end{array}$$

$$\begin{bmatrix} 1 & 0 & -1 & \vdots & -2 \\ 0 & 1 & 1 & \vdots & 3 \\ 0 & 0 & 1 & \vdots & 3 \end{bmatrix} -1 \times \text{Row 3}$$

$$\begin{bmatrix} 1 & 0 & 0 & \vdots & 1 \\ 0 & 1 & 0 & \vdots & 0 \\ 0 & 0 & 1 & \vdots & 3 \end{bmatrix} \begin{array}{l} \text{Row 1} + \text{Row 3} \\ \text{Row 2} + [-1 \times \text{Row 3}] \end{array}$$

The solution is (1, 0, 3). This same example was also solved in Exercise 9.2.

1. $x - 2y = 4$
   $x + 3y = -1$

2. $x + y = -1$
   $x - 4y = -14$

3. $3x - 2y = 13$
   $4x - y = 19$

4. $4x - 3y = 16$
   $2x + y = 8$

5. $x - 2y = 6$
   $3x + y = 25$

6. $x - y = -8$
   $x + 2y = 9$

7. $x + y - z = 0$
   $2x - y + z = -6$
   $x + 2y - 3z = 2$

8. $2x - y + 3z = 1$
   $x + 2y - z = -1$
   $3x + y + z = 2$

9. $2x - y = 0$
   $3y + z = 7$
   $2x + 3z = 1$

10. $3x - z = 7$
    $2x + y = 6$
    $3y - z = 7$

10.3 *Solution of Linear Systems by Using Row-Equivalent Matrices*

11. $2x - 5y + 3z = -1$
    $-3x - y + 2z = 11$
    $-2x + 7y + 5z = 9$

12. $2x + y + z = 4$
    $3x - z = 3$
    $2x + 3z = 13$

Use row transformations of the augmented matrix to solve each system of equations. Note the elementary transformations that are used. (The coefficient matrix may be singular.)

Example

$x - y + 2z = 3$

$2x + 3y - 6z = 1$

$4x + y - 2z = 7$

Solution

$$\begin{bmatrix} 1 & -1 & 2 & \vdots & 3 \\ 2 & 3 & -6 & \vdots & 1 \\ 4 & 1 & -2 & \vdots & 7 \end{bmatrix}$$ (augmented matrix)

$$\begin{bmatrix} 1 & -1 & 2 & \vdots & 3 \\ 0 & 5 & -10 & \vdots & -5 \\ 0 & 5 & -10 & \vdots & -5 \end{bmatrix}$$ Row 2 + [-2 × Row 1]
Row 3 + [-4 × Row 1]

$$\begin{bmatrix} 1 & -1 & 2 & \vdots & 3 \\ 0 & 1 & -2 & \vdots & -1 \\ 0 & 5 & -10 & \vdots & -5 \end{bmatrix}$$ $\frac{1}{5}$ × Row 2

$$\begin{bmatrix} 1 & 0 & 0 & \vdots & 2 \\ 0 & 1 & -2 & \vdots & -1 \\ 0 & 0 & 0 & \vdots & 0 \end{bmatrix}$$ Row 1 + Row 2
Row 3 + [-5 × Row 2]

The zero row indicates there are an infinite number of solutions. The final equivalent system is

$x = 2$
$y - 2z = -1$ or $x = 2$
$0x + 0y + 0z = 0$ $y = 2z - 1$.

For any real-number z the triple $(2, 2z - 1, z)$ is a solution of the system, and conversely. Thus the solution set may be described as the set of all triples of the form $(2, 2z - 1, z)$, $z \in R$.

13. $x + 2y + 2z = 3$
    $2x + 5y + 5z = 7$
    $y + z = 1$

14. $x - y + z = -1$
    $3x - 2y + 2z = 1$
    $2x - 4y + 4z = -10$

15. $x + 3y + 7z = 5$
    $-x + y + z = -1$
    $x + 11y + 22z = 14$

16. $x + 2y = -1$
    $3x + y - 5z = 5$
    $2x + 9y + 5z = -10$

17. $x + 2y + 2z = 5$
    $x + y + z = 4$
    $2x + 3y + 3z = 8$

18. $x - y + z = 4$
    $2x + 3y - z = 5$
    $x - 6y + 4z = 6$

19. $\begin{aligned} x + 2y - z &= 1 \\ 3x + y + z &= 3 \\ 2x - y + 2z &= -1 \end{aligned}$

20. $\begin{aligned} x + 3y + z &= 1 \\ 2x + 4y - z &= -1 \\ 2y + 3z &= 3 \end{aligned}$

B Show that each product is a matrix that is row-equivalent to $\begin{bmatrix} a & b \\ c & d \end{bmatrix}$ if $k \neq 0$.

21. $\begin{bmatrix} k & 0 \\ 0 & 1 \end{bmatrix} \begin{bmatrix} a & b \\ c & d \end{bmatrix}$

22. $\begin{bmatrix} 1 & 0 \\ 0 & k \end{bmatrix} \begin{bmatrix} a & b \\ c & d \end{bmatrix}$

23. $\begin{bmatrix} 0 & 1 \\ 1 & 0 \end{bmatrix} \begin{bmatrix} a & b \\ c & d \end{bmatrix}$

24. $\begin{bmatrix} 1 & k \\ 0 & 1 \end{bmatrix} \begin{bmatrix} a & b \\ c & d \end{bmatrix}$

25. $\begin{bmatrix} 1 & 0 \\ k & 1 \end{bmatrix} \begin{bmatrix} a & b \\ c & d \end{bmatrix}$

10.4 The Determinant Function

Associated with each square matrix A having real-number entries is a real number called the **determinant** of A and denoted by δA or $\delta(A)$ (read "the determinant of A"). Thus we have a function, δ (delta), with domain the set of all square matrices having real-number entries, and with range the set of all real numbers; $\delta(A_{n \times n})$ is called a determinant of **order** n.

Let us begin by examining δ over the set $S_{2 \times 2}$ of 2×2 matrices.

Definition 10.13   The *determinant* of the matrix

$$\begin{bmatrix} a_{11} & a_{12} \\ a_{21} & a_{22} \end{bmatrix}$$

is the number $a_{11}a_{22} - a_{12}a_{21}$.

The determinant of a square matrix is customarily displayed in the same form as the matrix, but with vertical bars in lieu of brackets. Thus,

$$\delta \begin{bmatrix} a_{11} & a_{12} \\ a_{21} & a_{22} \end{bmatrix} = \begin{vmatrix} a_{11} & a_{12} \\ a_{21} & a_{22} \end{vmatrix} = a_{11}a_{22} - a_{12}a_{21}.$$

Examples   Compute the determinant of each matrix.

a. $\begin{bmatrix} 2 & 4 \\ 1 & 0 \end{bmatrix}$

b. $\begin{bmatrix} 4 & -3 \\ 1 & 2 \end{bmatrix}$

10.4 The Determinant Function

Solutions

a. $\begin{vmatrix} 2 & 4 \\ 1 & 0 \end{vmatrix} = (2)(0) - (4)(1)$
$= -4$

b. $\begin{vmatrix} 4 & -3 \\ 1 & 2 \end{vmatrix} = (4)(2) - (-3)(1)$
$= 11$

Minors and cofactors

Before we define the determinant of a higher-order matrix, we need to define two terms. Throughout the following discussion, A is the $n \times n$ matrix

$$A = \begin{bmatrix} a_{11} & \cdots & a_{1n} \\ \vdots & & \vdots \\ a_{n1} & \cdots & a_{nn} \end{bmatrix}.$$

Definition 10.14 The **minor** M_{ij} of a_{ij} is the determinant of the $(n-1) \times (n-1)$ matrix obtained by deleting the ith row and the jth column of the matrix A.

Examples

Let $A = \begin{bmatrix} 1 & 4 & -2 \\ 2 & 3 & -4 \\ 0 & 2 & 3 \end{bmatrix}$. Compute **a.** M_{23} and **b.** M_{31}.

Solutions

a. Write the matrix A and cross out the second row and third column.

$$M_{23} = \delta \begin{vmatrix} 1 & 4 & -2 \\ \cancel{2} & \cancel{3} & \cancel{-4} \\ 0 & 2 & \cancel{3} \end{vmatrix} = \begin{vmatrix} 1 & 4 \\ 0 & 2 \end{vmatrix} = (1)(2) - (4)(0) = 2$$

b. Write the matrix A and cross out the third row and first column.

$$M_{31} = \delta \begin{vmatrix} 1 & 4 & -2 \\ 2 & 3 & 4 \\ \cancel{0} & \cancel{2} & \cancel{3} \end{vmatrix} = \begin{vmatrix} 4 & -2 \\ 3 & 4 \end{vmatrix} = (4)(4) - (-2)(3) = 22$$

Definition 10.15 The **cofactor** A_{ij} of the entry a_{ij} is
$$A_{ij} = (-1)^{i+j} M_{ij}.$$

Examples

Let $A = \begin{bmatrix} 1 & 4 & -2 \\ 2 & 3 & -4 \\ 0 & 2 & 3 \end{bmatrix}$. Compute **a.** A_{23} and **b.** A_{31}.

Solutions

The matrix A is the same one that was used in the foregoing example. Thus, from that example, we have $M_{23} = 2$ and $M_{31} - 22$. Hence by Definition 10.15

a. $A_{23} = (-1)^{2+3} M_{23}$
$= (-1)(2) = -2$

b. $A_{31} = (-1)^{3+1} M_{31}$
$= (1)(22) = 22$

Determinants of n × n matrices

We can use the cofactors of the entries of a square matrix A with order greater than 2×2 to define the determinant of A.

Definition 10.16 The **determinant** of the square matrix.

$$\begin{bmatrix} a_{11} & a_{12} & \cdots & a_{1n} \\ a_{21} & a_{22} & \cdots & a_{2n} \\ \vdots & & & \vdots \\ a_{n1} & a_{n2} & \cdots & a_{nn} \end{bmatrix}$$

is the sum of the n products formed by multiplying each entry in any single row (or any single column) by its cofactor.

In applying Definition 10.16 the determinant is said to be **expanded** about whatever row (or column) is chosen.

It can be shown, although we shall not do so here, that the value of the determinant is independent of the row or column about which the determinant is expanded.

Example Let $A = \begin{bmatrix} 1 & 0 & 1 \\ 2 & 3 & 0 \\ 3 & 0 & 4 \end{bmatrix}$. Compute $\delta(A)$.

Solution We shall expand about column 2.

$$\delta(A) = 0 \cdot A_{12} + 3 \cdot A_{22} + 0 \cdot A_{32}$$

$$= 3(-1)^{2+2} \begin{vmatrix} 1 & 1 \\ 3 & 4 \end{vmatrix} = 3(1)(1)$$

$$= 3$$

Choosing a row or column

Note that each term in the sum used to compute the determinant of a matrix is of the form $a_{ij}A_{ij}$. Thus, if $a_{ij} = 0$ then that term does not contribute to the sum and therefore the cofactor of that entry need not be computed. Therefore to minimize the number of operations necessary to compute the determinant of a matrix we expand about the row or column with the most zero entries. Thus, in the preceding example we expanded about the second column which has the largest number of zero entries.

Example Let $A = \begin{bmatrix} 1 & 0 & 1 & 1 \\ 4 & -2 & 0 & 0 \\ -3 & 1 & 1 & 0 \\ 0 & 2 & -4 & 1 \end{bmatrix}$. Compute $\delta(A)$.

10.4 The Determinant Function

Solution Observe that no row or column has more than two zero entries and both Row 2 and Column 4 have two zero entries. We therefore expand the determinant about either Row 2 or Column 4. We shall use Row 2.

$$\delta(A) = 4A_{21} + (-2)A_{22} + 0 \cdot A_{23} + 0 \cdot A_{24}$$

$$= 4(-1)^{2+1} \begin{vmatrix} 0 & 1 & 1 \\ 1 & 1 & 0 \\ 2 & -4 & 1 \end{vmatrix} + (-2)(-1)^{2+2} \begin{vmatrix} 1 & 1 & 1 \\ -3 & 1 & 0 \\ 0 & -4 & 1 \end{vmatrix}$$

Now, using cofactors we compute the two 3 × 3 determinants in the right-hand member above. We shall expand each of the determinants about the first column.

$$\begin{vmatrix} 0 & 1 & 1 \\ 1 & 1 & 0 \\ 2 & -4 & 1 \end{vmatrix} = 0 \cdot \begin{vmatrix} 1 & 0 \\ -4 & 1 \end{vmatrix} + 1(-1) \begin{vmatrix} 1 & 1 \\ -4 & 1 \end{vmatrix} + 2 \cdot \begin{vmatrix} 1 & 1 \\ 1 & 0 \end{vmatrix}$$

$$= 0(1) + (-1)(5) + 2(-1)$$

$$= -7$$

and

$$\begin{vmatrix} 1 & 1 & 1 \\ -3 & 1 & 0 \\ 0 & -4 & 1 \end{vmatrix} = 1 \cdot \begin{vmatrix} 1 & 0 \\ -4 & 1 \end{vmatrix} + (-3)(-1) \begin{vmatrix} 1 & 1 \\ -4 & 1 \end{vmatrix} + 0 \cdot \begin{vmatrix} 1 & 1 \\ 1 & 0 \end{vmatrix}$$

$$= 1(1) + 3(5) + 0(-1)$$

$$= 16$$

Thus,

$$\delta(A) = 4(-1)(-7) + (-2)(1)(16) = -4.$$

Exercise 10.4

A Compute the determinant of each matrix.

Examples

a. $\begin{bmatrix} -2 & \frac{1}{2} \\ 3 & 1 \end{bmatrix}$
b. $\begin{bmatrix} 3 & 2 \\ -1 & 0 \end{bmatrix}$

Solutions

a. $\begin{vmatrix} -2 & \frac{1}{2} \\ 3 & 1 \end{vmatrix} = (-2)(1) - \left(\frac{1}{2}\right)(3) = -\frac{7}{2}$
b. $\begin{vmatrix} 3 & 2 \\ -1 & 0 \end{vmatrix} = (3)(0) - (2)(-1) = 2$

1. $\begin{bmatrix} 3 & 0 \\ 0 & 1 \end{bmatrix}$
2. $\begin{bmatrix} 4 & 0 \\ 0 & -\frac{1}{2} \end{bmatrix}$
3. $\begin{bmatrix} 2 & -1 \\ 0 & 2 \end{bmatrix}$

4. $\begin{bmatrix} 3 & -5 \\ 0 & 10 \end{bmatrix}$ 5. $\begin{bmatrix} -2 & 4 \\ -1 & 2 \end{bmatrix}$ 6. $\begin{bmatrix} \frac{1}{2} & 1 \\ 4 & 8 \end{bmatrix}$

7. $\begin{bmatrix} 3 & 1 \\ 2 & -3 \end{bmatrix}$ 8. $\begin{bmatrix} 4 & -2 \\ 5 & -4 \end{bmatrix}$

Specify the minor and the cofactor (in determinant form) of the indicated entry of

$$A = \begin{bmatrix} 2 & 1 & -2 & 0 \\ 1 & 0 & 3 & -1 \\ -2 & 1 & 2 & 2 \\ 1 & -1 & 3 & 1 \end{bmatrix}.$$

Example a_{21}

Solution   The minor is $\begin{vmatrix} 1 & -2 & 0 \\ 1 & 2 & 2 \\ -1 & 3 & 1 \end{vmatrix}$ Since $2 + 1$ is odd, the cofactor is $-\begin{vmatrix} 1 & -2 & 0 \\ 1 & 2 & 2 \\ -1 & 3 & 1 \end{vmatrix}$.

9. a_{11} 10. a_{13} 11. a_{23} 12. a_{41}

13. a_{31} 14. a_{33} 15. a_{44} 16. a_{14}

Compute the determinant of the given matrix.

Example $\begin{bmatrix} 1 & 2 & 1 \\ 0 & -1 & 2 \\ 3 & 4 & 1 \end{bmatrix}$

Solution   Expand about Column 1.

$$\delta(A) = (1)A_{11} + (0)A_{21} + (3)A_{31}$$

Compute the necessary cofactors.

$$A_{11} = (-1)^{1+1} \begin{vmatrix} -1 & 2 \\ 4 & 1 \end{vmatrix} = (1)[(-1)(1) - (2)(4)] = -9$$

$$A_{31} = (-1)^{3+1} \begin{vmatrix} 2 & 1 \\ -1 & 2 \end{vmatrix} = (1)[(2)(2) - (1)(-1)] = 5$$

Therefore,

$$\delta(A) = (1)(-9) + (3)(5) = 6.$$

10.4 The Determinant Function

17. $\begin{bmatrix} 2 & 1 & 3 \\ 0 & 4 & 1 \\ 0 & 0 & 2 \end{bmatrix}$

18. $\begin{bmatrix} -3 & 0 & 0 \\ -1 & 4 & 0 \\ 5 & 6 & -\frac{1}{4} \end{bmatrix}$

19. $\begin{bmatrix} 0 & 1 & 1 \\ 1 & 2 & 1 \\ 3 & -1 & 0 \end{bmatrix}$

20. $\begin{bmatrix} 2 & -4 & 1 \\ 1 & 1 & 0 \\ 0 & 2 & 0 \end{bmatrix}$

21. $\begin{bmatrix} 2 & 1 & 1 \\ 3 & -4 & 2 \\ 4 & 2 & 2 \end{bmatrix}$

22. $\begin{bmatrix} 1 & -4 & 2 \\ 2 & -1 & 4 \\ -3 & 2 & -6 \end{bmatrix}$

Example

$\begin{bmatrix} 0 & 1 & 0 & 2 \\ 0 & 1 & 1 & 4 \\ 2 & 0 & 0 & 3 \\ 1 & 1 & 1 & 2 \end{bmatrix}$

Solution Expand about Column 3.

$$\delta(A) = (0)A_{13} + (1)A_{23} + (0)A_{33} + (1)A_{43}$$

Compute the necessary cofactors.

$$A_{23} = (-1)^{2+3} \begin{vmatrix} 0 & 1 & 2 \\ 2 & 0 & 3 \\ 1 & 1 & 2 \end{vmatrix} = -3$$

$$A_{43} = (-1)^{4+3} \begin{vmatrix} 0 & 1 & 2 \\ 0 & 1 & 4 \\ 2 & 0 & 3 \end{vmatrix} = -4$$

Thus,

$$\delta(A) = (1)(-3) + (1)(-4) = -7.$$

23. $\begin{bmatrix} 1 & 0 & 0 & 0 \\ 0 & 3 & 0 & 0 \\ 0 & 0 & -2 & 0 \\ 0 & 0 & 0 & 4 \end{bmatrix}$

24. $\begin{bmatrix} 1 & 0 & 2 & 0 \\ 0 & 1 & 3 & 1 \\ 0 & 0 & 2 & -1 \\ 0 & 0 & 0 & 1 \end{bmatrix}$

25. $\begin{bmatrix} 0 & 1 & 0 & 2 \\ 1 & 2 & 1 & 0 \\ 3 & -4 & 0 & 0 \\ 2 & 0 & 1 & 0 \end{bmatrix}$

26. $\begin{bmatrix} 1 & 4 & 1 & 1 \\ 0 & 2 & -3 & -4 \\ 2 & 0 & 0 & 4 \\ 1 & 1 & 1 & 2 \end{bmatrix}$

B 27. Show that $\begin{vmatrix} x & y & 1 \\ x_1 & y_1 & 1 \\ x_2 & y_2 & 1 \end{vmatrix} = 0$ represents an equation of the line through the points (x_1, y_1) and (x_2, y_2).

28. Use the results in Exercise 27 to find an equation of the line through the points $(3, -1)$ and $(-2, 5)$.

29. Show that if A is an $n \times n$ matrix with one row (or column) identically 0 then $\delta(A) = 0$.

30. In accordance with Definitions 10.14, 10.15, and 10.16, the determinant of an $n \times n$ matrix is the sum of a certain number of products of the entries. What is this number for $n = 2$? For $n = 3$? For $n = 4$?

31. Show that for any 2×2 matrix A, $\delta(aA) = a^2 \delta(A)$.

32. Show that for any 2×2 matrix A, $\delta(A^t) = \delta(A)$.

33. Show that for any 2×2 matrices A and B, $\delta(AB) = \delta(A)\delta(B)$.

10.5 Properties of Determinants

Determinants have some properties that are useful by virtue of the fact that they permit us to generate equivalent determinants (name the same number) with different and simpler configurations of entries. This, in turn, helps us find values for determinants. We shall state these properties without proof. Some of the proofs are left as exercises.

Theorem 10.7 *If each entry in any row, or each entry in any column, of a determinant is 0, then the determinant is equal to 0.*

Examples
a. $\begin{vmatrix} 0 & 0 \\ 1 & 2 \end{vmatrix} = 0$
b. $\begin{vmatrix} 1 & 1 & 0 \\ 3 & 5 & 0 \\ 2 & 7 & 0 \end{vmatrix} = 0$
c. $\begin{vmatrix} 0 & 1 & 0 & 0 \\ 1 & 0 & 0 & 0 \\ 0 & 0 & 0 & 1 \\ 0 & 0 & 0 & 1 \end{vmatrix} = 0$

Theorem 10.8 *If any two rows (or any two columns) of a determinant are interchanged, then the resulting determinant is the negative of the original determinant.*

Examples
a. $\begin{vmatrix} 1 & 2 \\ 3 & 4 \end{vmatrix} = -\begin{vmatrix} 3 & 4 \\ 1 & 2 \end{vmatrix}$
b. $\begin{vmatrix} 1 & 2 & 3 \\ 4 & 5 & 6 \\ 7 & 8 & 9 \end{vmatrix} = -\begin{vmatrix} 3 & 2 & 1 \\ 6 & 5 & 4 \\ 9 & 8 & 7 \end{vmatrix}$

In (**a**), Rows 1 and 2 were interchanged. In (**b**), Columns 1 and 3 were interchanged.

10.5 Properties of Determinants

Theorem 10.9 *If two rows (or two columns) in a determinant have corresponding entries that are equal, the determinant is equal to 0.*

Examples

a. $\begin{vmatrix} 1 & 1 \\ 3 & 3 \end{vmatrix} = 0$
b. $\begin{vmatrix} 1 & 2 & 1 \\ 3 & 1 & 0 \\ 1 & 2 & 1 \end{vmatrix} = 0$
c. $\begin{vmatrix} 1 & 2 & 3 & 4 \\ 5 & 6 & 7 & 8 \\ 0 & 0 & 1 & 0 \\ 1 & 2 & 3 & 4 \end{vmatrix} = 0$

In (**a**) Columns 1 and 2 are identical. In (**b**) Rows 1 and 3 are identical. In (**c**) Rows 1 and 4 are identical.

Theorem 10.10 *If each of the entries of one row (or column) of a determinant is multiplied by k, the determinant is multiplied by k.*

Examples

a. $\begin{vmatrix} 1 & 0 & 0 \\ 2 & 1 & 3 \\ 1 \times 2 & 3 \times 2 & 4 \times 2 \end{vmatrix} = 2\begin{vmatrix} 1 & 0 & 0 \\ 2 & 1 & 3 \\ 1 & 3 & 4 \end{vmatrix}$
b. $\begin{vmatrix} 4 & 5 & 8 \\ 1 & 1 & 2 \\ 3 & 1 & 6 \end{vmatrix} = 2\begin{vmatrix} 4 & 5 & 4 \\ 1 & 1 & 1 \\ 3 & 1 & 3 \end{vmatrix}$

Note that this process is different from that of the multiplication of a matrix by a real number. In the latter, each entry in the matrix is multiplied by the real number, rather than, as here, only the entries in a single row or column being so multiplied.

Theorem 10.11 *If each entry of one row (or column) of a determinant is multiplied by a real number k and the resulting product is added to the corresponding entry in another row (or column, respectively) in the determinant, the resulting determinant is equal to the original determinant.*

Examples

a. $\begin{vmatrix} 1 & 1 \\ 2 & 1 \end{vmatrix} = \begin{vmatrix} 1 & 1 \\ 2 + 3(1) & 1 + 3(1) \end{vmatrix} = \begin{vmatrix} 1 & 1 \\ 5 & 4 \end{vmatrix}$

b. $\begin{vmatrix} 1 & 2 & 3 \\ 4 & 5 & 6 \\ 7 & 8 & 9 \end{vmatrix} = \begin{vmatrix} 1 + 2(3) & 2 & 3 \\ 4 + 2(6) & 5 & 6 \\ 7 + 2(9) & 8 & 9 \end{vmatrix} = \begin{vmatrix} 7 & 2 & 3 \\ 16 & 5 & 6 \\ 25 & 8 & 9 \end{vmatrix}$

Evaluation of determinants The preceding theorems can be used to write sequences of equal determinants, leading from one form of a determinant to another and more useful form.

Example Evaluate

$$D = \begin{vmatrix} 2 & -1 & 1 & -3 \\ 1 & 3 & -4 & 2 \\ 1 & 0 & -2 & 1 \\ 3 & -1 & 5 & 2 \end{vmatrix}.$$

Solution overleaf

Solution As a step toward evaluating the determinant, we shall use Theorem 10.11 to produce an equal determinant with a row or a column containing zero entries in all but one place. Let us arbitrarily select the second column for this role, because one entry is already zero. Multiplying a_{1j} by 3 and adding the result to a_{2j}, we obtain

$$D = \begin{vmatrix} 2 & -1 & 1 & -3 \\ 1+3(2) & 3+3(-1) & -4+3(1) & 2+3(-3) \\ 1 & 0 & -2 & 1 \\ 3 & -1 & 5 & 2 \end{vmatrix}$$

$$= \begin{vmatrix} 2 & -1 & 1 & -3 \\ 7 & 0 & -1 & -7 \\ 1 & 0 & -2 & 1 \\ 3 & -1 & 5 & 2 \end{vmatrix}.$$

Next, multiplying a_{1j} by -1 and adding the result to a_{4j}, we find that

$$D = \begin{vmatrix} 2 & -1 & 1 & -3 \\ 7 & 0 & -1 & -7 \\ 1 & 0 & -2 & 1 \\ 3-1(2) & -1-1(-1) & 5-1(1) & 2-1(-3) \end{vmatrix}$$

$$= \begin{vmatrix} 2 & -1 & 1 & -3 \\ 7 & 0 & -1 & -7 \\ 1 & 0 & -2 & 1 \\ 1 & 0 & 4 & 5 \end{vmatrix}.$$

If we now expand the determinant about the second column, we have

$$D = \begin{vmatrix} 2 & -1 & 1 & -3 \\ 7 & 0 & -1 & -7 \\ 1 & 0 & -2 & 1 \\ 1 & 0 & 4 & 5 \end{vmatrix} = -(-1)\begin{vmatrix} 7 & -1 & -7 \\ 1 & -2 & 1 \\ 1 & 4 & 5 \end{vmatrix} + 0A_{22} + 0A_{32} + 0A_{42}.$$

From this point, we can reduce the third-order determinant to a second-order determinant by a similar procedure or, alternatively, expand directly about the elements in any row or column. Expanding about the elements of the first row, we obtain

$$D = \begin{vmatrix} 7 & -1 & -7 \\ 1 & -2 & 1 \\ 1 & 4 & 5 \end{vmatrix} = 7\begin{vmatrix} -2 & 1 \\ 4 & 5 \end{vmatrix} - (-1)\begin{vmatrix} 1 & 1 \\ 1 & 5 \end{vmatrix} + (-7)\begin{vmatrix} 1 & -2 \\ 1 & 4 \end{vmatrix},$$

from which

$$D = 7(-14) + (4) - 7(6)$$

$$= -98 + 4 - 42 = -136.$$

10.5 Properties of Determinants

Exercise 10.5

A Without evaluating, state why each statement is true. Verify selected examples by expansion.

Examples

a. $\begin{vmatrix} 1 & 0 & 3 \\ 2 & 1 & 7 \\ 1 & 0 & 2 \end{vmatrix} = \begin{vmatrix} 1 & 0 & 3 \\ 2 & 1 & 7 \\ 3 & 1 & 9 \end{vmatrix}$

b. $\begin{vmatrix} 1 & 2 & 1 \\ 0 & 0 & 2 \\ 1 & 2 & 1 \end{vmatrix} = 0$

Solutions

a. The right-hand determinant is obtained from the left by adding Row 2 to Row 3.

b. Rows 1 and 3 have corresponding entries that are equal.

1. $\begin{vmatrix} 2 & 3 & 1 \\ 0 & 0 & 0 \\ -1 & 2 & 0 \end{vmatrix} = 0$

2. $\begin{vmatrix} 3 & 1 & 3 \\ 0 & 1 & 0 \\ 1 & 2 & 1 \end{vmatrix} = 0$

3. $\begin{vmatrix} 2 & 3 & 1 & 1 \\ 2 & 0 & 1 & 2 \\ 2 & 3 & 1 & 1 \\ 0 & 1 & 2 & 0 \end{vmatrix} = 0$

4. $\begin{vmatrix} 7 & 3 & 2 & 0 \\ 2 & 1 & 2 & 0 \\ 4 & 1 & 1 & 0 \\ 0 & 2 & 1 & 0 \end{vmatrix} = 0$

5. $\begin{vmatrix} 4 & 2 & 1 \\ 0 & -1 & -2 \\ 1 & 0 & 2 \end{vmatrix} = -\begin{vmatrix} 4 & 2 & 1 \\ 0 & 1 & 2 \\ 1 & 0 & 2 \end{vmatrix}$

6. $\begin{vmatrix} -2 & 3 & 1 \\ -1 & 0 & 1 \\ -2 & 1 & 0 \end{vmatrix} = -\begin{vmatrix} 2 & 3 & 1 \\ 1 & 0 & 1 \\ 2 & 1 & 0 \end{vmatrix}$

7. $2\begin{vmatrix} 1 & 0 & 2 \\ -1 & 2 & 0 \\ 1 & 1 & 1 \end{vmatrix} = \begin{vmatrix} 1 & 0 & 2 \\ -1 & 2 & 0 \\ 2 & 2 & 2 \end{vmatrix}$

8. $\begin{vmatrix} 3 & -4 & 2 \\ 1 & -2 & 0 \\ 0 & 8 & 1 \end{vmatrix} = -2\begin{vmatrix} 3 & 2 & 2 \\ 1 & 1 & 0 \\ 0 & -4 & 1 \end{vmatrix}$

9. $\begin{vmatrix} 1 & 2 \\ 3 & 4 \end{vmatrix} = \begin{vmatrix} 1+2 & 2 \\ 3+4 & 4 \end{vmatrix}$

10. $\begin{vmatrix} 1 & 2 \\ 3 & 4 \end{vmatrix} = \begin{vmatrix} 1+4 & 2 \\ 3+8 & 4 \end{vmatrix}$

11. $\begin{vmatrix} 1 & 2 & 1 \\ 0 & 2 & 3 \\ 2 & -1 & 2 \end{vmatrix} = \begin{vmatrix} 1 & 2 & 1 \\ 0 & 2 & 3 \\ 0 & -5 & 0 \end{vmatrix}$

12. $\begin{vmatrix} -1 & 1 & 0 \\ 2 & 3 & -1 \\ 2 & 1 & 2 \end{vmatrix} = \begin{vmatrix} 0 & 1 & 0 \\ 5 & 3 & -1 \\ 3 & 1 & 2 \end{vmatrix}$

Theorem 10.11 was used on the left-hand member of each of the following equalities to produce the elements in the right-hand member. Complete the entries.

Example

$\begin{vmatrix} 1 & 5 \\ 4 & 3 \end{vmatrix} = \begin{vmatrix} 1 & 5 \\ 0 & \end{vmatrix}$

Solution

To obtain the 0 in the 2, 1 position using Theorem 10.11, it must be that -4 times row 1 was added to row 2. The missing entry is therefore -17.

13. $\begin{vmatrix} 1 & 3 \\ 2 & 2 \end{vmatrix} = \begin{vmatrix} 1 & 3 \\ 0 & \end{vmatrix}$

14. $\begin{vmatrix} 2 & -1 \\ 3 & 1 \end{vmatrix} = \begin{vmatrix} & 0 \\ 3 & 1 \end{vmatrix}$

15. $\begin{vmatrix} 1 & -2 & 1 \\ 3 & 1 & 4 \\ 0 & 2 & 1 \end{vmatrix} = \begin{vmatrix} 1 & -2 & 1 \\ 0 & 7 & \\ 0 & 2 & 1 \end{vmatrix}$

16. $\begin{vmatrix} 3 & -1 & 0 \\ 1 & 2 & 1 \\ 2 & 3 & 1 \end{vmatrix} = \begin{vmatrix} 3 & -1 & 0 \\ 1 & 2 & 1 \\ 1 & & 0 \end{vmatrix}$

17. $\begin{vmatrix} 2 & 3 & 1 & 4 \\ 0 & 2 & 1 & 2 \\ 1 & 1 & 2 & 3 \\ 0 & 1 & 1 & 1 \end{vmatrix} = \begin{vmatrix} 0 & 1 & & -2 \\ 0 & 2 & 1 & 2 \\ 1 & 1 & 2 & 3 \\ 0 & 1 & 1 & 1 \end{vmatrix}$

18. $\begin{vmatrix} 2 & 1 & 1 & 0 \\ 1 & 2 & 0 & 2 \\ 3 & 1 & 0 & 3 \\ 2 & 1 & 4 & 2 \end{vmatrix} = \begin{vmatrix} 2 & 1 & 1 & 0 \\ 1 & 2 & 0 & 2 \\ 3 & 1 & 0 & 3 \\ & -3 & 0 & 2 \end{vmatrix}$

First reduce each determinant to an equal 2 × 2 determinant and then evaluate.

Examples

a. $\begin{vmatrix} 0 & 3 & 2 \\ 1 & 7 & 8 \\ 0 & 5 & 4 \end{vmatrix}$

b. $\begin{vmatrix} 1 & 0 & 0 \\ 1 & 1 & 2 \\ 1 & -2 & 3 \end{vmatrix}$

Solutions

a. Expanding by Column 1,

$$-1 \cdot \begin{vmatrix} 3 & 2 \\ 5 & 4 \end{vmatrix} = -1(12 - 10) = -2.$$

b. Expanding by Row 1,

$$1 \cdot \begin{vmatrix} 1 & 2 \\ -2 & 3 \end{vmatrix} = 1(3 - (-4)) = 7.$$

19. $\begin{vmatrix} 2 & 1 & 0 \\ 3 & 2 & 1 \\ -1 & 2 & 0 \end{vmatrix}$

20. $\begin{vmatrix} 1 & 2 & 1 \\ 2 & -1 & 2 \\ 0 & 1 & 0 \end{vmatrix}$

21. $\begin{vmatrix} 1 & 0 & 3 \\ 2 & -1 & 1 \\ 1 & 2 & 1 \end{vmatrix}$

22. $\begin{vmatrix} 1 & 2 & -1 \\ 2 & 1 & 3 \\ 0 & 1 & 2 \end{vmatrix}$

23. $\begin{vmatrix} 1 & 2 & 1 \\ -1 & 2 & 3 \\ 2 & -1 & 1 \end{vmatrix}$

24. $\begin{vmatrix} 3 & -1 & 2 \\ 1 & 2 & 1 \\ -2 & 1 & 3 \end{vmatrix}$

25. $\begin{vmatrix} 0 & 0 & 1 & 2 \\ 6 & 0 & 0 & 1 \\ 6 & 1 & 0 & -1 \\ 6 & 1 & 0 & 2 \end{vmatrix}$

26. $\begin{vmatrix} 4 & 2 & 0 & 2 \\ -1 & 0 & 2 & 1 \\ 3 & 0 & -1 & 1 \\ 0 & 0 & 2 & 1 \end{vmatrix}$

27. $\begin{vmatrix} 0 & 1 & 0 & 2 \\ 0 & 2 & 0 & 3 \\ 2 & -1 & 1 & 0 \\ 0 & 0 & 8 & 8 \end{vmatrix}$

28. $\begin{vmatrix} 0 & 2 & -1 & 3 \\ 0 & 0 & 2 & 1 \\ 3 & 0 & 1 & 0 \\ -6 & 6 & 0 & 0 \end{vmatrix}$

29. $\begin{vmatrix} 1 & 2 & 3 & -1 \\ 0 & 4 & 8 & 4 \\ -2 & 0 & 1 & 1 \\ 2 & 1 & 0 & 1 \end{vmatrix}$

30. $\begin{vmatrix} 1 & 2 & 1 & 1 \\ 2 & -1 & 0 & 1 \\ 0 & 6 & 3 & 9 \\ 2 & 0 & -1 & 1 \end{vmatrix}$

B 31. Show that $\begin{vmatrix} 1 & a & a^2 \\ 1 & b & b^2 \\ 1 & c & c^2 \end{vmatrix} = (b-c)(c-a)(a-b)$.

32. Show that $\begin{vmatrix} a_{11} & a_{12} & a_{13} & a_{14} \\ a_{21} & a_{22} & a_{23} & a_{24} \\ 0 & 0 & a_{33} & a_{34} \\ 0 & 0 & a_{43} & a_{44} \end{vmatrix} = \begin{vmatrix} a_{11} & a_{12} \\ a_{21} & a_{22} \end{vmatrix} \cdot \begin{vmatrix} a_{33} & a_{34} \\ a_{43} & a_{44} \end{vmatrix}$.

33. Prove Theorem 10.9.

34. Prove Theorem 10.10.

10.6 The Inverse of a Square Matrix

In the field of real numbers, every element a except 0 has a multiplicative inverse $1/a$ with the property that $a \cdot 1/a = 1$. The question should (and does) arise, "Does every square matrix A have a multiplicative inverse A^{-1}?"

Definition 10.17 For a given square matrix A of order n, if there is a square matrix A^{-1} of order n such that

$$AA^{-1} = I \quad \text{and} \quad A^{-1}A = I,$$

where I is the multiplicative identity matrix of order n, then A^{-1} is the **multiplicative inverse** of A.

Inverse of a 2 × 2 matrix

To answer the question about the existence of a multiplicative inverse for a matrix, we shall begin by considering the simple case of 2 × 2 matrices. If we let

$$A = \begin{bmatrix} a_{11} & a_{12} \\ a_{21} & a_{22} \end{bmatrix},$$

we must see whether or not there exists a 2 × 2 matrix A^{-1} such that $AA^{-1} = I$. If so, let $A^{-1} = \begin{bmatrix} b & c \\ d & e \end{bmatrix}$. We wish to have

$$\begin{bmatrix} a_{11} & a_{12} \\ a_{21} & a_{22} \end{bmatrix} \begin{bmatrix} b & c \\ d & e \end{bmatrix} = \begin{bmatrix} 1 & 0 \\ 0 & 1 \end{bmatrix}.$$

This leads to
$$\begin{bmatrix} a_{11}b + a_{12}d & a_{11}c + a_{12}e \\ a_{21}b + a_{22}d & a_{21}c + a_{22}e \end{bmatrix} = \begin{bmatrix} 1 & 0 \\ 0 & 1 \end{bmatrix},$$
which is true if and only if
$$a_{11}b + a_{12}d = 1, \quad a_{11}c + a_{12}e = 0,$$
$$a_{21}b + a_{22}d = 0, \quad a_{21}c + a_{22}e = 1. \tag{1}$$
Solving these equations for $b, c, d,$ and e, we have
$$(a_{11}a_{22} - a_{12}a_{21})b = a_{22}, \quad (a_{11}a_{22} - a_{12}a_{21})c = -a_{12},$$
$$(a_{11}a_{22} - a_{12}a_{21})d = -a_{21}, \quad (a_{11}a_{22} - a_{12}a_{21})e = a_{11},$$
from which
$$b = \frac{a_{22}}{a_{11}a_{22} - a_{12}a_{21}}, \quad c = \frac{-a_{12}}{a_{11}a_{22} - a_{12}a_{21}},$$
$$d = \frac{-a_{21}}{a_{11}a_{22} - a_{12}a_{21}}, \quad e = \frac{a_{11}}{a_{11}a_{22} - a_{12}a_{21}},$$
provided $a_{11}a_{22} - a_{12}a_{21} \neq 0$. Now the denominator of each of these fractions is just $\delta(A)$, so that

$$A^{-1} = \begin{bmatrix} b & c \\ d & e \end{bmatrix} = \begin{bmatrix} \dfrac{a_{22}}{\delta(A)} & \dfrac{-a_{12}}{\delta(A)} \\ \dfrac{-a_{21}}{\delta(A)} & \dfrac{a_{11}}{\delta(A)} \end{bmatrix} = \frac{1}{\delta(A)} \begin{bmatrix} a_{22} & -a_{12} \\ -a_{21} & a_{11} \end{bmatrix}.$$

By direct multiplication, it can be verified not only that
$$AA^{-1} = I,$$
but also (surprisingly, since matrix multiplication is not always commutative) that
$$A^{-1}A = I.$$
Thus, to write the inverse of a 2×2 square matrix A for which $\delta(A) \neq 0$, we interchange the entries on the principal diagonal, replace each of the other two entries with its negative, and multiply the result by $1/\delta(A)$.

Example

If $A = \begin{bmatrix} 1 & 3 \\ 2 & -1 \end{bmatrix}$, find A^{-1}.

Solution

We first observe that $\delta(A) = (1)(-1) - (3)(2) = -7$. Hence,
$$A^{-1} = -\frac{1}{7}\begin{bmatrix} -1 & -3 \\ -2 & 1 \end{bmatrix} = \begin{bmatrix} \dfrac{1}{7} & \dfrac{3}{7} \\ \dfrac{2}{7} & -\dfrac{1}{7} \end{bmatrix}.$$

10.6 The Inverse of a Square Matrix

It is a good idea always to check the result when finding A^{-1}, because there is much room for blundering in the process of determining the inverse. In the present example, we have

$$A^{-1}A = -\frac{1}{7}\begin{bmatrix} -1 & -3 \\ -2 & 1 \end{bmatrix}\begin{bmatrix} 1 & 3 \\ 2 & -1 \end{bmatrix} = -\frac{1}{7}\begin{bmatrix} -7 & 0 \\ 0 & -7 \end{bmatrix} = \begin{bmatrix} 1 & 0 \\ 0 & 1 \end{bmatrix}.$$

Matrices with no inverse   We have now arrived at a position where we can answer the question, "Does every 2 × 2 square matrix A have an inverse?" The answer is "No," for if $\delta(A)$ is 0, then the foregoing Equations (1) for b, c, d, e would have no solution.

Example   The matrix $\begin{bmatrix} 3 & 5 \\ 6 & 10 \end{bmatrix}$ has no inverse because

$$\delta(A) = 3(10) - 6(5) = 0.$$

Inverse of an $n \times n$ matrix   More generally, and without proving it, we have the following result.

Theorem 10.12   If

$$A = \begin{bmatrix} a_{11} & a_{12} & \cdots & a_{1n} \\ a_{21} & a_{22} & \cdots & a_{2n} \\ \vdots & \vdots & & \vdots \\ a_{n1} & a_{n2} & \cdots & a_{nn} \end{bmatrix},$$

and if $\delta(A) \neq 0$, then A has an inverse A^{-1} given by

$$A^{-1} = \frac{1}{\delta(A)}\begin{bmatrix} A_{11} & A_{21} & \cdots & A_{n1} \\ A_{12} & A_{22} & \cdots & A_{n2} \\ \vdots & \vdots & & \vdots \\ A_{1n} & A_{2n} & \cdots & A_{nn} \end{bmatrix},$$

where A_{ij} is the cofactor of a_{ij} in A. If $\delta(A) = 0$, then A has no inverse.

Square matrices A for which $\delta(A) \neq 0$ are *nonsingular* (see page 349) for it can be shown that A is row-equivalent to the identity if and only if $\delta(A) \neq 0$. Thus by Theorem 10.12, A has an inverse if and only if A is nonsingular.

Observe that A^{-1} is $1/\delta(A)$ times the transpose of the matrix obtained by replacing each entry of A with its cofactor.

Example   If $A = \begin{bmatrix} 1 & 0 & 1 \\ 2 & 1 & 0 \\ 1 & -1 & 1 \end{bmatrix}$, find A^{-1}.

Solution overleaf

Solution We first observe that $\delta(A) = -2$, and since $\delta(A)$ is not zero, A has an inverse. Next, replacing each entry in A with its cofactor, we obtain the matrix

$$\begin{bmatrix} 1 & -2 & -3 \\ -1 & 0 & 1 \\ -1 & 2 & 1 \end{bmatrix}, \text{ whose transpose is } \begin{bmatrix} 1 & -1 & -1 \\ -2 & 0 & 2 \\ -3 & 1 & 1 \end{bmatrix},$$

so that

$$A^{-1} = -\frac{1}{2} \begin{bmatrix} 1 & -1 & -1 \\ -2 & 0 & 2 \\ -3 & 1 & 1 \end{bmatrix}.$$

As a check, we have

$$A^{-1}A = -\frac{1}{2} \begin{bmatrix} 1 & -1 & -1 \\ -2 & 0 & 2 \\ -3 & 1 & 1 \end{bmatrix} \begin{bmatrix} 1 & 0 & 1 \\ 2 & 1 & 0 \\ 1 & -1 & 1 \end{bmatrix}$$

$$= -\frac{1}{2} \begin{bmatrix} -2 & 0 & 0 \\ 0 & -2 & 0 \\ 0 & 0 & -2 \end{bmatrix} = \begin{bmatrix} 1 & 0 & 0 \\ 0 & 1 & 0 \\ 0 & 0 & 1 \end{bmatrix}.$$

Theorem 10.12 is applicable to $n \times n$ square matrices, although, clearly, the process of actually determining A^{-1} by the formula given in that theorem becomes very laborious for matrices much larger than 3×3.

Elementary transformations If the inverse of an $n \times n$ matrix A exists, it can be obtained by using elementary transformations. This is the method applied when using a computer to find the inverse of a matrix. To see this, we need the following result which we state without proof.

Theorem 10.13 If A is an $n \times n$ nonsingular matrix and if $[A \mid I]$ is the $n \times 2n$ matrix obtained by adjoining the $n \times n$ identity matrix to A, then

$$[A \mid I] \sim [I \mid A^{-1}].$$

The above theorem is used to compute A^{-1} by using elementary transformations to obtain $[I \mid A^{-1}]$ from $[A \mid I]$.

Example If $A = \begin{bmatrix} 1 & 2 \\ 2 & 0 \end{bmatrix}$, find A^{-1}.

10.6 The Inverse of a Square Matrix

Solution We first observe that $\delta(A) = -4$, and since $\delta(A)$ is not zero, A has an inverse.

$$[A \vdots I] = \begin{bmatrix} 1 & 2 & \vdots & 1 & 0 \\ 2 & 0 & \vdots & 0 & 1 \end{bmatrix}$$

$$\sim \begin{bmatrix} 0 & 2 & \vdots & 1 & -\frac{1}{2} \\ 2 & 0 & \vdots & 0 & 1 \end{bmatrix} \quad \text{Row } 1 + \left[-\frac{1}{2} \times \text{Row } 2 \right]$$

$$\sim \begin{bmatrix} 0 & 1 & \vdots & \frac{1}{2} & -\frac{1}{4} \\ 1 & 0 & \vdots & 0 & \frac{1}{2} \end{bmatrix} \quad \begin{array}{l} \frac{1}{2} \times \text{Row } 1 \\ \frac{1}{2} \times \text{Row } 2 \end{array}$$

$$\sim \begin{bmatrix} 1 & 0 & \vdots & 0 & \frac{1}{2} \\ 0 & 1 & \vdots & \frac{1}{2} & -\frac{1}{4} \end{bmatrix} \quad \begin{array}{l} \text{Interchange} \\ \text{Rows 1 and 2} \end{array}$$

$$= [I \vdots A^{-1}]$$

Thus, $A^{-1} = \begin{bmatrix} 0 & \frac{1}{2} \\ \frac{1}{2} & -\frac{1}{4} \end{bmatrix}$.

As a check, we have

$$A^{-1}A = \begin{bmatrix} 0 & \frac{1}{2} \\ \frac{1}{2} & -\frac{1}{4} \end{bmatrix} \begin{bmatrix} 1 & 2 \\ 2 & 0 \end{bmatrix} = \begin{bmatrix} 1 & 0 \\ 0 & 1 \end{bmatrix}.$$

Properties of matrices and their inverses

There are a number of useful properties associated with matrices and their inverses. For example, let A and B be $n \times n$ nonsingular matrices. If we right-multiply AB by $B^{-1}A^{-1}$, and apply Theorem 10.3, we have

$$AB \cdot B^{-1}A^{-1} = A \cdot I \cdot A^{-1} = A \cdot A^{-1} = I.$$

Moreover, if we left-multiply AB by $B^{-1}A^{-1}$, we have

$$B^{-1}A^{-1} \cdot AB = B^{-1} \cdot I \cdot B = B^{-1} \cdot B = I.$$

Thus, since $(AB)(B^{-1}A^{-1}) = (B^{-1}A^{-1})(AB) = I$, by the definition of the inverse of a matrix we have

$$(AB)^{-1} = B^{-1}A^{-1}.$$

This proves the following theorem.

Theorem 10.14 *If A and B are n × n nonsingular square matrices, then AB has an inverse, namely*

$$(AB)^{-1} = B^{-1}A^{-1}.$$

This theorem can be used to find the inverse of products of any number of nonsingular matrices. For example, if there are three factors A, B, and C in a product,

$$(ABC)^{-1} = [(AB)C]^{-1} = C^{-1}(AB)^{-1} = C^{-1}B^{-1}A^{-1}.$$

Exercise 10.6

A Find the inverse of each matrix if one exists.

Example

$$B = \begin{bmatrix} 1 & 0 & -1 \\ 1 & 3 & 1 \\ 0 & 1 & 2 \end{bmatrix}$$

Solution 1

The determinant $\delta(B)$ is given by

$$\delta \begin{bmatrix} 1 & 0 & -1 \\ 1 & 3 & 1 \\ 0 & 1 & 2 \end{bmatrix} = 1(5) - 0 - 1(1) = 4.$$

Replacing each entry of B with its cofactor gives

$$\begin{bmatrix} 5 & -2 & 1 \\ -1 & 2 & -1 \\ 3 & -2 & 3 \end{bmatrix};$$

$$B^{-1} = \frac{1}{\delta(B)} \begin{bmatrix} 5 & -2 & 1 \\ -1 & 2 & -1 \\ 3 & -2 & 3 \end{bmatrix}^t = \frac{1}{4} \begin{bmatrix} 5 & -1 & 3 \\ -2 & 2 & -2 \\ 1 & -1 & 3 \end{bmatrix} = \begin{bmatrix} \frac{5}{4} & -\frac{1}{4} & \frac{3}{4} \\ -\frac{2}{4} & \frac{2}{4} & -\frac{2}{4} \\ \frac{1}{4} & -\frac{1}{4} & \frac{3}{4} \end{bmatrix}.$$

10.6 The Inverse of a Square Matrix

Solution 2

$$\begin{bmatrix} 1 & 0 & -1 & \vdots & 1 & 0 & 0 \\ 1 & 3 & 1 & \vdots & 0 & 1 & 0 \\ 0 & 1 & 2 & \vdots & 0 & 0 & 1 \end{bmatrix}$$

$$\sim \begin{bmatrix} 1 & 0 & -1 & \vdots & 1 & 0 & 0 \\ 0 & 3 & 2 & \vdots & -1 & 1 & 0 \\ 0 & 1 & 2 & \vdots & 0 & 0 & 1 \end{bmatrix} \quad \text{Row 2} + [-1 \times \text{Row 1}]$$

$$\sim \begin{bmatrix} 1 & 0 & -1 & \vdots & 1 & 0 & 0 \\ 0 & 3 & 2 & \vdots & -1 & 1 & 0 \\ 0 & 0 & \frac{4}{3} & \vdots & \frac{1}{3} & -\frac{1}{3} & 1 \end{bmatrix} \quad \text{Row 3} + \left[-\frac{1}{3} \times \text{Row 2} \right]$$

$$\sim \begin{bmatrix} 1 & 0 & 0 & \vdots & \frac{5}{4} & -\frac{1}{4} & \frac{3}{4} \\ 0 & 3 & 0 & \vdots & -\frac{3}{2} & \frac{3}{2} & -\frac{3}{2} \\ 0 & 0 & \frac{4}{3} & \vdots & \frac{1}{3} & -\frac{1}{3} & 1 \end{bmatrix} \quad \begin{array}{l} \text{Row 1} + \left[\frac{3}{4} \times \text{Row 3} \right] \\ \text{Row 2} + \left[-\frac{3}{2} \times \text{Row 3} \right] \end{array}$$

$$\sim \begin{bmatrix} 1 & 0 & 0 & \vdots & \frac{5}{4} & -\frac{1}{4} & \frac{3}{4} \\ 0 & 1 & 0 & \vdots & -\frac{1}{2} & \frac{1}{2} & -\frac{1}{2} \\ 0 & 0 & 1 & \vdots & \frac{1}{4} & -\frac{1}{4} & \frac{3}{4} \end{bmatrix} \quad \begin{array}{l} \frac{1}{3} \times \text{Row 2} \\ \frac{3}{4} \times \text{Row 3} \end{array}$$

Thus, $A^{-1} = \begin{bmatrix} \frac{5}{4} & -\frac{1}{4} & \frac{3}{4} \\ -\frac{1}{2} & \frac{1}{2} & -\frac{1}{2} \\ \frac{1}{4} & -\frac{1}{4} & \frac{3}{4} \end{bmatrix}$.

1. $\begin{bmatrix} 1 & 2 \\ 1 & 3 \end{bmatrix}$
2. $\begin{bmatrix} 3 & 1 \\ 2 & -1 \end{bmatrix}$
3. $\begin{bmatrix} 2 & -3 \\ 1 & 1 \end{bmatrix}$
4. $\begin{bmatrix} 3 & -2 \\ 2 & 1 \end{bmatrix}$
5. $\begin{bmatrix} -2 & -1 \\ 4 & 2 \end{bmatrix}$
6. $\begin{bmatrix} 3 & 1 \\ 9 & 3 \end{bmatrix}$
7. $\begin{bmatrix} 5 & 7 \\ 3 & 4 \end{bmatrix}$
8. $\begin{bmatrix} 5 & -4 \\ 4 & -3 \end{bmatrix}$
9. $\begin{bmatrix} 7 & 4 \\ -4 & -2 \end{bmatrix}$
10. $\begin{bmatrix} -9 & 5 \\ -4 & 2 \end{bmatrix}$
11. $\begin{bmatrix} -2 & -6 \\ -3 & -9 \end{bmatrix}$
12. $\begin{bmatrix} 21 & 7 \\ 9 & 3 \end{bmatrix}$

13. $\begin{bmatrix} 1 & -1 & 2 \\ 2 & 1 & 3 \\ 0 & 0 & 2 \end{bmatrix}$ 14. $\begin{bmatrix} 0 & 4 & 2 \\ 1 & 0 & 2 \\ 0 & -1 & 1 \end{bmatrix}$ 15. $\begin{bmatrix} 2 & -1 & 1 \\ 3 & 0 & 1 \\ 2 & 2 & 1 \end{bmatrix}$

16. $\begin{bmatrix} 1 & 2 & 1 \\ 0 & 2 & 1 \\ -2 & 2 & 3 \end{bmatrix}$ 17. $\begin{bmatrix} 2 & 1 & 1 \\ 1 & 0 & 2 \\ 4 & 2 & 2 \end{bmatrix}$ 18. $\begin{bmatrix} -3 & 1 & -6 \\ 2 & 1 & 4 \\ 2 & 0 & 4 \end{bmatrix}$

19. $\begin{bmatrix} 1 & 2 & -3 \\ 3 & -1 & 0 \\ 5 & 3 & -6 \end{bmatrix}$ 20. $\begin{bmatrix} 2 & 4 & -1 \\ 1 & 6 & 2 \\ 5 & 14 & 0 \end{bmatrix}$ 21. $\begin{bmatrix} 2 & -1 & -5 \\ 1 & 3 & 4 \\ 0 & 1 & 2 \end{bmatrix}$

22. $\begin{bmatrix} 2 & 1 & -8 \\ 1 & 1 & -2 \\ 1 & 2 & 3 \end{bmatrix}$ 23. $\begin{bmatrix} 0 & 0 & 1 \\ 0 & 1 & 0 \\ 1 & 0 & 0 \end{bmatrix}$ 24. $\begin{bmatrix} 1 & 0 & 1 \\ 0 & 1 & 0 \\ 1 & 0 & 0 \end{bmatrix}$

25. Verify that

$$\left(\begin{bmatrix} 2 & 3 \\ 1 & -1 \end{bmatrix} \cdot \begin{bmatrix} 0 & 1 \\ 3 & 1 \end{bmatrix} \right)^{-1} = \begin{bmatrix} 0 & 1 \\ 3 & 1 \end{bmatrix}^{-1} \cdot \begin{bmatrix} 2 & 3 \\ 1 & -1 \end{bmatrix}^{-1}.$$

26. Verify that

$$\left(\begin{bmatrix} 1 & 2 \\ -1 & 0 \end{bmatrix} \cdot \begin{bmatrix} 1 & 1 \\ 2 & 0 \end{bmatrix} \cdot \begin{bmatrix} 2 & -1 \\ 0 & 1 \end{bmatrix} \right)^{-1} = \begin{bmatrix} 2 & -1 \\ 0 & 1 \end{bmatrix}^{-1} \cdot \begin{bmatrix} 1 & 1 \\ 2 & 0 \end{bmatrix}^{-1} \cdot \begin{bmatrix} 1 & 2 \\ -1 & 0 \end{bmatrix}^{-1}.$$

27. Verify that

$$\left(\begin{bmatrix} 3 & 0 & 1 \\ 2 & 1 & 0 \\ 0 & 1 & 2 \end{bmatrix} \cdot \begin{bmatrix} 2 & 1 & 0 \\ 1 & 1 & 2 \\ 0 & 1 & 0 \end{bmatrix} \right)^{-1} = \begin{bmatrix} 2 & 1 & 0 \\ 1 & 1 & 2 \\ 0 & 1 & 0 \end{bmatrix}^{-1} \cdot \begin{bmatrix} 3 & 0 & 1 \\ 2 & 1 & 0 \\ 0 & 1 & 2 \end{bmatrix}^{-1}.$$

B   28. Show that $[A^t]^{-1} = [A^{-1}]^t$ for each nonsingular 2×2 matrix.

29. Show that $\delta(A^{-1}) = 1/\delta(A)$ for each nonsingular 2×2 matrix.

30. Prove that if a and b are real numbers, then $\delta(aA^2 + bA) = \delta(aA + bI) \times \delta(A)$ for all 2×2 matrices A.

31. Prove that $\delta(B^{-1}AB) = \delta(A)$ for all nonsingular 2×2 matrices A and B.

32. Prove that if A is a 2×2 matrix and a, b, and c are real numbers, with $c \neq 0$, and if $aA^2 + bA + cI = 0$, then A has an inverse.

10.7   Solution of Linear Systems Using Matrix Inverses

In Section 10.3 we solved linear systems using row-equivalent matrices. The solution for a linear system can also be found by using the inverse of a matrix.

10.7 Solution of Linear Systems Using Matrix Inverses

We first verify the matrix product equation

$$\begin{bmatrix} a_{11} & a_{12} & \cdots & a_{1n} \\ \vdots & \vdots & & \vdots \\ a_{n1} & a_{n2} & \cdots & a_{nn} \end{bmatrix} \begin{bmatrix} x_1 \\ \vdots \\ x_n \end{bmatrix} = \begin{bmatrix} a_{11}x_1 + a_{12}x_2 + \cdots + a_{1n}x_n \\ \vdots \\ a_{n1}x_1 + a_{n2}x_2 + \cdots + a_{nn}x_n \end{bmatrix},$$

and hence note that the linear system

$$\begin{aligned} a_{11}x_1 + a_{12}x_2 + \cdots + a_{1n}x_n &= c_1 \\ a_{21}x_1 + a_{22}x_2 + \cdots + a_{2n}x_n &= c_2 \\ \vdots \quad \vdots \quad \vdots \quad \vdots & \\ a_{n1}x_1 + a_{n2}x_2 + \cdots + a_{nn}x_n &= c_n \end{aligned} \quad (1)$$

can be written as the matrix equation

$$\begin{bmatrix} a_{11} & a_{12} & \cdots & a_{1n} \\ \vdots & \vdots & & \vdots \\ a_{n1} & a_{n2} & \cdots & a_{nn} \end{bmatrix} \begin{bmatrix} x_1 \\ \vdots \\ x_n \end{bmatrix} = \begin{bmatrix} c_1 \\ \vdots \\ c_n \end{bmatrix},$$

where the first factor in the left-hand member is the coefficient matrix for the system. In more concise notation, this latter equation can be written

$$AX = B,$$

where A is an $n \times n$ square matrix, and X and B are $n \times 1$ column matrices.

Solution of systems

If A in the foregoing equation is nonsingular, we can left-multiply both members of this equation by A^{-1} to obtain the equivalent matrices

$$A^{-1}AX = A^{-1}B,$$
$$IX = A^{-1}B,$$
$$X = A^{-1}B,$$

where $A^{-1}B$ is an $n \times 1$ column matrix. Since X and $A^{-1}B$ are equal, each entry in X is equal to the corresponding entry in $A^{-1}B$, and hence these latter entries constitute the components of the solution of the given linear system. If A is a singular matrix, then of course it has no inverse, and either the system has no solution or the solution is not unique.

Example

Use matrices to find the solution set of

$$\begin{aligned} 2x + y + z &= 1 \\ x - 2y - 3z &= 1 \\ 3x + 2y + 4z &= 5. \end{aligned}$$

Solution overleaf

Solution   We first write this as a matrix equation of the form $AX = B$, thus:

$$\begin{bmatrix} 2 & 1 & 1 \\ 1 & -2 & -3 \\ 3 & 2 & 4 \end{bmatrix} \begin{bmatrix} x \\ y \\ z \end{bmatrix} = \begin{bmatrix} 1 \\ 1 \\ 5 \end{bmatrix}.$$

We next determine $\delta(A)$, obtaining

$$\delta \begin{bmatrix} 2 & 1 & 1 \\ 1 & -2 & -3 \\ 3 & 2 & 4 \end{bmatrix} = 2(-2) - 1(13) + 1(8) = -9.$$

Observing that A is nonsingular, we then find A^{-1} by either of the methods of Section 10.6. We shall use the first one discussed.

$$A^{-1} = \begin{bmatrix} 2 & 1 & 1 \\ 1 & -2 & -3 \\ 3 & 2 & 4 \end{bmatrix}^{-1} = -\frac{1}{9} \begin{bmatrix} -2 & -2 & -1 \\ -13 & 5 & 7 \\ 8 & -1 & -5 \end{bmatrix}$$

As a matter of routine, we check the latter by verifying that $A^{-1}A = I$.

$$A^{-1}A = -\frac{1}{9} \begin{bmatrix} -2 & -2 & -1 \\ -13 & 5 & 7 \\ 8 & -1 & -5 \end{bmatrix} \begin{bmatrix} 2 & 1 & 1 \\ 1 & -2 & -3 \\ 3 & 2 & 4 \end{bmatrix}$$

$$= -\frac{1}{9} \begin{bmatrix} -9 & 0 & 0 \\ 0 & -9 & 0 \\ 0 & 0 & -9 \end{bmatrix} = \begin{bmatrix} 1 & 0 & 0 \\ 0 & 1 & 0 \\ 0 & 0 & 1 \end{bmatrix}$$

Now, since $X = A^{-1}B$, we have

$$\begin{bmatrix} x \\ y \\ z \end{bmatrix} = -\frac{1}{9} \begin{bmatrix} -2 & -2 & -1 \\ -13 & 5 & 7 \\ 8 & -1 & -5 \end{bmatrix} \begin{bmatrix} 1 \\ 1 \\ 5 \end{bmatrix} = -\frac{1}{9} \begin{bmatrix} -9 \\ 27 \\ -18 \end{bmatrix} = \begin{bmatrix} 1 \\ -3 \\ 2 \end{bmatrix}.$$

Hence, $x = 1$, $y = -3$, and $z = 2$, and the solution set is $\{(1, -3, 2)\}$.

The computation of A^{-1} is laborious when A is a square matrix containing many rows and columns. The foregoing method is not always the easiest to use in solving systems. It is, however, very useful when solving several systems of the form $AX = B$ having the same coefficient matrix A.

Example   Solve.

a. $\begin{bmatrix} 2 & 1 & 1 \\ 1 & -2 & -3 \\ 3 & 2 & 4 \end{bmatrix} \begin{bmatrix} x \\ y \\ z \end{bmatrix} = \begin{bmatrix} 8 \\ 5 \\ 10 \end{bmatrix}$ b. $\begin{bmatrix} 2 & 1 & 1 \\ 1 & -2 & -3 \\ 3 & 2 & 4 \end{bmatrix} \begin{bmatrix} x \\ y \\ z \end{bmatrix} = \begin{bmatrix} 0 \\ 10 \\ -11 \end{bmatrix}$

10.7 Solution of Linear Systems Using Matrix Inverses

Solution In both examples we have $A = \begin{bmatrix} 2 & 1 & 1 \\ 1 & -2 & -3 \\ 3 & 2 & 4 \end{bmatrix}$, and from the previous example,

$$A^{-1} = -\frac{1}{9}\begin{bmatrix} -2 & -2 & -1 \\ -13 & 5 & 7 \\ 8 & -1 & -5 \end{bmatrix}.$$

a. The solution is

$$\begin{bmatrix} x \\ y \\ z \end{bmatrix} = A^{-1}\begin{bmatrix} 8 \\ 5 \\ 10 \end{bmatrix} = \begin{bmatrix} 4 \\ 1 \\ -1 \end{bmatrix}.$$

b. The solution is

$$\begin{bmatrix} x \\ y \\ z \end{bmatrix} = A^{-1}\begin{bmatrix} 0 \\ 10 \\ -11 \end{bmatrix} = \begin{bmatrix} 1 \\ 3 \\ -5 \end{bmatrix}.$$

Exercise 10.7

Find the solution set of the given system by using matrices. If the system has no unique solution, so state.

1. $2x - 3y = -1$
    $x + 4y = 5$

2. $3x - 4y = -2$
    $x - 2y = 0$

3. $3x + 6y = -2$
    $6x + 12y = 36$

4. $2x - 4y = 7$
    $x - 2y = 1$

5. $2x - 3y = 0$
    $2x + y = 16$

6. $2x + 3y = 3$
    $3x - 4y = 0$

7. $3x + y = -5$
    $2x - 4y = -16$

8. $2x - 3y = -8$
    $x - 4y = -9$

9. $3x + 9y = 2$
    $6x + 18y = 4$

10. $2x + y = 1$
    $4x + 2y = 2$

11. $x - 4y = -6$
    $4x - y = 6$

12. $2x + 3y = 3$
    $3x - 2y = -2$

13. $x + y = 2$
    $2x - z = 1$
    $2y - 3z = -1$

14. $2x - 6y + 3z = -12$
    $3x - 2y + 5z = -4$
    $4x + 5y - 2z = 10$

15. $x - 2y + z = -1$
    $3x + y - 2z = 4$
    $y - z = 1$

16. $2x + 5z = 9$
    $4x + 3y = -1$
    $3y - 4z = -13$

17. $2x + 2y + z = 1$
    $x - y + 6z = 21$
    $3x + 2y - z = -4$

18. $4x + 8y + z = -6$
    $2x - 3y + 2z = 0$
    $x + 7y - 3z = -8$

19. $x + y + z = 0$
    $2x - y - 4z = 15$
    $x - 2y - z = 7$

20. $x + y - 2z = 3$
    $3x - y + z = 5$
    $3x + 3y - 6z = 9$

21. Find the solution set of each system.

   a. $x + y + z = 3$
       $2x - y - 4z = 4$
       $x - 2y - z = -1$

   b. $x + y + z = -2$
       $2x - y - 4z = 1$
       $x - 2y - z = 0$

   c. $x + y + z = 1$
       $2x - y - 4z = 0$
       $x - 2y - z = 1$

22. Find the solution set of each system.

   a. $2x + 5z = 1$
       $4x + 3y = 1$
       $3y - 4z = 1$

   b. $2x + 5z = 2$
       $4x + 3y = -1$
       $3y - 4z = 1$

   c. $2x + 5z = 0$
       $4x + 3y = 2$
       $3y - 4z = 1$

10.8 Cramer's Rule

In Section 10.7, we obtained the solution set of the linear system (1) on page 375 with nonsingular coefficient matrix by first expressing the system in the matrix form $AX = B$ and then left-multiplying both members of the equation by A^{-1} to obtain

$$A^{-1}AX = X = A^{-1}B.$$

If, now, this technique is viewed in terms of determinants, we arrive at a general solution for such systems.

If the coefficient matrix A is nonsingular, then its inverse, A^{-1}, is

$$A^{-1} = \frac{1}{\delta(A)} \begin{bmatrix} A_{11} & A_{21} & \cdots & A_{n1} \\ \vdots & \vdots & & \vdots \\ A_{1n} & A_{2n} & \cdots & A_{nn} \end{bmatrix}.$$

Now, since $B = \begin{bmatrix} c_1 \\ c_2 \\ \vdots \\ c_n \end{bmatrix}$, we have

$$X = A^{-1}B = \frac{1}{\delta(A)} \begin{bmatrix} c_1 A_{11} + c_2 A_{21} + \cdots + c_n A_{n1} \\ c_1 A_{12} + c_2 A_{22} + \cdots + c_n A_{n2} \\ \vdots & \vdots & & \vdots \\ c_1 A_{1n} + c_2 A_{2n} + \cdots + c_n A_{nn} \end{bmatrix}.$$

Each entry in $X = A^{-1}B$ can be seen to be of the form

$$\frac{c_1 A_{1j} + c_2 A_{2j} + \cdots + c_n A_{nj}}{\delta(A)}.$$

10.8 Cramer's Rule

But $c_1 A_{1j} + c_2 A_{2j} + \cdots + c_n A_{nj}$ is just the expansion of the determinant

$$\begin{vmatrix} a_{11} & a_{12} & \cdots & c_1 & \cdots & a_{1n} \\ a_{21} & a_{22} & \cdots & c_2 & \cdots & a_{2n} \\ \vdots & \vdots & & \vdots & & \vdots \\ a_{n1} & a_{n2} & \cdots & c_n & \cdots & a_{nn} \end{vmatrix}$$

$\uparrow$ jth column

about the jth column, which has entries $c_1, c_2, \ldots, c_n$ in place of $a_{1j}, a_{2j}, \ldots, a_{nj}$.

Thus, each entry x_j in the matrix $X = \begin{bmatrix} x_1 \\ x_2 \\ \vdots \\ x_n \end{bmatrix} = A^{-1}B$ is

$$x_j = \frac{\delta(A_j)}{\delta(A)} = \frac{\begin{vmatrix} a_{11} & a_{12} & \cdots & c_1 & \cdots & a_{1n} \\ a_{21} & a_{22} & \cdots & c_2 & \cdots & a_{2n} \\ \vdots & \vdots & & \vdots & & \vdots \\ a_{n1} & a_{n2} & \cdots & c_n & \cdots & a_{nn} \end{vmatrix}}{\begin{vmatrix} a_{11} & a_{12} & \cdots & & & a_{1n} \\ a_{21} & a_{22} & \cdots & & & a_{2n} \\ \vdots & \vdots & & & & \vdots \\ a_{n1} & a_{n2} & \cdots & & & a_{nn} \end{vmatrix}}.$$

Application of Cramer's rule

This relationship expresses **Cramer's rule**. Cramer's rule is the assertion that if the determinant of the coefficient matrix of an $n \times n$ linear system *is not* 0, then the equations are consistent (the system has a solution) and have a unique solution which can be found as follows.

To find x_j in solving the matrix equation $AX = B$:

1. Write the determinant of the coefficient matrix for the system.
2. Replace each entry in the jth column of the coefficient matrix A with the corresponding entry from the column matrix B, and find the determinant of the resulting matrix.
3. Divide the result in Step 2 by the result in Step 1.

Hence, in a nonsingular 3×3 system:

$$x = \frac{\delta(A_x)}{\delta(A)}, \quad y = \frac{\delta(A_y)}{\delta(A)}, \quad \text{and} \quad z = \frac{\delta(A_z)}{\delta(A)}.$$

Example   Use Cramer's rule to solve the system
$$-4x + 2y - 9z = 2$$
$$3x + 4y + z = 5$$
$$x - 3y + 2z = 8.$$

Solution   By inspection,
$$\delta(A) = \begin{vmatrix} -4 & 2 & -9 \\ 3 & 4 & 1 \\ 1 & -3 & 2 \end{vmatrix}$$
$$= -4(11) - 2(5) - 9(-13)$$
$$= -44 - 10 + 117 = 63.$$

Replacing the entries in the first column of A with corresponding constants 2, 5, and 8, we have
$$\delta(A_x) = \begin{vmatrix} 2 & 2 & -9 \\ 5 & 4 & 1 \\ 8 & -3 & 2 \end{vmatrix}$$
$$= 2(11) - 2(2) - 9(-47)$$
$$= 22 - 4 + 423 = 441.$$

Hence, by Cramer's rule,
$$x = \frac{\delta(A_x)}{\delta(A)} = \frac{441}{63} = 7.$$

Similarly, by replacing, in turn, the entries of the second and third columns of A with the corresponding constants, 2, 5, and 8, we have
$$\delta(A_y) = \begin{vmatrix} -4 & 2 & -9 \\ 3 & 5 & 1 \\ 1 & 8 & 2 \end{vmatrix} \quad \text{and} \quad \delta(A_z) = \begin{vmatrix} -4 & 2 & 2 \\ 3 & 4 & 5 \\ 1 & -3 & 8 \end{vmatrix}.$$

Now,
$$\delta(A_y) = -4(2) - 2(5) - 9(19)$$
$$= -8 - 10 - 171 = -189$$

and
$$\delta(A_z) = -4(47) - 2(19) + 2(-13)$$
$$= -188 - 38 - 26 = -252,$$

so that
$$y = \frac{\delta(A_y)}{\delta(A)} = \frac{-189}{63} = -3$$

10.8 Cramer's Rule

and,

$$z = \frac{\delta(A_z)}{\delta(A)} = \frac{-252}{63} = -4,$$

and the solution set of the system is $\{(7, -3, -4)\}$.

If $\delta(A) = 0$ for a linear system, then the system either has infinitely many solutions (the equations are consistent and one of them can be obtained from the others by linear combinations) or has no solutions (the equations are inconsistent). The distinction can be determined using methods from Section 10.3.

Exercise 10.8

A *Find the solution set of each of the following systems by Cramer's rule. If $\delta(A) = 0$ in any of the systems, use the methods in Section 10.3 to determine whether or not the equations in the system are consistent.*

1. $x - y = 2$
   $x + 4y = 5$

2. $x + y = 4$
   $x - 2y = 0$

3. $3x - 4y = -2$
   $x + y = 6$

4. $2x - 4y = 7$
   $x - 2y = 1$

5. $\frac{1}{3}x - \frac{1}{2}y = 0$
   $\frac{1}{2}x + \frac{1}{4}y = 4$

6. $\frac{2}{3}x + y = 1$
   $x - \frac{4}{3}y = 0$

7. $x - 2y = 6$
   $\frac{2}{3}x - \frac{4}{3}y = 6$

8. $\frac{1}{2}x + y = 3$
   $-\frac{1}{4}x - y = -3$

9. $x - 3y = 1$
   $y = 1$

10. $2x - 3y = 12$
    $x = 4$

11. $ax + by = 1$
    $bx + ay = 1$

12. $x + y = a$
    $x - y = b$

13. $x - 2y + z = -1$
    $3x + y - 2z = 4$
    $y - z = 1$

14. $2x + 5z = 9$
    $4x + 3y = -1$
    $3y - 4z = -13$

15. $2x + 2y + z = 1$
    $x - y + 6z = 21$
    $3x + 2y - z = -4$

16. $4x + 8y + z = -6$
    $2x - 3y + 2z = 0$
    $x + 7y - 3z = -8$

17. $x + y + z = 0$
    $2x - y - 4z = 15$
    $x - 2y - z = 7$

18. $x + y - 2z = 2$
    $3x - y + z = 5$
    $3x + 3y - 6z = 6$

19. $x - 2y - 2z = 3$
    $2x - 4y + 4z = 1$
    $3x - 3y - 3z = 4$

20. $3x - 2y + 5z = 6$
    $4x - 4y + 3z = 0$
    $5x - 4y + z = -5$

21. $x - 4z = -1$
    $3x + 3y = 2$
    $3x + 4z = 5$

22. $2x - \frac{2}{3}y + z = 2$
    $6x - 4y - 3z = 0$
    $4x + 5y - 3z = -1$

23. $x + y + z = 0$
    $w + 2y - z = 4$
    $2w - y + 2z = 3$
    $-2w + 2y - z = -2$

24. $x + y + z = 0$
    $x + z + w = 0$
    $x + y + w = 0$
    $y + z + w = 0$

B 25. For the system

$$a_1 x + b_1 y + c_1 = 0$$
$$a_2 x + b_2 y + c_2 = 0,$$

show that if both $\delta(A_y) = 0$ and $\delta(A_x) = 0$, and if c_1 and c_2 are not both 0, then $\delta(A) = 0$, and the equations are consistent. *Hint*: Show that the first two determinant equations imply that $a_1 c_2 = a_2 c_1$ and $b_1 c_2 = b_2 c_1$ and that the rest follows from the formation of a proportion with these equations.

26. Show that if $\delta(A) = 0$ and $\delta(A_x) = 0$, and if a_1 and a_2 are not both 0, then $\delta(A_y) = 0$, where $\delta(A)$ is the determinant of the coefficient matrix of the system in Exercise 25.

Chapter Review

[10.1] *Write each sum or difference as a single matrix.*

1. $\begin{bmatrix} 4 & -7 \\ 2 & 1 \end{bmatrix} + \begin{bmatrix} -3 & 6 \\ -1 & 0 \end{bmatrix}$

2. $\begin{bmatrix} 3 & -1 & 7 \\ 6 & 2 & 5 \end{bmatrix} + \begin{bmatrix} -1 & 6 & -9 \\ 8 & -3 & 7 \end{bmatrix}$

3. $\begin{bmatrix} 2 & -4 & 3 \\ 6 & 1 & 7 \\ 2 & 8 & 0 \end{bmatrix} - \begin{bmatrix} 4 & -1 & 2 \\ 3 & 8 & 1 \\ 7 & 6 & -5 \end{bmatrix}$

4. $\begin{bmatrix} -11 & 2 & -6 \\ 7 & 1 & 2 \\ -3 & 4 & 8 \end{bmatrix} - \begin{bmatrix} -3 & 5 & 0 \\ 1 & 4 & 2 \\ 6 & -1 & 3 \end{bmatrix}$

[10.2] *Write each product as a single matrix.*

5. $-7 \begin{bmatrix} 3 & -1 \\ 2 & 0 \\ 1 & 1 \end{bmatrix}$

6. $\begin{bmatrix} 3 & -1 & 2 \end{bmatrix} \begin{bmatrix} 4 \\ -1 \\ 0 \end{bmatrix}$

7. $\begin{bmatrix} 3 & -1 \\ 6 & 5 \end{bmatrix} \cdot \begin{bmatrix} -4 & 2 \\ 1 & 3 \end{bmatrix}$

8. $\begin{bmatrix} -1 & 7 & 6 \\ 3 & 1 & 2 \\ 1 & 0 & 1 \end{bmatrix} \cdot \begin{bmatrix} 1 & -1 & 2 \\ 1 & 0 & 3 \\ 2 & 1 & 1 \end{bmatrix}$

Review Exercises

[10.3] Use row transformations on the augmented matrix to solve each system of equations.

9. $2x - y = 5$
   $x + 3y = -1$

10. $2x - y + z = 4$
    $x + 3y - z = 4$
    $x + 2y + z = 5$

[10.4] Evaluate each determinant.

11. $\begin{vmatrix} -3 & 0 \\ 2 & 1 \end{vmatrix}$

12. $\begin{vmatrix} 1 & 5 \\ -1 & 2 \end{vmatrix}$

13. $\begin{vmatrix} 3 & 1 & 0 \\ 2 & 0 & 1 \\ 1 & 2 & -1 \end{vmatrix}$

14. $\begin{vmatrix} 3 & 1 & -2 \\ -1 & 2 & 1 \\ 1 & -2 & 1 \end{vmatrix}$

15. $\begin{vmatrix} 3 & 0 & 1 & 1 \\ 0 & 2 & -1 & 0 \\ 0 & 1 & 0 & 2 \\ 1 & 0 & 0 & 1 \end{vmatrix}$

16. $\begin{vmatrix} -2 & 1 & 4 & 0 \\ 2 & 0 & 4 & 1 \\ 1 & 1 & 0 & 0 \\ 2 & -1 & 3 & 1 \end{vmatrix}$

[10.5] Reduce each determinant to an equal 2 × 2 determinant and evaluate.

17. $\begin{vmatrix} 3 & -1 & 2 \\ 1 & -2 & 0 \\ 2 & 1 & -1 \end{vmatrix}$

18. $\begin{vmatrix} 1 & 7 & -11 \\ 12 & 10 & 15 \\ 5 & 11 & 14 \end{vmatrix}$

[10.6] Find the inverse of each nonsingular matrix.

19. $\begin{bmatrix} -4 & 2 \\ 11 & 3 \end{bmatrix}$

20. $\begin{bmatrix} 1 & -1 & 2 \\ 3 & 1 & 0 \\ 2 & 1 & 1 \end{bmatrix}$

[10.7] Use matrices to solve each system.

21. $x - y = -3$
    $2x + 3y = -1$

22. $2x + z = 7$
    $y + 2z = 1$
    $3x + y + z = 9$

23. a. $x - y = 1$
       $3x + y = 1$
    b. $x - y = 2$
       $3x + y = -1$
    c. $x - y = 3$
       $3x + y = 1$

[10.8] Use Cramer's rule to solve each system.

24. $3x - y = -5$
    $x + 2y = -6$

25. $x - y + 2z = 3$
    $2x + y - z = 3$
    $x - 2y + 2z = 4$

11 Sequences and Series

A function with domain {1, 2, 3, ...} will relate some object with 1, some object with 2, and so forth. Thus, such a function puts objects in sequence: something goes first, second, and so on. The notion of a sequence is formalized in this chapter. The chapter concludes with a discussion of mathematical induction, a most useful technique for proving results about sequences.

11.1 Sequences

Let us consider a class of functions in which each function has as its domain either the set N of positive integers or a subset of successive members of N.

Definition 11.1 *A **sequence function** is a function having as its domain the set N of positive integers $1, 2, 3, \ldots$. A **finite-sequence function** has as its domain the set of positive integers $1, 2, 3, \ldots, n$, for some fixed n.*

For example, the function defined by
$$s(n) = n + 3, \quad n \in \{1, 2, 3, \ldots\}, \tag{1}$$
is a sequence function. The elements in the range of such a function, considered in the order
$$s(1), s(2), s(3), s(4), \ldots,$$
are said to form a **sequence**. Similarly, the elements of a finite-sequence function, considered in order, constitute a **finite sequence**.

11.1 Sequences

For example, the sequence associated with (1) is found by successively substituting the numbers 1, 2, 3, ..., for n:

$$s(1) = 1 + 3 = 4,$$
$$s(2) = 2 + 3 = 5,$$
$$s(3) = 3 + 3 = 6,$$
$$s(4) = 4 + 3 = 7,$$

and so on. Thus, the first four terms of (1) are 4, 5, 6, and 7. The nth term, commonly called the **general term**, is $n + 3$.

Examples Find the first three terms and the twenty-fifth term of the given sequence.

a. $s(n) = n^2$ **b.** $s(n) = \dfrac{1}{n}$

Solutions

a. $s(1) = 1^2 = 1$
$s(2) = 2^2 = 4$
$s(3) = 3^2 = 9$
$s(25) = 25^2 = 625$

b. $s(1) = \dfrac{1}{1} = 1$
$s(2) = \dfrac{1}{2}$
$s(3) = \dfrac{1}{3}$
$s(25) = \dfrac{1}{25}$

Given several terms in a sequence, we are often able to construct an expression for a general term of a sequence to which they belong. Such a sequence is not unique. Thus, if the first three terms in a sequence are 2, 4, 6, ..., we may *surmise* that the general term is $s(n) = 2n$. Note, however, that the sequences for both

$$s(n) = 2n$$

and

$$s(n) = 2n + (n-1)(n-2)(n-3)$$

start with 2, 4, 6, but that the two sequences differ for terms following the third.

Sequence notation

The notation ordinarily used for the terms in a sequence is not function notation as such. It is customary to denote a term in a sequence by means of a subscript. Thus, the sequence $s(1), s(2), s(3), s(4), \ldots$ would appear as $s_1, s_2, s_3, s_4, \ldots$. For example, instead of writing $s(n) = 1/n$, we write $s_n = 1/n$; the first three terms of this sequence are $s_1 = 1$, $s_2 = 1/2$, and $s_3 = 1/3$.

Arithmetic progressions

Let us next consider two special kinds of sequences that have many applications. The first kind can be defined as follows:

Definition 11.2 An **arithmetic progression** is a sequence defined by equations of the form

$$s_1 = a, \quad s_{n+1} = s_n + d,$$

where $a, d \in R$, and $n \in N$.

Since each term in such a sequence is obtained from the preceding term by adding d, d is called the **common difference**.

Examples Find the first four terms of the arithmetic progression.

a. $s_1 = 1, \ d = 3$
b. $s_1 = -3, \ d = 2$

Solutions

a. $s_1 = 1$
$s_2 = s_1 + 3 = 4$
$s_3 = s_2 + 3 = 7$
$s_4 = s_3 + 3 = 10;$
$1, 4, 7, 10$

b. $s_1 = -3$
$s_2 = s_1 + 2 = -1$
$s_3 = s_2 + 2 = 1$
$s_4 = s_3 + 2 = 3;$
$-3, -1, 1, 3$

In Definition 11.2, each term of the sequence, after the first term, is defined by its relation to previous terms. Such a formulation is called a **recursive** definition. To obtain the general term of an arithmetic progression with $s_1 = a$, observe that since the sequence progresses from term to term by adding the common difference, any term is obtained by adding an appropriate number of differences to a. The second term requires one difference, the third requires two differences and, in general, s_n requires $n - 1$ differences added to a. This indicates that the following theorem is true. A complete proof requires the technique of mathematical induction which is considered in Section 11.5.

Theorem 11.1 The nth term in the sequence defined by

$$s_1 = a, \quad s_{n+1} = s_n + d,$$

where $a, d \in R$, and $n \in N$, is

$$s_n = a + (n - 1)d. \tag{2}$$

Examples Find the general term and the one-hundredth term of the arithmetic progression.

a. $s_1 = 4, \ d = 5$
b. $s_1 = -3, \ d = 4$

Solutions

a. $s_n = 4 + 5(n - 1) = 5n - 1;$
$s_{100} = 5(100) - 1 = 499$

b. $s_n = -3 + 4(n - 1) = 4n - 7;$
$s_{100} = 4(100) - 7 = 393$

11.1 Sequences

Geometric progressions

The second kind of sequence we shall consider can be defined as follows:

Definition 11.3 A ***geometric progression*** is a sequence defined by equations of the form

$$s_1 = a, \quad s_{n+1} = rs_n,$$

where $a, r \in R$, $a, r \neq 0$, and $n \in N$.

Thus, 3, 9, 27, 81, ... is a geometric progression in which each term except the first is obtained by multiplying the preceding term by 3. Since the effect of multiplying the terms in this way is to produce a fixed ratio between any two successive terms, the multiplier, r, is called the **common ratio.**

Examples Find the first four terms in the geometric progression.

a. $s_1 = -3$, $r = 2$ **b.** $s_1 = 2$, $r = -1$

Solutions

a. $s_1 = -3$
$s_2 = rs_1 = -6$
$s_3 = rs_2 = -12$
$s_4 = rs_3 = -24;$
$-3, -6, -12, -24$

b. $s_1 = 2$
$s_2 = rs_1 = -2$
$s_3 = rs_2 = 2$
$s_4 = rs_3 = -2;$
$2, -2, 2, -2$

Definition 11.3 is a recursive definition of a geometric progression. To obtain the general term of a geometric progression with $s_1 = a$, observe that the sequence progresses from term to term by multiplying by the common ratio. Thus any term is obtained by multiplying by the common ratio an appropriate number of times. The second term requires one factor of the common ratio, the third requires two factors and in general the nth term, s_n, requires $n - 1$ factors of the common ratio multiplied by a. This indicates that the following theorem is true. The proof by mathematical induction is left as an exercise in Section 11.5.

Theorem 11.2 The nth term in the sequence defined by

$$s_1 = a, \quad s_{n+1} = rs_n,$$

where $a, r \in R$, $a \neq 0$, $r \neq 0$, and $n \in N$, is

$$s_n = ar^{n-1}.$$

Examples   Find the general term and the ninth term of the geometric progression.

a. $s_1 = 3$, $r = 2$
b. $s_1 = 1$, $r = -3$

Solutions   a. $s_n = 3(2)^{n-1}$;
$s_9 = 3(2^8) = 768$

b. $s_n = 1(-3)^{n-1}$;
$s_9 = (-3)^8 = 6561$

Exercise 11.1

A   Find the first four terms in the sequence with the general term as given.

Examples   a. $s_n = \dfrac{n(n+1)}{2}$
b. $s_n = (-1)^n 2^n$

Solutions   a. $s_1 = \dfrac{1(1+1)}{2} = 1$

$s_2 = \dfrac{2(2+1)}{2} = 3$

$s_3 = \dfrac{3(3+1)}{2} = 6$

$s_4 = \dfrac{4(4+1)}{2} = 10$;

1, 3, 6, 10

b. $s_1 = (-1)^1 2^1 = -2$
$s_2 = (-1)^2 2^2 = 4$
$s_3 = (-1)^3 2^3 = -8$
$s_4 = (-1)^4 2^4 = 16$;
$-2, 4, -8, 16$

1. $s_n = n - 5$
2. $s_n = 2n - 3$
3. $s_n = \dfrac{n^2 - 2}{2}$
4. $s_n = \dfrac{3}{n^2 + 1}$
5. $s_n = 1 + \dfrac{1}{n}$
6. $s_n = \dfrac{n}{2n - 1}$
7. $s_n = \dfrac{n(n-1)}{2}$
8. $s_n = \dfrac{5}{n(n+1)}$
9. $s_n = (-1)^n$
10. $s_n = (-1)^{n+1}$
11. $s_n = \dfrac{(-1)^n(n-2)}{n}$
12. $s_n = (-1)^{n-1} 3^{n+1}$

Write the next three terms in each of the following arithmetic progressions.

Examples   a. 5, 9, ...
b. $x, x - a, \ldots$

Solutions   Find the common difference and then continue the sequence.

a. $d = 9 - 5 = 4$;
13, 17, 21

b. $d = (x - a) - x = -a$;
$x - 2a, x - 3a, x - 4a$

11.1 Sequences

13. 3, 7, ...
14. −6, −1, ...
15. −4, −1, ...
16. −8, 5, ...
17. $x, x+1, \ldots$
18. $a, a+5, \ldots$
19. $2x+1, 2x+4, \ldots$
20. $3a, 5a, \ldots$

Write the next four terms in each of the following geometric progressions.

Examples
a. 3, 6, ...
b. $x, 2, \ldots$

Solutions   Find the common ratio, and then continue the sequence.

a. $r = \dfrac{6}{3} = 2$;

12, 24, 48, 96

b. $r = \dfrac{2}{x}$ $(x \neq 0)$;

$\dfrac{4}{x}, \dfrac{8}{x^2}, \dfrac{16}{x^3}, \dfrac{32}{x^4}$

21. 2, 8, ...
22. 4, 8, ...
23. $\dfrac{2}{3}, \dfrac{4}{3}, \ldots$
24. $\dfrac{1}{2}, -\dfrac{3}{2}, \ldots$
25. $\dfrac{a}{x}, -1, \ldots$
26. $\dfrac{a}{b}, \dfrac{a}{bc}, \ldots$
27. $-4, x, \ldots$
28. $-x, x^2, \ldots$

Example   Find the general term and the fourteenth term of the arithmetic progression $-6, -1, \ldots$.

Solution   Find the common difference.

$$d = -1 - (-6) = 5$$

Use $s_n = a + (n-1)d$.

$$s_n = -6 + (n-1)5 = 5n - 11;$$
$$s_{14} = 5(14) - 11 = 59$$

29. Find the general term and the seventh term in the arithmetic progression 7, 11, ...
30. Find the general term and the twelfth term in the arithmetic progression $2, \dfrac{5}{2}, \ldots$.
31. Find the general term and the twentieth term in the arithmetic progression 3, −2, ...
32. Find the general term and the ninth term in the arithmetic progression $\dfrac{3}{4}, 2, \ldots$.
33. Find the general term and the eighth term in the arithmetic progression $x, 5x, \ldots$.
34. Find the general term and the twelfth term in the arithmetic progression $2a, -2a, \ldots$.

Example Find the general term and the ninth term of the geometric progression $-24, 12, \ldots$.

Solution Find the common ratio.
$$r = \frac{12}{-24} = -\frac{1}{2}$$
Use $s_n = ar^{n-1}$.
$$s_n = -24\left(-\frac{1}{2}\right)^{n-1}; \quad s_9 = -24\left(-\frac{1}{2}\right)^8 = -\frac{3}{32}$$

35. Find the general term and the sixth term in the geometric progression $48, 96, \ldots$.

36. Find the general term and the eighth term in the geometric progression $-3, \dfrac{3}{2}, \ldots$.

37. Find the general term and the seventh term in the geometric progression $-\dfrac{1}{3}, 1, \ldots$.

38. Find the general term and the ninth term in the geometric progression $-81, 27, \ldots$.

B 39. Find the general term and the fifth term in the geometric progression $x, x^3, \ldots$.

40. Find the general term and the tenth term in the geometric progression $x, x(1+y), \ldots$.

41. Find the general term and the sixth term in the arithmetic progression $x, y, \ldots$.

42. Find the general term and the tenth term of the arithmetic progression $x, x^2, \ldots$.

43. Find the general term and the eighth term of the geometric progression $x, y, \ldots$.

44. Find the general term and the tenth term of the geometric progression $x, x+2, \ldots$.

Example Find the first term in an arithmetic progression in which the third term is 7 and the eleventh term is 55.

Solution Find the common difference by considering an arithmetic progression with first term 7 and with ninth term 55. Use $s_n = a + (n-1)d$.
$$s_9 = 7 + (9-1)d$$
$$55 = 7 + 8d$$
$$d = 6$$

Use the difference to find the first term in an arithmetic progression in which the third term is 7. Again use $s_n = a + (n-1)d$.
$$s_3 = a + (3-1)6$$
$$7 = a + 12$$
$$a = -5$$

11.2 Series

45. If the third term in an arithmetic progression is 7 and the eighth term is 17, find the common difference. What are the first and the twentieth terms?

46. If the fifth term of an arithmetic progression is -16 and the twentieth term is -46, what is the twelfth term?

47. Which term in the arithmetic progression $4, 1, \ldots$ is -77?

48. What is the twelfth term in an arithmetic progression in which the second term is x and the third term is y?

49. Find the first term of a geometric progression with fifth term 48 and ratio 2.

50. Find two different values for x so that $-\dfrac{3}{2}, x, -\dfrac{8}{27}$ will be in geometric progression.

11.2 Series

Associated with any sequence is a *series*.

Definition 11.4 A *series* is the indicated sum of the terms in a sequence.

We shall ordinarily denote a series of n terms by S_n. For example, with the finite sequence

$$4, 7, 10, \ldots, 3n + 1,$$

for a given positive integer n, there is associated the finite series

$$S_n = 4 + 7 + 10 + \cdots + (3n + 1);$$

similarly, with the finite sequence

$$x, x^2, x^3, x^4, \ldots, x^n,$$

there is associated the finite series

$$S_n = x + x^2 + x^3 + x^4 + \cdots + x^n.$$

Since the terms in the series are the same as those in the sequence, we can refer to the first term or the second term or the general term of a series in the same manner as we do for a sequence.

Sum of the first n terms of an arithmetic progression

Consider the series S_n of the first n terms of the general arithmetic progression,

$$S_n = a + (a + d) + (a + 2d) + \cdots + [a + (n - 1)d], \quad (1)$$

and then consider the same series written as

$$S_n = s_n + (s_n - d) + (s_n - 2d) + \cdots + [s_n - (n - 1)d], \quad (2)$$

where the terms are displayed in reverse order. Adding (1) and (2) term by term, we have

$$S_n + S_n = (a + s_n) + (a + s_n) + (a + s_n) + \cdots + (a + s_n),$$

where the term $(a + s_n)$ occurs n times. Then

$$2S_n = n(a + s_n),$$

$$S_n = \frac{n}{2}(a + s_n). \tag{3}$$

If (3) is rewritten as

$$S_n = n\left(\frac{a + s_n}{2}\right),$$

we observe that the sum is given by the product of the number of terms in the series and the average of the first and last terms. The validity of (3) can be established by mathematical induction and is deferred until Section 11.5.

An alternative form for (3) is obtained by substituting $a + (n - 1)d$ for s_n in (3) to obtain

$$S_n = \frac{n}{2}(a + [a + (n - 1)d]),$$

$$S_n = \frac{n}{2}[2a + (n - 1)d], \tag{3'}$$

where the sum is now expressed in terms of a, n, and d.

Example Compute $1 + 3 + \cdots + 99$.

Solution The sequence $1, 3, \ldots$ is an arithmetic progression with $s_1 = 1$, $d = 2$. We have $s_n = 99$, $a = 1$, and $d = 2$. Thus, from Equation (2), page 386, we have

$$99 = 1 + (n - 1)2.$$

Solving for n, we obtain $n = 50$. Now we use Equation (3') to compute

$$S_{50} = \frac{50}{2}[2a + (50 - 1)d]$$

$$= \frac{50}{2}[2(1) + (49)(2)] = 2500.$$

Sum of the first n terms of a geometric progression

To find an explicit representation for the sum of a given number of terms in a geometric progression in terms of a, r, and n, we employ a device somewhat similar to the one used in finding the sum in an arithmetic progression. Consider the geometric series (4) containing n terms, and the series (5) obtained by multiplying both members of (4) by r:

$$S_n = a + ar + ar^2 + ar^3 + \cdots + ar^{n-2} + ar^{n-1}, \tag{4}$$

$$rS_n = ar + ar^2 + ar^3 + ar^4 + \cdots + ar^{n-1} + ar^n. \tag{5}$$

11.2 Series

When we subtract (5) from (4), all terms in the right-hand members except the first term in (4) and the last term in (5) vanish, yielding

$$S_n - rS_n = a - ar^n.$$

Factoring S_n from the left-hand member gives

$$(1 - r)S_n = a - ar^n,$$

$$S_n = \frac{a - ar_n}{1 - r}, \tag{6}$$

if $r \neq 1$, and we have a formula for the sum of the first n terms of a geometric progression. Establishing the validity of Equation (6), which can be accomplished by mathematical induction, is left to the exercises in Section 11.5.

Examples Find the sum of the first ten terms in the geometric progression.

a. $a = 2$, $r = \dfrac{1}{2}$ 　　　　**b.** $a = \dfrac{1}{2}$, $r = -2$

Solutions **a.** $S_{10} = \dfrac{a - ar^{10}}{1 - r}$ 　　　　**b.** $S_{10} = \dfrac{a - ar^{10}}{1 - r}$

$= \dfrac{2 - 2(1/2)^{10}}{1 - (1/2)} = \dfrac{1023}{256}$ 　　　　$= \dfrac{(1/2) - (1/2)(-2)^{10}}{1 - (-2)} = \dfrac{1023}{6}$

An alternative expression for (6) can be obtained by first writing

$$S_n = \frac{a - r(ar^{n-1})}{1 - r},$$

and then, since $s_n = ar^{n-1}$, expressing this as

$$S_n = \frac{a - rs_n}{1 - r}, \tag{7}$$

where the sum is now given in terms of a, s_n, and r.

Examples Find the sum of the terms in the geometric progression up to the *n*th term.

a. $a = 2$, $r = \dfrac{1}{2}$, $s_n = \dfrac{1}{8}$ 　　　　**b.** $a = \dfrac{1}{2}$, $r = -3$, $s_n = \dfrac{81}{2}$

Solutions **a.** $S_n = \dfrac{a - rs_n}{1 - r}$ 　　　　**b.** $S_n = \dfrac{a - rs_n}{1 - r}$

$= \dfrac{2 - (1/2)(1/8)}{1 - (1/2)} = \dfrac{31}{8}$ 　　　　$= \dfrac{(1/2) - (-3)(81/2)}{1 - (-3)} = \dfrac{61}{2}$

Sigma notation

A series for which the general term is known can be represented in a very convenient, compact way by means of what is called **sigma**, or **summation**, **notation**. The Greek letter $\sum$ (sigma) is used to denote a sum. For example,

$$S_n = 4 + 7 + 10 + \cdots + (3n + 1)$$

can be written

$$S_n = \sum_{j=1}^{n} (3j + 1),$$

where we understand that S_n is the series having terms obtained by replacing j in the expression $3j + 1$ with the numbers $1, 2, 3, \ldots, n$, successively. Similarly,

$$S = \sum_{j=3}^{6} j^2$$

appears in expanded form as

$$S = 3^2 + 4^2 + 5^2 + 6^2,$$

where the first value for j is 3 and the last is 6.

The variable used in conjunction with summation notation is called the **index of summation**, and the set of integers over which we sum (in this case, $\{3, 4, 5, 6\}$) is called the **range of summation**.

Notation for an infinite sum

To indicate that a series has an infinite number of terms, we cannot use the notation S_n for the sum, because there is no value to substitute for n. We therefore adopt notation such as

$$S_\infty = \sum_{j=1}^{\infty} \frac{1}{2^j} \tag{8}$$

to indicate that there is no last term in a series. In expanded form, the infinite series (8) is given by

$$S_\infty = \frac{1}{2} + \frac{1}{4} + \frac{1}{8} + \cdots.$$

The meaning of such an infinite sum will be discussed in Section 11.3.

Exercise 11.2

A *Write each series in expanded form.*

Examples

a. $\sum_{j=2}^{5} (j^2 + 1)$

b. $\sum_{k=1}^{\infty} (-1)^k 2^{k+1}$

11.2 Series

Solutions

a. $j = 2, \ 2^2 + 1 = 5;$
$j = 3, \ 3^2 + 1 = 10;$
$j = 4, \ 4^2 + 1 = 17;$
$j = 5, \ 5^2 + 1 = 26.$

$$\sum_{j=2}^{5} (j^2 + 1) = 5 + 10 + 17 + 26$$

b. $k = 1, \ (-1)^1 2^{1+1} = (-1)(4) = -4;$
$k = 2, \ (-1)^2 2^{2+1} = (1)(8) = 8;$
$k = 3, \ (-1)^3 2^{3+1} = (-1)(16) = -16.$

$$\sum_{k=1}^{\infty} (-1)^k 2^{k+1} = -4 + 8 - 16 + \cdots$$

1. $\sum_{j=1}^{4} j^2$

2. $\sum_{j=1}^{4} (3j - 2)$

3. $\sum_{j=1}^{3} \frac{(-1)^j}{2^j}$

4. $\sum_{j=3}^{5} \frac{(-1)^{j+1}}{j - 2}$

5. $\sum_{j=-3}^{4} (2j + 1)$

6. $\sum_{j=-2}^{2} j^3$

7. $\sum_{j=0}^{\infty} \frac{1}{2^j}$

8. $\sum_{j=0}^{\infty} \frac{j}{1 + j}$

Write each series in sigma notation with the range of summation starting with 1.

Examples

a. $5 + 8 + 11 + 14$

b. $x^2 + x^4 + x^6$

c. $\frac{3}{5} + \frac{5}{7} + \frac{7}{9} + \cdots$

Solutions

Find an expression for the general term and write in sigma notation.

a. $3j + 2$

$\sum_{j=1}^{4} (3j + 2)$

b. x^{2j}

$\sum_{j=1}^{3} x^{2j}$

c. $\frac{2j + 1}{2j + 3}$

$\sum_{j=1}^{\infty} \frac{2j + 1}{2j + 3}$

9. $x + x^3 + x^5 + x^7$

10. $x^3 + x^5 + x^7 + x^9 + x^{11}$

11. $1 + 4 + 9 + 16 + 25$

12. $\frac{1}{3} + \frac{1}{9} + \frac{1}{27} + \frac{1}{81}$

13. $1 \cdot 2 + 2 \cdot 3 + 3 \cdot 4 + 4 \cdot 5 + \cdots$

14. $\frac{1}{2} + \frac{2}{3} + \frac{3}{4} + \frac{4}{5} + \cdots$

15. $\frac{2}{1} + \frac{3}{2} + \frac{4}{3} + \frac{5}{4} + \cdots$

16. $\frac{1}{1} + \frac{2}{3} + \frac{3}{5} + \frac{4}{7} + \cdots$

Find each of the following sums.

Example

$$\sum_{j=1}^{12} (4j + 1)$$

Solution overleaf

Solution Write the first two or three terms in expanded form:

$$5 + 9 + 13 + \cdots.$$

This is an arithmetic series. The first term is 5 and the common difference is 4. Therefore we can use

$$S_n = \frac{n}{2}[2a + (n-1)d]$$

to obtain

$$S_{12} = \frac{12}{2}[2(5) + (12-1)4] = 324.$$

17. $\sum_{j=1}^{7}(2j+1)$ 18. $\sum_{j=1}^{21}(3j-2)$ 19. $\sum_{j=3}^{15}(7j-1)$

20. $\sum_{j=10}^{20}(2j-3)$ 21. $\sum_{k=1}^{8}\left(\frac{1}{2}k - 3\right)$ 22. $\sum_{k=1}^{100} k$

23. $\sum_{k=1}^{100}(2k-1)$ 24. $\sum_{k=1}^{100}(10k-2)$

Example $\sum_{j=2}^{5}\left(\frac{1}{3}\right)^j$

Solution Write the first two or three terms in expanded form:

$$\left(\frac{1}{3}\right)^2 + \left(\frac{1}{3}\right)^3 + \cdots.$$

This is a geometric series in which the first term is $\frac{1}{9}$, the ratio is $\frac{1}{3}$, and $n = 4$. Therefore, we can use $S_n = \dfrac{a - ar^n}{1 - r}$ to obtain

$$S_4 = \frac{\frac{1}{9} - \frac{1}{9}\left(\frac{1}{3}\right)^4}{1 - \frac{1}{3}} = \frac{40}{243}.$$

25. $\sum_{j=1}^{6} 3^j$ 26. $\sum_{j=1}^{4} 2^j$ 27. $\sum_{k=3}^{7}\left(\frac{1}{2}\right)^{k-2}$

28. $\sum_{j=3}^{12} 2^{j-5}$ 29. $\sum_{j=1}^{6}\left(\frac{1}{3}\right)^{j+1}$ 30. $\sum_{j=1}^{4} 2^{j+3}$

31. $\sum_{j=4}^{8}(-2)^{j+4}$ 32. $\sum_{j=5}^{10}\left(-\frac{1}{2}\right)^{j+2}$

B 33. Find the sum of all even integers n, for $13 < n < 29$.

34. Find the sum of all integral multiples of 7 between 8 and 110.

35. Find the sum of the powers of 2 between 5 and 2000.

36. Find the sum of the powers of $1/2$ between $1/300$ and 1.

37. How many bricks will there be in a wall one brick in thickness if there are 27 bricks in the bottom row, 25 in the second row, and so forth, to the top row, which has one brick?

38. If there are a total of 256 bricks in a wall arranged in the manner of those in Exercise 37, how many bricks are there in the tenth row from the bottom?

39. If ten dollars is deposited each month in an account which draws 12% annual interest and which is compounded monthly, how much will be in the account at the end of ten years?

40. How much has to be deposited monthly in an account earning 9% annual interest compounded monthly to guarantee that the account will be worth one million dollars at the end of thirty years?

41. Find $\sum_{j=1}^{n} \left(\frac{1}{2}\right)^j$ for $n = 2, 3, 4,$ and 5. What value do you think $\sum_{j=1}^{n} \left(\frac{1}{2}\right)^j$ approximates as n becomes greater and greater?

42. Find p if $\sum_{j=1}^{5} pj = 14$.

43. Find p and q if $\sum_{j=1}^{4} (pj + q) = 28$ and $\sum_{j=2}^{5} (pj + q) = 44$.

44. Consider $S_n = \sum_{j=1}^{n} f(j)$. Explain why this equation defines a sequence function. What is the variable denoting an element in the domain? The range?

11.3 Limits of Sequences and Series

A sequence that is strictly increasing but bounded

Consider the sequence function defined by

$$S_n = \frac{n}{n+1}, \quad n \in N. \tag{1}$$

If we write the range of (1) in the form

$$\frac{1}{2}, \frac{2}{3}, \frac{3}{4}, \frac{4}{5}, \ldots, \frac{n}{n+1}, \ldots,$$

then it is clear that each of the terms is greater than the preceding term; indeed, the difference of consecutive terms is

$$\frac{n+1}{n+2} - \frac{n}{n+1} = \frac{(n^2 + 2n + 1) - (n^2 + 2n)}{(n+1)(n+2)} = \frac{1}{(n+1)(n+2)} > 0.$$

Such a sequence is said to be **strictly increasing**. On the other hand, it is also clear that, no matter how large a value is assigned to *n*, we have

$$\frac{n}{n+1} < 1,$$

because the denominator is one larger than the numerator; in fact, we have

$$1 - \frac{n}{n+1} = \frac{(n+1) - n}{n+1} = \frac{1}{n+1} > 0.$$

Thus we have a sequence in which each term is greater than the preceding term and yet no term is equal to or greater than 1. We note, however—and this is a very basic consideration—that the value of $n/(n+1)$ is as close to 1 as we please if *n* is large enough.

Example

The value of $\dfrac{n}{n+1}$ is within 1/1000 of 1 if $n > 999$, because

$$1 - \frac{n}{n+1} = \frac{1}{n+1}, \quad \text{and we have} \quad \frac{1}{n+1} < \frac{1}{1000}$$

provided $n + 1 > 1000$—that is, $n > 999$.

Limit of a sequence

If it is true that the *n*th term in a sequence differs from the number *L* by as little as we please for all sufficiently large *n*, we say that **the sequence approaches the number *L* as a limit**. The symbolism

$$\lim_{n \to \infty} s_n = L$$

(read "the limit, as *n* increases without bound, of s_n is *L*") is used to denote this situation. A thorough discussion of the notion of a limit is included in courses in calculus, and will not be attempted here. A few elementary ideas, however, are in order.

A sequence in which the *n*th term approaches a number *L* as $n \to \infty$ is said to be a **convergent sequence**, and the sequence is said to **converge** to *L*. It is not necessary for convergence that a sequence be strictly increasing. For example,

$$1, \frac{1}{2}, \frac{1}{3}, \frac{1}{4}, \ldots, \frac{1}{n}, \ldots$$

converges to 0, but each term in the sequence is less than, instead of greater than, the term that precedes it. Again, the sequence

$$-1, \frac{1}{2}, -\frac{1}{3}, \frac{1}{4}, \ldots, \frac{(-1)^n}{n}, \ldots$$

converges to 0 but is neither increasing nor decreasing. We can rephrase the definition of convergence of a sequence as follows:

11.3 Limits of Sequences and Series

Definition 11.5 A sequence $s_1, s_2, \ldots, s_n, \ldots$ **converges** to the number L,

$$\lim_{n \to \infty} s_n = L,$$

if and only if the absolute value of the difference between the nth term in the sequence and the number L is as small as we please for all sufficiently large n. Thus the sequence converges to the number L if and only if

$$\lim_{n \to \infty} |L - s_n| = 0.$$

For example, the **alternating** (because the signs alternate) **sequence**

$$-\frac{1}{3}, \frac{1}{9}, -\frac{1}{27}, \ldots, \frac{(-1)^n}{3^n}, \ldots$$

converges to 0 since the absolute value of the difference between $\frac{(-1)^n}{3^n}$ and 0 is as small as we please for n large enough. We express this by writing

$$\lim_{n \to \infty} \left(\frac{-1}{3}\right)^n = 0.$$

On the other hand, the alternating sequence

$$\frac{1}{2}, -\frac{2}{3}, \frac{3}{4}, -\frac{4}{5}, \ldots, (-1)^{n+1} \frac{n}{n+1}, \ldots$$

does not converge. As n increases, the nth term oscillates back and forth from the neighborhood of $+1$ to the neighborhood of -1, and we cannot find a number L such that $\lim_{n \to \infty} |L - s_n| = 0$. Such a sequence is said to **diverge**. A sequence such as

$$1, 2, 3, \ldots, n, \ldots$$

also is said to diverge. An answer to the logical question of what we mean by "enough" when we say "n large enough" requires a more precise definition of limit than we have given here. As remarked earlier, a course in the calculus will treat this in detail.

Sequence of partial sums of a series

For an infinite series,

$$S_\infty = \sum_{j=1}^{\infty} s_j,$$

we can consider the infinite sequence of **partial sums**:

$$S_1 = s_1$$
$$S_2 = s_1 + s_2$$
$$\cdot \ \cdot \ \cdot$$
$$S_n = s_1 + s_2 + \cdots + s_n$$
$$\cdot \ \cdot \ \cdot$$

Definition 11.6 An infinite series

$$S_\infty = \sum_{j=1}^{\infty} s_j$$

converges if and only if $S_1, S_2, \ldots, S_n, \ldots$, the corresponding sequence of partial sums, converges.

If the sequence of partial sums converges to the number L,

$$\lim_{n \to \infty} S_n = L,$$

then L is said to be the **sum** of the infinite series, and we write

$$S_\infty = \sum_{j=1}^{\infty} s_j = L.$$

If the sequence of partial sums diverges, then the series is said to **diverge**.

Sum of an infinite geometric progression

We recall from Section 11.2 that the sum of n terms (the nth partial sum) of a geometric progression is given, for $r \neq 1$, by

$$S_n = \frac{a - ar^n}{1 - r}. \tag{2}$$

If $|r| < 1$, that is, if $-1 < r < 1$, then $|r|^n$ becomes smaller and smaller for increasingly large n. For example, if $r = 1/2$, then

$$r^2 = \frac{1}{4}, \quad r^3 = \frac{1}{8}, \quad r^4 = \frac{1}{16},$$

and so forth; and $(1/2)^n$ is as small as we please if n is sufficiently large. Writing (2) as

$$S_n = \frac{a}{1 - r}(1 - r^n), \tag{3}$$

we see that the value of the factor $(1 - r^n)$ is as close as we please to 1 provided $|r| < 1$ and n is taken large enough. Since this argument shows that the sequence of partial sums (3) converges to

$$\frac{a}{1 - r},$$

we have the following result.

Theorem 11.3 The sum of an infinite geometric progression,

$$a + ar + ar^2 + \cdots + ar^n + \cdots,$$

Theorem continued

11.3 Limits of Sequences and Series

with $|r| < 1$, is

$$S_\infty = \lim_{n \to \infty} S_n = \frac{a}{1-r}.$$

An interesting application of this sum arises in connection with repeating decimals—that is, decimal numerals that, after a finite number of decimal places, have endlessly repeating groups of digits. For example,

$$0.2121\overline{21},$$

$$0.138512512\overline{512}$$

are repeating decimals. The bar denotes that the numerals appearing under it are repeated endlessly. Consider the problem of expressing such a decimal fraction as an arithmetic fraction. We illustrate the process involved with the first example above. The decimal $0.2121\overline{21}$ can be written as

$$0.21 + 0.0021 + 0.000021 + \cdots, \tag{4}$$

which is a geometric progression with ratio $r = 0.01$. Since the ratio is less than 1 in absolute value, we can use Theorem 11.3 to find the sum of the infinite series (4). Thus,

$$S_\infty = \frac{a}{1-r} = \frac{0.21}{1 - 0.01} = \frac{21}{99} = \frac{7}{33},$$

and the given decimal fraction is equivalent to 7/33.

Exercise 11.3

A Discuss the limiting behavior of each expression as $n \to \infty$.

Example $\quad \dfrac{n^2 + 3}{n^2}$

Solution $\quad$ By writing $\dfrac{n^2 + 3}{n^2}$ as $\dfrac{n^2}{n^2} + \dfrac{3}{n^2}$ and then as $1 + \dfrac{3}{n^2}$, we observe that

$$\lim_{n \to \infty} \frac{n^2 + 3}{n^2} = \lim_{n \to \infty} \left(1 + \frac{3}{n^2}\right) = 1 + 0 = 1.$$

1. $\dfrac{1}{n}$
2. $1 + \dfrac{1}{n^2}$
3. $\dfrac{n+1}{n}$
4. $\dfrac{n+3}{n^2}$

5. $2n$
6. $(-1)^n$
7. $\dfrac{1}{2^n}$
8. $(-1)^n \dfrac{1}{n}$

State which of the following sequences are convergent.

Example

$$1, \frac{3}{2}, \frac{7}{4}, \frac{15}{8}, \ldots, \frac{2^n - 1}{2^{n-1}}, \ldots$$

Solution

Writing the general term as $\frac{2^n}{2^{n-1}} - \frac{1}{2^{n-1}}$, or $2 - \frac{1}{2^{n-1}}$, we observe that

$$\lim_{n \to \infty} \frac{2^n - 1}{2^{n-1}} = \lim_{n \to \infty} \left(2 - \frac{1}{2^{n-1}}\right) = 2 - 0 = 2.$$

The sequence is convergent.

9. $\frac{1}{2}, \frac{1}{4}, \frac{1}{8}, \frac{1}{16}, \ldots, \frac{1}{2^n}, \ldots$

10. $2, \frac{3}{2}, \frac{4}{3}, \frac{5}{4}, \ldots, \frac{n+1}{n}, \ldots$

11. $1, 2, 3, 4, 5, \ldots, n, \ldots$

12. $2, 4, 6, 8, \ldots, 2n, \ldots$

13. $1, -\frac{1}{2}, \frac{1}{4}, -\frac{1}{8}, \ldots, (-1)^{n-1}\frac{1}{2^{n-1}}, \ldots$

14. $1, -1, 1, -1, \ldots, (-1)^{n+1}, \ldots$

Find the sum of each of the following infinite geometric series. If the series has no sum, so state.

Examples

a. $3 + 2 + \cdots$

b. $\frac{1}{81} - \frac{1}{54} + \cdots$

Solutions

a. $r = \frac{2}{3}$; series has a sum since $|r| < 1$. Using Theorem 11.3, we have

$$S_\infty = \frac{a}{1 - r} = \frac{3}{1 - \frac{2}{3}} = 9.$$

b. $r = -\frac{1}{54} \div \frac{1}{81} = -\frac{3}{2}$; series does not have a sum since $|r| > 1$.

15. $12 + 6 + \cdots$

16. $2 + 1 + \cdots$

17. $\frac{1}{36} + \frac{1}{30} + \cdots$

18. $\frac{1}{16} - \frac{1}{8} + \cdots$

19. $\sum_{j=1}^{\infty} \left(\frac{2}{3}\right)^j$

20. $\sum_{j=1}^{\infty} \left(-\frac{1}{4}\right)^j$

21. $\sum_{j=2}^{\infty} \left(\frac{2}{3}\right)^{j+1}$

22. $\sum_{j=-1}^{\infty} \left(\frac{-1}{3}\right)^{j+2}$

Find an arithmetic fraction equal to each of the given decimal numerals.

Example $0.81818\overline{1}$

Solution Rewrite as a series: $0.81 + 0.0081 + 0.000081 + \cdots$.

Find the common ratio: $r = 0.01$. Use $S_\infty = \dfrac{a}{1-r}$.

$$S_\infty = \frac{0.81}{1-0.01} = \frac{81}{99} = \frac{9}{11}.$$

23. $0.31313\overline{1}$ 24. $0.45454\overline{5}$ 25. $2.41041\overline{0}$

26. $3.02702\overline{7}$ 27. $0.12888\overline{8}$ 28. $0.8333\overline{3}$

B 29. A force is applied to a particle moving in a straight line in such a fashion that each second it moves only one half of the distance it moved the preceding second. If the particle moves ten centimeters the first second, approximately how far will it move before coming to rest?

30. The arc length through which the bob on a pendulum moves is nine-tenths of its preceding arc length. Approximately how far will the bob move before coming to rest if the first arc length is 12 inches?

11.4 The Binomial Theorem

Factorial notation

There are situations in which it is necessary to write the product of several consecutive positive integers. To facilitate writing products of this type, we use a special symbol $n!$ (read "n factorial" or "factorial n"), which is defined by

$$n! = n(n-1)(n-2)\cdots(3)(2)(1).$$

Thus,

$$5! = 5 \cdot 4 \cdot 3 \cdot 2 \cdot 1 \quad \text{(read ``five factorial''),}$$

and

$$8! = 8 \cdot 7 \cdot 6 \cdot 5 \cdot 4 \cdot 3 \cdot 2 \cdot 1 \quad \text{(read ``eight factorial'').}$$

Factorial notation can also be used to represent products of consecutive positive integers, beginning with integers different from 1. For example,

$$8 \cdot 7 \cdot 6 \cdot 5 = \frac{8!}{4!},$$

because

$$\frac{8!}{4!} = \frac{8 \cdot 7 \cdot 6 \cdot 5 \cdot 4 \cdot 3 \cdot 2 \cdot 1}{4 \cdot 3 \cdot 2 \cdot 1} = 8 \cdot 7 \cdot 6 \cdot 5.$$

Since

$$n! = n(n-1)(n-2)(n-3) \cdots 5 \cdot 4 \cdot 3 \cdot 2 \cdot 1$$

and

$$(n-1)! = (n-1)(n-2)(n-3) \cdots 5 \cdot 4 \cdot 3 \cdot 2 \cdot 1,$$

for $n > 1$ we can write the recursive relationship

$$\boldsymbol{n! = n(n-1)!.} \tag{1}$$

For example,

$$7! = 7 \cdot 6!,$$
$$27! = 27 \cdot 26!,$$
$$(n+2)! = (n+2)(n+1)!.$$

If (1) is to hold also for $n = 1$, then we must have

$$1! = 1 \cdot (1-1)!$$

or

$$1! = 1 \cdot 0!.$$

Therefore, for consistency, we define $0!$ by

$$\boldsymbol{0! = 1.}$$

A special case of the use of factorial notation occurs in the binomial expansion that follows. We make the following definition:

$$\binom{n}{r} = \frac{n!}{r!(n-r)!}. \tag{2}$$

The symbol $\binom{n}{r}$ is read "the binomial coefficient n, r," or simply, "n, r." Some examples are

$$\binom{5}{3} = \frac{5!}{3!(5-3)!} = \frac{5!}{3!2!} = \frac{5 \cdot 4 \cdot 3!}{3!2 \cdot 1} = 10,$$

$$\binom{5}{1} = \frac{5!}{1!(5-1)!} = \frac{5!}{4!} = \frac{5 \cdot 4!}{4!} = 5,$$

and

$$\binom{5}{0} = \frac{5!}{0!(5-0)!} = \frac{5!}{5!} = 1.$$

11.4 The Binomial Theorem

Binomial expansions

The series obtained by expanding a binomial of the form

$$(a + b)^n$$

is particularly useful in certain branches of mathematics. Starting with familiar examples, where n takes the values 1, 2, 3, 4, and 5 in turn, we can show by direct multiplication that

$$(a + b)^1 = a + b$$
$$(a + b)^2 = a^2 + 2ab + b^2$$
$$(a + b)^3 = a^3 + 3a^2b + 3ab^2 + b^3$$
$$(a + b)^4 = a^4 + 4a^3b + 6a^2b^2 + 4ab^3 + b^4$$
$$(a + b)^5 = a^5 + 5a^4b + 10a^3b^2 + 10a^2b^3 + 5ab^4 + b^5.$$

We observe that in each case:

1. The first term is a^n.

2. The variable factors of the second term are $a^{n-1}b^1$, and the coefficient is n, which can be written in the form

$$\frac{n}{1!}.$$

3. The variable factors of the third term are $a^{n-2}b^2$, and the coefficient can be written in the form

$$\frac{n(n-1)}{2!}.$$

4. The variable factors of the fourth term are $a^{n-3}b^3$, and the coefficient can be written in the form

$$\frac{n(n-1)(n-2)}{3!}.$$

The foregoing expansions suggest the following result, known as the **binomial theorem**. Its proof, which requires the use of mathematical induction, is omitted.

Theorem 11.4 *For each natural number n,*

$$(a + b)^n = a^n + \frac{n}{1!}a^{n-1}b + \frac{n(n-1)}{2!}a^{n-2}b^2 + \frac{n(n-1)(n-2)}{3!}a^{n-3}b^3$$

$$+ \cdots + \frac{n(n-1)(n-2)\cdots(n-r+2)}{(r-1)!}a^{n-r+1}b^{r-1} + \cdots + b^n, \quad (3)$$

where r is the number of the term.

For example,

$$(x-2)^4 = x^4 + \frac{4}{1!}x^3(-2)^1 + \frac{4\cdot 3}{2!}x^2(-2)^2 + \frac{4\cdot 3\cdot 2}{3!}x(-2)^3$$
$$+ \frac{4\cdot 3\cdot 2\cdot 1}{4!}(-2)^4$$
$$= x^4 - 8x^3 + 24x^2 - 32x + 16.$$

In this case, $a = x$ and $b = -2$ in the binomial expansion.

Observe that the coefficients of the terms in the binomial expansion (3) can be represented as follows.

1st term: $\quad 1 = \dfrac{n!}{0!n!} = \dbinom{n}{0}$,

2nd term: $\quad \dfrac{n}{1!} = \dfrac{n\cdot(n-1)!}{1!(n-1)!} = \dfrac{n!}{1!(n-1)!} = \dbinom{n}{1}$,

3rd term: $\quad \dfrac{n(n-1)}{2!} = \dfrac{n(n-1)(n-2)!}{2!(n-2)!} = \dfrac{n!}{2!(n-2)!} = \dbinom{n}{2}$,

rth term: $\quad \dfrac{n(n-1)(n-2)\cdots(n-r+2)}{(r-1)!}$

$$= \frac{n(n-1)(n-2)\cdots(n-r+2)(n-r+1)!}{(r-1)!(n-r+1)!}$$

$$= \frac{n!}{(r-1)!(n-r+1)!} = \binom{n}{r-1}.$$

Hence, the binomial expansion (3) can be represented by the expression

$$(a+b)^n = \binom{n}{0}a^n + \binom{n}{1}a^{n-1}b + \binom{n}{2}a^{n-2}b^2 + \binom{n}{3}a^{n-3}b^3 + \cdots$$
$$+ \binom{n}{r-1}a^{n-r+1}b^{r-1} + \cdots + \binom{n}{n}b^n. \quad (4)$$

For example,

$$(x-2)^4 = \binom{4}{0}x^4 + \binom{4}{1}x^3(-2)^1 + \binom{4}{2}x^2(-2)^2 + \binom{4}{3}x(-2)^3 + \binom{4}{4}(-2)^4.$$

Using Formula (2) for $\dbinom{n}{r}$ to find the coefficients, we obtain the same result as above,

$$(x-2)^4 = x^4 - 8x^3 + 24x^2 - 32x + 16.$$

11.4 The Binomial Theorem

rth term in a binomial expansion

Note that the rth term in a binomial expansion is given by

$$\binom{n}{r-1}a^{n-r+1}b^{r-1} = \frac{n!}{(r-1)!(n-r+1)!}a^{n-r+1}b^{r-1}$$

$$= \frac{n(n-1)(n-2)\cdots(n-r+2)}{(r-1)!}a^{n-r+1}b^{r-1}. \quad (5)$$

For example, by the left-hand member of (5), the seventh term of $(x-2)^{10}$ is

$$\binom{10}{6}x^4(-2)^6 = \frac{10!}{6!4!}x^4(-2)^6 = \frac{10\cdot 9\cdot 8\cdot 7\cdot 6!}{6!\cdot 4\cdot 3\cdot 2\cdot 1}x^4(64)$$

$$= 13440x^4,$$

while, by the right-hand member of (5), we have

$$\frac{10\cdot 9\cdot 8\cdot 7\cdot 6\cdot 5}{6\cdot 5\cdot 4\cdot 3\cdot 2\cdot 1}x^4(64) = 13440x^4.$$

Exercise 11.4

A 1. Write $(2n)!$ in expanded form for $n = 4$.
2. Write $2n!$ in expanded form for $n = 4$.
3. Write $n(n-1)!$ in expanded form for $n = 6$.
4. Write $2n(2n-1)!$ in expanded form for $n = 2$.

Write in expanded form and simplify.

Examples a. $\dfrac{7!}{4!}$ b. $\dfrac{4!6!}{8!}$

Solutions a. $\dfrac{7\cdot 6\cdot 5\cdot 4!}{4!}$ b. $\dfrac{4\cdot 3\cdot 2\cdot 1\cdot 6!}{8\cdot 7\cdot 6!}$

210

$\dfrac{3}{7}$

5. $5!$ 6. $7!$ 7. $\dfrac{9!}{7!}$ 8. $\dfrac{12!}{11!}$

9. $\dfrac{5!7!}{8!}$ 10. $\dfrac{12!8!}{16!}$ 11. $\dfrac{8!}{2!(8-2)!}$ 12. $\dfrac{10!}{4!(10-4)!}$

Write each product in factorial notation.

Examples   **a.** $1 \cdot 2 \cdot 3 \cdot 4 \cdot 5 \cdot 6$ **b.** $11 \cdot 12 \cdot 13 \cdot 14$ **c.** 150

Solutions   **a.** $6!$ **b.** $\dfrac{14!}{10!}$ **c.** $\dfrac{150!}{149!}$

13. $1 \cdot 2 \cdot 3$ **14.** $1 \cdot 2 \cdot 3 \cdot 4 \cdot 5$ **15.** $3 \cdot 4 \cdot 5 \cdot 6$

16. 7 **17.** $8 \cdot 7 \cdot 6$ **18.** $28 \cdot 27 \cdot 26 \cdot 25 \cdot 24$

Write each expression in factorial notation and simplify.

Examples   **a.** $\binom{6}{2}$ **b.** $\binom{4}{4}$

Solutions   **a.** $\binom{6}{2} = \dfrac{6!}{2!(6-2)!} = \dfrac{6!}{2!4!}$ **b.** $\binom{4}{4} = \dfrac{4!}{4!(4-4)!}$

$= \dfrac{6 \cdot 5 \cdot 4!}{2 \cdot 1 \cdot 4!} = 15$ $= \dfrac{4!}{4!0!} = 1$

19. $\binom{6}{5}$ **20.** $\binom{4}{2}$ **21.** $\binom{3}{3}$ **22.** $\binom{5}{5}$

23. $\binom{7}{0}$ **24.** $\binom{2}{0}$ **25.** $\binom{5}{2}$ **26.** $\binom{5}{3}$

Write each expression in factored form and show the first three factors and the last three factors.

Example $(2n + 1)!$

Solution $(2n + 1)(2n)(2n - 1) \cdot \ldots \cdot 3 \cdot 2 \cdot 1$

27. $n!$ **28.** $(n + 4)!$ **29.** $(3n)!$

30. $3n!$ **31.** $(n - 2)!$ **32.** $(3n - 2)!$

Simplify each expression.

Examples   **a.** $\dfrac{(n-1)!}{(n-3)!}$ **b.** $\dfrac{(n-1)!(2n)!}{2n!(2n-2)!}$

11.4 The Binomial Theorem

Solutions

a. $\dfrac{(n-1)(n-2)(n-3)!}{(n-3)!}$

$(n-1)(n-2)$

b. $\dfrac{(n-1)!(2n)(2n-1)(2n-2)!}{2(n)(n-1)!(2n-2)!}$

$2n - 1$

33. $\dfrac{(n+2)!}{n!}$

34. $\dfrac{(n+2)!}{(n-1)!}$

35. $\dfrac{(n+1)(n+2)!}{(n+3)!}$

36. $\dfrac{(2n+4)!}{(2n+2)!}$

37. $\dfrac{(2n)!(n-2)!}{4(2n-2)!(n)!}$

38. $\dfrac{(2n+1)!(2n-1)!}{[(2n)!]^2}$

Expand.

Example $(a - 3b)^4$

Solution From the binomial expansion (3) on page 405,

$$(a-3b)^4 = a^4 + \frac{4}{1!}a^3(-3b) + \frac{4\cdot 3}{2!}a^2(-3b)^2 + \frac{4\cdot 3\cdot 2}{3!}a(-3b)^3$$

$$+ \frac{4\cdot 3\cdot 2\cdot 1}{4!}(-3b)^4$$

$$= a^4 - 12a^3b + 54a^2b^2 - 108ab^3 + 81b^4.$$

Alternatively, from the abbreviation of the binomial expansion (4) on page 406, we have

$$(a-3b)^4 = \binom{4}{0}a^4 + \binom{4}{1}a^3(-3b) + \binom{4}{2}a^2(-3b)^2 + \binom{4}{3}a(-3b)^3 + \binom{4}{4}(-3b)^4,$$

which also simplifies to the expression obtained above.

39. $(x+3)^5$
40. $(2x+y)^4$
41. $(x-3)^4$
42. $(2x-1)^5$

43. $\left(2x - \dfrac{y}{2}\right)^3$
44. $\left(\dfrac{x}{3} + 3\right)^5$
45. $\left(\dfrac{x}{2} + 2\right)^6$
46. $\left(\dfrac{2}{3} - a^2\right)^4$

Write the first four terms in each expansion. Do not simplify the terms.

Example $(x + 2y)^{15}$

Solution From the binomial expansion (3), we have

$$x^{15} + \frac{15}{1!}x^{14}(2y) + \frac{15\cdot 14}{2!}x^{13}(2y)^2 + \frac{15\cdot 14\cdot 13}{3!}x^{12}(2y)^3,$$

or alternatively from the form (4),

$$\binom{15}{0}x^{15} + \binom{15}{1}x^{14}(2y) + \binom{15}{2}x^{13}(2y)^2 + \binom{15}{3}x^{12}(2y)^3.$$

47. $(x + y)^{20}$ 48. $(x - y)^{15}$ 49. $(a - 2b)^{12}$

50. $(2a - b)^{12}$ 51. $(x - \sqrt{2})^{10}$ 52. $\left(\dfrac{x}{2} + 2\right)^8$

B *Find each power to the nearest hundredth.*

Example $(0.97)^7$

Solution Either form of the binomial expansion (3) or (4) can be used. We shall use (3).

$$(0.97)^7 = (1 - 0.03)^7$$

$$= 1^7 + \frac{7}{1!}(1)^6(-0.03)^1 + \frac{7 \cdot 6}{2!}(1)^5(-0.03)^2 + \frac{7 \cdot 6 \cdot 5}{3!}(1)^4(-0.03)^3 + \cdots$$

$$= 1 - 0.21 + 0.0189 - 0.000945 + \cdots$$

$$= 0.807955^+$$

Hence, to the nearest hundredth, $(0.97)^7 = 0.81$.

53. $(1.02)^{10}$ *Hint*: $1.02 = (1 + 0.02)$ 54. $(1.01)^{15}$

55. If an amount of money (P) is invested at 4% compounded annually, the amount (A) present at the end of (n) years is given by $A = P(1 + 0.04)^n$. Find the amount A (to the nearest dollar) if $1000 was invested for 10 years.

56. In Exercise 55, find the amount present at the end of 20 years.

Find each specified term.

Example $(x - 2y)^{12}$, the seventh term.

Solution In Equation (5), page 407, use $n = 12$ and $r = 7$.

$$\frac{12!}{6!(12 - 6)!} x^6(-2y)^6 = \frac{12 \cdot 11 \cdot 10 \cdot 9 \cdot 8 \cdot 7}{6!} x^6(-2y)^6 = 59{,}136 x^6 y^6$$

57. $(a - b)^{15}$, the sixth term 58. $(x + 2)^{12}$, the fifth term
59. $(x - 2y)^{10}$, the fifth term 60. $(a^3 - b)^9$, the seventh term

61. Given that the binomial formula holds for $(1 + x)^n$ where n is a negative integer:
   a. Write the first four terms of $(1 + x)^{-1}$.
   b. Find the first four terms of the quotient $1/(1 + x)$, by dividing 1 by $(1 + x)$. Compare the results of (a) and (b).

11.5 Mathematical Induction

62. Given that the binomial formula holds as an infinite "sum" for $(1 + x)^n$, where n is a rational number and $|x| < 1$, find to two decimal places.

    a. $\sqrt{1.02}$ **b.** $\sqrt{0.99}$

63. Show that
$$\binom{k}{r} + \binom{k}{r-1} = \binom{k+1}{r}.$$

64. Show that
$$n\binom{n-1}{k} = (n-k)\binom{n}{k}.$$

11.5 Mathematical Induction

The material in the present section depends on a special property of the set of natural numbers, or positive integers, $N = \{1, 2, 3, \ldots\}$:

    a. $1 \in N$
    b.   If $k \in N$, then $k + 1 \in N$.

In addition, N contains no elements not implied by properties a and b. These properties underlie the following theorem, called the **principle of mathematical induction**, which we state without proof.

Theorem 11.5   *If a given sentence involving natural numbers n is true for $n = 1$, and if its truth for $n = k$ implies its truth for $n = k + 1$, then it is true for every natural number n.*

Requirements of a proof by mathematical induction

We can exploit Theorem 11.5 to prove a number of assertions. Although the technique we shall use is called **proof by mathematical induction**, the argument we shall employ is deductive, as have been all of the other arguments in this book. Proofs by mathematical induction require two things:

    a. A demonstration that the assertion to be proved is true for the natural number 1.
    b. A demonstration that the truth of the assertion for a natural number k implies its truth for $k + 1$.

When these two demonstrations have been made, the principle of mathematical induction assures us that the assertion is true for every natural number.

Example Prove that the sum of the first n natural numbers is $\dfrac{n(n+1)}{2}$.

Solution We wish to show that
$$1 + 2 + 3 + \cdots + n = \frac{n(n+1)}{2}.$$

We must do two things:

1. We must first show that the assertion is true for $n = 1$, that is, that
$$1 = \frac{1(1+1)}{2},$$
which is true.

2. We must next show that the truth of the assertion for $n = k$ implies its truth for $n = k + 1$. That is, we must show that the truth of
$$1 + 2 + 3 + \cdots + k = \frac{k(k+1)}{2} \qquad (1)$$
implies the truth of
$$1 + 2 + 3 + \cdots + k + (k+1) = \frac{(k+1)[(k+1)+1]}{2}. \qquad (2)$$

Assuming the truth of (1), we add $k + 1$ to each member of this equation, obtaining
$$1 + 2 + 3 + \cdots + k + (k+1) = \frac{k(k+1)}{2} + (k+1)$$
$$= (k+1)\left(\frac{k}{2} + 1\right)$$
$$= (k+1)\left(\frac{k+2}{2}\right),$$
from which we obtain
$$1 + 2 + 3 + \cdots + k + (k+1) = \frac{(k+1)[(k+1)+1]}{2},$$
which is (2).

Thus the second fact necessary for our proof is established. By the principle of mathematical induction, the assertion is true for every natural number n.

Example Prove that for any arithmetic progression,
$$S_n = \frac{n}{2}[2a + (n-1)d].$$

11.5 Mathematical Induction

Solution

For $n = 1$, the assertion is that

$$S_1 = \frac{1}{2}(2a + 0 \cdot d) = a,$$

which is true. We next must show that

$$S_k = \frac{k}{2}[2a + (k-1)d]$$

implies that

$$S_{k+1} = \frac{k+1}{2}[2a + (k+1-1)d] = \frac{k+1}{2}(2a + kd).$$

Now since S_{k+1} is the sum of the first $k + 1$ terms, S_{k+1} is the sum of the first k terms and the $k + 1$st term. That is,

$$S_{k+1} = S_k + s_{k+1}.$$

Therefore

$$S_{k+1} = \frac{k}{2}[2a + (k-1)d] + a + kd$$

$$= ka + a + \frac{k(k-1)}{2}d + kd$$

$$= (k+1)a + \frac{k(k+1)}{2} \cdot d = \frac{k+1}{2}(2a + kd).$$

Thus, by the principle of mathematical induction, the assertion is true for every natural number n.

This method of proof is often compared to lining up a row of dominoes, with the assumption that whenever one domino is toppled, the one following will topple. One then needs only to topple the first domino ($n = 1$) and the whole row behind it will topple, as indicated in Figure 11.1.

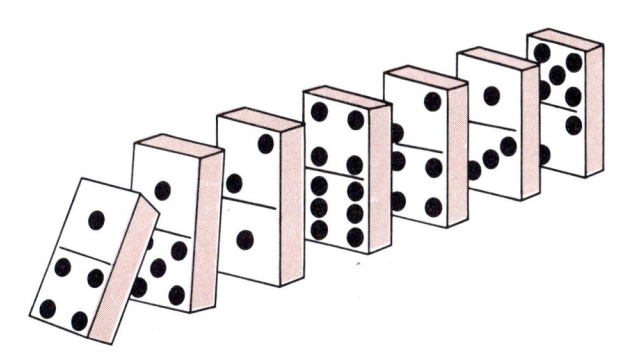

Figure 11.1

Exercise 11.5

A By mathematical induction, prove the validity of the formulas in Exercises 1–14 for all positive integral values of n.

1. $\dfrac{1}{2} + \dfrac{2}{2} + \dfrac{3}{2} + \cdots + \dfrac{n}{2} = \dfrac{n(n+1)}{4}$

2. $1 + 3 + 5 + \cdots + (2n - 1) = n^2$

3. $2 + 4 + 6 + \cdots + 2n = n(n + 1)$

4. $2 + 6 + 10 + \cdots + (4n - 2) = 2n^2$

5. $2 + 8 + 14 + \cdots + (6n - 4) = n(3n - 1)$

6. $1 + 5 + 9 + \cdots + (4n - 3) = n(2n - 1)$

7. $7 + 10 + 13 + \cdots + (3n + 4) = \dfrac{n(3n + 11)}{2}$

8. $6 + 11 + 16 + \cdots + (5n + 1) = \dfrac{n(5n + 7)}{2}$

9. $1^2 + 2^2 + 3^2 + \cdots + n^2 = \dfrac{n(n + 1)(2n + 1)}{6}$

10. $2 + 2^2 + 2^3 + \cdots + 2^n = 2^{n+1} - 2$

11. $1^3 + 3^3 + 5^3 + \cdots + (2n - 1)^3 = n^2(2n^2 - 1)$

12. $\dfrac{1}{1 \cdot 2} + \dfrac{1}{2 \cdot 3} + \dfrac{1}{3 \cdot 4} + \cdots + \dfrac{1}{n(n + 1)} = \dfrac{n}{n + 1}$

13. $1 \cdot 2 + 2 \cdot 3 + 3 \cdot 4 + \cdots + n(n + 1) = \dfrac{n(n + 1)(n + 2)}{3}$

14. $1 \cdot 4 + 2 \cdot 9 + 3 \cdot 16 + \cdots + n(n + 1)^2 = \dfrac{1}{12} n(n + 1)(n + 2)(3n + 5)$

B

15. Show that $n^3 + 2n$ is divisible by 3 for every $n \in N$.

16. Show that $3^{2n} - 1$ is divisible by 8 for every $n \in N$.

17. Show that if $2 + 4 + 6 + \cdots + 2n = n(n + 1) + 2$ is true for $n = k$, then it is true for $n = k + 1$. Is it true for every $n \in N$?

18. Show that $n^3 + 11n = 6(n^2 + 1)$ is true for $n = 1, 2,$ and 3. Is it true for every $n \in N$?

19. Use mathematical induction to prove Theorem 11.1.

20. Use mathematical induction to prove Theorem 11.2.

21. Use mathematical induction to prove that for a geometric progression, $S_n = \dfrac{a - ar^n}{1 - r}$ for all $n \in N$, provided $r \neq 1$.

Chapter Review

[11.1] *Write the first three terms of the sequence with the general term as given.*

1. $s_n = n^2 + 1$
2. $s_n = \dfrac{1}{n+1}$

Write the next three terms of each of the following arithmetic progressions.

3. $7, 10, \ldots$
4. $a, a - 2, \ldots$

Write the next three terms in each of the following geometric progressions.

5. $-2, 6, \ldots$
6. $\dfrac{2}{3}, 1, \ldots$

7. Find the general term and the seventh term of the arithmetic progression $-3, 2, \ldots$.

8. Find the general term and the fifth term of the geometric progression $-2, \dfrac{2}{3}, \ldots$.

9. If the fourth term of an arithmetic progression is 13 and the ninth term is 33, find the seventh term.

10. Which term in a geometric progression $-\dfrac{2}{9}, \dfrac{2}{3}, \ldots$ is 54?

[11.2] 11. Write $\displaystyle\sum_{k=2}^{5} k(k-1)$ in expanded form.

12. Write $x^2 + x^3 + x^4 + \cdots$ in sigma notation.

13. Find the value for $\displaystyle\sum_{j=3}^{9} (3j - 1)$.

14. Find the value for $\displaystyle\sum_{j=1}^{5} \left(\dfrac{1}{3}\right)^j$.

[11.3] 15. Specify the limit of $\dfrac{3n^2 - 1}{n^2}$ as $n \to \infty$.

16. Find the value of the geometric series $4 - 2 + 1 - \dfrac{1}{2} + \cdots$.

17. Find the value of $\displaystyle\sum_{i=1}^{\infty} \left(\dfrac{1}{3}\right)^i$.

18. Find a fraction equivalent to $0.44\overline{4}$.

19. Write $n(n-3)!$ in expanded form for $n = 5$.

[11.4] *Write each of the following expressions in expanded form and simplify.*

20. $\dfrac{8!3!}{7!}$
21. $\dbinom{7}{2}$
22. $\dfrac{(n-1)!}{n!(n+1)!}$

23. Write the first four terms of the binomial expansion of $(x - 2y)^{10}$.
24. Find the eighth term in the expansion of $(x - 2y)^{10}$.

[11.5] *By mathematical induction, prove each formula for all positive integral values of n.*

25. $3 + 6 + 9 + \cdots + 3n = \dfrac{3n(n+1)}{2}$
26. $\dfrac{1}{2} + \dfrac{1}{4} + \dfrac{1}{8} + \cdots + \dfrac{1}{2^n} = 1 - \dfrac{1}{2^n}$

Supplemental Exercises for Chapters 9–11

The problems in this supplement are significantly more difficult than those in the exercise sets at the end of each section and the chapter reviews. They are designed to challenge the student, and a solution may draw on a combination of ideas and techniques from any of the preceding chapters. Many of these problems contain results that are used in more advanced classes in mathematics.

1. Find constants a, b, c, and d so that
$$\frac{3x^3 - x - 2}{x^4 + x^3 + x^2} = \frac{a}{x} + \frac{b}{x^2} + \frac{cx + d}{x^2 + x + 1}.$$

2. Find constants a, b, c, d, and e so that
$$\frac{5x^4 - 19x^3 + 36x^2 - 33x + 13}{(x - 1)(x^2 - 2x + 2)^2} = \frac{a}{x - 1} + \frac{bx + c}{x^2 - 2x + 2} + \frac{dx + e}{(x^2 - 2x + 2)^2}.$$

Solve each system of equations.

3. $\begin{aligned} 2x + 3y + z + w &= 2 \\ x \phantom{{}+3y} + 2z - w &= -4 \\ x + y + 2z + w &= 0 \\ 2x - y + 2z + 2w &= -24 \end{aligned}$

4. $\begin{aligned} x + 3y + z - 2w &= 4 \\ 2x - y + 2z + w &= 10 \\ - y + 2z - 3w &= 10 \\ x \phantom{{}+3y} + 2z + w &= 7 \end{aligned}$

5. $\begin{aligned} \frac{1}{x^2} + \frac{3}{y} + z &= 2 \\ \frac{2}{x^2} - \frac{1}{y} + z &= 7 \\ \frac{2}{x^2} + \frac{1}{y} + 2z &= 9 \end{aligned}$

6. $\begin{aligned} \frac{2}{x^2} - y^2 + 2z &= 0 \\ \frac{4}{x^2} + 4y^2 - 8z &= 3 \\ \frac{1}{x^2} + 2y^2 - 3z &= \frac{5}{2} \end{aligned}$

417

7. $2x + 3y - z = 1$
   $x + 2y - 3z = 1$

8. $-x + 2y = 4$
   $2y - z = 1$

9. Show that the determinant of a 4×4 diagonal matrix is the product of the entries on the principal diagonal. Does this result generalize to $n \times n$ diagonal matrices?

10. We say that a square matrix A is **upper triangular** if $a_{ij} = 0$ for $i > j$. Show that the determinant of a 4×4 upper-triangular matrix is the product of the entries on the principal diagonal. Does this result generalize to $n \times n$ upper-triangular matrices?

11. Show that if A is a 3×3 nonsingular upper-triangular matrix, then A^{-1} is an upper-triangular matrix. Does this result generalize to $n \times n$ matrices?

12. State conditions under which

$$\begin{bmatrix} a_1 & 0 & 0 & 0 \\ 0 & a_2 & 0 & 0 \\ 0 & 0 & a_3 & 0 \\ 0 & 0 & 0 & a_4 \end{bmatrix}$$

is nonsingular, and find the inverse assuming those conditions are satisfied. Does this result generalize to $n \times n$ diagonal matrices?

13. Show that if A and B are 4×4 matrices with $\delta(AB) = \delta(A)\delta(B)$ and $AB = I$, then $BA = I$.

14. Assume $\delta(AB) = \delta(A) \cdot \delta(B)$ and show that if $A_{4 \times 4}$ is nonsingular, then

$$\delta(A_{4 \times 4}^{-1}) = \frac{1}{\delta(A_{4 \times 4})}.$$

Find the limit of each of the following sequences if it exists.

15. $s_n = \dfrac{2n^2 + 1}{3n^2 - 2n + 1}$

16. $s_n = \dfrac{n^{1/2} + 1}{2n^{3/2} + n^{1/2} - 1}$

17. $s_n = \dfrac{2n}{\sqrt{n^2 + 1}}$

18. $s_n = \dfrac{\sqrt{n^2 + 2n + 4}}{n}$

Compute the value of each series. Hint: Use partial fractions to rewrite each term in the sum and then try to find a general expression for each partial sum.

19. $\sum\limits_{n=1}^{\infty} \dfrac{2}{n^2 + n}$

20. $\sum\limits_{n=1}^{\infty} \dfrac{1/3}{n^2 + 7n + 12}$

Using the principle of mathematical induction, prove each of the following.

21. For any natural number n, 3 is a factor of $n(n + 1)(n + 2)$.

22. If I is the $n \times n$ identity matrix, then $\delta(I) = 1$.

23. If A is a square matrix and $X_0 \neq 0$ satisfies $AX = \lambda X$, then X_0 satisfies $A^n X = \lambda^n X$ for every $n \in N$.

24. The binomial theorem (Theorem 11.4).

12 Conic Sections

In previous chapters we introduced some basic notions of analytic geometry that related ordered pairs of real numbers with points in a plane. In this chapter we shall continue our study of analytic geometry with particular emphasis on graphing conic sections.

12.1 Relations Whose Graphs Are Given

Let us begin our discussion by describing a method of finding an equation or inequality for the coordinates of the points on a graph when a geometric description of the graph is given. Any set of points in the plane can be called a *graph* or a **locus** (plural **loci**). The words *graph* and *locus* are ordinarily used, however, to refer to the set of points meeting one or more predesignated conditions. The following examples illustrate ways of finding equations (or inequalities) for given loci.

Example Find an equation of the locus of points at a distance of 7 units from the point $P_1(3, -2)$.

Solution We first make a sketch showing a representative point $P(x, y)$ in the locus. By the distance formula, the distance d from P_1 to P is

$$d = \sqrt{(x - 3)^2 + (y + 2)^2}.$$

Since this distance is given to be 7, we have

$$\sqrt{(x - 3)^2 + (y + 2)^2} = 7.$$

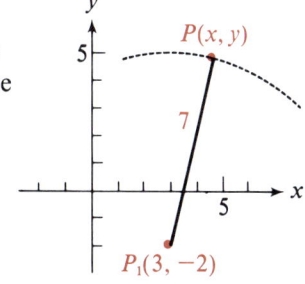

Squaring both members produces
$$(x - 3)^2 + (y + 2)^2 = 49,$$
$$x^2 - 6x + 9 + y^2 + 4y + 4 = 49,$$
or, finally,
$$x^2 + y^2 - 6x + 4y - 36 = 0. \tag{1}$$

Conversely, since the steps are reversible, any point whose coordinates satisfy (1) is on the graph. Hence (1) is an equation of the locus.

In a similar way, on replacing the sign $=$ with $\leq$ in the example above, it could be shown that the set of points in the plane at distance not more than 7 from P_1, is the graph of the relation
$$x^2 + y^2 - 6x + 4y - 36 \leq 0.$$

Example Find an equation of the locus of all points that are twice as far from the point $(3, -2)$ as they are from $(-2, 1)$.

Solution We first make a sketch. We then note that, by the distance formula, any point $P(x, y)$ in the plane located at a distance d_1 from $(-2, 1)$ has coordinates that satisfy
$$d_1 = \sqrt{(x + 2)^2 + (y - 1)^2},$$
and any point $P(x, y)$ at a distance d_2 from $(3, -2)$ has coordinates that satisfy
$$d_2 = \sqrt{(x - 3)^2 + (y + 2)^2}.$$

We now observe that, in order for $P(x, y)$ to lie in the described locus, it is necessary and sufficient that
$$d_2 = 2d_1,$$
or
$$\sqrt{(x - 3)^2 + (y + 2)^2} = 2\sqrt{(x + 2)^2 + (y - 1)^2}.$$

Then, equating squares of both members, we have
$$(x - 3)^2 + (y + 2)^2 = 4[(x + 2)^2 + (y - 1)^2],$$
$$x^2 - 6x + 9 + y^2 + 4y + 4 = 4x^2 + 16x + 16 + 4y^2 - 8y + 4,$$
from which we obtain the simplified equation
$$3x^2 + 3y^2 + 22x - 12y + 7 = 0. \tag{2}$$

Again the steps are reversible, so that (2) is an equation of the given locus.

Loci as paths Another way of considering a locus intuitively is to think of it as a path that has been traced in the plane by a moving point (or particle).

12.1 Relations Whose Graphs Are Given

Example Find an equation of the locus of a point that moves in the plane so that it is always at the same distance from the line with equation $x = 3$ as it is from the point $(8, 0)$.

Solution We first make a sketch as shown. We let $P(x, y)$ be any point on the locus and let P_1 be the point with coordinates $(3, y)$. Since segment $P_1 P$ is parallel to the x-axis, the distance d_1 from $P_1(3, y)$ to $P(x, y)$ is given by

$$d_1 = |x - 3|,$$

while d_2, the distance from $P_2(8, 0)$ to $P(x, y)$, is given by

$$d_2 = \sqrt{(x - 8)^2 + (y - 0)^2}.$$

The condition imposed on $P(x, y)$ is that

$$d_1 = d_2.$$

Thus we have

$$|x - 3| = \sqrt{(x - 8)^2 + (y - 0)^2}.$$

Squaring each member, we obtain

$$(x - 3)^2 = (x - 8)^2 + y^2,$$
$$x^2 - 6x + 9 = x^2 - 16x + 64 + y^2,$$

from which we have the simplified equation

$$y^2 - 10x + 55 = 0. \tag{3}$$

Once more, the steps are reversible, so that (3) is an equation of the given locus.

Exercise 12.1

A Find an equation of the locus of a point that moves in the plane in such a way that it meets the given condition.

1. It is a distance of 5 units from the point $(4, 2)$.
2. It is a distance of 3 units from the point $(-1, 3)$.
3. It is equidistant from $(3, 2)$ and $(5, 0)$.
4. It is equidistant from $(-3, -4)$ and $(1, 2)$.
5. Its distance from $(1, -4)$ is twice its distance from $(6, -1)$.
6. Its distance from $(5, 3)$ is three times its distance from $(5, 9)$.
7. Its distance from the line with equation $x = -2$ is equal to its distance from $(2, 0)$.
8. Its distance from the line with equation $y = 4$ is equal to its distance from the origin.
9. It is equidistant from the x-axis and the point $(-3, -2)$.

10. It is equidistant from the y-axis and the point (4, 3).
11. The sum of its distances from (2, 0) and (−2, 0) is 8.
12. The sum of its distances from (0, 4) and (0, −4) is 12.
13. The difference of its distances from (2, 0) and (−2, 0) is 2.
14. The difference of its distances from (0, 4) and (0, −4) is 6.
15. Its distance from (0, p) equals its distance from the x-axis.
16. Its distance from (p, 0) equals its distance from the y-axis.
17. Find an inequality specifying the set of points at distance greater than or equal to 3 from the point (−4, 1).
18. Find a continued inequality (of the form $a < b < c$) specifying the set of points at distance from (3, −4) that is greater than 5 but less than 7.

B 19. Show that the perpendicular distance d from the point $P(x_0, y_0)$ to the line $y = kx$, $k \neq 0$, is given by

$$d = \frac{|y_0 - kx_0|}{\sqrt{k^2 + 1}}.$$

Hint: Find an equation of the line perpendicular to the given line through the point P and then use the two equations to find the point P' of intersection of the two lines.

20. Find an equation of the locus of a point that moves in such a way that it is equidistant from the line $y = 4x$ and the point (5, 1). *Hint*: Use the result of Exercise 19.
21. Find an equation of the locus of a point that moves in such a way that its distance to the line $y = 2x$ is always the same as the sum of its distance to (1, 3) and its distance to (−1, 1).
22. Find a formula for the perpendicular distance from $P(x_0, y_0)$ to the line $Ax + By = 0$, where $A^2 + B^2 \neq 0$.

12.2 Parabolas

Here and in Sections 12.3 and 12.4 we shall use the methods developed in Section 12.1 to derive equations of special curves that occur frequently in applications. The first of these curves is the *parabola*.

Definition 12.1 *A **parabola** is the set of points in the plane that are equidistant from a given point F and a given line l in the plane.*

The given point F is called the **focus** of the parabola, and the given line l is called the **directrix** of the parabola.

12.2 Parabolas

Equation of a parabola

To derive an equation of a parabola, let us make some simplifying assumptions about its location in the plane. First, let us assume that the focus $F(0, p)$ is on the positive y-axis and that the directrix is the line $y = -p$ (see Figure 12.1). By the distance formula, any point $P(x, y)$ that lies at a distance d_1 from F has coordinates that satisfy

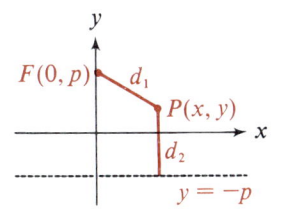

$$d_1 = \sqrt{(x - 0)^2 + (y - p)^2}.$$

Figure 12.1

Also, any point $P(x, y)$ that lies at a distance d_2 from the line $y = -p$ has coordinates that satisfy

$$d_2 = |y + p|.$$

In order for a point P to lie on the parabola, by Definition 12.1 it must satisfy

$$d_1 = d_2.$$

Thus we have

$$\sqrt{(x - 0)^2 + (y - p)^2} = |y + p|.$$

Squaring both sides of this equation, we obtain

$$x^2 + (y - p)^2 = |y + p|^2.$$

This is equivalent to

$$x^2 + y^2 - 2py + p^2 = y^2 + 2py + p^2,$$

which upon simplification yields

$$x^2 = 4py. \quad (1)$$

These steps are reversible, and thus any point whose coordinates satisfy Equation (1) lies on the parabola. Hence Equation (1) is an equation of the parabola. The graph of Equation (1) is shown in Figure 12.2.

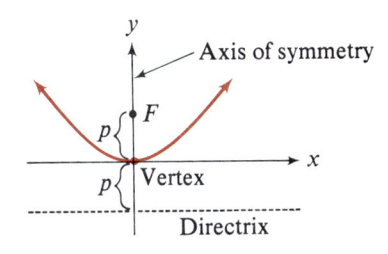

Figure 12.2

Axis of symmetry, vertex

The line that passes through the focus and is perpendicular to the directrix is called the **axis of symmetry** of the parabola, and the midpoint of the segment of the axis of symmetry joining the focus and the directrix is called the **vertex** of the parabola.

Example

Find an equation of the parabola with focus $F(0, 4)$ and directrix $y = -4$.

Solution

Since the focus is on the positive y-axis and the directrix is parallel to the x-axis, we can use Equation (1) with $p = 4$. Thus the desired equation is

$$x^2 = 4(4)y$$
$$= 16y.$$

Standard position

The simplifying assumptions that were made in the above derivations were the following:

1. The focus of the parabola is on the positive y-axis.
2. The directrix of the parabola is parallel to the x-axis.
3. The focus and directrix are on opposite sides of the x-axis and equidistant from it.

Similar assumptions are made for three other cases. The equations and the corresponding graphs are shown in Figure 12.3. In each case, assume $p > 0$. The derivations are similar to the one above and are left as exercises.

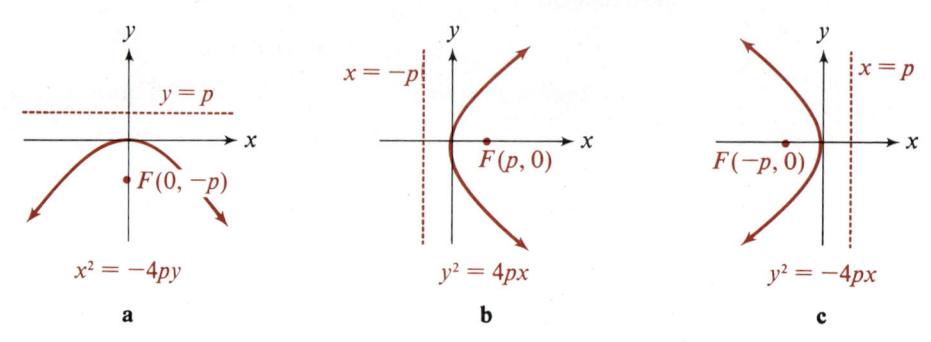

$x^2 = -4py$
a

$y^2 = 4px$
b

$y^2 = -4px$
c

Figure 12.3

The parabolas in Figures 12.2 and 12.3 are in **standard position**, since in each case the vertex is at the origin and the axis of symmetry coincides with one of the coordinate axes. Equation (1) and the equations in Figure 12.3 are the **standard forms** of the equation of a parabola. Note that Equation (1) and the equation in Figure 12.3-a define quadratic functions. These functions were considered in Section 2.6.

The parabola is one of a group of curves called conic sections. **Conic sections** result from the intersection of a plane with a right circular cone. In the case of the parabola, the plane is parallel to one of the elements of the cone, as illustrated by Figure 12.4.

Parabola

Figure 12.4

Exercise 12.2

A *Find an equation of the parabola in standard position satisfying the given condition or conditions and sketch the graph. Let F and d denote the focus and the directrix, respectively.*

12.2 Parabolas

Example $F(1, 0)$

Solution   Because the focus is on the positive x-axis, the equation is of the form (see Figure 12.3-b)

$$y^2 = 4px.$$

Since the focus is $(1, 0)$, we have $p = 1$. Thus the equation is

$$y^2 = 4(1)x$$
$$= 4x.$$

The graph is shown in the figure.

1. $F(0, 4)$
2. $d: y = 3$
3. $d: x = 2$
4. $F(4, 0)$
5. $F(2, 0)$
6. $d: y = -1$
7. The directrix is parallel to the x-axis, the vertex is at the origin, and the parabola contains the point $(2, 4)$.
8. The directrix is parallel to the y-axis, the vertex is at the origin, and the parabola contains the point $(1, 3)$.
9. The directrix is parallel to the y-axis, the vertex is at the origin, and the parabola contains the point $(-1, 1)$.
10. The directrix is parallel to the x-axis, the vertex is at the origin, and the parabola contains the point $(2, 8)$.

Find the focus and the directrix of the given parabola.

Example $y = x^2$

Solution   This equation is of the form $x^2 = 4py$. Thus the focus is on the positive y-axis and the directrix is $y = -p$. Since $4p = 1$, we have $p = 1/4$ and the focus is $(0, 1/4)$. Hence the directrix is $y = -1/4$.

11. $x^2 = 2y$
12. $y^2 = 4x$
13. $y^2 = -16x$
14. $x^2 = -18y$
15. $y^2 = 12x$
16. $4x^2 = 3y$
17. $y^2 + 4x = 0$
18. $4y^2 - 2x = 0$
19. $2y^2 - 3x = 0$
20. $2x^2 = 8y$

B   *Derive an equation of the parabola satisfying the given condition.*

21. Vertex $(0, 0)$, focus $(0, -p)$ $(p > 0)$
22. Vertex $(0, 0)$, focus $(p, 0)$ $(p > 0)$
23. Vertex $(0, 0)$, focus $(-p, 0)$ $(p > 0)$

24. The **latus rectum** of a parabola is the line segment perpendicular to the axis of symmetry, passing through the focus, and with endpoints on the parabola. Show that the length of the latus rectum of any parabola in standard position is 4*p*.

12.3 Circles and Ellipses

Like the parabola, the *circle* and the *ellipse* are conic sections. A circle is the intersection of a right circular cone with a plane perpendicular to the axis of the cone (see Figure 12.5-a), and an ellipse is the intersection of a right circular cone with a plane that is not perpendicular to the axis of the cone but that intersects only one nappe of the cone (see Figure 12.5-b).

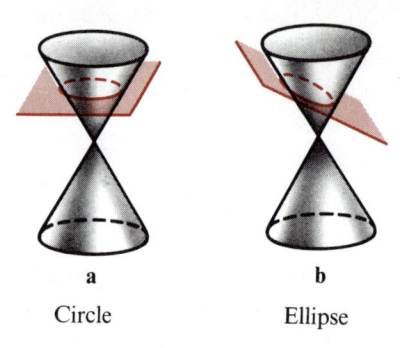

a
Circle

b
Ellipse

Figure 12.5

Circle

Recall from elementary geometry the following definition of a circle.

Definition 12.2 *A circle is the set of all points in the plane at a given distance from a fixed point C in the plane.*

The fixed point C in the above definition is called the **center** of the circle, and the common distance of all points on the circle from the center is called the **radius** of the circle.

To derive an equation of a circle with center at $(0, 0)$, we employ the techniques we used in Section 12.1. If r is the radius, then by the distance formula in order for a point $P(x, y)$ to lie on the circle (see Figure 12.6), we must have

$$\sqrt{x^2 + y^2} = r. \tag{1}$$

Squaring both sides of Equation (1) yields

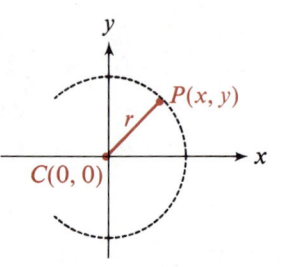

$$x^2 + y^2 = r^2. \tag{2}$$

Figure 12.6

12.3 Circles and Ellipses

These steps are reversible. Thus any point whose coordinates satisfy Equation (2) lie on the circle. Equation (2) is the **standard form** of the equation of a circle with center at the origin and radius r.

Example Find an equation of the circle with center $(0, 0)$ and passing through the point $(1, 4)$.

Solution Since we know that the center of the circle is the origin, we need only compute the radius in order to use Equation (2). The radius is the distance from the center to any point on the circle. Thus we have

$$r = \sqrt{(1 - 0)^2 + (4 - 0)^2} = \sqrt{17}.$$

Hence the desired equation is

$$x^2 + y^2 = 17.$$

Ellipse The ellipse is defined as follows.

Definition 12.3 An **ellipse** is the set of all points in the plane the sum of whose distances to two fixed points F_1 and F_2 in the plane is constant.

The points F_1 and F_2 are called the **foci** of the ellipse, and the midpoint of the line segment joining the foci is called the **center** of the ellipse.

Equation of an ellipse In order to simplify the derivation of the equation of the ellipse, we shall make the following assumptions:

1. The center of the ellipse is at the origin.
2. The line passing through the foci is the x-axis.
3. The foci are the points $F_1(-c, 0)$ and $F_2(c, 0)$, $c > 0$, and the constant sum of the distances from a point on the ellipse to the foci is $2a$, where $a > 0$.

In order for a point $P(x, y)$ to be on the ellipse, it must satisfy

$$d_1 + d_2 = 2a, \quad (3)$$

where d_1 and d_2 are the distances from P to F_1 and F_2, respectively (see Figure 12.7). Thus

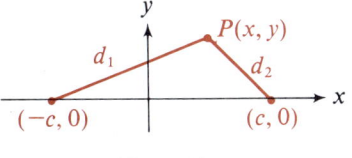

$$\sqrt{(x - c)^2 + (y - 0)^2} + \sqrt{(x + c)^2 + (y - 0)^2} = 2a. \quad (4)$$

Figure 12.7

Rearranging terms in Equation (4), we obtain

$$\sqrt{(x - c)^2 + y^2} = 2a - \sqrt{(x + c)^2 + y^2}; \quad (5)$$

and squaring both sides of (5), we find

$$(x - c)^2 + y^2 = 4a^2 - 4a\sqrt{(x + c)^2 + y^2} + (x + c)^2 + y^2.$$

Performing the squaring operations and simplifying yields

$$cx + a^2 = a\sqrt{(x + c)^2 + y^2}. \tag{6}$$

Squaring both sides of (6), we obtain

$$c^2x^2 + 2ca^2x + a^4 = a^2[(x + c)^2 + y^2],$$

which upon simplification yields

$$(a^2 - c^2)x^2 + a^2y^2 = a^2(a^2 - c^2). \tag{7}$$

Since $a^2 - c^2 > 0$ (see Exercise 24 on page 430), we can simplify the notation by letting

$$b^2 = a^2 - c^2. \tag{8}$$

Then dividing both sides of Equation (7) by a^2b^2, we obtain

$$\frac{x^2}{a^2} + \frac{y^2}{b^2} = 1. \tag{9}$$

The steps in this derivation are reversible. Thus any point whose coordinates satisfy Equation (9) is on the ellipse. Hence Equation (9) is an equation of the ellipse. The graph of the ellipse is shown in Figure 12.8-a.

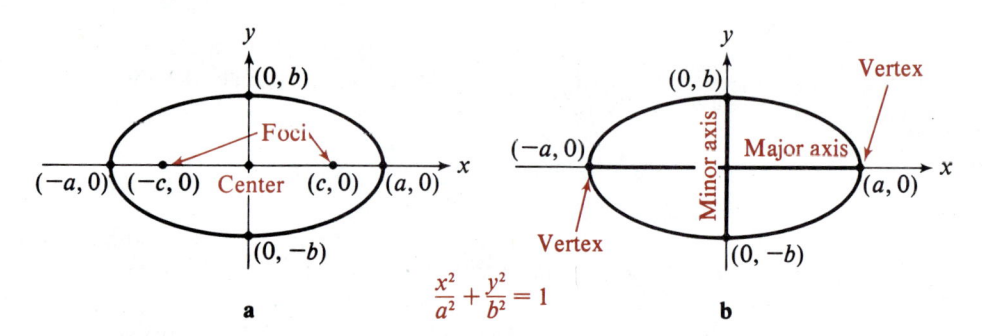

Figure 12.8

Axes and vertices

The line segment that passes through the foci with endpoints on the ellipse is called the **major axis** of the ellipse, and the line segment passing through the center of the ellipse, perpendicular to the major axis and with endpoints on the ellipse, is called the **minor axis**. The endpoints of the major axis are called the **vertices** of the ellipse (see Figure 12.8-b).

In the preceding derivation, it was assumed that the major axis lay along the x-axis. If the major axis lies along the y-axis, the derivation is essentially the same, and the equation takes the form

$$\frac{y^2}{a^2} + \frac{x^2}{b^2} = 1. \tag{10}$$

12.3 Circles and Ellipses

The graph of this equation is given in Figure 12.9.

In the derivations of Equations (9) and (10), we let $b^2 = a^2 - c^2$. Thus we have

$$a^2 = b^2 + c^2,$$

and since $c^2 > 0$, it follows that $a^2 > b^2$.

An ellipse centered at the origin with axes lying along the coordinate axes is said to be in **standard position**, and Equations (9) and (10) are the **standard forms** of the equation of an ellipse in standard position.

$$\frac{y^2}{a^2} + \frac{x^2}{b^2} = 1$$

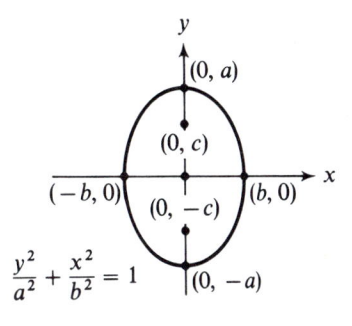

Figure 12.9

Exercise 12.3

A *State whether the graph of the given equation is a circle or an ellipse and give the value of r or the values of a and b.*

Example $\quad x^2 + 3y^2 = 9$

Solution $\quad$ Divide both sides of the equation by 9.

$$\frac{x^2}{9} + \frac{y^2}{3} = 1$$

This is an equation of an ellipse with $a = 3$ and $b = \sqrt{3}$.

1. $2x^2 + 2y^2 = 4$
2. $3x^2 + y^2 = 3$
3. $\dfrac{x^2}{3} + \dfrac{y^2}{2} = 4$
4. $\dfrac{x^2}{2} + \dfrac{y^2}{2} = 1$

Find an equation of the circle or ellipse in standard position satisfying the given condition, and sketch the graph. Assume that F_1, F_2 denote foci in the case of an ellipse and that r denotes the radius in the case of a circle.

Example $\quad$ Ellipse with $F_2(1, 0)$ and $a = 2$.

Solution $\quad$ Since the foci are on the x-axis, the equation of the ellipse is of the form

$$\frac{x^2}{a^2} + \frac{y^2}{b^2} = 1.$$

Since $a = 2$ and $c = 1$,

$$b^2 = a^2 - c^2 = 3.$$

The equation is

$$\frac{x^2}{4} + \frac{y^2}{3} = 1.$$

The graph is shown in the figure.

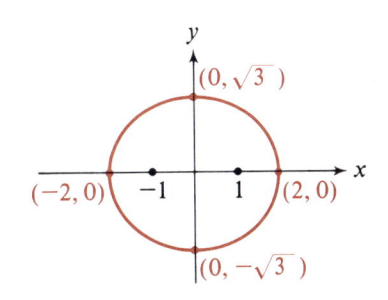

5. Circle centered at the origin with radius 5.
6. Circle centered at the origin and containing the point $(-1, -3)$.
7. Ellipse with $F_1(0, -1)$ and $b = 2$.
8. Ellipse with vertices at $(0, \pm 4)$ and $b = 1$.
9. Circle centered at the origin with radius 4.
10. Circle centered at the origin with radius 1/2.
11. Ellipse with $F_2(0, 2)$ and a vertex at $(0, 4)$.
12. Ellipse with $F_1(-4, 0)$ and $b = 4$.
13. Ellipse with major axis of length 4 along the x-axis and minor axis of length 1.
14. Ellipse with $F_2(3, 0)$ and major axis of length 12.

Find the foci and vertices of the given ellipse.

Example

$$\frac{x^2}{4} + \frac{y^2}{9} = 1$$

Solution

Since $a^2 > b^2$, $a^2 = 9$ and $b^2 = 4$; the equation is of the form

$$\frac{y^2}{a^2} + \frac{x^2}{b^2} = 1,$$

and the major axis lies along the y-axis. Furthermore,

$$c^2 = a^2 - b^2 = 9 - 4 = 5.$$

Thus the foci are $F_1(0, -\sqrt{5})$ and $F_2(0, \sqrt{5})$, and the vertices are $V_1(0, -3)$ and $V_2(0, 3)$.

15. $\dfrac{x^2}{9} + \dfrac{y^2}{4} = 1$
16. $\dfrac{x^2}{16} + \dfrac{y^2}{25} = 1$
17. $\dfrac{x^2}{144} + \dfrac{y^2}{16} = 1$

18. $\dfrac{x^2}{9} + \dfrac{y^2}{16} = 1$
19. $3y^2 + 4x^2 = 12$
20. $2y^2 + x^2 = 1$

B
21. Derive Equation (10) on page 428.
22. Assume that $a = b$ in Equation (10). What figure does the resulting equation represent?
23. What effect does decreasing the value of c (i.e., moving the foci closer to the origin) have on the shape of the ellipse? How does this relate to Exercise 22?
24. Show that $a^2 - c^2$ in Equation (7) is positive.
25. Show that if the major axis of an ellipse in standard position lies along the x-axis, and if $2a$ is the constant sum of the distances from the foci to the points on the ellipse, then the endpoints of the major axis are $(\pm a, 0)$.
26. Show that the endpoints of the minor axis of the ellipse in Exercise 25 are $(0, \pm b)$, where $b^2 = a^2 - c^2$ and the foci are $(\pm c, 0)$.

12.4 Hyperbolas

The figure obtained by intersecting a right circular cone with a plane cutting both nappes, as illustrated in Figure 12.10, is called a *hyperbola* and is defined as follows.

Definition 12.4 *A **hyperbola** is the set of all points in the plane such that the absolute value of the difference of the distances to two fixed points F_1 and F_2 in the plane is constant.*

The two fixed points F_1 and F_2 are called the **foci** of the hyperbola. The midpoint of the line segment joining the foci of a hyperbola is called the **center** of the hyperbola.

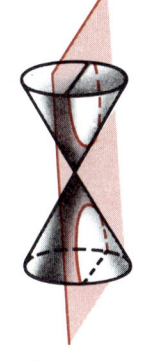

Equation of a hyperbola

In order to simplify the derivation of an equation for a hyperbola, let us make the following assumptions:

Hyperbola

Figure 12.10

1. The center is at the origin.
2. The line segment joining the foci lies along the x-axis.
3. The foci are $F_1(-c, 0)$ and $F_2(c, 0)$, $c > 0$, and the absolute value of the difference of the distances from the foci to the points on the ellipse is $2a$, $a > 0$.

By Definition 12.4, any point $P(x, y)$ on the hyperbola must have coordinates that satisfy

$$|d_2 - d_1| = 2a, \qquad (1)$$

where d_1 and d_2 are the distances from P to F_1 and F_2, respectively (see Figure 12.11). Equation (1) can be rewritten equivalently as

$$d_2 - d_1 = \pm 2a. \qquad (2)$$

Figure 12.11

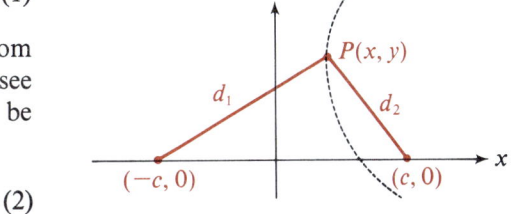

Thus, since

$$d_1 = \sqrt{(x - c)^2 + y^2} \quad \text{and} \quad d_2 = \sqrt{(x + c)^2 + y^2},$$

Equation (2) is equivalent to

$$\sqrt{(x - c)^2 + y^2} = \pm 2a + \sqrt{(x + c)^2 + y^2}. \qquad (3)$$

Squaring both sides of (3), we obtain

$$(x - c)^2 + y^2 = 4a^2 \pm 4a\sqrt{(x + c)^2 + y^2} + (x + c)^2 + y^2,$$

which upon simplification yields

$$cx + a^2 = \pm a\sqrt{(x + c)^2 + y^2}. \qquad (4)$$

Then, squaring both sides of (4), we have

$$c^2x^2 + 2a^2cx + a^4 = a^2[(x + c)^2 + y^2],$$

which is equivalent to

$$(c^2 - a^2)x^2 - a^2y^2 = a^2(c^2 - a^2). \tag{5}$$

Since $c^2 - a^2 > 0$ (see Exercise 17), we let

$$b^2 = c^2 - a^2.$$

Dividing both sides of (5) by a^2b^2 yields

$$\frac{x^2}{a^2} - \frac{y^2}{b^2} = 1. \tag{6}$$

The graph of this equation is shown in Figure 12.12-a.

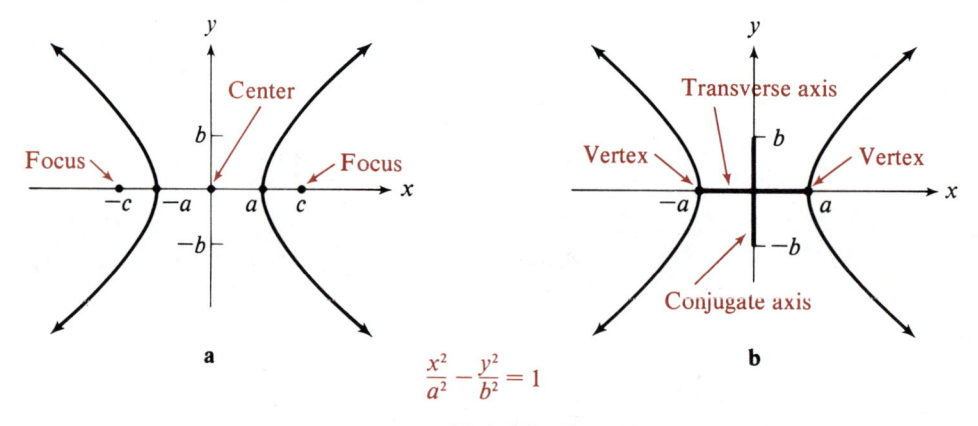

Figure 12.12

Vertices and axes

The points where the hyperbola intersects the line joining the foci are called the **vertices** of the hyperbola, and the line segment joining the vertices is called the **transverse axis**. The line segment of length $2b$ through the center that is bisected by and perpendicular to the transverse axis is called the **conjugate axis** (see Figure 12.12-b).

In the derivation of Equation (6), it was assumed that the transverse axis lay along the x-axis. Precisely the same steps can be used to derive an equation of a hyperbola with transverse axis along the y-axis (see Exercise 18). The resulting equation is

$$\frac{y^2}{a^2} - \frac{x^2}{b^2} = 1. \tag{7}$$

The graph of Equation (7) is given in Figure 12.13.

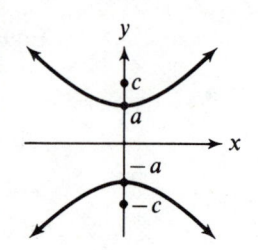

Figure 12.13

12.4 Hyperbolas

A hyperbola centered at the origin with **axes lying along the coordinate axes** is said to be in **standard position**, and Equations (6) and (7) are the **standard forms** of the equation of a hyperbola.

Asymptotes

It was observed in Section 3.5 that some curves approach asymptotes. It can be shown that each hyperbola has two asymptotes, and if the hyperbola is in standard position, the equations of the asymptotes can be obtained by setting the right-hand member of (6) or (7) equal to 0. It follows that

1. $y = \pm \dfrac{b}{a} x$ are asymptotes if Equation (6) is an equation of the hyperbola (see Figure 12.14-a);

2. $y = \pm \dfrac{a}{b} x$ are asymptotes if Equation (7) is an equation of the hyperbola (see Figure 12.14-b).

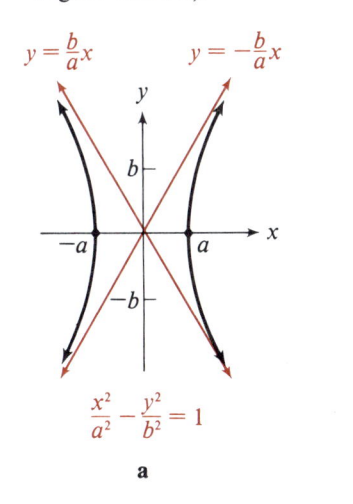

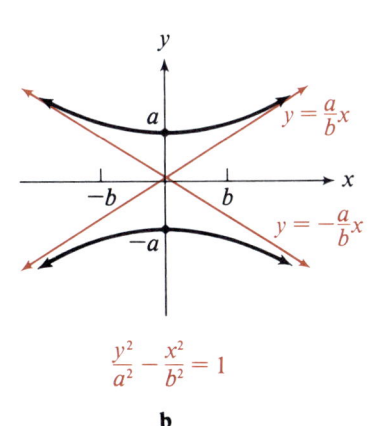

Figure 12.14

Exercise 12.4

A *Find an equation and sketch the graph of the hyperbola in standard position with the given property. The letters F and V stand for focus and vertex, respectively.*

Example $V_1(0, 1), F_1(0, 2)$

Solution   In this case $a = 1$ and $c = 2$. Thus

$$b^2 = c^2 - a^2$$
$$= 4 - 1 = 3.$$

Since the foci lie on the y-axis, the equation is

$$\frac{y^2}{1} - \frac{x^2}{3} = 1,$$

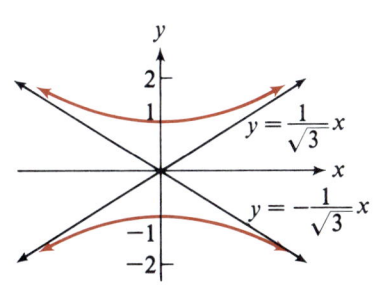

Solution continued overleaf

and the asymptotes are

$$y = \pm \frac{a}{b} x = \pm \frac{1}{\sqrt{3}} x.$$

1. $F_1(0, 3)$, $b = 2$
2. $F_2(4, 0)$, $a = 2$
3. $F_1(0, 4)$, $V_2(0, -2)$
4. $F_1\left(0, \frac{1}{2}\right)$, $V_1\left(0, \frac{1}{4}\right)$
5. $a = 4$, $F_1(5, 0)$
6. $b = 2$, $V_1(1, 0)$
7. $F_2(0, 1)$, asymptotes $y = \pm \frac{1}{2} x$
8. $V_1(6, 0)$, asymptotes $y = \pm x$

Find the foci, vertices, and the equations of the asymptotes of each of the following hyperbolas.

Example $x^2 - y^2 = 1$

Solution This is an equation of the form $x^2/a^2 - y^2/b^2 = 1$, and hence the transverse axis lies along the x-axis (see Figure 12.12-b). Since $a = b = 1$,

$$c^2 = a^2 + b^2 = 2.$$

Thus the foci are $F_1(\sqrt{2}, 0)$ and $F_2(-\sqrt{2}, 0)$, the vertices are $V_1(1, 0)$ and $V_2(-1, 0)$, and the asymptotes are

$$y = \pm \frac{b}{a} x = \pm x.$$

9. $\dfrac{x^2}{9} - \dfrac{y^2}{16} = 1$
10. $\dfrac{y^2}{4} - \dfrac{x^2}{9} = 1$
11. $x^2 - \dfrac{y^2}{2} = 1$
12. $\dfrac{x^2}{4} - \dfrac{y^2}{2} = 1$
13. $4y^2 - 2x^2 = 16$
14. $9x^2 - 4y^2 = 1$
15. $25y^2 - 16x^2 = 1$
16. $25y^2 - 9x^2 = 225$

B 17. Show that $c^2 - a^2 > 0$ in Equation (5) on page 432.

18. Derive Equation (7).

19. Find an equation of the hyperbola with $a = \sqrt{2}$ and with foci $F_1(\sqrt{2}, \sqrt{2})$ and $F_2(-\sqrt{2}, -\sqrt{2})$. Sketch the graph.

20. Give an intuitive argument for the fact that the hyperbola

$$\frac{x^2}{a^2} - \frac{y^2}{b^2} = 1$$

has asymptotes $y = \pm \dfrac{b}{a} x$. *Hint*: Solve for y in terms of x and then factor x^2 from the quantity under the radical.

12.5 More About Graphs of Quadratic Equations

In Sections 12.2, 12.3, and 12.4 we derived equations of the conic sections of the form

$$y = ax^2, \quad a \neq 0,$$

or

$$x = ay^2, \quad a \neq 0,$$

or

$$Ax^2 + By^2 = C, \quad A^2 + B^2 \neq 0.$$

These forms are frequently encountered in mathematics, and you should be able to sketch their graphs quickly. The following summary should be used as an aid.

1. A quadratic equation of the form

$$y = ax^2, \quad a \neq 0,$$

has a graph that is a parabola, opening upward if $a > 0$ and downward if $a < 0$.

Similarly, any equation of the form

$$x = ay^2, \quad a \neq 0,$$

has a graph that is a parabola, opening to the right if $a > 0$ and to the left if $a < 0$.

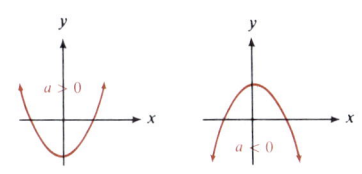

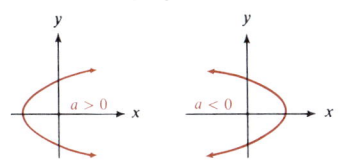

2. A quadratic equation of the form

$$Ax^2 + By^2 = C, \quad A^2 + B^2 \neq 0,$$

has a graph that is

   a. a **circle** if $A = B$ and A, B, and C have like signs;

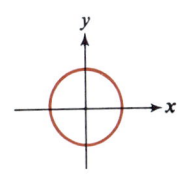

   b. an **ellipse** if $A \neq B$ and A, B, and C have like signs;

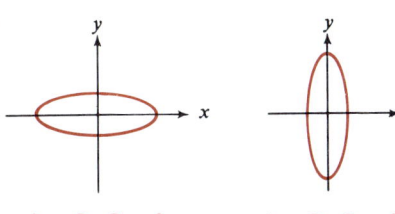

   $A < B, C > 0$ \quad\quad $A > B, C > 0$

c. a **hyperbola** if A and B are opposite in sign and $C \neq 0$;

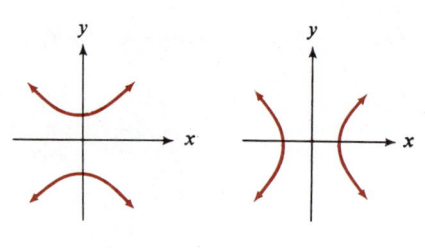

$A < 0, B, C > 0$ $A, C > 0, B < 0$

d. **two distinct lines** through the origin if A and B are opposite in sign and $C = 0$ (see Exercise 25);

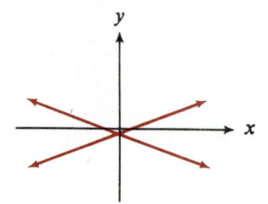

e. **two distinct parallel lines** if one of A and $B = 0$ and the other has the same sign as C;

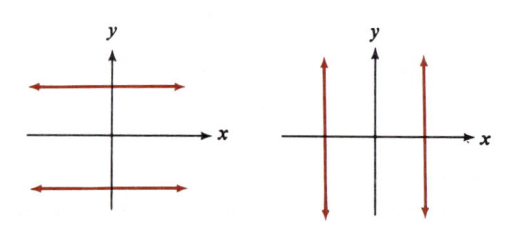

f. **two coincident parallel lines** (one line) through the origin if either $A = 0$ or $B = 0$ but not both and also $C = 0$ (see Exercise 27);

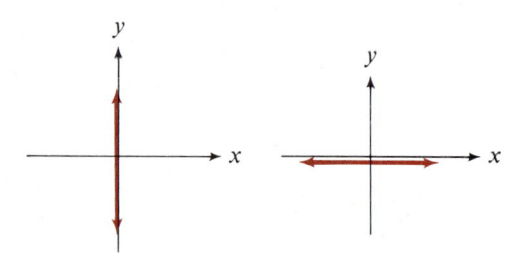

g. **a point** if A and B are both greater than 0 or both less than 0 and $C = 0$;

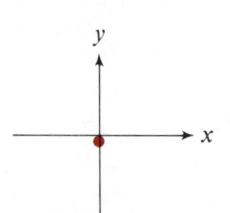

h. the **null set**, $\emptyset$, if A and B are both greater than or equal to 0 and $C < 0$, or if both A and B are less than or equal to 0 and $C > 0$ (see Exercise 28).

Note that we did not consider the special cases 2d to 2h in the preceding sections of this chapter. These cases are called **degenerate cases** of the conic sections.

Review Exercises

Exercise 12.5

A Name and sketch the graph of each of the following relations. Specify the vertices (or intercepts), and for the hyperbolas give equations of the asymptotes.

Example $4x^2 - 36 = -9y^2$

Solution Rewrite the defining equation equivalently in standard form,

$$4x^2 + 9y^2 = 36.$$

By inspection, the graph is an ellipse; x-intercepts are -3 and 3; y-intercepts are -2 and 2. Sketch the graph.

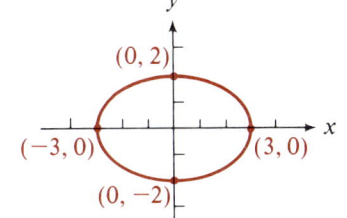

1. $y = 4x^2$
2. $y = 9x^2$
3. $y = -8x^2$
4. $y = -3x^2$
5. $x^2 = 16y$
6. $x^2 = 4y$
7. $x^2 = -y$
8. $x^2 = -4y$
9. $x^2 + y^2 = 49$
10. $x^2 + y^2 = 64$
11. $4x^2 + 25y^2 = 100$
12. $x^2 + 2y^2 = 8$
13. $4x^2 = 4y^2$
14. $x^2 - 9y^2 = 0$
15. $x^2 = 9 + y^2$
16. $x^2 = 2y^2 + 8$
17. $4x^2 + 4y^2 = 1$
18. $9x^2 + 9y^2 = 2$
19. $3x^2 - 12 = -4y^2$
20. $12 - 3y^2 = 4x^2$
21. $4x^2 - y^2 = 0$
22. $4x^2 + y^2 = 0$
23. $y^2 = 0$
24. $x^2 = 4$

B 25. Discuss the graph of an equation of the form $Ax^2 - By^2 = 0$, $A, B > 0$.

26. Discuss the graph of an equation of the form $Ax^2 + By^2 = 0$, $A, B > 0$.

27. Discuss the graph of an equation of the form $Ax^2 = 0$, $A \neq 0$.

28. Explain why the graph of $x^2 + y^2 = -1$ is the null set. Generalize from the result and discuss the graph of any equation of the form

$$Ax^2 + By^2 = C, \quad A^2 + B^2 \neq 0, \quad A, B \geq 0, \quad C < 0.$$

Chapter Review

[12.1] 1. Find an equation of the locus of a point in the plane that moves so that its distance from $(2, -1)$ is twice its distance from $(0, 4)$.

2. Find an equation of the locus of a point in the plane that moves so that its distance from $(3, 4)$ is 5.

3. Find an equation of the locus of a point in the plane that moves so that its distance from the x-axis is 3.

4. Find an equation of the locus of a point in the plane that moves so that the sum of its distances from $(-2, 6)$ and $(0, 4)$ is 8.

[12.2] Find an equation of the parabola in standard position satisfying the given condition, and sketch the graph.

5. Focus $(0, 6)$
6. Directrix $x = -3$
7. Directrix $y = 2$
8. Focus $(-4, 0)$

Find the focus and the directrix of the given parabola.

9. $y^2 = \dfrac{1}{2}x$
10. $x^2 = -y$
11. $x^2 - 20y = 0$
12. $y^2 = -x$

[12.3] Find an equation and sketch the graph of the circle or ellipse in standard position satisfying the given condition.

13. Center $(0, 0)$, $r = 16$
14. Focus $(0, 4)$, $a = 8$
15. Vertex $(9, 0)$, $b = 3$
16. Focus $\left(-\dfrac{1}{2}, 0\right)$, $b = 1$

Find the foci and vertices of the given ellipse.

17. $\dfrac{x^2}{4} + \dfrac{y^2}{16} = 1$
18. $\dfrac{x^2}{8} + y^2 = 1$
19. $\dfrac{x^2}{9} + \dfrac{y^2}{8} = 1$
20. $\dfrac{y^2}{9} + \dfrac{x^2}{4} = 1$

[12.4] Find an equation and sketch the graph of the hyperbola in standard position satisfying the given condition.

21. Focus $(0, 2)$, $b = 1$
22. Focus $(0, 4)$, vertex $(0, 2)$
23. Vertex $(1, 0)$, asymptotes $y = \pm 2x$
24. Focus $(2, 0)$, $a = 1$

Find the foci, vertices, and asymptotes of the given hyperbola.

25. $\dfrac{x^2}{2} - y^2 = 1$
26. $y^2 - x^2 = 1$
27. $\dfrac{y^2}{4} - \dfrac{x^2}{2} = 1$
28. $\dfrac{x^2}{10} - \dfrac{y^2}{4} = 1$

[12.5] *Name and sketch the graph of each of the following equations.*

29. $x^2 = 12y$
30. $x^2 = -4y$
31. $y^2 = 16x$
32. $y^2 = 8x$
33. $\dfrac{x^2}{2} - y^2 = 1$
34. $\dfrac{x^2}{2} + \dfrac{y^2}{4} = 3$
35. $x^2 - y^2 = 0$
36. $y^2 = -x^2$

13 Additional Topics in Analytic Geometry

In Chapter 12 we found the equations of the conic sections in standard position. Each of these equations was a quadratic equation in two variables. In this chapter we shall consider two processes that will allow us to determine the graph of the general quadratic equation in two variables.

We shall also consider several other methods for representing curves in the plane.

13.1   Translation of Axes

The equations of conic sections discussed in Sections 12.2, 12.3, and 12.4 were developed for the conic sections in standard position. By using a process called **translation of axes**, it is possible to discuss conic sections with axes parallel to, but distinct from, the coordinate axes. Figure 13.1 shows a point P in the plane together with two sets of coordinate axes, an xy-system and an $x'y'$-system, whose corresponding axes are parallel. Each set of axes can be viewed as the result of "sliding" the other set across the plane while the axes remain parallel to their original position, suggesting the terminology "translation of axes."

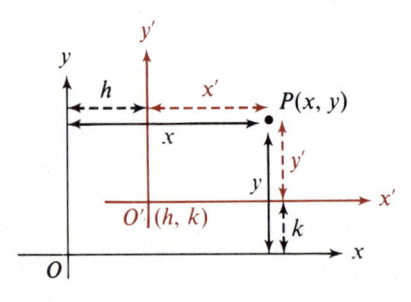

Figure 13.1

If the origin O' of the $x'y'$-system shown in Figure 13.1 corresponds to the point (h, k) of the xy-system, then the $x'y'$-coordinates of the point P in the plane are related to the xy-coordinates of P by the equations

$$x = x' + h,$$
$$y = y' + k. \tag{1}$$

13.1 Translation of Axes

These equations can be solved for x' and y' to obtain the equivalent equations

$$x' = x - h,$$
$$y' = y - k. \qquad (2)$$

Examples

The origin in an xy-system of coordinates is translated into the point $(4, -3)$, which becomes the origin in an $x'y'$-system. Under this translation,

a. what are the $x'y'$-coordinates of the point whose xy-coordinates are $(5, 8)$?
b. what is the $x'y'$-equation for $x + 5y = 3$?

Solutions

a. Using Equations (2) with $h = 4$ and $k = -3$, we have

$$x' = x - 4,$$
$$y' = y + 3,$$

from which, for $(x, y) = (5, 8)$, we obtain

$$x' = 5 - 4 = 1,$$
$$y' = 8 + 3 = 11.$$

Therefore, the $x'y'$-coordinates are $(1, 11)$.

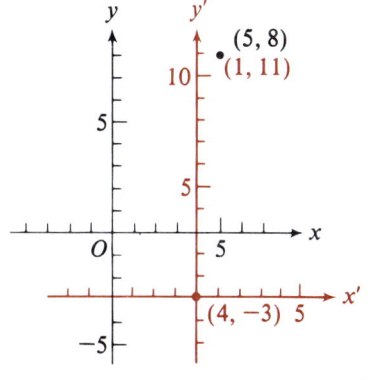

b. Using Equation (1), with $h = 4$ and $k = -3$, we have

$$x = x' + 4, \qquad y = y' - 3.$$

Replacing x and y in $x + 5y = 3$ with the right-hand members of these latter equations, we have

$$(x' + 4) + 5(y' - 3) = 3,$$

from which we obtain the desired equation,

$$x' + 5y' = 14.$$

Translated conic sections

Now consider a parabola with vertex at the point $P(h, k)$ in an xy-coordinate system and with its axis of symmetry parallel to the x-axis, as shown in Figure 13.2. If an $x'y'$-coordinate system had origin at P, then the $x'y'$-equation of the parabola would have the form

$$y'^2 = 4px'.$$

As before, the constant p represents the distance between the focus and the vertex. By Equations (2), the xy-equation for the parabola would then have the form

$$(y - k)^2 = 4p(x - h). \qquad (3)$$

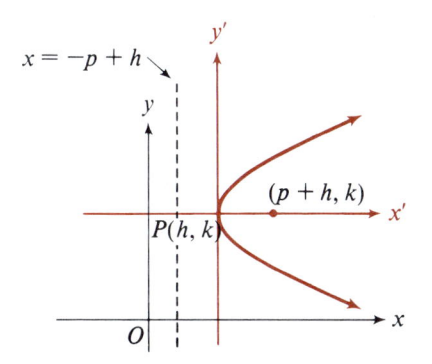

Figure 13.2

Since in Figure 13.2, $p > 0$, the coordinates of the focus must then be $(p + h, k)$, while an equation for the directrix is $x = -p + h$.

Similar considerations lead to the conclusions given in Table 13.1 regarding parabolas, circles, ellipses, and hyperbolas with center (or vertex for parabolas) at

Table 13.1

Curve	xy-equation	Major axis parallel to	Foci	Vertices
Parabola	$(y - k)^2 = 4p(x - h)$ $(y - k)^2 = -4p(x - h)$	x-axis	$(h + p, k)$ $(h - p, k)$	(h, k)
	$(x - h)^2 = 4p(y - k)$ $(x - h)^2 = -4p(y - k)$	y-axis	$(h, k + p)$ $(h, k - p)$	(h, k)
Circle	$(x - h)^2 + (y - k)^2 = r^2$	—	—	—
Ellipse $c^2 = a^2 - b^2$	$\dfrac{(x - h)^2}{a^2} + \dfrac{(y - k)^2}{b^2} = 1$	x-axis	$(h \pm c, k)$	$(h \pm a, k)$
	$\dfrac{(y - k)^2}{a^2} + \dfrac{(x - h)^2}{b^2} = 1$	y-axis	$(h, k \pm c)$	$(h, k \pm a)$
Hyperbola $c^2 = a^2 + b^2$	$\dfrac{(x - h)^2}{a^2} - \dfrac{(y - k)^2}{b^2} = 1$	x-axis	$(h \pm c, k)$	$(h \pm a, k)$
	$\dfrac{(y - k)^2}{a^2} - \dfrac{(x - h)^2}{b^2} = 1$	y-axis	$(h, k \pm c)$	$(h, k \pm a)$

the point (h, k) in the xy-plane. The data in the table can be used to help find an equation when certain information about a graph is given.

Example Find an equation for the ellipse with major axis of length 12 and foci at $(-1, -1)$ and $(-1, 7)$. Sketch the curve.

Solution The equal x-coordinates of the foci tells us that the major axis of the ellipse is parallel to the y-axis. Since the center is the midpoint of the segment with the foci as endpoints, the coordinates of the center are

$$\left(-1, \frac{7 - 1}{2}\right) = (-1, 3),$$

and we see that c, the distance from the center to each focus, is 4. Because the major axis has length 12, it follows that a, the

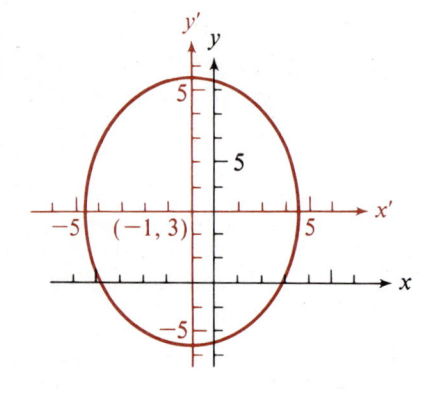

13.1 Translation of Axes

distance from the center to each vertex, is 6. Hence, $a^2 = 36$. From the fact that $b^2 = a^2 - c^2$, we have

$$b^2 = 36 - 16 = 20.$$

The desired equation, then, is

$$\frac{(y-3)^2}{36} + \frac{(x+1)^2}{20} = 1.$$

Using $a = 6$ and $b = 2\sqrt{5}$, and starting with the center at $(-1, 3)$, we sketch the curve as shown on page 442.

We can use the equations of translation (2) together with the process of completing the square to identify and sketch the graph of an equation of the form

$$Ax^2 + By^2 + Dx + Ey + F = 0 \quad (A \neq 0, B \neq 0). \tag{4}$$

We do this by first completing the square in x and y in Equation (4) to obtain an equation of the form

$$A(x-h)^2 + B(y-k)^2 = C.$$

Then using Equations (2) and substituting x' for $x - h$ and y' for $y - k$, we transform this to

$$Ax'^2 + By'^2 = C.$$

The graph of this equation is one of those discussed in Section 12.5. In standard form, we would have

$$\frac{x'^2}{C/A} + \frac{y'^2}{C/B} = 1 \quad (A, B, \text{ and } C \neq 0).$$

If in Equation (4) either $A = 0$ or $B = 0$ but not both, then we can use translation of axes to transform Equation (4) into the standard form of the equation of a parabola.

Example Find an equation in standard form in the $x'y'$-system for the graph of

$$8x^2 + 4y^2 + 24x - 4y + 1 = 0.$$

Solution We first rewrite the given equation equivalently as

$$8(x^2 + 3x \quad) + 4(y^2 - y \quad) = -1$$

and complete the square in x and y to obtain

$$8\left(x^2 + 3x + \frac{9}{4}\right) + 4\left(y^2 - y + \frac{1}{4}\right) = -1 + 18 + 1,$$

$$8\left(x + \frac{3}{2}\right)^2 + 4\left(y - \frac{1}{2}\right)^2 = 18.$$

Solution continued overleaf

If now we take $h = -3/2$ and $k = 1/2$ in Equations (1), we find upon substituting $x' - 3/2$ for x and $y' + 1/2$ for y that this latter equation is equivalent to

$$4x'^2 + 2y'^2 = 9.$$

In standard form, we have

$$\frac{y'^2}{9/2} + \frac{x'^2}{9/4} = 1,$$

the equation of an ellipse, with $a = 3/\sqrt{2}$ and $b = 3/2$.

It is sometimes convenient to use an auxiliary coordinate system as an aid in sketching graphs. For example, to graph

$$8x^2 + 4y^2 + 24x - 4y + 1 = 0$$

in the foregoing example, we can use the $x'y'$-coordinate system with origin at $(-3/2, 1/2)$ in the xy-system and then graph

$$\frac{x'^2}{9/4} + \frac{y'^2}{9/2} = 1.$$

First, we sketch the $x'y'$-axes and identify the x'- and y'-intercepts, $\pm 3/2$ and $\pm 3/\sqrt{2}$, and then we complete the graph as shown in Figure 13.3.

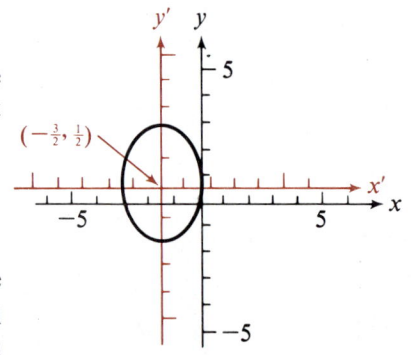

Figure 13.3

Exercise 13.1

A For Exercises 1–8, assume that the origin in an xy-system of coordinates is translated into the first point given, which then becomes the origin of an $x'y'$-system.

1. $(4, 4)$; find the $x'y'$-coordinates of the point whose xy-coordinates are $(-3, 1)$.
2. $(-3, 6)$; find the $x'y'$-coordinates of the point whose xy-coordinates are $(7, 0)$.
3. $(-5, -2)$; find the $x'y'$-coordinates of the point whose xy-coordinates are $(0, -3)$.
4. $(7, 0)$; find the $x'y'$-coordinates of the point whose xy-coordinates are $(0, 7)$.
5. $(3, 4)$; find the $x'y'$-equation for $x - 2y = 8$.
6. $(-1, 6)$; find the $x'y'$-equation for $3x + y = 1$.
7. $(2, -5)$; find the $x'y'$-equation for $x^2 - 2y^2 = 6$.
8. $(-6, -1)$; find the $x'y'$-equation for $x^2 + 2y = -2$.

Find an equation in the xy-system for each curve. Sketch the curve.

9. Circle with center at $(-5, 1)$ and radius 3.
10. Ellipse with major axis of length 10 and foci at $(3, 7)$ and $(3, -1)$.
11. Parabola with vertex at $(3, 3)$ and directrix with equation $y = 1$.
12. Hyperbola with foci at $(-1, 2)$ and $(-7, 2)$ and a vertex at $(-3, 2)$.
13. Ellipse with vertices at $(3, 2)$ and $(-7, 2)$ and length of minor axis 6.
14. Parabola with focus at $(3, 2)$ and vertex at $(3, 4)$.
15. Hyperbola with center at $(3, 2)$, one focus at $(8, 2)$, and one vertex at $(0, 2)$.
16. Hyperbola with center $(-1, 5)$, length of transverse axis 8, and length of conjugate axis 6.

Find an equation in the x'y'-system and in standard form for the graph of the given equation in x and y. Use the auxiliary x'y'-system to graph the equation.

17. $y^2 - 4y = 12x - 52$
18. $x^2 + 4x + 8y = 4$
19. $x^2 - 6x + 3 = -y^2 - y$
20. $x^2 + y^2 = 6x - y + 3$
21. $3x^2 + 2y^2 + 12x - 4y = -2$
22. $16x^2 + 25y^2 - 32x + 100y - 284 = 0$
23. $9x^2 - 16y^2 - 90x + 64y + 17 = 0$
24. $9x^2 - 4y^2 = -61 - 54x - 16y$

B *Find conditions on the coefficients of the quadratic equation*

$$Ax^2 + By^2 + Cx + Dy + F = 0$$

so that the graph of the equation is the given curve.

25. Parabola with the axis of symmetry parallel to the y-axis.
26. Ellipse
27. Circle
28. Hyperbola

13.2 Rotation of Axes

In Section 13.1 we used translation of axes to discuss conic sections with axes that are parallel to the coordinate axes. By using a process called **rotation of axes**, we can similarly discuss conic sections with axes that are not parallel to the coordinate axes.

If we visualize all points on the plane as remaining fixed while we rigidly rotate the x- and y-coordinate axes through an angle α into a new set of x'- and y'-coordinate axes, then we have the situation depicted in Figure 13.4. Using right triangles we can write the coordinates of a point P relative to the xy-coordinate axes as

$$x = r \cos \beta \quad \text{and} \quad y = r \sin \beta$$

where β is $\angle POA$, and relative to the x'y'-coordinate axes as

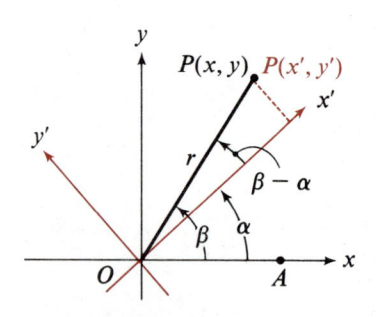

$$x' = r \cos (\beta - \alpha) \quad \text{and} \quad y' = r \sin (\beta - \alpha). \tag{1}$$

Figure 13.4

Using the identities

$$\cos (\beta - \alpha) = \cos \beta \cos \alpha + \sin \beta \sin \alpha$$

and

$$\sin (\beta - \alpha) = \sin \beta \cos \alpha - \cos \beta \sin \alpha$$

(Sections 7.2, 7.3), we can rewrite (1) equivalently as

$$x' = r \cos \beta \cos \alpha + r \sin \beta \sin \alpha$$

and

$$y' = r \sin \beta \cos \alpha - r \cos \beta \sin \alpha.$$

Since $x = r \cos \beta$ and $y = r \sin \beta$, the latter equations become

$$x' = x \cos \alpha + y \sin \alpha \quad \text{and} \quad y' = -x \sin \alpha + y \cos \alpha. \tag{2}$$

If Equations (2) are solved for x and y in terms of x' and y', we then have

$$x = x' \cos \alpha - y' \sin \alpha \quad \text{and} \quad y = x' \sin \alpha + y' \cos \alpha. \tag{3}$$

Equations (2) and (3) relate the xy-coordinates and the x'y'-coordinates of a point in the plane under a rotation of axes through an angle α.

Example

Find the x'y'-coordinates of the point P with xy-coordinates $(5, -2)$ under a rotation of axes through an angle measuring $60°$.

Solution

We have $\alpha = 60°$, so that, from (2) above,

$$x' = 5 \cos 60° + (-2) \sin 60° \quad \text{and} \quad y' = -5 \sin 60° + (-2) \cos 60°.$$

Then, since $\cos 60° = 1/2$ and $\sin 60° = \sqrt{3}/2$, we have

$$x' = 5\left(\frac{1}{2}\right) + (-2)\frac{\sqrt{3}}{2} = \frac{5}{2} - \sqrt{3}$$

13.2 Rotation of Axes

and
$$y' = -5\left(\frac{\sqrt{3}}{2}\right) + (-2)\left(\frac{1}{2}\right) = \frac{-5\sqrt{3}}{2} - 1.$$

Thus, the new coordinates of P are $\left(\dfrac{5 - 2\sqrt{3}}{2}, \dfrac{-5\sqrt{3} - 2}{2}\right)$.

General second-degree equation

Now consider the general second-degree equation in two variables,
$$Ax^2 + Bxy + Cy^2 + Dx + Ey + F = 0. \tag{4}$$

If $B \ne 0$, we can use a rotation of axes to remove the $x'y'$ term from the transformed equation and produce an equivalent one of the form (4) on page 443. Under a rotation through an angle α, we can replace x and y in (4) above with the right-hand members of Equation (3) to obtain

$$A(x' \cos \alpha - y' \sin \alpha)^2 + B(x' \cos \alpha - y' \sin \alpha)(x' \sin \alpha + y' \cos \alpha)$$
$$+ C(x' \sin \alpha + y' \cos \alpha)^2 + D(x' \cos \alpha - y' \sin \alpha)$$
$$+ E(x' \sin \alpha + y' \cos \alpha) + F$$
$$= 0.$$

If this equation is simplified, we arrive at one of the form
$$A'x'^2 + B'x'y' + C'y'^2 + D'x' + E'y' + F' = 0,$$
where
$$A' = A \cos^2 \alpha + B \cos \alpha \sin \alpha + C \sin^2 \alpha, \tag{5}$$
$$B' = 2(C - A) \sin \alpha \cos \alpha + B(\cos^2 \alpha - \sin^2 \alpha),$$
$$C' = A \sin^2 \alpha - B \cos \alpha \sin \alpha + C \cos^2 \alpha, \tag{6}$$
$$D' = D \cos \alpha + E \sin \alpha,$$
$$E' = -D \sin \alpha + E \cos \alpha,$$
$$F' = F.$$

If, now, we recall that
$$\sin 2\alpha = 2 \sin \alpha \cos \alpha \quad \text{and} \quad \cos 2\alpha = \cos^2 \alpha - \sin^2 \alpha$$
(Sections 7.2, 7.3), the equation for B' becomes
$$B' = (C - A) \sin 2\alpha + B \cos 2\alpha.$$

If we wish $B' = 0$, we have two cases to consider. On the one hand, if $C = A$, then
$$B' = B \cos 2\alpha = 0,$$
provided that
$$\cos 2\alpha = 0.$$

This will occur when $\alpha = 45° + k(90°)$, for $k \in J$. On the other hand, if $C \ne A$, then
$$B' = (C - A) \sin 2\alpha + B \cos 2\alpha = 0,$$

provided that
$$(A - C) \sin 2\alpha = B \cos 2\alpha,$$
from which
$$\tan 2\alpha = \frac{B}{A - C}. \qquad (7)$$

Clearly, there are infinitely many angles of rotation we can select, but, in particular, we can restrict the angle of rotation to a positive acute angle.

Example Use a rotation of axes to identify the graph of $3x^2 + 4\sqrt{3}xy - y^2 - 15 = 0$ and sketch the graph.

Solution We first eliminate the $x'y'$ term, using
$$\tan 2\alpha = \frac{B}{A - C} = \frac{4\sqrt{3}}{3 - (-1)} = \frac{4\sqrt{3}}{4} = \sqrt{3}.$$

From Table 6.1, $2\alpha = 60°$, from which $\alpha = 30°$, so that
$$\sin \alpha = \frac{1}{2} \quad \text{and} \quad \cos \alpha = \frac{\sqrt{3}}{2}.$$

From the equations on page 447, we have
$$A' = 3\left(\frac{\sqrt{3}}{2}\right)^2 + 4\sqrt{3}\left(\frac{\sqrt{3}}{2}\right)\left(\frac{1}{2}\right) + (-1)\left(\frac{1}{2}\right)^2 = 5,$$
$$B' = 0 \quad \left(\text{because } \tan 2\alpha = \frac{B}{A - C} = \sqrt{3}\right),$$
$$C' = 3\left(\frac{1}{2}\right)^2 - 4\sqrt{3}\left(\frac{\sqrt{3}}{2}\right)\left(\frac{1}{2}\right) + (-1)\left(\frac{\sqrt{3}}{2}\right)^2 = -3,$$
$$F' = -15.$$

Hence, the new equation is
$$5x'^2 - 3y'^2 = 15,$$
which is the equation of a hyperbola. Its graph is shown at the right.

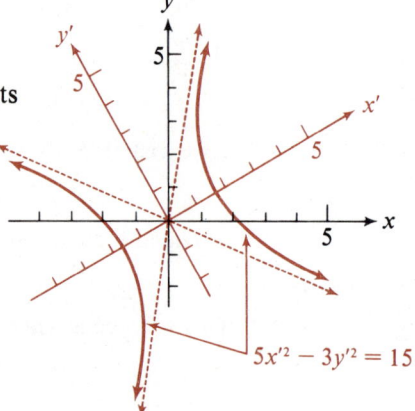

Exercise 13.2

A Find the $x'y'$-coordinates of a point whose coordinates are given under a rotation of axes through an angle with the given measure.

1. $(-3, 5)$; $30°$
2. $(7, -1)$; $45°$
3. $(8, 6)$; $60°$
4. $(-2, -2)$; $90°$
5. $(-6, 8)$; $120°$
6. $(4, -3)$; $135°$
7. $(6, -5)$; $150°$
8. $(-3, 0)$; $225°$

In Exercises 9–18, use a rotation of axes to eliminate the $x'y'$ term; then, when necessary, use a translation to eliminate the first-degree terms from the resulting equation. Identify the graph of the original equation.

9. $3x^2 - 10xy + 3y^2 = -32$
10. $2x^2 + 3xy + 2y^2 = 25$
11. $x^2 + \sqrt{3}xy + 2y^2 = 18$
12. $2x^2 - 2\sqrt{3}xy + 4y^2 = 9$
13. $x^2 + 3xy + 5y^2 = 22$
14. $19x^2 + 6xy + 11y^2 = 20$
15. $3x^2 + 10xy + 3y^2 - 2x - 14y = 15$
16. $3x^2 + 2\sqrt{3}xy + y^2 - 8x + 8\sqrt{3}y = -4$
17. $x^2 + 2xy + y^2 + 2x - 4y + 5 = 0$
18. $x^2 + 4xy + 5y^2 - 8x + 3y + 12 = 0$

B
19. Prove that $A + C$ is invariant (does not change) under a rotation of axes.
20. Prove that $B^2 - 4AC$ is invariant under a rotation of axes.
21. Find an equation in the xy-coordinate system for the parabola with vertex at the origin and with focus having xy-coordinates $(1/\sqrt{2}, 1/\sqrt{2})$. *Hint*: Consider a new coordinate system with x'-axis containing the focus; use Equations (2).
22. Find an equation in the xy-coordinate system for the ellipse centered at the origin, with $a = 4$ and one focus having xy-coordinates $(2, 1)$.
23. Find an equation in the xy-coordinate system of the hyperbola centered at the origin, with $a = \sqrt{2}$ and foci having xy-coordinates $(\sqrt{2}, \sqrt{2})$ and $(-\sqrt{2}, -\sqrt{2})$.
24. Find an equation in the xy-coordinate system of the hyperbola with foci having xy-coordinates $(\sqrt{3}, 1)$ and $(-\sqrt{3}, -1)$, and one vertex having xy-coordinates $(\sqrt{15}/2, \sqrt{5}/2)$.

13.3 Parametric Equations

In this section we shall consider an alternative method for algebraically representing curves in a plane. If the x- and y-coordinates of the points on the curve are each related to a third variable t whose replacement set (an interval of real numbers) is contained in the common domain of the functions $x = g(t)$ and $y = h(t)$, the variable t is called a **parameter** and the equations.

$$x = g(t) \quad \text{and} \quad y = h(t)$$

are called **parametric equations** for the curve.

Graphing The graph of a curve can be obtained from its parametric equations by assigning values to the parameter, plotting the points corresponding to the ordered pairs obtained, and joining them with a smooth curve.

Example Graph the curve

$$x = t \quad \text{and} \quad y = 1 - t, \quad t \in R$$

in an *xy*-coordinate system.

Solution By assigning some values to t (as shown in the table below) and obtaining the corresponding values for x and y and then plotting those points, we obtain the figure shown.

t	$x = t$	$y = 1 - t$	(x, y)
-3	-3	4	$(-3, 4)$
-2	-2	3	$(-2, 3)$
-1	-1	2	$(-1, 2)$
0	0	1	$(0, 1)$
1	1	0	$(1, 0)$
2	2	-1	$(2, -1)$
3	3	-2	$(3, -2)$

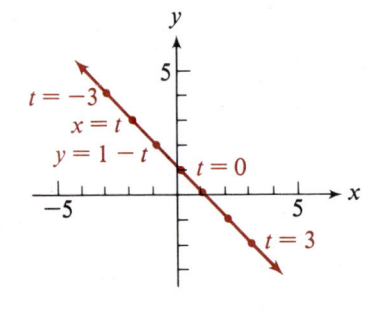

Parametric to Cartesian form Sometimes a Cartesian equation for a locus can be obtained if we are given a set of parametric equations. Thus, if $x = t$ and $y = 1 - t, t \in R$, as in the foregoing example, we can eliminate the parameter by substituting x for t in $y = 1 - t$, to obtain

$$y = 1 - x, \quad \text{or} \quad x + y = 1,$$

the Cartesian form of a linear equation.

When eliminating a parameter from a set of parametric equations, care must be taken to ensure that all restrictions on the variables x and y are noted. For example, for the curve

$$x = \cos t \quad \text{and} \quad y = \cos 2t, \quad t \in [0, \pi]$$

13.3 Parametric Equations

we can eliminate the parameter by using the trigonometric identity

$$\cos 2t = 2\cos^2 t - 1$$

and writing

$$y = \cos 2t = 2\cos^2 t - 1,$$

from which, because $x = \cos t$ and $|\cos t| \le 1$, we have

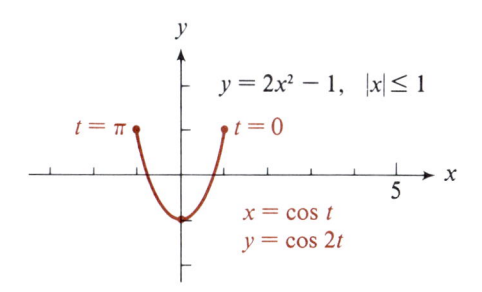

$$y = 2x^2 - 1, \quad |x| \le 1.$$

Figure 13.5

The graph of this relation is the section of the parabola shown in Figure 13.5.

It is sometimes convenient to describe the curve represented by the parametric equations $x = g(t)$ and $y = h(t)$, for t in some interval as a path that has been traced out in the plane by the point $(g(t), h(t))$ as t varies through the interval.

Examples Describe the curve with the given parametric representation.

a. $x = \cos t$ and $y = \sin t$, $t \in [0, 2\pi]$
b. $x = \cos t$ and $y = -\sin t$, $t \in [0, 2\pi]$

Solutions **a.** We observe that the parametric equations

$$x = \cos t \quad \text{and} \quad y = \sin t$$

satisfy the equation $x^2 + y^2 = 1$. Thus these are parametric equations for the circle centered at the origin of radius 1. Further, by plotting several points as in Figure a, we see that as t varies from 0 to 2π, the point $(\cos t, \sin t)$ traverses the circle one time in a counterclockwise direction beginning at the point $(1, 0)$.

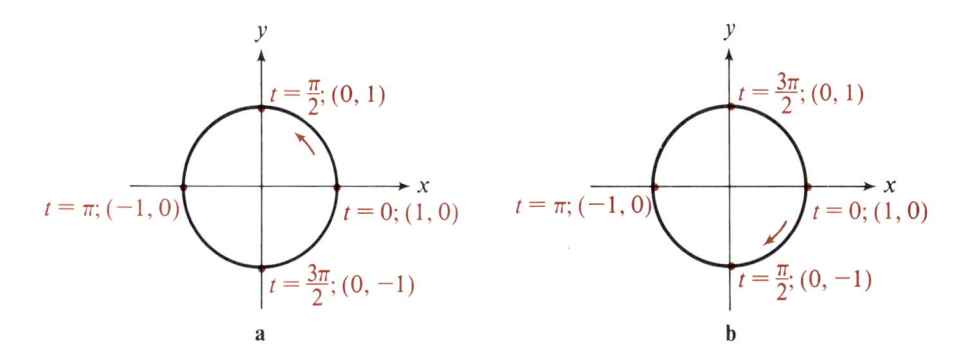

b. As in Example a, the parametric equations satisfy the equation $x^2 + y^2 = 1$. However, plotting several points as in Figure b indicates that in this case the point $(\cos t, -\sin t)$ traverses the circle one time in a clockwise direction beginning at the point $(1, 0)$.

Exercise 13.3

A **a.** Sketch the graph of the given parametric equations.
b. Eliminate the parameter and give a Cartesian equation for the graph.

Example $x = t^2 - 3; \quad y = t + 1, \quad t \in R$

Solution **a.** Make a table of values using appropriate replacements for t.

t	-4	-3	-2	-1	0	1	2	3	4
x	13	6	1	-2	-3	-2	1	6	13
y	-3	-2	-1	0	1	2	3	4	5

Locate the points associated with the ordered pairs (x, y), connect them in order of increasing t, and sketch the graph. (It appears to be a parabola.)

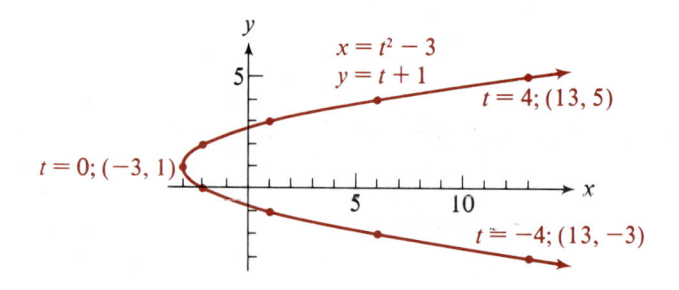

b. Eliminate the parameter t from the original equations by first writing the equation $y = t + 1$ equivalently as

$$t = y - 1$$

and then substituting $y - 1$ for t in the equation $x = t^2 - 3$.

$$x = (y - 1)^2 - 3$$
$$x = y^2 - 2y + 1 - 3$$
$$x = y^2 - 2y - 2$$

The graph is indeed a parabola.

1. $x = 2 - 3t; \quad y = 3t - 1; \quad t \in R$
2. $x = 4t + 6; \quad y = 2t - 3; \quad t \in R$
3. $x = 3t - 2; \quad y = 3 - 2t; \quad t \in R$
4. $x = t + 2; \quad y = 2t - 3; \quad t \in R$
5. $x = t^2; \quad y = 1 - t; \quad t \in R$
6. $x = t + 1; \quad y = t^2 - 1; \quad t \in R$
7. $x = 2 + t^3; \quad y = t^3 + 1; \quad t \in R$
8. $x = 2t^3 + 1; \quad y = 3t^3 - 1; \quad t \in R$

9. $x = \sin t;\quad y = \cos^2 t;\quad t \in [0, \pi/2]$

10. $x = \cos 2t - 1;\quad y = \sin t;\quad t \in [0, \pi/2]$

11. $x = \cos \alpha;\quad y = \sin \alpha;\quad \alpha \in [0, 2\pi]$

12. $x = 2 \cos \alpha;\quad y = 2 \sin \alpha;\quad \alpha \in [-2\pi, 0]$

13. $x = 1 - \cos t;\quad y = 1 + \sin t;\quad t \in [0, 2\pi]$

14. $x = 2 + \cos t;\quad y = \cos 2t;\quad t \in [0, \pi]$

15. $x = 4 \cos \alpha;\quad y = 3 \sin \alpha;\quad \alpha \in [0, 2\pi]$

16. $x = 2 \cos \alpha;\quad y = 5 \sin \alpha;\quad \alpha \in [0, 2\pi]$

Describe the curve with the given parametric representation.

Example $x = 1 - \cos t;\quad y = 1 + \sin t;\quad t \in [0, 4\pi]$

Solution Rewrite the equations as

$$\cos t = 1 - x \quad \text{and} \quad \sin t = y - 1.$$

Square both members of each equation and use the relation $\sin^2 t + \cos^2 t = 1$ to obtain

$$x^2 + y^2 - 2x - 2y + 1 = 0.$$

The curve is a circle of radius 1 centered at the point (1, 1). Plot selected points to determine the direction the curve is traversed. The circle is traversed two times in a counterclockwise direction starting from the point (0, 1).

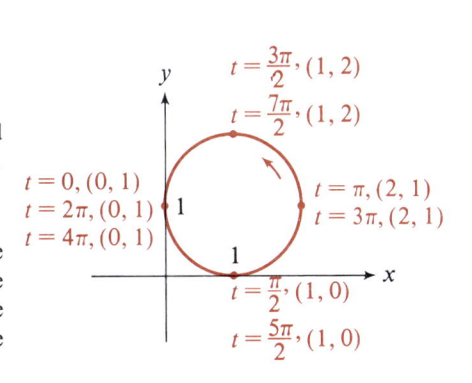

17. $x = \sin t;\quad y = \cos t;\quad t \in [0, 2\pi]$

18. $x = -\sin t;\quad y = -\cos t;\quad t \in [0, 2\pi]$

19. $x = \cos t;\quad y = \cos 2t;\quad t \in [0, 2\pi]$

20. $x = \sin t;\quad y = \cos 2t;\quad t \in [0, 2\pi]$

21. $x = \sin t;\quad y = \cos^2 t;\quad t \in [0, 2\pi]$

22. $x = \cos t;\quad y = \sin^2 t;\quad t \in [0, 2\pi]$

B 23. Show that the equations

$$x = x_1 + at,\quad y = y_1 + bt \quad (a \neq 0)$$

are parametric equations for the line passing through the point (x_1, y_1) with slope b/a.

24. Show that the equations

$$x = (1 - t)x_1 + tx_2,\quad y = (1 - t)y_1 + ty_2$$

are parametric equations for the line passing through the points (x_1, y_1) and (x_2, y_2).

25. Show that the equations

$$x = x_1 + r \cos t,\quad y = y_1 + r \sin t \quad (r \neq 0)$$

are parametric equations for the circle of radius r centered at the point (x_1, y_1).

26. Show that the equations

$$x = kt, \quad y = mt^2 - nt \quad (m \neq 0)$$

are parametric equations for the parabola with x intercepts $(0, 0)$ and $(kn/m, 0)$ and vertex at

$$\left(\frac{kn}{2m}, -\frac{n^2}{4m}\right).$$

Find parametric equations for the given curve.

27. The graph of $x^2 + y^2 = 1$. *Hint*: Use circular functions.

28. The line segment joining the points $(0, 0)$ and $(2, 2)$. *Hint*: Use Exercise 24 and restrict the values of t.

13.4 Polar Coordinates

In many areas of plane geometry, the Cartesian coordinate system is a convenient tool for analytically describing figures in the plane. Sometimes, however, we deal with figures that, due to their particular geometric properties, can be more readily described using alternative coordinate systems. For example, a curve that spirals out from the origin (Figure 13.6) does not lend itself to a simple description using Cartesian coordinates. In this section we shall discuss a different method for coordinatizing the plane.

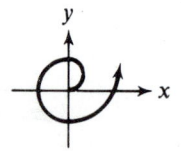

Figure 13.6

In the plane, consider a fixed ray $\overline{PB}$ and any point A. We can describe the location of A by giving the distance r from P to A and specifying the angle BPA (Figure 13.7), which is customarily designated by θ. We ordinarily write the pair (r, θ), where θ is the measure of the angle in either degrees or radians. The components of such an ordered pair are called **polar coordinates** of A. The fixed ray $\overline{PB}$ is called the **polar axis**, and the initial point P of the polar axis is called the **pole** of the system.

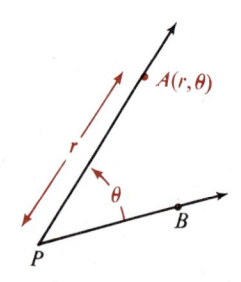

Figure 13.7

Notice that while there is a one-to-one correspondence between the set of ordered pairs in a Cartesian coordinate system and the points in the geometric plane, each point in the plane has infinitely many pairs of polar coordinates. In the first place, if (r, θ) are polar coordinates of A, then for any integer k so is

$$(r, \theta + k \cdot 360°)$$

(see Figure 13.8-a). In the second place, if we let $\overline{PA'}$ denote the ray in a direction opposite that of $\overline{PA}$ (Figure 13.8-b), then we see that $-r \leq 0$ denotes the *directed*

13.4 Polar Coordinates

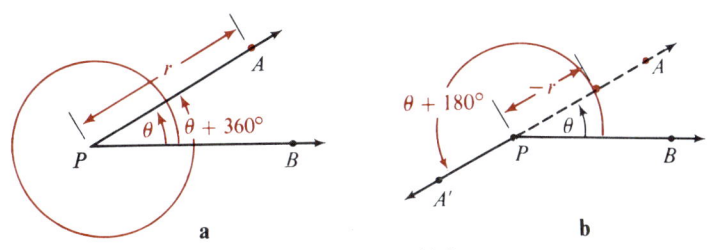

Figure 13.8

distance from P to A along the negative extension of $\overline{PA'}$. Thus we see that also $(-r, \theta + 180°)$, and more generally, for any integer k,

$$(-r, \theta + 180° + k \cdot 360°)$$

are polar coordinates of A. The pole P itself is represented by $(0, \theta)$ for any θ.

Example   Write four additional sets of polar coordinates for the point having coordinates $(3, 30°)$.

Solution   With positive values for r, two more pairs of polar coordinates for $(3, 30°)$ are $(3, 390°)$ and $(3, -330°)$. Using negative values for r, we have $(-3, 210°)$ and $(-3, -150°)$. The figures show these cases. Of course there are infinitely many other possible polar coordinates for the same point.

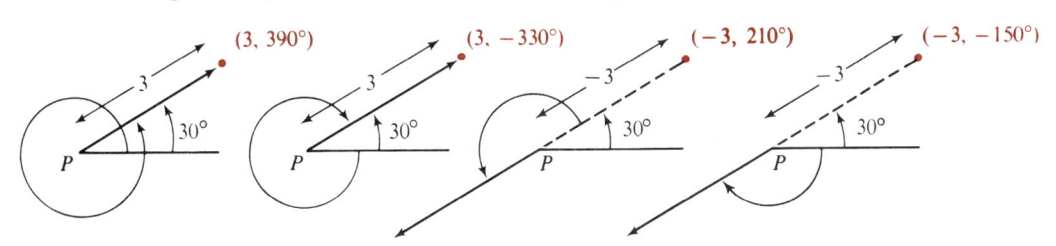

Conversion equations   Cartesian and polar coordinates of a point can be related by means of the trigonometric functions. If the pole P in a polar coordinate system is also the origin O in a Cartesian coordinate system, and if the polar axis coincides with the positive x-axis of the Cartesian system, as in Figure 13.9, then the coordinates (x, y) can be expressed in terms of the polar coordinates (r, θ) by the following equations:

$$x = r \cos \theta \quad \text{and} \quad y = r \sin \theta. \tag{1}$$

Conversely, we have

$$r = \pm \sqrt{x^2 + y^2}, \quad \cos \theta = \frac{x}{\pm\sqrt{x^2 + y^2}},$$

$$\sin \theta = \frac{y}{\pm\sqrt{x^2 + y^2}} \quad [(x, y) \neq (0, 0)]. \tag{2}$$

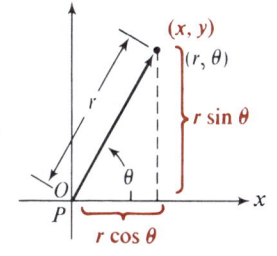

Figure 13.9

The sets of Equations (1) and (2) enable us to find rectangular coordinates for a point with a given pair of polar coordinates, and vice versa. Often, in determining θ, it is simplest first to determine the quadrant from the signs of x and y and then to use the relation

$$\tan \theta = \frac{y}{x}.$$

Example Find the Cartesian coordinates of the point with polar coordinates (4, 30°).

Solution Using Equations (1), on page 455 we obtain

$$x = 4 \cos 30° = 4 \cdot \frac{\sqrt{3}}{2} = 2\sqrt{3},$$

$$y = 4 \sin 30° = 4 \cdot \frac{1}{2} = 2.$$

The rectangular coordinates are $(2\sqrt{3}, 2)$.

Example Find a pair of polar coordinates for the point with Cartesian coordinates $(7, -2)$.

Solution By (2), $r = \pm\sqrt{x^2 + y^2} = \pm\sqrt{49 + 4} = \pm\sqrt{53}$. Choosing the positive sign and noting that the point $(7, -2)$ is in the fourth quadrant, we have

$$\tan \theta = -\frac{2}{7},$$

from which

$$\theta \approx -15.9°.$$

A pair of polar coordinates is therefore $(\sqrt{53}, -15.9°)$, where the given angle measure is an approximation.

Equations (1) on page 455 can be used to transform Cartesian equations to polar form, and Equations (2) can be used to transform polar equations to Cartesian form.

Example Transform $x^2 + y^2 = 4$ into polar form.

Solution By Equations (1) we have

$$x^2 + y^2 = r^2 \cos^2 \theta + r^2 \sin^2 \theta$$
$$= r^2 (\cos^2 \theta + \sin^2 \theta)$$
$$= r^2.$$

Thus the polar form of $x^2 + y^2 = 4$ is

$$r^2 = 4.$$

Example Transform $r \sin \theta + 1 = 0$ into Cartesian form.

Solution By Equations (2) on page 455, we have

$$r \sin \theta + 1 = \pm\sqrt{x^2 + y^2}\left(\frac{y}{\pm\sqrt{x^2 + y^2}}\right) + 1$$
$$= y + 1.$$

Thus the Cartesian form of $r \sin \theta + 1 = 0$ is

$$y + 1 = 0.$$

Exercise 13.4

A *Find four additional sets of polar coordinates* $(-360° < \theta \le 360°)$ *for the point with polar coordinates as given.*

1. $(6, 485°)$
2. $(-3, 518°)$
3. $(-2, -450°)$
4. $(5, 720°)$
5. $(6, -600°)$
6. $(3, -395°)$

Find the Cartesian coordinates of a point with polar coordinates as given.

7. $(5, 45°)$
8. $(-3, 30°)$
9. $(1/2, 330°)$
10. $(3/4, 225°)$
11. $(10, -135°)$
12. $(-6, -240°)$

Find two sets of polar coordinates, one involving an angle of positive measure and one an angle of negative measure, for the point with Cartesian coordinates as given.

13. $(3\sqrt{2}, 3\sqrt{2})$
14. $(-\sqrt{3}/2, 1/2)$
15. $(-1, -\sqrt{3})$
16. $(0, -4)$
17. $(0, 0)$
18. $(-6, 0)$

Transform the given equation into an equation in polar form.

Example $x^2 + y^2 - 2x + 3 = 0$

Solution From Equations (1) on page 455, $x^2 = r^2 \cos^2 \theta$ and $y^2 = r^2 \sin^2 \theta$. Thus

$$r^2 \cos^2 \theta + r^2 \sin^2 \theta - 2r \cos \theta + 3 = 0,$$

from which

$$r^2(\cos^2 \theta + \sin^2 \theta) - 2r \cos \theta + 3 = 0,$$
$$r^2 - 2r \cos \theta + 3 = 0.$$

19. $x^2 + y^2 = 25$ 20. $x = 3$ 21. $y = -4$
22. $x^2 + y^2 - 4y = 0$ 23. $x^2 + 9y^2 = 9$ 24. $x^2 - 4y^2 = 4$

Transform the given equation into an equation in Cartesian form free of radicals.

Example $r(1 - 2\cos\theta) = 3$

Solution   Use Equations (2) on page 455 to obtain

$$\pm\sqrt{x^2 + y^2}\left[1 - 2\left(\frac{x}{\pm\sqrt{x^2+y^2}}\right)\right] = 3,$$

from which

$$\pm\sqrt{x^2 + y^2} - 2x = 3,$$
$$\pm\sqrt{x^2 + y^2} = 2x + 3.$$

Square each member to obtain

$$x^2 + y^2 = 4x^2 + 12x + 9,$$

from which

$$3x^2 - y^2 + 12x + 9 = 0.$$

25. $r = 5$ 26. $r = 4\sin\theta$ 27. $r = 9\cos\theta$
28. $r\cos\theta = 3$ 29. $r(1 - \cos\theta) = 2$ 30. $r(1 + \sin\theta) = 2$

B   31. Show by transformation of coordinates that the graph of $r = \sec^2(\theta/2)$ is a parabola.
32. Show that the graph of $r = \csc^2(\theta/2)$ is a parabola.
33. Show that the graph of $r^2(1 + 2\sin^2\theta) = 1$ is an ellipse.
34. Show that the graph of $r^2(1 - 2\sin^2\theta) = 1$ is a hyperbola.
35. Show that the graph of $\cos 2\theta = 0$ is two intersecting straight lines.
36. Show that the graph of $\sin 2\theta = 0$ is two intersecting straight lines.

13.5   Graphing in Polar Coordinates

The graph of an equation in polar coordinates is the set of all points in the plane having *at least* one pair of polar coordinates that satisfy the given equation.

Plotting points   We can graph an equation in polar form by following the same procedure that was introduced at the beginning of Section 2.6; that is, we find a set of ordered pairs that are solutions of the polar equation, graph these ordered pairs, see if a pattern develops, and then join the plotted points with a smooth curve.

13.5 Graphing in Polar Coordinates

Note that although either degree or radian measure can be used when graphing in polar coordinates, for consistency we shall use radian measure throughout this section.

Example Graph $r = 1 + \cos \theta$.

Solution We first construct a table of values that satisfy the equation.

θ	0	$\frac{\pi^R}{6}$	$\frac{\pi^R}{4}$	$\frac{\pi^R}{3}$	$\frac{\pi^R}{2}$	$\frac{2\pi^R}{3}$	$\frac{3\pi^R}{4}$	$\frac{5\pi^R}{6}$	π^R	$\frac{7\pi^R}{6}$	$\frac{5\pi^R}{4}$	$\frac{4\pi^R}{3}$	$\frac{3\pi^R}{2}$	$\frac{5\pi^R}{3}$	$\frac{7\pi^R}{4}$	$\frac{11\pi^R}{6}$
r	2.0	1.9	1.7	1.5	1.0	0.5	0.3	0.1	0	0.1	0.3	0.5	1.0	1.5	1.7	1.9

We then plot these points on a polar grid and join them with a smooth curve to obtain the figure shown.

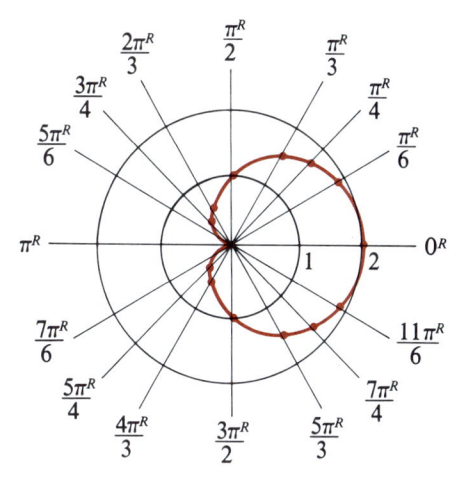

Symmetry The process of plotting points used in the above example can be simplified if the curve is symmetric with respect to either the x-axis, the y-axis, or the pole. For a given curve, we can determine symmetry by using Table 13.2 on page 460 whose entries are determined by the geometric conditions illustrated in Figure 13.10. The curve has

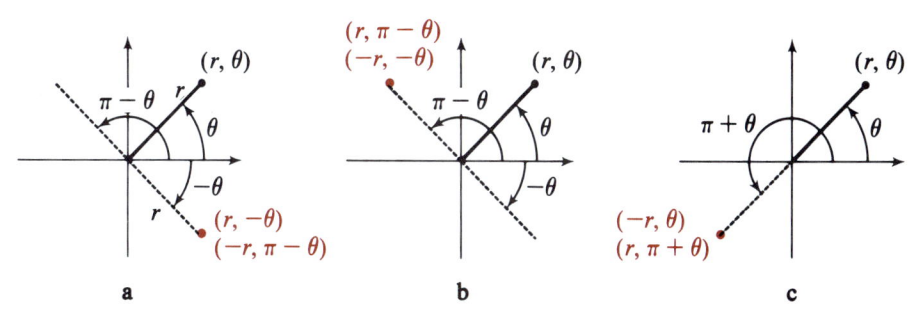

Figure 13.10

Table 13.2

	Symmetry with respect to	Replacement	
a	Polar axis	$(r, -\theta)$	1
	$(\theta = 0)$	$(-r, \pi - \theta)$	2
b	Vertical axis	$(r, \pi - \theta)$	3
	$\left(\theta = \dfrac{\pi}{2}\right)$	$(-r, -\theta)$	4
c	Pole	$(r, \pi + \theta)$	5
	$(r = 0)$	$(-r, \theta)$	6

the symmetry indicated if the corresponding replacement is a solution when (r, θ) is a solution.

Example   Show that the graph of $r = 1 + \cos \theta$ is symmetric with respect to the polar axis.

Solution   If (r, θ) is a solution of $r = 1 + \cos \theta$, then

$$1 + \cos(-\theta) = 1 + \cos \theta = r.$$

Hence $(r, -\theta)$ is also a solution to $r = 1 + \cos \theta$. Thus, by entry 1 of Table 13.2, the graph of $r = 1 + \cos \theta$ is symmetric with respect to the polar axis.

Example   Show that the graph of $r = \cos 2\theta$ is symmetric with respect to the vertical axis.

Solution   If (r, θ) is a solution of $r = \cos 2\theta$, then

$$\cos 2(\pi - \theta) = \cos(2\pi - 2\theta)$$
$$= \cos 2\theta$$
$$= r.$$

Hence $(r, \pi - \theta)$ is also a solution of $r = \cos 2\theta$. Thus, by entry 3 of Table 13.2, the graph of $r = \cos 2\theta$ is symmetric with respect to the vertical axis.

Example   Show that the graph of $r = \cos 2\theta$ is symmetric with respect to the pole.

Solution   If (r, θ) is a solution to $r = \cos 2\theta$, then

$$\cos 2(\pi + \theta) = \cos(2\pi + 2\theta)$$
$$= \cos 2\theta$$
$$= r.$$

13.5 Graphing in Polar Coordinates

Hence $(r, \pi + \theta)$ is also a solution of $r = \cos 2\theta$. Thus, by entry 5 of Table 13.2, the graph of $r = \cos 2\theta$ is symmetric with respect to the pole.

Special curves There are several curves with simple equations in polar coordinates that have special names. The graph of $r = 1 + \cos \theta$ given on page 459 is called a **cardioid** because of its heartlike shape. Some other polar curves are shown in Figure 13.11.

$r^2 = \sin \theta$
(Lemniscate)
a

$r = \cos 2\theta$
(4-petalled rose)
b

$r = \theta, \theta > 0$
(Spiral of Archimedes)
c

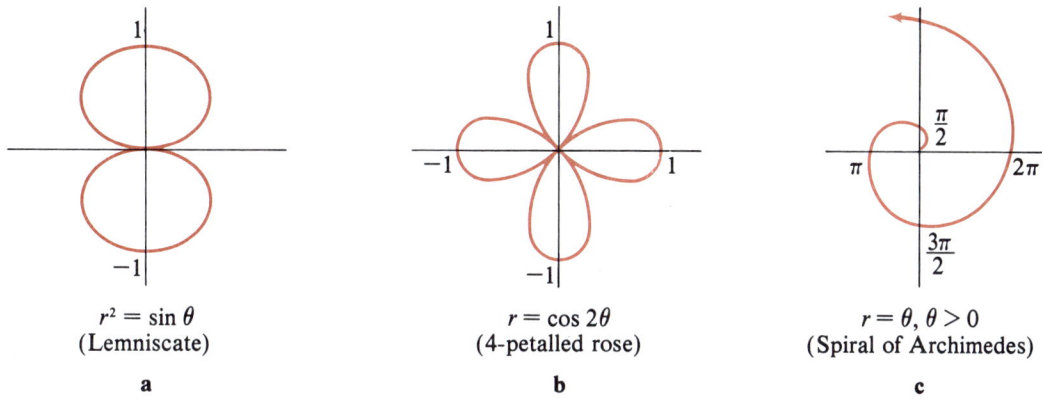

Figure 13.11

Exercise 13.5

A Graph each of the following equations.

Example $r = \cot \theta$

Solution Checking for symmetry, we find that

$$\cot(-\theta) = -\cot \theta = -r.$$

Thus $(-r, -\theta)$ is a solution when (r, θ) is a solution, and by entry 4 of Table 13.2 the curve is symmetric with respect to the vertical axis. Further, since the cotangent has period π, we have

$$\cot \theta = \cot(\theta + \pi),$$

and thus if (r, θ) is a solution of $r = \cot \theta$, so is $(r, \theta + \pi)$. Thus by entry 5 of Table 13.2, the curve is symmetric with respect to the pole. Due to these symmetries, we need only plot the curve in the first quadrant. We then use the symmetries to finish the sketch.

θ	0^R	$\dfrac{\pi^R}{6}$	$\dfrac{\pi^R}{3}$	$\dfrac{\pi^R}{2}$
r	Und.	1.7	0.6	0

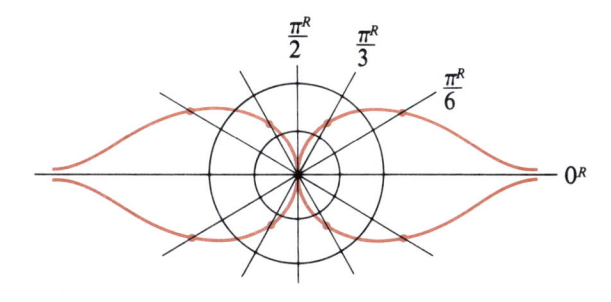

1. $r = 4 \sin \theta$
2. $r = 9 \cos \theta$
3. $r = 1 + \cos \theta$
4. $r = 1 - \sin \theta$
5. $r \cos \theta = 3$
6. $r \sin \theta = -3$
7. $r = 4 \sin 2\theta$
8. $r = 4 \cos 2\theta$
9. $r = \tan \theta$
10. $r^2 = \cos \theta$
11. $r = \cos 3\theta$
12. $r = \sin 3\theta$
13. $r = 2 \cos \dfrac{\theta}{2}$
14. $r(2 - \cos \theta) = 4$
15. $r = \sin^2 \theta$
16. $r = \sec \theta$
17. $r = \cos^2 \dfrac{\theta}{2}$
18. $r^2 = \tan \theta$
19. $r = 3$ Hint: All solutions are of the form $(3, \theta)$.
20. $\theta = \dfrac{\pi^R}{6}$ Hint: All solutions are of the form $\left(r, \dfrac{\pi^R}{6}\right)$.

B Graph each of the following equations.

21. $r = \theta, \quad \theta < 0$
22. $r = \theta, \quad \theta > 0$
23. $r = \theta + \dfrac{\pi}{2}, \quad \theta < 0$
24. $r = \theta + \dfrac{\pi}{2}, \quad \theta > 0$

Chapter Review

[13.1]   For Exercises 1 and 2, assume that the origin in an xy-system of coordinates is translated into the first point given, which then becomes the origin of an $x'y'$-system.

1. $(2, -1)$; find the $x'y'$-coordinates of the point whose xy-coordinates are $(-5, 3)$.
2. $(-3, 0)$; find the $x'y'$-equation for $2x^2 - y = 1$.

Find an equation in the xy-system for each curve. Sketch the curve.

3. Circle with center at $(1, -3)$ and radius 4.
4. Parabola with focus at $(5, 0)$ and vertex at $(7, 0)$.

Find an equation in standard form for the graph of the given equation in x and y.

5. $x^2 - 2x = 6y + 14$
6. $2x^2 + y^2 - x + 4y - 5 = 0$
7. Use an auxiliary system to graph the equation in Exercise 5.
8. Use an auxiliary system to graph the equation in Exercise 6.

Review Exercises

[13.2] *Find the $x'y'$-coordinates of a point whose coordinates are given under a rotation of axes through an angle with the given measure.*

9. $(2, -1)$; $210°$
10. $(0, 4)$; $315°$

Use a rotation of axes to eliminate the $x'y'$ term; then, when necessary, use a translation to eliminate the first-degree terms from the resulting equation. Identify the graph of the equation.

11. $x^2 + 4xy + y^2 = 20$
12. $5x^2 - \sqrt{3}xy + 4y^2 - \sqrt{3}x + y = 7$

[13.3] **a.** *Sketch the graph of the given parametric equations.*
b. *Eliminate the parameter and give a Cartesian equation for the graph.*

13. $x = 2 + t$; $y = 3t^2$; $t \in R$
14. $x = 2 \sin t$; $y = 3 \cos t$; $t \in [0, 2\pi]$

Describe how the curve with the given parametric representation is traversed.

15. $x = 2 \cos t$; $y = 2 \sin t$; $t \in [0, \pi]$
16. $x = 2 \cos t$; $y = \sin t$; $t \in [0, \pi]$

[13.4]
17. Find four additional sets of polar coordinates $(-360° < \theta \leq 360°)$ for the point with polar coordinates $(4, -420°)$.
18. Find the Cartesian coordinates of the point with polar coordinates $(6, 135°)$.
19. Transform $x^2 + y^2 - 9x = 0$ to an equation in polar form.
20. Transform $r = 4 \cos \theta$ to an equation in Cartesian form.

[13.5] *Graph each equation.*

21. $r = \sin \theta$
22. $r = 1 + \sin \theta$
23. $r = 2 \sin 2\theta$
24. $r = 4\theta$

Supplemental Exercises for Chapters 12–13

The problems in this supplement are significantly more difficult than those in the exercise sets at the end of each section and the chapter reviews. They are designed to challenge the student, and a solution may draw on a combination of ideas and techniques from any of the preceding chapters. Many of these problems contain results that are used in more advanced classes in mathematics.

1. Find an equation in the xy-coordinate system for the ellipse with vertices $(0, 0)$ and $(2, 2/\sqrt{3})$ and with minor axis of length $1/\sqrt{3}$.

2. Find an equation in the xy-coordinate system for the hyperbola with vertices $(-1, -\sqrt{3})$ and $(2, 2\sqrt{3})$ and one focus at $(7/2, 7\sqrt{3}/2)$. What are the equations in the xy-coordinate system of the asymptotes?

For Exercises 3–6, assume the graph of

$$Ax^2 + Bxy + Cy^2 + Dx + Ey + F = 0 \qquad (1)$$

is not a degenerate conic section.

3. Show that if the $x'y'$-coordinate system is obtained from the xy-coordinate system by rotation through an angle α satisfying $\tan 2\alpha = B/(A - C)$ and if Equation (1) transforms into $A'x'^2 + C'y'^2 + D'x' + E'y' + F' = 0$, then $B^2 - 4AC = -4A'C'$. (*Hint:* See Exercise 20 on page 449).

4. Show that if $B^2 - 4AC = 0$, then the graph of Equation (1) is a parabola. *Hint:* Use Exercise 3.

5. Show that if $B^2 - 4AC < 0$ and $B \neq 0$ or $A \neq C$, then the graph of Equation (1) is an ellipse. *Hint:* Use Exercise 3.

6. Show that if $B^2 - 4AC > 0$, then the graph of Equation (1) is a hyperbola. *Hint:* Use Exercise 3.

464

Supplemental Exercises for Chapters 12–13 465

7. Under what conditions on the coefficients is the graph of Equation (1) a circle?

8. Show that if the graph of an equation in polar coordinates is symmetric with respect to the polar axis and the vertical axis, then it must also be symmetric with respect to the pole.

9. Show that if the graph of an equation in polar coordinates is symmetric with respect to the vertical axis and the pole, then it must be symmetric with the polar axis.

10. Show that if the graph of an equation in polar coordinates is symmetric with respect to the pole and the polar axis, then it is symmetric with respect to the vertical axis.

Graph each equation.

11. $r = e^\theta$, $\theta \geq 0$

12. $r = e^{-\theta}$, $\theta \geq 0$

13. Let f be an increasing function on $[0, \infty)$. Describe the graph in polar coordinates of $r = f(\theta)$, $\theta \geq 0$.

14. Let f be a decreasing function on $[0, \infty)$ with $y = 0$ as a horizontal asymptote. Describe the graph in polar coordinates of $r = f(\theta)$, $\theta \geq 0$, near the pole.

Appendices

A Reference Outline

This reference outline contains all of the basic facts presented in the text. For easy reference the facts are listed by section number. The outline is an excellent tool for the student preparing for examinations.

[1.1] If $P(x)$, $Q(x)$, and $R(x)$ are expressions, then for all values of x for which $P(x)$, $Q(x)$, and $R(x)$ are real numbers, the sentence

$$P(x) = Q(x)$$

is equivalent to each of the following:

I. $P(x) + R(x) = Q(x) + R(x)$;

II. $P(x) \cdot R(x) = Q(x) \cdot R(x)$ for $R(x) \neq 0$.

If $a, b \in R$, $a \neq 0$, then $ax + b = 0$ is equivalent to

$$x = -\frac{b}{a}.$$

[1.2] If $P(x)$ and $Q(x)$ are algebraic expressions, then $P(x)Q(x) = 0$ if and only if $P(x) = 0$ or $Q(x) = 0$ or both.

If $a, b, c \in R$, $a \neq 0$, then $ax^2 + bx + c = 0$ is equivalent to

$$x = \frac{-b \pm \sqrt{b^2 - 4ac}}{2a}.$$

[1.3] The solution set of $U(x) = V(x)$ is a subset of but not necessarily equal to the solution set of $[U(x)]^n = [V(x)]^n$, for each natural number n.

[1.4] If $P(x)$, $Q(x)$, and $R(x)$ are expressions, then for all values of x for which $P(x)$, $Q(x)$, and $R(x)$ are real numbers, the sentence

$$P(x) < Q(x)$$

is equivalent to each of the following:

I. $P(x) + R(x) < Q(x) + R(x)$;

II. $P(x) \cdot R(x) < Q(x) \cdot R(x)$ for $R(x) > 0$;

III. $P(x) \cdot R(x) > Q(x) \cdot R(x)$ for $R(x) < 0$.

Similarly, the sentence

$$P(x) \leq Q(x)$$

is equivalent to sentences of the form I–III, with $<$ (or $>$) replaced by $\leq$ (or $\geq$) under the same conditions, $R(x) > 0$ and $R(x) < 0$, as above.

[1.5] $|x| = a$ is equivalent to $x = a$ or $x = -a$.

$|x| < a$ is equivalent to $-a < x$ and $x < a$, i.e., $-a < x < a$.

$|x| > a$ is equivalent to $x < -a$ or $x > a$.

$|x| \leq a$ is equivalent to $-a \leq x$ and $x \leq a$, i.e., $-a \leq x \leq a$.

$|x| \geq a$ is equivalent to $x \leq -a$ or $x \geq a$.

[2.3] The length of the line segment between two points (x_1, y_1) and (x_2, y_2) is given by

$$d = \sqrt{(x_2 - x_1)^2 + (y_2 - y_1)^2},$$

and the slope is given by

$$m = \frac{y_2 - y_1}{x_2 - x_1} \quad (x_2 \neq x_1).$$

[2.4] Forms for linear equations:

$y - y_1 = m(x - x_1)$ *Point-slope form,*

$y = mx + b$ *Slope-intercept form,*

$\dfrac{x}{a} + \dfrac{y}{b} = 1$ *Intercept form.*

[2.5] Two lines are parallel if and only if they have the same slope.

Two lines with slopes m_1 and m_2 are perpendicular if and only if $m_1 m_2 = -1$.

[3.1] If $P(x)$ is a real polynomial, then for every real number c there exists a unique real polynomial $Q(x)$ such that

$$P(x) = (x - c)Q(x) + P(c).$$

If $P(x)$ is a polynomial with real-number coefficients and $P(r) = 0$, then $(x - r)$ is a factor of $P(x)$.

[3.2] If $P(x) = a_n x^n + a_{n-1} x^{n-1} + \cdots + a_0$ $(a_n \neq 0)$, and if k is between $P(x_1)$ and $P(x_2)$, then there exists at least one c between x_1 and x_2 such that $P(c) = k$.

Let $P(x)$ be a polynomial over the field R of real numbers. If $P(x_1)$ and $P(x_2)$ are opposite in sign, then there exists at least one value of c between x_1 and x_2 such that $P(c) = 0$.

Let $P(x)$ be a polynomial with real-number coefficients.

I. If $r_1 \geq 0$ and the coefficients of the terms in $Q(x)$ and the term $P(r_1)$ are all of the same sign in the right-hand member of

$$P(x) = (x - r_1)Q(x) + P(r_1),$$

then $P(x) = 0$ can have no solution greater than r_1.

II. If $r_2 \leq 0$ and the coefficients of the terms in $Q(x)$ and the term $P(r_2)$ alternate in sign (zero suitably denoted by $+0$ and -0) in the right-hand member of

$$P(x) = (x - r_2)Q(x) + P(r_2),$$

then $P(x) = 0$ can have no solution less than r_2.

If $P(x)$ is a polynomial over the field R of real numbers, then the number of positive real solutions of $P(x) = 0$ either is equal to the number of variations of sign occurring in $P(x)$ or is less than that number by an even natural number. Moreover, the number of negative real solutions of $P(x) = 0$ either is equal to the number of variations in sign occurring in $P(-x)$ or is less than that number by an even natural number.

[3.3] If a is a composite number, then a is the product of only one set of prime factors; that is, the prime factorization of a is unique except for the ordering of the factors.

If the rational number p/q, in lowest terms, is a solution of

$$P(x) = a_n x^n + a_{n-1} x^{n-1} + \cdots + a_0 = 0, \quad a_n \neq 0,$$

where $a_j \in J$, then p is an integral factor of a_0 and q is an integral factor of a_n.

[3.4] If $P(x) = a_n x^n + a_{n-1} x^{n-1} + \cdots + a_0$ is a real polynomial equation of degree n, then the graph of

$$P(x) = a_n x^n + a_{n-1} x^{n-1} + \cdots + a_0, \quad a_n \neq 0,$$

is a smooth curve that has at most $n - 1$ turning points.

The first components of the turning points of the graph of
$$P(x) = a_n x^n + a_{n-1} x^{n-1} + \cdots + a_1 x + a_0$$
are solutions of the equation
$$P'(x) = n a_n x^{n-1} + (n-1) a_{n-1} x^{n-2} + \cdots + 2 a_2 x + a_1 = 0.$$

[3.5] The graph of the rational function over R defined by $y = P(x)/Q(x)$ has a vertical asymptote at $x = a$ for each value a at which $Q(x)$ vanishes and $P(x)$ does not vanish.

The graph of the rational function over R defined by
$$y = \frac{a_n x^n + a_{n-1} x^{n-1} + \cdots + a_0}{b_m x^m + b_{m-1} x^{m-1} + \cdots + b_0},$$
where $a_n, b_m \neq 0$ and n, m are nonnegative integers, has

  I. a horizontal asymptote at $y = 0$ if $n < m$,

  II. a horizontal asymptote at $y = a_n/b_m$ if $n = m$,

  III. no horizontal asymptotes if $n > m$.

[4.1] Let $x, y \in Q$, and let $x > y > 0$. Then
$$b^x > b^y \text{ if } b > 1, \quad b^x = b^y \text{ if } b = 1, \quad \text{and} \quad b^x < b^y \text{ if } 0 < b < 1.$$

[4.2] Assume x_1 and x_2 are values for which all expressions are defined. Then
$$\log_b (x_1 x_2) = \log_b x_1 + \log_b x_2,$$
$$\log_b \frac{x_2}{x_1} = \log_b x_2 - \log_b x_1,$$
$$\log_b (x_1)^m = m \log_b x_1.$$

[5.1] When measuring arcs on the unit circle, the counterclockwise direction is considered positive and the clockwise direction is considered negative.

If s is any real number, then
$$\sin^2 s + \cos^2 s = 1$$

[5.3] For any arc A on the unit circle, with initial point $(1, 0)$ and with measure x, the reference arc is the arc on the unit circle with least nonnegative measure between the terminal point of A and the horizontal axis. The length of the reference arc is denoted by $\bar{x}$.

A Reference Outline 473

For any arc of measure x,

$$|\cos x| = \cos \bar{x} \quad \text{and} \quad |\sin x| = \sin \bar{x}.$$

[5.4] If f is a function such that for some $p \in R$, $p \neq 0$,

$$f(x + p) = f(x - p) = f(x)$$

for all x in the domain of f, then f is *periodic* with period p.

If p is the least positive period of a periodic function f, then p is called the *fundamental period of f*.

The fundamental period of the cosine function and of the sine function is 2π.

[5.5] The fundamental period of the tangent function is π.

For any arc of measure x with $\tan x$ defined,

$$|\tan x| = \tan \bar{x}.$$

[6.1] $$1^R = \left(\frac{180}{\pi}\right)^\circ \quad \text{and} \quad 1^\circ = \left(\frac{\pi}{180}\right)^R$$

If α is the radian measure of a central angle of a circle of radius r, then the length s of the intercepted arc of the circle is

$$s = r\alpha.$$

[6.2] For any angle α in standard position, the reference angle $\bar{\alpha}$ is the least nonnegative angle that the terminal ray of α forms with the x-axis.

If T is any trigonometric function and $\bar{\alpha}$ is the reference angle for the angle α, then

$$|T(\alpha)| = T(\bar{\alpha}).$$

[6.3] If α is an angle of a right triangle, then

$$\cos \alpha = \frac{\text{length of side adjacent to } \alpha}{\text{length of hypotenuse}},$$

$$\sin \alpha = \frac{\text{length of side opposite } \alpha}{\text{length of hypotenuse}},$$

$$\tan \alpha = \frac{\text{length of side opposite } \alpha}{\text{length of side adjacent to } \alpha}.$$

[6.4] If α, β, and γ are the angles of a triangle, and a, b, and c are the lengths of the sides opposite α, β, and γ, respectively, then

$$\frac{\sin \alpha}{a} = \frac{\sin \beta}{b} = \frac{\sin \gamma}{c}. \quad \textbf{Law of sines}$$

[6.5] If α, β, and γ are the angles of a triangle, and a, b, and c are the lengths of the sides opposite α, β, and γ, respectively, then

$$c^2 = a^2 + b^2 - 2ab \cos \gamma,$$
$$b^2 = a^2 + c^2 - 2ac \cos \beta, \quad \textbf{Law of cosines}$$
$$a^2 = b^2 + c^2 - 2bc \cos \alpha.$$

[7.1] The following basic identities are useful:

$$\tan^2 x + 1 = \sec^2 x \qquad 1 + \cot^2 x = \csc^2 x$$
$$\sin(-x) = -\sin x \qquad \cos(-x) = \cos x \qquad \tan(-x) = -\tan x$$

[7.2]
$$\cos(x_1 + x_2) = \cos x_1 \cos x_2 - \sin x_1 \sin x_2$$
$$\cos(x_1 - x_2) = \cos x_1 \cos x_2 + \sin x_1 \sin x_2$$
$$\cos 2x = \cos^2 x - \sin^2 x$$
$$= 1 - 2 \sin^2 x$$
$$= 2 \cos^2 x - 1$$

$$\cos \frac{x}{2} = \begin{cases} \sqrt{\dfrac{1 + \cos x}{2}} & \text{when } \dfrac{x}{2} \text{ terminates in Quadrant I or IV;} \\ -\sqrt{\dfrac{1 + \cos x}{2}} & \text{when } \dfrac{x}{2} \text{ terminates in Quadrant II or III.} \end{cases}$$

[7.3]
$$\sin(x_1 + x_2) = \sin x_1 \cos x_2 + \cos x_1 \sin x_2$$
$$\sin(x_1 - x_2) = \sin x_1 \cos x_2 - \cos x_1 \sin x_2$$
$$\sin 2x = 2 \sin x \cos x$$

$$\sin \frac{x}{2} = \begin{cases} \sqrt{\dfrac{1 - \cos x}{2}} & \text{when } \dfrac{x}{2} \text{ terminates in Quadrant I or II;} \\ -\sqrt{\dfrac{1 - \cos x}{2}} & \text{when } \dfrac{x}{2} \text{ terminates in Quadrant III or IV.} \end{cases}$$

[7.4] $\tan(x_1 + x_2) = \dfrac{\tan x_1 + \tan x_2}{1 - \tan x_1 \tan x_2} \qquad \tan(x_1 - x_2) = \dfrac{\tan x_1 - \tan x_2}{1 + \tan x_1 \tan x_2}$

$$\tan 2x = \frac{2 \tan x}{1 - \tan^2 x} \qquad\qquad \tan \frac{x}{2} = \frac{1 - \cos x}{\sin x}$$

A Reference Outline

[8.1] The complex numbers satisfy the field axioms (See Preliminary Concepts).

[8.2] If $P(z)$ is a polynomial over the real numbers, and $P(z) = 0$ for some $z \in C$, then $P(\bar{z}) = 0$.

Every polynomial function of degree $n \geq 1$ over the complex numbers has at least one complex zero.

If $P(z)$ is a polynomial of degree $n \geq 1$ over the complex numbers, then $P(z)$ can be expressed as a product of a constant and n linear factors of the form $(z - z_k)$.

[8.3] If r is the modulus and θ is an argument of the complex number $a + bi$, then

$$a + bi = r[\cos(\theta + k \cdot 360°) + i \sin(\theta + k \cdot 360°)], \quad k \in J.$$

If $z_1 = r_1 \operatorname{cis} \theta_1$ and $z_2 = r_2 \operatorname{cis} \theta_2$, then

I. $z_1 \cdot z_2 = r_1 r_2 \operatorname{cis}(\theta_1 + \theta_2)$;

II. $\dfrac{z_1}{z_2} = \dfrac{r_1}{r_2} \operatorname{cis}(\theta_1 - \theta_2) \quad (z_2 \neq 0 + 0i)$.

[8.4] If $z \in C$, $z = r \operatorname{cis} \theta$ and $n \in J$, then

$$z^n = r^n \operatorname{cis} n\theta. \quad \textbf{De Moivre's theorem}$$

[9.1] Any ordered pair that satisfies both the equations

$$f(x, y) = 0 \quad \text{and} \quad g(x, y) = 0$$

will also satisfy the equation

$$a \cdot f(x, y) + b \cdot g(x, y) = 0,$$

for all real numbers a and b.

[9.2] If any equation in the system

$$f(x, y, z) = 0, \quad g(x, y, z) = 0, \quad \text{and} \quad h(x, y, z) = 0$$

is replaced by a linear combination, with nonzero coefficients, of itself and any one of the other equations in the system, then the result is an equivalent system.

[9.3] If $r_1 \neq r_2$, then there exist constants $c_1, c_2 \in R$ such that

$$\frac{ax + b}{(x - r_1)(x - r_2)} = \frac{c_1}{x - r_1} + \frac{c_2}{x - r_2}.$$

If $P(x)$ is of degree less than $Q(x)$, and if $Q(x) = (x - r_1)(x - r_2) \cdots (x - r_n)$, where no two factors are identical, then there exist constants $c_1, c_2, \ldots, c_n \in R$ such that

$$\frac{P(x)}{Q(x)} = \frac{c_1}{x - r_1} + \frac{c_2}{x - r_2} + \cdots + \frac{c_n}{x - r_n}.$$

If $P(x)$ is of degree less than $Q(x)$, and if $Q(x) = (x - r_1)^n$, then there exist constants $c_1, c_2, \ldots, c_n \in R$ such that

$$\frac{P(x)}{Q(x)} = \frac{c_1}{x - r_1} + \frac{c_2}{(x - r_1)^2} + \cdots + \frac{c_n}{(x - r_1)^n}.$$

[9.5] Any half-plane is a convex set.

The intersection of two convex sets is a convex set.

[10.1] If A, B, and C are $m \times n$ matrices with real-number entries, then

 I. $(A + B)_{m \times n}$ is a matrix with real-number entries;
 II. $(A + B) + C = A + (B + C)$;
 III. the matrix $\mathbf{0}_{m \times n}$ has the property that, for every matrix $A_{m \times n}$,
$$A + 0 = A \quad \text{and} \quad 0 + A = A;$$
 IV. for every matrix $A_{m \times n}$, the matrix $-A_{m \times n}$ has the property that
$$A + (-A) = 0 \quad \text{and} \quad (-A) + A = 0,$$
 V. $A + B = B + A$.

[10.2] If A and B are $m \times n$ matrices and $c, d \in R$, then

 I. cA is an $m \times n$ matrix; V. $1A = A$;
 II. $c(dA) = (cd)A$; VI. $(-1)A = -A$;
 III. $(c + d)A = cA + dA$; VII. $0A = \mathbf{0}$;
 IV. $c(A + B) = cA + cB$; VIII. $c\mathbf{0} = \mathbf{0}$.

If A, B, and C are $n \times n$ square matrices, then
$$(AB)C = A(BC).$$

If A, B, and C are $n \times n$ square matrices, then
$$A(B + C) = AB + AC \quad \text{and} \quad (B + C)A = BA + CA.$$

For each matrix $A_{n \times n}$, we have

$$A_{n \times n} I_{n \times n} = I_{n \times n} A_{n \times n} = A_{n \times n}.$$

Furthermore, $I_{n \times n}$ is the unique matrix having this property for all matrices $A_{n \times n}$.

If A and B are $n \times n$ square matrices, and a is a real number, then

$$a(AB) = (aA)B = A(aB).$$

[10.5] If each entry in any row, or each entry in any column, of a determinant is 0, then the determinant is equal to 0.

If any two rows (or any two columns) of a determinant are interchanged, the resulting determinant is the negative of the original determinant.

If two rows (or two columns) in a determinant have corresponding entries that are equal, the determinant is equal to 0.

If each of the entries of one row (or column) of a determinant is multiplied by k, the determinant is multiplied by k.

If each entry of one row (or column) of a determinant is multiplied by a real number k and the resulting product is added to the corresponding entry in another row (or column, respectively) in the determinant, the resulting determinant is equal to the original determinant.

[10.6] If

$$A = \begin{bmatrix} a_{11} & a_{12} & \cdots & a_{1n} \\ a_{21} & a_{22} & \cdots & a_{2n} \\ \vdots & \vdots & & \vdots \\ a_{n1} & a_{n2} & \cdots & a_{nn} \end{bmatrix},$$

and if $\delta(A) \neq 0$, then

$$A^{-1} = \frac{1}{\delta(A)} \begin{bmatrix} A_{11} & A_{21} & \cdots & A_{n1} \\ A_{12} & A_{22} & \cdots & A_{n2} \\ \vdots & \vdots & & \vdots \\ A_{1m} & A_{2n} & \cdots & A_{nn} \end{bmatrix},$$

where A_{ij} is the cofactor of a_{ij} in A. If $\delta(A) = 0$, then A has no inverse.

If A is an $n \times n$ nonsingular matrix and if $[A \mid I]$ is the $n \times 2n$ matrix obtained by adjoining the $n \times n$ identity matrix to A, then

$$[A \mid I] \sim [I \mid A^{-1}].$$

If A and B are $n \times n$ nonsingular square matrices, then AB has an inverse, namely
$$(AB)^{-1} = B^{-1}A^{-1}.$$

[10.8] Cramer's rule:
$$x = \frac{\delta(A_x)}{\delta(A)}, \quad y = \frac{\delta(A_y)}{\delta(A)}, \quad \text{and} \quad z = \frac{\delta(A_z)}{\delta(A)}, \quad (\delta(A) \neq 0).$$

[11.1] The nth term in the sequence defined by
$$s_1 = a, \quad s_{n+1} = s_n + d,$$
where $a, d \in R$, and $n \in N$, is
$$s_n = a + (n-1)d.$$
The nth term in the sequence defined by
$$s_1 = a, \quad s_{n+1} = rs_n,$$
where $a, r \in R$, $a \neq 0$, $r \neq 0$, and $n \in N$, is
$$s_n = ar^{n-1}.$$

[11.2] The sum of the first n terms of an arithmetic progression is
$$S_n = \frac{n}{2}(a + s_n), \quad \text{or} \quad S_n = \frac{n}{2}[2a + (n-1)d].$$
The sum of the first n terms of a geometric progression is
$$S_n = \frac{a - ar^n}{1 - r}, \quad \text{or} \quad S_n = \frac{a - rs_n}{1 - r} \quad (r \neq 1).$$

[11.3] The sum of an infinite geometric progression, $a + ar + ar^2 + \cdots + ar^n + \cdots$, with $|r| < 1$, is
$$S_\infty = \lim_{n \to \infty} S_n = \frac{a}{1 - r}.$$

[11.4] For each natural number n,
$$(a + b)^n = a^n + \frac{n}{1!}a^{n-1}b + \frac{n(n-1)}{2!}a^{n-2}b^2 + \frac{n(n-1)(n-2)}{3!}a^{n-3}b^3 + \cdots$$
$$+ \frac{n(n-1)(n-2)\cdots(n-r+2)}{(r-1)!}a^{n-r+1}b^{r-1} + \cdots + b^n,$$

where r is the number of the term. Alternatively,

$$(a + b)^n = \binom{n}{0}a^n + \binom{n}{1}a^{n-1}b + \binom{n}{2}a^{n-2}b^2 + \binom{n}{3}a^{n-3}b^3 + \cdots$$
$$+ \binom{n}{r-1}a^{n-r+1}b^{r-1} + \cdots + \binom{n}{n}b^n.$$

The rth term in a binomial expansion is given by

$$\binom{n}{r-1}a^{n-r+1}b^{r-1} = \frac{n!}{(r-1)!(n-r+1)!} a^{n-r+1}b^{r-1}$$
$$= \frac{n(n-1)(n-2)(n-r+2)}{(r-1)!} a^{n-r+1}b^{r-1}.$$

[11.5] If a given statement involving n is true for $n = 1$, and if its truth for $n = k$ implies its truth for $n = k + 1$, then it is true for every natural number n.

[12.2] The standard forms of the equation of a parabola in standard position are as follows:

Equation	Axis of Symmetry	Focus
$x^2 = 4py$	$x = 0$	$(0, p)$
$x^2 = -4py$	$x = 0$	$(0, -p)$
$y^2 = 4px$	$y = 0$	$(p, 0)$
$y^2 = -4px$	$y = 0$	$(-p, 0)$

[12.3] The standard form of the equation of a circle of radius r centered at $(0, 0)$ is

$$x^2 + y^2 = r^2.$$

The standard forms of the equation of an ellipse in standard position centered at $(0, 0)$ are as follows:

Equation	Vertices	Foci
$\dfrac{x^2}{a^2} + \dfrac{y^2}{b^2} = 1$	$(\pm a, 0)$	$(\pm c, 0)$
$\dfrac{y^2}{a^2} + \dfrac{x^2}{b^2} = 1$	$(0, \pm a)$	$(0, \pm c)$

where $c^2 = a^2 - b^2$.

[12.4] The standard forms of the equation of a hyperbola in standard position centered at (0, 0) are as follows:

Equation	Vertices	Foci	Asymptotes
$\dfrac{x^2}{a^2} - \dfrac{y^2}{b^2} = 1$	$(\pm a, 0)$	$(\pm c, 0)$	$y = \pm \dfrac{b}{a} x$
$\dfrac{y^2}{a^2} - \dfrac{x^2}{b^2} = 1$	$(0, \pm a)$	$(0, \pm c)$	$y = \pm \dfrac{a}{b} x$

where $c^2 = a^2 + b^2$.

[13.1] If the origin in the $x'y'$-coordinate system has xy-coordinates (h, k), then the xy-coordinates and $x'y'$-coordinates of a point are related by the equations

$$x' = x - h,$$
$$y' = y - k.$$

[13.2] If the $x'y'$-coordinate axes are obtained by rotating the xy-axes through an angle α, then the xy-coordinates and the $x'y'$-coordinates of a point are related by the equations

$$x' = x(\cos \alpha) + y(\sin \alpha),$$
$$y' = -x(\sin \alpha) + y(\cos \alpha).$$

[13.4] If (r, θ) are polar coordinates of the point with Cartesian coordinates (x, y), then

$$x = r \cos \theta \quad \text{and} \quad y = r \sin \theta$$

and

$$r = \pm\sqrt{x^2 + y^2} \quad \text{and} \quad \tan \theta = \frac{y}{x}.$$

B Tables

Table I
Exponential Functions, Base e

x	e^x	e^{-x}	x	e^x	e^{-x}
0.00	1.0000	1.0000	1.5	4.4817	0.2231
0.01	1.0101	0.9901	1.6	4.9530	0.2019
0.02	1.0202	0.9802	1.7	5.4739	0.1827
0.03	1.0305	0.9705	1.8	6.0496	0.1653
0.04	1.0408	0.9608	1.9	6.6859	0.1496
0.05	1.0513	0.9512	2.0	7.3891	0.1353
0.06	1.0618	0.9418	2.1	8.1662	0.1225
0.07	1.0725	0.9324	2.2	9.0250	0.1108
0.08	1.0833	0.9331	2.3	9.9742	0.1003
0.09	1.0942	0.9139	2.4	11.023	0.0907
0.10	1.1052	0.9048	2.5	12.182	0.0821
0.11	1.1163	0.8958	2.6	13.464	0.0743
0.12	1.1275	0.8869	2.7	14.880	0.0672
0.13	1.1388	0.8781	2.8	16.445	0.0608
0.14	1.1503	0.8694	2.9	18.174	0.0550
0.15	1.1618	0.8607	3.0	20.086	0.0498
0.16	1.1735	0.8521	3.1	22.198	0.0450
0.17	1.1853	0.8437	3.2	24.533	0.0408
0.18	1.1972	0.8353	3.3	27.113	0.0369
0.19	1.2092	0.8270	3.4	29.964	0.0334
0.20	1.2214	0.8187	3.5	33.115	0.0302
0.21	1.2337	0.8106	3.6	36.598	0.0273
0.22	1.2461	0.8025	3.7	40.447	0.0247
0.23	1.2586	0.7945	3.8	44.701	0.0224
0.24	1.2712	0.7866	3.9	49.402	0.0202
0.25	1.2840	0.7788	4.0	54.598	0.0183
0.30	1.3499	0.7408	4.1	60.340	0.0166
0.35	1.4191	0.7047	4.2	66.686	0.0150
0.40	1.4918	0.6703	4.3	73.700	0.0136
0.45	1.5683	0.6376	4.4	81.451	0.0123
0.50	1.6487	0.6065	4.5	90.017	0.0111
0.55	1.7333	0.5769	4.6	99.484	0.0101
0.60	1.8221	0.5488	4.7	109.95	0.0091
0.65	1.9155	0.5220	4.8	121.51	0.0082
0.70	2.0138	0.4966	4.9	134.29	0.0074
0.75	2.1170	0.4724	5.0	148.41	0.0067
0.80	2.2255	0.4493	5.5	244.69	0.0041
0.85	2.3396	0.4274	6.0	403.43	0.0025
0.90	2.4596	0.4066	6.5	665.14	0.0015
0.95	2.5857	0.3867	7.0	1096.6	0.0009
1.0	2.7183	0.3679	7.5	1808.0	0.0006
1.1	3.0042	0.3329	8.0	2981.0	0.0003
1.2	3.3201	0.3012	8.5	4914.8	0.0002
1.3	3.6693	0.2725	9.0	8103.1	0.0001
1.4	4.0552	0.2466	10.0	22026	0.00005

Table II
Common Logarithms

x	0	1	2	3	4	5	6	7	8	9
1.0	0.0000	0.0043	0.0086	0.0128	0.0170	0.0212	0.0253	0.0294	0.0334	0.0374
1.1	0.0414	0.0453	0.0492	0.0531	0.0569	0.0607	0.0645	0.0682	0.0719	0.0755
1.2	0.0792	0.0828	0.0864	0.0899	0.0934	0.0969	0.1004	0.1038	0.1072	0.1106
1.3	0.1139	0.1173	0.1206	0.1239	0.1271	0.1303	0.1335	0.1367	0.1399	0.1430
1.4	0.1461	0.1492	0.1523	0.1553	0.1584	0.1614	0.1644	0.1673	0.1703	0.1732
1.5	0.1761	0.1790	0.1818	0.1847	0.1875	0.1903	0.1931	0.1959	0.1987	0.2014
1.6	0.2041	0.2068	0.2095	0.2122	0.2148	0.2175	0.2201	0.2227	0.2253	0.2279
1.7	0.2304	0.2330	0.2355	0.2380	0.2405	0.2430	0.2455	0.2480	0.2504	0.2529
1.8	0.2553	0.2577	0.2601	0.2625	0.2648	0.2672	0.2695	0.2718	0.2742	0.2765
1.9	0.2788	0.2810	0.2833	0.2856	0.2878	0.2900	0.2923	0.2945	0.2967	0.2989
2.0	0.3010	0.3032	0.3054	0.3075	0.3096	0.3118	0.3139	0.3160	0.3181	0.3201
2.1	0.3222	0.3243	0.3263	0.3284	0.3304	0.3324	0.3345	0.3365	0.3385	0.3404
2.2	0.3424	0.3444	0.3464	0.3483	0.3502	0.3522	0.3541	0.3560	0.3579	0.3598
2.3	0.3617	0.3636	0.3655	0.3674	0.3692	0.3711	0.3729	0.3747	0.3766	0.3784
2.4	0.3802	0.3820	0.3838	0.3856	0.3874	0.3892	0.3909	0.3927	0.3945	0.3962
2.5	0.3979	0.3997	0.4014	0.4031	0.4048	0.4065	0.4082	0.4099	0.4116	0.4133
2.6	0.4150	0.4166	0.4183	0.4200	0.4216	0.4232	0.4249	0.4265	0.4281	0.4298
2.7	0.4314	0.4330	0.4346	0.4362	0.4378	0.4393	0.4409	0.4425	0.4440	0.4456
2.8	0.4472	0.4487	0.4502	0.4518	0.4533	0.4548	0.4564	0.4579	0.4594	0.4609
2.9	0.4624	0.4639	0.4654	0.4669	0.4683	0.4698	0.4713	0.4728	0.4742	0.4757
3.0	0.4771	0.4786	0.4800	0.4814	0.4829	0.4843	0.4857	0.4871	0.4886	0.4900
3.1	0.4914	0.4928	0.4942	0.4955	0.4969	0.4983	0.4997	0.5011	0.5024	0.5038
3.2	0.5051	0.5065	0.5079	0.5092	0.5105	0.5119	0.5132	0.5145	0.5159	0.5172
3.3	0.5185	0.5198	0.5211	0.5224	0.5237	0.5250	0.5263	0.5276	0.5289	0.5302
3.4	0.5315	0.5328	0.5340	0.5353	0.5366	0.5378	0.5391	0.5403	0.5416	0.5428
3.5	0.5441	0.5453	0.5465	0.5478	0.5490	0.5502	0.5514	0.5527	0.5539	0.5551
3.6	0.5563	0.5575	0.5587	0.5599	0.5611	0.5623	0.5635	0.5647	0.5658	0.5670
3.7	0.5682	0.5694	0.5705	0.5717	0.5729	0.5740	0.5752	0.5763	0.5775	0.5786
3.8	0.5798	0.5809	0.5821	0.5832	0.5843	0.5855	0.5866	0.5877	0.5888	0.5899
3.9	0.5911	0.5922	0.5933	0.5944	0.5955	0.5966	0.5977	0.5988	0.5999	0.6010
4.0	0.6021	0.6031	0.6042	0.6053	0.6064	0.6075	0.6085	0.6096	0.6107	0.6117
4.1	0.6128	0.6138	0.6149	0.6160	0.6170	0.6180	0.6191	0.6201	0.6212	0.6222
4.2	0.6232	0.6243	0.6253	0.6263	0.6274	0.6284	0.6294	0.6304	0.6314	0.6325
4.3	0.6335	0.6345	0.6355	0.6365	0.6375	0.6385	0.6395	0.6405	0.6415	0.6425
4.4	0.6435	0.6444	0.6454	0.6464	0.6474	0.6484	0.6493	0.6503	0.6513	0.6522
4.5	0.6532	0.6542	0.6551	0.6561	0.6571	0.6580	0.6590	0.6599	0.6609	0.6618
4.6	0.6628	0.6637	0.6646	0.6656	0.6665	0.6675	0.6684	0.6693	0.6702	0.6712
4.7	0.6721	0.6730	0.6739	0.6749	0.6758	0.6767	0.6776	0.6785	0.6794	0.6803
4.8	0.6812	0.6821	0.6830	0.6839	0.6848	0.6857	0.6866	0.6875	0.6884	0.6893
4.9	0.6902	0.6911	0.6920	0.6928	0.6937	0.6946	0.6955	0.6964	0.6972	0.6981
5.0	0.6990	0.6998	0.7007	0.7016	0.7024	0.7033	0.7042	0.7050	0.7059	0.7067
5.1	0.7076	0.7084	0.7093	0.7101	0.7110	0.7118	0.7126	0.7135	0.7143	0.7152
5.2	0.7160	0.7168	0.7177	0.7185	0.7193	0.7202	0.7210	0.7218	0.7226	0.7235
5.3	0.7243	0.7251	0.7259	0.7267	0.7275	0.7284	0.7292	0.7300	0.7308	0.7316
5.4	0.7324	0.7332	0.7340	0.7348	0.7356	0.7364	0.7372	0.7380	0.7388	0.7396
x	0	1	2	3	4	5	6	7	8	9

Table II (*continued*)

x	0	1	2	3	4	5	6	7	8	9
5.5	0.7404	0.7412	0.7419	0.7427	0.7435	0.7443	0.7451	0.7459	0.7466	0.7474
5.6	0.7482	0.7490	0.7497	0.7505	0.7513	0.7520	0.7528	0.7536	0.7543	0.7551
5.7	0.7559	0.7566	0.7574	0.7582	0.7589	0.7597	0.7604	0.7612	0.7619	0.7627
5.8	0.7634	0.7642	0.7649	0.7657	0.7664	0.7672	0.7679	0.7686	0.7694	0.7701
5.9	0.7709	0.7716	0.7723	0.7731	0.7738	0.7745	0.7752	0.7760	0.7767	0.7774
6.0	0.7782	0.7789	0.7796	0.7803	0.7810	0.7818	0.7825	0.7832	0.7839	0.7846
6.1	0.7853	0.7860	0.7868	0.7875	0.7882	0.7889	0.7896	0.7903	0.7910	0.7917
6.2	0.7924	0.7931	0.7938	0.7945	0.7952	0.7959	0.7966	0.7973	0.7980	0.7987
6.3	0.7993	0.8000	0.8007	0.8014	0.8021	0.8028	0.8035	0.8041	0.8048	0.8055
6.4	0.8062	0.8069	0.8075	0.8082	0.8089	0.8096	0.8102	0.8109	0.8116	0.8122
6.5	0.8129	0.8136	0.8142	0.8149	0.8156	0.8162	0.8169	0.8176	0.8182	0.8189
6.6	0.8195	0.8202	0.8209	0.8215	0.8222	0.8228	0.8235	0.8241	0.8248	0.8254
6.7	0.8261	0.8267	0.8274	0.8280	0.8287	0.8293	0.8299	0.8306	0.8312	0.8319
6.8	0.8325	0.8331	0.8338	0.8344	0.8351	0.8357	0.8363	0.8370	0.8376	0.8382
6.9	0.8388	0.8395	0.8401	0.8407	0.8414	0.8420	0.8426	0.8432	0.8439	0.8445
7.0	0.8451	0.8457	0.8463	0.8470	0.8476	0.8482	0.8488	0.8494	0.8500	0.8506
7.1	0.8513	0.8519	0.8525	0.8531	0.8537	0.8543	0.8549	0.8555	0.8561	0.8567
7.2	0.8573	0.8579	0.8585	0.8591	0.8597	0.8603	0.8609	0.8615	0.8621	0.8627
7.3	0.8633	0.8639	0.8645	0.8651	0.8657	0.8663	0.8669	0.8675	0.8681	0.8686
7.4	0.8692	0.8698	0.8704	0.8710	0.8716	0.8722	0.8727	0.8733	0.8739	0.8745
7.5	0.8751	0.8756	0.8762	0.8768	0.8774	0.8779	0.8785	0.8791	0.8797	0.8802
7.6	0.8808	0.8814	0.8820	0.8825	0.8831	0.8837	0.8842	0.8848	0.8854	0.8859
7.7	0.8865	0.8871	0.8876	0.8882	0.8887	0.8893	0.8899	0.8904	0.8910	0.8915
7.8	0.8921	0.8927	0.8932	0.8938	0.8943	0.8949	0.8954	0.8960	0.8965	0.8971
7.9	0.8976	0.8982	0.8987	0.8993	0.8998	0.9004	0.9009	0.9015	0.9020	0.9025
8.0	0.9031	0.9036	0.9042	0.9047	0.9053	0.9058	0.9063	0.9069	0.9074	0.9079
8.1	0.9085	0.9090	0.9096	0.9101	0.9106	0.9112	0.9117	0.9122	0.9128	0.9133
8.2	0.9138	0.9143	0.9149	0.9154	0.9159	0.9165	0.9170	0.9175	0.9180	0.9186
8.3	0.9191	0.9196	0.9201	0.9206	0.9212	0.9217	0.9222	0.9227	0.9232	0.9238
8.4	0.9243	0.9248	0.9253	0.9258	0.9263	0.9269	0.9274	0.9279	0.9284	0.9289
8.5	0.9294	0.9299	0.9304	0.9309	0.9315	0.9320	0.9325	0.9330	0.9335	0.9340
8.6	0.9345	0.9350	0.9355	0.9360	0.9365	0.9370	0.9375	0.9380	0.9385	0.9390
8.7	0.9395	0.9400	0.9405	0.9410	0.9415	0.9420	0.9425	0.9430	0.9435	0.9440
8.8	0.9445	0.9450	0.9455	0.9460	0.9465	0.9469	0.9474	0.9479	0.9484	0.9489
8.9	0.9494	0.9499	0.9504	0.9509	0.9513	0.9518	0.9523	0.9528	0.9533	0.9538
9.0	0.9542	0.9547	0.9552	0.9557	0.9562	0.9566	0.9571	0.9576	0.9581	0.9586
9.1	0.9590	0.9595	0.9600	0.9605	0.9609	0.9614	0.9619	0.9624	0.9628	0.9633
9.2	0.9638	0.9643	0.9647	0.9652	0.9657	0.9661	0.9666	0.9671	0.9675	0.9680
9.3	0.9685	0.9689	0.9694	0.9699	0.9703	0.9708	0.9713	0.9717	0.9722	0.9727
9.4	0.9731	0.9736	0.9741	0.9745	0.9750	0.9754	0.9759	0.9763	0.9768	0.9773
9.5	0.9777	0.9782	0.9786	0.9791	0.9795	0.9800	0.9805	0.9809	0.9814	0.9818
9.6	0.9823	0.9827	0.9832	0.9836	0.9841	0.9845	0.9850	0.9854	0.9859	0.9863
9.7	0.9868	0.9872	0.9877	0.9881	0.9886	0.9890	0.9894	0.9899	0.9903	0.9908
9.8	0.9912	0.9917	0.9921	0.9926	0.9930	0.9934	0.9939	0.9943	0.9948	0.9952
9.9	0.9956	0.9961	0.9965	0.9969	0.9974	0.9978	0.9983	0.9987	0.9991	0.9996
x	0	1	2	3	4	5	6	7	8	9

Table III
Natural Logarithms

n	$\log_e n$	n	$\log_e n$	n	$\log_e n$
		4.5	1.5041	9.0	2.1972
0.1	7.6974 †	4.6	1.5261	9.1	2.2083
0.2	8.3906	4.7	1.5476	9.2	2.2192
0.3	8.7960	4.8	1.5686	9.3	2.2300
0.4	9.0837	4.9	1.5892	9.4	2.2407
0.5	9.3069	5.0	1.6094	9.5	2.2513
0.6	9.4892	5.1	1.6292	9.6	2.2618
0.7	9.6433	5.2	1.6487	9.7	2.2721
0.8	9.7769	5.3	1.6677	9.8	2.2824
0.9	9.8946	5.4	1.6864	9.9	2.2925
1.0	0.0000	5.5	1.7047	10	2.3026
1.1	0.0953	5.6	1.7228	11	2.3979
1.2	0.1823	5.7	1.7405	12	2.4849
1.3	0.2624	5.8	1.7579	13	2.5649
1.4	0.3365	5.9	1.7750	14	2.6391
1.5	0.4055	6.0	1.7918	15	2.7081
1.6	0.4700	6.1	1.8083	16	2.7726
1.7	0.5306	6.2	1.8245	17	2.8332
1.8	0.5878	6.3	1.8405	18	2.8904
1.9	0.6419	6.4	1.8563	19	2.9444
2.0	0.6931	6.5	1.8718	20	2.9957
2.1	0.7419	6.6	1.8871	25	3.2189
2.2	0.7885	6.7	1.9021	30	3.4012
2.3	0.8329	6.8	1.9169	35	3.5553
2.4	0.8755	6.9	1.9315	40	3.6889
2.5	0.9163	7.0	1.9459	45	3.8067
2.6	0.9555	7.1	1.9601	50	3.9120
2.7	0.9933	7.2	1.9741	55	4.0073
2.8	1.0296	7.3	1.9879	60	4.0943
2.9	1.0647	7.4	2.0015	65	4.1744
3.0	1.0986	7.5	2.0149	70	4.2485
3.1	1.1314	7.6	2.0281	75	4.3175
3.2	1.1632	7.7	2.0412	80	4.3820
3.3	1.1939	7.8	2.0541	85	4.4427
3.4	1.2238	7.9	2.0669	90	4.4998
3.5	1.2528	8.0	2.0794	100	4.6052
3.6	1.2809	8.1	2.0919	110	4.7005
3.7	1.3083	8.2	2.1041	120	4.7875
3.8	1.3350	8.3	2.1163	130	4.8676
3.9	1.3610	8.4	2.1282	140	4.9416
4.0	1.3863	8.5	2.1401	150	5.0106
4.1	1.4110	8.6	2.1518	160	5.0752
4.2	1.4351	8.7	2.1633	170	5.1358
4.3	1.4586	8.8	2.1748	180	5.1930
4.4	1.4816	8.9	2.1861	190	5.2470

† Subtract 10 for $n < 1$.
Thus, $\log_e 0.1 = 7.6974 - 10 = -2.3026$.

Table IV
Values of Circular Functions

Real Number x or θ radians	θ degrees	sin x or sin θ	csc x or csc θ	tan x or tan θ	cot x or cot θ	sec x or sec θ	cos x or cos θ
0.00	0° 00′	0.0000	No value	0.0000	No value	1.000	1.000
.01	0° 34′	.0100	100.0	.0100	100.0	1.000	1.000
.02	1° 09′	.0200	50.00	.0200	49.99	1.000	0.9998
.03	1° 43′	.0300	33.34	.0300	33.32	1.000	0.9996
.04	2° 18′	.0400	25.01	.0400	24.99	1.001	0.9992
0.05	2° 52′	0.0500	20.01	0.0500	19.98	1.001	0.9988
.06	3° 26′	.0600	16.68	.0601	16.65	1.002	.9982
.07	4° 01′	.0699	14.30	.0701	14.26	1.002	.9976
.08	4° 35′	.0799	12.51	.0802	12.47	1.003	.9968
.09	5° 09′	.0899	11.13	.0902	11.08	1.004	.9960
0.10	5° 44′	0.0998	10.02	0.1003	9.967	1.005	0.9950
.11	6° 18′	.1098	9.109	.1104	9.054	1.006	.9940
.12	6° 53′	.1197	8.353	.1206	8.293	1.007	.9928
.13	7° 27′	.1296	7.714	.1307	7.649	1.009	.9916
.14	8° 01′	.1395	7.166	.1409	7.096	1.010	.9902
0.15	8° 36′	0.1494	6.692	0.1511	6.617	1.011	0.9888
.16	9° 10′	.1593	6.277	.1614	6.197	1.013	.9872
.17	9° 44′	.1692	5.911	.1717	5.826	1.015	.9856
.18	10° 19′	.1790	5.586	.1820	5.495	1.016	.9838
.19	10° 53′	.1889	5.295	.1923	5.200	1.018	.9820
0.20	11° 28′	0.1987	5.033	0.2027	4.933	1.020	0.9801
.21	12° 02′	.2085	4.797	.2131	4.692	1.022	.9780
.22	12° 36′	.2182	4.582	.2236	4.472	1.025	.9759
.23	13° 11′	.2280	4.386	.2341	4.271	1.027	.9737
.24	13° 45′	.2377	4.207	.2447	4.086	1.030	.9713
0.25	14° 19′	0.2474	4.042	0.2553	3.916	1.032	0.9689
.26	14° 54′	.2571	3.890	.2660	3.759	1.035	.9664
.27	15° 28′	.2667	3.749	.2768	3.613	1.038	.9638
.28	16° 03′	.2764	3.619	.2876	3.478	1.041	.9611
.29	16° 37′	.2860	3.497	.2984	3.351	1.044	.9582
0.30	17° 11′	0.2955	3.384	0.3093	3.233	1.047	0.9553
.31	17° 46′	.3051	3.278	.3203	3.122	1.050	.9523
.32	18° 20′	.3146	3.179	.3314	3.018	1.053	.9492
.33	18° 54′	.3240	3.086	.3425	2.920	1.057	.9460
.34	19° 29′	.3335	2.999	.3537	2.827	1.061	.9428
0.35	20° 03′	0.3429	2.916	0.3650	2.740	1.065	0.9394
.36	20° 38′	.3523	2.839	.3764	2.657	1.068	.9359
.37	21° 12′	.3616	2.765	.3879	2.578	1.073	.9323
.38	21° 46′	.3709	2.696	.3994	2.504	1.077	.9287
.39	22° 21′	.3802	2.630	.4111	2.433	1.081	.9249
0.40	22° 55′	0.3894	2.568	0.4228	2.365	1.086	0.9211
.41	23° 29′	.3986	2.509	.4346	2.301	1.090	.9171
.42	24° 04′	.4078	2.452	.4466	2.239	1.095	.9131
.43	24° 38′	.4169	2.399	.4586	2.180	1.100	.9090
.44	25° 13′	.4259	2.348	.4708	2.124	1.105	.9048
0.45	25° 47′	0.4350	2.299	0.4831	2.070	1.111	0.9004

Table IV (continued)

Real Number x or θ radians	θ degrees	sin x or sin θ	csc x or csc θ	tan x or tan θ	cot x or cot θ	sec x or sec θ	cos x or cos θ
0.45	25° 47′	0.4350	2.299	0.4831	2.070	1.111	0.9004
.46	26° 21′	.4439	2.253	.4954	2.018	1.116	.8961
.47	26° 56′	.4529	2.208	.5080	1.969	1.122	.8916
.48	27° 30′	.4618	2.166	.5206	1.921	1.127	.8870
.49	28° 04′	.4706	2.125	.5334	1.875	1.133	.8823
0.50	28° 39′	0.4794	2.086	0.5463	1.830	1.139	0.8776
.51	29° 13′	.4882	2.048	.5594	1.788	1.146	.8727
.52	29° 48′	.4969	2.013	.5726	1.747	1.152	.8678
.53	30° 22′	.5055	1.978	.5859	1.707	1.159	.8628
.54	30° 56′	.5141	1.945	.5994	1.668	1.166	.8577
0.55	31° 31′	0.5227	1.913	0.6131	1.631	1.173	0.8525
.56	32° 05′	.5312	1.883	.6269	1.595	1.180	.8473
.57	32° 40′	.5396	1.853	.6410	1.560	1.188	.8419
.58	33° 14′	.5480	1.825	.6552	1.526	1.196	.8365
.59	33° 48′	.5564	1.797	.6696	1.494	1.203	.8309
0.60	34° 23′	0.5646	1.771	0.6841	1.462	1.212	0.8253
.61	34° 57′	.5729	1.746	.6989	1.431	1.220	.8196
.62	35° 31′	.5810	1.721	.7139	1.401	1.229	.8139
.63	36° 06′	.5891	1.697	.7291	1.372	1.238	.8080
.64	36° 40′	.5972	1.674	.7445	1.343	1.247	.8021
0.65	37° 15′	0.6052	1.652	0.7602	1.315	1.256	0.7961
.66	37° 49′	.6131	1.631	.7761	1.288	1.266	.7900
.67	38° 23′	.6210	1.610	.7923	1.262	1.276	.7838
.68	38° 58′	.6288	1.590	.8087	1.237	1.286	.7776
.69	39° 32′	.6365	1.571	.8253	1.212	1.297	.7712
0.70	40° 06′	0.6442	1.552	0.8423	1.187	1.307	0.7648
.71	40° 41′	.6518	1.534	.8595	1.163	1.319	.7584
.72	41° 15′	.6594	1.517	.8771	1.140	1.330	.7518
.73	41° 50′	.6669	1.500	.8949	1.117	1.342	.7452
.74	42° 24′	.6743	1.483	.9131	1.095	1.354	.7385
0.75	42° 58′	0.6816	1.467	0.9316	1.073	1.367	0.7317
.76	43° 33′	.6889	1.452	.9505	1.052	1.380	.7248
.77	44° 07′	.6961	1.436	.9697	1.031	1.393	.7179
.78	44° 41′	.7033	1.422	.9893	1.011	1.407	.7109
.79	45° 16′	.7104	1.408	1.009	.9908	1.421	.7038
0.80	45° 50′	0.7174	1.394	1.030	0.9712	1.435	0.6967
.81	46° 25′	.7243	1.381	1.050	.9520	1.450	.6895
.82	46° 59′	.7311	1.368	1.072	.9331	1.466	.6822
.83	47° 33′	.7379	1.355	1.093	.9146	1.482	.6749
.84	48° 08′	.7446	1.343	1.116	.8964	1.498	.6675
0.85	48° 42′	0.7513	1.331	1.138	0.8785	1.515	0.6600
.86	49° 16′	.7578	1.320	1.162	.8609	1.533	.6524
.87	49° 51′	.7643	1.308	1.185	.8437	1.551	.6448
.88	50° 25′	.7707	1.297	1.210	.8267	1.569	.6372
.89	51° 00′	.7771	1.287	1.235	.8100	1.589	.6294
0.90	51° 34′	0.7833	1.277	1.260	0.7936	1.609	0.6216
.91	52° 08′	.7895	1.267	1.286	.7774	1.629	.6137
.92	52° 43′	.7956	1.257	1.313	.7615	1.651	.6058
.93	53° 17′	.8016	1.247	1.341	.7458	1.673	.5978
.94	53° 51′	.8076	1.238	1.369	.7303	1.696	.5898
0.95	54° 26′	0.8134	1.229	1.398	0.7151	1.719	0.5817

Table IV (*continued*)

Real Number x or θ radians	θ degrees	sin x or sin θ	csc x or csc θ	tan x or tan θ	cot x or cot θ	sec x or sec θ	cos x or cos θ
0.95	54° 26′	0.8134	1.229	1.398	0.7151	1.719	0.5817
.96	55° 00′	.8192	1.221	1.428	.7001	1.744	.5735
.97	55° 35′	.8249	1.212	1.459	.6853	1.769	.5653
.98	56° 09′	.8305	1.204	1.491	.6707	1.795	.5570
.99	56° 43′	.8360	1.196	1.524	.6563	1.823	.5487
1.00	57° 18′	0.8415	1.188	1.557	0.6421	1.851	0.5403
1.01	57° 52′	.8468	1.181	1.592	.6281	1.880	.5319
1.02	58° 27′	.8521	1.174	1.628	.6142	1.911	.5234
1.03	59° 01′	.8573	1.166	1.665	.6005	1.942	.5148
1.04	59° 35′	.8624	1.160	1.704	.5870	1.975	.5062
1.05	60° 10′	0.8674	1.153	1.743	0.5736	2.010	0.4976
1.06	60° 44′	.8724	1.146	1.784	.5604	2.046	.4889
1.07	61° 18′	.8772	1.140	1.827	.5473	2.083	.4801
1.08	61° 53′	.8820	1.134	1.871	.5344	2.122	.4713
1.09	62° 27′	.8866	1.128	1.917	.5216	2.162	.4625
1.10	63° 02′	0.8912	1.122	1.965	0.5090	2.205	0.4536
1.11	63° 36′	.8957	1.116	2.014	.4964	2.249	.4447
1.12	64° 10′	.9001	1.111	2.066	.4840	2.295	.4357
1.13	64° 45′	.9044	1.106	2.120	.4718	2.344	.4267
1.14	65° 19′	.9086	1.101	2.176	.4596	2.395	.4176
1.15	65° 53′	0.9128	1.096	2.234	0.4475	2.448	0.4085
1.16	66° 28′	.9168	1.091	2.296	.4356	2.504	.3993
1.17	67° 02′	.9208	1.086	2.360	.4237	2.563	.3902
1.18	67° 37′	.9246	1.082	2.427	.4120	2.625	.3809
1.19	68° 11′	.9284	1.077	2.498	.4003	2.691	.3717
1.20	68° 45′	0.9320	1.073	2.572	0.3888	2.760	0.3624
1.21	69° 20′	.9356	1.069	2.650	.3773	2.833	.3530
1.22	69° 54′	.9391	1.065	2.733	.3659	2.910	.3436
1.23	70° 28′	.9425	1.061	2.820	.3546	2.992	.3342
1.24	71° 03′	.9458	1.057	2.912	.3434	3.079	.3248
1.25	71° 37′	0.9490	1.054	3.010	0.3323	3.171	0.3153
1.26	72° 12′	.9521	1.050	3.113	.3212	3.270	.3058
1.27	72° 46′	.9551	1.047	3.224	.3102	3.375	.2963
1.28	72° 20′	.9580	1.044	3.341	.2993	3.488	.2867
1.29	73° 55′	.9608	1.041	3.467	.2884	3.609	.2771
1.30	74° 29′	0.9636	1.038	3.602	0.2776	3.738	0.2675
1.31	75° 03′	.9662	1.035	3.747	.2669	3.878	.2579
1.32	75° 38′	.9687	1.032	3.903	.2562	4.029	.2482
1.33	76° 12′	.9711	1.030	4.072	.2456	4.193	.2385
1.34	76° 47′	.9735	1.027	4.256	.2350	4.372	.2288
1.35	77° 21′	0.9757	1.025	4.455	0.2245	4.566	0.2190
1.36	77° 55′	.9779	1.023	4.673	.2140	4.779	.2092
1.37	78° 30′	.9799	1.021	4.913	.2035	5.014	.1994
1.38	79° 04′	.9819	1.018	5.177	.1931	5.273	.1896
1.39	79° 38′	.9837	1.017	5.471	.1828	5.561	.1798
1.40	80° 13′	0.9854	1.015	5.798	0.1725	5.883	0.1700
1.41	80° 47′	.9871	1.013	6.165	.1622	6.246	.1601
1.42	81° 22′	.9887	1.011	6.581	.1519	6.657	.1502
1.43	81° 56′	.9901	1.010	7.055	.1417	7.126	.1403
1.44	82° 30′	.9915	1.009	7.602	.1315	7.667	.1304
1.45	83° 05′	0.9927	1.007	8.238	0.1214	8.299	0.1205

Table IV (continued)

Real Number x or θ radians	θ degrees	sin x or sin θ	csc x or csc θ	tan x or tan θ	cot x or cot θ	sec x or sec θ	cos x or cos θ
1.45	83° 05′	0.9927	1.007	8.238	0.1214	8.299	0.1205
1.46	83° 39′	.9939	1.006	8.989	.1113	9.044	.1106
1.47	84° 13′	.9949	1.005	9.887	.1011	9.938	.1006
1.48	84° 48′	.9959	1.004	10.98	.0910	11.03	.0907
1.49	85° 22′	.9967	1.003	12.35	.0810	12.39	.0807
1.50	85° 57′	0.9975	1.003	14.10	0.0709	14.14	0.0707
1.51	86° 31′	.9982	1.002	16.43	.0609	16.46	.0608
1.52	87° 05′	.9987	1.001	19.67	.0508	19.69	.0508
1.53	87° 40′	.9992	1.001	24.50	.0408	24.52	.0408
1.54	88° 14′	.9995	1.000	32.46	.0308	32.48	.0308
1.55	88° 49′	0.9998	1.000	48.08	0.0208	48.09	0.0208
1.56	89° 23′	.9999	1.000	92.62	.0108	92.63	.0108
1.57	89° 57′	1.000	1.000	1256	.0008	1256	.0008

Table V
Values of Trigonometric Functions

θ deg	θ deg-min	sin θ	cos θ	tan θ	csc θ	sec θ	cot θ		
0.0°	0°00′	0.0000	1.0000	0.0000	no value	1.0000	no value	90°0′	90.0°
0.1	0 06	0.0017	1.0000	0.0017	572.96	1.0000	572.96	89 54	89.9
0.2	0 12	0.0035	1.0000	0.0035	286.48	1.0000	286.48	89 48	89.8
0.3	0 18	0.0052	1.0000	0.0052	190.99	1.0000	190.98	89 42	89.7
0.4	0 24	0.0070	1.0000	0.0070	143.24	1.0000	143.24	89 36	89.6
0.5	0 30	0.0087	1.0000	0.0087	114.59	1.0000	114.59	89 30	89.5
0.6	0 36	0.0105	0.9999	0.0105	95.495	1.0001	95.490	89 24	89.4
0.7	0 42	0.0122	0.9999	0.0122	81.853	1.0001	81.847	89 18	89.3
0.8	0 48	0.0140	0.9999	0.0140	71.622	1.0001	71.615	89 12	89.2
0.9	0 54	0.0157	0.9999	0.0157	63.665	1.0001	63.657	89 06	89.1
1.0°	1°00′	0.0175	0.9998	0.0175	57.299	1.0002	57.290	89°00′	89.0°
1.1	1 06	0.0192	0.9998	0.0192	52.090	1.0002	52.081	88 54	88.9
1.2	1 12	0.0209	0.9998	0.0209	47.750	1.0002	47.740	88 48	88.8
1.3	1 18	0.0227	0.9997	0.0227	44.077	1.0003	44.066	88 42	88.7
1.4	1 24	0.0244	0.9997	0.0244	40.930	1.0003	40.917	88 36	88.6
1.5	1 30	0.0262	0.9997	0.0262	38.202	1.0003	38.188	88 30	88.5
1.6	1 36	0.0279	0.9996	0.0279	35.815	1.0004	35.801	88 24	88.4
1.7	1 42	0.0297	0.9996	0.0297	33.708	1.0004	33.694	88 18	88.3
1.8	1 48	0.0314	0.9995	0.0314	31.836	1.0005	31.821	88 12	88.2
1.9	1 54	0.0332	0.9995	0.0332	30.161	1.0005	30.145	88 06	88.1
2.0°	2°00′	0.0349	0.9994	0.0349	28.654	1.0006	28.636	88°00′	88.0°
2.1	2 06	0.0366	0.9993	0.0367	27.290	1.0007	27.271	87 54	87.9
2.2	2 12	0.0384	0.9993	0.0384	26.050	1.0007	26.031	87 48	87.8
2.3	2 18	0.0401	0.9992	0.0402	24.918	1.0008	24.898	87 42	87.7
2.4	2 24	0.0419	0.9991	0.0419	23.880	1.0009	23.859	87 36	87.6
2.5	2 30	0.0436	0.9990	0.0437	22.926	1.0010	22.904	87 30	87.5
2.6	2 36	0.0454	0.9990	0.0454	22.044	1.0010	22.022	87 24	87.4
2.7	2 42	0.0471	0.9989	0.0472	21.229	1.0011	21.205	87 18	87.3
2.8	2 48	0.0488	0.9988	0.0489	20.471	1.0012	20.446	87 12	87.2
2.9	2 54	0.0506	0.9987	0.0507	19.766	1.0013	19.740	87 06	87.1
3.0°	3°00′	0.0523	0.9986	0.0524	19.107	1.0014	19.081	87°00′	87.0°
3.1	3 06	0.0541	0.9985	0.0542	18.492	1.0015	18.464	86 54	86.9
3.2	3 12	0.0558	0.9984	0.0559	17.914	1.0016	17.886	86 48	86.8
3.3	3 18	0.0576	0.9983	0.0577	17.372	1.0017	17.343	86 42	86.7
3.4	3 24	0.0593	0.9982	0.0594	16.862	1.0018	16.832	86 36	86.6
3.5	3 30	0.0610	0.9981	0.0612	16.380	1.0019	16.350	86 30	86.5
3.6	3 36	0.0628	0.9980	0.0629	15.926	1.0020	15.895	86 24	86.4
3.7	3 42	0.0645	0.9979	0.0647	15.496	1.0021	15.464	86 18	86.3
3.8	3 48	0.0663	0.9978	0.0664	15.089	1.0022	15.056	86 12	86.2
3.9	3 54	0.0680	0.9977	0.0682	14.703	1.0023	14.669	86 06	86.1
4.0°	4°00′	0.0698	0.9976	0.0699	14.336	1.0024	14.301	86°00′	86.0°
4.1	4 06	0.0715	0.9974	0.0717	13.987	1.0026	13.951	85 54	85.9
4.2	4 12	0.0732	0.9973	0.0734	13.654	1.0027	13.617	85 48	85.8
4.3	4 18	0.0750	0.9972	0.0752	13.337	1.0028	13.300	85 42	85.7
4.4	4 24	0.0767	0.9971	0.0769	13.035	1.0030	12.996	85 36	85.6
4.5	4 30	0.0785	0.9969	0.0787	12.746	1.0031	12.706	85 30	85.5
4.6	4 36	0.0802	0.9968	0.0805	12.469	1.0032	12.429	85 24	85.4
4.7	4 42	0.0819	0.9966	0.0822	12.204	1.0034	12.163	85 18	85.3
4.8	4 48	0.0837	0.9965	0.0840	11.951	1.0035	11.909	85 12	85.2
4.9	4 54	0.0854	0.9963	0.0857	11.707	1.0037	11.665	85°06′	85.1°
		cos θ	sin θ	cot θ	sec θ	csc θ	tan θ	θ deg-min	θ deg

Table V (continued)

θ deg	θ deg-min	sin θ	cos θ	tan θ	csc θ	sec θ	cot θ		
5.0°	5°00′	0.0872	0.9962	0.0875	11.474	1.0038	11.430	85°00′	85.0°
5.1	5 06	0.0889	0.9960	0.0892	11.249	1.0040	11.205	84 54	84.9
5.2	5 12	0.0906	0.9959	0.0910	11.034	1.0041	10.988	84 48	84.8
5.3	5 18	0.0924	0.9957	0.0928	10.826	1.0043	10.780	84 42	84.7
5.4	5 24	0.0941	0.9956	0.0945	10.626	1.0045	10.579	84 36	84.6
5.5	5 30	0.0958	0.9954	0.0963	10.433	1.0046	10.385	84 30	84.5
5.6	5 36	0.0976	0.9952	0.0981	10.248	1.0048	10.199	84 24	84.4
5.7	5 42	0.0993	0.9951	0.0998	10.069	1.0050	10.019	84 18	84.3
5.8	5 48	0.1011	0.9949	0.1016	9.8955	1.0051	9.8448	84 12	84.2
5.9	5 54	0.1028	0.9947	0.1033	9.7283	1.0053	9.6768	84 06	84.1
6.0°	6°00′	0.1045	0.9945	0.1051	9.5668	1.0055	9.5144	84°00′	84.0°
6.1	6 06	0.1063	0.9943	0.1069	9.4105	1.0057	9.3573	83 54	83.9
6.2	6 12	0.1080	0.9942	0.1086	9.2593	1.0059	9.2052	83 48	83.8
6.3	6 18	0.1097	0.9940	0.1104	9.1129	1.0061	9.0579	83 42	83.7
6.4	6 24	0.1115	0.9938	0.1122	8.9711	1.0063	8.9152	83 36	83.6
6.5	6 30	0.1132	0.9936	0.1139	8.8337	1.0065	8.7769	83 30	83.5
6.6	6 36	0.1149	0.9934	0.1157	8.7004	1.0067	8.6428	83 24	83.4
6.7	6 42	0.1167	0.9932	0.1175	8.5711	1.0069	8.5126	83 18	83.3
6.8	6 48	0.1184	0.9930	0.1192	8.4457	1.0071	8.3863	83 12	83.2
6.9	6 54	0.1201	0.9928	0.1210	8.3238	1.0073	8.2636	83 06	83.1
7.0°	7°00′	0.1219	0.9925	0.1228	8.2055	1.0075	8.1444	83°00′	83.0°
7.1	7 06	0.1236	0.9923	0.1246	8.0905	1.0077	8.0285	82 54	82.9
7.2	7 12	0.1253	0.9921	0.1263	7.9787	1.0079	7.9158	82 48	82.8
7.3	7 18	0.1271	0.9919	0.1281	7.8700	1.0082	7.8062	82 42	82.7
7.4	7 24	0.1288	0.9917	0.1299	7.7642	1.0084	7.6996	82 36	82.6
7.5	7 30	0.1305	0.9914	0.1317	7.6613	1.0086	7.5958	82 30	82.5
7.6	7 36	0.1323	0.9912	0.1334	7.5611	1.0089	7.4947	82 24	82.4
7.7	7 42	0.1340	0.9910	0.1352	7.4635	1.0091	7.3962	82 18	82.3
7.8	7 48	0.1357	0.9907	0.1370	7.3684	1.0093	7.3002	82 12	82.2
7.9	7 54	0.1374	0.9905	0.1388	7.2757	1.0096	7.2066	82 06	82.1
8.0°	8°00′	0.1392	0.9903	0.1405	7.1853	1.0098	7.1154	82°00′	82.0°
8.1	8 06	0.1409	0.9900	0.1423	7.0972	1.0101	7.0264	81 54	81.9
8.2	8 12	0.1426	0.9898	0.1441	7.0112	1.0103	6.9395	81 48	81.8
8.3	8 18	0.1444	0.9895	0.1459	6.9273	1.0106	6.8548	81 42	81.7
8.4	8 24	0.1461	0.9893	0.1477	6.8454	1.0108	6.7720	81 36	81.6
8.5	8 30	0.1478	0.9890	0.1495	6.7655	1.0111	6.6912	81 30	81.5
8.6	8 36	0.1495	0.9888	0.1512	6.6874	1.0114	6.6122	81 24	81.4
8.7	8 42	0.1513	0.9885	0.1530	6.6111	1.0116	6.5350	81 18	81.3
8.8	8 48	0.1530	0.9882	0.1548	6.5366	1.0119	6.4596	81 12	81.2
8.9	8 54	0.1547	0.9880	0.1566	6.4637	1.0122	6.3859	81 06	81.1
9.0°	9°00′	0.1564	0.9877	0.1584	6.3925	1.0125	6.3138	81°00′	81.0°
9.1	9 06	0.1582	0.9874	0.1602	6.3228	1.0127	6.2432	80 54	80.9
9.2	9 12	0.1599	0.9871	0.1620	6.2547	1.0130	6.1742	80 48	80.8
9.3	9 18	0.1616	0.9869	0.1638	6.1880	1.0133	6.1066	80 42	80.7
9.4	9 24	0.1633	0.9866	0.1655	6.1227	1.0136	6.0405	80 36	80.6
9.5	9 30	0.1650	0.9863	0.1673	6.0589	1.0139	5.9758	80 30	80.5
9.6	9 36	0.1668	0.9860	0.1691	5.9963	1.0142	5.9124	80 24	80.4
9.7	9 42	0.1685	0.9857	0.1709	5.9351	1.0145	5.8502	80 18	80.3
9.8	9 48	0.1702	0.9854	0.1727	5.8751	1.0148	5.7894	80 12	80.2
9.9	9 54	0.1719	0.9851	0.1745	5.8164	1.0151	5.7297	80°06′	80.1°
		cos θ	sin θ	cot θ	sec θ	csc θ	tan θ	θ deg-min	θ deg

Table V (*continued*)

θ deg	θ deg-min	sin θ	cos θ	tan θ	csc θ	sec θ	cot θ		
10.0°	10°00′	0.1736	0.9848	0.1763	5.7588	1.0154	5.6713	80°00′	80.0°
10.1	10 06	0.1754	0.9845	0.1781	5.7023	1.0157	5.6140	79 54	79.9
10.2	10 12	0.1771	0.9842	0.1799	5.6470	1.0161	5.5578	79 48	79.8
10.3	10 18	0.1788	0.9839	0.1817	5.5928	1.0164	5.5027	79 42	79.7
10.4	10 24	0.1805	0.9836	0.1835	5.5396	1.0167	5.4486	79 36	79.6
10.5	10 30	0.1822	0.9833	0.1853	5.4874	1.0170	5.3955	79 30	79.5
10.6	10 36	0.1840	0.9829	0.1871	5.4362	1.0174	5.3435	79 24	79.4
10.7	10 42	0.1857	0.9826	0.1890	5.3860	1.0177	5.2924	79 18	79.3
10.8	10 48	0.1874	0.9823	0.1908	5.3367	1.0180	5.2422	79 12	79.2
10.9	10 54	0.1891	0.9820	0.1926	5.2883	1.0184	5.1929	79 06	79.1
11.0°	11°00′	0.1908	0.9816	0.1944	5.2408	1.0187	5.1446	79°00′	79.0°
11.1	11 06	0.1925	0.9813	0.1962	5.1942	1.0191	5.0970	78 54	78.9
11.2	11 12	0.1942	9.9810	0.1980	5.1484	1.0194	5.0504	78 48	78.8
11.3	11 18	0.1959	0.9806	0.1998	5.1034	1.0198	5.0045	78 42	78.7
11.4	11 24	0.1977	0.9803	0.2016	5.0593	1.0201	4.9595	78 36	78.6
11.5	11 30	0.1994	0.9799	0.2035	5.0159	1.0205	4.9152	78 30	78.5
11.6	11 36	0.2011	0.9796	0.2053	4.9732	1.0209	4.8716	78 24	78.4
11.7	11 42	0.2028	0.9792	0.2071	4.9313	1.0212	4.8288	78 18	78.3
11.8	11 48	0.2045	0.9789	0.2089	4.8901	1.0216	4.7867	78 12	78.2
11.9	11 54	0.2062	0.9785	0.2107	4.8496	1.0220	4.7453	78 06	78.1
12.0°	12°00′	0.2079	0.9781	0.2126	4.8097	1.0223	4.7046	78°00′	78.0°
12.1	12 06	0.2096	0.9778	0.2144	4.7706	1.0227	4.6646	77 54	77.9
12.2	12 12	0.2113	0.9774	0.2162	4.7321	1.0231	4.6252	77 48	77.8
12.3	12 18	0.2130	0.9770	0.2180	4.6942	1.0235	4.5864	77 42	77.7
12.4	12 24	0.2147	0.9767	0.2199	4.6569	1.0239	4.5483	77 36	77.6
12.5	12 30	0.2164	0.9763	0.2217	4.6202	1.0243	4.5107	77 30	77.5
12.6	12 36	0.2181	0.9759	0.2235	4.5841	1.0247	4.4737	77 24	77.4
12.7	12 42	0.2198	0.9755	0.2254	4.5486	1.0251	4.4374	77 18	77.3
12.8	12 48	0.2215	0.9751	0.2272	4.5137	1.0255	4.4015	77 12	77.2
12.9	12 54	0.2232	0.9748	0.2290	4.4793	1.0259	4.3662	77 06	77.1
13.0°	13°00′	0.2250	0.9744	0.2309	4.4454	1.0263	4.3315	77°00′	77.0°
13.1	13 06	0.2267	0.9740	0.2327	4.4121	1.0267	4.2972	76 54	76.9
13.2	13 12	0.2284	0.9736	0.2345	4.3792	1.0271	4.2635	76 48	76.8
13.3	13 18	0.2300	0.9732	0.2364	4.3469	1.0276	4.2303	76 42	76.7
13.4	13 24	0.2317	0.9728	0.2382	4.3150	1.0280	4.1976	76 36	76.6
13.5	13 30	0.2334	0.9724	0.2401	4.2837	1.0284	4.1653	76 30	76.5
13.6	13 36	0.2351	0.9720	0.2419	4.2528	1.0288	4.1335	76 24	76.4
13.7	13 42	0.2368	0.9715	0.2438	4.2223	1.0293	4.1022	76 18	76.3
13.8	13 48	0.2385	0.9711	0.2456	4.1923	1.0297	4.0713	76 12	76.2
13.9	13 54	0.2402	0.9707	0.2475	4.1627	1.0302	4.0408	76 06	76.1
14.0°	14°00′	0.2419	0.9703	0.2493	4.1336	1.0306	4.0108	76°00′	76.0°
14.1	14 06	0.2436	0.9699	0.2512	4.1048	1.0311	3.9812	75 54	75.9
14.2	14 12	0.2453	0.9694	0.2530	4.0765	1.0315	3.9520	75 48	75.8
14.3	14 18	0.2470	0.9690	0.2549	4.0486	1.0320	3.9232	75 42	75.7
14.4	14 24	0.2487	0.9686	0.2568	4.0211	1.0324	3.8947	75 36	75.6
14.5	14 30	0.2504	0.9681	0.2586	3.9939	1.0329	3.8667	75 30	75.5
14.6	14 36	0.2521	0.9677	0.2605	3.9672	1.0334	3.8391	75 24	75.4
14.7	14 42	0.2538	0.9673	0.2623	3.9408	1.0338	3.8118	75 18	75.3
14.8	14 48	0.2554	0.9668	0.2642	3.9147	1.0343	3.7849	75 12	75.2
14.9	14 54	0.2571	0.9664	0.2661	3.8890	1.0348	3.7583	75°06′	75.1°
		cos θ	sin θ	cot θ	sec θ	csc θ	tan θ	θ deg-min	θ deg

Table V (continued)

θ deg	θ deg-min	sin θ	cos θ	tan θ	csc θ	sec θ	cot θ		
15.0°	15°00′	0.2588	0.9659	0.2679	3.8637	1.0353	3.7321	75°00′	75.0°
15.1	15 06	0.2605	0.9655	0.2698	3.8387	1.0358	3.7062	74 54	74.9
15.2	15 12	0.2622	0.9650	0.2717	3.8140	1.0363	3.6806	74 48	74.8
15.3	15 18	0.2639	0.9646	0.2736	3.7897	1.0367	3.6554	74 42	74.7
15.4	15 24	0.2656	0.9641	0.2754	3.7657	1.0372	3.6305	74 36	74.6
15.5	15 30	0.2672	0.9636	0.2773	3.7420	1.0377	3.6059	74 30	74.5
15.6	15 36	0.2689	0.9632	0.2792	3.7186	1.0382	3.5816	74 24	74.4
15.7	15 42	0.2706	0.9627	0.2811	3.6955	1.0388	3.5576	74 18	74.3
15.8	15 48	0.2723	0.9622	0.2830	3.6727	1.0393	3.5339	74 12	74.2
15.9	15 54	0.2740	0.9617	0.2849	3.6502	1.0398	3.5105	74 06	74.1
16.0°	16°00′	0.2756	0.9613	0.2867	3.6280	1.0403	3.4874	74°00′	74.0°
16.1	16 06	0.2773	0.9608	0.2886	3.6060	1.0408	3.4646	73 54	73.9
16.2	16 12	0.2790	0.9603	0.2905	3.5843	1.0413	3.4420	73 48	73.8
16.3	16 18	0.2807	0.9598	0.2924	3.5629	1.0419	3.4197	73 42	73.7
16.4	16 24	0.2823	0.9593	0.2943	3.5418	1.0424	3.3977	73 36	73.6
16.5	16 30	0.2840	0.9588	0.2962	3.5209	1.0429	3.3759	73 30	73.5
16.6	16 36	0.2857	0.9583	0.2981	3.5003	1.0435	3.3544	73 24	74.4
16.7	16 42	0.2874	0.9578	0.3000	3.4800	1.0440	3.3332	73 18	73.3
16.8	16 48	0.2890	0.9573	0.3019	3.4598	1.0446	3.3122	73 12	73.2
16.9	16 54	0.2907	0.9568	0.3038	3.4399	1.0451	3.2914	73 06	73.1
17.0°	17 00′	0.2924	0.9563	0.3057	3.4203	1.0457	3.2709	73°00′	73.0°
17.1	17 06	0.2940	0.9558	0.3076	3.4009	1.0463	3.2506	72 54	72.9
17.2	17 12	0.2957	0.9553	0.3096	3.3817	1.0468	3.2305	72 48	72.8
17.3	17 18	0.2974	0.9548	0.3115	3.3628	1.0474	3.2106	72 42	72.7
17.4	17 24	0.2990	0.9542	0.3134	3.3440	1.0480	3.1910	72 36	72.6
17.5	17 30	0.3007	0.9537	0.3153	3.3255	1.0485	3.1716	72 30	72.5
17.6	17 36	0.3024	0.9532	0.3172	3.3072	1.0491	3.1524	72 24	72.4
17.7	17 42	0.3040	0.9527	0.3191	3.2891	1.0497	3.1334	72 18	72.3
17.8	17 48	0.3057	0.9521	0.3211	3.2712	1.0503	3.1146	72 12	72.2
17.9	17 54	0.3074	0.9516	0.3230	3.2536	1.0509	3.0961	72 06	72.1
18.0°	18°00′	0.3090	0.9511	0.3249	3.2361	1.0515	3.0777	72°00′	72.0°
18.1	18 06	0.3107	0.9505	0.3268	3.2188	1.0521	3.0595	71 54	71.9
18.2	18 12	0.3123	0.9500	0.3288	3.2017	1.0527	3.0415	71 48	71.8
18.3	18 18	0.3140	0.9494	0.3307	3.1848	1.0533	3.0237	71 42	71.7
18.4	18 24	0.3156	0.9489	0.3327	3.1681	1.0539	3.0061	71 36	71.6
18.5	18 30	0.3173	0.9483	0.3346	3.1515	1.0545	2.9887	71 30	71.5
18.6	18 36	0.3190	0.9478	0.3365	3.1352	1.0551	2.9714	71 24	71.4
18.7	18 42	0.3206	0.9472	0.3385	3.1190	1.0557	2.9544	71 18	71.3
18.8	18 48	0.3223	0.9466	0.3404	3.1030	1.0564	2.9375	71 12	71.2
18.9	18 54	0.3239	0.9461	0.3424	3.0872	1.0570	2.9208	71 06	71.1
19.0°	19°00′	0.3256	0.9455	0.3443	3.0716	1.0576	2.9042	71°00′	71.0°
19.1	19 06	0.3272	0.9449	0.3463	3.0561	1.0583	2.8878	70 54	70.9
19.2	19 12	0.3289	0.9444	0.3482	3.0407	1.0589	2.8716	70 48	70.8
19.3	19 18	0.3305	0.9438	0.3502	3.0256	1.0595	2.8556	70 42	70.7
19.4	19 24	0.3322	0.9432	0.3522	3.0106	1.0602	2.8397	70 36	70.6
19.5	19 30	0.3338	0.9426	0.3541	2.9957	1.0608	2.8239	70 30	70.5
19.6	19 36	0.3355	0.9421	0.3561	2.9811	1.0615	2.8083	70 24	70.4
19.7	19 42	0.3371	0.9415	0.3581	2.9665	1.0622	2.7929	70 18	70.3
19.8	19 48	0.3387	0.9409	0.3600	2.9521	1.0628	2.7776	70 12	70.2
19.9	19 54	0.3404	0.9403	0.3620	2.9379	1.0635	2.7625	70°06′	70.1°
		cos θ	sin θ	cot θ	sec θ	csc θ	tan θ	θ deg-min	θ deg

Table V (*continued*)

θ deg	θ deg-min	$\sin\theta$	$\cos\theta$	$\tan\theta$	$\csc\theta$	$\sec\theta$	$\cot\theta$		
20.0°	20°00′	0.3420	0.9397	0.3640	2.9238	1.0642	2.7475	70°00′	70.0°
20.1	20 06	0.3437	0.9391	0.3659	2.9099	1.0649	2.7326	69 54	69.9
20.2	20 12	0.3453	0.9385	0.3679	2.8960	1.0655	2.7179	69 48	69.8
20.3	20 18	0.3469	0.9379	0.3699	2.8824	1.0662	2.7034	69 42	69.7
20.4	20 24	0.3486	0.9373	0.3719	2.8688	1.0669	2.6889	69 36	69.6
20.5	20 30	0.3502	0.9367	0.3739	2.8555	1.0676	2.6746	69 30	69.5
20.6	20 36	0.3518	0.9361	0.3759	2.8422	1.0683	2.6605	69 24	69.4
20.7	20 42	0.3535	0.9354	0.3779	2.8291	1.0690	2.6464	69 18	69.3
20.8	20 48	0.3551	0.9348	0.3799	2.8161	1.0697	2.6325	69 12	69.2
20.9	20 54	0.3567	0.9342	0.3819	2.8032	1.0704	2.6187	69 06	69.1
21.0°	21°00′	0.3584	0.9336	0.3839	2.7904	1.0711	2.6051	69°00′	69.0°
21.1	21 06	0.3600	0.9330	0.3859	2.7778	1.0719	2.5916	68 54	68.9
21.2	21 12	0.3616	0.9323	0.3879	2.7653	1.0726	2.5782	68 48	68.8
21.3	21 18	0.3633	0.9317	0.3899	2.7529	1.0733	2.5649	68 42	68.7
21.4	21 24	0.3649	0.9311	0.3919	2.7407	1.0740	2.5517	68 36	68.6
21.5	21 30	0.3665	0.9304	0.3939	2.7285	1.0748	2.5386	68 30	68.5
21.6	21 36	0.3681	0.9298	0.3959	2.7165	1.0755	2.5257	68 24	68.4
21.7	21 42	0.3697	0.9291	0.3979	2.7046	1.0763	2.5129	68 18	68.3
21.8	21 48	0.3714	0.9285	0.4000	2.6927	1.0770	2.5002	68 12	68.2
21.9	21 54	0.3730	0.9278	0.4020	2.6811	1.0778	2.4876	68 06	68.1
22.0°	22°00′	0.3746	0.9272	0.4040	2.6695	1.0785	2.4751	68°00′	68.0°
22.1	22 06	0.3762	0.9265	0.4061	2.6580	1.0793	2.4627	67 54	67.9
22.2	22 12	0.3778	0.9259	0.4081	2.6466	1.0801	2.4504	67 48	67.8
22.3	22 18	0.3795	0.9252	0.4101	2.6354	1.0808	2.4383	67 42	67.7
22.4	22 24	0.3811	0.9245	0.4122	2.6242	1.0816	2.4262	67 36	67.6
22.5	22 30	0.3827	0.9239	0.4142	2.6131	1.0824	2.4142	67 30	67.5
22.6	22 36	0.3843	0.9232	0.4163	2.6022	1.0832	2.4023	67 24	67.4
22.7	22 42	0.3859	0.9225	0.4183	2.5913	1.0840	2.3906	67 18	67.3
22.8	22 48	0.3875	0.9219	0.4204	2.5805	1.0848	2.3789	67 12	67.2
22.9	22 54	0.3891	0.9212	0.4224	2.5699	1.0856	2.3673	67 06	67.1
23.0°	23°00′	0.3907	0.9205	0.4245	2.5593	1.0864	2.3559	67°00′	67.0°
23.1	23 06	0.3923	0.9198	0.4265	2.5488	1.0872	2.3445	66 54	66.9
23.2	23 12	0.3939	0.9191	0.4286	2.5384	1.0880	2.3332	66 48	66.8
23.3	23 18	0.3955	0.9184	0.4307	2.5282	1.0888	2.3220	66 42	66.7
23.4	23 24	0.3971	0.9178	0.4327	2.5180	1.0896	2.3109	66 36	66.6
23.5	23 30	0.3987	0.9171	0.4348	2.5078	1.0904	2.2998	66 30	66.5
23.6	23 36	0.4003	0.9164	0.4369	2.4978	1.0913	2.2889	66 24	66.4
23.7	23 42	0.4019	0.9157	0.4390	2.4879	1.0921	2.2781	66 18	66.3
23.8	23 48	0.4035	0.9150	0.4411	2.4780	1.0929	2.2673	66 12	66.2
23.9	23 54	0.4051	0.9143	0.4431	2.4683	1.0938	2.2566	66 06	66.1
24.0°	24°00′	0.4067	0.9135	0.4452	2.4586	1.0946	2.2460	66°00′	66.0°
24.1	24 06	0.4083	0.9128	0.4473	2.4490	1.0955	2.2355	65 54	65.9
24.2	24 12	0.4099	0.9121	0.4494	2.4395	1.0963	2.2251	65 48	65.8
24.3	24 18	0.4115	0.9114	0.4515	2.4301	1.0972	2.2148	65 42	65.7
24.4	24 24	0.4131	0.9107	0.4536	2.4207	1.0981	2.2045	65 36	65.6
24.5	24 30	0.4147	0.9100	0.4557	2.4114	1.0989	2.1943	65 30	65.5
24.6	24 36	0.4163	0.9092	0.4578	2.4022	1.0998	2.1842	65 24	65.4
24.7	24 42	0.4179	0.9085	0.4599	2.3931	1.1007	2.1742	65 18	65.3
24.8	24 48	0.4195	0.9078	0.4621	2.3841	1.1016	2.1642	65 12	65.2
24.9	24 54	0.4210	0.9070	0.4642	2.3751	1.1025	2.1543	65°06′	65.1°
		$\cos\theta$	$\sin\theta$	$\cot\theta$	$\sec\theta$	$\csc\theta$	$\tan\theta$	θ deg-min	θ deg

Table V (continued)

θ deg	deg-min	sin θ	cos θ	tan θ	csc θ	sec θ	cot θ		
25.0°	25°00′	0.4226	0.9063	0.4663	2.3662	1.1034	2.1445	65°00′	65.0°
25.1	25 06	0.4242	0.9056	0.4684	2.3574	1.1043	2.1348	64 54	64.9
25.2	25 12	0.4258	0.9048	0.4706	2.3486	1.1052	2.1251	64 48	64.8
25.3	25 18	0.4274	0.9041	0.4727	2.3400	1.1061	2.1155	64 42	64.7
25.4	25 24	0.4289	0.9033	0.4748	2.3314	1.1070	2.1060	64 36	64.6
25.5	25 30	0.4305	0.9026	0.4770	2.3228	1.1079	2.0965	64 30	64.5
25.6	25 36	0.4321	0.9018	0.4791	2.3144	1.1089	2.0872	64 24	64.4
25.7	25 42	0.4337	0.9011	0.4813	2.3060	1.1098	2.0778	64 18	64.3
25.8	25 48	0.4352	0.9003	0.4834	2.2976	1.1107	2.0686	64 12	64.2
25.9	25 54	0.4368	0.8996	0.4856	2.2894	1.1117	2.0594	64 06	64.1
26.0°	26°00′	0.4384	0.8988	0.4877	2.2812	1.1126	2.0503	64°00′	64.0°
26.1	26 06	0.4399	0.8980	0.4899	2.2730	1.1136	2.0413	63 54	63.9
26.2	26 12	0.4415	0.8973	0.4921	2.2650	1.1145	2.0323	63 48	63.8
26.3	26 18	0.4431	0.8965	0.4942	2.2570	1.1155	2.0233	63 42	63.7
26.4	26 24	0.4446	0.8957	0.4964	2.2490	1.1164	2.0145	63 36	63.6
26.5	26 30	0.4462	0.8949	0.4986	2.2412	1.1174	2.0057	63 30	63.5
26.6	26 36	0.4478	0.8942	0.5008	2.2333	1.1184	1.9970	63 24	63.4
26.7	26 42	0.4493	0.8934	0.5029	2.2256	1.1194	1.9883	63 18	63.3
26.8	26 48	0.4509	0.8926	0.5051	2.2179	1.1203	1.9797	63 12	63.2
26.9	26 54	0.4524	0.8918	0.5073	2.2103	1.1213	1.9711	63 06	63.1
27.0°	27°00′	0.4540	0.8910	0.5095	2.2027	1.1223	1.9626	63°00′	63.0°
27.1	27 06	0.4555	0.8902	0.5117	2.1952	1.1233	1.9542	62 54	62.9
27.2	27 12	0.4571	0.8894	0.5139	2.1877	1.1243	1.9458	62 48	62.8
27.3	27 18	0.4586	0.8886	0.5161	2.1803	1.1253	1.9375	62 42	62.7
27.4	27 24	0.4602	0.8878	0.5184	2.1730	1.1264	1.9292	62 36	62.6
27.5	27 30	0.4617	0.8870	0.5206	2.1657	1.1274	1.9210	62 30	62.5
27.6	27 36	0.4633	0.8862	0.5228	2.1584	1.1284	1.9128	62 24	62.4
27.7	27 42	0.4648	0.8854	0.5250	2.1513	1.1294	1.9047	62 18	62.3
27.8	27 48	0.4664	0.8846	0.5272	2.1441	1.1305	1.8967	62 12	62.2
27.9	27 54	0.4679	0.8838	0.5295	2.1371	1.1315	1.8887	62 06	62.1
28.0°	28°00′	0.4695	0.8829	0.5317	2.1301	1.1326	1.8807	62°00′	62.0°
28.1	28 06	0.4710	0.8821	0.5339	2.1231	1.1336	1.8728	61 54	61.9
28.2	28 12	0.4726	0.8813	0.5362	2.1162	1.1347	1.8650	61 48	61.8
28.3	28 18	0.4741	0.8805	0.5384	2.1093	1.1357	1.8572	61 42	61.7
28.4	28 24	0.4756	0.8796	0.5407	2.1025	1.1368	1.8495	61 36	61.6
28.5	28 30	0.4772	0.8788	0.5430	2.0957	1.1379	1.8418	61 30	61.5
28.6	28 36	0.4787	0.8780	0.5452	2.0890	1.1390	1.8341	61 24	61.4
28.7	28 42	0.4802	0.8771	0.5475	2.0824	1.1401	1.8265	61 18	61.3
28.8	28 48	0.4818	0.8763	0.5498	2.0758	1.1412	1.8190	61 12	61.2
28.9	28 54	0.4833	0.8755	0.5520	2.0692	1.1423	1.8115	61 06	61.1
29.0°	29°00′	0.4848	0.8746	0.5543	2.0627	1.1434	1.8040	61°00′	61.0°
29.1	29 06	0.4863	0.8738	0.5566	2.0562	1.1445	1.7966	60 54	60.9
29.2	29 12	0.4879	0.8729	0.5589	2.0598	1.1456	1.7893	60 48	60.8
29.3	29 18	0.4894	0.8721	0.5612	2.0434	1.1467	1.7820	60 42	60.7
29.4	29 24	0.4909	0.8712	0.5635	2.0371	1.1478	1.7747	60 36	60.6
29.5	29 30	0.4924	0.8704	0.5658	2.0308	1.1490	1.7675	60 30	60.5
29.6	29 36	0.4939	0.8695	0.5681	2.0245	1.1501	1.7603	60 24	60.4
29.7	29 42	0.4955	0.8686	0.5704	2.0183	1.1512	1.7532	60 18	60.3
29.8	29 48	0.4970	0.8678	0.5727	2.0122	1.1524	1.7461	60 12	60.2
29.9	29 54	0.4985	0.8669	0.5750	2.0061	1.1535	1.7391	60°06′	60.1°
		cos θ	sin θ	cot θ	sec θ	csc θ	tan θ	θ deg-min	θ deg

Table V (continued)

θ deg	θ deg-min	sin θ	cos θ	tan θ	csc θ	sec θ	cot θ		
30.0°	30°00′	0.5000	0.8660	0.5774	2.0000	1.1547	1.7321	60°00′	60.0°
30.1	30 06	0.5015	0.8652	0.5797	1.9940	1.1559	1.7251	59 54	59.9
30.2	30 12	0.5030	0.8643	0.5820	1.9880	1.1570	1.7182	59 48	59.8
30.3	30 18	0.5045	0.8634	0.5844	1.9821	1.1582	1.7113	59 42	59.7
30.4	30 24	0.5060	0.8625	0.5867	1.9762	1.1594	1.7045	59 36	59.6
30.5	30 30	0.5075	0.8616	0.5890	1.9703	1.1606	1.6977	59 30	59.5
30.6	30 36	0.5090	0.8607	0.5914	1.9645	1.1618	1.6909	59 24	59.4
30.7	30 42	0.5105	0.8599	0.5938	1.9587	1.1630	1.6842	59 18	59.3
30.8	30 48	0.5120	0.8590	0.5961	1.9530	1.1642	1.6775	59 12	59.2
30.9	30 54	0.5135	0.8581	0.5985	1.9473	1.1654	1.6709	59 06	59.1
31.0°	31°00′	0.5150	0.8572	0.6009	1.9416	1.1666	1.6643	59°00′	59.0°
31.1	31 06	0.5165	0.8563	0.6032	1.9360	1.1679	1.6577	58 54	58.9
31.2	31 12	0.5180	0.8554	0.6056	1.9304	1.1691	1.6512	58 48	58.8
31.3	31 18	0.5195	0.8545	0.6080	1.9249	1.1703	1.6447	58 42	58.7
31.4	31 24	0.5210	0.8536	0.6104	1.9194	1.1716	1.6383	58 36	58.6
31.5	31 30	0.5225	0.8526	0.6128	1.9139	1.1728	1.6319	58 30	58.5
31.6	31 36	0.5240	0.8517	0.6152	1.9084	1.1741	1.6255	58 24	58.4
31.7	31 42	0.5255	0.8508	0.6176	1.9031	1.1753	1.6191	58 18	58.3
31.8	31 48	0.5270	0.8499	0.6200	1.8977	1.1766	1.6128	58 12	58.2
31.9	31 54	0.5284	0.8490	0.6224	1.8924	1.1779	1.6066	58 06	58.1
32.0°	32°00′	0.5299	0.8480	0.6249	1.8871	1.1792	1.6003	58°00′	58.0°
32.1	32 06	0.5314	0.8471	0.6273	1.8818	1.1805	1.5941	57 54	57.9
32.2	32 12	0.5329	0.8462	0.6297	1.8766	1.1818	1.5880	57 48	57.8
32.3	32 18	0.5344	0.8453	0.6322	1.8714	1.1831	1.5818	57 42	57.7
32.4	32 24	0.5358	0.8443	0.6346	1.8663	1.1844	1.5757	57 36	57.6
32.5	32 30	0.5373	0.8434	0.6371	1.8612	1.1857	1.5697	57 30	57.5
32.6	32 36	0.5388	0.8425	0.6395	1.8561	1.1870	1.5637	57 24	57.4
32.7	32 42	0.5402	0.8415	0.6420	1.8510	1.1883	1.5577	57 18	57.3
32.8	32 48	0.5417	0.8406	0.6445	1.8460	1.1897	1.5517	57 12	57.2
32.9	32 54	0.5432	0.8396	0.6469	1.8410	1.1910	1.5458	57 06	57.1
33.0°	33°00′	0.5446	0.8387	0.6494	1.8361	1.1924	1.5399	57°00′	57.0°
33.1	33 06	0.5461	0.8377	0.6519	1.8312	1.1937	1.5340	56 54	56.9
33.2	33 12	0.5476	0.8368	0.6544	1.8263	1.1951	1.5282	56 48	56.8
33.3	33 18	0.5490	0.8358	0.6569	1.8214	1.1964	1.5224	56 42	56.7
33.4	33 24	0.5505	0.8348	0.6594	1.8166	1.1978	1.5166	56 36	56.6
33.5	33 30	0.5519	0.8339	0.6619	1.8118	1.1992	1.5108	56 30	56.5
33.6	33 36	0.5534	0.8329	0.6644	1.8070	1.2006	1.5051	56 24	56.4
33.7	33 42	0.5548	0.8320	0.6669	1.8023	1.2020	1.4994	56 18	56.3
33.8	33 48	0.5563	0.8310	0.6694	1.7976	1.2034	1.4938	56 12	56.2
33.9	33 54	0.5577	0.8300	0.6720	1.7929	1.2048	1.4882	56 06	56.1
34.0°	34°00′	0.5592	0.8290	0.6745	1.7883	1.2062	1.4826	56°00′	56.0°
34.1	34 06	0.5606	0.8281	0.6771	1.7837	1.2076	1.4770	55 54	55.9
34.2	34 12	0.5621	0.8271	0.6796	1.7791	1.2091	1.4715	55 48	55.8
34.3	34 18	0.5635	0.8261	0.6822	1.7745	1.2105	1.4659	55 42	55.7
34.4	34 24	0.5650	0.8251	0.6847	1.7700	1.2120	1.4605	55 36	55.6
34.5	34 30	0.5664	0.8241	0.6873	1.7655	1.2134	1.4550	55 30	55.5
34.6	34 36	0.5678	0.8231	0.6899	1.7610	1.2149	1.4496	55 24	55.4
34.7	34 42	0.5693	0.8221	0.6924	1.7566	1.2163	1.4442	55 18	55.3
34.8	34 48	0.5707	0.8211	0.6950	1.7522	1.2178	1.4388	55 12	55.2
34.9	34 54	0.5721	0.8202	0.6976	1.7478	1.2193	1.4335	55°06′	55.1°
		cos θ	sin θ	cot θ	sec θ	csc θ	tan θ	θ deg-min	θ deg

Table V (continued)

θ deg	θ deg-min	sin θ	cos θ	tan θ	csc θ	sec θ	cot θ		
35.0°	35°00′	0.5736	0.8192	0.7002	1.7434	1.2208	1.4281	55°00′	55.0°
35.1	35 06	0.5750	0.8181	0.7028	1.7391	1.2223	1.4229	54 54	54.9
35.2	35 12	0.5764	0.8171	0.7054	1.7348	1.2238	1.4176	54 48	54.8
35.3	35 18	0.5779	0.8161	0.7080	1.7305	1.2253	1.4124	54 42	54.7
35.4	35 24	0.5793	0.8151	0.7107	1.7263	1.2268	1.4071	54 36	54.6
35.5	35 30	0.5807	0.8141	0.7133	1.7221	1.2283	1.4019	54 30	54.5
35.6	35 36	0.5821	0.8131	0.7159	1.7179	1.2299	1.3968	54 24	54.4
35.7	35 42	0.5835	0.8121	0.7186	1.7137	1.2314	1.3916	54 18	54.3
35.8	35 48	0.5850	0.8111	0.7212	1.7095	1.2329	1.3865	54 12	54.2
35.9	35 54	0.5864	0.8100	0.7239	1.7054	1.2345	1.3814	54 06	54.1
36.0°	36°00′	0.5878	0.8090	0.7265	1.7013	1.2361	1.3764	54°00′	54.0°
36.1	36 06	0.5892	0.8080	0.7292	1.6972	1.2376	1.3713	53 54	53.9
36.2	36 12	0.5906	0.8070	0.7319	1.6932	1.2392	1.3663	53 48	53.8
36.3	36 18	0.5920	0.8059	0.7346	1.6892	1.2408	1.3613	53 42	53.7
36.4	36 24	0.5934	0.8049	0.7373	1.6852	1.2424	1.3564	53 36	53.6
36.5	36 30	0.5948	0.8039	0.7400	1.6812	1.2440	1.3514	53 30	53.5
36.6	36 36	0.5962	0.8028	0.7427	1.6772	1.2456	1.3465	53 24	53.4
36.7	36 42	0.5976	0.8018	0.7454	1.6733	1.2472	1.3416	53 18	53.3
36.8	36 48	0.5990	0.8007	0.7481	1.6694	1.2489	1.3367	53 12	53.2
36.9	36 54	0.6004	0.7997	0.7508	1.6655	1.2505	1.3319	53 06	53.1
37.0°	37°00′	0.6018	0.7986	0.7536	1.6616	1.2521	1.3270	53°00′	53.0°
37.1	37 06	0.6032	0.7976	0.7563	1.6578	1.2538	1.3222	52 54	52.9
37.2	37 12	0.6046	0.7965	0.7590	1.6540	1.2554	1.3175	52 48	52.8
37.3	37 18	0.6060	0.7955	0.7618	1.6502	1.2571	1.3127	52 42	52.7
37.4	37 24	0.6074	0.7944	0.7646	1.6464	1.2588	1.3079	52 36	52.6
37.5	37 30	0.6088	0.7934	0.7673	1.6427	1.2605	1.3032	52 30	52.5
37.6	37 36	0.6101	0.7923	0.7701	1.6390	1.2622	1.2985	52 24	52.4
37.7	37 42	0.6115	0.7912	0.7729	1.6353	1.2639	1.2938	52 18	52.3
37.8	37 48	0.6129	0.7902	0.7757	1.6316	1.2656	1.2892	52 12	52.2
37.9	37 54	0.6143	0.7891	0.7785	1.6279	1.2673	1.2846	52 06	52.1
38.0°	38°00′	0.6157	0.7880	0.7813	1.6243	1.2690	1.2799	52°00′	52.0°
38.1	38 06	0.6170	0.7869	0.7841	1.6207	1.2708	1.2753	51 54	51.9
38.2	38 12	0.6184	0.7859	0.7869	1.6171	1.2725	1.2708	51 48	51.8
38.3	38 18	0.6198	0.7848	0.7898	1.6135	1.2742	1.2662	51 42	51.7
38.4	38 24	0.6211	0.7837	0.7926	1.6099	1.2760	1.2617	51 36	51.6
38.5	38 30	0.6225	0.7826	0.7954	1.6064	1.2778	1.2572	51 30	51.5
38.6	38 36	0.6239	0.7815	0.7983	1.6029	1.2796	1.2527	51 24	51.4
38.7	38 42	0.6252	0.7804	0.8012	1.5994	1.2813	1.2482	51 18	51.3
38.8	38 48	0.6266	0.7793	0.8040	1.5959	1.2831	1.2437	51 12	51.2
38.9	38 54	0.6280	0.7782	0.8069	1.5925	1.2849	1.2393	51 06	51.1
39.0°	39°00′	0.6293	0.7771	0.8098	1.5890	1.2868	1.2349	51°00′	51.0°
39.1	39 06	0.6307	0.7760	0.8127	1.5856	1.2886	1.2305	50 54	50.9
39.2	39 12	0.6320	0.7749	0.8156	1.5822	1.2904	1.2261	50 48	50.8
39.3	39 18	0.6334	0.7738	0.8185	1.5788	1.2923	1.2218	50 42	50.7
39.4	39 24	0.6347	0.7727	0.8214	1.5755	1.2941	1.2174	50 36	50.6
39.5	39 30	0.6361	0.7716	0.8243	1.5721	1.2960	1.2131	50 30	50.5
39.6	39 36	0.6374	0.7705	0.8273	1.5688	1.2978	1.2088	50 24	50.4
39.7	39 42	0.6388	0.7694	0.8302	1.5655	1.2997	1.2045	50 18	50.3
39.8	39 48	0.6401	0.7683	0.8332	1.5622	1.3016	1.2002	50 12	50.2
39.9	39 54	0.6414	0.7672	0.8361	1.5590	1.3035	1.1960	50°06′	50.1°
		cos θ	sin θ	cot θ	sec θ	csc θ	tan θ	θ deg-min	θ deg

Table V (continued)

θ deg	θ deg-min	sin θ	cos θ	tan θ	csc θ	sec θ	cot θ		
40.0°	40°00′	0.6428	0.7660	0.8391	1.5557	1.3054	1.1918	50°00′	50.0°
40.1	40 06	0.6441	0.7649	0.8421	1.5525	1.3073	1.1875	49 54	49.9
40.2	40 12	0.6455	0.7638	0.8451	1.5493	1.3092	1.1833	49 48	49.8
40.3	40 18	0.6468	0.7627	0.8481	1.5461	1.3112	1.1792	49 42	49.7
40.4	40 24	0.6481	0.7615	0.8511	1.5429	1.3131	1.1750	49 36	49.6
40.5	40 30	0.6494	0.7604	0.8541	1.5398	1.3151	1.1708	49 30	49.5
40.6	40 36	0.6508	0.7593	0.8571	1.5366	1.3171	1.1667	49 24	49.4
40.7	40 42	0.6521	0.7581	0.8601	1.5335	1.3190	1.1626	49 18	49.3
40.8	40 48	0.6534	0.7570	0.8632	1.5304	1.3210	1.1585	49 12	49.2
40.9	40 54	0.6547	0.7559	0.8662	1.5273	1.3230	1.1544	49 06	49.1
41.0°	41°00′	0.6561	0.7547	0.8693	1.5243	1.3250	1.1504	49°00′	49.0°
41.1	41 06	0.6574	0.7536	0.8724	1.5212	1.3270	1.1463	48 54	48.9
41.2	41 12	0.6587	0.7524	0.8754	1.5182	1.3291	1.1423	48 48	48.8
41.3	41.18	0.6600	0.7513	0.8785	1.5151	1.3311	1.1383	48 42	48.7
41.4	41 24	0.6613	0.7501	0.8816	1.5121	1.3331	1.1343	48 36	48.6
41.5	41 30	0.6626	0.7490	0.8847	1.5092	1.3352	1.1303	48 30	48.5
41.6	41 36	0.6639	0.7478	0.8878	1.5062	1.3373	1.1263	48 24	48.4
41.7	41 42	0.6652	0.7466	0.8910	1.5032	1.3393	1.1224	48 18	48.3
41.8	41 48	0.6665	0.7455	0.8941	1.5003	1.3414	1.1184	48 12	48.2
41.9	41 54	0.6678	0.7443	0.8972	1.4974	1.3435	1.1145	48 06	48.1
42.0°	42°00′	0.6691	0.7431	0.9004	1.4945	1.3456	1.1106	48°00′	48.0°
42.1	42 06	0.6704	0.7420	0.9036	1.4916	1.3478	1.1067	47 54	47.9
42.2	42 12	0.6717	0.7408	0.9067	1.4887	1.3499	1.1028	47 48	47.8
42.3	42 18	0.6730	0.7396	0.9099	1.4859	1.3520	1.0990	47 42	47.7
42.4	42 24	0.6743	0.7385	0.9131	1.4830	1.3542	1.0951	47 36	47.6
42.5	42 30	0.6756	0.7373	0.9163	1.4802	1.3563	1.0913	47 30	47.5
42.6	42 36	0.6769	0.7361	0.9195	1.4774	1.3585	1.0875	47 24	47.4
42.7	42 42	0.6782	0.7349	0.9228	1.4746	1.3607	1.0837	47 18	47.3
42.8	42 48	0.6794	0.7337	0.9260	1.4718	1.3629	1.0799	47 12	47.2
42.9	42 54	0.6807	0.7325	0.9293	1.4690	1.3651	1.0761	47 06	47.1
43.0°	43°00′	0.6820	0.7314	0.9325	1.4663	1.3673	1.0724	47°00′	47.0°
43.1	43 06	0.6833	0.7302	0.9358	1.4635	1.3696	1.0686	46 54	46.9
43.2	43 12	0.6845	0.7290	0.9391	1.4608	1.3718	1.0649	46 48	46.8
43.3	43 18	0.6858	0.7278	0.9424	1.4581	1.3741	1.0612	46 42	46.7
43.4	43 24	0.6871	0.7266	0.9457	1.4554	1.3763	1.0575	46 36	46.6
43.5	43 30	0.6884	0.7254	0.9490	1.4527	1.3786	1.0538	46 30	46.5
43.6	43 36	0.6896	0.7242	0.9523	1.4501	1.3809	1.0501	46 24	46.4
43.7	43 42	0.6909	0.7230	0.9556	1.4474	1.3832	1.0464	46 18	46.3
43.8	43 48	0.6921	0.7218	0.9590	1.4448	1.3855	1.0428	46 12	46.2
43.9	43 54	0.6934	0.7206	0.9623	1.4422	1.3878	1.0392	46 06	46.1
44.0°	44°00′	0.6947	0.7193	0.9657	1.4396	1.3902	1.0355	46°00′	46.0°
44.1	44 06	0.6959	0.7181	0.9691	1.4370	1.3925	1.0319	45 54	45.9
44.2	44 12	0.6972	0.7169	0.9725	1.4344	1.3949	1.0283	45 48	45.8
44.3	44 18	0.6984	0.7157	0.9759	1.4318	1.3972	1.0247	45 42	45.7
44.4	44 24	0.6997	0.7145	0.9793	1.4293	1.3996	1.0212	45 36	45.6
44.5	44 30	0.7009	0.7133	0.9827	1.4267	1.4020	1.0176	45 30	45.5
44.6	44 36	0.7022	0.7120	0.9861	1.4242	1.4044	1.0141	45 24	45.4
44.7	44 42	0.7034	0.7108	0.9896	1.4217	1.4069	1.0105	45 18	45.3
44.8	44 48	0.7046	0.7096	0.9930	1.4192	1.4093	1.0070	45 12	45.2
44.9	44 54	0.7059	0.7083	0.9965	1.4167	1.4118	1.0035	45 06	45.1
45.0°	45°00′	0.7071	0.7071	1.0000	1.4142	1.4142	1.0000	45°00′	45.0°
		cos θ	sin θ	cot θ	sec θ	csc θ	tan θ	θ deg-min	θ deg

Table VI
Squares, Square Roots, and Prime Factors

No.	Sq.	Sq. Root	Prime Factors	No.	Sq.	Sq. Root	Prime Factors
1	1	1.000		51	2,601	7.141	$3 \cdot 17$
2	4	1.414	2	52	2,704	7.211	$2^2 \cdot 13$
3	9	1.732	3	53	2,809	7.280	53
4	16	2.000	2^2	54	2,916	7.348	$2 \cdot 3^3$
5	25	2.236	5	55	3,025	7.416	$5 \cdot 11$
6	36	2.449	$2 \cdot 3$	56	3,136	7.483	$2^3 \cdot 7$
7	49	2.646	7	57	3,249	7.550	$3 \cdot 19$
8	64	2.828	2^3	58	3,364	7.616	$2 \cdot 29$
9	81	3.000	3^2	59	3,481	7.681	59
10	100	3.162	$2 \cdot 5$	60	3,600	7.746	$2^2 \cdot 3 \cdot 5$
11	121	3.317	11	61	3,721	7.810	61
12	144	3.464	$2^2 \cdot 3$	62	3,844	7.874	$2 \cdot 31$
13	169	3.606	13	63	3,969	7.937	$3^2 \cdot 7$
14	196	3.742	$2 \cdot 7$	64	4,096	8.000	2^6
15	225	3.873	$3 \cdot 5$	65	4,225	8.062	$5 \cdot 13$
16	256	4.000	2^4	66	4,356	8.124	$2 \cdot 3 \cdot 11$
17	289	4.123	17	67	4,489	8.185	67
18	324	4.243	$2 \cdot 3^2$	68	4,624	8.246	$2^2 \cdot 17$
19	361	4.359	19	69	4,761	8.307	$3 \cdot 23$
20	400	4.472	$2^2 \cdot 5$	70	4,900	8.367	$2 \cdot 5 \cdot 7$
21	441	4.583	$3 \cdot 7$	71	5,041	8.426	71
22	484	4.690	$2 \cdot 11$	72	5,184	8.485	$2^3 \cdot 3^2$
23	529	4.796	23	73	5,329	8.544	73
24	576	4.899	$2^3 \cdot 3$	74	5,476	8.602	$2 \cdot 37$
25	625	5.000	5^2	75	5,625	8.660	$3 \cdot 5^2$
26	676	5.099	$2 \cdot 13$	76	5,776	8.718	$2^2 \cdot 19$
27	729	5.196	3^3	77	5,929	8.775	$7 \cdot 11$
28	784	5.292	$2^2 \cdot 7$	78	6,084	8.832	$2 \cdot 3 \cdot 13$
29	841	5.385	29	79	6,241	8.888	79
30	900	5.477	$2 \cdot 3 \cdot 5$	80	6,400	8.944	$2^4 \cdot 5$
31	961	5.568	31	81	6,561	9.000	3^4
32	1,024	5.657	2^5	82	6,724	9.055	$2 \cdot 41$
33	1,089	5.745	$3 \cdot 11$	83	6,889	9.110	83
34	1,156	5.831	$2 \cdot 17$	84	7,056	9.165	$2^2 \cdot 3 \cdot 7$
35	1,225	5.916	$5 \cdot 7$	85	7,225	9.220	$5 \cdot 17$
36	1,296	6.000	$2^2 \cdot 3^2$	86	7,396	9.274	$2 \cdot 43$
37	1,369	6.083	37	87	7,569	9.327	$3 \cdot 29$
38	1,444	6.164	$2 \cdot 19$	88	7,744	9.381	$2^3 \cdot 11$
39	1,521	6.245	$3 \cdot 13$	89	7,921	9.434	89
40	1,600	6.325	$2^3 \cdot 5$	90	8,100	9.487	$2 \cdot 3^2 \cdot 5$
41	1,681	6.403	41	91	8,281	9.539	$7 \cdot 13$
42	1,764	6.481	$2 \cdot 3 \cdot 7$	92	8,464	9.592	$2^2 \cdot 23$
43	1,849	6.557	43	93	8,649	9.644	$3 \cdot 31$
44	1,936	6.633	$2^2 \cdot 11$	94	8,836	9.695	$2 \cdot 47$
45	2,025	6.708	$3^2 \cdot 5$	95	9,025	9.747	$5 \cdot 19$
46	2,116	6.782	$2 \cdot 23$	96	9,216	9.798	$2^5 \cdot 3$
47	2,209	6.856	47	97	9,409	9.849	97
48	2,304	6.928	$2^4 \cdot 3$	98	9,604	9.899	$2 \cdot 7^2$
49	2,401	7.000	7^2	99	9,801	9.950	$3^2 \cdot 11$
50	2,500	7.071	$2 \cdot 5^2$	100	10,000	10.000	$2^2 \cdot 5^2$

Answers to Odd-Numbered Exercises

Exercise 1.1 (page 13) **1.** $\{11\}$ **3.** $\{5\}$ **5.** $\{2\}$ **7.** $\left\{\dfrac{1}{4}\right\}$ **9.** The empty set **11.** $\left\{-\dfrac{1}{3}\right\}$ **13.** $\{9\}$ **15.** $\{0\}$ **17.** $\{1\}$ **19.** $\left\{\dfrac{1}{2}\right\}$ **21.** The empty set **23.** $\left\{\dfrac{2}{5}\right\}$ **25.** Any x; $x \neq 1, -1$ **27.** $\{0\}$ **29.** The empty set **31.** Any x; $x \neq -2, 4$ **33.** $r = \dfrac{S}{2\pi h}, h \neq 0$ **35.** $k = v - gt$ **37.** $c = \dfrac{2A - bh}{h}, h \neq 0$ **39.** $n = \dfrac{l - a + d}{d}, d \neq 0$ **41.** $r^2 = \dfrac{V + \pi R^2 h}{-\pi h}, h \neq 0$ **43.** $y' = -\dfrac{x}{y}, y \neq 0$ **45.** $y' = \dfrac{1}{x}, x \neq 0, y \neq 0$ **47.** $y' = \dfrac{1 + 3x}{x^2 - 2y^3}, x^2 \neq 2y^3$ **49.** $x_1 = \dfrac{x_4}{x_2 - 2x_3}, x_2 \neq 2x_3$ **51.** $y = 6x - 6x_1 + y_1$ **53.** $k = \dfrac{-3}{5}$ **55.** $k = -13$ **57.** $k = 11$

Exercise 1.2 (page 19) **1.** $\{-2, 1\}$ **3.** $\{0, 5\}$ **5.** $\{1\}$ **7.** $\left\{-\dfrac{1}{2}, -\dfrac{1}{3}\right\}$ **9.** $\left\{\dfrac{1}{2}, 6\right\}$ **11.** $\left\{-\dfrac{1}{2}, -\dfrac{2}{3}\right\}$ **13.** $\left\{-\dfrac{3}{2}, 2\right\}$ **15.** $\left\{-\dfrac{3}{2}, -3\right\}$ **17.** $\left\{\dfrac{3}{5}, -2\right\}$ **19.** $\{2, -2\}$ **21.** $\{\sqrt{3}, -\sqrt{3}\}$ **23.** $\{\sqrt{8}, -\sqrt{8}\}$ **25.** $\{1, -3\}$ **27.** $\{1 + \sqrt{3}, 1 - \sqrt{3}\}$ **29.** $\left\{\dfrac{1}{2}, \dfrac{3}{2}\right\}$ **31.** $\{1, -9\}$ **33.** $\{-6, -5\}$ **35.** $\left\{2, -\dfrac{1}{3}\right\}$ **37.** $\left\{\dfrac{1}{2} + \dfrac{\sqrt{17}}{2}, \dfrac{1}{2} - \dfrac{\sqrt{17}}{2}\right\}$ **39.** $\left\{\dfrac{3}{2} + \dfrac{\sqrt{21}}{2}, \dfrac{3}{2} - \dfrac{\sqrt{21}}{2}\right\}$ **41.** $\left\{3 - \dfrac{\sqrt{88}}{2}, 3 + \dfrac{\sqrt{88}}{2}\right\}$ **43.** $\{-2, 1\}$ **45.** $\left\{0, \dfrac{2}{3}\right\}$ **47.** $\left\{-\dfrac{13}{12}\right\}$ **49.** $\{2 + \sqrt{5}, 2 - \sqrt{5}\}$ **51.** $\{\sqrt{2}, -\sqrt{2}\}$ **53.** $k = 1$ **55.** $k = \pm 6$ **57.** $k = 4$ **59.** $x = \pm\sqrt{\dfrac{1}{k}}$ **61.** $x = 0, x = \dfrac{1}{k}$ **63.** $x = -k$ **65.** $x = \dfrac{1 \pm \sqrt{5}}{2} k$

67. If $ax^2 + bx + c = 0$ and $a \neq 0$, then completing the square we have

$$x^2 + \frac{b}{a}x + \frac{b^2}{4a^2} = \frac{-c}{a} + \frac{b^2}{4a^2},$$

$$\left(x + \frac{b}{2a}\right)^2 = \frac{b^2 - 4ac}{4a^2}.$$

Extracting roots, we obtain

$$x + \frac{b}{2a} = \frac{\pm\sqrt{b^2 - 4ac}}{2a}.$$

Thus,

$$x = \frac{-b \pm \sqrt{b^2 - 4ac}}{2a}.$$

Exercise 1.3 (page 24) **1.** $\{64\}$ **3.** $\{-7\}$ **5.** $\{4\}$ **7.** $\{-25\}$ **9.** $\{16\}$ **11.** $\{1\}$ **13.** $\{5\}$ **15.** $\left\{-\frac{7}{4}\right\}$ **17.** $\{4\}$ **19.** $\{1, 3\}$ **21.** $A = \pi r^2$ **23.** $y = \frac{1}{x^3}$ **25.** $y = \pm\sqrt{a^2 - x^2}$ **27.** $V = \pi h(R^2 - r^2)$ **29.** $\{1\}$ **31.** $\{216, -27\}$ **33.** $\left\{\frac{1}{9}, -\frac{1}{2}\right\}$ **35.** $\left\{-\frac{6}{7}, -\frac{5}{4}\right\}$ **37.** $\{16, 256\}$ **39.** $\{87\}$ **41.** $\left\{-\frac{25}{8}, -\frac{11}{4}\right\}$ **43.** $\{16\}$ **45.** $\left\{\frac{1}{9}\right\}$ **47.** $\{1, -1\}$ **49.** $\{64\}$ **51.** $\left\{-1, -\frac{35}{16}, \frac{19}{16}\right\}$

Exercise 1.4 (page 33) **1.** $(-6, +\infty)$ **3.** $(-\infty, 3]$ **5.** $[-6, +\infty)$ **7.** $\left[\frac{5}{2}, +\infty\right)$ **9.** $(-\infty, -1)$ **11.** $[-2, +\infty)$ **13.** $[-3, +\infty)$ **15.** $\left(-\infty, \frac{8}{3}\right)$ **17.** $\left[-\frac{6}{5}, +\infty\right)$ **19.** $[-6, 1]$ **21.** $\left(-\frac{22}{3}, -4\right]$ **23.** $[-25, -16]$ **25.** $(-\infty, -1) \cup (5, +\infty)$ **27.** $\left[-\frac{1}{2}, \frac{8}{3}\right]$ **29.** $(-\infty, -3) \cup (2, +\infty)$ **31.** $(-\infty, -2) \cup (4, +\infty)$ **33.** $(-\infty, -5) \cup (4, +\infty)$ **35.** $\left[-\frac{1}{3}, 2\right]$ **37.** $\left(-\infty, \frac{-3 - \sqrt{33}}{8}\right] \cup \left[\frac{-3 + \sqrt{33}}{8}, +\infty\right)$ **39.** No solution

41. $(-\infty, 2 - \sqrt{3}] \cup [2 + \sqrt{3}, +\infty)$ **43.** No solution   **45.** $\left(-\infty, -\dfrac{1}{2}\right] \cup (0, +\infty)$

47. $(-\infty, 1)$ **49.** $(-2, 0)$ **51.** $(-1, 1) \cup (3, +\infty)$ **53.** $(-\infty, -1) \cup (0, 2)$

55. $[-3, -1] \cup [1, +\infty)$ **57.** $(-\infty, -4) \cup (-2, 0) \cup (2, +\infty)$ **59.** $\left(\dfrac{1}{4}(5 - \varepsilon), +\infty\right)$

61. $\left(\dfrac{1}{3}(-2 - \varepsilon), \dfrac{1}{3}(\varepsilon - 2)\right)$ **63.** $\left(\dfrac{1}{3}(7 - \varepsilon), \dfrac{1}{3}(7 + \varepsilon)\right)$ **65.** $(-\sqrt{\varepsilon}, \sqrt{\varepsilon})$

67. $(-\sqrt{1 + \varepsilon}, \sqrt{1 + \varepsilon})$

Exercise 1.5 (page 40)   **1.** $\{-5, 1\}$ **3.** $\{-1, 7\}$ **5.** $\{1, -2\}$ **7.** $(-3, 7)$ **9.** $\left(-\dfrac{1}{6}, \dfrac{7}{6}\right)$

11. $(-\infty, 3) \cup (5, +\infty)$ **13.** $(-\infty, -10) \cup (4, +\infty)$ **15.** $[3, 5]$ **17.** $[0, 3]$

19. $\left(-\infty, -\dfrac{3}{2}\right] \cup \left[\dfrac{5}{2}, +\infty\right)$ **21.** $\left(-\infty, -\dfrac{2}{3}\right] \cup \left[\dfrac{4}{3}, +\infty\right)$ **23.** $[-3, 3]$ **25.** $|x| \le 3$

27. $|x - 2| < 4$ **29.** $|x + 5| \le 3$ **31.** $(-1, 0) \cup (0, 1)$

33. $(-\infty, 2 - 2\sqrt{2}] \cup \{2\} \cup [2 + 2\sqrt{2}, +\infty)$ **35.** $\left(-\infty, -\dfrac{3}{5}\right] \cup \left[-\dfrac{3}{7}, +\infty\right)$

Chapter 1 Review (page 41)   **1.** $\left\{-\dfrac{23}{2}\right\}$ **2.** $\{3\}$ **3.** $\left\{-\dfrac{5}{2}\right\}$ **4.** $\left\{\dfrac{3}{5}\right\}$ **5.** $y = \dfrac{x}{4}$

6. $x = 4y$ **7.** $\{3, -2\}$ **8.** $\{-7, 3\}$ **9.** $\{\sqrt{5}, -\sqrt{5}\}$ **10.** $\{5 + \sqrt{2}, 5 - \sqrt{2}\}$

11. $\left\{\dfrac{-5}{2} + \dfrac{\sqrt{33}}{2}, \dfrac{-5}{2} - \dfrac{\sqrt{33}}{2}\right\}$ **12.** $\left\{\dfrac{1}{2}, -1\right\}$ **13.** $\left\{\dfrac{1 \pm \sqrt{15}}{4}\right\}$ **14.** $\{-1 \pm \sqrt{5}\}$

15. $\left\{\dfrac{3 \pm \sqrt{9 - 4k}}{2k}\right\}$ **16.** $\left\{\dfrac{-k \pm \sqrt{k^2 + 16}}{2}\right\}$ **17.** $\{4, 9\}$ **18.** $\{8\}$ **19.** $\{1, 4\}$

20. $\left\{\dfrac{1}{7}, -\dfrac{1}{6}\right\}$ **21.** $[32, +\infty)$ **22.** $\left(-\infty, -\dfrac{6}{5}\right)$ **23.** $(-5, 2)$ **24.** $\left(\dfrac{1}{3}, 1\right)$ **25.** $\left\{-2, \dfrac{4}{3}\right\}$

26. $\left\{-1, \dfrac{7}{3}\right\}$ **27.** $[-1, 0]$ **28.** $\left(-\infty, -\dfrac{1}{2}\right] \cup \left[\dfrac{7}{2}, +\infty\right)$ **29.** $(-\sqrt{3}, -1) \cup (1, \sqrt{3})$

30. $\left(-\infty, -\dfrac{5}{3}\right) \cup \left(-\dfrac{1}{3}, +\infty\right)$ **31.** $|x + 1| \le 4$ **32.** $|x + 2| < 4$

Exercise 2.1 (page 47)   **1.** $\left\{(0, 6), (1, 4), (2, 2), (-3, 12), \left(\dfrac{2}{3}, \dfrac{14}{3}\right)\right\}$

3. $\left\{(0, 0), (1, -3), (2, 3), \left(-3, -\dfrac{9}{7}\right), \left(\dfrac{2}{3}, -\dfrac{9}{7}\right)\right\}$

5. $\left\{(0, \sqrt{11}), (1, \sqrt{14}), (2, \sqrt{17}), (-3, \sqrt{2}), \left(\dfrac{2}{3}, \sqrt{13}\right)\right\}$

7. a. Domain: $\{2, 5, 7\}$; range: $\{3, 7, 8\}$ **b.** A function
9. a. Domain: $\{2, 3\}$; range: $\{-1, 4, 6\}$ **b.** Not a function
11. a. Domain: $\{5, 6, 7\}$; range: $\{5, 6, 7\}$ **b.** A function

13. Domain: $\{x \mid x \in R\}$ **15.** Domain: $\{x \mid x \in R\}$ **17.** Domain: $\{x \mid x \in R, x \neq 2\}$
19. Domain: $\{x \mid x \in R, x \geq 0\}$ **21.** Domain: $\{x \mid x \in R, -2 \leq x \leq 2\}$
23. Domain: $\{x \mid x \in R, x \neq 0, 1\}$ **25.** Yes **27.** Yes **29.** No **31.** No

Exercise 2.2 (page 50) **1.** 5 **3.** 0 **5.** -1 **7.** $-\dfrac{1}{3}$ **9.** $\sqrt{3}$ **11.** $-2a + 2$

13. $\dfrac{1}{a^2} + 1$ **15.** $(a-2)^4 - 2(a-2)^2 + 1$ **17. a.** 13 **b.** 49 **19. a.** $\dfrac{1}{a}$ **b.** $\dfrac{1}{a}$

21. 3 **23.** 1 **25.** $2h$ **27.** $-3h$ **29.** $-2ah - h^2$ **31.** $2x + h - 1$ **33.** $\dfrac{-1}{x(x+h)}$

35. $\dfrac{1}{\sqrt{x} + \sqrt{x+h}}$

Exercise 2.3 (page 57) **1.** **3.** **5.**

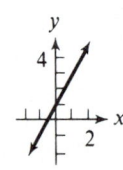

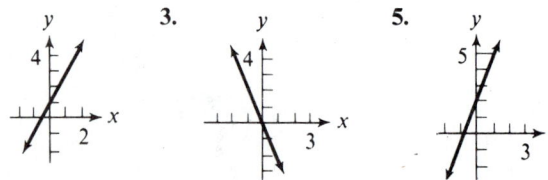

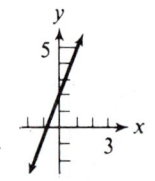

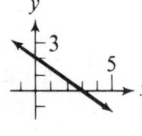

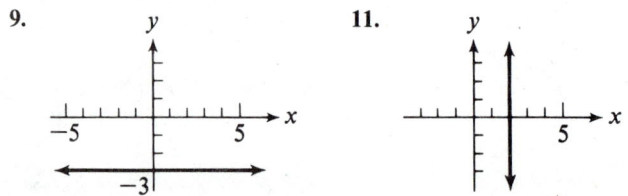

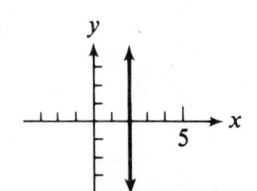

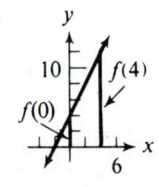

7. **9.** **11.** **13.**

15. Distance, 5; slope, $\dfrac{4}{3}$ **17.** Distance, 13; slope, $\dfrac{12}{5}$ **19.** Distance, $\sqrt{2}$; slope, 1

21. Distance, $3\sqrt{5}$; slope, $\dfrac{1}{2}$ **23.** Distance, 5; slope, 0 **25.** Distance, 10; slope, undefined

27. 7, $\sqrt{68}$, $\sqrt{89}$ **29.** 10, 21, 17

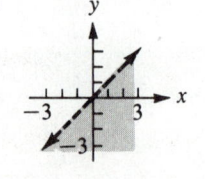

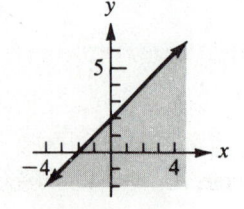

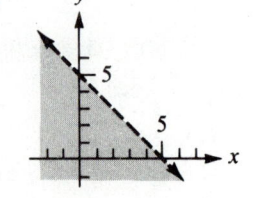

31. **33.** **35.**

37. From Exercise 28, the lengths of the sides of the triangle are 15, $3\sqrt{5}$, and $6\sqrt{5}$, respectively; since $15^2 = 225$, $(3\sqrt{5})^2 = 45$, and $(6\sqrt{5})^2 = 180$, it follows that $15^2 = (3\sqrt{5})^2 + (6\sqrt{5})^2$, and the converse of the Pythagorean theorem applies. Hence, the triangle is a right triangle.

39. $x - 2y = 8$ **41.** 2, slope of the line segment joining $(x, f(x))$ with $(a, f(a))$.

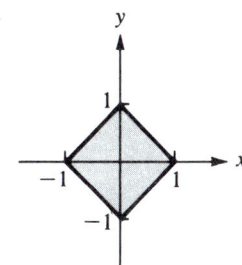

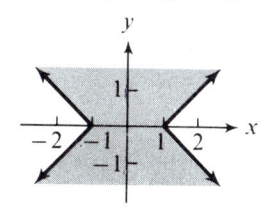

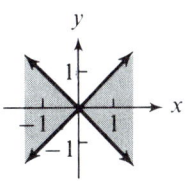

43.   **45.**   **47.**

Exercise 2.4 (page 62)   **1.** $4x - y - 7 = 0$ **3.** $x + y - 10 = 0$ **5.** $3x - y = 0$
7. $x + 2y + 2 = 0$ **9.** $3x + 4y + 14 = 0$ **11.** $y - 2 = 0$
13. $y = -x + 3$; slope, -1; intercept, 3   **15.** $y = -\frac{3}{2}x + \frac{1}{2}$; slope, $-\frac{3}{2}$; intercept, $\frac{1}{2}$
17. $y = \frac{1}{3}x - \frac{2}{3}$; slope, $\frac{1}{3}$; intercept, $-\frac{2}{3}$ **19.** $y = \frac{4}{3}x - \frac{7}{3}$; slope, $\frac{4}{3}$; intercept, $-\frac{7}{3}$
21. $y = 6$; slope, 0; intercept, 6   **23.** $x = 5$; slope, undefined; no y-intercept
25. $3x + 2y - 6 = 0$ **27.** $5x + 2y + 10 = 0$ **29.** $6x - 2y + 3 = 0$
31. Substituting $\dfrac{y_2 - y_1}{x_2 - x_1}$ for m in the formula $y - y_1 = m(x - x_1)$ gives the desired result.
33. $F(x) = -\frac{1}{3}x + \frac{11}{3}$

Exercise 2.5 (page 65)   **1. a.** $3x + y = 16$ **b.** $x - 3y = 2$
3. a. $3x - 2y = 0$ **b.** $2x + 3y = 0$ **5. a.** $2x + 3y = -7$ **b.** $3x - 2y = -1$
7. a. $x + 2y = 5$ **b.** $2x - y = -10$ **9.** $x + y = 12$ **11.** $7x - 3y = -8$
13. $ax - by = \frac{1}{2}(a^2 - b^2)$

15. Slope of $\overline{AB}$: $m_1 = -1$; slope of $\overline{AC}$: $m_2 = 1$; $m_1 m_2 = -1$; thus, by Theorem 2.3, $\overline{AB} \perp \overline{AC}$. Hence, $\triangle ABC$ is a right triangle.

17. Slope of $\overline{PQ}$: $m_4 = \frac{1}{3}$; slope of $\overline{QR}$: $m_2 = -\frac{2}{3}$; slope of $\overline{RS}$: $m_3 = \frac{1}{3}$; slope of $\overline{SP}$: $m_4 = -\frac{2}{3}$; $m_1 = m_3$ and $m_2 = m_4$; thus, by Theorem 2.2, $\overline{PQ}$ is parallel to $\overline{RS}$, and $\overline{QR}$ is parallel to $\overline{SP}$. Hence, $PQRS$ is a parallelogram.

19. From geometry, if L_1 is parallel to L_2, then $\angle P_3 P_1 Q_1$ is congruent to $\angle P_4 P_2 Q_2$. Also, $\angle P_1 Q_1 P_3$ is congruent to $\angle P_2 Q_2 P_4$, since each is a right angle. Hence, $\triangle P_1 Q_1 P_3$ is similar to $\triangle P_2 Q_2 P_4$. Because the lengths of corresponding sides of similar triangles are proportional, it follows that $\dfrac{Q_1 P_3}{P_1 Q_1} = \dfrac{Q_2 P_4}{P_2 Q_2}$.

But $\dfrac{Q_1 P_3}{P_1 Q_1} = \dfrac{y_3}{x_3 - x_1} = m_1$ and $\dfrac{Q_2 P_4}{P_2 Q_2} = \dfrac{y_4}{x_4 - x_2} = m_2$. Hence, $m_1 = m_2$.

504   *Answers to Odd-Numbered Exercises*

21. $\dfrac{y_1 - y_2}{x_2 - x_1} = -\dfrac{y_2 - y_1}{x_2 - x_1} = -m_2$; $\dfrac{x_2 - x_1}{y_3 - y_1} = \dfrac{1}{m_1}$. Hence, $-m_2 = \dfrac{1}{m_1}$, or $m_1 m_2 = -1$.

23. From geometry, if $\overline{PQ}$ is parallel to $\overline{P_2 Q_2}$ and if P is the midpoint of $\overline{P_1 P_2}$, then Q is the midpoint of $\overline{P_1 Q_2}$, or $x_2 - x_1 = 2(x - x_1)$. Then

$$x_2 - x_1 = 2x - 2x_1,$$
$$2x = x_1 + x_2,$$
$$x = \frac{x_2 + x_1}{2}.$$

Also, from geometry, if $\overline{PQ}$ is parallel to $\overline{P_2 Q_2}$ and if P is the midpoint of $\overline{P_1 P_2}$, then $\overline{PQ}$ is one-half of $\overline{P_2 Q_2}$, or $y_2 - y_1 = 2(y - y_1)$. Then

$$y_2 - y_1 = 2y - 2y_1,$$
$$2y = y_1 + y_2,$$
$$y = \frac{y_1 + y_2}{2}.$$

Exercise 2.6 (page 72)   **1.** $y, 4$; $x, 1,$ and 4; $\left(\dfrac{5}{2}, -\dfrac{9}{4}\right)$, minimum

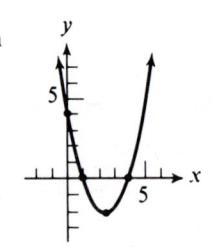

3. $y, -7$; $x, -1,$ and 7; $(3, -16)$, minimum 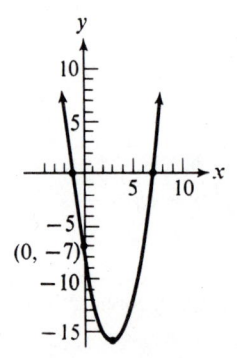   **5.** $y, -4$; $x, 1,$ and 4; $\left(\dfrac{5}{2}, \dfrac{9}{4}\right)$, maximum

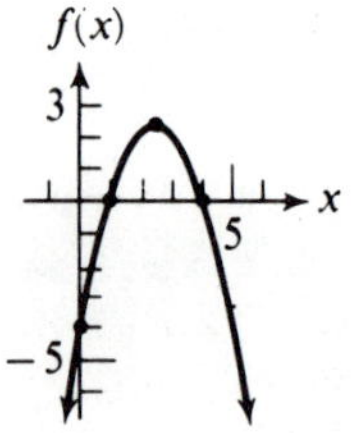

7. $y, 2$; no x; $(0, 2)$, minimum 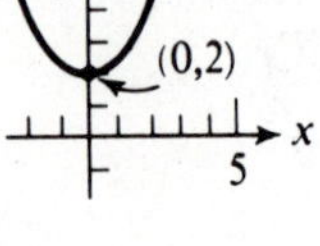   **9.** 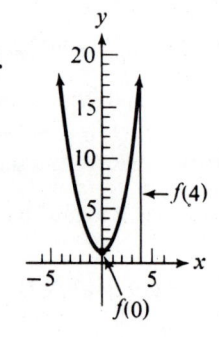   **11.** $\dfrac{16}{3}$, slope of secant line

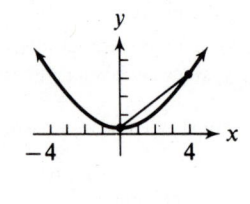

Answers to Odd-Numbered Exercises

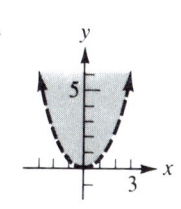

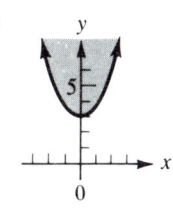

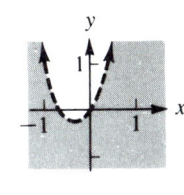

13. [graph] **15.** [graph] **17.** [graph] **19.** [graph] **21.** 4; 4

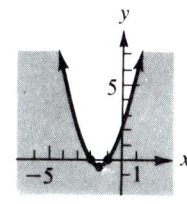

23. Varying k has the effect of translating the graph along the y-axis.

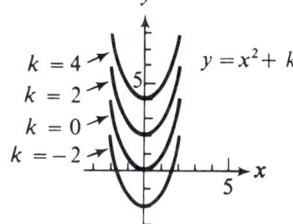

25. Varying k has the effect of translating the axis of symmetry along the x-axis and thus changes the vertex.

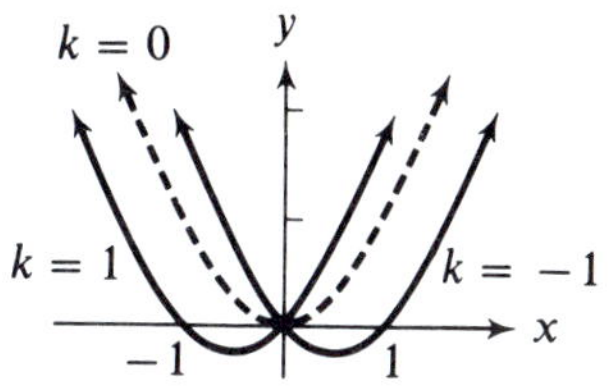

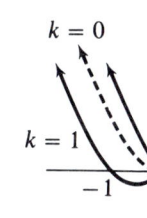

27. [graph with points $(-4,0)$, $(0,2)$, $(0,-2)$]

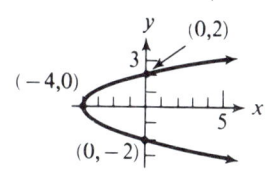

29. [graph with points $(0,2)$, $(4,0)$]

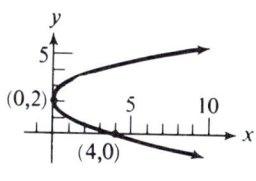

31. [graph]

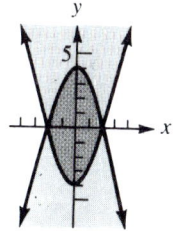

33. $2(x + a) - 1$, slope of the line joining $(x, f(x))$ and $(a, f(a))$.

Exercise 2.7 (page 78)

1. [graph with points $(-2,3)$, $(5,9)$, $(4,7)$, $(9,5)$, $(7,4)$, $(3,-2)$]

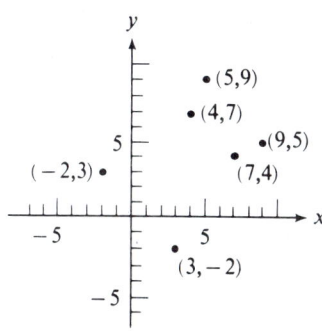

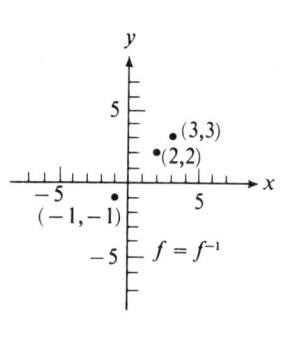

3. [graph with points $(3,3)$, $(2,2)$, $(-1,-1)$; $f = f^{-1}$]

5. [graph of $y = 2x + 6$ and $x = 2y + 6$]

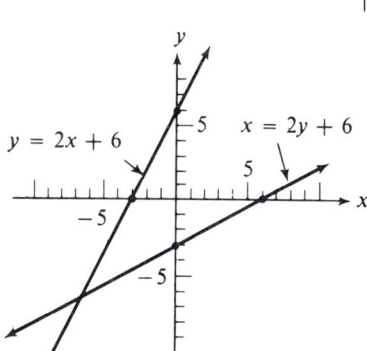

7. [graph of $y = 4 - 2x$ and $x = 4 - 2y$]

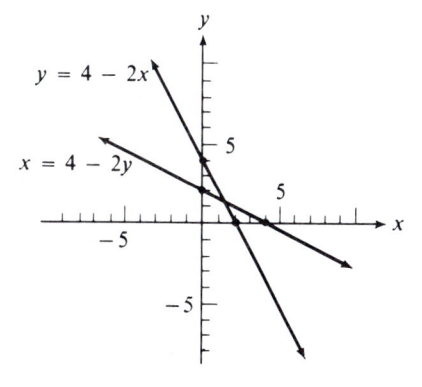

9.

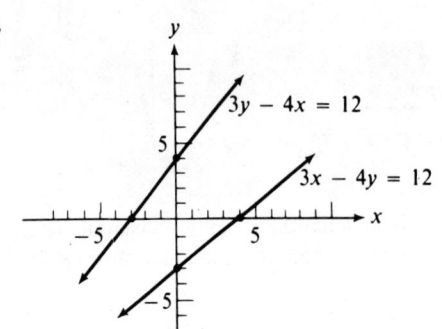

11.

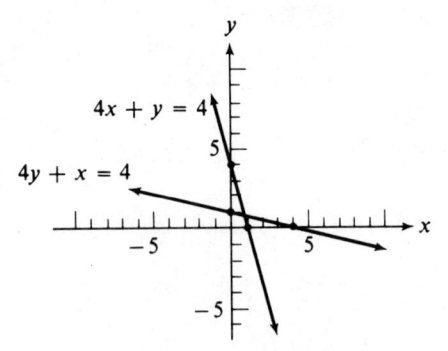

13.

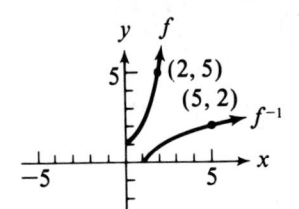

15.

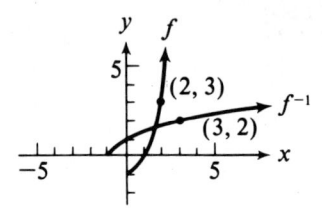

17.

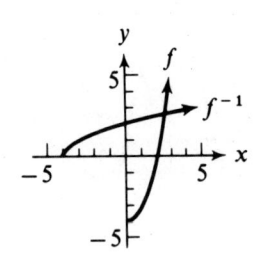

19.

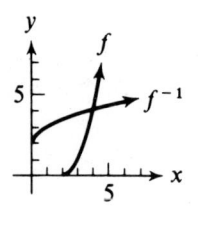

21. $f(x) = x$, $f^{-1}(x) = x$; $f[f^{-1}(x)] = f(x) = x$, $f^{-1}[f(x)] = f^{-1}(x) = x$

23. $f(x) = 4 - 2x$, $f^{-1}(x) = \dfrac{4-x}{2}$; $f[f^{-1}(x)] = f\left[\dfrac{4-x}{2}\right] = 4 - 2\left(\dfrac{4-x}{2}\right) = 4 - 4 + x = x$,

$f^{-1}[f(x)] = f^{-1}[4 - 2x] = \dfrac{4 - (4 - 2x)}{2} = \dfrac{4 - 4 + 2x}{2} = \dfrac{2x}{2} = x$

25. $f(x) = \dfrac{3x - 12}{4}$, $f^{-1}(x) = \dfrac{4x + 12}{3}$; $f[f^{-1}(x)] = f\left[\dfrac{4x + 12}{3}\right] = \dfrac{(3)\left(\dfrac{4x + 12}{3}\right) - 12}{4} =$

$\dfrac{4x + 12 - 12}{4} = \dfrac{4x}{4} = x$, $f^{-1}[f(x)] = f^{-1}\left[\dfrac{3x - 12}{4}\right] = \dfrac{(4)\left(\dfrac{3x - 12}{4}\right) + 12}{3} = \dfrac{3x - 12 + 12}{3} = \dfrac{3x}{3} = x$

27. $f(x) = \dfrac{1}{x}$, $f^{-1}(x) = \dfrac{1}{x}$; $f[f^{-1}(x)] = \dfrac{1}{f^{-1}(x)} = \dfrac{1}{1/x} = x$, $f^{-1}[f(x)] = \dfrac{1}{f(x)} = \dfrac{1}{1/x} = x$ 29. 4

31. 3

33. The line joining the points (b, a) and (a, b) has slope -1 and hence is perpendicular to the line $y = x$. The midpoint of the line segment joining the points (a, b) and (b, a) is $\left(\dfrac{a + b}{2}, \dfrac{a + b}{2}\right)$, which is on the line $y = x$. Thus the line $y = x$ is the perpendicular bisector of the line segment joining the points (b, a) and (a, b).

Exercise 2.8 (page 83)   **1.** 1   **3.** 8

5.

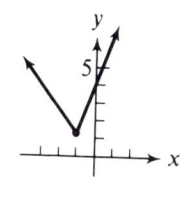

7.

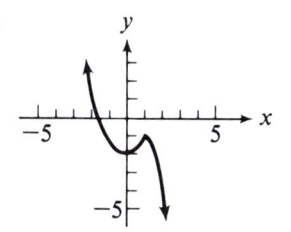

9.

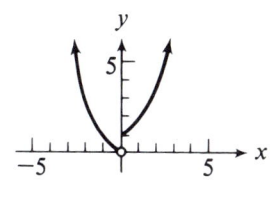

11.

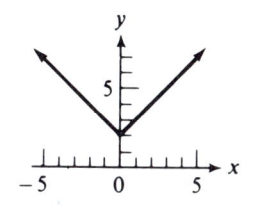

13.

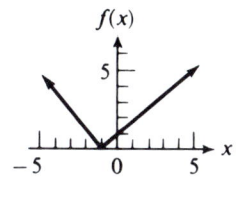

15.

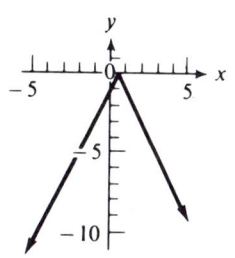

17.

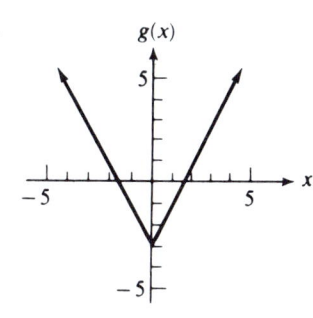

19.

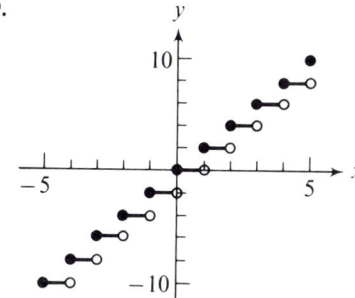

21.

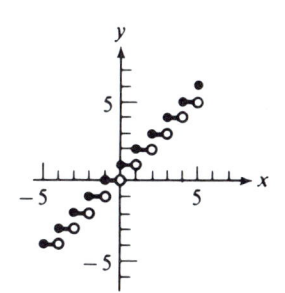

23.

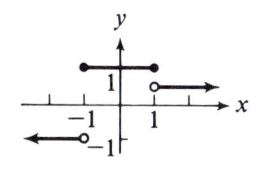

25.

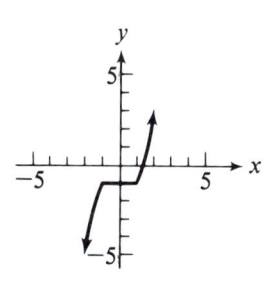

27.

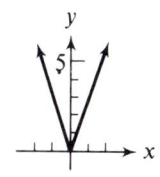

29.

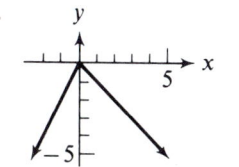

31.

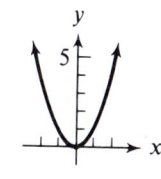

33.

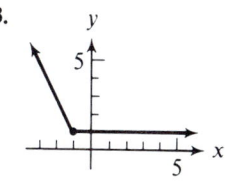

35. [graph of F(x)]

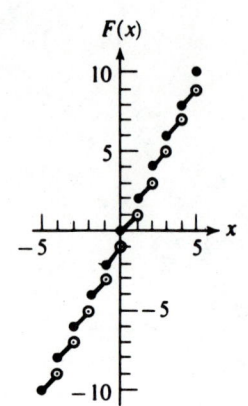

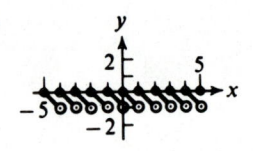

37. [graph]

39. Cost in cents: $C = -c[-x]$, $x > 0$.

Chapter 2 Review (page 84) **1.** $\{4, 2, 3\}$ **2.** $\{1\}$ **3.** $\{x \mid x \neq -4\}$ **4.** $\{x \mid x \geq 6\}$
5. 13 **6.** -5 **7.** $a - h - 3$ **8.** $a^2 + 2ah + h^2 + 4$
9. $f(g(x)) = x - 2\sqrt{x}$, $g(f(x)) = \sqrt{x^2 - 2x}$ **10.** $f[g(x)] = x$, $g[f(x)] = x$ **11.** $2h$
12. $2ah + h^2$ **13.** 12 **14.** $\dfrac{\sqrt{3}}{3} - 1$ **15.** [graph] **16.** [graph]

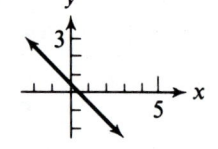

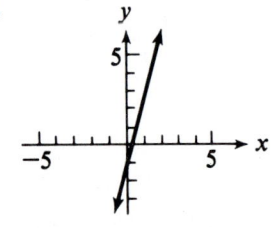

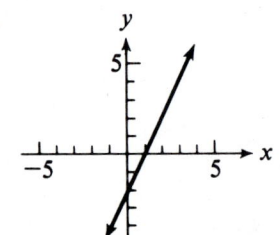

17. [graph] **18.** [graph]

19. $d = \sqrt{41}$, $m = -\dfrac{4}{5}$ **20.** $d = \sqrt{5}$, $m = -\dfrac{1}{2}$

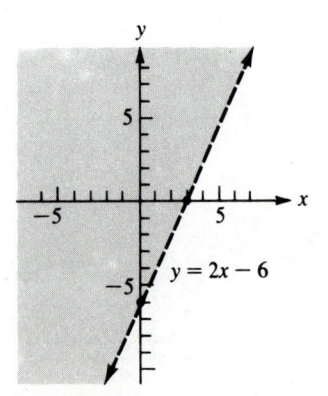

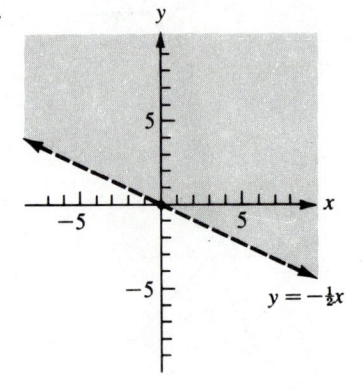

21. [graph, $y = 2x - 6$]

22. [graph, $y = -\tfrac{1}{2}x$]

Answers to Odd-Numbered Exercises 509

23. [graph showing $y = 3$ and $y = -2$]

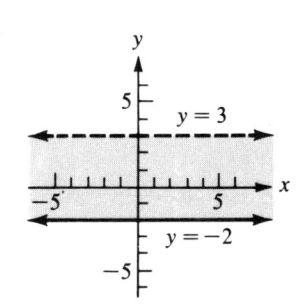

24. [graph showing $x = 3$ and $x = 5$]

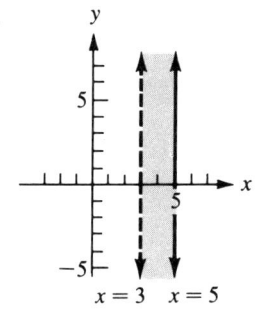

25. $-4x + y + 15 = 0$ **26.** $2y - x - 12 = 0$ **27.** $y = -4x + 6$, $m = -4$, $b = 6$

28. $y = \frac{3}{2}x - 8$, $m = \frac{3}{2}$, $b = -8$ **29.** $-\frac{2}{3}x + y - 2 = 0$ **30.** $y = +12x - 4$

31. a. $y - 1 = \frac{3}{2}(x - 1)$ **b.** $y - 1 = -\frac{2}{3}(x - 1)$

32. a. $y - 1 = x + 1$ **b.** $y - 1 = -(x + 1)$ **33.** $\left(0, \frac{3}{2}\right)$ **34.** $\left(\frac{1}{4}, \frac{1}{6}\right)$

35. x-intercepts $3, -2$; axis of symmetry $x = \frac{1}{2}$; minimum point $\left(\frac{1}{2}, -\frac{25}{4}\right)$

36. x-intercepts $5, 2$; axis of symmetry $x = \frac{7}{2}$; maximum point $\left(\frac{7}{2}, \frac{9}{4}\right)$

37. [parabola with point $(4,6)$, $f(4)$, and minimum $\left(\frac{1}{2}, -\frac{25}{4}\right)$]

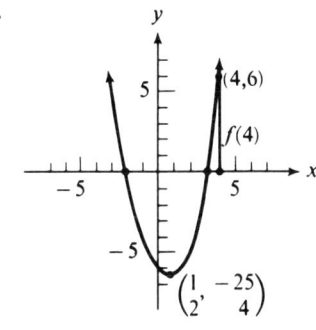

38. [downward parabola with point $(3,2)$, $f(3)$]

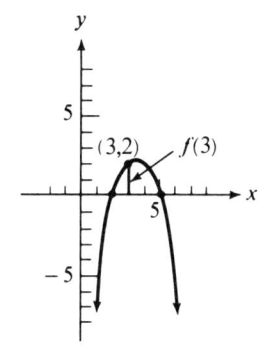

39. [parabola graph, dashed]

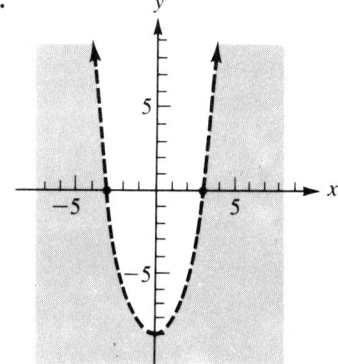

40. [parabola graph with vertex $(-3,-4)$]

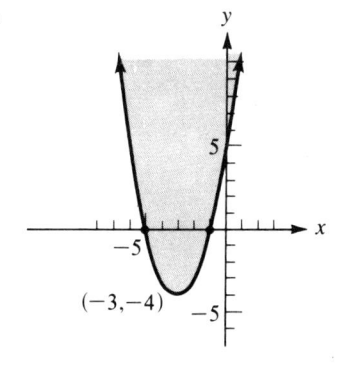

41.

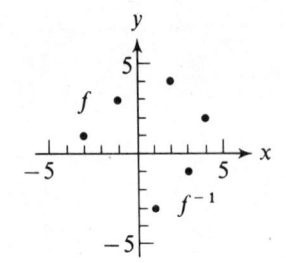

42.

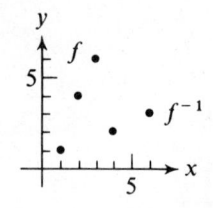

43.

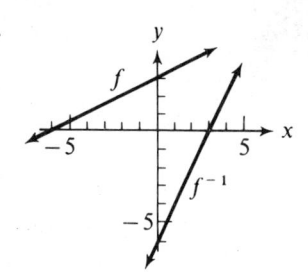

44.

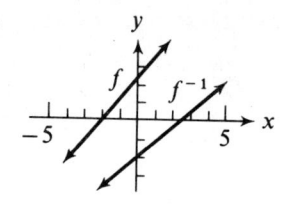

45. $y = x^2 - 9, \quad x \geq 0$; $x = y^2 - 9, \quad y \geq 0$

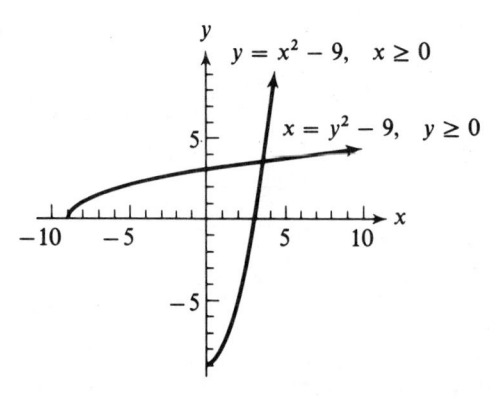

46. $y = x^2 - 4x - 5, \quad x \leq 2$; $(-9, 2)$; $x = y^2 - 4y - 5, \quad y \leq 2$; $(2, -9)$

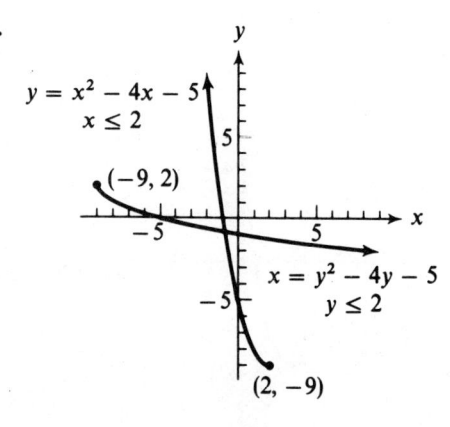

47. 4   **48.** -7 **49.**

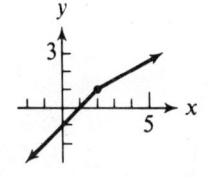

50.

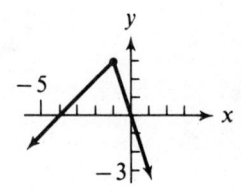

51.

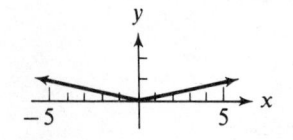

52. $\left(\frac{1}{4}, 0\right)$

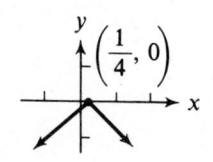

Answers to Odd-Numbered Exercises

53.

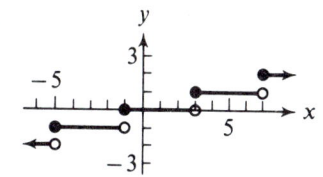

54.

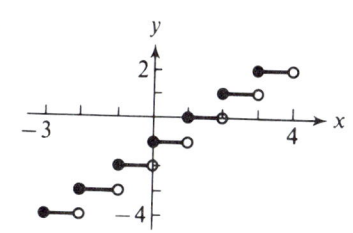

Exercise 3.1 (page 92)
1. $2y^2 - y + \dfrac{13}{2} + \dfrac{-3/2}{2y+1}, \left(y \neq -\dfrac{1}{2}\right)$
3. $2y^3 - y^2 - \dfrac{1}{2}y - \dfrac{5}{4} + \dfrac{(-13y/4) + (9/4)}{2y^2 + y + 1}$
5. $x^3 - x^2 - \dfrac{1}{x-2}, (x \neq 2)$
7. $2x^2 - 2x + 3 + \dfrac{-8}{x+1}, (x \neq -1)$
9. $2x^3 + 10x^2 + 50x + 249 + \dfrac{1251}{x-5}, (x \neq 5)$
11. $x^2 + 2x - 3 + \dfrac{4}{x+2}, (x \neq -2)$
13. $x^5 + x^4 + 2x^3 + 2x^2 + 2x + 1 + \dfrac{1}{x-1}, (x \neq 1)$
15. $x^4 + x^3 + x^2 + x + 1, (x \neq 1)$
17. 3, 17, 47
19. 56, 12, 326
21. $-685, 95, 719$
23. $(x-2)(x+1)^2$
25. $(x+1)(x+3)(x-2)$
27. $(2x-1)(x-2)(x+2)$
29. $x(x+1)(x-2)(x-1)$
31. $x(x-1)(2x+1)(x+2)$
33. $k = 80$
35. $k = -22$

Exercise 3.2 (page 98)
1. $f(0) = 1$ and $f(1) = -1$
3. $g(-3) = 10$ and $g(-2) = -33$
5. $P(-3) = 2$ and $P(-2) = -4$, $P(0) = -4$, and $P(1) = 2$
7. 0 or 2 positive, 0 or 2 negative
9. 0 positive, 1 negative
11. 0 positive, 0 negative
13. Zero within one-half unit of 3/2 and 2
15. Zero within one-half unit of -3 and $-5/2$
17. Upper bound 3; lower bound -4
19. Upper bound 4; lower bound -3
21. Upper bound 2; lower bound -3
23. Upper bound 1; lower bound -2
25. Upper bound 6; lower bound -4
27. There is a zero between $-3/2$ and $-11/8$.
29. There is a zero between -1 and $-7/8$.

Exercise 3.3 (page 102)
1. $\{4\}$
3. $\{1, -2\}$
5. None
7. $\left\{-\dfrac{3}{2}\right\}$
9. $\left\{2, -\dfrac{7}{4}\right\}$
11. $\left\{\dfrac{3}{2}\right\}$
13. $\left\{-\dfrac{1}{3}, 1+\sqrt{5}, 1-\sqrt{5}\right\}$
15. $\left\{-1, \dfrac{1}{2}, \dfrac{-1+\sqrt{3}}{2}, \dfrac{-1-\sqrt{3}}{2}\right\}$
17. $\left\{\dfrac{3}{4}, -\dfrac{4}{3}, \sqrt{2}, -\sqrt{2}\right\}$
19. $(2x+3)(x-1)(x+1)$
21. Consider $x^2 - 3 = 0$. If the equation has real roots, then they are either rational or irrational. If rational, by Theorem 3.9, the only possibilities are ± 1 and ± 3; however, direct substitution shows none of these satisfies the equation. Now, $\sqrt{3}$ is a real number, and since $(\sqrt{3})^2 - 3 = 3 - 3 = 0$, $\sqrt{3}$ is a real zero of the equation, and since it is not among the rational zeros, it must be irrational.
23. By Theorem 3.9, the only possible rational roots of the equation $x^2 - p = 0$ are ± 1 and $\pm p$. However, since p is prime, $p \neq 1$ and thus $(\pm 1)^2 - p = 1 - p \neq 0$ and neither 1 nor -1 is a solution. If either p or $-p$ is a solution, then $p^2 - p = 0$, and $p = 0$ or $p = 1$. Since p is prime, $p \neq 0$ and $p \neq 1$. Thus, neither p nor $-p$ is a solution. Finally, $\sqrt{p}$ is a real solution; since it is not rational, it must be irrational.

Exercise 3.4 (page 108)
1. (1, 15) is a local maximum; (2, 14) is a local minimum.
3. $(-1, 0)$ and $(1, 0)$ are local minima; $(0, 1)$ is a local maximum.
5. (0, 10) is neither a local maximum nor a local minimum; (1, 8) is a local minimum.

7. Graph with points $(-1, 2)$ and $(1, -2)$.

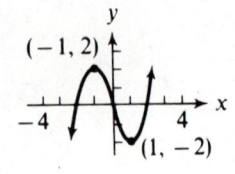

9. Graph with points $(\frac{1}{2}, 2)$ and $(\frac{3}{2}, 0)$.

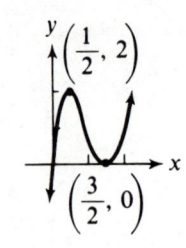

11. Graph with points $(-2, \frac{22}{3})$ and $(3, -\frac{27}{2})$.

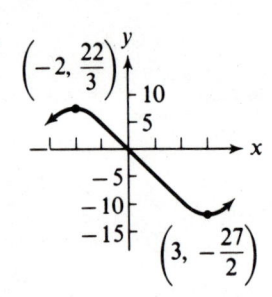

13. Graph with points $(-\frac{1}{3}, \frac{14}{27})$ and $(3, -18)$.

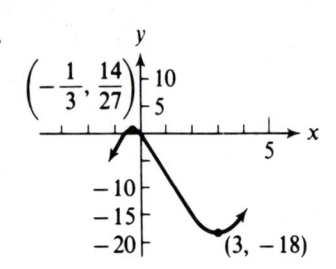

15. Graph with points $(\frac{4}{3}, \frac{5}{27})$ and $(0, -1)$.

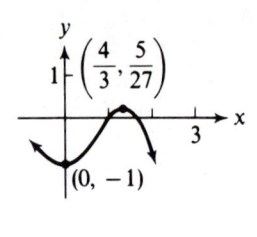

17. Graph with points $(\frac{7}{3}, \frac{299}{9})$ and $(-\frac{1}{3}, \frac{43}{9})$.

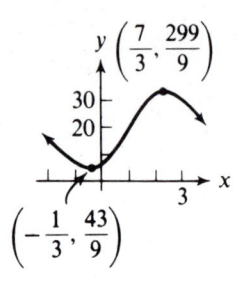

19. Graph with points $(-\frac{\sqrt{2}}{2}, \frac{1}{4})$, $(\frac{\sqrt{2}}{2}, \frac{1}{4})$, and $(0, 0)$.

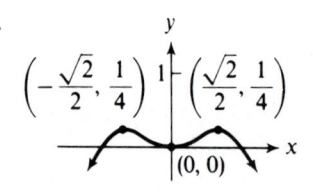

21. Graph with points $(0, 9)$, $(-\sqrt{5}, -16)$, and $(\sqrt{5}, -16)$.

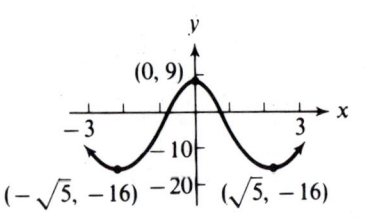

23. Graph.

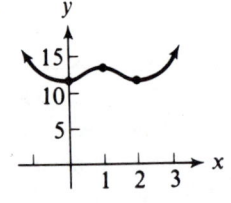

25. Graph.

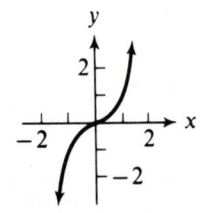

27. The vertex of the graph of $P(x) = ax^2 + bx + c$ is the only turning point. The only turning point occurs where $P'(x) = 2ax + b = 0$, or $x = \frac{-b}{2a}$. The y-coordinate is $P\left(-\frac{b}{2a}\right) = \frac{4ac - b^2}{4a}$.

Exercise 3.5 (page 114) **1.** $x = 3$ **3.** $x = 3$; $x = -2$ **5.** $x = -1$; $x = -4$ **7.** $x = -1$

9. Graph with points $(1, 1)$ and $(-1, -1)$.

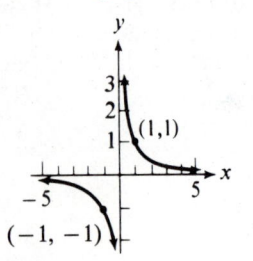

11. Graph with points $(4, 1)$, $(5, \frac{1}{2})$, $(-1, -\frac{1}{4})$, and $(1, -\frac{1}{2})$.

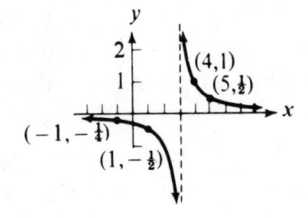

13. [graph]

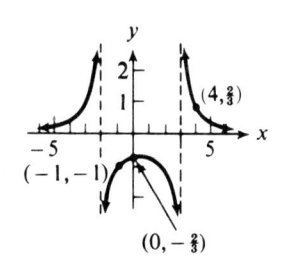

15. [graph]

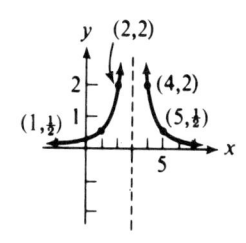

17. $x = 2$; $x = -2$; $y = 0$ **19.** $x = 4$; $y = x + 4$ **21.** $x = 4$; $x = -1$; $y = 1$

23. [graph]

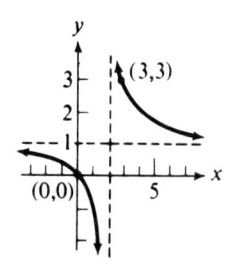

25. [graph]

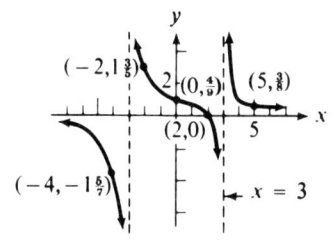

27. [graph]

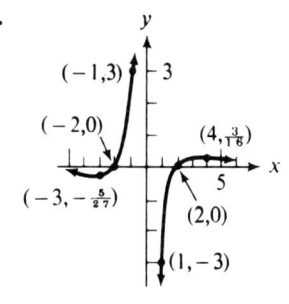

29. [graph]

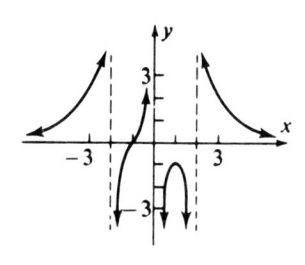

31. [graph]

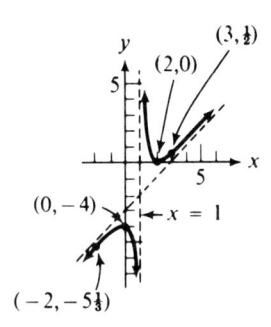

33. [graph]

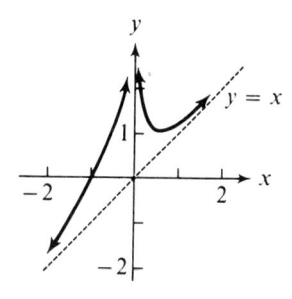

35. [graph]

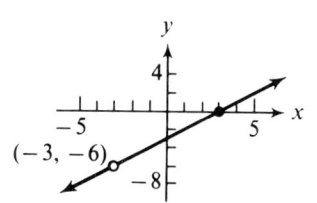

37. [graph]

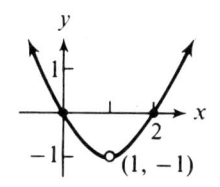

39. [graph]

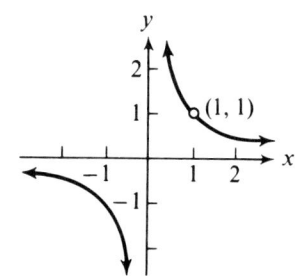

41. Behaves like $y = x^2$ as $|x|$ becomes large.

43. No, because if $x = a$ is a vertical asymptote, then the function is not defined at $x = a$; thus there is no y so that the point (a, y) is on the graph.

45. $(1, 3)$

Chapter 3 Review (page 116)
1. $2n + 1$
2. $r + 1$
3. $2x^2 - 3x + 1$
4. $3x^2 - 2x + 3 + \dfrac{1}{x + 3}$
5. $(x - 1)(x - 2)(x + 1)$
6. $(x + 2)(2x + 1)(x - 3)$
7. $P(1) = -3$ and $P(2) = 10$
8. $P\left(\dfrac{1}{2}\right) = -\dfrac{7}{16}$ and $P\left(\dfrac{3}{2}\right) = \dfrac{57}{16}$
9. Upper bound 2; lower bound -2
10. Upper bound 3; lower bound -4
11. 1 positive, 0 or 2 negative
12. 0 or 2 positive, 0 or 2 negative
13. $\{4\}$
14. $\left\{-\dfrac{1}{2}, 3\right\}$
15. $\{-2, \sqrt{2}, -\sqrt{2}\}$
16. $\left\{\dfrac{1}{2}, -\dfrac{1}{3}, \sqrt{\dfrac{2}{3}}, -\sqrt{\dfrac{2}{3}}\right\}$
17. $(1, 5)$ is a local maximum; $(2, 4)$ is a local minimum.
18. $(0, 0)$ is a local maximum; $\left(\dfrac{2}{3}, \dfrac{-4}{27}\right)$ is a local minimum.

19.

20.

21.

22.

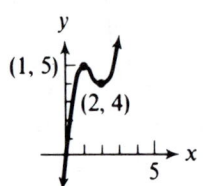

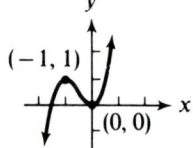

23. Vertical asymptote: $x = -2$
Horizontal asymptote: $y = 0$

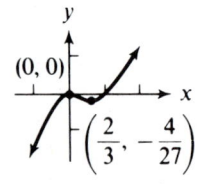

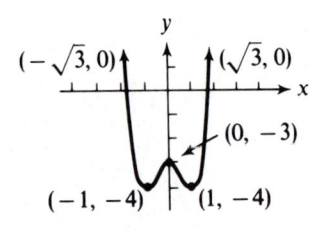

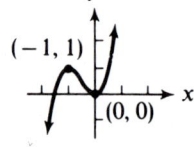

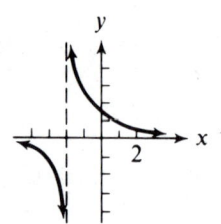

24. Vertical asymptotes: $x = -3$; $x = 4$
Horizontal asymptote: $y = 0$

25. Vertical asymptote: $x = 3$
Horizontal asymptote: $y = 1$

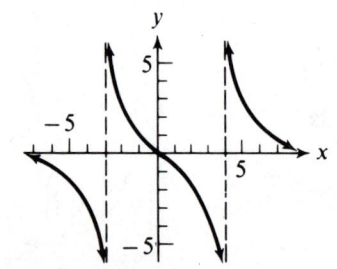

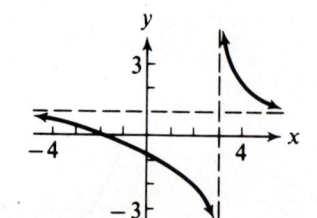

26. Vertical asymptotes: $x = 1$; $x = 2$
Horizontal asymptote: $y = 0$

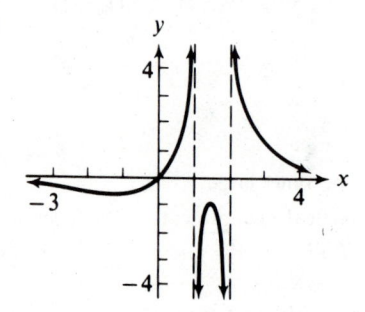

Exercise 4.1 (page 120) **1.** (0, 1), (1, 3), (2, 9) **3.** (0, −1), (1, −5), (2, −25) **5.** $(-3, 8), (0, 1), \left(3, \frac{1}{8}\right)$
7. $\left(-2, \frac{1}{100}\right), (1, 10), (0, 1)$

9.

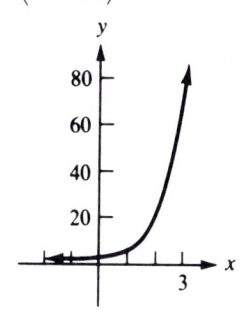

11.

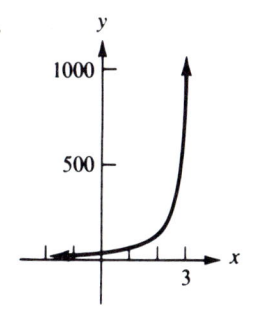

13.

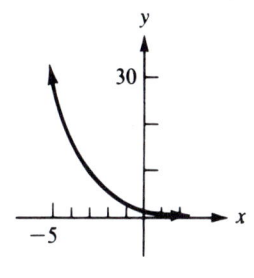

15.

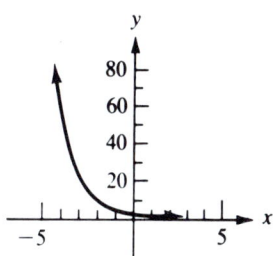

17.

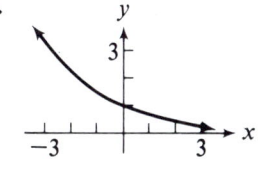

19.

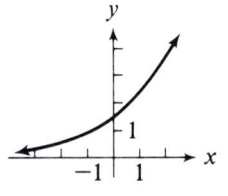

21. $y = 10^x$, (1, 10), (0, 1), (10, 1), (1, 0), $x = 10^y$

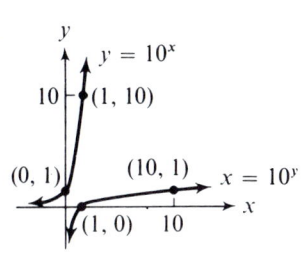

23. {−2}   **25.** $\left\{\frac{3}{4}\right\}$ **27.** 2   **29.** −2

31. $b = 4$, $b = 3$, $b = 2$

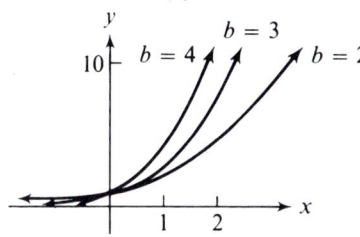

As $b > 1$ increases, the rate of increase of each curve becomes larger.

Exercise 4.2 (page 124)   **1.** $y = \log_2 x$ **3.** $y = \log_2 4x$ **5.** $y = \log_2 (x + 3)$

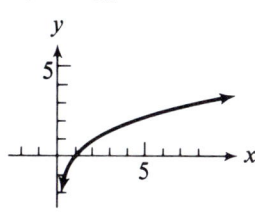

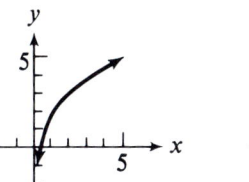

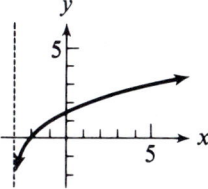

7. $\log_4 16 = 2$ **9.** $\log_3 27 = 3$ **11.** $\log_{1/2} \frac{1}{4} = 2$ **13.** $\log_8 \frac{1}{2} = -\frac{1}{3}$ **15.** $\log_{10} 100 = 2$

17. $\log_{10} (0.1) = -1$ **19.** $2^6 = 64$ **21.** $3^2 = 9$ **23.** $\left(\frac{1}{3}\right)^{-2} = 9$ **25.** $10^3 = 1000$

27. 2   **29.** 3   **31.** −1   **33.** 1   **35.** 2   **37.** −1   **39.** {2}   **41.** {2}   **43.** {64}

45. $\{-3\}$ **47.** $\{100\}$ **49.** $\{4\}$ **51.** $\log_b x + \log_b y$ **53.** $\log_b x - \log_b y$ **55.** $5 \log_b x$
57. $\frac{1}{3} \log_b x$ **59.** $\frac{1}{2}(\log_b x - \log_b z)$ **61.** $\frac{1}{3}(\log_{10} x + 2 \log_{10} y - \log_{10} z)$ **63.** $\log_b 2xy^3$
65. $\log_b x^{1/2} y^{2/3}$ **67.** $\log_b \frac{x^3 y}{z^2}$ **69.** $\log_{10} \frac{x(x-2)}{z^2}$ **71.** $\{500\}$ **73.** $\{4\}$ **75.** $\{3\}$
77. By definition, $\log_b 1$ is a number such that $b^{\log_b 1} = 1$; therefore $\log_b 1 = 0$.
79. Since $x_1 = b^{\log_b x_1}$ and $x_2 = b^{\log_b x_2}$, it follows that

$$\frac{x_2}{x_1} = \frac{b^{\log_b x_2}}{b^{\log_b x_1}} = b^{\log_b x_2 - \log_b x_1},$$

and by definition $\log_b (x_2/x_1) = \log_b x_2 - \log_b x_1$.

81.

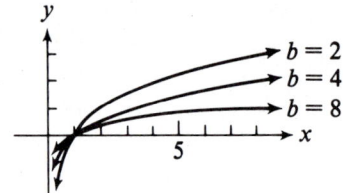

As $b > 1$ increases, $\log_b x$ increases more slowly.

(Graph showing curves for $b=2$, $b=4$, $b=8$)

Exercise 4.3 (page 133) *Note*: The authors have used a calculator to obtain these answers. If the tables are used, slightly different values may be obtained.

1. 2 **3.** -3 **5.** -4 **7.** 4 **9.** 0.8280 **11.** $9.9101 - 10$, or -0.0899
13. $8.9031 - 10$, or -1.0969 **15.** 2.3945 **17.** $6.5328 - 10$, or -3.4672 **19.** 2.4330
21. 4.10 **23.** 3.67 **25.** 703 **27.** 2060 **29.** 0.0065 **31.** 0.0041 **33.** 1.0986
35. 2.8332 **37.** 6.1092 **39.** -2.7334 **41.** 1.649 **43.** 29.96 **45.** 1.259 **47.** 0.8607
49. 0.0821 **51.** 0.7788 **53.** 1.13 **55.** 0.923 **57.** 1.27
59. Since $e^N = x$, we have $\log_{10} e^N = \log_{10} x$; thus, by Theorem 4.2-III, $N \log_{10} e = \log_{10} x$. Hence,

$$N = \frac{\log_{10} x}{\log_{10} e}.$$

61. 0.2057, 0.0450, 0.0069, 0.0009 **63.** $-0.2303, -0.0461, -0.0069, -0.0009$

Exercise 4.4 (page 139) *Note*: The authors have used a calculator to obtain these answers. If tables are used, slightly different values may result.

1. 8.80 **3.** 220.26 **5.** 1.08 **7.** $\frac{\log_{10} 6}{3} = 0.2594$ **9.** $-\frac{1}{3} + \frac{1}{3} \log_{10} 9 = -0.0152$
11. $\pm\sqrt{\log_{10} 150} = \pm 1.4752$ **13.** $\frac{1}{3} \ln 5 = 0.5365$ **15.** $-\frac{1}{2} + \frac{1}{2} \ln 25 = 1.1094$
17. $\pm\sqrt{2 \ln 15} = \pm 2.3273$ **19.** $\frac{\log_{10} 7}{\log_{10} 2} = 2.8074$ **21.** $\frac{1}{2}\left(\frac{\log_{10} 3}{\log_{10} 7} + 1\right) = 0.7823$
23. $\pm\sqrt{\frac{\log_{10} 16}{\log_{10} 4}} = \pm 1.4142$ **25.** $t = \frac{1}{k} \log_{10} A$ **27.** $t = \frac{1}{k} \ln y$ **29.** $n = \frac{\log_{10} y}{\log_{10} x}$
31. 30.0 inches; 16.1 inches **33.** 12.05 grams **35.** 7.7 **37.** 2.5×10^{-6} **39.** 7%
41. $12,391.14; $12,648.84 **43.** $767.15 **45.** $27,782 **47.** 6.93% **49.** $k = 0.23026$

Answers to Odd-Numbered Exercises   517

Chapter 4 Review (page 144) *Note*: Where possible, the authors have used a calculator to obtain these answers. If tables are used, slightly different values may result.

1.

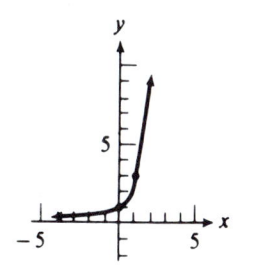

2.

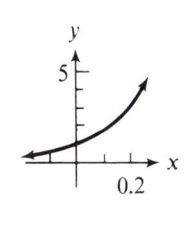

3. $\log_{16} \frac{1}{4} = -\frac{1}{2}$ **4.** $\log_7 343 = 3$ **5.** $2^3 = 8$ **6.** $10^{-4} = 0.0001$ **7.** $y = 4$

8. $x = 1000$ **9.** $\frac{1}{3} \log_{10} x + \frac{2}{3} \log_{10} y$ **10.** $\log_{10} 2 + 3 \log_{10} R - \frac{1}{2} \log_{10} P - \frac{1}{2} \log_{10} Q$

11. $\log_b \frac{x^2}{\sqrt[3]{y}}$ **12.** $\log_{10} \frac{\sqrt[3]{x^2 y}}{z^3}$ **13.** 1.6232   **14.** 7.4969 − 10, or −2.5031   **15.** 2.8340

16. 8.6172 − 10, or −1.3828   **17.** 67.4   **18.** 2.65   **19.** 0.0466   **20.** 0.6644   **21.** 1.9459
22. 3.1355   **23.** 5.4848   **24.** 6.2344   **25.** 403.4   **26.** 6.0496   **27.** 0.0302   **28.** 0.6570
29. 1.43   **30.** 0.696   **31.** $\frac{\log_{10} 8}{-4} = -0.4515$ **32.** $-2 - \log_{10} 25 = -3.3979$

33. $\frac{\ln 8}{4} = 0.5199$ **34.** $\frac{1}{2}(2 - \ln 16) = -0.3863$ **35.** $\frac{\log_{10} 2}{\log_{10} 3} = 0.6309$

36. $x = \frac{\log_{10} 80}{\log_{10} 3} - 1 = 2.9887$ **37.** $t = -\ln\left(\frac{y - A}{A}\right)$ **38.** $t = -c - \ln\left(\frac{y - A}{A}\right)$ **39.** 8%

40. 6.31×10^{-7} **41.** 7.13%   **42.** 0.65   **43.** $634.84   **44.** A little more than 223 months.

Supplemental Exercises, Chapters 1-4 (page 147)   **1.** $\left\{-1, \frac{5}{7}, 2\right\}$ **2.** $\left\{0, -1, \frac{-13}{12} + \frac{\sqrt{73}}{12}, \frac{-13}{12} - \frac{\sqrt{73}}{12}\right\}$

3. $\left\{\frac{7}{6}, \sqrt{\frac{41}{168}}, -\sqrt{\frac{41}{168}}\right\}$ **4.** $\left\{\frac{1}{\sqrt[3]{3}}\right\}$

5. If $R = 0$, then every $y \neq 0$ is a solution. If $R \neq 0$, then $y = -R$ and $y = R/3$ are solutions.

6. $x = \frac{L\sqrt[3]{A/B}}{1 + \sqrt[3]{A/B}}$ **7.** $x = \frac{V_0}{C}$ or $x = \frac{V_0}{2 + C}$ **8.** $x = \frac{B}{A}, \frac{A}{B}$ **9.** $(-1, 0) \cup (0, +\infty)$

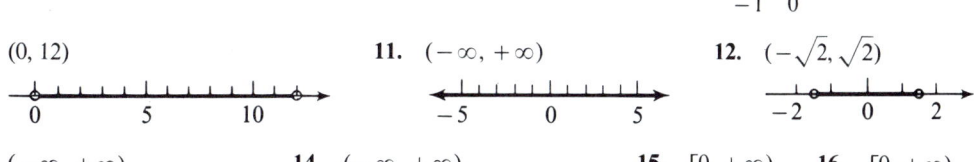

10. (0, 12)   **11.** $(-\infty, +\infty)$ **12.** $(-\sqrt{2}, \sqrt{2})$

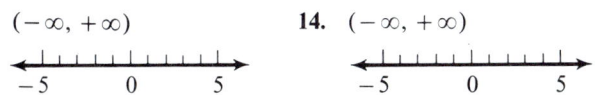

13. $(-\infty, +\infty)$ **14.** $(-\infty, +\infty)$ **15.** $[0, +\infty)$ **16.** $[0, +\infty)$

17.

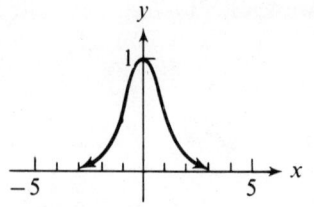

18.

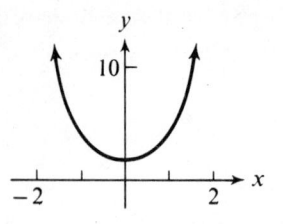

19.

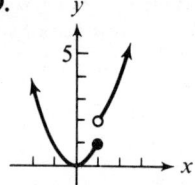

20.

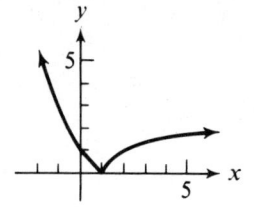

21.

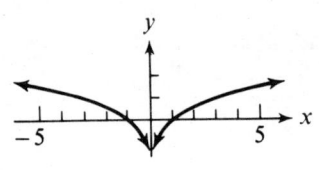

22.

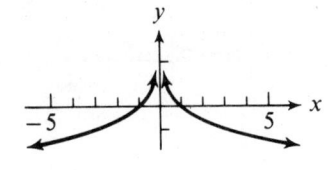

23.

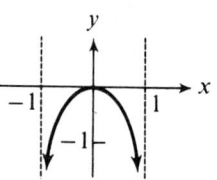

24.

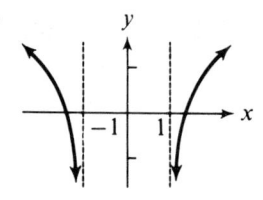

25. $(1, \frac{97}{3})$, $(3, 11)$, $(-2, -31)$

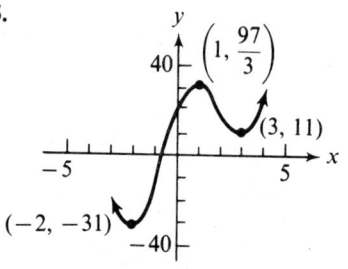

26. $(1, 1)$

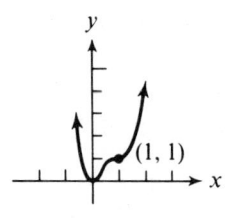

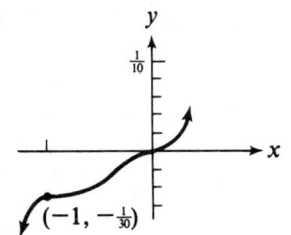

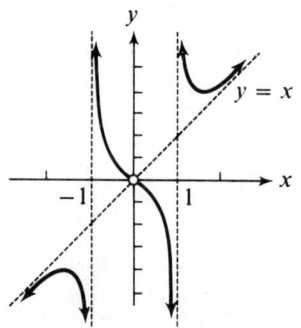

27. $(-1, -\frac{1}{30})$

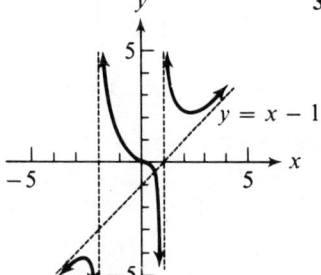

28. $y = x$

29. $y = x - 1$

30. $y = x^3$

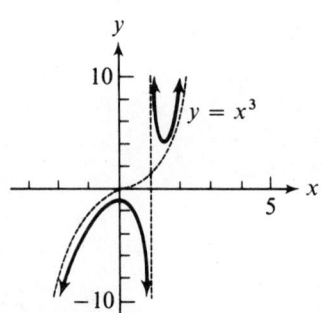

31.

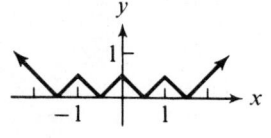

32.

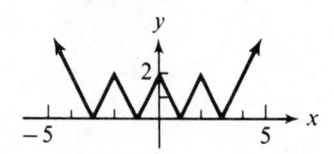

33. $\{y | y > -1\}$ **34.** $\{y | y > 1\}$ **35.** $\{y | 0 < y \le 1\}$ **36.** $\{y | y \le 0\}$

37. $y = \ln\left(\dfrac{x + \sqrt{x^2 + 4}}{2}\right)$ **38.** $y = \ln\left(\dfrac{x + \sqrt{x^2 - 4}}{2}\right)$

39. Assume that $P_1P = m_2$ and $PP_2 = m_1$, so that $P_1P_2 = m_1 + m_2$. Note that $\triangle P_1 QP$ is similar to $\triangle P_1 A P_2$ and therefore

$$\frac{P_1 Q}{P_1 P} = \frac{P_1 A}{P_1 P_2}.$$

But $P_1 Q = x - x_1$ and $P_1 A = x_2 - x_1$. Thus we obtain

$$\frac{x - x_1}{m_2} = \frac{x_2 - x_1}{m_1 + m_2},$$

or

$$x = x_1 + \frac{m_2}{m_1 + m_2}(x_2 - x_1)$$

$$= \left(1 - \frac{m_1}{m_1 + m_2}\right)x_1 + \frac{m_2}{m_1 + m_2}x_2$$

$$= \frac{m_1}{m_1 + m_2}x_1 + \frac{m_2}{m_1 + m_2}x_2.$$

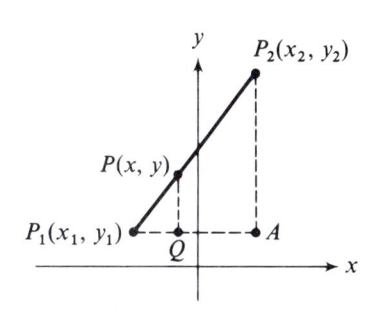

A similar argument gives

$$y = \frac{m_1}{m_1 + m_2}y_1 + \frac{m_2}{m_1 + m_2}y_2.$$

40. $\log_a x = \dfrac{\log_b x}{\log_b a}$

Exercise 5.1 (page 153)   **1.** I   **3.** C   **5.** H   **7.** K   **9.** I   **11.** K   **13.** M   **15.** K
17. Positive   **19.** Positive   **21.** Positive   **23.** Negative   **25.** Positive   **27.** Positive
29. $\dfrac{4}{5}$, I   **31.** $-\dfrac{12}{13}$, IV   **33.** $\dfrac{\sqrt{5}}{3}$, IV   **35.** $-\dfrac{1}{2}$, III   **37.** $\dfrac{4}{5}$, IV

Exercise 5.2 (page 158)   **1.** $\dfrac{1}{2}$ **3.** $\dfrac{\sqrt{2}}{2}$ **5.** 1   **7.** 1   **9.** $\dfrac{2 + \sqrt{2}}{2}$ **11.** $\dfrac{1}{2}$ **13.** 0.9460
15. 0.6816   **17.** 0.1601   **19.** 0.9735   **21.** 0.9940   **23.** 0.9490   **25.** 0.5000   **27.** 0.0045
29. 2.4572   **31.** 1.9015   **33.** 1.3104   **35.** 1.4265   **37.** $K = 25$ **39.** $d = 0$
41. $E = 1.2$ **43.** $E = 0.6$ **45.** $E = 0$
47. 0.9983, 1.0000, 1.0000, 1.0000; $(\sin x)/x$ approaches 1 as x approaches 0.
49. 0.0997, 0.0100, 0.0010, 0.0001; $(\sin^2 x)/x$ approaches 0 as x approaches 0.
51. $-0.0500, -0.0050, -0.0005, -0.0001$; $(\cos x - 1)/x$ approaches 0 as x approaches 0.
53. 0.0899, 0.0099, 0.0010, 0.0001; $(1 - x)\sin x$ approaches 0 as x approaches 0.

Exercise 5.3 (page 164) *Note*: A calculator in conjunction with a reference arc was used to compute the values in Exercises 13–24, 31–36, and 43–48. If you used Table IV, your answers may differ slightly from those given below.

1. $-\dfrac{\sqrt{3}}{2}$ 3. $-\dfrac{1}{2}$ 5. $\dfrac{\sqrt{2}}{2}$ 7. $-\dfrac{\sqrt{3}}{2}$ 9. $-\dfrac{\sqrt{2}}{2}$ 11. $\dfrac{\sqrt{2}}{2}$
13. -0.5148 15. -0.9323 17. 0.9916 19. 0.9999
21. -0.0200 23. -0.9992 25. $\dfrac{1}{2}$ 27. $-\dfrac{\sqrt{2}}{2}$ 29. $-\dfrac{1}{2}$
31. -0.9608 33. 0.4169 35. 0.9208 37. $\dfrac{\sqrt{3}}{2}$ 39. $\dfrac{\sqrt{2}}{2}$
41. $\dfrac{1}{2}$ 43. 0.1197 45. 0.9927 47. -0.5396
49. $\dfrac{1}{2}(\sqrt{3}+\sqrt{2})$ 51. 2.2750 53. 0.0093
55. **a.** -0.6594 **b.** -0.7149 **c.** The answer in (b) is more accurate because the calculator uses a more accurate approximation of π than 3.14.
57. **a.** -0.9287 **b.** -0.8839 **c.** Same as Answer 55c.

Exercise 5.4 (page 174)

1. $A = 2, \; p = 2\pi$

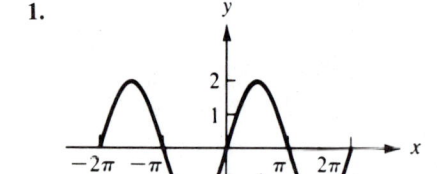

3. $A = \dfrac{1}{2}, \; p = 2\pi$

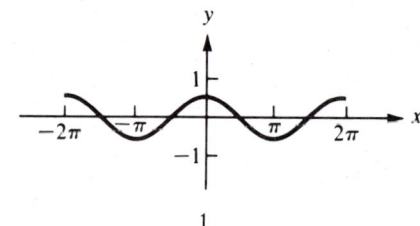

5. $A = 4, \; p = 2\pi$

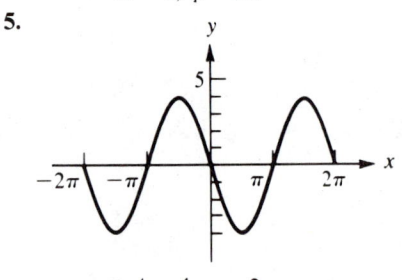

7. $A = 1, \; p = \pi$

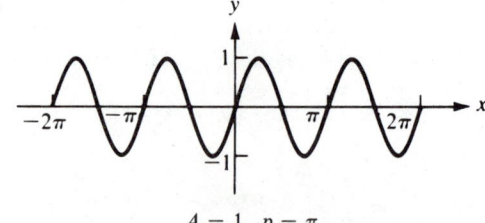

9. $A = 1, \; p = 6\pi$

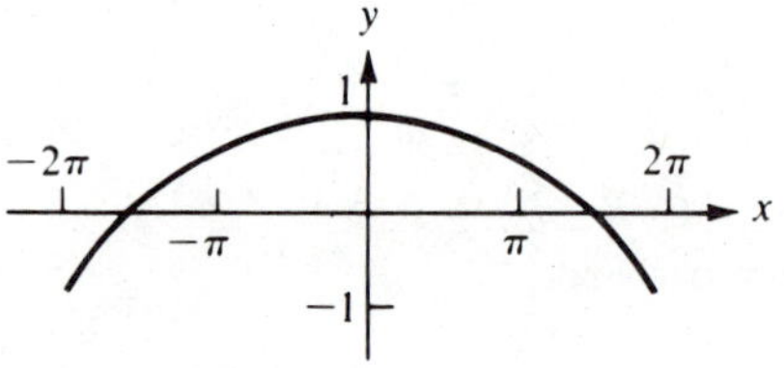

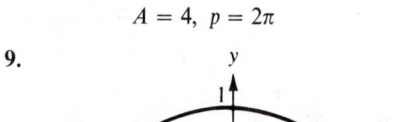

11. $A = 3, \; p = \pi$

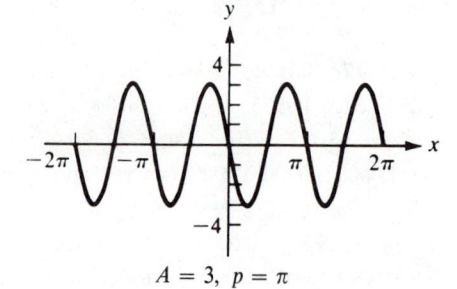

Answers to Odd-Numbered Exercises

13.

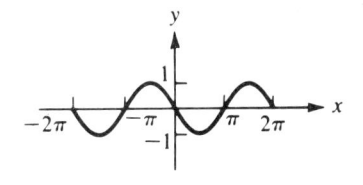

$A = 1$, $p = 2\pi$, leads $y = \sin x$ by π

15.

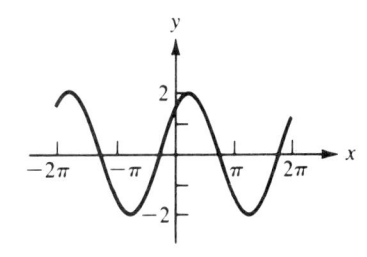

$A = 2$, $p = 2\pi$, lags $y = 2\cos x$ by $\dfrac{\pi}{4}$

17.

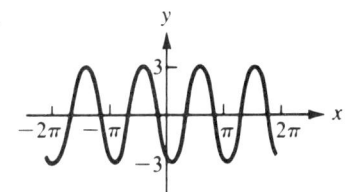

$A = 3$, $p = \pi$, lags $y = 3\sin 2x$ by $\dfrac{2\pi}{3}$

19.

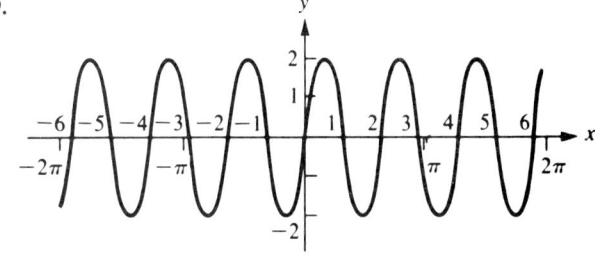

$A = 2$, $p = 2$

21.

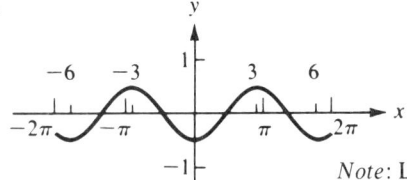

$A = 1/2$, $p = 6$

23. $\left\{-2\pi, -\dfrac{3\pi}{2}, -\pi, -\dfrac{\pi}{2}, 0, \dfrac{\pi}{2}, \pi, \dfrac{3\pi}{2}, 2\pi\right\}$

Note: Long marks, π units; short marks, integers

25. $\left\{-\dfrac{3\pi}{2}, \dfrac{3\pi}{2}\right\}$ **27.** $\{-6, -5, -4, -3, -2, -1, 0, 1, 2, 3, 4, 5, 6\}$

29.

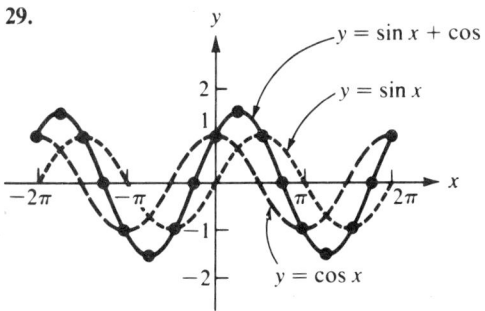

$y = \sin x + \cos x$
$y = \sin x$
$y = \cos x$

31.

$y = \sin 2x + \tfrac{1}{2}\cos x$
$y = \sin 2x$
$y = \tfrac{1}{2}\cos x$

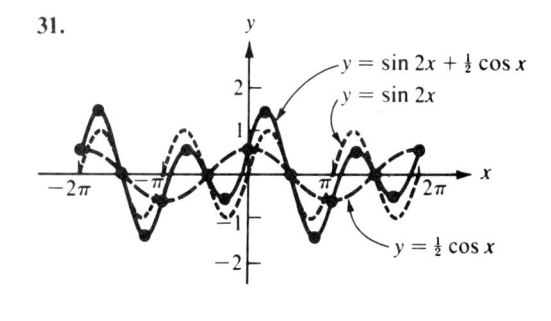

33.

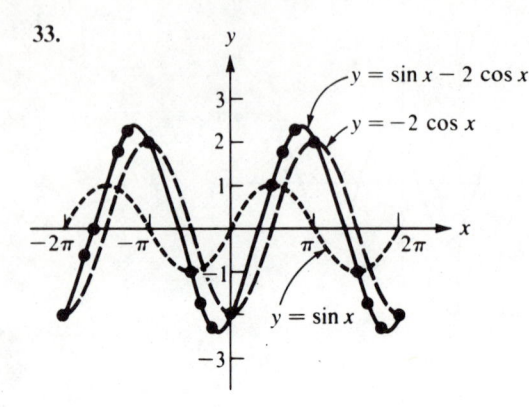

35.

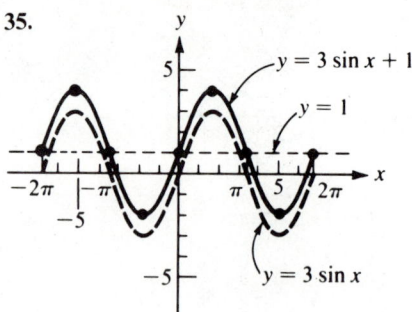

37.

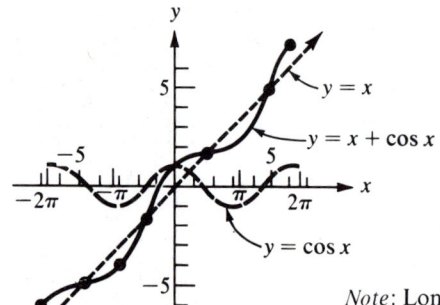

Note: Long marks, π units; short marks, integers

39.

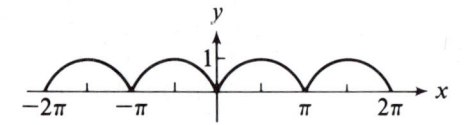

41.

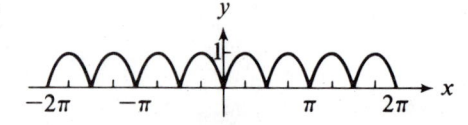

43.

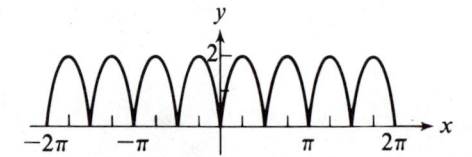

45.

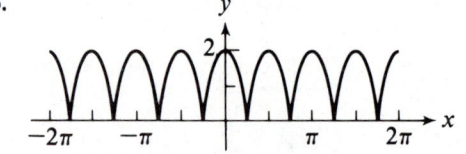

47. Since the circumference of the unit circle is exactly 2π, any arc of length $x + 2\pi$ along the unit circle from the point (1, 0) has the same terminal point as an arc of length x along the unit circle from the point (1, 0). Thus,

$$\sin(x + 2\pi) = \sin x.$$

Exercise 5.5 (page 180) *Note*: A calculator in conjunction with a reference arc was used to compute the values in Exercises 17–24. If you used Table IV, your answers may differ slightly from those given below.

1. 0   **3.** 1   **5.** Undefined   **7.** Undefined   **9.** $-\sqrt{3}$ **11.** $\dfrac{\sqrt{3}}{3}$

13. $-\dfrac{\sqrt{3}}{3}$ **15.** Undefined   **17.** 5.1774   **19.** 2.3600   **21.** -0.0601 **23.** 0.1003

25. $-\dfrac{7}{\sqrt{15}}$ **27.** $-\sqrt{63}$

Answers to Odd-Numbered Exercises

29. Vertical asymptotes:
$x = \pm\pi$
Zeros:
$x = 0, \pm 2\pi$

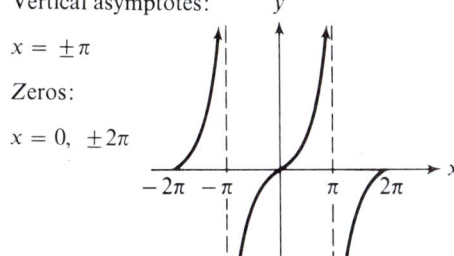

31. Vertical asymptotes:
$x = 0, \pm\pi, \pm 2\pi$
Zeros:
$x = \pm\dfrac{\pi}{2}, \pm\dfrac{3\pi}{2}$

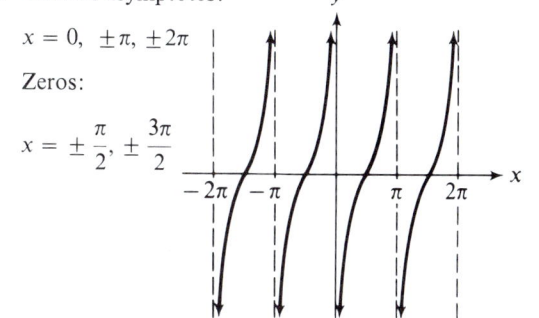

33. 1.0033, 1.0000, 1.0000, 1.0000; (tan x)/x approaches 1 as x approaches 0.

35. 0.0100, 0.0001, 0.0000, 0.0000; x tan x approaches 0 as x approaches 0.

Exercise 5.6 (page 186) *Note*: A calculator in conjunction with a reference arc was used to compute the values in Exercises 21–32. If you used Table IV, your answers may differ slightly from those given below.

1. 0   **3.** $\sqrt{2}$ **5.** $\dfrac{2\sqrt{3}}{3}$ **7.** Undefined   **9.** $\sqrt{3}$ **11.** $\sqrt{2}$ **13.** 2

15. Undefined   **17.** $\dfrac{2\sqrt{3}}{3}$ **19.** -1 **21.** 0.4237   **23.** 2.0098

25. 1.0033   **27.** 0.5870   **29.** -1.1395 **31.** -7.1662

33. Vertical asymptotes:
$x = 0, \dfrac{\pi}{3}, \dfrac{2\pi}{3}, \pi$
Zeros:
$x = \dfrac{\pi}{6}, \dfrac{\pi}{2}, \dfrac{5\pi}{6}$

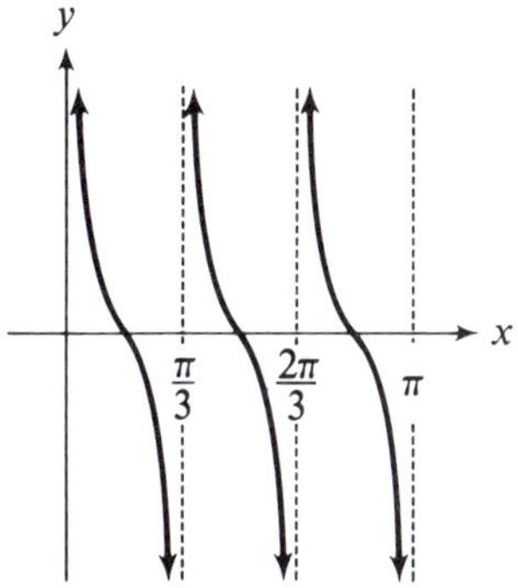

35. Vertical asymptotes:
$x = \dfrac{\pi}{6}, \dfrac{\pi}{2}, \dfrac{5\pi}{6}$
No zeros.

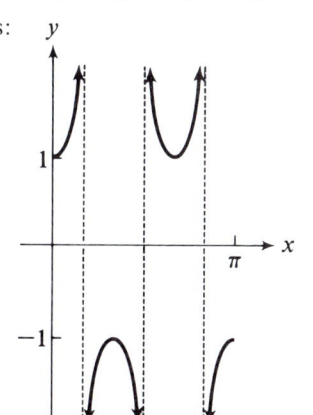

37. Vertical asymptotes:
$x = \dfrac{\pi}{3}, \dfrac{2\pi}{3}, \pi$
No zeros

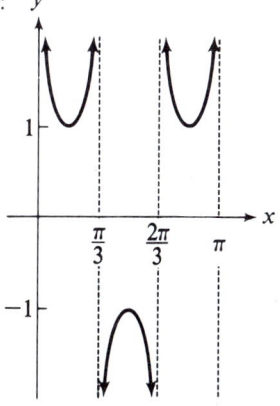

Exercise 5.7 (page 192)

1. $\dfrac{\pi}{6}$ 3. $\dfrac{\pi}{4}$ 5. $\dfrac{\pi}{3}$ 7. $-\dfrac{\pi}{6}$ 9. 0.10 11. 0.39 13. -1.04 15. 0.38 17. 0.32
19. 0.53 21. $\dfrac{\pi}{4}$ 23. $\dfrac{\pi}{3}$ 25. $\dfrac{\sqrt{3}}{2}$ 27. $\dfrac{\sqrt{3}}{2}$ 29. 0.4472 31. $\dfrac{3\pi}{4}$
33. $x = \dfrac{1}{2}\cos\dfrac{y}{3},\ 0 \le y \le 3\pi$ 35. $x = -\pi + \tan 2y,\ -\dfrac{\pi}{4} < y < \dfrac{\pi}{4}$
37. $x = \dfrac{1}{2}(-\pi + \cos 2y),\ 0 \le y \le \dfrac{\pi}{2}$ 39. $x = \dfrac{1}{\pi}\left[-\dfrac{\pi}{2} + \tan(-3y)\right],\ -\dfrac{\pi}{6} < y < \dfrac{\pi}{6}$

Chapter 5 Review (page 195)

Note: A calculator (in conjunction with a reference arc where appropriate) was used to compute the values in Exercises 11–16, 21–28, 37–40, 50–52, and 59–64. If you used Table IV, your answers may differ slightly from those given below.

1. Negative 2. Negative 3. Positive 4. Negative
5. $-\dfrac{\sqrt{5}}{3}$, IV 6. $\dfrac{\sqrt{55}}{8}$, I 7. $\dfrac{\sqrt{3}}{2}$ 8. 0 9. 0 10. $\dfrac{\sqrt{3}}{3}$
11. 0.9581 12. -0.2160 13. 0.0962 14. 3.2842 15. $K = -25$
16. $d = 71.4$ 17. $-\dfrac{\sqrt{2}}{2}$ 18. $\dfrac{\sqrt{3}}{2}$ 19. $-\dfrac{\sqrt{2}}{2}$ 20. $\dfrac{1}{2}$
21. -0.9902 22. 0.9856 23. 0.5891 24. -0.2085 25. 0.0408
26. 0.9611 27. -0.9967 28. 0.8016 29. $A = 2,\ p = \dfrac{2\pi}{3}$; no phase shift
30. $A = \dfrac{3}{2},\ p = 2\pi$; lags $y = \dfrac{3}{2}\cos x$ by $\dfrac{\pi}{4}$. 31. $A = 4,\ p = \pi$; lags $y = 4\sin 2x$ by $\dfrac{\pi}{2}$.

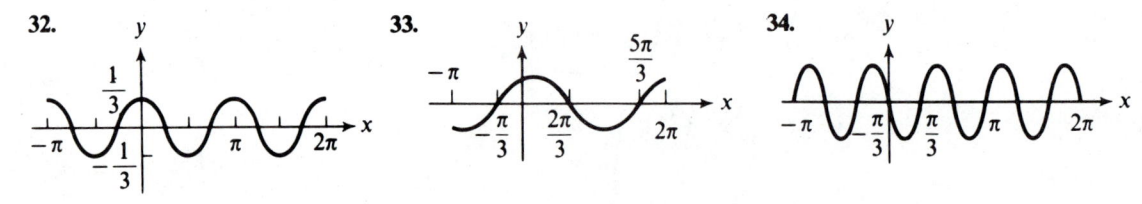

32.

33.

34.

35. 1 36. $\sqrt{3}$ 37. -2.4979 38. 0.5594 39. 1.4284 40. 1.4592

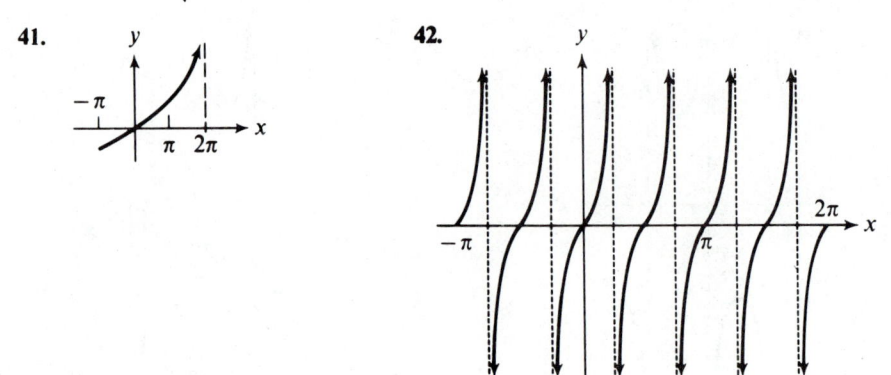

41.

42.

43. $\sqrt{3}$ **44.** $\dfrac{2\sqrt{3}}{3}$ **45.** 1 **46.** Undefined **47.** -1 **48.** $\sqrt{2}$
49. -2 **50.** 1.1634 **51.** -1.0099 **52.** -1.0033 **53.** $\dfrac{\pi}{4}$ **54.** 0
55. $\dfrac{\pi}{6}$ **56.** $\dfrac{\pi}{6}$ **57.** $\dfrac{2\pi}{3}$ **58.** $-\dfrac{\pi}{6}$ **59.** 0.95 **60.** 1.20 **61.** 0.30 **62.** 0.90
63. 0.60 **64.** 0.80 **65.** 0.25 **66.** 0 **67.** 1 **68.** $\dfrac{\sqrt{3}}{2}$

Exercise 6.1 (page 202) **1. a.** $0°$ **b.** $90°$ **c.** $180°$ **d.** $270°$ **e.** $360°$ **3.** $40°$
5. $252°$ **7.** $17.2°$ **9.** $207.5°$ **11.** 0.35^R **13.** 2.27^R **15.** 7.33^R **17.** 13.08^R **19.** $57.32°$
21. $s \approx 2.09$ cm **23.** $s \approx 0.72$ m **25.** $s \approx 4.71$ m **27.** $90°$ **29.** $32.7°$ **31.** $495.3°$
33. $390°, -330°$; $30° + 360°k$, $k \in J$ **35.** $120°, 480°$; $-240° + 360°k$, $k \in J$
37. $60°, -300°$; $420° + 360°k$, $k \in J$ **39.** $30°, 390°$; $-330° + 360°k$, $k \in J$

Exercise 6.2 (page 212) *Note*: A calculator (in conjunction with a reference angle where appropriate) was used to compute the values in Exercises 27–56. If you used Table IV or Table V, your answers may differ slightly from those given below.

1. $\sin \alpha = \dfrac{4}{5}$, $\cos \alpha = \dfrac{3}{5}$, $\tan \alpha = \dfrac{4}{3}$, $\cot \alpha = \dfrac{3}{4}$, $\sec \alpha = \dfrac{5}{3}$, $\csc \alpha = \dfrac{5}{4}$

3. $\sin \alpha = -\dfrac{\sqrt{2}}{2}$, $\cos \alpha = \dfrac{\sqrt{2}}{2}$, $\tan \alpha = -1$, $\cot \alpha = -1$, $\sec \alpha = \sqrt{2}$, $\csc \alpha = -\sqrt{2}$

5. $\sin \alpha = -\dfrac{4}{5}$, $\cos \alpha = -\dfrac{3}{5}$, $\tan \alpha = \dfrac{4}{3}$, $\cot \alpha = \dfrac{3}{4}$, $\sec \alpha = -\dfrac{5}{3}$, $\csc \alpha = -\dfrac{5}{4}$

7. $\sin \alpha = 1$, $\cos \alpha = 0$, $\tan \alpha$ not defined, $\cot \alpha = 0$, $\sec \alpha$ not defined, $\csc \alpha = 1$

9. IV **11.** I **13.** IV **15.** $-\dfrac{1}{\sqrt{3}}$ **17.** $-\dfrac{1}{\sqrt{2}}$ **19.** $-\dfrac{1}{2}$ **21.** $-\dfrac{1}{2}$ **23.** $\dfrac{\sqrt{3}}{2}$
25. $\sqrt{3}$ **27.** 0.5299 **29.** 0.6494 **31.** 1.0457 **33.** 0.8080 **35.** 0.1519 **37.** 1.0145
39. 0.7431 **41.** -0.8391 **43.** -0.6428 **45.** 0.8772 **47.** 1.8270 **49.** -0.0608
51. a. $\dfrac{\pi^R}{2}$ **b.** $90°$ **53. a.** $\dfrac{\pi^R}{4}$ **b.** $45°$ **55. a.** 0.19^R **b.** $11.0°$
57. a. 0.66^R **b.** $37.8°$ **59. a.** 0.66^R **b.** $37.7°$ **61. a.** 0.84^R **b.** $48.4°$
63. If the coordinates are associated with lengths of line segments as indicated in the figure, then by the Pythagorean theorem, $OP_1 = \sqrt{x_1^2 + y_1^2}$ and $OP_2 = \sqrt{x_2^2 + y_2^2}$. Since $\triangle OAP_1$ and $\triangle OBP_2$ are similar, their corresponding sides are proportional and the desired results follow by substituting in the following ratios:

$$\dfrac{OA}{AP_1} = \dfrac{OB}{BP_2}; \quad \dfrac{OA}{OP_1} = \dfrac{OB}{OP_2}; \quad \text{and} \quad \dfrac{AP_1}{OP_1} = \dfrac{BP_2}{OP_2}$$

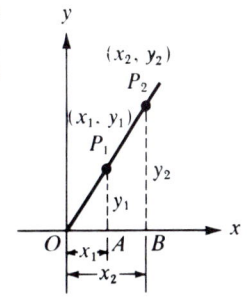

Exercise 6.3 (page 217) **1.** $c = 10$, $A = 36.9°$, $B = 53.1°$
3. $c = 148.3$, $a = 87.2$, $A = 36°$ **5.** $a = 6.2$, $b = 14.8$, $B = 67.3°$ **7.** 16
9. $\cos \theta = \dfrac{-\sqrt{3}}{2}$, $\tan \theta = \dfrac{1}{\sqrt{3}}$, $\cot \theta = \sqrt{3}$, $\csc \theta = -2$, $\sec \theta = \dfrac{-2}{\sqrt{3}}$
11. $\sin \theta = \dfrac{2\sqrt{10}}{7}$, $\tan \theta = -\dfrac{2\sqrt{10}}{3}$, $\cot \theta = -\dfrac{3}{2\sqrt{10}}$, $\sec \theta = -\dfrac{7}{3}$, $\csc \theta = \dfrac{7}{2\sqrt{10}}$
13. $\dfrac{\sqrt{2}}{4}$ **15.** $\dfrac{3\sqrt{13}}{13}$ **17.** $\sqrt{1-x^2}$ **19.** $\dfrac{y}{\sqrt{1-y^2}}$ **21.** $1/x$

23. Let $\alpha = \text{Arcsin}\left(\dfrac{2}{5}\right)$. Then from the illustrated triangle we read $\tan \alpha = \dfrac{2}{\sqrt{21}}$. Thus, $\text{Arctan}\left(\dfrac{2}{\sqrt{21}}\right) = \alpha = \text{Arcsin}\left(\dfrac{2}{5}\right)$.

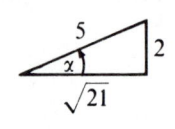

25. 20.5 **27.** 38.7° and 51.3° **29.** 24.0 meters **31.** 927.2 meters
33. 10,585.6 meters

Exercise 6.4 (page 223) **1.** $\gamma = 70°$, $c = 18.8$, $a = 19.7$ **3.** $\beta = 58.7°$, $b = 74.8$, $a = 74.3$
5. $B = 47.9°$, $b = 66.5$, $c = 33.7$ **7.** One triangle possible **9.** One triangle possible
11. Two triangles possible **13.** A right triangle **15.** $\gamma = 48.4°$, $\beta = 18.6°$, $b = 1.7$
17. $\gamma = 28.6°$, $\alpha = 108.7°$, $a = 8.7$
19. $\beta = 139.9°$, $\alpha = 7.6°$, $a = 0.4$ and $\beta = 40.1°$, $\alpha = 107.4°$, $a = 2.7$
21. No solution **23.** $B = 25.7°$, $C = 94.3°$, $c = 15.9$
25. $C = 53.1°$, $A = 93.5°$, $a = 767.8$ and $C = 126.9°$, $A = 19.7°$, $a = 259.3$
27. 197 feet **29.** 2235.5 meters **31.** 6.9

Exercise 6.5 (page 226) **1.** $c = 6.9$, $\alpha = 131.9°$, $\beta = 17.5°$ **3.** $b = 10.3$, $\alpha = 23.8°$, $\gamma = 33.8°$
5. $a = 14.9$, $B = 75°$, $C = 24.3°$ **7.** $C = 60°$, $B = 81.8°$, $A = 38.2°$ **9.** 71.7° **11.** 77.7
13. 279.3 yards **15.** 23.2° west of north or on a heading of 336.8°
17. Set the triangle on a coordinate system as shown in the adjoining figure; then $a^2 = (b - x)^2 + y^2$. But notice $c^2 = x^2 + y^2$ and thus
$$a^2 = (b - x)^2 + c^2 - x^2$$
$$= b^2 - 2bx + x^2 + c^2 - x^2$$
$$= b^2 + c^2 - 2bx.$$
Since $\cos \alpha = x/c$, we have $a^2 = b^2 + c^2 - 2bc \cos \alpha$.

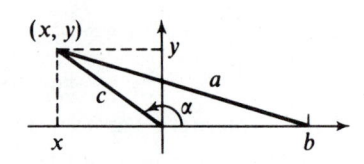

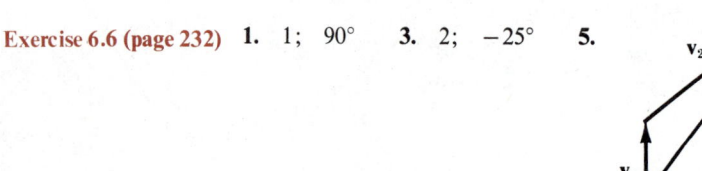

Exercise 6.6 (page 232) **1.** 1; 90° **3.** 2; −25° **5.**

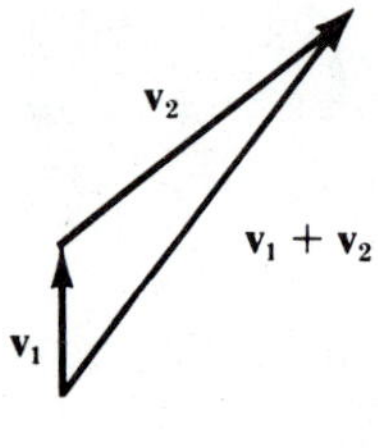

7.

2**v**₃

9.

v₄

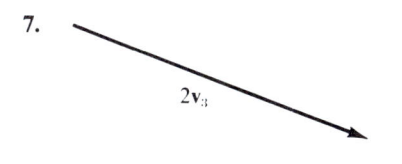

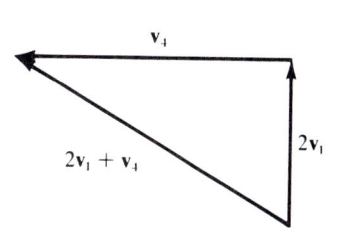

2**v**₁

2**v**₁ + **v**₄

11. $\|\mathbf{v}_x\| = 3$, $\|\mathbf{v}_y\| = 3$ **13.** $\|\mathbf{v}_x\| = \dfrac{7\sqrt{3}}{2}$, $\|\mathbf{v}_y\| = \dfrac{7}{2}$ **15.** $R \approx 8.60$ pounds; $\alpha \approx 54.5°$

17. $R \approx 7.2$ pounds; $\alpha \approx 106.1°$ **19.** 721.1 miles **21.** 80 pounds

Chapter 6 Review (page 233) *Note:* A calculator (in conjunction with a reference angle where appropriate) was used to calculate the values in Exercises 17–24 and 26. If you used Table IV or Table V, your answers may differ slightly from those given below.

1. 67.5° **2.** 499.9° **3.** 0.70^R **4.** 10.64^R **5.** 5.0″ **6.** 4.2′ **7.** 30°, 390°
8. −202°, 158° **9.** III **10.** III **11.** $-\dfrac{\sqrt{3}}{2}$ **12.** $-\dfrac{1}{2}$ **13.** 1
14. $-\dfrac{1}{2}$ **15.** $\dfrac{1}{2}$ **16.** $\dfrac{1}{\sqrt{2}}$ **17.** 0.9205 **18.** 2.9119 **19.** 0.8660 **20.** −0.7071
21. −0.2079 **22.** 1.8807 **23.** −0.8577 **24.** 0.9168 **25. a.** $\dfrac{\pi R}{6}$ **b.** 30°
26. a. 0.09^R **b.** 4.9° **27.** $c = 15$, $A = 36.9°$, $B = 53.1°$ **28.** $a = 7.2$, $b = 8.3$, $\beta = 49°$
29. $c = 24$, $b = 23.5$, $B = 78°$ **30.** $b = 9.8$, $B = 63°$, $A = 27°$
31. $\cos\theta = \dfrac{4}{5}$, $\tan\theta = \dfrac{3}{4}$, $\cot\theta = \dfrac{4}{3}$, $\sec\theta = \dfrac{5}{4}$, $\csc\theta = \dfrac{5}{3}$
32. $\sin\theta = \dfrac{5}{13}$, $\cos\theta = \dfrac{12}{13}$, $\cot\theta = \dfrac{12}{5}$, $\sec\theta = \dfrac{13}{12}$, $\csc\theta = \dfrac{13}{5}$ **33.** 101.45 meters
34. 4.3 meters
35. $a = 45.7$, $c = 46.6$, $\gamma = 80°$ **36.** $b = 22.4$, $\beta = 111.3°$, $\gamma = 38.7°$, $b' = 3.6$, $\beta' = 8.7°$, $\gamma' = 141.3°$
37. $a = 15.2$, $c = 19.8$, $A = 50°$ **38.** $a = 36$, $b = 31.7$, $B = 60°$ **39.** 538.2 feet
40. 4.2 kilometers **41.** $c = 9.4$, $\beta = 7.8°$, $\alpha = 147.2°$
42. $a = 14.1$, $\beta = 63.6°$, $\gamma = 35.7°$ **43.** $C = 58.8°$, $B = 71.8°$, $A = 49.5°$
44. $C = 36.9°$, $B = 53.1°$, $A = 90°$ **45.** $A = 40.2°$, $B = 19.6°$, $C = 120.3°$
46. 329.3 meters **47.** $\|\mathbf{v}_x\| = 6\sqrt{3}$, $\|\mathbf{v}_y\| = 6$ **48.** $\|\mathbf{v}_x\| = 11.88$, $\|\mathbf{v}_y\| = 11.88$
49. $\|\mathbf{v}_3\| \approx 13.16$; $\alpha \approx 46.3°$ **50.** 2.56 mph

Exercise 7.1 (page 241) **1.** Recall $\tan x = \dfrac{\sin x}{\cos x}$; hence $\cos x \tan x = \cos x \left(\dfrac{\sin x}{\cos x}\right) = \sin x$.

3. Recall $\cot x = \dfrac{\cos x}{\sin x}$; then $\cot^2 x = \dfrac{\cos^2 x}{\sin^2 x}$; hence $\sin^2 x \cot^2 x = \sin^2 x \left(\dfrac{\cos^2 x}{\sin^2 x}\right) = \cos^2 x$.

5. By Equation (2), $1 + \tan^2 \alpha = \sec^2 \alpha$; also $\sec \alpha = \dfrac{1}{\cos \alpha}$. Then $\sec^2 \alpha = \dfrac{1}{\cos^2 \alpha}$;

hence $(\cos^2 \alpha)(1 + \tan^2 \alpha) = (\cos^2 \alpha)(\sec^2 \alpha) = (\cos^2 \alpha)\left(\dfrac{1}{\cos^2 \alpha}\right) = 1$.

7. By Equation (1), $\sin^2 \alpha + \cos^2 \alpha = 1$; hence $1 - \cos^2 \alpha = \sin^2 \alpha$. Using the definitions of secant, cosecant, and tangent yields $\sec \alpha \csc \alpha - \cot \alpha = \left(\dfrac{1}{\cos \alpha}\right)\left(\dfrac{1}{\sin \alpha}\right) - \dfrac{\cos \alpha}{\sin \alpha} = \dfrac{1 - \cos^2 \alpha}{\cos \alpha \sin \alpha} = \dfrac{\sin^2 \alpha}{\cos \alpha \sin \alpha} = \dfrac{\sin \alpha}{\cos \alpha} = \tan \alpha$.

9. By the definitions of secant and tangent, $\dfrac{\sin \theta \sec \theta}{\tan \theta} = \dfrac{(\sin \theta)(1/\cos \theta)}{(\sin \theta/\cos \theta)} = \dfrac{\sin \theta}{\sin \theta} = 1$.

11. By Equation (2), $\tan^2 \theta + 1 = \sec^2 \theta$; hence $\sec^2 \theta - 1 = \tan^2 \theta$; also, by Equation (3), $1 + \cot^2 \theta = \csc^2 \theta$; hence $\csc^2 \theta - 1 = \cot^2 \theta$. Using these results with the fact that $\cot x = \dfrac{1}{\tan x}$ yields $(\sec^2 \theta - 1)(\csc^2 \theta - 1) = (\tan^2 \theta)(\cot^2 \theta) = (\tan^2 \theta)\left(\dfrac{1}{\tan^2 \theta}\right) = 1$.

13. From Equation (1), since $\sin^2 \theta + \cos^2 \theta = 1$, $1 - \sin^2 \theta = \cos^2 \theta$. From this result and the definition of secant, $\dfrac{1}{1 + \sin \theta} + \dfrac{1}{1 - \sin \theta} = \dfrac{1 - \sin \theta + 1 + \sin \theta}{(1 + \sin \theta)(1 - \sin \theta)} = \dfrac{2}{1 - \sin^2 \theta} = \dfrac{2}{\cos^2 \theta} = 2\left(\dfrac{1}{\cos \theta}\right)^2 = 2 \sec^2 \theta$.

15. From the definition of cotangent, $\sin x \cot x = \sin x \dfrac{\cos x}{\sin x} = \cos x$.

17. Recall that $\tan \theta = \dfrac{\sin \theta}{\cos \theta}$; hence $\dfrac{\sin \theta}{\tan \theta} = \dfrac{\sin \theta}{\sin \theta/\cos \theta} = \cos \theta$. By Equation (5), $\cos(-\theta) = \cos \theta$; hence $\dfrac{\sin \theta}{\tan \theta} = \cos(-\theta)$.

19. Recall that $\tan x = \dfrac{\sin x}{\cos x}$; hence, $\tan^2 x \cos^2 x = \dfrac{\sin^2 x}{\cos^2 x} \cos^2 x = \sin^2 x$.

21. By Equation (1), $1 - \sin^2 \theta = \cos^2 \theta$; hence
$$\dfrac{(1 + \sin \theta)(1 - \sin \theta)}{\cos \theta} = \dfrac{1 - \sin^2 \theta}{\cos \theta} = \dfrac{\cos^2 \theta}{\cos \theta} = \cos \theta.$$

23. From the definition of secant and tangent,
$$\sec x - \cos x = \dfrac{1}{\cos x} - \cos x = \dfrac{1 - \cos^2 x}{\cos x} = \dfrac{\sin^2 x}{\cos x} = \sin x \left(\dfrac{\sin x}{\cos x}\right) = \sin x \tan x.$$

25. By Equation (3), $1 + \cot^2 x = \csc^2 x$; also, $\cot x = \dfrac{1}{\tan x}$; hence $\tan x \cot x = 1$ and $\tan x = \dfrac{1}{\cot x}$, $\cot x \neq 0$. Then
$$\dfrac{1 + \tan^2 x}{\tan^2 x} = \dfrac{1}{\tan^2 x} + 1 = \cot^2 x + 1 = \csc^2 x.$$

27. From the definition of tangent and Equation (1),
$$\tan^2 \alpha - \sin^2 \alpha = \dfrac{\sin^2 \alpha}{\cos^2 \alpha} - \sin^2 \alpha = \dfrac{\sin^2 \alpha - \sin^2 \alpha \cos^2 \alpha}{\cos^2 \alpha} = \dfrac{\sin^2 \alpha(1 - \cos^2 \alpha)}{\cos^2 \alpha}$$
$$= \left(\dfrac{\sin \alpha}{\cos \alpha}\right)^2 (1 - \cos^2 \alpha) = \sin^2 \alpha \tan^2 \alpha.$$

29. From Equation (2),
$$\dfrac{1}{\sec \alpha - \tan \alpha} \cdot \dfrac{\sec \alpha + \tan \alpha}{\sec \alpha + \tan \alpha} = \dfrac{\sec \alpha + \tan \alpha}{\sec^2 \alpha - \tan^2 \alpha} = \dfrac{\sec \alpha + \tan \alpha}{(\tan^2 \alpha + 1) - \tan^2 \alpha} = \sec \alpha + \tan \alpha.$$

31. Recall $\cot x = \dfrac{\cos x}{\sin x}$; hence $\dfrac{\cos x + \sin x}{\sin x} = \dfrac{\cos x}{\sin x} + \dfrac{\sin x}{\sin x} = \cot x + 1$.

33. Adding the fractions in the left-hand member, we have $\dfrac{1}{1 + \sin x} + \dfrac{1}{1 - \sin x} = \dfrac{2}{1 - \sin^2 x}$. By the definition of secant and Equation (1), $\dfrac{1}{1 + \sin x} + \dfrac{1}{1 - \sin x} = \dfrac{2}{1 - \sin^2 x} = \dfrac{2}{\cos^2 x} = 2\sec^2 x$.

35. Recall $\sec \theta = \dfrac{1}{\cos \theta}$; hence $\dfrac{1 + \sec \theta}{\sec \theta} = \dfrac{1 + (1/\cos \theta)}{1/\cos \theta} = \dfrac{(\cos \theta + 1)/\cos \theta}{1/\cos \theta} = \cos \theta + 1$.

Now, multiplying by $\dfrac{1 - \cos \theta}{1 - \cos \theta}$, we have $\dfrac{1 + \sec \theta}{\sec \theta} = \dfrac{(1 + \cos \theta)(1 - \cos \theta)}{1 - \cos \theta} = \dfrac{1 - \cos^2 \theta}{1 - \cos \theta}$.

By Equation (1), $1 - \cos^2 \theta = \sin^2 \theta$; hence, $\dfrac{1 + \sec \theta}{\sec \theta} = \dfrac{\sin^2 \theta}{1 - \cos \theta}$.

37. Multiplying the left-hand member by $\dfrac{1 - \sin \alpha}{1 - \sin \alpha}$, we have

$$\dfrac{1 + \sin \alpha}{\cos \alpha} = \dfrac{(1 + \sin \alpha)(1 - \sin \alpha)}{\cos \alpha (1 - \sin \alpha)} = \dfrac{1 - \sin^2 \alpha}{\cos \alpha (1 - \sin \alpha)}.$$

By Equation (1), $1 - \sin^2 \alpha = \cos^2 \alpha$; hence $\dfrac{1 + \sin \alpha}{\cos \alpha} = \dfrac{\cos^2 \alpha}{\cos \alpha (1 - \sin \alpha)} = \dfrac{\cos \alpha}{1 - \sin \alpha}$.

39. Multiplying the left-hand member by $\dfrac{\sec \alpha}{\sec \alpha}$, we have

$$\dfrac{1 - \cos \alpha}{1 + \cos \alpha} = \dfrac{(1 - \cos \alpha)\sec \alpha}{(1 + \cos \alpha)\sec \alpha} = \dfrac{\sec \alpha - 1}{\sec \alpha + 1}.$$

41. By Equation (2), $\tan^2 \theta = \sec^2 \theta - 1$;
hence $\tan^4 \theta + \tan^2 \theta = (\sec^2 \theta - 1)^2 + (\sec^2 \theta - 1) = \sec^4 \theta - 2\sec^2 \theta + 1 + \sec^2 \theta - 1 = \sec^4 \theta - \sec^2 \theta$.

43. $\tan^2 45° + \sec^2 45° = 1 + 2 = 3 \neq 1$ **45.** $\sin 45° + \tan 45° = 1 + \dfrac{\sqrt{2}}{2} \neq \dfrac{\sqrt{2}}{2} = \cos 45°$.

47. Expanding the terms of the left-hand member, we have $(a \sin \alpha + b \cos \alpha)^2 + (a \cos \alpha - b \sin \alpha)^2$
$= a^2 \sin^2 \alpha + 2ab \sin \alpha \cos \alpha + b^2 \cos^2 \alpha + a^2 \cos^2 \alpha - 2ab \sin \alpha \cos \alpha + b^2 \sin^2 \alpha$
$= a^2 (\sin^2 \alpha + \cos^2 \alpha) + b^2 (\sin^2 \alpha + \cos^2 \alpha)$. By Equation (1), $\sin^2 \alpha + \cos^2 \alpha = 1$;
hence $(a \sin \alpha + b \cos \alpha)^2 + (a \cos \alpha - b \sin \alpha)^2 = a^2 + b^2$.

49. Convert the terms in the left-hand member into terms involving only sines and cosines and simplify to get

$\left(\dfrac{\sec \theta - 1}{\sec \theta + 1}\right)\sin^2 \theta + \sin^2 \theta - 2 = \left(\dfrac{1/\cos \theta - 1}{1/\cos \theta + 1}\right)\sin^2 \theta + \sin^2 \theta - 2 = \left(\dfrac{1 - \cos \theta}{1 + \cos \theta}\right)\sin^2 \theta + \sin^2 \theta - 2$.

By Equation (1), $\sin^2 \theta = 1 - \cos^2 \theta$; hence $\left(\dfrac{\sec \theta - 1}{\sec \theta + 1}\right)\sin^2 \theta + \sin^2 \theta - 2 =$

$\dfrac{(1 - \cos \theta)(1 - \cos^2 \theta)}{1 + \cos \theta} + (1 - \cos^2 \theta) - 2 = (1 - \cos \theta)(1 - \cos \theta) + 1 - \cos^2 \theta - 2 = -2 \cos \theta$.

Exercise 7.2 (page 245) 1. $\dfrac{-\sqrt{2}(1+\sqrt{3})}{4}$ 3. $\dfrac{\sqrt{2}(1-\sqrt{3})}{4}$ 5. $\dfrac{\sqrt{2}(\sqrt{3}-1)}{4}$

7. a. $\dfrac{-7}{25}$ b. $\dfrac{\sqrt{5}}{5}$ 9. $\cos(\pi+x) = \cos\pi\cos x - \sin\pi\sin x = -\cos x$

11. $\cos\left(\dfrac{\pi}{2}+x\right) = \cos\dfrac{\pi}{2}\cos x - \sin\dfrac{\pi}{2}\sin x = -\sin x$

13. $\cos\left(\dfrac{3\pi}{2}+x\right) = \cos\dfrac{3\pi}{2}\cos x - \sin\dfrac{3\pi}{2}\sin x = \sin x$

15. By Equation (4), $\cos 2x = 1 - 2\sin^2 x$; hence $\dfrac{1-\cos 2x}{2} = \dfrac{1-(1-2\sin^2 x)}{2} = \sin^2 x.$

17. Write the terms of the right-hand member in terms of sine and cosine to obtain

$$\dfrac{1-\tan^2\theta}{1+\tan^2\theta} = \dfrac{1-(\sin^2\theta/\cos^2\theta)}{1+(\sin^2\theta/\cos^2\theta)} = \dfrac{[(\cos^2\theta-\sin^2\theta)/\cos^2\theta]}{[(\cos^2\theta+\sin^2\theta)/\cos^2\theta]} = \dfrac{\cos^2\theta-\sin^2\theta}{\cos^2\theta+\sin^2\theta}.$$

By Equation (1) of Section 7.1 and Equation (3), we have

$$\dfrac{1-\tan^2\theta}{1+\tan^2\theta} = \cos 2\theta.$$

19. $\cos 3x = \cos(x+2x) = \cos x\cos 2x - \sin x\sin 2x$
$= \cos x(2\cos^2 x - 1) - \sqrt{1-\cos^2 x}\sqrt{1-\cos^2 2x}$
$= \cos x(2\cos^2 x - 1) - \sqrt{1-\cos^2 x}\sqrt{1-(2\cos^2 x - 1)^2}$
$= \cos x(2\cos^2 x - 1) - \sqrt{1-\cos^2 x}\sqrt{1-4\cos^4 x + 4\cos^2 x - 1}$
$= \cos x(2\cos^2 x - 1) - \sqrt{1-\cos^2 x}\sqrt{4\cos^2 x(1-\cos^2 x)}$
$= \cos x(2\cos^2 x - 1) - 2\cos x(1-\cos^2 x)$
$= 4\cos^3 x - 3\cos x$

Note that we use the positive square roots for $\sin x$ and $\sin 2x$ because x was assumed to be in the interval $[0, \pi/2]$.

21. $\dfrac{\cos(x+h)-\cos x}{h} = \dfrac{\cos x\cos h - \sin x\sin h - \cos x}{h} = \cos x\left(\dfrac{\cos h - 1}{h}\right) - \dfrac{\sin h}{h}\sin x$

Exercise 7.3 (page 250) 1. $\dfrac{\sqrt{2}(\sqrt{3}-1)}{4}$ 3. $\dfrac{-\sqrt{2}(1+\sqrt{3})}{4}$

5. $\dfrac{\sqrt{2}(-\sqrt{3}+1)}{4}$ 7. a. $\dfrac{24}{25}$ b. $\dfrac{-2\sqrt{5}}{5}$

9. $\sin(\pi+x) = \sin\pi\cos x + \sin x\cos\pi = -\sin x$

11. $\sin\left(\dfrac{\pi}{2}+x\right) = \sin\dfrac{\pi}{2}\cos x + \sin x\cos\dfrac{\pi}{2} = \cos x$

13. $\sin\left(\dfrac{3\pi}{2}+x\right) = \sin\dfrac{3\pi}{2}\cos x + \sin x\cos\dfrac{3\pi}{2} = -\cos x$

15. By Equation (5), $\sin 2\alpha = (2\sin\alpha\cos\alpha)\left(\dfrac{\cos\alpha}{\cos\alpha}\right) = 2\tan\alpha\cos^2\alpha = \dfrac{2\tan\alpha}{\sec^2\alpha} = \dfrac{2\tan\alpha}{1+\tan^2\alpha}.$

17. By Equation (5) of Section 7.2 and Equation (5),

$$\frac{1+\cos 2x}{\sin 2x} = \frac{1+2\cos^2 x - 1}{2\sin x \cos x} = \frac{\cos^2 x}{\sin x \cos x} = \frac{\cos x}{\sin x} = \cot x.$$

19. $\sin 3x = \sin(2x + x) = \sin 2x \cos x + \sin x \cos 2x$
$= 2 \sin x \cos^2 x + \sin x (1 - 2\sin^2 x) = 2 \sin x - 2 \sin^3 x + \sin x - 2 \sin^3 x = 3 \sin x - 4 \sin^3 x$

21. $\dfrac{\sin(x+h) - \sin x}{h} = \dfrac{\sin x \cos h + \cos x \sin h - \sin x}{h} = \sin x \left(\dfrac{\cos h - 1}{h}\right) + \cos x \dfrac{\sin h}{h}$

Exercise 7.4 (page 255) **1.** $\dfrac{\sqrt{3}+1}{1-\sqrt{3}}$ **3.** $\sqrt{2}-1$ **5.** $\sqrt{2}+1$ **7. a.** $\dfrac{24}{7}$ **b.** $\dfrac{1}{3}$

9. $\tan(\pi - x) = \dfrac{\tan \pi - \tan x}{1 + \tan \pi \tan x} = -\tan x$ **11.** $\tan(2\pi - x) = \dfrac{\tan 2\pi - \tan x}{1 + \tan 2\pi \tan x} = -\tan x$

13. By Identity 19, $\cot \theta = \dfrac{1}{\tan \theta}$ and $\cot 2\theta = \dfrac{1}{\tan 2\theta}$, by Identity 22, $\tan 2\theta = \dfrac{2 \tan \theta}{1 - \tan^2 \theta}$;

hence $\cot \theta - \cot 2\theta = \dfrac{1}{\tan \theta} - \dfrac{1 - \tan^2 \theta}{2 \tan \theta} = \dfrac{2 - 1 + \tan^2 \theta}{2 \tan \theta} = \dfrac{1 + \tan^2 \theta}{2 \tan \theta}$. By Identity 14,

$1 + \tan^2 \theta = \sec^2 \theta$; hence $\cot \theta - \cot 2\theta = \dfrac{\sec^2 \theta}{2 \tan \theta}$.

15. $\tan 3x = \tan(2x + x) = \dfrac{\tan 2x + \tan x}{1 - \tan 2x \tan x}$

$$= \dfrac{\dfrac{2 \tan x}{1 - \tan^2 x} + \tan x}{1 - \dfrac{2 \tan x}{1 - \tan^2 x} \cdot \tan x} = \dfrac{\dfrac{2 \tan x + \tan x - \tan^3 x}{1 - \tan^2 x}}{\dfrac{1 - \tan^2 x - 2 \tan^2 x}{1 - \tan^2 x}} = \dfrac{3 \tan x - \tan^3 x}{1 - 3 \tan^2 x}$$

17. $2 \tan 4x = 2 \tan 2(2x) = 2 \dfrac{2 \tan 2x}{1 - \tan^2 2x} = \dfrac{4 \tan 2x}{1 - \tan^2 2x}$

19. By Exercise 19 of Sections 7.2 and 7.3, $\cos 3x = 4 \cos^3 x - 3 \cos x$ and

$\sin 3x = 3 \sin x - 4 \sin^3 x$; hence $\tan \dfrac{3x}{2} = \dfrac{1 - \cos 3x}{\sin 3x} = \dfrac{1 - 4 \cos^3 x + 3 \cos x}{3 \sin x - 4 \sin^3 x}$.

21. $\tan\left(\dfrac{\pi}{2} - x\right) = \dfrac{\sin\left(\dfrac{\pi}{2} - x\right)}{\cos\left(\dfrac{\pi}{2} - x\right)} = \dfrac{\cos x}{\sin x} = \cot x$ **23.** $\tan\left(\dfrac{3\pi}{2} - x\right) = \dfrac{\sin\left(\dfrac{3\pi}{2} - x\right)}{\cos\left(\dfrac{3\pi}{2} - x\right)} = \dfrac{-\cos x}{-\sin x} = \cot x$

Exercise 7.5 (page 260)

1. a. $\{(60 + 360k)°\} \cup \{(300 + 360k)°\}, k \in J$ **b.** $\left\{\left(\dfrac{\pi}{3} + k \cdot 2\pi\right)^R\right\} \cup \left\{\left(\dfrac{5\pi}{3} + k \cdot 2\pi\right)^R\right\}, k \in J$

3. a. $\{(60 + 180k)°\}, k \in J$ **b.** $\left\{\left(\dfrac{\pi}{3} + k\pi\right)^R\right\}, k \in J$

5. a. $\{(45 + 360k)°\} \cup \{(315 + 360k)°\}, k \in J$ **b.** $\left\{\left(\dfrac{\pi}{4} + k \cdot 2\pi\right)^R\right\} \cup \left\{\left(\dfrac{7\pi}{4} + k \cdot 2\pi\right)^R\right\}, k \in J$

7. a. $\{14.5° + 360°k\} \cup \{163.5° + 360°k\},\ k \in J$ **b.** $\{(0.25 + k \cdot 2\pi)^R\} \cup \{(2.89 + k \cdot 2\pi)^R\},\ k \in J$
9. a. $\{71.6° + 180°k\},\ k \in J$ **b.** $\{(1.25 + k\pi)^R\},\ k \in J$
11. a. $\{66.4° + 360°k\} \cup \{293.6° + 360°k\},\ k \in J$ **b.** $\{(1.16 + k \cdot 2\pi)^R\} \cup \{(5.12 + k \cdot 2\pi)^R\},\ k \in J$
13. $\left\{x \mid x = \dfrac{\pi}{3} + k\pi\right\},\ k \in J$ **15.** $\left\{x \mid x = \dfrac{\pi}{4} + \dfrac{k\pi}{2}\right\},\ k \in J$ **17.** $\left\{x \mid x = \dfrac{\pi}{2} + k\pi\right\},\ k \in J$
19. $\{30°, 45°, 135°, 150°, 225°, 315°\}$ **21.** $\{0°, 120°, 180°, 240°\}$ **23.** $\left\{\dfrac{\pi^R}{4}, \dfrac{5\pi^R}{4}\right\}$
25. $\{0^R\}$ **27.** $\{1.11^R, 1.77^R, 4.25^R, 4.91^R\}$ **29.** $\{0.67^R, 2.47^R\}$
31. $\{1.9, 0, -1.9\}$

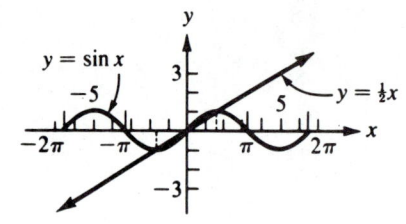

Note: Long marks, π units; short marks, integers

Exercise 7.6 (page 263) **1.** $\{22.5°, 157.5°, 202.5°, 337.5°\}$ **3.** $\{60°, 300°\}$
5. $\{0°, 60°, 120°, 180°, 240°, 300°\}$ **7.** $\{45°, 225°\}$ **9.** $\{67.5°, 157.5°, 247.5°, 337.5°\}$
11. $\{30°, 90°, 150°, 210°, 270°, 330°\}$ **13.** $\left\{\dfrac{\pi}{2} + k\pi, \dfrac{\pi}{6} + 2k\pi, \dfrac{5\pi}{6} + 2k\pi\right\},\ k \in J$
15. $\left\{\dfrac{\pi}{2} + k\pi, \dfrac{\pi}{3} + 2k\pi, \dfrac{5\pi}{3} + 2k\pi\right\},\ k \in J$ **17.** $\left\{k\pi, \dfrac{\pi}{2} + k\pi\right\},\ k \in J$ **19.** $\left\{\dfrac{k\pi}{2}\right\},\ k \in J$

Chapter 7 Review (page 264) **1.** $\tan x = \dfrac{\sin x}{\cos x} = \sin x \sec x$

2. $\cos x - \sin x = (\cos x - \sin x)\dfrac{\cos x}{\cos x} = \left(\dfrac{\cos x}{\cos x} - \dfrac{\sin x}{\cos x}\right)\cos x = (1 - \tan x)\cos x$

3. $\dfrac{1 - \tan^2 \alpha}{\tan \alpha} = \dfrac{1}{\tan \alpha} - \dfrac{\tan^2 \alpha}{\tan \alpha} = \cot \alpha - \tan \alpha$

4. $\dfrac{\cos^2 \alpha}{1 - \sin \alpha} = \dfrac{\cos \alpha \cdot \dfrac{1}{\sec \alpha}}{1 - \sin \alpha} = \dfrac{\cos \alpha}{(1 - \sin \alpha)(\sec \alpha)} = \dfrac{\cos \alpha}{\sec \alpha - \dfrac{\sin \alpha}{\cos \alpha}} = \dfrac{\cos \alpha}{\sec \alpha - \tan \alpha}$

5. $\dfrac{1}{\cot^2 \theta} - \dfrac{1}{\csc^2 \theta} = \dfrac{\csc^2 \theta - \cot^2 \theta}{\csc^2 \theta \cot^2 \theta} = \dfrac{1}{\csc^2 \theta \cot^2 \theta}$

6. $\dfrac{2}{1 - \sin \theta} + \dfrac{2}{1 + \sin \theta} = \dfrac{2(1 + \sin \theta) + 2(1 - \sin \theta)}{1 - \sin^2 \theta} = \dfrac{4}{\cos^2 \theta}$

7. $\dfrac{1 + \sin x}{\cos x} = \dfrac{1}{\cos x} + \dfrac{\sin x}{\cos x} = \sec x + \tan x = \dfrac{\sec^2 x - \tan^2 x}{\sec x - \tan x} = \dfrac{1}{\sec x - \tan x}$

8. $\dfrac{1}{\cos^2 x} + \dfrac{1}{\sin^2 x} = \dfrac{\sin^2 x + \cos^2 x}{\sin^2 x \cos^2 x} = \dfrac{1}{\sin^2 x \cos^2 x}$

Answers to Odd-Numbered Exercises

9. $\dfrac{-2\sqrt{3}-\sqrt{5}}{6}$ 10. $\dfrac{-2-\sqrt{15}}{6}$ 11. $\dfrac{-2\sqrt{3}+\sqrt{5}}{6}$ 12. $\dfrac{-2+\sqrt{15}}{6}$ 13. $-\dfrac{1}{9}$

14. $\dfrac{\sqrt{6}}{6}$ 15. $\dfrac{-\sqrt{5}+2\sqrt{3}}{6}$ 16. $\dfrac{-\sqrt{15}+2}{6}$ 17. $\dfrac{-\sqrt{5}-2\sqrt{3}}{6}$ 18. $\dfrac{-\sqrt{15}-2}{6}$

19. $\dfrac{-4\sqrt{5}}{9}$ 20. $\sqrt{\dfrac{3+\sqrt{5}}{6}}$ 21. $\dfrac{-6\sqrt{5}+5\sqrt{3}}{15+2\sqrt{15}}$ 22. $\dfrac{5\sqrt{3}-2\sqrt{5}}{5+2\sqrt{15}}$ 23. $\dfrac{5\sqrt{3}+6\sqrt{5}}{15-2\sqrt{15}}$

24. $\dfrac{5\sqrt{3}+2\sqrt{5}}{5-2\sqrt{15}}$ 25. $-4\sqrt{5}$ 26. $\dfrac{3+\sqrt{5}}{2}$ 27. $\{0°, 180°, 360°\}$ 28. $\{30°, 90°, 150°, 270°\}$

29. $\left\{\dfrac{\pi^R}{2}\right\}$ 30. $\left\{\dfrac{\pi^R}{6}+k\pi^R\right\},\ k\in J$ 31. $\{20°, 40°\}$ 32. $\{90°\}$ 33. $\left\{\left(\dfrac{\pi}{8}+\dfrac{k}{2}\pi\right)\right\},\ k\in J$

34. $\left\{\left(\dfrac{\pi}{6}+\dfrac{k}{2}\pi\right)\right\}\cup\left\{\left(\dfrac{\pi}{3}+\dfrac{k}{2}\pi\right)\right\},\ k\in J$

Exercise 8.1 (page 273) 1. $1-i$ 3. $9-i$ 5. $4-3i$ 7. 6 9. $7+i$ 11. $6+2i$
13. $-1+i$ 15. $6+7i$ 17. $2i$ 19. 5 21. $-2i$ 23. $3i$ 25. $-2+6i$
27. $\dfrac{1}{5}-\dfrac{2}{5}i$ 29. $\dfrac{1}{2}+\dfrac{1}{2}i$ 31. $-\dfrac{1}{4}i$ 33. $\dfrac{1}{2}+\dfrac{1}{2}i$ 35. $\dfrac{2}{5}-\dfrac{9}{5}i$ 37. $\dfrac{1}{10}+\dfrac{3}{10}i$
39. $\dfrac{3}{2}+\dfrac{3}{2}i$ 41. $-\dfrac{1}{2}-i$ 43. $-1-i$ 45. $4+(3-\sqrt{2})i$ 47. $(3+3\sqrt{2})+(9-\sqrt{2})i$
49. $\dfrac{3-3\sqrt{2}}{11}+\dfrac{9+\sqrt{2}}{11}i$ 51. $(4+2\sqrt{3})i$ 53. 1 55. $-i$ 57. 1 59. $-i$
61. $-2+2i$ 63. $-1+i$ 65. $-\dfrac{4}{25}-\dfrac{3}{25}i$ 67. $-1+i, -1-i$
69. $-\dfrac{1}{4}+i\dfrac{\sqrt{7}}{4}, -\dfrac{1}{4}-i\dfrac{\sqrt{7}}{4}$ 71. $1+\sqrt{3}i, 1-\sqrt{3}i$ 73. $-i$ 75. $2i, -4i$

77. Let $z_1 = a_1 + b_1 i$ and $z_2 = a_2 + b_2 i$. Then
$$\overline{z_1 \pm z_2} = \overline{(a_1 \pm a_2)+(b_1 \pm b_2)i}$$
$$= (a_1 \pm a_2) - (b_1 \pm b_2)i$$
$$= (a_1 - b_1 i) \pm (a_2 - b_2 i)$$
$$= \bar{z}_1 \pm \bar{z}_2.$$

79. Repeated applications of Exercise 78 with $z = z_1 = z_2$ yields this result.
81. Let $z = a + bi$, then $\bar{z} = a - bi$ and
$$z + \bar{z} = (a+bi)+(a-bi)$$
$$= 2a + 0i$$
$$= 2a.$$

Exercise 8.2 (page 278) 1. $-2i$ 3. $-i$ 5. $i, \dfrac{1}{2}$ 7. $i, -2i$ 9. $3i, \dfrac{-3+\sqrt{29}}{2}, \dfrac{-3-\sqrt{29}}{2}$
11. $-2i$ (multiplicity 2) and 2 13. $P(x) = (2x-3)(x-2+2i)(x-2-2i)$ 15. $4+i, 1+i, 1-i$
17. $1-i,\ x^3 - 2x + 4 = 0$

19. Let $P(x) = 0$ be a polynomial equation with real coefficients of degree n, where n is odd; then, $P(x) = 0$ must have at least n zeros and, since $P(x)$ has only real coefficients, any complex zeros must occur as conjugate pairs; since n is odd there must always be at least one real zero, then $P(x) = 0$ must have at least one real root.

21. $-1, \dfrac{1+\sqrt{3}i}{2}, \dfrac{1-\sqrt{3}i}{2}$

23. Let $P(x) = a_n x^n + \cdots + a_0$ and suppose $P(z) = 0$; then, by Exercises 77–79, Section 8.1, we can write

$$P(\bar{z}) = a_n \bar{z}^n + \cdots + a_0$$
$$= \overline{a_n z^n} + \cdots + \overline{a_0} \quad \text{(since } a_i \in R\text{)}$$
$$= \overline{a_n z^n + \cdots + a_0}$$
$$= \overline{P(z)} = \bar{0} = 0.$$

25. Let $S(x) = P(x) - Q(x)$; then the degree of S is less than or equal to n. But there are more than n values of x for which $S(x) = 0$ [namely the values where $P(x) = Q(x)$]. But the number of roots of $S(x) = 0$ must be the same as the degree of S if the degree of S is positive. Since the number of roots of $S(x) = 0$ is greater than the degree of S, it must be the case that the degree of S is 0 and $S(x)$ must be a constant. Since the equation $S(x) = 0$ has solutions, it must be the case that $P(x) - Q(x) = S(x) = 0$, or $P(x) = Q(x)$ for all x.

Exercise 8.3 (page 283)

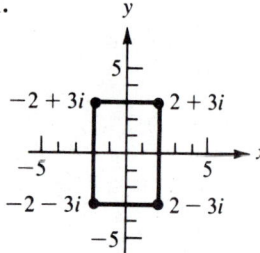

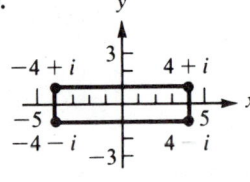

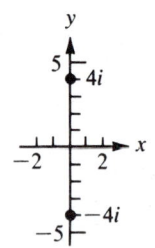

1.

3.

5.

7. 4 **9.** $\sqrt{13}$ **11.** 2 **13.** $\sqrt{5}$ **15.** $3\sqrt{2}$ cis 45° **17.** $5\sqrt{2}$ cis 135° **19.** 4 cis 330°
21. $2\sqrt{2}$ cis 225° **23.** 3 cis 0° **25.** 2 cis 90° **27.** $-2 - 2\sqrt{3}i$ **29.** $3\sqrt{3} - 3i$
31. $6 + 6\sqrt{3}i$ **33.** $6i$ **35.** $1.1490 + 0.9642i$ **37.** $1.2856 - 1.5321i$
39. a. $-3 + 3i$ **b.** $\dfrac{3}{2} + \dfrac{3}{2}i$ **41. a.** 108 **b.** $\dfrac{1}{6} - \dfrac{\sqrt{3}}{6}i$
43. a. $4.5963 + 3.8567i$ **b.** $0.6566 - 0.1158i$
45. Let $z = a + bi$; then $\bar{z} = a - bi$. $(a + bi) + (a - bi) = (a + a) + (b - b)i = 2a + 0i = 2a \in R$; $(a + bi) \cdot (a - bi) = (a^2 + b^2) + (ab - ab)i = (a^2 + b^2) + 0i = (a^2 + b^2) \in R$.
47. From the result of Exercise 46, and from Theorem 8.8-I,

$$(a + bi)^3 = (a + bi)^2(a + bi) = (r^2 \text{ cis } 2\theta)(r \text{ cis } \theta) = r^3 \text{ cis } (2\theta + \theta) = r^3 \text{ cis } 3\theta.$$

Exercise 8.4 (page 288)
1. $-64\sqrt{3} + 64i$ **3.** $1 + 0i$ **5.** $\dfrac{729}{2} + \dfrac{729\sqrt{3}}{2}i$ **7.** $\dfrac{-\sqrt{3}}{64} + \dfrac{1}{64}i$
9. $\dfrac{\sqrt{3}-1}{64} - \dfrac{\sqrt{3}+1}{64}i$ **11.** $-\dfrac{1}{2} - \dfrac{1}{2}i$
13. 2 cis 9°; 2 cis 81°; 2 cis 153°; 2 cis 225°; 2 cis 297°

Answers to Odd-Numbered Exercises   535

15. 2 cis 30°;   2 cis 102°;   2 cis 174°;   2 cis 246°;   2 cis 318°
17. cis 45°;   cis 105°;   cis 165°;   cis 225°;   cis 285°;   cis 345°
19. 2 cis 60°;   2 cis 132°;   2 cis 204°;   2 cis 276°;   2 cis 348°
21. cis $25\frac{5}{7}°$;   cis $77\frac{1}{7}°$;   cis $128\frac{4}{7}°$;   cis 180°;   cis $231\frac{3}{7}°$;   cis $282\frac{6}{7}°$;   cis $334\frac{2}{7}°$
23. $(x - 2 \text{ cis } 45°)(x - 2 \text{ cis } 135°)(x - 2 \text{ cis } 225°)(x - 2 \text{ cis } 315°)$
25. The four roots are $-1, 1, i, -i$; hence their sum is $0 + 0i$.
27. Theorem 8.9 guarantees $z^n = r^n$ cis $n\theta$ for $n > 0$. For $n = 0$ we have $z^0 = 1 + 0i$ by Definition 8.10-I. But

$$1 + 0i = 1 \cdot \text{cis } 0° = r^0 (\text{cis } 0 \cdot \theta).$$

Thus, for $n = 0$, $z^n = r^n$ cis $n\theta$. Now if $n < 0$, then $-n > 0$ and by Definition 8.10-II and Theorem 8.9 we have

$$z^n = \frac{1}{z^{-n}} = \frac{1}{r^{-n} \text{ cis } (-n\theta)}$$

$$= r^n \left(\frac{1}{\cos(-n\theta) + i \sin(-n\theta)} \right)$$

$$= r^n [\cos(-n\theta) - i \sin(-n\theta)]$$

$$= r^n \text{ cis } (n\theta)$$

Chapter 8 Review (page 289) 1. $4 - 2i$ 2. $6 - 8i$ 3. $2 + 4i$ 4. i 5. $2i$ 6. $8 + 2i$
7. i 8. i 9. $2, 2i, -2i$ 10. $\frac{3}{2}, 2 + 2i, 2 - 2i$ 11. $-2, 1, 2i, -2i,$
12. $-2, 1, 2 + 3i, 2 - 3i$ 13. $2\sqrt{2}$ cis 45°   14. 10 cis 300°   15. $-\frac{3\sqrt{3}}{2} + \frac{3}{2}i$ 16. $2 - 2\sqrt{3}i$
17. a. $-9\sqrt{3} + 9i$ b. $0 - \frac{2}{9}i$ 18. a. $3 + 3\sqrt{3}i$ b. $\frac{3\sqrt{3}}{4} - \frac{3}{4}i$ 19. $-\frac{243\sqrt{2}}{2} - \frac{243\sqrt{2}}{2}i$
20. $-1 - \sqrt{3}i$ 21. $2^{1/6}$ cis 15°, $2^{1/6}$ cis 135°, $2^{1/6}$ cis 255°
22. $6^{1/4}$ cis 82.5°, $6^{1/4}$ cis 172.5°, $6^{1/4}$ cis 262.5°, $6^{1/4}$ cis 352.5°
23. $\{\text{cis } 45°, \text{cis } 135°, \text{cis } 225°, \text{cis } 315°\}$
24. $\{\text{cis } 0°, \text{cis } 72°, \text{cis } 144°, \text{cis } 216°, \text{cis } 288°\}$

Supplemental Exercises, Chapters 5-8 (page 291)
1. $-0.506366, 0.826879, -0.305613, 0.035762, -0.350116$; $\sin(1/x)$ does not seem to approach any value as x approaches 0.
2. $0.862319, 0.562379, -0.952156, -0.99360, 0.936706$; $\cos(1/x)$ does not seem to approach any value as x approaches 0.
3. $-0.005064, 0.000827, -0.000031, 0.000000, 0.000000$; $x \sin(1/x)$ approaches 0 as x approaches 0.
4. $0.008623, 0.000562, -0.000095, -0.000010, 0.000001$; $x \cos(1/x)$ approaches 0 as x approaches 0.
5. $-0.050637, 0.026148, -0.003056, 0.000113, -0.000350$; $\sqrt{x} \sin(1/x)$ approaches 0 as x approaches 0.
6. $0.086232, 0.017784, -0.009522, -0.003160, 0.000937$; $\sqrt{x} \cos(1/x)$ approaches 0 as x approaches 0.
7.   8.

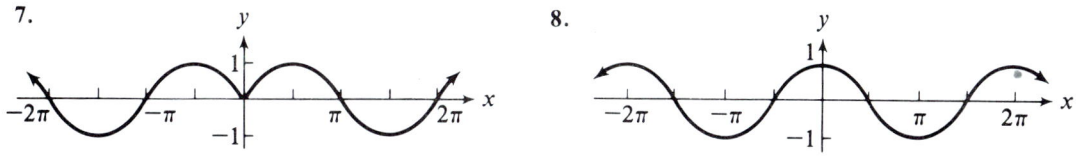

9.
10.
11.
12.
13.
14.
15.
16.
17.
18.

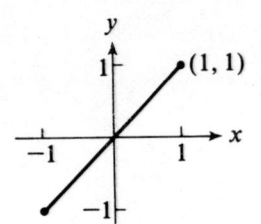

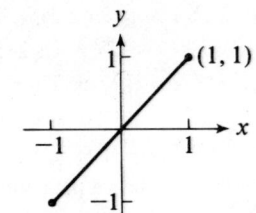

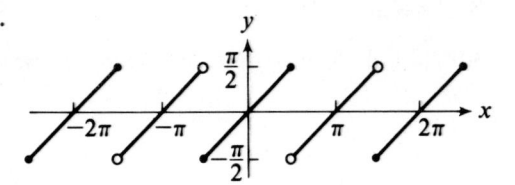

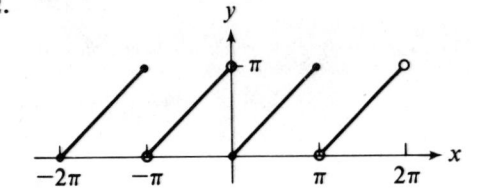

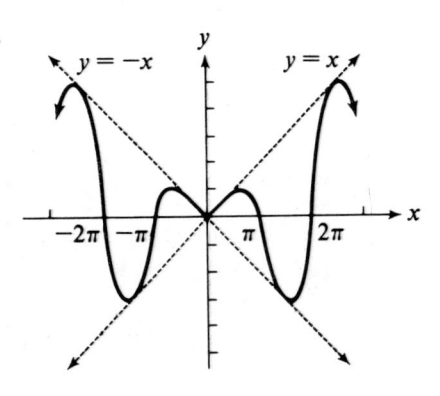

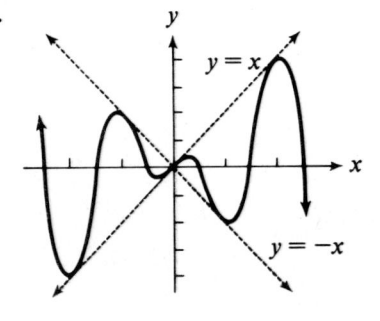

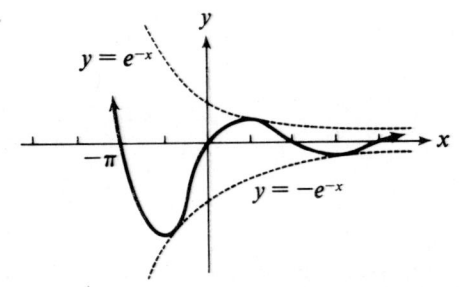

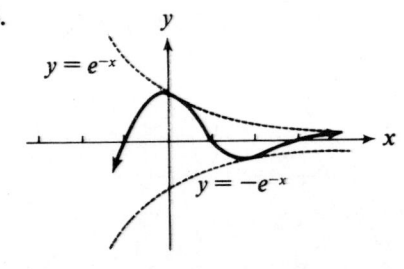

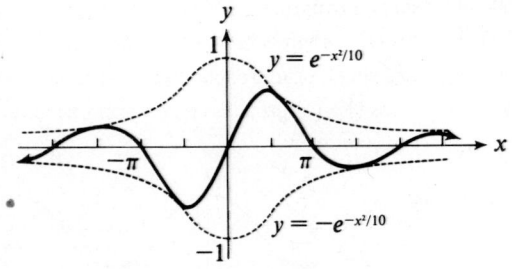

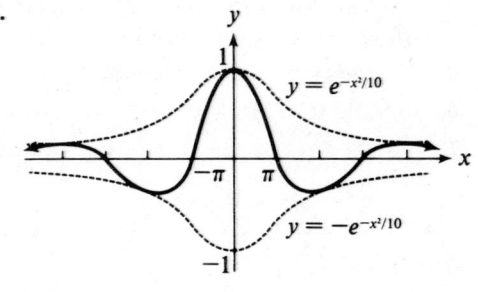

Answers to Odd-Numbered Exercises

19.

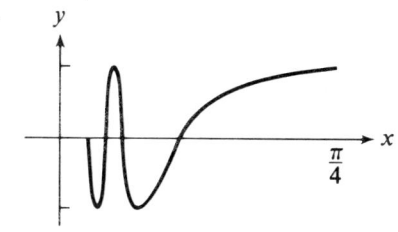

The curve oscillates between −1 and 1 infinitely often near 0. It is undefined at 0.

20.

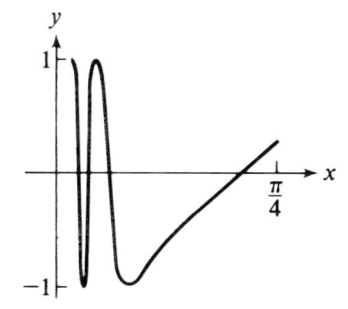

The curve oscillates between −1 and 1 infinitely often near 0. It is undefined at 0.

21.

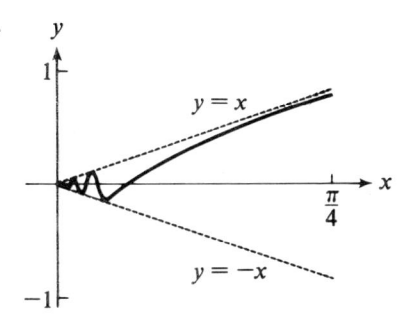

The curve oscillates but the oscillations are damped to 0 near $x = 0$. See Exercise 3.

22.

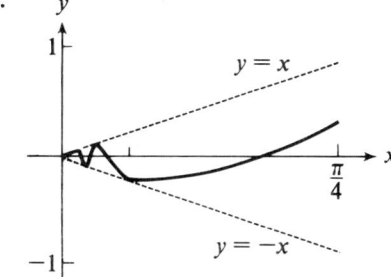

The curve oscillates but the oscillations are damped to 0 near $x = 0$. See Exercise 4.

23. $\cos|x| = \begin{cases} \cos x & \text{if } x \geq 0 \\ \cos(-x) & \text{if } x < 0 \end{cases} = \begin{cases} \cos x & \text{if } x \geq 0 \\ \cos x & \text{if } x < 0 \end{cases} = \cos x$

24. $\sin^2|x| = (\sin|x|)^2 = \begin{cases} (\sin x)^2 & \text{if } x \geq 0 \\ [\sin(-x)]^2 & \text{if } x < 0 \end{cases} = \begin{cases} (\sin x)^2 & \text{if } x \geq 0 \\ (-\sin x)^2 & \text{if } x < 0 \end{cases} = \sin^2 x$

25. Note that if $z = a + bi$ and $w \in C$, then

(1) $|a| < |z|$,

(2) $|zw| = |z||w|$ (by Theorem 8.8-I),

(3) $|\bar{z}| = |z|$ (by Exercise 48, page 285).

Now let $z, w \in C$, then by Exercise 45 (page 285) we have

$$\begin{aligned}
|z \pm w|^2 &= (z \pm w)\overline{(z \pm w)} \\
&= (z \pm w)(\bar{z} \pm \bar{w}) \\
&= z\bar{z} \pm w\bar{z} \pm \bar{w}z + w\bar{w} \\
&= |z|^2 \pm (w\bar{z} + \overline{w\bar{z}}) + |w|^2 \\
&= |z|^2 \pm 2a + |w|^2 \quad \text{(where } w\bar{z} = a + bi) \\
&\leq |z|^2 + 2|a| + |w|^2 \\
&\leq |z|^2 + 2|w\bar{z}| + |w|^2 \\
&= |z|^2 + 2|w||z| + |w|^2 \\
&= (|z| + |w|)^2.
\end{aligned}$$

Thus $|z \pm w| \leq |z| + |w|$.

26. Let $z, w \in C$. Since $z = (z + w) - w$ by Exercise 25, we have
$$|z| = |(z + w) - w|$$
$$\leq |z + w| + |w|.$$
Thus $|z| - |w| \leq |z + w|$. Similarly $|w| - |z| \leq |z + w|$. Thus $||z| - |w|| \leq |z + w|$.

27. Let $z_0, z_1, \ldots, z_{n-1}$ be the n nth roots of the complex number z. Then the solutions of the equation $x^n - z = 0$ are $z_0, \ldots, z_{n-1}$. Thus we can factor $x^n - z$ as
$$x^n - z = (x - z_0)(x - z_1) \cdots (x - z_{n-1}).$$
Multiplying the right-hand member, we have
$$x^n - z = x^n - (z_0 + z_1 + \cdots + z_{n-1})x^{n-1} + \cdots + (-1)^n z_0 z_1 \cdots z_{n-1}.$$
Equating coefficients of x^{n-1} yields
$$z_0 + z_1 + \cdots + z_{n-1} = 0.$$

28. Let $n = 2j$; then the distinct nth roots of unity are
$$\operatorname{cis}\left(\frac{k \cdot 360°}{2j}\right), \quad k = 0, 1, \ldots, 2j - 1.$$
Thus the squares of the distinct $n$ths roots of unity are
$$\operatorname{cis}\left(\frac{k \cdot 360°}{j}\right), \quad k = 0, 1, \ldots, 2j - 1.$$
Note that
$$\operatorname{cis}\left(\frac{k \cdot 360°}{j}\right) = \operatorname{cis}\left(\frac{(k + j)360°}{j}\right), \quad k = 0, \ldots, j - 1.$$
Thus if S denotes the sum of the squares of the nth roots of unity, then
$$S = \operatorname{cis}\left(\frac{0 \cdot 360°}{j}\right) + \cdots + \operatorname{cis}\left(\frac{(2j - 1) \cdot 360°}{j}\right)$$
$$= 2\left[\operatorname{cis}\left(\frac{0 \cdot 360°}{j}\right) + \cdots + \operatorname{cis}\left(\frac{(j - 1) \cdot 360°}{j}\right)\right].$$
But the last sum on the right is the sum of the jth roots of unity, which by Exercise 27 is 0.

29. By the law of cosines,
$$(a - c)^2 + (b - d)^2 = a^2 + b^2 + c^2 + d^2 - 2\|\mathbf{v}_1\|\|\mathbf{v}_2\| \cos \theta$$
or
$$a^2 - 2ac + c^2 + b^2 - 2bd + d^2 = a^2 + b^2 + c^2 + d^2 - 2\|\mathbf{v}_1\|\|\mathbf{v}_2\| \cos \theta.$$
Thus
$$\frac{ac + bd}{\|\mathbf{v}_1\|\|\mathbf{v}_2\|} = \cos \theta.$$

30. By Exercise 29,
$$ac + bd = \|\mathbf{v}_1\|\|\mathbf{v}_2\| \cos \theta.$$
Since $\cos \theta \leq 1$, we have
$$ac + bd \leq \|\mathbf{v}_1\|\|\mathbf{v}_2\|.$$

Answers to Odd-Numbered Exercises 539

31. If v_1 is perpendicular to v_2, then by Exercise 29

$$ac + bd = \|v_1\|\|v_2\| \cos 90° = 0.$$

If $ac + bd = 0$, then by Exercise 29

$$\|v_1\|\|v_2\| \cos \theta = 0.$$

Since $\|v_1\| \neq 0$ and $\|v_2\| \neq 0$, we have $\cos \theta = 0$. Thus v_1 is perpendicular to v_2.

32. If v_1 is perpendicular to v_2, then $v_1 + v_2$ is the hypotenuse of a right triangle with legs v_1 and v_2. Thus

$$\|v_1 + v_2\|^2 = \|v_1\|^2 + \|v_2\|^2.$$

If $\|v_1 + v_2\|^2 = \|v_1\|^2 + \|v_2\|^2$, then

$$(a + c)^2 + (b + d)^2 = a^2 + b^2 + c^2 + d^2.$$

Hence,

$$ac + bd = 0.$$

Thus, by Exercise 31, v_1 is perpendicular to v_2.

33. $\left[-2\pi, -\dfrac{11\pi}{6}\right] \cup \left[-\dfrac{7\pi}{6}, \dfrac{\pi}{6}\right] \cup \left[\dfrac{5\pi}{6}, 2\pi\right]$ **34.** $\left[-\dfrac{5\pi}{3}, -\dfrac{\pi}{3}\right] \cup \left[\dfrac{\pi}{3}, \dfrac{5\pi}{3}\right]$

35. $\left[-2\pi, -\dfrac{5\pi}{3}\right] \cup \left[-\dfrac{4\pi}{3}, -\dfrac{2\pi}{3}\right] \cup \left[-\dfrac{\pi}{3}, \dfrac{\pi}{3}\right] \cup \left[\dfrac{2\pi}{3}, \dfrac{4\pi}{3}\right] \cup \left[\dfrac{5\pi}{3}, 2\pi\right]$

36. $\left[-\dfrac{11\pi}{6}, -\dfrac{7\pi}{6}\right] \cup \left[-\dfrac{5\pi}{6}, -\dfrac{\pi}{6}\right] \cup \left[\dfrac{\pi}{6}, \dfrac{5\pi}{6}\right] \cup \left[\dfrac{7\pi}{6}, \dfrac{11\pi}{6}\right]$

Exercise 9.1 (page 298) **1.** $\{(3, 2)\}$ **3.** $\{(2, 1)\}$ **5.** $\{(-5, 4)\}$ **7.** $\left\{\left(0, \dfrac{3}{2}\right)\right\}$ **9.** $\left\{\left(\dfrac{2}{3}, -1\right)\right\}$

11. $\{(1, 2)\}$ **13.** $\left\{\left(-\dfrac{19}{5}, -\dfrac{18}{5}\right)\right\}$ **15.** $\left(x, -\dfrac{3}{2}x + 3\right)$, $x \in R$ **17.** No solution

19. $a = 1$, $b = -1$ **21.** $a = 0$, $b = 2$ **23.** $y = -\dfrac{10}{3}x + 2$ **25.** $C = \dfrac{5}{9}(F - 32)$

27. 32 pounds **29.** $14,400 at 9% and $4,800 at 8% **31.** 32 and 64 mph **33.** $\{(-1, 1)\}$
35. $\{(2, 1)\}$

37. a. From $\dfrac{a_1}{a_2} = \dfrac{b_1}{b_2}$, it follows that $\dfrac{a_1}{b_1} = \dfrac{a_2}{b_2}$. Hence the slopes are equal; the two lines are the same or they are parallel. The equations are either consistent or inconsistent, respectively. Assume that the equations are consistent and the two lines are the same. Then, the y-intercepts $\dfrac{-c_1}{b_1} = \dfrac{-c_2}{b_2}$, from which $\dfrac{b_1}{b_2} = \dfrac{c_1}{c_2}$, but this contradicts the hypothesis $\dfrac{b_1}{b_2} \neq \dfrac{c_1}{c_2}$; hence the equations are inconsistent.

b. From (a), $\dfrac{a_1}{b_1} = \dfrac{a_2}{b_2}$, and it follows that $\dfrac{a_1}{a_2} = \dfrac{b_1}{b_2}$. Now, either $\dfrac{b_1}{b_2} = \dfrac{c_1}{c_2}$ or $\dfrac{b_1}{b_2} \neq \dfrac{c_1}{c_2}$. Assume $\dfrac{b_1}{b_2} = \dfrac{c_1}{c_2}$; then $\dfrac{c_1}{b_1} = \dfrac{c_2}{b_2}$; the graphs would be the same straight line, and the equations would be consistent. This contradicts the hypothesis that states that the equations are inconsistent; hence $\dfrac{b_1}{b_2} \neq \dfrac{c_1}{c_2}$.

Answers to Odd-Numbered Exercises

Exercise 9.2 (page 305)   **1.** $\{(1, 2, -1)\}$ **3.** $\{(2, -2, 0)\}$ **5.** $\{(2, 2, 1)\}$ **7.** $\{(0, 1, 2)\}$
9. Dependent   **11.** $\{(4, -2, 2)\}$ **13.** 3, 6, 6   **15.** 60 nickels, 20 dimes, 5 quarters
17. $x = 40$ cm, $y = 60$ cm, $z = 55$ cm   **19.** $a = -6$, $b = -8$, $c = 0$ **21.** $a = 3$, $b = 1$, $c = -2$
23. $a = \frac{1}{4}$, $b = \frac{1}{2}$, $c = 0$ **25.** $\{(1, 1, 1)\}$ **27.** $\{(0, 1, 2)\}$
29. $\left(\frac{5}{3} - \frac{1}{3}z, -\frac{5}{3}z + \frac{1}{3}, z\right)$, $z \in R$; $\left\{\left(\frac{5}{3}, \frac{1}{3}, 0\right), (2, 2, -1)\right\}$

Exercise 9.3 (page 312)   **1.** $\frac{4}{x} - \frac{4}{x + 1}$ **3.** $\frac{-1}{x + 1} + \frac{2}{x + 2}$ **5.** $\frac{-4}{x + 2} + \frac{5}{x + 3}$
7. $\frac{1/2}{x} + \frac{1}{x + 1} + \frac{-3/2}{x + 2}$ **9.** $\frac{1/2}{x} + \frac{-3}{x + 1} + \frac{7/2}{x + 2}$ **11.** $\frac{1/2}{x} + \frac{-1}{x + 1} + \frac{1/2}{x + 2}$
13. $\frac{-3/4}{x} + \frac{-1/2}{x^2} + \frac{3/4}{x - 2}$ **15.** $\frac{1/2}{x - 1} + \frac{1/4}{(x - 1)^2} + \frac{1/2}{x + 1} + \frac{-1/4}{(x + 1)^2}$ **17.** $\frac{-1}{x} + \frac{-1}{x^2} + \frac{1}{x - 1}$
19. $x^2 + \frac{2}{x} + \frac{-1}{x + 1}$ **21.** $x + 2 + \frac{3}{x - 1} + \frac{1}{(x - 1)^2}$ **23.** $a = 1$, $b = -1$, $c = 0$
25. $a = -\frac{3}{4}$, $b = \frac{3}{4}$, $c = \frac{1}{2}$

Exercise 9.4 (page 320)   **1.** $\{(-1, -4), (5, 20)\}$ **3.** $\{(2, 3), (3, 2)\}$ **5.** $\{(4, -3), (-3, 4)\}$
7. $\{(-1, 3), (-1, -3), (1, 3), (1, -3)\}$ **9.** $\{(-3, \sqrt{2}), (-3, -\sqrt{2}), (3, \sqrt{2}), (3, -\sqrt{2})\}$
11. $\{(\sqrt{3}, 4), (\sqrt{3}, -4), (-\sqrt{3}, 4), (-\sqrt{3}, -4)\}$ **13.** $\{(1, -2), (-1, 2), (2, -1), (-2, 1)\}$
15. $\{(3, 1), (-3, -1), (-2\sqrt{7}, \sqrt{7}), (2\sqrt{7}, -\sqrt{7})\}$
17. $\left\{\left(2\sqrt{\frac{1}{3}}, \sqrt{\frac{1}{3}}\right), \left(-2\sqrt{\frac{1}{3}}, -\sqrt{\frac{1}{3}}\right), (-6i, 4i), (6i, -4i)\right\}$ **19.** $\{(i\sqrt{7}, 4), (-i\sqrt{7}, 4)\}$
21. $\{(2i, -2i), (-2i, 2i), (2\sqrt{2}, \sqrt{2}), (-2\sqrt{2}, -\sqrt{2})\}$ **23.** $\left\{\left(-1, \frac{1}{10}\right)\right\}$ **25.** $\{(1, 0), (10, -1)\}$
27. $\{(0, 2)\}$ **29.** $\left\{(1, 0), \left(\frac{3}{2}, \frac{1}{2}\right)\right\}$

31. $x = 2$, $y = 3$ **33.** $800 at 4%   **35. a.** 1   **b.** 2   **c.** 4   **37.** $\{(2, 2), (-2, -2)\}$

Exercise 9.5 (page 325)

1. **3.** **5.** **7.**

Answers to Odd-Numbered Exercises 541

9. **11.** 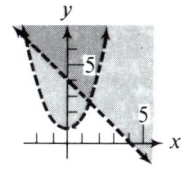 **13.**

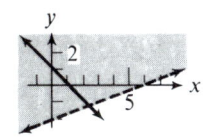

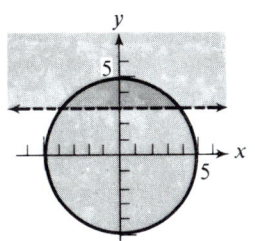

15. 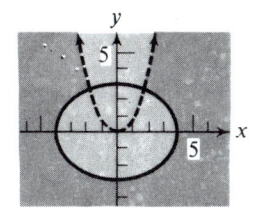 **17.** 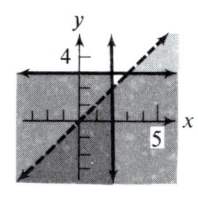 **19.**

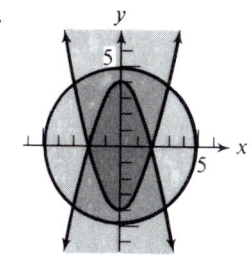

21. 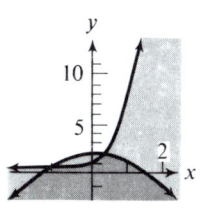 **23.**

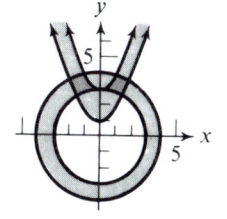

Exercise 9.6 (page 328)

1. **3.**

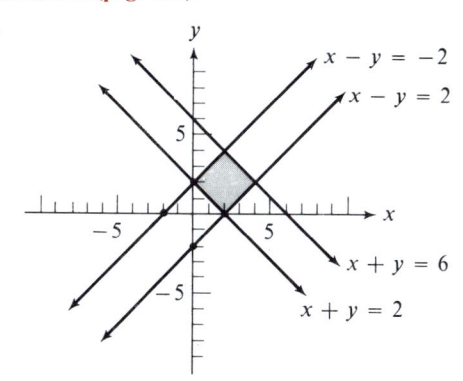

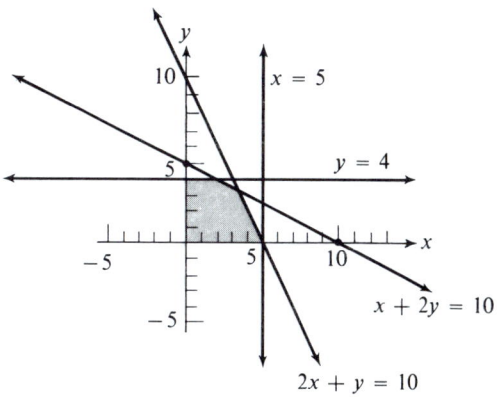

5. 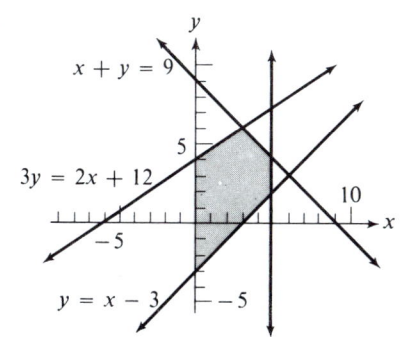 **7.** No **9.** Yes

Exercise 9.7 (page 330) **1.** Maximum: 14; minimum: 2 **3.** Maximum: 80/3; minimum: 0
5. Maximum: 17; minimum: -16 **7.** Maximum profit: \$1,860 **9.** Maximum profit: \$790
11. 200 standard models and 400 deluxe models

Chapter 9 Review (page 332) **1.** $\left\{\left(\frac{1}{2}, \frac{7}{2}\right)\right\}$ **2.** $\{(1, 2)\}$ **3.** $\left\{\left(\frac{37}{13}, -\frac{10}{13}\right)\right\}$ **4.** $\{(-2, 1)\}$
5. $\left(x, \frac{2}{3}x - \frac{1}{3}\right)$, $x \in R$ **6.** No solution **7.** $a = 2$, $b = 5$ **8.** $a = \frac{2}{3}$, $b = \frac{5}{3}$ **9.** $\{(2, 0, -1)\}$
10. $\{(2, 1, -1)\}$ **11.** $\{(2, -1, 3)\}$ **12.** $a = 2$, $b = -3$, $c = 4$ **13.** $\frac{2}{2x-3} - \frac{1}{x+1}$
14. $\frac{-1}{x} + \frac{1}{x-1} + \frac{1}{x+1}$ **15.** $\{(1, 2), (4, -13)\}$ **16.** $\{(1, 1), (-1, -1)\}$
17. $\{(2, 3), (2, -3), (-2, 3), (-2, -3)\}$
18. $\left\{\left(\frac{1}{\sqrt{5}}, -\frac{1}{\sqrt{5}}\right), \left(-\frac{1}{\sqrt{5}}, \frac{1}{\sqrt{5}}\right), \left(\frac{1}{\sqrt{11}}, \frac{-2}{\sqrt{11}}\right), \left(-\frac{1}{\sqrt{11}}, \frac{2}{\sqrt{11}}\right)\right\}$ **19.** $\left\{\left(-\frac{1}{3}, 10^{-2/3}\right)\right\}$
20. $\left\{\left(\frac{1}{20}, -1\right)\right\}$ **21.** $\left\{\left(\frac{7}{8}, \frac{17}{8}\right)\right\}$

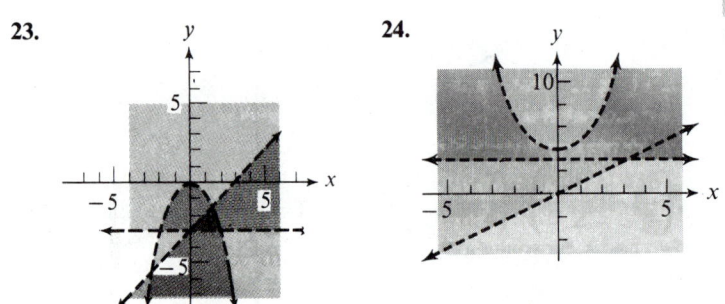

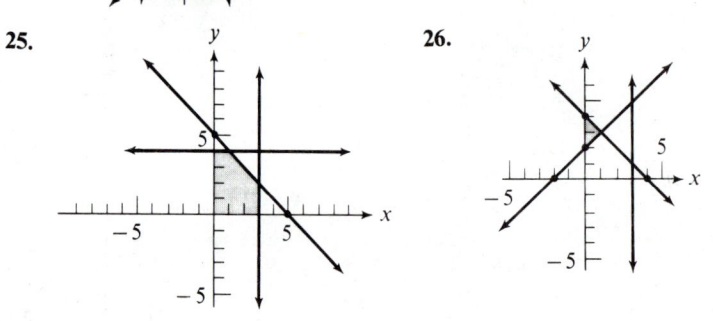

27. Minimum: 0; maximum: 8 **28.** Minimum: 0; maximum: 11
29. Maximum profit: \$1,100 **30.** Maximum profit: \$210

Answers to Odd-Numbered Exercises 543

Exercise 10.1 (page 339) 1. 2×2, $\begin{bmatrix} 6 & 2 \\ -1 & 3 \end{bmatrix}$ 3. 2×3, $\begin{bmatrix} 2 & 1 \\ -7 & 4 \\ 3 & 0 \end{bmatrix}$ 5. 3×3, $\begin{bmatrix} 2 & 4 & -2 \\ 3 & 0 & 3 \\ -1 & 1 & 1 \end{bmatrix}$

7. 2×4, $\begin{bmatrix} 4 & 2 \\ -3 & 1 \\ -1 & 1 \\ 0 & 6 \end{bmatrix}$ 9. $\begin{bmatrix} 3 & 1 \\ 3 & 9 \end{bmatrix}$ 11. $\begin{bmatrix} 9 & -1 & -1 \\ 2 & 3 & 6 \end{bmatrix}$ 13. $\begin{bmatrix} -2 & -3 \\ 4 & 0 \end{bmatrix}$ 15. $\begin{bmatrix} 2 & -9 & -13 \\ 7 & -4 & 1 \\ 9 & 0 & 0 \end{bmatrix}$

17. $\begin{bmatrix} 10 \\ 3 \\ -3 \end{bmatrix}$ 19. $\begin{bmatrix} 2 & 3 & 4 \\ -1 & 6 & 2 \\ 1 & 0 & 3 \end{bmatrix}$ 21. $\begin{bmatrix} 2 & -1 \\ 2 & 1 \end{bmatrix}$ 23. $\begin{bmatrix} 1 & 0 \\ -5 & 3 \end{bmatrix}$

25. Let $A = \begin{bmatrix} a_{11} & a_{12} \\ a_{21} & a_{22} \end{bmatrix}$; then $A^t = \begin{bmatrix} a_{11} & a_{21} \\ a_{12} & a_{22} \end{bmatrix}$ and $[A^t]^t = \begin{bmatrix} a_{11} & a_{12} \\ a_{21} & a_{22} \end{bmatrix} = A$. This result holds for both $n \times n$ and $m \times n$ matrices.

27. The i,j entry of $A + B$ is $a_{ij} + b_{ij}$, which is a real number. Hence $A + B$ is a matrix with real-number entries.

28. The i,j entry of $A + B$ is $a_{ij} + b_{ij}$ and the i,j entry of $B + A$ is $b_{ij} + a_{ij}$. By the commutative law for addition of real-numbers, $a_{ij} + b_{ij} = b_{ij} + a_{ij}$. Thus $A + B = B + A$.

Exercise 10.2 (page 345) 1. $\begin{bmatrix} 0 & -5 & 5 \\ -15 & 5 & -10 \end{bmatrix}$ 3. $[-1]$ 5. $\begin{bmatrix} 1 & -13 \\ 4 & -7 \end{bmatrix}$ 7. $\begin{bmatrix} 30 & -39 \\ 29 & 14 \end{bmatrix}$

9. $\begin{bmatrix} -5 & -1 \\ 8 & -1 \end{bmatrix}$ 11. $\begin{bmatrix} -1 & 0 & -2 \\ 1 & 2 & 8 \\ 0 & 1 & 3 \end{bmatrix}$ 13. $\begin{bmatrix} 1 & 0 & 0 \\ 0 & 1 & 0 \\ 0 & 0 & 1 \end{bmatrix}$ 15. $\begin{bmatrix} 1 & 0 \\ -1 & 2 \end{bmatrix}$ 17. $\begin{bmatrix} 1 & -2 \\ 1 & 2 \end{bmatrix}$

19. $\begin{bmatrix} -2 & 3 \\ 2 & -4 \end{bmatrix}$ 21. $\begin{bmatrix} -1 & 1 \\ -1 & -1 \end{bmatrix}$ 23. $\begin{bmatrix} 1 & -1 \\ 1 & 0 \end{bmatrix}$

25. Since $A + B = \begin{bmatrix} 0 & 2 \\ -1 & 3 \end{bmatrix}$, $A - B = \begin{bmatrix} -2 & 2 \\ 1 & -1 \end{bmatrix}$, $A^2 = \begin{bmatrix} 1 & 0 \\ 0 & 1 \end{bmatrix}$, $B^2 = \begin{bmatrix} 1 & 0 \\ -3 & 4 \end{bmatrix}$,

and $AB = \begin{bmatrix} -3 & 4 \\ -1 & 2 \end{bmatrix}$, we have **a.** $(A + B)(A + B) = \begin{bmatrix} 0 & 2 \\ -1 & 3 \end{bmatrix} \cdot \begin{bmatrix} 0 & 2 \\ -1 & 3 \end{bmatrix} = \begin{bmatrix} -2 & 6 \\ -3 & 7 \end{bmatrix}$

and $A^2 + 2AB + B^2 = \begin{bmatrix} 1 & 0 \\ 0 & 1 \end{bmatrix} + 2\begin{bmatrix} -3 & 4 \\ -1 & 2 \end{bmatrix} + \begin{bmatrix} 1 & 0 \\ -3 & 4 \end{bmatrix} = \begin{bmatrix} -4 & 8 \\ -5 & 9 \end{bmatrix}$;

hence $(A + B)(A + B) \neq A^2 + 2AB + B^2$. **b.** $(A + B)(A - B) = \begin{bmatrix} 0 & 2 \\ -1 & 4 \end{bmatrix} \cdot \begin{bmatrix} -2 & 2 \\ 1 & -1 \end{bmatrix} = \begin{bmatrix} 2 & -2 \\ 6 & -6 \end{bmatrix}$

and $A^2 - B^2 = \begin{bmatrix} 1 & 0 \\ 0 & 1 \end{bmatrix} - \begin{bmatrix} 1 & 0 \\ -3 & 4 \end{bmatrix} = \begin{bmatrix} 0 & 0 \\ 3 & -3 \end{bmatrix}$; hence $(A + B)(A - B) \neq A^2 - B^2$.

For Exercises 27 and 29, let $A = \begin{bmatrix} a_{11} & a_{12} \\ a_{21} & a_{22} \end{bmatrix}$, $B = \begin{bmatrix} b_{11} & b_{12} \\ b_{21} & b_{22} \end{bmatrix}$, and $C = \begin{bmatrix} c_{11} & c_{12} \\ c_{21} & c_{22} \end{bmatrix}$.

27. Since $AB = \begin{bmatrix} a_{11}b_{11} + a_{12}b_{21} & a_{11}b_{12} + a_{12}b_{22} \\ a_{21}b_{11} + a_{22}b_{21} & a_{21}b_{12} + a_{22}b_{22} \end{bmatrix}$,

$(AB)C = \begin{bmatrix} (a_{11}b_{11} + a_{12}b_{21})c_{11} + (a_{11}b_{12} + a_{12}b_{22})c_{21} & (a_{11}b_{11} + a_{12}b_{21})c_{12} + (a_{11}b_{12} + a_{12}b_{22})c_{22} \\ (a_{21}b_{11} + a_{22}b_{21})c_{11} + (a_{21}b_{12} + a_{22}b_{22})c_{21} & (a_{21}b_{11} + a_{22}b_{22})c_{12} + (a_{21}b_{12} + a_{22}b_{22})c_{22} \end{bmatrix}$.

The element in the first row, first column is given by $a_{11}b_{11}c_{11} + a_{12}b_{21}c_{11} + a_{11}b_{12}c_{21} + a_{12}b_{22}c_{21} = (a_{11}b_{11}c_{11} + a_{11}b_{12}c_{21}) + (a_{12}b_{21}c_{11} + a_{12}b_{22}c_{21}) = a_{11}(b_{11}c_{11} + b_{12}c_{21}) + a_{12}(b_{21}c_{11} + b_{22}c_{21})$.

When the remaining elements are treated in a similar manner, we get

$$(AB)C = \begin{bmatrix} a_{11}(b_{11}c_{11} + b_{12}c_{21}) + a_{12}(b_{21}c_{11} + b_{22}c_{21}) & a_{11}(b_{11}c_{12} + b_{12}c_{22}) + a_{12}(b_{21}c_{12} + b_{22}c_{22}) \\ a_{21}(b_{11}c_{11} + b_{12}c_{21}) + a_{22}(b_{21}c_{11} + b_{22}c_{21}) & a_{21}(b_{11}c_{12} + b_{12}c_{22}) + a_{22}(b_{21}c_{12} + b_{22}c_{22}) \end{bmatrix}$$

$$= \begin{bmatrix} a_{11} & a_{12} \\ a_{21} & a_{22} \end{bmatrix} \cdot \begin{bmatrix} b_{11}c_{11} + b_{12}c_{21} & b_{11}c_{12} + b_{12}c_{22} \\ b_{21}c_{11} + b_{22}c_{21} & b_{21}c_{12} + b_{22}c_{22} \end{bmatrix} = \begin{bmatrix} a_{11} & a_{12} \\ a_{21} & a_{22} \end{bmatrix} \cdot \left(\begin{bmatrix} b_{11} & b_{12} \\ b_{21} & b_{22} \end{bmatrix} \cdot \begin{bmatrix} c_{11} & c_{12} \\ c_{21} & c_{22} \end{bmatrix} \right)$$

$$= A(BC).$$

29. $B + C = \begin{bmatrix} b_{11} + c_{11} & b_{12} + c_{12} \\ b_{21} + c_{21} & b_{22} + c_{22} \end{bmatrix}$;

hence

$$(B + C)A = \begin{bmatrix} (b_{11} + c_{11})a_{11} + (b_{12} + c_{12})a_{21} & (b_{11} + c_{11})a_{12} + (b_{12} + c_{12})a_{22} \\ (b_{21} + c_{22})a_{11} + (b_{22} + c_{22})a_{21} & (b_{21} + c_{22})a_{12} + (b_{22} + c_{22})a_{22} \end{bmatrix}.$$

The element in the first row, first column is given by

$$b_{11}a_{11} + c_{11}a_{21} + b_{12}a_{21} + c_{12}a_{21} = (b_{11}a_{11} + b_{12}a_{21}) + (c_{11}a_{21} + c_{12}a_{21}).$$

When the remaining elements are treated in a similar manner, we get

$$(B + C)A = \begin{bmatrix} (b_{11}a_{11} + b_{12}a_{21}) + (c_{11}a_{11} + c_{12}a_{21}) & (b_{11}a_{12} + b_{12}a_{22}) + (c_{11}a_{12} + c_{12}a_{22}) \\ (b_{21}a_{11} + b_{22}a_{21}) + (c_{21}a_{11} + c_{22}a_{21}) & (b_{21}a_{12} + b_{22}a_{22}) + (c_{21}a_{12} + c_{22}a_{22}) \end{bmatrix}$$

$$= \begin{bmatrix} b_{11}a_{11} + b_{12}a_{21} & b_{11}a_{12} + b_{12}a_{22} \\ b_{21}a_{11} + b_{22}a_{21} & b_{21}a_{12} + b_{22}a_{22} \end{bmatrix} + \begin{bmatrix} c_{11}a_{11} + c_{12}a_{21} & c_{11}a_{12} + c_{12}a_{22} \\ c_{21}a_{11} + c_{22}a_{21} & c_{21}a_{12} + c_{22}a_{22} \end{bmatrix}$$

$$= \begin{bmatrix} b_{11} & b_{12} \\ b_{21} & b_{22} \end{bmatrix} \cdot \begin{bmatrix} a_{11} & a_{12} \\ a_{21} & a_{22} \end{bmatrix} + \begin{bmatrix} c_{11} & c_{12} \\ c_{21} & c_{22} \end{bmatrix} \cdot \begin{bmatrix} a_{11} & a_{12} \\ a_{21} & a_{22} \end{bmatrix} = BA + CA.$$

31.
$$(A_{2 \times 2} \cdot B_{2 \times 2}) = \begin{bmatrix} a_{11}b_{11} + a_{12}b_{21} & a_{11}b_{12} + a_{12}b_{22} \\ a_{21}b_{11} + a_{22}b_{21} & a_{21}b_{12} + a_{22}b_{22} \end{bmatrix}$$

Hence,

$$(A_{2 \times 2} \cdot B_{2 \times 2})^t = \begin{bmatrix} a_{11}b_{11} + a_{12}b_{21} & a_{21}b_{11} + a_{22}b_{21} \\ a_{11}b_{12} + a_{12}b_{22} & a_{21}b_{12} + a_{22}b_{22} \end{bmatrix}.$$

By the commutative property of multiplication for real numbers,

$$(A_{2 \times 2} \cdot B_{2 \times 2})^t = \begin{bmatrix} b_{11}a_{11} + b_{21}a_{12} & b_{11}a_{21} + b_{21}a_{22} \\ b_{12}a_{11} + b_{22}a_{12} & b_{12}a_{21} + b_{22}a_{22} \end{bmatrix} = \begin{bmatrix} b_{11} & b_{21} \\ b_{12} & b_{22} \end{bmatrix} \cdot \begin{bmatrix} a_{11} & a_{21} \\ a_{12} & a_{22} \end{bmatrix} = B^t_{2 \times 2} \cdot A^t_{2 \times 2}.$$

33. Let $A = \begin{bmatrix} a_{11} & a_{12} \\ a_{21} & a_{22} \end{bmatrix}$ and c be a scalar; then by Definition 10.6 and the closure property of multiplication

of real numbers, $c\begin{bmatrix} a_{11} & a_{12} \\ a_{21} & a_{22} \end{bmatrix} = \begin{bmatrix} ca_{11} & ca_{12} \\ ca_{21} & ca_{22} \end{bmatrix}$, which is a 2×2 matrix.

35. By Definition 10.6 and the distributive property for real numbers and by Definition 10.3,

$$(c + d)\begin{bmatrix} a_{11} & a_{12} \\ a_{21} & a_{22} \end{bmatrix} = \begin{bmatrix} (c + d)a_{11} & (c + d)a_{12} \\ (c + d)a_{21} & (c + d)a_{22} \end{bmatrix} = \begin{bmatrix} ca_{11} + da_{11} & ca_{12} + da_{12} \\ ca_{21} + da_{21} & ca_{22} + da_{22} \end{bmatrix}$$

$$= \begin{bmatrix} ca_{11} & ca_{12} \\ ca_{21} & ca_{22} \end{bmatrix} + \begin{bmatrix} da_{11} & da_{12} \\ da_{21} & da_{22} \end{bmatrix} = c\begin{bmatrix} a_{11} & a_{12} \\ a_{21} & a_{22} \end{bmatrix} + d\begin{bmatrix} a_{11} & a_{12} \\ a_{21} & a_{22} \end{bmatrix} = cA + dA.$$

37. By Definitions 10.6 and 10.5, $(-1)A = -1\begin{bmatrix} a_{11} & a_{12} \\ a_{21} & a_{22} \end{bmatrix} = \begin{bmatrix} -a_{11} & -a_{12} \\ -a_{21} & -a_{22} \end{bmatrix} = -A.$

Exercise 10.3 (page 354)   **1.** $\{(2, -1)\}$ **3.** $\{(5, 1)\}$ **5.** $\{(8, 1)\}$ **7.** $\{(-2, 2, 0)\}$

9. $\left\{\left(\dfrac{5}{4}, \dfrac{5}{2}, -\dfrac{1}{2}\right)\right\}$ **11.** $\left\{\left(-\dfrac{77}{27}, -\dfrac{8}{27}, \dfrac{29}{27}\right)\right\}$

13. $\{(1, 1 - z, z)\}$ **15.** $(3, 3, -1)$ **17.** No solution   **19.** No solution

21. $\begin{bmatrix} k & 0 \\ 0 & 1 \end{bmatrix}\begin{bmatrix} a & b \\ c & d \end{bmatrix} = \begin{bmatrix} k\cdot a + 0\cdot c & k\cdot b + 0\cdot d \\ 0\cdot a + 1\cdot c & 0\cdot b + 1\cdot d \end{bmatrix} = \begin{bmatrix} ka & kb \\ c & d \end{bmatrix}$

23. $\begin{bmatrix} 0 & 1 \\ 1 & 0 \end{bmatrix}\begin{bmatrix} a & b \\ c & d \end{bmatrix} = \begin{bmatrix} 0\cdot a + 1\cdot c & 0\cdot b + 1\cdot d \\ 1\cdot a + 0\cdot c & 1\cdot b + 0\cdot d \end{bmatrix} = \begin{bmatrix} c & d \\ a & b \end{bmatrix}$

25. $\begin{bmatrix} 1 & 0 \\ k & 1 \end{bmatrix}\begin{bmatrix} a & b \\ c & d \end{bmatrix} = \begin{bmatrix} 1\cdot a + 0\cdot c & 1\cdot b + 0\cdot d \\ k\cdot a + 1\cdot c & k\cdot b + 1\cdot d \end{bmatrix} = \begin{bmatrix} a & b \\ ka + c & kb + d \end{bmatrix}$

Exercise 10.4 (page 359)   **1.** 3   **3.** 4   **5.** 0   **7.** -11

9. $M_{11} = \begin{vmatrix} 0 & 3 & -1 \\ 1 & 2 & 2 \\ -1 & 3 & 1 \end{vmatrix}$, $A_{11} = \begin{vmatrix} 0 & 3 & -1 \\ 1 & 2 & 2 \\ -1 & 3 & 1 \end{vmatrix}$

11. $M_{23} = \begin{vmatrix} 2 & 1 & 0 \\ -2 & 1 & 2 \\ 1 & -1 & 1 \end{vmatrix}$, $A_{23} = -\begin{vmatrix} 2 & 1 & 0 \\ -2 & 1 & 2 \\ 1 & -1 & 1 \end{vmatrix}$

13. $M'_{31} = \begin{vmatrix} 1 & -2 & 0 \\ 0 & 3 & -1 \\ -1 & 3 & 1 \end{vmatrix}$, $A_{31} = \begin{vmatrix} 1 & -2 & 0 \\ 0 & 3 & -1 \\ -1 & 3 & 1 \end{vmatrix}$

15. $M_{44} = \begin{vmatrix} 2 & 1 & -2 \\ 1 & 0 & 3 \\ -2 & 1 & 2 \end{vmatrix}$, $A_{44} = \begin{vmatrix} 2 & 1 & -2 \\ 1 & 0 & 3 \\ -2 & 1 & 2 \end{vmatrix}$

17. 16   **19.** -4 **21.** 0   **23.** -24 **25.** 4

27. Let $A = \begin{vmatrix} x & y & 1 \\ x_1 & y_1 & 1 \\ x_2 & y_2 & 1 \end{vmatrix} = 0$.

Expanding about the first row gives $\delta(A) = x\begin{vmatrix} y_1 & 1 \\ y_2 & 1 \end{vmatrix} - y\begin{vmatrix} x_1 & 1 \\ x_2 & 1 \end{vmatrix} + 1\begin{vmatrix} x_1 & y_1 \\ x_2 & y_2 \end{vmatrix} = 0$; hence $(y_1 - y_2)x +$ $(x_2 - x_1)y + (x_1 y_2 - y_1 x_2) = 0$. Further, y_1, y_2, x_1, x_2 are real numbers; hence there exist real numbers a, b, c such that $(y_1 - y_2) = a, (x_2 - x_1) = b$ and $(x_1 y_2 - y_1 x_2) = c$ with a and b not both 0 since $(x_1, y_1) \neq (x_2, y_2)$. Substituting yields $ax + by + c = 0$, which is the equation of a straight line. By Theorem 10.9, the line passes through (x_1, y_1) and (x_2, y_2).

29. If the ith row of A is identically 0, then expanding about the ith row we have

$$\delta(A) = 0\cdot A_{i1} + 0\cdot A_{i2} + \cdots + 0\cdot A_{in} = 0.$$

Similarly if the ith column is identically 0, then $\delta(A) = 0$.

31. Let $A = \begin{bmatrix} a_{11} & a_{12} \\ a_{21} & a_{22} \end{bmatrix}$; by Definition 10.6 and the fact that $\delta(A) = a_{11}a_{22} - a_{12}a_{21}$, it follows that

$aA = \begin{bmatrix} aa_{11} & aa_{12} \\ aa_{21} & aa_{22} \end{bmatrix}$ and $\delta(aA) = a^2 a_{11}a_{22} - a^2 a_{12}a_{21} = a^2(a_{11}a_{22} - a_{12}a_{21}) = a^2 \delta(A)$.

33. Let $A = \begin{bmatrix} a_{11} & a_{12} \\ a_{21} & a_{22} \end{bmatrix}$, $B = \begin{bmatrix} b_{11} & b_{12} \\ b_{21} & b_{22} \end{bmatrix}$; then

$$AB = \begin{bmatrix} a_{11}b_{11} + a_{12}b_{21} & a_{11}b_{12} + a_{12}b_{22} \\ a_{21}b_{11} + a_{22}b_{21} & a_{21}b_{12} + a_{22}b_{22} \end{bmatrix}. \text{ Thus}$$

$$\delta(AB) = (a_{11}b_{11} + a_{12}b_{21})(a_{21}b_{12} + b_{22}a_{22}) - (a_{11}b_{12} + a_{12}b_{22})(a_{21}b_{11} + a_{22}b_{21})$$
$$= a_{11}b_{11}a_{21}b_{12} + a_{12}b_{21}a_{21}b_{12} + a_{11}b_{11}a_{22}b_{22} + a_{12}b_{21}a_{22}b_{22}$$
$$\quad - a_{11}b_{12}a_{21}b_{11} - a_{12}b_{22}a_{21}b_{11} - a_{11}b_{12}a_{22}b_{21} - a_{12}b_{22}a_{22}b_{21}$$
$$= a_{11}b_{11}a_{22}b_{22} - a_{12}a_{21}b_{11}b_{22} - a_{11}a_{22}b_{21}b_{12} + a_{21}a_{12}b_{21}b_{12}$$
$$= b_{11}b_{22}(a_{11}a_{22} - a_{12}a_{21}) - b_{21}b_{12}(a_{11}a_{22} - a_{21}a_{12}) = (a_{11}a_{22} - a_{21}a_{12})(b_{11}b_{22} - b_{21}b_{12})$$
$$= \delta(A)\,\delta(B).$$

Exercise 10.5 (page 365)   **1.** Theorem 10.7   **3.** Theorem 10.9   **5.** Theorem 10.10
7. Theorem 10.10   **9.** Theorem 10.11   **11.** Theorem 10.11

13. $\begin{vmatrix} 1 & 3 \\ 0 & -4 \end{vmatrix}$ **15.** $\begin{vmatrix} 1 & -2 & 1 \\ 0 & 7 & 1 \\ 0 & 2 & 1 \end{vmatrix}$ **17.** $\begin{vmatrix} 0 & 1 & -3 & -2 \\ 0 & 2 & 1 & 2 \\ 1 & 1 & 2 & 3 \\ 0 & 1 & 1 & 1 \end{vmatrix}$ **19.** $-1 \begin{vmatrix} 2 & 1 \\ -1 & 2 \end{vmatrix} = -5$

21. $\begin{vmatrix} -1 & -5 \\ 2 & -2 \end{vmatrix} = 12$ **23.** $\begin{vmatrix} 4 & 4 \\ 3 & 7 \end{vmatrix} = 16$ **25.** $\begin{vmatrix} 6 & 1 \\ 0 & 3 \end{vmatrix} = 18$ **27.** $-16 \begin{vmatrix} 1 & 2 \\ 2 & 3 \end{vmatrix} = 16$

29. $\begin{vmatrix} 4 & -4 \\ 3 & -9 \end{vmatrix} = -24$

31. Multiply column 1 by $(-a)$ and add result to column 2; also, multiply column 1 by $(-a^2)$ and add result to column 3, to obtain $\begin{vmatrix} 1 & a & a^2 \\ 1 & b & b^2 \\ 1 & c & c^2 \end{vmatrix} = \begin{vmatrix} 1 & 0 & 0 \\ 1 & b-a & b^2 - a^2 \\ 1 & c-a & c^2 - a^2 \end{vmatrix}$. Expand about the first row to obtain

$$1 \begin{vmatrix} b-a & b^2 - a^2 \\ c-a & c^2 - a^2 \end{vmatrix} = (b-a)(c^2 - a^2) - (c-a)(b^2 - a^2)$$
$$= (b-a)[(c-a)(c+a)] - (c-a)[(b-a)(b+a)]$$
$$= -(a-b)[(c-a)(c+a)] + (c-a)[(a-b)(a+b)]$$
$$= (a-b)(c-a)[-(c+a) + (a+b)]$$
$$= (a-b)(c-a)(b-c) = (b-c)(c-a)(a-b).$$

33. If Rows i and j of the matrix A are identical, then interchanging the two rows results in the same matrix. Thus, by Theorem 10.8, $\delta(A) = -\delta(A)$ and hence $\delta(A) = 0$. A similar argument shows $\delta(A) = 0$ if two columns are identical.

Answers to Odd-Numbered Exercises

Exercise 10.6 (page 372) 1. $\begin{bmatrix} 3 & -2 \\ -1 & 1 \end{bmatrix}$ 3. $\frac{1}{5}\begin{bmatrix} 1 & 3 \\ -1 & 2 \end{bmatrix}$ 5. $|A| = 0$; no inverse

7. $-1\begin{bmatrix} 4 & -7 \\ -3 & 5 \end{bmatrix}$ 9. $\frac{1}{2}\begin{bmatrix} -2 & -4 \\ 4 & 7 \end{bmatrix}$ 11. $|A| = 0$; no inverse

13. $\frac{1}{6}\begin{bmatrix} 2 & 2 & -5 \\ -4 & 2 & 1 \\ 0 & 0 & 3 \end{bmatrix}$ 15. $\frac{1}{3}\begin{bmatrix} -2 & 3 & -1 \\ -1 & 0 & 1 \\ 6 & -6 & 3 \end{bmatrix}$ 17. $|A| = 0$; no inverse

19. $|A| = 0$; no inverse 21. $\begin{bmatrix} 2 & -3 & 11 \\ -2 & 4 & -13 \\ 1 & -2 & 7 \end{bmatrix}$ 23. $-1\begin{bmatrix} 0 & 0 & -1 \\ 0 & -1 & 0 \\ -1 & 0 & 0 \end{bmatrix}$

25. Let $A \cdot B = \begin{bmatrix} 2 & 3 \\ 1 & -1 \end{bmatrix} \cdot \begin{bmatrix} 0 & 1 \\ 3 & 1 \end{bmatrix}$; $A \cdot B = \begin{bmatrix} 9 & 5 \\ -3 & 0 \end{bmatrix}$ and $\delta(AB) = 15$,

so $(A \cdot B)^{-1} = \frac{1}{15}\begin{bmatrix} 0 & -5 \\ 3 & 9 \end{bmatrix}$; also, since $\delta(A) = -5$ and $\delta(B) = -3$,

$B^{-1} = -\frac{1}{3}\begin{bmatrix} 1 & -1 \\ -3 & 0 \end{bmatrix}$ and $A^{-1} = \frac{1}{5}\begin{bmatrix} -1 & -3 \\ -1 & 2 \end{bmatrix}$; $B^{-1} \cdot A^{-1} = \frac{1}{15}\begin{bmatrix} 0 & -5 \\ 3 & 9 \end{bmatrix}$; hence $(A \cdot B)^{-1} = B^{-1} \cdot A^{-1}$.

27. Let $A = \begin{bmatrix} 3 & 0 & 1 \\ 2 & 1 & 0 \\ 0 & 1 & 2 \end{bmatrix}$; then $\delta(A) = 8$ and $A^{-1} = \frac{1}{8}\begin{bmatrix} 2 & 1 & -1 \\ -4 & 6 & 2 \\ 2 & -3 & 3 \end{bmatrix}$.

Let $B = \begin{bmatrix} 2 & 1 & 0 \\ 1 & 1 & 2 \\ 0 & 1 & 0 \end{bmatrix}$; then $\delta(B) = -\frac{1}{4}$ and $B^{-1} = -\frac{1}{4}\begin{bmatrix} -2 & 0 & 2 \\ 0 & 0 & -4 \\ 1 & -2 & 1 \end{bmatrix}$;

$A \cdot B = \begin{bmatrix} 6 & 4 & 0 \\ 5 & 3 & 2 \\ 1 & 3 & 2 \end{bmatrix}$, $\delta(A \cdot B) = -32$, and $(A \cdot B)^{-1} = -\frac{1}{32}\begin{bmatrix} 0 & -8 & 8 \\ -8 & 12 & -12 \\ 12 & -14 & -2 \end{bmatrix}$;

$B^{-1} \cdot A^{-1} = -\frac{1}{4}\begin{bmatrix} -2 & 0 & 2 \\ 0 & 0 & -4 \\ 1 & -2 & 1 \end{bmatrix} \cdot \frac{1}{8}\begin{bmatrix} 2 & 1 & -1 \\ -4 & 6 & 2 \\ 2 & -3 & 3 \end{bmatrix} = -\frac{1}{32}\begin{bmatrix} 0 & -8 & 8 \\ -8 & 12 & -12 \\ 12 & -14 & -2 \end{bmatrix}$;

hence $(A \cdot B)^{-1} = B^{-1} \cdot A^{-1}$.

29. Let $A = \begin{bmatrix} a_{11} & a_{12} \\ a_{21} & a_{22} \end{bmatrix}$; then $\delta(A) = (a_{11}a_{22} - a_{12}a_{21}) \neq 0$, since A is nonsingular, and hence $1/\delta(A)$ is defined and A^{-1} exists.

$$A^{-1} = \frac{1}{\delta(A)}\begin{bmatrix} a_{22} & -a_{12} \\ -a_{21} & a_{11} \end{bmatrix} = \begin{bmatrix} \frac{a_{22}}{\delta(A)} & \frac{-a_{12}}{\delta(A)} \\ \frac{-a_{21}}{\delta(A)} & \frac{a_{11}}{\delta(A)} \end{bmatrix}$$

and $\delta(A^{-1}) = \frac{a_{22}a_{11}}{[\delta(A)]^2} - \frac{a_{12}a_{21}}{[\delta(A)]^2} = \frac{a_{22}a_{11} - a_{12}a_{21}}{[\delta(A)]^2} = \frac{\delta(A)}{[\delta(A)]^2} = \frac{1}{\delta(A)}$.

31. From the results of Exercise 33 on page 362, and Exercise 29 above,

$$\delta[B^{-1}AB] = \delta[B^{-1}(AB)] = \delta(B^{-1}) \cdot \delta(AB) = \delta(B^{-1}) \cdot \delta(A) \cdot \delta(B) = \frac{1}{\delta(B)} \cdot \delta(A) \cdot \delta(B) = \delta(A).$$

Exercise 10.7 (page 377) 1. $\{(1, 1)\}$ 3. No solution   5. $\{(6, 4)\}$ 7. $\left\{\left(-\dfrac{18}{7}, \dfrac{19}{7}\right)\right\}$
9. $\left(x, -\dfrac{1}{3}x + \dfrac{2}{9}\right)$, $x \in R$ 11. $\{(2, 2)\}$ 13. $\{(1, 1, 1)\}$ 15. $\{(1, 1, 0)\}$ 17. $\{(1, -2, 3)\}$
19. $\{(3, -1, -2)\}$ 21. a. $\left\{\left(\dfrac{11}{6}, \dfrac{5}{3}, -\dfrac{1}{2}\right)\right\}$ b. $\left\{\left(-\dfrac{13}{12}, -\dfrac{1}{6}, -\dfrac{3}{4}\right)\right\}$ c. $\left\{\left(\dfrac{5}{6}, -\dfrac{1}{3}, \dfrac{1}{2}\right)\right\}$

Exercise 10.8 (page 381) 1. $\left\{\left(\dfrac{13}{5}, \dfrac{3}{5}\right)\right\}$ 3. $\left\{\left(\dfrac{22}{7}, \dfrac{20}{7}\right)\right\}$ 5. $\{(6, 4)\}$ 7. Inconsistent
9. $\{(4, 1)\}$ 11. $\left\{\left(\dfrac{1}{a+b}, \dfrac{1}{a+b}\right)\right\}$ $(a \neq -b)$ 13. $\{(1, 1, 0)\}$ 15. $\{(1, -2, 3)\}$
17. $\{(3, -1, -2)\}$ 19. $\left\{\left(-\dfrac{1}{3}, -\dfrac{25}{24}, -\dfrac{5}{8}\right)\right\}$ 21. $\left\{\left(1, -\dfrac{1}{3}, \dfrac{1}{2}\right)\right\}$
23. $\{(w, x, y, z)\} = \{(2, -1, 1, 0)\}$
25. $|A| = \begin{vmatrix} a_1 & b_1 \\ a_2 & b_2 \end{vmatrix}$ and $\delta(A) = a_1 b_2 - a_2 b_1$;

$|A_y| = \begin{vmatrix} a_1 & c_1 \\ a_2 & c_2 \end{vmatrix}$ and $\delta(A_y) = a_1 c_2 - a_2 c_1 = 0$, so $a_1 c_2 = a_2 c_1$;

$|A_x| = \begin{vmatrix} c_1 & b_1 \\ c_2 & b_2 \end{vmatrix}$ and $\delta(A_x) = b_2 c_1 - b_1 c_2 = 0$, so $b_1 c_2 = b_2 c_1$;

hence, $\dfrac{a_1 c_2}{b_1 c_2} = \dfrac{a_2 c_1}{b_2 c_1}$; $\dfrac{a_1}{b_1} = \dfrac{a_2}{b_2}$ and $a_1 b_2 = a_2 b_1$, so $a_1 b_2 - a_2 b_1 = 0$; therefore $\delta(A) = 0$.

Chapter 10 Review (page 382) 1. $\begin{bmatrix} 1 & -1 \\ 1 & 1 \end{bmatrix}$ 2. $\begin{bmatrix} 2 & 5 & -2 \\ 14 & -1 & 12 \end{bmatrix}$ 3. $\begin{bmatrix} -2 & -3 & 1 \\ 3 & -7 & 6 \\ -5 & 2 & 5 \end{bmatrix}$
4. $\begin{bmatrix} -8 & -3 & -6 \\ 6 & -3 & 0 \\ -9 & 5 & 5 \end{bmatrix}$ 5. $\begin{bmatrix} -21 & 7 \\ -14 & 0 \\ -7 & -7 \end{bmatrix}$ 6. $[13]$ 7. $\begin{bmatrix} -13 & 3 \\ -19 & 27 \end{bmatrix}$ 8. $\begin{bmatrix} 18 & 7 & 25 \\ 8 & -1 & 11 \\ 3 & 0 & 3 \end{bmatrix}$
9. $\{(2, -1)\}$ 10. $\{(2, 1, 1)\}$ 11. -3 12. 7   13. -3 14. 14   15. 6   16. -1
17. 15   18. $-1{,}578$ 19. $\dfrac{1}{34}\begin{bmatrix} -3 & 2 \\ 11 & 4 \end{bmatrix}$ 20. $\dfrac{1}{6}\begin{bmatrix} 1 & 3 & -2 \\ -3 & -3 & 6 \\ 1 & -3 & 4 \end{bmatrix}$ 21. $\{(-2, 1)\}$
22. $\{(3, -1, 1)\}$ 23. a. $\left\{\left(\dfrac{1}{2}, -\dfrac{1}{2}\right)\right\}$ b. $\left\{\left(\dfrac{1}{4}, -\dfrac{7}{4}\right)\right\}$ c. $\{(1, -2)\}$ 24. $\left\{\left(-\dfrac{16}{7}, -\dfrac{13}{7}\right)\right\}$
25. $\{(2, -1, 0)\}$

Exercise 11.1 (page 388) 1. $-4, -3, -2, -1$ 3. $-\dfrac{1}{2}, 1, \dfrac{7}{2}, 7$ 5. $2, \dfrac{3}{2}, \dfrac{4}{3}, \dfrac{5}{4}$ 7. $0, 1, 3, 6$
9. $-1, 1, -1, 1$ 11. $1, 0, -\dfrac{1}{3}, \dfrac{1}{2}$ 13. $11, 15, 19$ 15. $2, 5, 8$ 17. $x + 2, x + 3, x + 4$

19. $2x + 7, 2x + 10, 2x + 13$ **21.** $32, 128, 512, 2{,}048$ **23.** $\dfrac{8}{3}, \dfrac{16}{3}, \dfrac{32}{3}, \dfrac{64}{3}$ **25.** $\dfrac{x}{a}, -\dfrac{x^2}{a^2}, \dfrac{x^3}{a^3}, -\dfrac{x^4}{a^4}$

27. $\dfrac{x^2}{-4}, \dfrac{x^3}{16}, \dfrac{x^4}{-64}$ **29.** $4n + 3, 31$ **31.** $8 - 5n, -92$ **33.** $(4n + 1)x - 4x, 29x$

35. $48(2)^{n-1}, 1{,}536$ **37.** $-\dfrac{1}{3}(-3)^{n-1}, -243$ **39.** $x + (n-1)(x^3 - x); -3x + 4x^3$

41. $x + (n-1)(y - x); 5y - 4x$ **43.** $\dfrac{y^{n-1}}{x^{n-2}}; \dfrac{y^7}{x^6}$ **45.** $2; 3; 41$ **47.** 28th   **49.** 3

Exercise 11.2 (page 394)   **1.** $1 + 4 + 9 + 16$ **3.** $-\dfrac{1}{2} + \dfrac{1}{4} - \dfrac{1}{8}$ **5.** $-5 - 3 - 1 + 1 + 3 + 5 + 7$

7. $1 + \dfrac{1}{2} + \dfrac{1}{4} + \cdots$ **9.** $\sum_{j=1}^{4} x^{2j-1}$ **11.** $\sum_{j=1}^{5} j^2$ **13.** $\sum_{j=1}^{\infty} j(j+1)$ **15.** $\sum_{j=1}^{\infty} \dfrac{j+1}{j}$ **17.** 63

19. 806   **21.** -6 **23.** 10,000   **25.** 1,092   **27.** $\dfrac{31}{32}$ **29.** $\dfrac{364}{2{,}187}$ **31.** 2,816   **33.** 168

35. 2,040   **37.** 196   **39.** $\$2{,}333.39$ **41.** $\dfrac{3}{4}, \dfrac{7}{8}, \dfrac{15}{16}, \dfrac{31}{32}; 1$ **43.** $p = 4, q = -3$

Exercise 11.3 (page 401)   **1.** $\lim_{n \to \infty} s_n = 0$ **3.** $\lim_{n \to \infty} s_n = 1$ **5.** $\lim_{n \to \infty} s_n$ is undefined   **7.** $\lim_{n \to \infty} s_n = 0$

9. Convergent, $\lim_{n \to \infty} \left|0 - \dfrac{1}{2^n}\right| = 0$ **11.** Divergent, $\lim_{n \to \infty} n$ is undefined

13. Convergent, $\lim_{n \to \infty} \left|0 - (-1)^{n+1}\dfrac{1}{2^{n-1}}\right| = 0$ **15.** 24   **17.** No sum   **19.** 2   **21.** $\dfrac{8}{9}$

23. $\dfrac{31}{99}$ **25.** $2\dfrac{410}{999}$ **27.** $\dfrac{29}{225}$ **29.** 20 cm

Exercise 11.4 (page 407)   **1.** $8 \cdot 7 \cdot 6 \cdot 5 \cdot 4 \cdot 3 \cdot 2 \cdot 1$ **3.** $6 \cdot 5 \cdot 4 \cdot 3 \cdot 2 \cdot 1$ **5.** $5 \cdot 4 \cdot 3 \cdot 2 \cdot 1 = 120$

7. $\dfrac{9 \cdot 8 \cdot 7!}{7!} = 72$ **9.** $\dfrac{5 \cdot 4 \cdot 3 \cdot 2 \cdot 1 \cdot 7!}{8 \cdot 7!} = 15$ **11.** $\dfrac{8 \cdot 7 \cdot 6!}{2 \cdot 1 \cdot 6!} = 28$ **13.** $3!$ **15.** $\dfrac{6!}{2!}$ **17.** $\dfrac{8!}{5!}$

19. $\dfrac{6!}{5!1!} = 6$ **21.** $\dfrac{3!}{3!0!} = 1$ **23.** $\dfrac{7!}{0!7!} = 1$ **25.** $\dfrac{5!}{2!3!} = 10$

27. $(n)(n-1)(n-2) \cdot \cdots \cdot 3 \cdot 2 \cdot 1$ **29.** $(3n)(3n-1)(3n-2) \cdot \cdots \cdot 3 \cdot 2 \cdot 1$

31. $(n-2)(n-3)(n-4) \cdot \cdots \cdot 3 \cdot 2 \cdot 1$ **33.** $(n+2)(n+1)$ **35.** $\dfrac{n+1}{n+3}$ **37.** $\dfrac{2n-1}{2n-2}$

39. $x^5 + 15x^4 + 90x^3 + 270x^2 + 405x + 243$ **41.** $x^4 - 12x^3 + 54x^2 - 108x + 81$

43. $8x^3 - 6x^2y + \dfrac{3}{2}xy^2 - \dfrac{1}{8}y^3$ **45.** $\dfrac{1}{64}x^6 + \dfrac{3}{8}x^5 + \dfrac{15}{4}x^4 + 20x^3 + 60x^2 + 96x + 64$

47. $x^{20} + 20x^{19}y + \dfrac{20 \cdot 19}{2!}x^{18}y^2 + \dfrac{20 \cdot 19 \cdot 18}{3!}x^{17}y^3$, or $\binom{20}{0}x^{20} + \binom{20}{1}x^{19}y + \binom{20}{2}x^{18}y^2 + \binom{20}{3}x^{17}y^3$

49. $a^{12} + 12a^{11}(-2b) + \dfrac{12 \cdot 11}{2!} a^{10}(-2b)^2 + \dfrac{12 \cdot 11 \cdot 10}{3!} a^9(-2b)^3$, or

$\binom{12}{0}a^{12} + \binom{12}{1}a^{11}(-2b) + \binom{12}{2}a^{10}(-2b)^2 + \binom{12}{3}a^9(-2b)^3$

51. $x^{10} + 10x^9(-\sqrt{2}) + \dfrac{10 \cdot 9}{2!} x^8(-\sqrt{2})^2 + \dfrac{10 \cdot 9 \cdot 8}{3!} x^7(-\sqrt{2})^3$, or

$\binom{10}{0}x^{10} + \binom{10}{1}x^9(-\sqrt{2}) + \binom{10}{2}x^8(-\sqrt{2})^2 + \binom{10}{3}x^7(-\sqrt{2})^3$ **53.** 1.22   **55.** $1,480

57. $-3003a^{10}b^5$ **59.** $3360x^6y^4$ **61.** **a.** $1 - x + x^2 - x^3 + \cdots$ **b.** $1 - x + x^2 - x^3 + \cdots$

63. $\binom{k}{r} + \binom{k}{r-1} = \dfrac{k!}{r!(k-r)!} + \dfrac{k!}{(r-1)!(k-r+1)!}$

$= \dfrac{k!(k-r+1) + k!r}{r!(k-r+1)!} = \dfrac{k!(k+1)}{r!(k-r+1)!}$

$= \dfrac{(k+1)!}{r![(k+1)-r]!} = \binom{k+1}{r}$

Exercise 11.5 (page 414)   **1.**   **a.** For $n = 1$, $\dfrac{n}{2} = \dfrac{1}{2}$; $\dfrac{n(n+1)}{4} = \dfrac{1(1+1)}{4} = \dfrac{1}{2}$.

b. For $n = k$, $\dfrac{1}{2} + \dfrac{2}{2} + \dfrac{3}{2} + \cdots + \dfrac{k}{2} = \dfrac{k(k+1)}{4}$ and $(k+1)$st term $= \dfrac{k+1}{2}$;

hence $\dfrac{1}{2} + \dfrac{2}{2} + \dfrac{3}{2} + \cdots + \dfrac{k}{2} + \dfrac{k+1}{2} = \dfrac{k(k+1)}{4} + \dfrac{k+1}{2} = \dfrac{k^2 + k + 2k + 2}{4}$

$= \dfrac{k^2 + 3k + 2}{4} = \dfrac{(k+1)(k+2)}{4}$.

3.   **a.** For $n = 1$, $2n = 2(1) = 2$; $n(n+1) = 1(1+1) = 2$.

b. For $n = k$, $2 + 4 + 6 + \cdots + 2k = k(k+1)$ and $(k+1)$st term is $2(k+1)$;

hence $2 + 4 + 6 + \cdots + 2k + 2(k+1) = k(k+1) + 2(k+1)$

$= (k+1)(k+2)$.

5.   **a.** For $n = 1$, $6n - 4 = 6(1) - 4 = 2$; $n(3n - 1) = 1[3(1) - 1] = 2$.

b. For $n = k$, $2 + 8 + 14 + \cdots + (6k - 4) = k(3k - 1)$ and $(k+1)$st term is $6(k+1) - 4$;

hence $2 + 8 + 14 + \cdots + (6k - 4) + [6(k+1) - 4] = k(3k - 1) + 6k + 2 = 3k^2 + 5k + 2$

$= (k+1)(3k+2) = (k+1)[3(k+1) - 1]$.

7.   **a.** For $n = 1$, $3n + 4 = 3(1) + 4 = 7$; $\dfrac{n(3n+11)}{2} = \dfrac{1[3(1) + 11]}{2} = 7$.

b. For $n = k$, $7 + 10 + 13 + \cdots + (3k + 4) = \dfrac{k(3k+11)}{2}$ and $(k+1)$st term is $3(k+1) + 4$;

hence $7 + 10 + 13 + \cdots + (3k + 4) + [3(k+1) + 4] = \dfrac{k(3k+11)}{2} + 3k + 7 = \dfrac{3k^2 + 17k + 14}{2}$

$= \dfrac{(k+1)(3k+14)}{2} = \dfrac{(k+1)[3(k+1) + 11]}{2}$.

Answers to Odd-Numbered Exercises 551

9. a. For $n = 1$, $n^2 = 1^2 = 1$; $\dfrac{n(n + 1)(2n + 1)}{6} = \dfrac{1(2)(3)}{6} = 1$.

   b. For $n = k$, $1^2 + 2^2 + 3^2 + \cdots + k^2 = \dfrac{k(k + 1)(2k + 1)}{6}$ and $(k + 1)$st term is $(k + 1)^2$;

hence $1^2 + 2^2 + 3^2 + \cdots + k^2 + (k + 1)^2 = \dfrac{k(k + 1)(2k + 1)}{6} + (k + 1)^2$

$= \dfrac{k(k + 1)(2k + 1) + 6(k + 1)^2}{6} = \dfrac{(k + 1)[k(2k + 1) + 6(k + 1)]}{6} = \dfrac{(k + 1)(2k^2 + 7k + 6)}{6}$

$= \dfrac{(k + 1)(k + 2)(2k + 3)}{6} = \dfrac{(k + 1)[(k + 1) + 1][2(k + 1) + 1]}{6}.$

11. a. For $n = 1$, $(2n - 1)^3 = (2 - 1)^3 = 1^3 = 1$; $n^2(2n^2 - 1) = 1(2 - 1) = 1(1) = 1$.

   b. For $n = k$, $1^3 + 3^3 + 5^3 + \cdots + (2k - 1)^3 = k^2(2k^2 - 1)$ and the $(k + 1)$st term is $[2(k + 1) - 1]^3 = (2k + 1)^3$; hence

$1^3 + 3^3 + 5^3 + \cdots + (2k - 1)^3 + (2k + 1)^3 = k^2(2k^2 - 1) + (2k - 1)^3 = 2k^4 + 8k^3 + 11k^2 + 6k + 1;$

by use of the factor theorem and synthetic division;

$$2k^4 + 8k^3 + 11k^2 + 6k + 1 = (k + 1)(k + 1)(2k^2 + 4k + 1);$$

also, $2k^2 + 4k + 1 = 2(k^2 + 2k + 1) - 2 + 1 = 2(k + 1)^2 - 1$;
hence, $2k^4 + 8k^3 + 11k^2 + 6k + 1 = (k + 1)^2[2(k + 1)^2 - 1]$.

13. a. For $n = 1$, $n(n + 1) = 1(2) = 2$; $\dfrac{n(n + 1)(n + 2)}{3} = \dfrac{1(2)(3)}{3} = 2$.

   b. For $n = k$, $1 \cdot 2 + 2 \cdot 3 + 3 \cdot 4 + \cdots + k(k + 1) = \dfrac{k(k + 1)(k + 2)}{3}$ and the $(k + 1)$st term is

$(k + 1)[(k + 1) + 1](k + 1)(k + 2)$; hence

$1 \cdot 2 + 2 \cdot 3 + 3 \cdot 4 + \cdots + k(k + 1) + [(k + 1)(k + 2)]$

$= \dfrac{k(k + 1)(k + 2)}{3} + (k + 1)(k + 2) = \dfrac{[k(k + 1)(k + 2)] + [3(k + 1)(k + 2)]}{3}$

$= \dfrac{(k + 1)(k + 2)(k + 3)}{3} = \dfrac{(k + 1)[(k + 1) + 1][(k + 1) + 2]}{3}.$

15. a. For $n = 1$, $1^3 + 2 \cdot 1 = 3$; 3 is divisible by 3.
   b. For $n = k$, assume $k^3 + 2k$ is divisible by 3; for $n = k + 1$,

$$(k + 1)^3 + 2(k + 1) = k^3 + 3k^2 + 3k + 1 + 2k + 2$$
$$= (k^3 + 2k) + (3k^2 + 3k + 3).$$

Since by hypothesis $k^3 + 2k$ is divisible by 3 and since $3k^2 + 3k + 3$ is divisible by 3, $(k + 1)^3 + 2(k + 1)$ is divisible by 3.

17. For $n = k$, $2 + 4 + 6 + \cdots + 2k = k(k + 1) + 2$ and $(k + 1)$st term is $2(k + 1)$;
hence

$2 + 4 + 6 + \cdots + 2k + 2(k + 1) = k(k + 1) + 2 + 2(k + 1) = (k^2 + 3k + 2) + 2$

$= (k + 1)(k + 2) + 2 = (k + 1)[(k + 1) + 1] + 2.$

However, for $n = 1$, $2n = 2(1) = 2$; $n(n + 1) + 2 = 1(2) + 2 = 4$. Hence it is not true for every $n \in N$.

19. Since $s_1 = a$ and $s_{n+1} = s_n + d$, it follows that
  a. For $n = 1$, $s_n = s_1 = a = a + 0 \cdot d = a + (n-1)d$.
  b. For $n = k$, $s_k = a + (k-1)d$; now
  $$s_{k+1} = s_k + d = a + (k-1)d + d = a + kd = a + [(k+1)-1]d.$$

21. Since $s_n = ar^{n-1}$ and $S_{n+1} = S_n + s_{n+1}$, it follows that
  a. For $n = 1$, $S_1 = s_1 = a = a \dfrac{1-r}{1-r} = \dfrac{a - ar^1}{1-r}$.
  b. For $n = k$, $S_k = \dfrac{a - ar^k}{1-r}$ and $s_{k+1} = ar^k$; now
  $$S_{k+1} = S_k + s_{k+1} = \dfrac{a - ar^k}{1-r} + ar^k = \dfrac{a - ar^k + ar^k - ar^{k+1}}{1-r} = \dfrac{a - ar^{k+1}}{1-r}.$$

Chapter 11 Review (page 415)   **1.** 2, 5, 10   **2.** $\dfrac{1}{2}, \dfrac{1}{3}, \dfrac{1}{4}$ **3.** 13, 16, 19   **4.** $a-4, a-6, a-8$
5. $-18, 54, -162$ **6.** $\dfrac{3}{2}, \dfrac{9}{4}, \dfrac{27}{8}$ **7.** $s_n = 5n - 8$; $s_7 = 27$ **8.** $s_n = (-2)\left(\dfrac{-1}{3}\right)^{n-1}$; $s_5 = \dfrac{-2}{81}$
9. 25   **10.** Sixth term   **11.** $2 + 6 + 12 + 20$ **12.** $\displaystyle\sum_{k=1}^{\infty} x^{k+1}$ **13.** 119   **14.** $\dfrac{121}{243}$ **15.** 3
16. $\dfrac{8}{3}$ **17.** $\dfrac{1}{2}$ **18.** $\dfrac{4}{9}$ **19.** $5 \cdot (2 \cdot 1)$ **20.** 48   **21.** 21   **22.** $\dfrac{1}{n(n+1)!}$
23. $x^{10} - 20x^9 y + 180x^8 y^2 - 960x^7 y^3$ **24.** $-15{,}360x^3 y^7$
25. Formula holds for 1. Assume formula holds for n; test for $n+1$:
$$3 + 6 + 9 + \cdots + 3n + 3(n+1) = \dfrac{3n(n+1)}{2} + 3(n+1).$$
Right-hand member is equivalent to
$$\dfrac{3n^2 + 3n}{2} + \dfrac{6(n+1)}{2} = \dfrac{3n^2 + 9n + 6}{2} = \dfrac{3(n^2 + 3n + 2)}{2} = \dfrac{3(n+1)(n+2)}{2} = \dfrac{3(n+1)[(n+1)+1]}{2}.$$
26. Formula holds for 1. Assume formula holds for n; test for $n+1$:
$$\dfrac{1}{2} + \dfrac{1}{4} + \dfrac{1}{8} + \cdots + \dfrac{1}{2^n} + \dfrac{1}{2^{n+1}} = 1 - \dfrac{1}{2^n} + \dfrac{1}{2^{n+1}}.$$
Right-hand member is equivalent to $\dfrac{2^{n+1} - 2 + 1}{2^{n+1}} = \dfrac{2^{n+1} - 1}{2^{n+1}} = 1 - \dfrac{1}{2^{n+1}}$.

Supplemental Exercises, Chapters 9–11 (page 417)   **1.** $a=1, b=-2, c=2, d=1$
2. $a=2, b=3, c=-2, d=2, e=-1$ **3.** $\left\{\left(-\dfrac{28}{3}, \dfrac{20}{3}, 2, -\dfrac{4}{3}\right)\right\}$ **4.** $\{(2, -1, 3, -1)\}$
5. $\{(1, -1, 4), (-1, -1, 4)\}$ **6.** $\left\{\left(2, \sqrt{3}, \dfrac{5}{4}\right), \left(-2, \sqrt{3}, \dfrac{5}{4}\right), \left(2, -\sqrt{3}, \dfrac{5}{4}\right), \left(-2, -\sqrt{3}, \dfrac{5}{4}\right)\right\}$
7. $(-7z-1, 5z+1, z),\ z \in R$ **8.** $(2y-4, y, 2y-1),\ y \in R$

9. Let $A = \begin{bmatrix} a_1 & 0 & 0 & 0 \\ 0 & a_2 & 0 & 0 \\ 0 & 0 & a_3 & 0 \\ 0 & 0 & 0 & a_4 \end{bmatrix}$; then expanding all minors about the first row we have

$$\delta(A) = a_1 \begin{vmatrix} a_2 & 0 & 0 \\ 0 & a_3 & 0 \\ 0 & 0 & a_4 \end{vmatrix} = a_1 a_2 \begin{vmatrix} a_3 & 0 \\ 0 & a_4 \end{vmatrix} = a_1 a_2 a_3 a_4.$$

This result generalizes to $n \times n$ diagonal matrices.

10. This proof is identical to Exercise 9 except we expand all minors about the first column instead of the first row. The result generalizes to $n \times n$ upper-triangular matrices.

11. Let $A = \begin{bmatrix} a_{11} & a_{12} & a_{13} \\ 0 & a_{22} & a_{23} \\ 0 & 0 & a_{33} \end{bmatrix}$; then $A_{12} = A_{13} = A_{23} = 0$ because each of the minors used to compute these cofactors has a zero row or zero column. Now

$$A^{-1} = \frac{1}{\delta(A)} \begin{bmatrix} A_{11} & A_{12} & A_{13} \\ A_{21} & A_{22} & A_{23} \\ A_{31} & A_{32} & A_{33} \end{bmatrix}^t = \frac{1}{\delta(A)} \begin{bmatrix} A_{11} & 0 & 0 \\ A_{21} & A_{22} & 0 \\ A_{31} & A_{32} & A_{33} \end{bmatrix}^t$$

$$= \frac{1}{\delta(A)} \begin{bmatrix} A_{11} & A_{21} & A_{31} \\ 0 & A_{22} & A_{32} \\ 0 & 0 & A_{33} \end{bmatrix},$$

which is upper triangular. The result generalizes to nonsingular, $n \times n$, upper-triangular matrices.

12. $a_1 \neq 0$, $a_2 \neq 0$, $a_3 \neq 0$, $a_4 \neq 0$; $A^{-1} = \begin{bmatrix} \frac{1}{a_1} & 0 & 0 & 0 \\ 0 & \frac{1}{a_2} & 0 & 0 \\ 0 & 0 & \frac{1}{a_3} & 0 \\ 0 & 0 & 0 & \frac{1}{a_4} \end{bmatrix}.$

This result generalizes to $n \times n$ diagonal matrices.

13. Note $\delta(A)\delta(B) = \delta(AB) = \delta(I) = 1$, thus $\delta(A) \neq 0$ and A is nonsingular. Thus, $B = A^{-1}$ and $BA = I$.

14. Note $1 = \delta(I_{4 \times 4}) = \delta(A_{4 \times 4}^{-1} A_{4 \times 4}) = \delta(A_{4 \times 4}^{-1})\delta(A_{4 \times 4})$. Hence, $\delta(A_{4 \times 4}^{-1}) = \dfrac{1}{\delta(A_{4 \times 4})}$.

15. $\dfrac{2}{3}$ **16.** 0 **17.** 2 **18.** 1 **19.** 2 **20.** $\dfrac{1}{12}$

21. a. For $n = 1$, $n(n + 1)(n + 2) = 6$ and 3 is a factor of 6. **b.** For $n = k$, 3 is a factor of $k(k + 1)(k + 2)$; i.e., there is $p \in N$ so that $k(k + 1)(k + 2) = 3p$. For $n = k + 1$,
$(k + 1)(k + 1 + 1)(k + 1 + 2) = (k + 1)(k + 2)(k + 3) = k(k + 1)(k + 2) + 3(k + 1)(k + 2)$
$= 3p + 3(k + 1)(k + 2) = 3[p + (k + 1)(k + 2)].$
Since $p + (k + 1)(k + 2) \in N$, 3 is a factor of $(k + 1)(k + 2)(k + 3)$.

22. a. For $n = 2$, $I_{2 \times 2} = \begin{bmatrix} 1 & 0 \\ 0 & 1 \end{bmatrix}$ and $\delta(I_{2 \times 2}) = (1)(1) - (0)(0) = 1$.

b. For $n = k$, $\delta(I_{k \times k}) = 1$. For $n = k + 1$, we expand the determinant of $I_{(k+1) \times (k+1)}$ about the first row and note that $\delta(I_{(k+1) \times (k+1)}) = (1)(-1)^{1+1}\delta(I_k) = 1$.

23. a. For $n = 1$, X_0 satisfies $AX = \lambda X$ by hypothesis.

b. For $n = k$, X_0 satisfies $A^k X = \lambda^k X$, i.e., $A^k X_0 = \lambda^k X_0$. For $n = k + 1$,

$$A^{k+1} X_0 = A(A^k X_0) = A(\lambda^k X_0) = \lambda^k A X_0 = \lambda^k (\lambda X_0) = \lambda^{k+1} X_0.$$

Thus, X_0 satisfies $A^{k+1} X = \lambda^{k+1} X$.

24. a. For $n = 1$, $(x + y)^1 = \sum_{j=0}^{1} \binom{1}{j} x^j y^{1-j}$.

b. For $n = k$, $(x + y)^k = \sum_{j=0}^{k} \binom{k}{j} x^j y^{k-j}$. For $n = k + 1$,

$$(x + y)^{k+1} = (x + y)(x + y)^k$$

$$= (x + y) \sum_{j=0}^{k} \binom{k}{j} x^j y^{k-j}$$

$$= x \sum_{j=0}^{k} \binom{k}{j} x^j y^{k-j} + y \sum_{j=0}^{k} \binom{k}{j} x^j y^{k-j}$$

$$= x \left[x^k + \sum_{j=0}^{k-1} \binom{k}{j} x^j y^{k-j} \right] + y \left[y^k + \sum_{j=1}^{k} \binom{k}{j} x^j y^{k-j} \right]$$

$$= x^{k+1} + x \sum_{j=0}^{k-1} \binom{k}{j} x^j y^{k-j} + y \sum_{j=1}^{k} \binom{k}{j} x^j y^{k-j} + y^{k+1}$$

$$= x^{k+1} + \sum_{j=0}^{k-1} \binom{k}{j} x^{j+1} y^{k-j} + \sum_{j=1}^{k} \binom{k}{j} x^j y^{k+1-j} + y^{k+1}$$

$$= x^{k+1} + \sum_{r=1}^{k} \binom{k}{r-1} x^r y^{k+1-r} + \sum_{r=1}^{k} \binom{k}{r} x^r y^{k+1-r} + y^{k+1}$$

$$= x^{k+1} + \sum_{r=1}^{k} \left[\binom{k}{r-1} + \binom{k}{r} \right] x^r y^{k+1-r} + y^{k+1}$$

$$= x^{k+1} + \sum_{r=1}^{k} \binom{k+1}{r} x^r y^{k+1-r} + y^{k+1} \text{ (by Exercise 63 on page 411)}$$

$$= \sum_{r=0}^{k+1} \binom{k+1}{r} x^r y^{k+1-r}.$$

Exercise 12.1 (page 421) **1.** $x^2 + y^2 - 8x - 4y - 5 = 0$ **3.** $x - y - 3 = 0$
5. $3x^2 + 3y^2 - 46x + 131 = 0$ **7.** $y^2 - 8x = 0$ **9.** $x^2 + 6x + 4y + 13 = 0$
11. $3x^2 + 4y^2 - 48 = 0$ **13.** $3x^2 - y^2 - 3 = 0$ **15.** $x^2 - 2py + p^2 = 0$
17. $(x + 4)^2 + (y - 1)^2 \geq 9$
19. The equation of the line perpendicular to the line $y = kx$ passing through the point $P(x_0, y_0)$ is
$(y - y_0) = -\dfrac{1}{k}(x - x_0)$. The point of intersection of these two lines is $Q\left(\dfrac{ky_0 + x_0}{k^2 + 1}, \dfrac{k^2 y_0 + k x_0}{k^2 + 1}\right)$. The distance

Answers to Odd-Numbered Exercises   555

from P to the line is the distance from P to Q. Thus,

$$d = \sqrt{\left(x_0 - \frac{ky_0 + x_0}{k^2 + 1}\right)^2 + \left(y_0 - \frac{k^2 y_0 + kx_0}{k^2 + 1}\right)^2} = \sqrt{\left(\frac{k^2 x_0 - ky_0}{k^2 + 1}\right)^2 + \left(\frac{y_0 - kx_0}{k^2 + 1}\right)^2}$$

$$= \sqrt{\frac{k^2(y_0 - kx_0)^2}{(k^2 + 1)^2} + \frac{(y_0 - kx_0)^2}{(k^2 + 1)^2}} = \sqrt{\frac{(y_0 - kx_0)^2}{k^2 + 1}} = \frac{|y_0 - kx_0|}{\sqrt{k^2 + 1}}.$$

21. $\dfrac{|y - 2x|}{\sqrt{5}} = \sqrt{(x - 1)^2 + (y - 3)^2} + \sqrt{(x + 1)^2 + (y - 1)^2}$

Exercise 12.2 (page 424)   1. $x^2 = 16y$ 3. $y^2 = -8x$

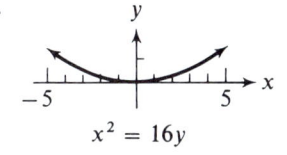

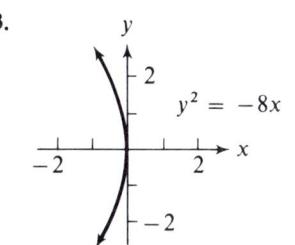

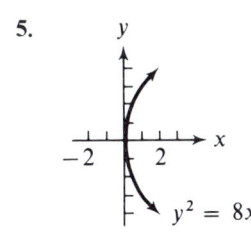

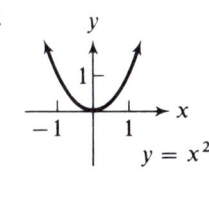

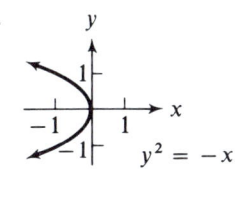

5. $y^2 = 8x$ 7. $y = x^2$ 9. $y^2 = -x$

11. $F\left(0, \dfrac{1}{2}\right)$, d: $y = -\dfrac{1}{2}$ 13. $F(-4, 0)$, d: $x = 4$ 15. $F(3, 0)$, d: $x = -3$

17. $F(-1, 0)$, d: $x = 1$ 19. $F\left(\dfrac{3}{8}, 0\right)$, d: $x = -\dfrac{3}{8}$

21. The point $P(x, y)$ is on the parabola with vertex $(0, 0)$ and focus $(0, -p)$ if and only if
$p - y = \sqrt{(x - 0)^2 + (y + p)^2}$. By the same procedure used in the text, this equation simplifies to $x^2 = -4py$.
23. The point $P(x, y)$ is on the parabola with vertex $(0, 0)$ and focus $(-p, 0)$ if and only if
$p - x = \sqrt{(x + p)^2 + (y - 0)^2}$. By the same procedure used in the text, this equation simplifies to $y^2 = -4px$.

Exercise 12.3 (page 429)   1. Circle, $r = \sqrt{2}$ 3. Ellipse, $a = \sqrt{12}$, $b = \sqrt{8}$

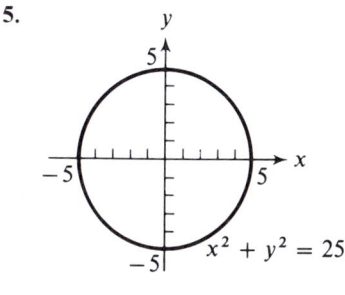

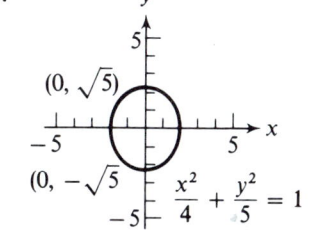

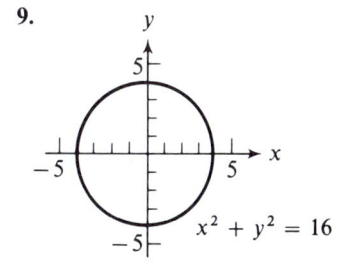

5. $x^2 + y^2 = 25$ 7. $\dfrac{x^2}{4} + \dfrac{y^2}{5} = 1$ 9. $x^2 + y^2 = 16$

11.

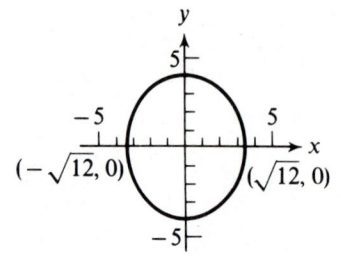

$$\frac{x^2}{12} + \frac{y^2}{16} = 1$$

13.

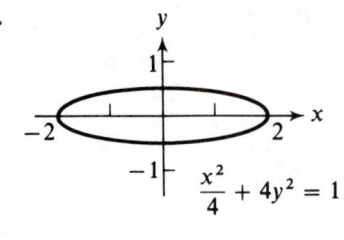

$$\frac{x^2}{4} + 4y^2 = 1$$

15. $F_1(-\sqrt{5}, 0)$, $F_2(\sqrt{5}, 0)$, $V_1(-3, 0)$, $V_2(3, 0)$

17. $F_1(-8\sqrt{2}, 0)$, $F_2(8\sqrt{2}, 0)$, $V_1(-12, 0)$, $V_2(12, 0)$

19. $F_1(0, -1)$, $F_2(0, 1)$, $V_1(0, -2)$, $V_2(0, 2)$

21. The steps in the derivation are the same as those in the derivation of Equation (9) except that instead of Equation (4), we start with $\sqrt{(x-0)^2 + (y-c)^2} + \sqrt{(x-0)^2 + (y+c)^2} = 2a$.

23. If $c \approx 0$, then $a \approx b$ and the ellipse is more nearly circular, as one would expect from Exercise 22.

25. The endpoints of the major axis are the x-intercepts of the curve $\frac{x^2}{a^2} + \frac{y^2}{b^2} = 1$. Setting $y = 0$, we have $\frac{x^2}{a^2} = 1$, or $x = \pm a$.

Exercise 12.4 (page 433)

1.

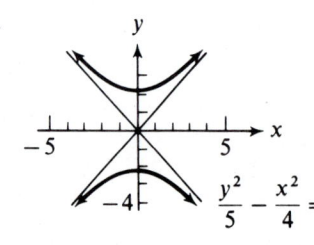

$$\frac{y^2}{5} - \frac{x^2}{4} = 1$$

3.

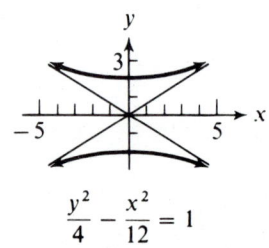

$$\frac{y^2}{4} - \frac{x^2}{12} = 1$$

5.

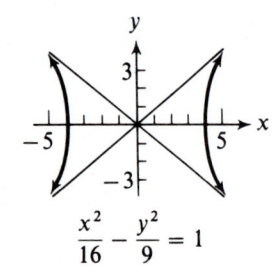

$$\frac{x^2}{16} - \frac{y^2}{9} = 1$$

7.

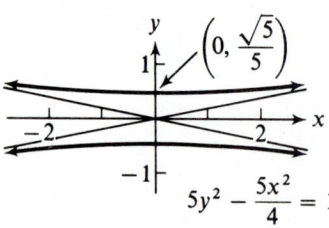

$$5y^2 - \frac{5x^2}{4} = 1$$

9. $F_1(5, 0)$, $F_2(-5, 0)$, $V_1(3, 0)$, $V_2(-3, 0)$, $y = \pm \frac{4}{3}x$

11. $F_1(\sqrt{3}, 0)$, $F_2(-\sqrt{3}, 0)$, $V_1(1, 0)$, $V_2(-1, 0)$, $y = \pm\sqrt{2}x$

13. $F_1(0, 2\sqrt{3})$, $F_2(0, -2\sqrt{3})$, $V_1(0, 2)$, $V_2(0, -2)$, $y = \pm \frac{\sqrt{2}}{2}x$

15. $F_1\left(0, \frac{\sqrt{41}}{20}\right)$, $F_2\left(0, -\frac{\sqrt{41}}{20}\right)$, $V_1\left(0, \frac{1}{5}\right)$, $V_2\left(0, -\frac{1}{5}\right)$, $y = \pm \frac{4}{5}x$

Answers to Odd-Numbered Exercises

17. By the law of cosines and Figure 12.11,

$$4c^2 = d_1^2 + d_2^2 - 2d_1 d_2 \cos \alpha,$$

where α is the angle opposite the line segment joining the foci. We are assuming $|d_1 - d_2| = 2a$; thus $d_1^2 - 2d_1 d_2 + d_2^2 = 4a^2$. Hence $d_1^2 + d_2^2 = 4a^2 + 2d_1 d_2$. Thus,

$$\begin{aligned} 4c^2 &= 4a^2 + 2d_1 d_2 - 2d_1 d_2 \cos \alpha \\ &= 4a^2 + 2d_1 d_2 (1 - \cos \alpha). \end{aligned}$$

Since $d_1 > 0$, $d_2 > 0$, and $1 - \cos \alpha > 0$, we have $4c^2 > 4a^2$. Thus $c^2 - a^2 > 0$.

19. The equation is $|\sqrt{(x + \sqrt{2})^2 + (y + \sqrt{2})^2} - \sqrt{(x - \sqrt{2})^2 + (y - \sqrt{2})^2}| = 2\sqrt{2}$, which simplifies to $xy = 1$.

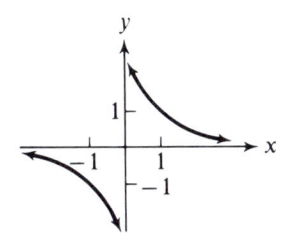

Exercise 12.5 (page 437)

1. Parabola, intercepts (0, 0)

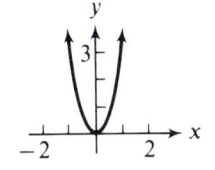

3. Parabola, intercepts (0, 0)

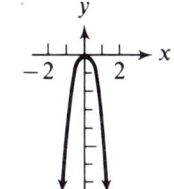

5. Parabola, intercepts (0, 0)

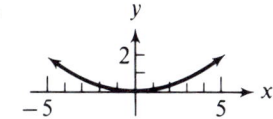

7. Parabola, intercepts (0, 0)

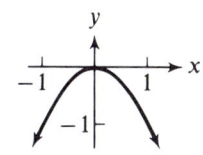

9. Circle, x-intercepts (± 7, 0), y-intercepts (0, ± 7)

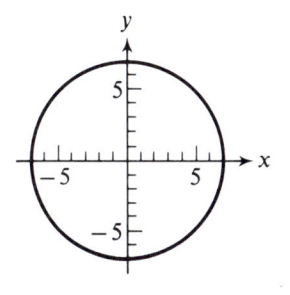

11. Ellipse, x-intercepts (± 5, 0), y-intercepts (0, ± 2)

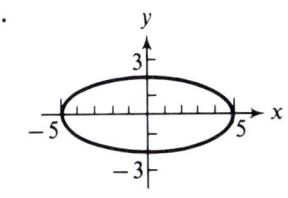

13.

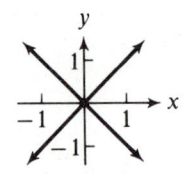

Two intersecting lines, intercepts (0, 0)

15.

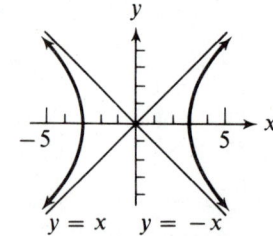

Hyperbola, x-intercepts (± 3, 0), no y-intercepts

$y = x \quad y = -x$

17.

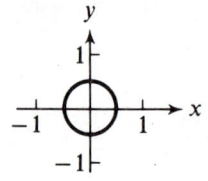

Circle, x-intercepts $\left(\pm\dfrac{1}{2}, 0\right)$, y-intercepts $\left(0, \pm\dfrac{1}{2}\right)$

19.

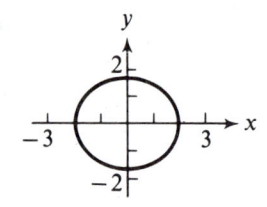

Ellipse, x-intercepts (± 2, 0), y-intercepts ($0, \pm\sqrt{3}$)

21.

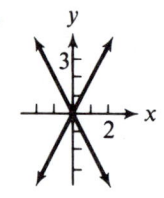

Two intersecting lines, intercepts (0, 0)

23.

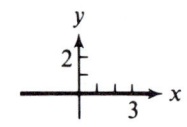

Straight line, every real number is an x-intercept, y-intercept (0, 0)

25. Since A and B are positive, $Ax^2 - By^2 = 0$ is equivalent to $y = \pm\sqrt{A/B}\, x$. Thus the graph of the relation is a pair of straight lines with slopes $\pm\sqrt{A/B}$ that intersect at (0, 0).

27. Since $A \neq 0$, $Ax^2 = 0$ is equivalent to $x^2 = 0$, which is equivalent to $x = 0$. Thus the graph of $Ax^2 = 0$ is a vertical straight line passing through the point (0, 0).

Chapter 12 Review (page 437)   **1.** $3x^2 + 3y^2 + 4x - 34y + 59 = 0$ **2.** $(x - 3)^2 + (y - 4)^2 = 25$

3. $|y| = 3$ **4.** $\sqrt{(x+2)^2 + (y-6)^2} + \sqrt{x^2 + (y-4)^2} = 8$

5.

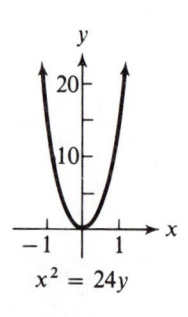

$x^2 = 24y$

6.

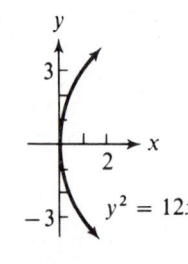

$y^2 = 12x$

7.

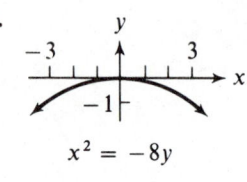

$x^2 = -8y$

8.

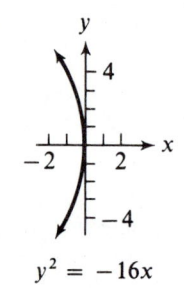

$y^2 = -16x$

9. $F\left(\dfrac{1}{8}, 0\right)$, $x = -\dfrac{1}{8}$ **10.** $F\left(0, -\dfrac{1}{4}\right)$, $y = \dfrac{1}{4}$ **11.** $F(0, 5)$, $y = -5$ **12.** $F\left(-\dfrac{1}{4}, 0\right)$, $x = \dfrac{1}{4}$

13.

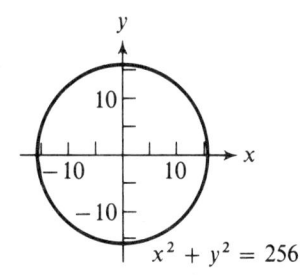

[Graph: circle $x^2 + y^2 = 256$, shown on axes from -10 to 10]

14.

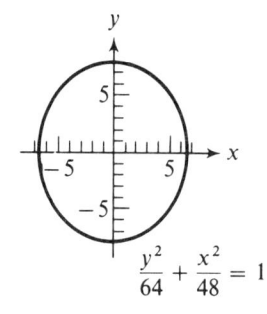

[Graph: ellipse $\dfrac{y^2}{64} + \dfrac{x^2}{48} = 1$, shown on axes from -5 to 5]

15.

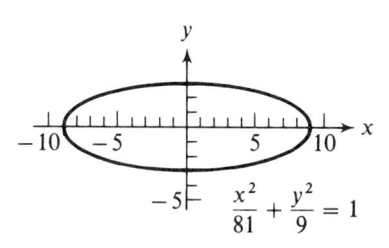

[Graph: ellipse $\dfrac{x^2}{81} + \dfrac{y^2}{9} = 1$, shown on axes from -10 to 10]

16.

[Graph: ellipse $\dfrac{4x^2}{5} + y^2 = 1$, shown on axes from -1 to 1]

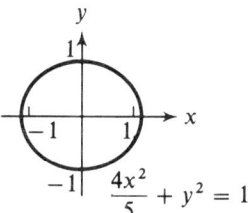

17. $F_1(0, -2\sqrt{3})$, $F_2(0, 2\sqrt{3})$, $V_1(0, -4)$, $V_2(0, 4)$

18. $F_1(-\sqrt{7}, 0)$, $F_2(\sqrt{7}, 0)$, $V_1(-\sqrt{8}, 0)$, $V_2(\sqrt{8}, 0)$

19. $F_1(-1, 0)$, $F_2(1, 0)$, $V_1(-3, 0)$, $V_2(3, 0)$

20. $F_1(0, -\sqrt{5})$, $F_2(0, \sqrt{5})$, $V_1(0, -3)$, $V_2(0, 3)$

21.

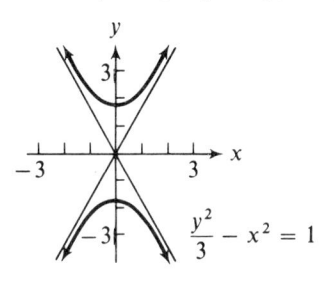

[Graph: hyperbola $\dfrac{y^2}{3} - x^2 = 1$]

22.

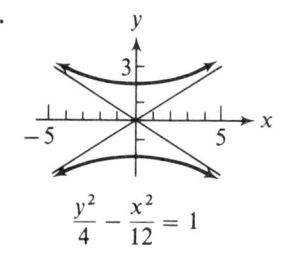

[Graph: hyperbola $\dfrac{y^2}{4} - \dfrac{x^2}{12} = 1$]

23.

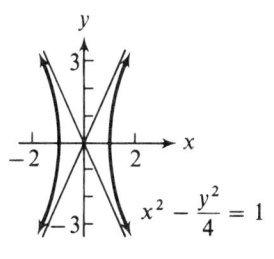

[Graph: hyperbola $x^2 - \dfrac{y^2}{4} = 1$]

24.

[Graph: hyperbola $x^2 - \dfrac{y^2}{3} = 1$]

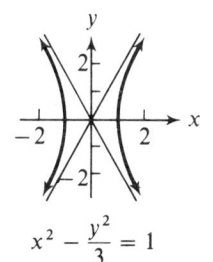

25. $F_1(\sqrt{3}, 0)$, $F_2(-\sqrt{3}, 0)$, $V_1(\sqrt{2}, 0)$, $V_2(-\sqrt{2}, 0)$, $y = \pm \dfrac{\sqrt{2}}{2} x$

26. $F_1(0, \sqrt{2})$, $F_2(0, -\sqrt{2})$, $V_1(0, 1)$, $V_2(0, -1)$, $y = \pm x$

27. $F_1(0, \sqrt{6})$, $F_2(0, -\sqrt{6})$, $V_1(0, 2)$, $V_2(0, -2)$, $y = \pm \sqrt{2} x$

28. $F_1(\sqrt{14}, 0)$, $F_2(-\sqrt{14}, 0)$, $V_1(\sqrt{10}, 0)$, $V_2(-\sqrt{10}, 0)$, $y = \pm \dfrac{\sqrt{10}}{5} x$

29. Parabola

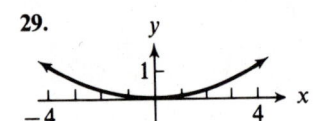

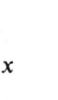

30. Parabola

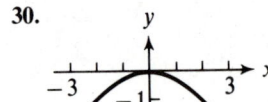

31. Parabola

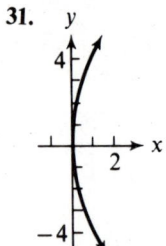

32. Parabola

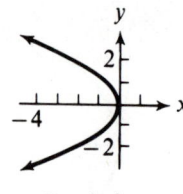

33. Hyperbola

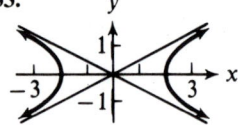

34. Ellipse

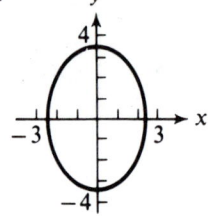

35. Two intersecting lines

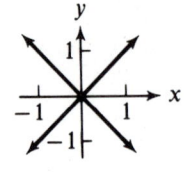

36. Single point

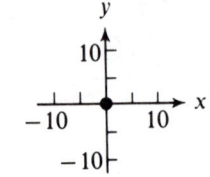

Exercise 13.1 (page 444) **1.** $(-7, -3)$ **3.** $(5, -1)$ **5.** $x' - 2y' = 13$ **7.** $x'^2 - 2y'^2 + 4x' + 20y' - 52 = 0$ **9.** $(x + 5)^2 + (y - 1)^2 = 9$ **11.** $(x - 3)^2 = 8(y - 3)$

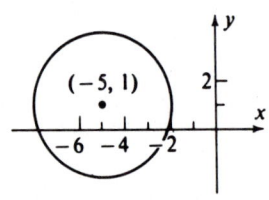

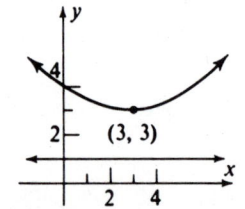

13. $\dfrac{(x + 2)^2}{25} + \dfrac{(y - 2)^2}{9} = 1$

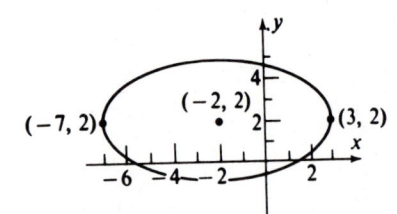

15. $\dfrac{(x - 3)^2}{9} - \dfrac{(y - 2)^2}{16} = 1$

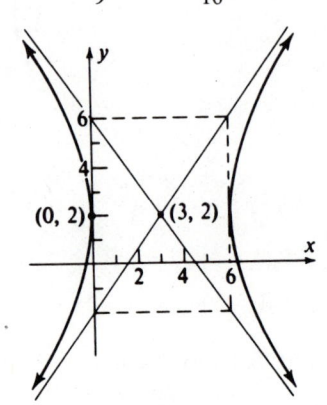

17. 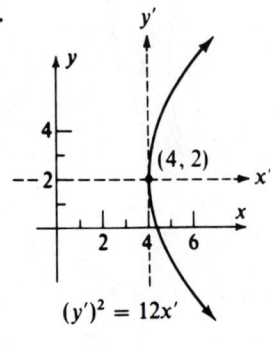 $(y')^2 = 12x'$

Answers to Odd-Numbered Exercises

19. [graph showing circle with center $(3, -\frac{1}{2})$ in rotated coordinates]

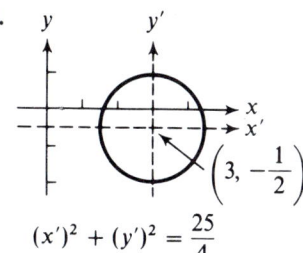

$$(x')^2 + (y')^2 = \frac{25}{4}$$

21. [graph showing ellipse with center $(-2, 1)$]

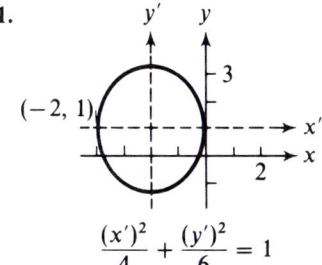

$$\frac{(x')^2}{4} + \frac{(y')^2}{6} = 1$$

23. [graph showing hyperbola with center $(5, 2)$]

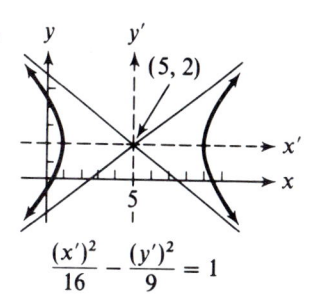

$$\frac{(x')^2}{16} - \frac{(y')^2}{9} = 1$$

25. $A \neq 0$, $B = 0$, $D \neq 0$ **27.** $A \neq 0$, $B \neq 0$, $A = B$, $\dfrac{C^2}{4A} + \dfrac{D^2}{4B} - F > 0$

Exercise 13.2 (page 449)

1. $x' = \dfrac{-3\sqrt{3} + 5}{2}$, $y' = \dfrac{3 + 5\sqrt{3}}{2}$ **3.** $x' = 4 + 3\sqrt{3}$, $y' = -4\sqrt{3} + 3$

5. $x' = 3 + 4\sqrt{3}$, $y' = 3\sqrt{3} - 4$ **7.** $x' = -3\sqrt{3} - \dfrac{5}{2}$, $y' = -3 + \dfrac{5}{2}\sqrt{3}$

9. $-x'^2 + 4y'^2 = -16$; hyperbola   **11.** $5x'^2 + y'^2 = 36$; ellipse

13. $11x'^2 + y'^2 = 44$; ellipse

15. $8x'^2 - 2y'^2 - 8\sqrt{2}x' - 6\sqrt{2}y' = 15$; eliminating the first-degree term gives $4x''^2 - y''^2 = 5$; hyperbola

17. $2x'^2 - \sqrt{2}x' - 3\sqrt{2}y' + 5 = 0$; eliminating the first-degree terms gives
$2x''^2 - 3\sqrt{2}y'' + \dfrac{19}{4} = 0$; parabola

19. Using the equations on page 447, we have

$$A' + C' = A\cos^2\alpha + B\sin\alpha\cos\alpha + C\sin^2\alpha + A\sin^2\alpha - B\sin\alpha\cos\alpha + C\cos^2\alpha$$
$$= A(\cos^2\alpha + \sin^2\alpha) + C(\sin^2\alpha + \cos^2\alpha)$$

and, because $\cos^2\alpha + \sin^2\alpha = 1$, $A' + C' = A + C$.

21. $y^2 - 2xy + x^2 - 4\sqrt{2}x - 4\sqrt{2}y = 0$ **23.** $xy = 1$

Exercise 13.3 (page 452)

1. a. [graph of line with $x = 2 - 3t$, $y = 3t - 1$, showing $t = 1$ and $t = 0$]   **b.** $x + y - 1 = 0$

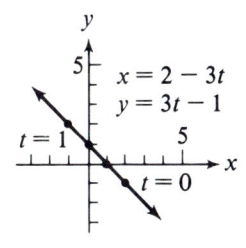

3. a. [graph of line with $x = 3t - 2$, $y = 3 - 2t$, showing $t = 0$ and $t = \frac{3}{2}$]   **b.** $2x + 3y - 5 = 0$

5. a. [graph of parabola with $x = t^2$, $y = 1 - t$, showing $t = -2, -1, 0, 1, 2$]   **b.** $x = 1 - 2y + y^2$

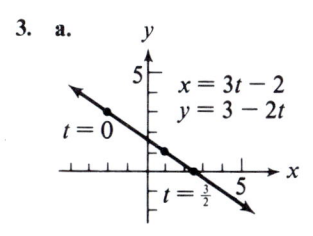

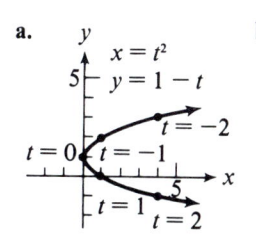

7. a.

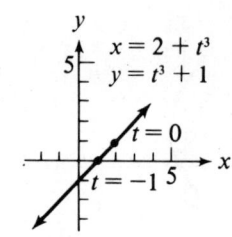

$x = 2 + t^3$
$y = t^3 + 1$

b. $x - y - 1 = 0$

9. a.

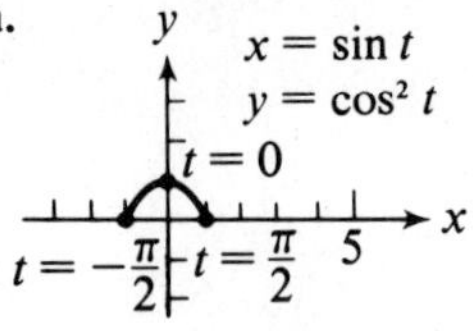

$x = \sin t$
$y = \cos^2 t$

b. $x^2 + y = 1$, $|x| \le 1$

11. a.

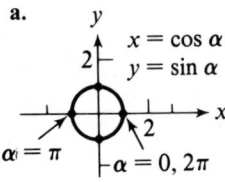

$x = \cos \alpha$
$y = \sin \alpha$

b. $x^2 + y^2 = 1$

13. a.

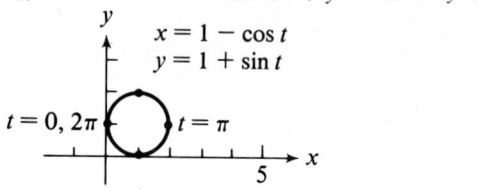

$x = 1 - \cos t$
$y = 1 + \sin t$

b. $x^2 + y^2 - 2x - 2y + 1 = 0$

15. a.

$x = 4 \cos \alpha$
$y = 3 \sin \alpha$

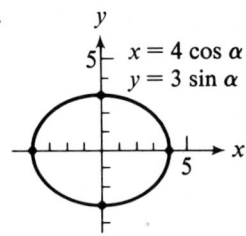

b. $\dfrac{x^2}{16} + \dfrac{y^2}{9} = 1$

17.

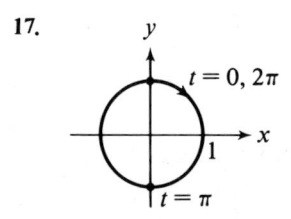

The circle is traversed once in the clockwise direction beginning from the point $(0, 1)$.

19.

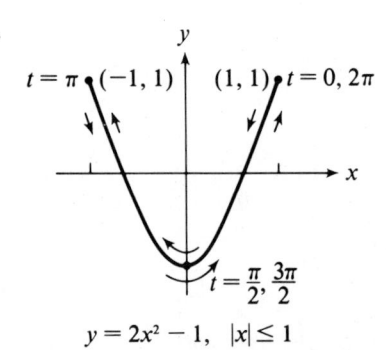

$y = 2x^2 - 1$, $|x| \le 1$

The curve is traversed once from $(1, 1)$ to $(-1, 1)$ as t takes values from 0 to π and then is traversed from $(-1, 1)$ back to $(1, 1)$ as t takes values from π to 2π.

21.

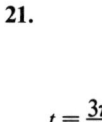

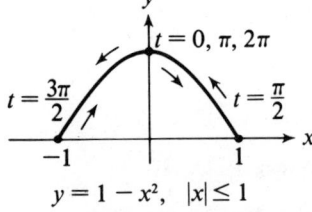

$y = 1 - x^2$, $|x| \le 1$

The curve is traversed from $(0, 1)$ to $(1, 0)$ as t takes values from 0 to $\pi/2$; it then is traversed from $(1, 0)$ to $(-1, 0)$ as t takes values from $\pi/2$ to $3\pi/2$; finally it is traversed from $(-1, 0)$ back to $(0, 1)$ as t takes values from $3\pi/2$ to 2π.

23. Note that $(y_1 + bt - y_1) = b/a \, (x_1 + at - x_1)$. Thus $x = x_1 + at$ and $y = y_1 + bt$ satisfy $y - y_1 = b/a \, (x - x_1)$.

25. Note that $(x_1 + r \cos t - x_1)^2 + (y_1 + r \sin t - y_1)^2 = r^2$. Thus $x = x_1 + r \cos t$ and $y = y_1 + r \sin t$ satisfy, $(x - x_1)^2 + (y - y_1)^2 = r^2$.

27. $x = \cos t$, $y = \sin t$, $t \in [0, 2\pi]$

Exercise 13.4 (page 457) **1.** $(6, 125°), (6, -235°), (-6, 305°), (-6, -55°)$
3. $(-2, -90°), (-2, 270°), (2, 90°), (2, -270°)$ **5.** $(6, -240°), (6, 120°), (-6, -60°), (-6, 300°)$
7. $\left(\dfrac{5}{\sqrt{2}}, \dfrac{5}{\sqrt{2}}\right)$ **9.** $\left(\dfrac{\sqrt{3}}{4}, -\dfrac{1}{4}\right)$ **11.** $\left(\dfrac{-10}{\sqrt{2}}, \dfrac{-10}{\sqrt{2}}\right)$ **13.** $(6, 45°), (6, -315°)$
15. $(2, 240°), (2, -120°)$ **17.** $(0, 0°), (0, -0°)$ for any $\theta > 0$ **19.** $r = 5$ **21.** $r \sin \theta = -4$
23. $r^2(\cos^2 \theta + 9 \sin^2 \theta) = 9$ **25.** $x^2 + y^2 = 25$ **27.** $x^2 + y^2 - 9x = 0$ **29.** $y^2 = 4x + 4$
31. $\sec \dfrac{\theta}{2} = \dfrac{1}{\cos(\theta/2)}$ and $\cos^2 \dfrac{\theta}{2} = \dfrac{1 + \cos \theta}{2}$; hence $r = \sec^2 \dfrac{\theta}{2} = \dfrac{1}{\cos^2(\theta/2)} = \dfrac{2}{1 + \cos \theta}$.

Now $\cos \theta = \dfrac{x}{r}$ and $r^2 = x^2 + y^2$; so $r = \dfrac{2}{1 + \dfrac{x}{r}}$ and $r\left(1 + \dfrac{x}{r}\right) = 2$;

hence $r + x = 2$, or $r = 2 - x$. Squaring each member gives $r^2 = 4 - 4x + x^2$, and substituting for r^2 gives $x = 1 - \dfrac{1}{4}y^2$, whose graph is a parabola.

33. $r^2(1 + 2\sin^2 \theta) = 1$ is equivalent to $r^2 + 2r^2 \sin^2 \theta = 1$. Since $r^2 = x^2 + y^2$ and $y = r \sin \theta$, we have $x^2 + y^2 + 2y^2 = 1$ or $x^2 + 3y^2 = 1$, which is an equation of an ellipse.
35. $\cos 2\theta = 0$ if and only if $2\theta = (\pi/2) + k\pi$, $k \in J$ or $\theta = (\pi/4) + k(\pi/2)$, $k \in J$. The graph of this equation is the two intersecting lines $y = x$ and $y = -x$.

Exercise 13.5 (page 461)

1.

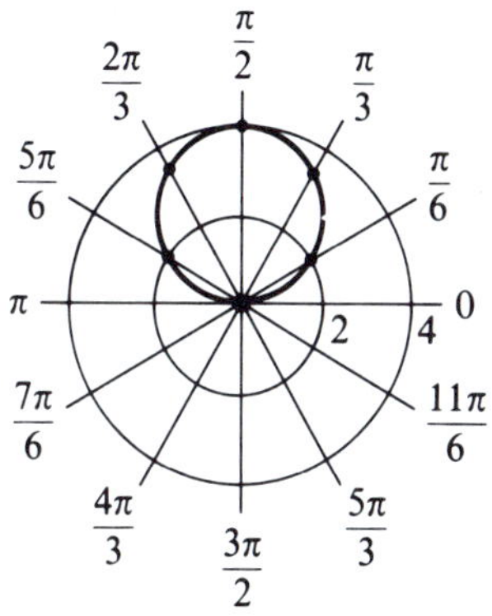

3.

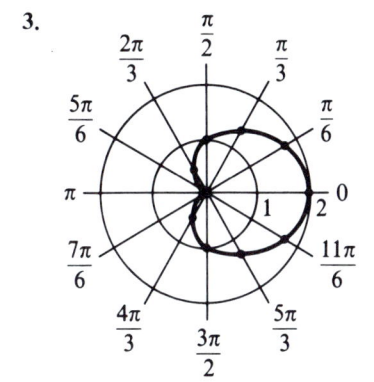

5.

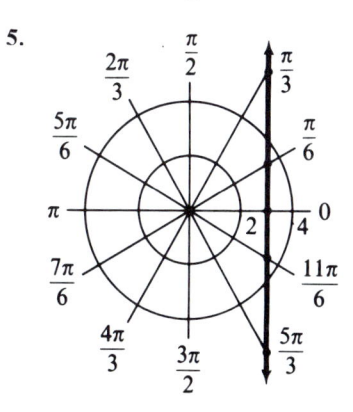

7.

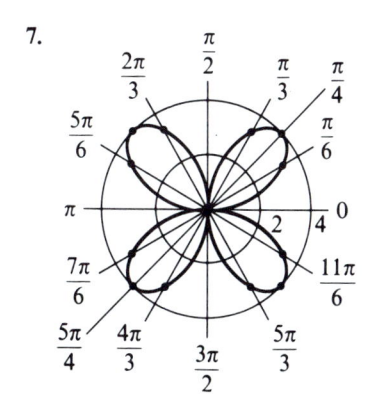

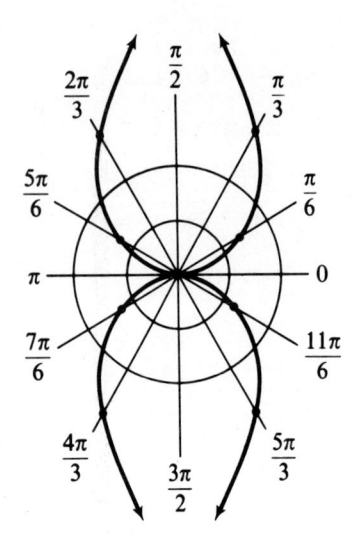

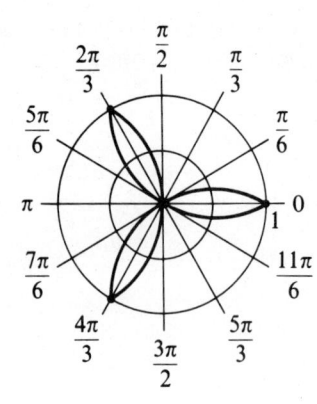

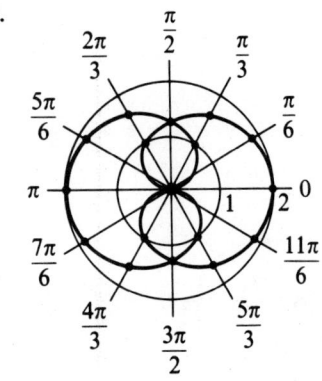

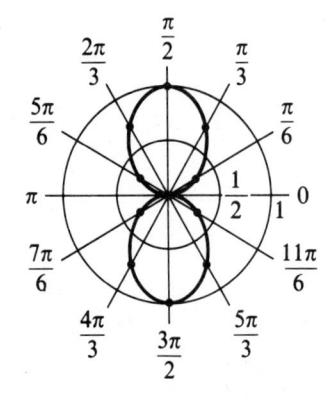

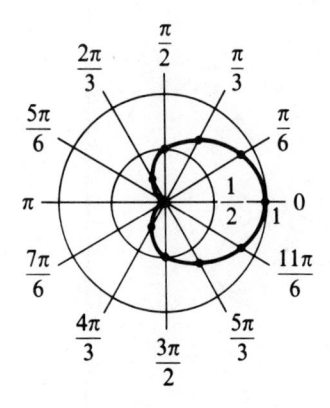

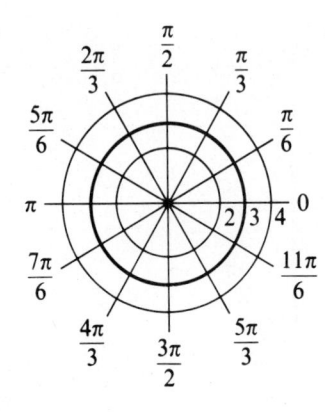

9.

11.

13.

15.

17.

19.

Answers to Odd-Numbered Exercises 565

21.

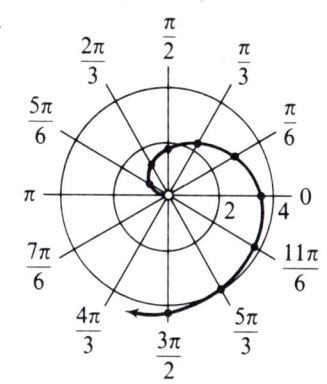

23.

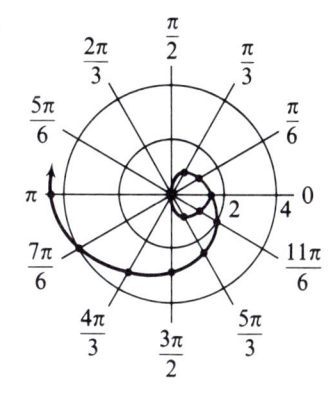

Chapter 13 Review (page 462)   **1.** $(-7, 4)$ **2.** $2x'^2 - 12x' - y' + 17 = 0$

3. $(x - 1)^2 + (y + 3)^2 = 16$ **4.** $y^2 = -8(x - 7)$

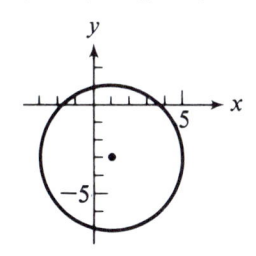

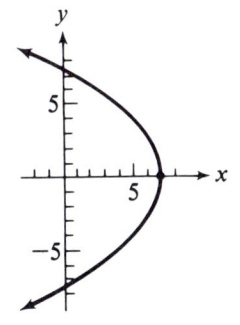

5. $x'^2 = 6y'$ **6.** $2x'^2 + y'^2 = \dfrac{73}{8}$

7.

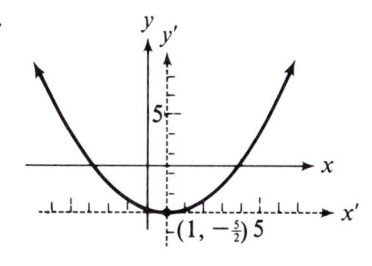

8.

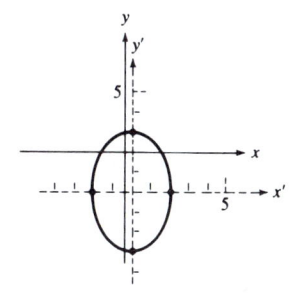

9. $\left(-\sqrt{3} + \dfrac{1}{2}, 1 + \dfrac{\sqrt{3}}{2}\right)$ **10.** $(-2\sqrt{2}, 2\sqrt{2})$ **11.** $3x'^2 - y'^2 = 2$, hyperbola

12. $14x''^2 + 22y''^2 = \dfrac{316}{11}$, ellipse

13. a.

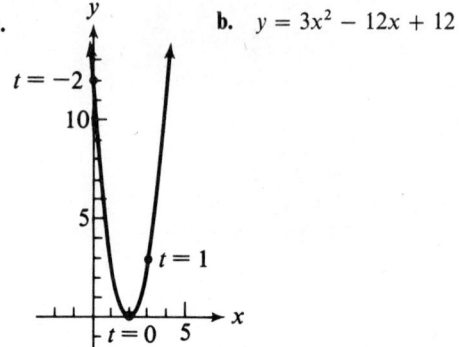

[Graph showing parabola with points labeled $t=-2$, $t=0$, $t=1$, passing through y-axis around 10]

b. $y = 3x^2 - 12x + 12$

14. a.

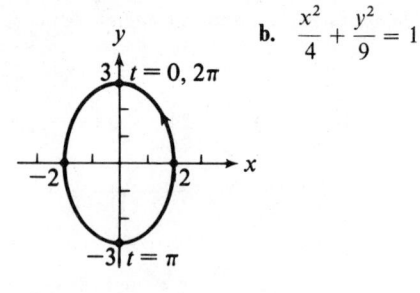

[Ellipse graph with $t=0, 2\pi$ at top (0,3) and $t=\pi$ at bottom (0,-3), extending to ±2 on x-axis]

b. $\dfrac{x^2}{4} + \dfrac{y^2}{9} = 1$

15.

[Upper semicircle from (-2,0) to (2,0) with $t=\pi/2$ at top, $t=\pi$ at (-2,0), $t=0$ at (2,0)]

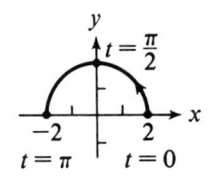

The semicircle is traversed once from (2, 0) to (−2, 0).

16.

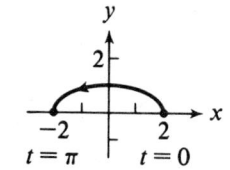

[Curve from (2,0) to (-2,0) passing through approximately (0,2)]

The curve is traversed once from (2, 0) to (−2, 0).

17. $(-4, 120°), (4, 300°), (4, -60°), (-4, -240°)$
18. $(-3\sqrt{2}, 3\sqrt{2})$
19. $r = 9\cos\theta$
20. $x^2 + y^2 - 4x = 0$

21.

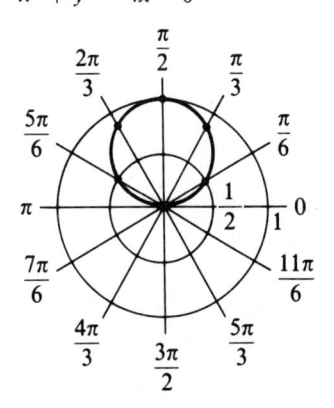

[Polar graph showing a circle in the upper half passing through origin]

22.

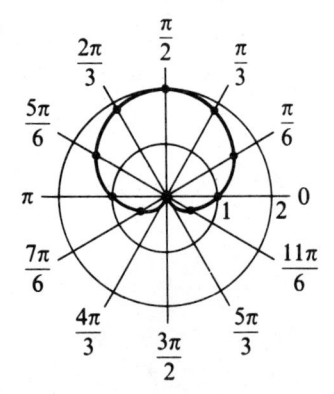

[Polar graph showing a circle centered on positive x-axis]

23.

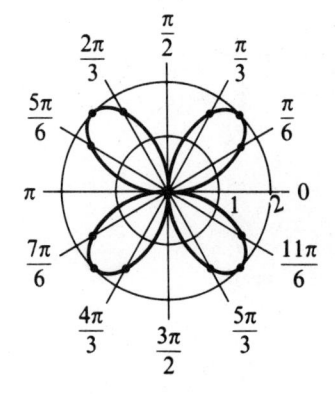

[Polar graph showing a four-petaled rose]

24.

[Polar graph showing a cardioid-like curve opening to the right]

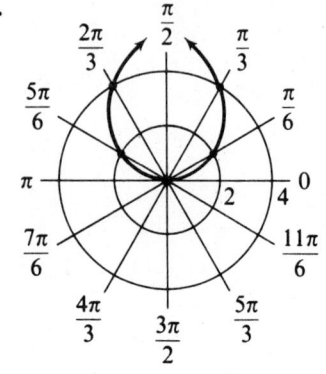

Supplemental Exercises, Chapters 12–13 (page 464)

1. $3\left(\dfrac{1}{2}x + \dfrac{\sqrt{3}}{2}y - 2\right)^2 + \dfrac{\left(-\dfrac{\sqrt{3}}{2}x + \dfrac{1}{2}y\right)^2}{8} = 1$

2. $\left(\dfrac{1}{2}x + \dfrac{\sqrt{3}}{2}y - 1\right)^2 + \left(-\dfrac{\sqrt{3}}{2}x + \dfrac{1}{2}y\right)^2 = 18$

3. By Exercise 20 on page 449, $B^2 - 4AC = B'^2 - 4A'C'$ but $B' = 0$. Thus $B^2 - 4AC = -4A'C'$.

For Exercises 4–6, assume that the coordinate axes have been rotated through an angle α that transforms the equation

$$Ax^2 + Bxy + Cy^2 + Dx + Ey + F = 0 \qquad (1)$$

into the equation

$$A'x'^2 + C'y'^2 + D'x' + E'y' + F' = 0; \qquad (2)$$

i.e., $B' = 0$.

4. By Exercise 3, if $B^2 - 4AC = 0$, then $-4A'C' = 0$. Thus either $A' = 0$ or $C' = 0$ or both.

Case I: $A' = 0$, $C' \neq 0$ Then the equation is linear in x' and quadratic in y'. Thus after completing the square in y' and translating axes we arrive at an equation of the form

$$y''^2 = 4px'' \text{ or } y''^2 = -4px'',$$

which is an equation of a parabola.

Case II: If $C' = 0$, $A' \neq 0$, use an argument similar to the one used for Case I.

Case III: If $A' = C' = 0$, the equation is linear in both variables and has a degenerate conic section as its graph.

5. Note that if $B = 0$, then $-4AC < 0$ implies A and C have the same sign. Assume without loss of generality that $A > 0$ and $C > 0$. Then completing the square on Equation (1) and translating axes yields the equation

$$Ax''^2 + Cy''^2 = K.$$

Since the conic is not degenerate, $K > 0$, and since for $B = 0$ we are assuming $A \neq C$, this is an equation of an ellipse.

Now if $B \neq 0$, then, by Exercise 3. $-4A'C' < 0$. Thus A' and C' must have the same sign.

Now if $A' = C'$, then by Equations (5) and (6) on page 447 we have

$$A \cos^2 \alpha + B \cos \alpha \sin \alpha + C \sin^2 \alpha = A \sin^2 \alpha - B \cos \alpha \sin \alpha + C \cos^2 \alpha$$

or equivalently

$$A(\cos^2 \alpha - \sin^2 \alpha) + 2B \cos \alpha \sin \alpha + C(\sin^2 \alpha - \cos^2 \alpha) = 0.$$

Hence

$$(A - C) \cos 2\alpha + B \sin 2\alpha = 0$$

or

$$\tan 2\alpha = \dfrac{C - A}{B}.$$

But the angle of rotation α satisfies

$$\tan 2\alpha = \dfrac{B}{A - C}.$$

Thus we have

$$B^2 = -(C - A)^2.$$

Since this implies $B = 0$, it must be the case that our assumption $A' = C'$ is false. Thus $A' \neq C'$. Now, using the same argument as in the case $B = 0$, $A \neq C$, we see that the equation reduces to that of an ellipse.

6. If $B^2 - 4AC > 0$, then $-4A'C' > 0$, and hence A' and C' have different signs, say $A' > 0$ and $C' < 0$. Then after completing the square and translating axes our equation transforms to

$$A'x''^2 + C'y''^2 = K$$

where $K \neq 0$ since the conic is not degenerate. Since $A' > 0$ and $C' < 0$, this is an equation of a hyperbola.

7. $B = 0$, $A = C$, $\dfrac{4AF + D^2 + E^2}{4A} < 0$

For Exercises 8–10, assume (r, θ) is a set of polar coordinates for a point in the graph.

8. Symmetry with respect to the polar axis guarantees that the point with polar coordinates $(r, -\theta)$ is in the graph. Then symmetry with respect to the vertical axis guarantees that the point with polar coordinates given by $(-r, -(-\theta)) = (-r, \theta)$ is in the graph. Thus the graph is symmetric with respect to the pole.

9. Symmetry with respect to the vertical axis guarantees that the point with polar coordinates $(-r, -\theta)$ is in the graph. Then symmetry with respect to the pole guarantees that the point with polar coordinates $(-(-r), -\theta) = (r, -\theta)$ is in the graph. Thus the graph is symmetric with respect to the polar axis.

10. Symmetry with respect to the polar axis guarantees that the point with polar coordinates $(r, -\theta)$ is in the graph. Then symmetry with respect to the pole guarantees that the point with polar coordinates $(-r, -\theta)$ is in the graph. Thus the graph is symmetric with respect to the vertical axis.

11.

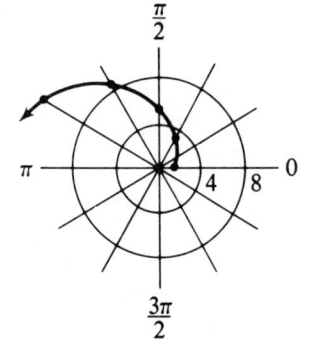

12.

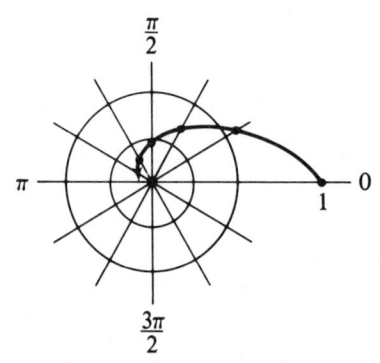

13. The graph spirals out away from the pole.

14. The graph spirals in toward the pole but never reaches it.

Index

a^0, 3
a^{-n}, 3
Abscissa, 44
Absolute value:
  in equation, 37 *ff*
  in inequalities, 37 *ff*
  of a complex number, 280
  of a real number, 5
Addition:
  associative law of:
    for matrices, 336
    for real numbers, 6
  closure for, of matrices, 336
    of real numbers, 6
  commutative law of:
    for matrices, 337
    for real numbers, 6
  identity element for:
    matrices, 336
    real numbers, 6
  law of:
    for equality, 6
    of matrices, 336
    of quotients, 7
    of vectors, 230
  parallelogram law for, of vectors, 230
Additive inverse:
  of a matrix, 337
  of a real number, 6
Algebraic expression, 2
Alternating sequence, 399
Ambiguous case for law of sines, 221
Amplitude:
  of a complex number, 280
  of a sine wave, 169
Angle(s):
  coterminal, 201
  definition of, 198
  in a quadrant, 199
  in standard position, 198
  initial side of, 198
  measure of an, using a circle, 199
  reference, 208

Angle(s) *(continued)*:
  sides of, 198
  symbols for measure, 199
  terminal side of, 198
  vertex of, 198
Antilogarithm, 130
Arc(s):
  reference, 161
Arccosecant function, 190
Arccosine function, 190
Arccosine relation, 190
Arccotangent function, 190
Arcsecant function, 190
Arcsine:
  function, 190
  relation, 188
Arctangent function, 190
Arctangent relation, 189
Argand plane, 280
Argument of a complex number, 280
Arithmetic progression:
  common difference of, 386
  meaning of, 386
  nth term of, 386
  sum of n terms, 391
Associative law:
  of addition of matrices, 336
  of addition of real numbers, 6
  of multiplication of a matrix by a real number, 340
  of multiplication of real numbers, 7
Asymptote(s):
  horizontal, 110
  oblique, 111
  of graph of a rational function, 100 *ff*
  vertical, 110
Augmented matrix, 350
Axioms, 6
Axes, rotation of, 445
  translation of, 440
Axis:
  conjugate, of a hyperbola, 432
  imaginary, 280

Axis *(continued)*:
   major, of an ellipse, 428
   minor of an ellipse, 428
   polar, 454
   real, 279
   transverse, of a hyperbola, 432
Axis of symmetry, 69, 423

Base of a logarithm, 122
Base of a power, 2
Binomial:
   coefficient of rth term in expansion of, 407
   expansion of, 405
   theorem, 403 *ff*

Cartesian coordinate system, 44
   graph of an ordered pair on, 44
   in three dimensions, 304
Cartesian product, 43
Characteristic of a logarithm, 129
Circle, 150, 426
   center of a, 426
   equation of a, 426
   radius of a, 426
Circular functions, 150 *ff*
   inverse, 190
Closure law:
   for addition of matrices, 336
   for addition of positive real numbers, 5
   for addition of real numbers, 6
   for multiplication of positive real numbers, 8
   for multiplication of real numbers, 6
Coefficient(s):
   detached, 90
   leading, of a polynomial, 3
   numerical, 2
   of terms in a binomial expansion, 407
Coefficient matrix of a system of linear equations, 350
Cofactor, of an element in a determinant, 357
Column matrix, 334
Column vector, 334
Common difference of an arithmetic progression, 386
Common logarithms (*see also* Logarithms), 127
Common ratio of a geometric progression, 387
Commutative law:
   of addition of matrices, 337
   of addition of real numbers, 6
   of multiplication of real numbers, 7
Completely factored polynomial, 100
Completing the square, 18
Complex number(s): 266 *ff*
   absolute value of, 280
   addition of, 267
   amplitude of, 280
   argument of, 280
   as zeros of a polynomial function, 277
   conjugate of, 270
   definition of set of, 266
   division of, 271, 282

Complex number(s) *(continued)*:
   equality of, 266
   factor theorem over, 276
   graph of, 279
   imaginary part of, 279
   modulus of, 280
   multiplication of, 267
   notation for, 266
   polar form of, 281
   powers of, 285
   product of, 267, 282
   pure imaginary, 268
   real numbers related to, 268
   real part of, 279
   roots of, 287
   subtraction of, 269
   sum of, 267
   trigonometric form of, 279 *ff*
Complex plane, 280
Components of an ordered pair, 43
Composite number, 100
Conditional equation(s), 257 *ff*
Conformable matrices, 343
Conics, 424
Conic sections, 424
Conjugate(s):
   definition, 270
   of a complex number, 270
Conjugate complex roots, 276
Consistent equations, 293, 301
Constant, 2
Constant function, 53
Convergent sequence, 398
Conversion formulas for angle measure, 200
Convex polygon, 326
Convex sets, 326
   intersection of, 326
Coordinate(s):
   of a point in the plane, 43
   of points on unit circle, 150
Cosecant:
   definition, 185
   domain of function, 185
   graph of function, 185
   period, 185
   range of function, 185
Cosine(s):
   definition, 150
   domain of function, 151
   double-angle formula for, 244
   function, 150
   graph of function, 167 *ff*
   half-angle formula for, 245
   law of, 224 *ff*
   period, 168
   range of function, 151
   sum and difference formulas, 242
   table of values of, 157
Cotangent:
   definition, 182
   domain of function, 182
   graph of function, 183
   period, 183

Index

Cotangent *(continued)*:
   range of function, 182
Coterminal angles, 201
Counterexample, 240
Cramer's rule, 378 *ff*
Critical numbers, 31
Cube root, 4
Cycle of a periodic function, 169
Cycle of a sine wave, 169

Decreasing function, 105
Degenerate cases, conic sections, 436
Degree as angle measure, 199
Degree of a monomial, 2
Degree of a polynomial, 2
De Moivre's theorem, 285
Denominator of a fraction, 3
Descartes' rule of signs, 97
Detached coefficients, 90
Determinant(s):
   cofactor of an element of, 357
   definition of, 356, 358
   expansion of, 358
   function, 356 *ff*
   minor of an element of, 357
   notation, 356
   properties of, 356 *ff*, 362 *ff*
   use of, in solutions of systems, 379
   value of, 356
Diagonal matrix, 344
Difference:
   common, 386
   definition of, 5
   of complex numbers, 269
   of matrices, 336
   of quotients, 7
   of real numbers, 5
Difference formula:
   for cosine function, 243
   for sine function, 248
   for tangent function, 252
Dimension of a matrix, 334
Direction angle of a vector, 229
Directrix of a parabola, 422
Discriminant of a quadratic equation, 19
Distance formula, 54
Distance, directed, 55, 216
Distributive law:
   for $n \times n$ square matrices, 343
   for real numbers, 7
   for scalar multiplication over matrix addition, 343
Divergent sequence, 399
Divergent series, 400
Division:
   of complex numbers, 271, 282
   of polynomials, 88 *ff*
   of quotients, 8
   of real numbers, 7
   synthetic, 89
Domain:
   of a logarithmic function, 121

Domain *(continued)*:
   of a relation or function, 45
   of cosecant function, 185
   of cosine function, 151
   of cotangent function, 182
   of secant function, 184
   of sine function, 151
   of tangent function, 176
Double-angle formula:
   for cosine function, 244
   for sine function, 248
   for tangent function, 252

e, 120
Element:
   of a matrix, 334
   of a set, 1
Elementary transformation:
   in terms of matrices, 370
   of an equation, 11
   of an inequality, 26
Ellipse, 427
   center of an, 427
   equation of an, 428
   focus of an, 427
   major axis of an, 428
   minor axis of an, 428
   standard equation of an, 429
   vertex of an, 428
Entry of a matrix, 334
Equality:
   addition law for, 8
   multiplication law for, 8
   of complex numbers, 266
   of like powers, 21
   of matrices, 335
   of quotients, 7
   of sets, 2
   of vectors, 229
   properties of, 6
Equation(s):
   of a circle, 426
   conditional, 257 *ff*
   consistent, 293
   elementary transformation of an, 11
   of an ellipse, 428
   equivalent, 10 *ff*
   exponential, solution of, 136
   first-degree, in one variable, 12
   first-degree, in two variables, 52
   graph of a linear, in two variables, 52
   of a hyperbola, 431
   in one variable, 10 *ff*
   inconsistent, 293
   intercept form, 61
   in two variables, 44
   involving absolute value, 37 *ff*
   linear, in one variable, 12
   linear, in two variables, 52
   linearly dependent, 293
   linearly independent, 293
   logarithmic, 123

Equation(s) *(continued)*:
   nonlinear, 68
   of a parabola, 423
   parametric, 450
   point-slope form, 60
   polar form, 455
   quadratic, in one variable, 16
   quadratic, in two variables, 68
   radical, 22
   root of an, 17
   second-degree, in one variable, 16
   second-degree, in two variables, 68
   slope-intercept form, 61
   solution of an, 10 *ff*
   solution set of an, 10
   systems of, 293 *ff*
   two-point form, 63
Equivalence symbol, 1
Equivalent expressions, 2
Equivalent rational expressions, 3
Equivalent systems, 294
Expansion:
   binomial, 405
   of a determinant, 358
Exponent(s):
   definition, 2
   laws of integral, 4
   laws of natural number, 8
   laws of rational, 4
   logarithms as, 122
   properties of, 8
Exponential function(s):
   definition of an, 119
   graph of an, 119
   inverse of, 180
Extraction of roots, 17
Extraneous solutions, 21

Factor(s):
   of quadratics, 16
   zero as a, 16
Factor theorem, 92, 276
Factorial notation, 403
Factoring polynomials, 92
Finite sequences, 384
First-degree equation(s):
   in one variable, 10
   in two variables, 52
   intercept form of a, in two variables, 61
   point-slope form for, in two variables, 60
   slope-intercept form for, in two variables, 61
   solution of systems of, by Cramer's rule, 378 *ff*
   standard form for, in two variables, 52
Focus:
   of an ellipse, 427
   of a hyperbola, 431
   of a parabola, 422
Fractions:
   definition of, 3
   equivalent, 3
   fundamental principle of, 7
   lowest terms, 7

Fractions *(continued)*:
   partial, 307
   reducing, 7
Function(s):
   absolute value, 80
   bracket, 82
   circular, 150
   composition, 49
   constant, 53
   cosecant, 185
   cosine, 150
   cotangent, 182
   decreasing, 105
   definition of a, 46
   determinant, 356 *ff*
   exponential, 118 *ff*
   greatest integer, 82
   increasing, 105
   inverse, 74 *ff*, 188 *ff*
   inverse of an exponential, 121
   inverse of a trigonometric, 210
   linear, 52 *ff*
   logarithmic, 121 *ff*
   nonlinear, 68
   notation, 49 *ff*
   one-to-one, 75
   period of, 168
   periodic, 167 *ff*
   polynomial, 88 *ff*
   quadratic, 68 *ff*
   rational, 88 *ff*, 109 *ff*
   real-valued, 46
   secant, 184
   sequence, 384
   sine, 150
   tangent, 176
   trigonometric, 198 *ff*, 205 *ff*
   zeros of, 71, 94, 100
Fundamental period, 167
   of cosecant function, 185
   of cosine function, 168
   of cotangent function, 183
   of secant function, 184
   of sine function, 168
   of tangent function, 176
Fundamental principle of fractions, 7
Fundamental theorem of algebra, 277
Fundamental theorem of arithmetic, 100

Gauss plane, 280
Geometric plane, 43
Geometric progression(s):
   common ratio of, 387
   definition of, 387
   infinite, 400
   nth term of, 387
   sum of an infinite, 400
   sum of n terms of, 392
Geometric vector, 227 *ff*
Graph(s):
   in Argand plane, 280
   in three dimensions, 304

Graph(s) *(continued)*:
  line, 29
  maximum point of, 105
  minimum point of, 105
  of an absolute value function, 37
  of a conic section, 422
  of an exponential function, 119
  of a first-degree equation, 52
  of a first-degree relation, 57
  of an inequality in two variables, 57
  of inverse circular function, 188
  of a logarithmic function, 121
  of a point on the number line, 27
  in polar coordinates, 458
  of a polynomial function, 104
  of a quadratic function, 68
  of a rational function, 110
  of cosecant function, 185
  of cosine function, 167 *ff*
  of cotangent function, 183
  of secant function, 184
  of sine function, 167 *ff*
  of tangent function, 178
Greater than, 5
Greatest integer function, 82

Half-angle formula:
  for cosine function, 245
  for sine function, 249
  for tangent function, 253
Half-plane, 57, 324
Horizontal:
  asymptote, 110
Hyperbola, 431
  center of a, 431
  conjugate axis of a, 432
  equation of a, 431
  focus of a, 431
  standard position of a, 433
  transverse axis of a, 432
  vertex of a, 432

i, 266
Identities:
  definition of, 238
  proof of, 238
  summary of, 474
Identity element:
  for addition of complex numbers, 269
  for addition of matrices, 336
  for addition of real numbers, 6
  for multiplication of complex numbers, 269
  for multiplication of matrices, 345
  for multiplication of real numbers, 7
Imaginary:
  number, 268
  part of complex number, 268
  pure, 268
Inclination, 55
Inconsistent equations, 381
Increasing function, 105

Index of a radical, 4
Index of summation, 394
Induction, mathematical, 411
Inequality (Inequalities):
  absolute, 39
  elementary tranformations of, 26
  equivalent, 26
  graph of, in one variable, 28
  graph of, in two variables, 57
  in one variable, 28
  in two variables, 52
  involving absolute values, 37 *ff*
  second-degree, in one variable, 29 *ff*
  second-degree, in two variables, 68 *ff*
  solution of, 26
  solution set, 26
  systems of, 324
Infinite series, 400
Initial side of an angle, 98
Integers, relatively prime, 100
Integers, set of, 5
Intercept form for a linear equation, 61
Intercept of a graph, 52
Intersection of sets, 1
Interval notation, 27
Inverse:
  additive, 6
  multiplicative, 7, 270
  of a circular function, 188
  of a function, 74 *ff*
  of a square matrix, 367 *ff*
  of a trigonometric function, 210
  of an exponential function, 121
Inverse function, 74 *ff*, 188 *ff*
Inverse operations, 7, 8
Inverse relation, 74
Irrational numbers, set of, 5

Lag, in phase of periodic function, 173
Latus rectum, 426
Law of cosines, 224 *ff*
Law of sines, 219 *ff*
Lead, in phase of periodic function, 173
Length of a line segment, 54
Less than, 5
Limit:
  as sum of an infinite series, 400
  of a sequence, 398
Line segment:
  inclination of a, 54
  length of a, 54
  slope of a, 56
Linear combination, 295
Linear equation (*See* First-degree equation)
Linear function, 52 *ff*
Linear programming, 329
Linearly dependent equations, 293
Linearly independent equations, 293
Location theorem, 95
Locus, 419
Logarithm(s):
  applications of, 136

Logarithm(s) *(continued)*:
  base e, 130
  base 10, 127 *ff*
  characteristic of, 129
  common, 127 *ff*
  computations with, 136 *ff*
  laws of, 123
  mantissa of, 129
  natural, 130
  reading tables of, 127
Logarithmic equation, 123
Logarithmic function, 121 *ff*
  graph of, 180
Long-division algorithm, 88
Lower bound:
  for zeros of real polynomial function, 95
Lowest terms, 100

Magnitude of vector, 229
Mantissa of a logarithm, 129
Mathematical induction:
  principle of, 411
  proof by, 411 *ff*
Matrix (Matrices):
  addition of, 335
  additive inverse of a, 337
  associative law of addition for, 336
  augmented, 350
  coefficient, of a system of linear equations, 350
  column, 334
  commutative law of addition for, 337
  conformable, for multiplication, 343
  diagonal, 344
  difference of, 336
  dimension of, 334
  elements of, 334
  entries of, 334
  equality of, 335
  identity element for addition of, 336
  identity element for multiplication of, 345
  inverse of a square, 367 *ff*
  meaning of, 334
  negative of, 336
  noncommutativity of, products, 343
  nonsingular, 349
  notation, 334
  order of, 334
  principal diagonal of, 344
  product of, and a scalar, 340
  product of two, 341
  properties of products, 341
  properties of sums, 336
  row, 334
  row-equivalent, 348
  singular, 349
  solution of linear systems, 347, 374
  square, 343
  sum of two, 335
  transpose of a, 335
  zero, 336
Maximum point:
  on a graph, 69

Maximum point *(continued)*:
  on a parabola, 69
Member of a set, 1
Midpoint, 65
Minimum point:
  on a graph, 69
  on a parabola, 69
Minor of an element in a determinant, 357
Modulus of a complex number, 280
Monomial, 2
Multiplication:
  associative law of, for matrices, 336
  associative law of, for real numbers, 7
  closure under, of set of real numbers, 6
  commutative law of, for real numbers, 7
  identity element for, in set of matrices, 336, 345
  identity element for, in set of real numbers, 7
  law of, for equality, 8
  of complex numbers, 267
  of a matrix by a scalar, 340
  of a vector by a scalar, 230
  of matrices, 341
Mutliplicative identity, 7
Multiplicative inverse, 7

Natural logarithm, 130
Natural number, 5
Negative:
  of a matrix, 336
  of a real number, 5
  of a vector, 229
Negative number(s):
  real, 6
Norm of a vector, 229
Nth root, 4
Number(s):
  lowest terms for a rational, 7
  negative of a, 5
Numerator of a fraction, 3

Oblique asymptote, 111
One-to-one function, 75
Order:
  axioms for real numbers, 5
  of a matrix, 334
  properties of, 5
  trichotomy law of, 5
Ordered field:
  rational numbers as, 5
  real numbers as, 5
Ordered pair(s):
  components of, 43
  definition of, 43
  in Cartesian products, 43
Ordinate, 44

Parabola, 69, 422
  axis of a, 69
  axis of symmetry of a, 423
  directrix of a, 422

Parabola *(continued)*:
  equation of a, 423
  focus of a, 422
  latus rectum of a, 426
  lowest or highest point of a, 69
  standard position of a, 424
  vertex of a, 423
Parallel lines, 63
Parallelogram law for vector sums, 230
Parameter, 450
Parametric equations, 450
Partial fractions, 307
Partial sums, 399
Period:
  fundamental, of a function, 167
  of a function, 167
Periodic function, 167 *ff*
Perpendicular lines, 64
Phase shift, 173
Point-slope form of a linear equation, 60
Polar axis, 454
Polar coordinates, 454 *ff*
Pole, 454
Polygon, convex, 326
Polygonal regions, 326
Polynomial(s):
  definition of, 2
  degree of, 2
  Descartes' rule of signs for, 97
  division of, 88 *ff*
  exactly divisible, 88
  function, 88
  graph of, function, 104
  in a real variable, 3
  leading coefficient of, 3
  leading term, 3
  location theorem for zeros of, function, 95
  monic, 3
  over R, 3
  rational zeros, 100 *ff*
  real, 3
  real zeros of, function, 94
  remainder theorem for, 91, 276
  simplification of, 27
  standard form of, 2
  value of, 91
Polynomial equation(s) *(see also* Polynomial function):
  rational roots of, 100
Polynomial function(s), 88 *ff*
  continuity of, 94
  graph of, 68
  isolating zeros of real, 95
  rational zeros of, 100 *ff*
  real zeros of, 94 *ff*
  zeros of, 95 *ff*
Positive real number(s):
  closure of set of, 5
Power(s):
  definition, 2
  integral, 3
  of complex numbers, 285
  rational, of a real number, 8

Power(s) *(continued)*:
  real, of a real number, 8
Prime number, 100
Principal diagonal of a matrix, 344
Product(s):
  Cartesian, 43
  of complex numbers, 267, 282
  of matrices, 341
  of matrix and scalar, 340
  of monomials, 8
  of polynomials, 8
  of powers, 8
  of quotients, 7
  of vector and scalar, 230
Progression:
  arithmetic, 386
  geometric, 387
Proof:
  by mathematical induction, 411
Pure imaginary number, 268

Quadrantal values, 155
Quadratic equation(s) *(see* Second-degree equations)
Quadratic formula, 18
  discriminant of, 19
Quadratic function(s), 68 *ff*
  graph of, 68
Quadratic inequalities, 72
Quotient(s):
  addition of, 7
  difference of two, 7
  equality of, 7
  of complex numbers, 271, 282
  of polynomials, 88 *ff*
  of powers, 8
  of real numbers, 7
  products of, 7
  subtraction of, 7
  sums of, 7

Radian measure of an angle, 199
Radical expressions:
  definition of, 4
  products of, 9
  properties of, 9
Radicand, 4
Radius of a circle, 426
Range:
  of an exponential function, 120
  of a logarithmic function, 122
  of a relation or function, 44
  of cosecant function, 185
  of cosine function, 151
  of cotangent function, 182
  of secant function, 184
  of sine function, 151
  of summation, 394
  of tangent function, 176
Rational expressions:
  definition of, 3

Rational expressions *(continued)*:
  differences of, 7
  products of, 7
  quotients of, 8
  sums of, 7
Rational functions, 88 *ff*, 109 *ff*
Rational numbers:
  set of, 5
Rational powers, properties of, 8
Ratios, trigonometric, 205, 215
Real number(s):
  absolute value of a, 5
  axioms for, 5
  negative, 6
  negative of, 5
  order in set of, 5
  positive, 6
  set of, 5
Real part of a complex number, 279
Real polynomial, 3
Rectangular coordinate system, 44
Recursive property of a factorial, 404
Reference angle, 208
Reference arc, 161
Reference outline, 469 *ff*
Regions, polygonal, 326
Relation(s):
  definition of, 44
  domain of, 45
  inverse, 74
  nonlinear, 68
  range of, 45
Remainder theorem, 91, 276
Resultant of vectors, 229
Right triangles, solution of, 215
Root(s):
  extraction of, 17
  of a complex number, 287
  nth, 4
Rotation of axes, 445
Row-equivalent matrices, 348
Row matrix, 334
Row vector, 334

Scalar:
  in vector algebra, 230
  product of a matrix by, 340
  product of a vector by, 230
Scalar product of vectors, 230
Scientific notation, 4
Secant:
  definition, 184
  domain of, 184
  fundamental period of, 184
  graph of, 184
  period, 184
  range of, 184
Second-degree equation(s):
  definition of, 16
  discriminant of, 19
  in one variable, 16 *ff*
  number of solutions, 17

Second-degree equation(s) *(continued)*:
  solution of, by completing the square, 18
  solution of, by extraction of roots, 17
  solution of, by factoring, 16
  solution of, by formula, 18
  standard form, 16
Second-degree inequalities, 29 *ff*
Sequence:
  alternating, 399
  convergent, 398
  definition of, 384
  divergent, 399
  finite, 384
  function, 384
  general term of, 385
  limit of, 397
  notation, 385
  nth term, 386, 387
Series:
  definition of, 391
  divergent, 400
  infinite, 400
  sigma notation for, 394
Set(s):
  Cartesian product of, 43
  closed convex polygonal, 326
  convex, 326
  definition, 1
  designation of a, 1
  element of a, 1
  equality of, 2
  equivalence of, 2
  intersection, 1
  member of a, 1
  notation, 1
  of complex numbers, 5
  of imaginary numbers, 268
  of integers, 5
  of irrational numbers, 5
  of natural numbers, 4
  of rational numbers, 5
  of real numbers, 5
  polygonal, 326
  union of, 1
Set-builder notation, 1
Sigma notation, 394
Sine:
  definition, 150
  domain of function, 151
  double-angle formula for, 248
  graph of, 167 *ff*
  half-angle formula for, 249
  period, 168
  range of, 151
  sum and difference formulas, 248
  table of values of, 157
Sine waves, 169
Sines, law of, 219 *ff*
Sinusoid, 169
Slope:
  of a line, 56
  of a line segment, 56
Slope-intercept form for a linear equation, 61

Solution(s):
　extraneous, 21
　matrix, of linear systems, 347, 374
　of a system, 293 *ff*
　of an equation in one variable, 10
　of an equation in two variables, 44
　of linear inequalities, 26
　of multiplicity two, 17
　of nonlinear inequalities, 29
　of triangles, 215
Solution set, 10, 293
Square matrix (*see* Matrix)
Standard form:
　for first-degree equations, 52
　for polynomials, 2
　for quadratic equations, 16
Standard position of an angle, 198
Subset:
　definition of, 1
Substitution in solving equations, 23, 297
Subtraction:
　of complex numbers, 269
　of quotients, 7
　of real numbers, 5
Sum(s):
　of a geometric progression, 392
　of an arithmetic progression, 391
　of an infinite geometric progression, 400
　of an infinite series, 400
　of complex numbers, 267
　of matrices, 335
　of quotients, 7
　of real numbers, 5
　of vectors, 229
　partial, 399
Sum formula:
　for cosine function, 243
　for sine function, 248
　for tangent function, 252
Summation notation, 394
　index of, 394
　range of, 394
Symmetry, 459
　axis of, 69, 423
Synthetic division, 89
Systems, equivalent, 294
Systems of inequalities, 324
Systems of linear equations:
　in three variables, 301 *ff*
　in two variables, 293 *ff*
Systems of linear equations, solution by:
　determinants, 379
　linear combinations, 301
　matrices, 347 *ff*
　substitution, 297
Systems of nonlinear equations, 314
　graphs of, 315

Tangent:
　definition, 176
　domain of, 176
　double-angle, formula for, 252

Tangent *(continued)*:
　graph of, 178
　half-angle, formula for, 253
　period, 178
　range of, 176
　sum and difference formulas for, 252
Term, 2
　leading, 3
Terminal side of an angle, 198
Transformation(s):
　elementary, 11, 347
Translation of axes, 440
Transpose of a matrix, 335
Triangle:
　right, 214 *ff*
　solution of, 215 *ff*
Trichotomy law, 5
Trigonometric functions, 198 *ff*, 205 *ff*
Trigonometric ratios, 205, 210
Turning point of a graph, 105
Two-dimensional vector, 228

Union of sets, 1
Unique-factorization theorem, 100
Unit circle, 150, 206
Upper bound:
　for zeros of real polynomial function, 95

Value:
　absolute, 5
　of a determinant, 356
Variable, 3
Variation in sign of coefficients, 97
Vector(s): 227 *ff*
　applications, 230 *ff*
　column, 334
　direction angle of, 229
　equality of, 229
　equivalence of, 229
　geometric, 227 *ff*
　magnitude of, 229
　negative of, 229
　norm of, 229
　notation for, 228
　parallelogram law of addition for, 230
　resultant, 229
　row, 334
　scalar product of, 230
　sum, 229
　triangle law of addition for, 229
　two-dimensional, 228
　with opposite direction, 230
　with same direction, 230
　zero, 230
Vertex:
　of an angle, 198
　of an ellipse, 428
　of a hyperbola, 432
　of a parabola, 423
Vertical:
　asymptote, 110

Zero(s):
  as a factor in a product, 16
  bounds for, of real polynomial function, 95
  complex, of a polynomial function, 277
  location theorem for, 94
  of a function, 71, 94

Zero(s) *(continued)*:
  rational, of polynomial function, 100 *ff*
  real, of polynomial function, 95
  test for rational, 101
Zero matrix, 336
Zero vector, 230

STUDENT QUESTIONNAIRE

Your chance to rate **Functions and Graphs** *(Beckenbach/Grady/Drooyan)*

In order to keep this text responsive to your needs, it would help us to know what you, the student, thought of this text. We would appreciate it if you would answer the following questions. Then cut out the page, fold, seal, and mail it; no postage is required. Thank you for your help.

Which chapters did you cover?   (circle)   1   2   3   4   5   6   7   8   9   10   11   12   13   All

Does the book have enough worked-out examples?   Yes_____   No_____

Enough exercises?   Yes_____   No_____

Which helped most?

Explanations_____   Examples_____   Exercises_____   All three_____   Other _____
(fill in)

Were the answers at the back of the book helpful?   Yes_____   No_____

Did the answers have any typos or misprints? If so, where?

For you, was the course elective? _____   Required? _____

Do you plan to take more mathematics courses?   Yes_____   No_____

If yes, which ones?

How much algebra did you have before this course?   Terms in high school   (circle)   1   2   3   4

Courses in college   1   2   3

If you had algebra before, how long ago?

Last 2 years_____   3–5 years ago_____   5 years or longer_____

What is your major or your career goal? _____   Your age? _____

What did you like the most about *Functions and Graphs*?

May we quote you?   Yes_____   No_____

What did you like least about the book?

College _____   State _____

-- FOLD HERE --

| | First Class
PERMIT NO. 34
Belmont Ca. |

BUSINESS REPLY MAIL
No postage necessary if mailed in United States

Postage will be paid by
WADSWORTH PUBLISHING COMPANY, INC.
10 Davis Drive
Belmont, California 94002

   ATTN: Rich Jones, Mathematics editor